T0122297

Studies in Fuzziness and Soft Computing

Volume 392

Series Editor

Janusz Kacprzyk, Systems Research Institute, Polish Academy of Sciences, Warsaw, Poland

The series "Studies in Fuzziness and Soft Computing" contains publications on various topics in the area of soft computing, which include fuzzy sets, rough sets, neural networks, evolutionary computation, probabilistic and evidential reasoning, multi-valued logic, and related fields. The publications within "Studies in Fuzziness and Soft Computing" are primarily monographs and edited volumes. They cover significant recent developments in the field, both of a foundational and applicable character. An important feature of the series is its short publication time and world-wide distribution. This permits a rapid and broad dissemination of research results.

Indexed by ISI, DBLP and Ulrichs, SCOPUS, Zentralblatt Math, GeoRef, Current Mathematical Publications, IngentaConnect, MetaPress and Springerlink. The books of the series are submitted for indexing to Web of Science.

More information about this series at http://www.springer.com/series/2941

Cengiz Kahraman · Fatma Kutlu Gündoğdu
Editors

Decision Making with Spherical Fuzzy Sets

Theory and Applications

Editors
Cengiz Kahraman
Department of Industrial Engineering
Faculty of Management
Istanbul Technical University
Istanbul, Turkey

Fatma Kutlu Gündoğdu
Turkish Air Force Academy
National Defence University
Istanbul, Turkey

ISSN 1434-9922 ISSN 1860-0808 (electronic)
Studies in Fuzziness and Soft Computing
ISBN 978-3-030-45463-0 ISBN 978-3-030-45461-6 (eBook)
https://doi.org/10.1007/978-3-030-45461-6

© Springer Nature Switzerland AG 2021
This work is subject to copyright. All rights are reserved by the Publisher, whether the whole or part of the material is concerned, specifically the rights of translation, reprinting, reuse of illustrations, recitation, broadcasting, reproduction on microfilms or in any other physical way, and transmission or information storage and retrieval, electronic adaptation, computer software, or by similar or dissimilar methodology now known or hereafter developed.
The use of general descriptive names, registered names, trademarks, service marks, etc. in this publication does not imply, even in the absence of a specific statement, that such names are exempt from the relevant protective laws and regulations and therefore free for general use.
The publisher, the authors and the editors are safe to assume that the advice and information in this book are believed to be true and accurate at the date of publication. Neither the publisher nor the authors or the editors give a warranty, expressed or implied, with respect to the material contained herein or for any errors or omissions that may have been made. The publisher remains neutral with regard to jurisdictional claims in published maps and institutional affiliations.

This Springer imprint is published by the registered company Springer Nature Switzerland AG
The registered company address is: Gewerbestrasse 11, 6330 Cham, Switzerland

Preface

The fuzzy set theory, introduced by Lotfi A. Zadeh in 1965, had enclosed a lot of ground with excellent achievements in almost all branches of science. It found many application areas in both theoretical and practical studies from engineering area to arts and humanities, from computer science to health sciences, and from life sciences to physical sciences. In this book, a new extension of fuzzy sets, entitled as spherical fuzzy sets, is introduced by pioneer researchers with different points of view and different research areas. This book consists of an Introduction and three parts. The first part involves seven chapters presenting valuable contributions on the elements of spherical fuzzy theory such as spherical fuzzy numbers, operational laws, new aggregation operators, similarity measures, and spherical fuzzy graphs. The second part contains 11 chapters. These chapters include spherical fuzzy decision-making methods and different applications to the real-life problems. Finally, the last part which contains four chapters is on mathematical programming with spherical fuzzy sets.

Introduction discusses the emergence of fuzzy sets from a historical perspective. In this Introduction, a comprehensive literature review on the fuzzy set theory is realized. In the recent years, ordinary fuzzy sets have been extended to new types, and these extensions have been employed in many areas such as mathematics, energy, environment, medicine, economics, and decision sciences. This literature review also analyzes the chronological development of these extensions.

The first chapter in the first part of the book deals with the mathematics of spherical fuzzy theory. In this chapter, single-valued spherical fuzzy sets and interval-valued spherical fuzzy sets are introduced with their score and accuracy functions; arithmetic and aggregation operations such as spherical fuzzy weighted arithmetic mean operator and interval-valued spherical fuzzy geometric mean operator. The second chapter is on preliminaries of interval-valued spherical fuzzy sets. The interval-valued spherical fuzzy sets, operational laws, and aggregation operators with their properties are introduced in this chapter. The proposed aggregation operators are also applied to a decision-making problem to choose the best station which examines the quality of air. The third chapter introduces spherical trapezoidal fuzzy numbers (STF numbers) and spherical triangular fuzzy numbers (STrF numbers) with laws of their arithmetic operations and their properties. By utilizing the STF numbers and STrF numbers, two multi-criteria decision-making methods are developed. The fourth chapter introduces spherical fuzzy Bonferroni mean (SFBM) and spherical fuzzy normalized weighted Bonferroni mean (SFNWBM). Based on the proposed aggregation operator (SFNWBM), it presents an approach for multi-criteria group decision-making problems under the spherical fuzzy environment. The fifth chapter includes novel Dice similarity measures of spherical fuzzy sets. This chapter also presents the generalized Dice similarity measure-based multiple attribute group decision-making models under spherical fuzzy environment. The sixth chapter describes spherical fuzzy soft sets (SFSSs) with their properties. This chapter shows that DeMorgan's laws are valid in SFSS theory. Also, it gives an algorithm to solve decision-making problems based on adjustable soft discernibility matrix. The seventh chapter presents certain concepts of spherical fuzzy graphs and describes various methods of their construction. It computes degree and total degree of spherical fuzzy graphs with their important properties. In addition, this chapter shows some applications of spherical fuzzy graphs to decision making.

The first chapter of the second part of the book is on spherical fuzzy TOPSIS method, and this method is used in solving a multiple criteria selection problem for optimal site selection of electric vehicle charging station to verify the developed approach. The second chapter proposes a novel spherical fuzzy VIKOR method. Supplier selection problem is solved by using spherical fuzzy VIKOR with the four implementations, and the results are compared with the results of the spherical fuzzy TOPSIS. The third chapter presents simple additive weighting (SAW) and weighted product methods (WPM) to their spherical fuzzy versions. Scoring methods are the most frequently used multi-attribute decision-making methods because of their easiness and effectiveness. In this chapter, single-valued and interval-valued spherical SAW and WPM methods are applied to the selection of insurance options. The fourth chapter utilizes the spherical fuzzy sets (SFSs) for the applicability of the available data for the WASPAS method. In this study, a multi-criteria decision-making method, WASPAS, based on single-valued spherical fuzzy sets is applied for the prioritization of the manufacturing challenges of a contract manufacturing company by considering evaluations of a group of experts. The fifth chapter is on livability indices which help to understand how a place is livable. In this chapter, COmbinative Distance-based ASsessment (CODAS) method is extended to its Spherical CODAS version for handling the impreciseness and vagueness in human thoughts. The applicability of the methodology is illustrated through the assessment of livability index of suburban districts. Sensitivity analysis demonstrates the robustness of the decision-making methodology. The sixth chapter is on Industry 4.0, which connects new technologies providing flexibility in manufacturing where the conditions change rapidly. Evaluating companies' performance based on Industry 4.0 is a complex multi-criteria problem including both quantitative and qualitative factors. A novel fuzzy MULTIMOORA method based on interval-valued spherical fuzzy sets is proposed to evaluate the performances of companies using Industry 4.0 technologies. The seventh chapter extends the analytic hierarchy process (AHP) using SFSs. In the proposed method, SFSs are used to construct the pairwise comparison matrices. The proposed method is used to solve a case study in global supplier selection based on SFSs and intuitionistic fuzzy sets (IFSs) for comparative purposes. The eighth chapter is on generalized three-dimensional spherical fuzzy sets introduced by Kutlu Gündoğdu and Kahraman. The authors propose spherical fuzzy analytic hierarchy process method, which is extended to interval-valued spherical fuzzy AHP method. The proposed method is used to compare the service performances of several hospitals. The method is designed to analyze the service quality in the healthcare industry based on SERVQUAL dimensions. The ninth chapter extends the conventional PROMETHEE method into its spherical fuzzy version. In the proposed method, both the weights of the criteria and the preference relations are SFSs. The relative degree of closeness is used to determine the preference indices. Two examples are solved to illustrate the applicability and efficiency of the proposed method in solving MCDM problems. The tenth chapter is on the usage of spherical fuzzy quality function deployment for the design of delivery drones. The important ratings and global weights of customer requirements and improvement directions of design requirements are represented by SFSs. Spherical fuzzy aggregation operators are used to aggregate the opinions of different decision-makers. The eleventh chapter is on the design of mobile phone applications based on mobile Internet usage. The fact that these investments should be utilized by many users has become an important agenda of business plans for many brands. A new method is proposed to decide which design parameters are affective for the related mobile application and to determine the importance degree of the design parameters. The methods including Kano model, quality function deployment, and spherical fuzzy sets are integrated into the proposed method.

The first chapter of the third and last part of the book deals with spherical fuzzy linear programming problem (SFLPP) in which the different parameters are represented by spherical fuzzy numbers. The crisp version of the SFLPP is obtained with the aid of positive, neutral, and negative membership degrees. Furthermore, the spherical fuzzy optimization model is presented to solve the SFLPP. The second chapter in this part proposes a new algorithm based on spherical fuzzy sets called spherical fuzzy multi-objective programming problem (SFMOLPP). The SFMOLPP inevitably involves the degree of neutrality along with positive and negative membership degrees of the element into the feasible solution set. It also generalizes the decision set by imposing the restriction that the sum of squares of each membership function must be less than or equal to one. The attainment of achievement function is determined by maximizing the positive membership function and minimization of neutral and negative membership function of each objective function under the spherical fuzzy decision set. The third chapter is on spherical fuzzy goal programming problem (SFGP). The SFGP unavoidably involves the degree of neutrality along with truth and a falsity membership degree of the element into the feasible decision set. The fourth chapter presents a new algorithm based on spherical fuzzy sets called spherical fuzzy geometric programming problem (SFGPP) together with several numerical examples.

We hope that this book will provide a useful resource of ideas, techniques, and methods for the development of the spherical fuzzy set theory. We are grateful to the referees whose valuable and highly appreciated works contributed to select the high-quality chapters published in this book. We would like to also thank Prof. Janusz Kacprzyk, the editor of *Studies in Fuzziness and Soft Computing* at Springer for his supportive role in this process.

Istanbul, Turkey

Cengiz Kahraman
Fatma Kutlu Gündoğdu

From Ordinary Fuzzy Sets to Spherical Fuzzy Sets

Abstract Fuzzy sets have a great progress in every scientific research area. It found many application areas in both theoretical and practical studies from engineering area to arts and humanities, from computer science to health sciences, and from life sciences to physical sciences. In this paper, a comprehensive literature review on the fuzzy set theory is realized. In the recent years, ordinary fuzzy sets have been extended to new types of fuzzy sets, and these extensions have been used in many areas such as energy, medicine, material, economics, and pharmacology sciences. This literature review also analyzes the chronological development of these extensions. In the last section of the paper, we present our new theory entitled spherical fuzzy sets theory.

Introduction

The pioneer of the fuzzy set theory, Lotfi Zadeh published the first paper on his new theory that is a way of handling uncertainty by representing every element in the set together with a membership degree in 1965. In the real world, we encounter situations that we cannot determine whether the state is true or false. The fuzzy logic provides a very valuable tool for reasoning in these situations. In this way, we can represent the truthiness and falsity of any situation. In Boolean system, truth value 1.0 represents absolute truth value and 0.0 represents absolute false value. However, in fuzzy logic, there is an intermediate value too present which is partially true and partially false. Between 1965 and 1975, Zadeh extended the foundation of the fuzzy set theory by launching fuzzy similarity relations, fuzzy decision making, and linguistic hedges. In 1970s, some research groups in Japan started to study the fuzzy set theory. The success of fuzzy logic was observed in Japan at the beginning of 1980s, and this led to revitalization in fuzzy logic in the USA at the end of 1980s. Along with the emergence of the fuzzy set theory, many objections to the theory have appeared. Nevertheless, the fuzzy logic showed its influence through the real technology applications. There are many technology applications in the real world by using fuzzy logic and sets. Fuzzy logic is used in the aerospace field for altitude control of spacecraft and satellite. It has employed in the automotive system for speed control and traffic control. It is also used for decision-making support systems and personal evaluation in the large company business. It has many applications in chemical industry for controlling the pH, drying, and chemical distillation process. Fuzzy logic is used in natural language processing and various intensive applications in artificial intelligence. For example, it has been applied in 3D animation systems for generating crowds as it was used extensively in the making of the Lord of the Rings. It is also extensively operated in modern control systems such as expert systems and neural networks as it mimics how a person would make decisions (Kahraman et al. 2016). Briefly, fuzzy set theory has been employed to obtain efficient solutions under uncertainty for the research areas engineering, mathematics, computer sciences, medical sciences, business and economics, social sciences, and human behaviors.

In this chapter, a comprehensive literature review for the fuzzy set theory is realized. It aims at presenting the expansion areas of the fuzzy set theory in its 60th year and exhibits the results of literature review with graphical illustrations and a classification with respect to the extensions of fuzzy sets.

The rest of the chapter is organized as follows. In section "Fuzzy Set Theory with Graphical Illustrations", some graphical illustrations on the most productive authors writing fuzzy papers and their affiliations; the frequencies, sources, and countries of these papers are presented. Section "Fuzzy Set Theory with Graphical Illustrations" includes the subject areas of fuzzy papers and some related statistics. Section "Advanced Fuzzy Extensions" gives the classification of extensions of fuzzy sets including the brief definitions on these sets. Section "Spherical Fuzzy Sets" presents the literature review on spherical fuzzy set theory. Finally, section "Conclusions" concludes the chapter.

Fuzzy Set Theory with Graphical Illustrations

In this section, the whole publications including the word *fuzzy* in their titles are classified with respect to their publication years, publication sources, authors, the affiliations of authors, the countries of the authors, document types, and subject areas. Each classification type is enhanced by the graphical and tabular illustrations. The search was made in December 25, 2019, based on Scopus database. Figure 1 demonstrates the frequencies of the papers using the fuzzy set theory with respect to years. It shows that there is an increasing trend starting at the beginning of 1990s. After 2002, the numbers of published papers using the fuzzy set theory have nearly an arithmetic increase. The number of these papers reaches to a saddle point of 11,059 papers per year.

Table 1 illustrates the sources publishing fuzzy papers and the number of fuzzy papers published over 500 papers so far. According to this table, "Fuzzy Sets and Systems" and "IEEE International Conference on Fuzzy Systems" are the sources most publishing fuzzy papers.

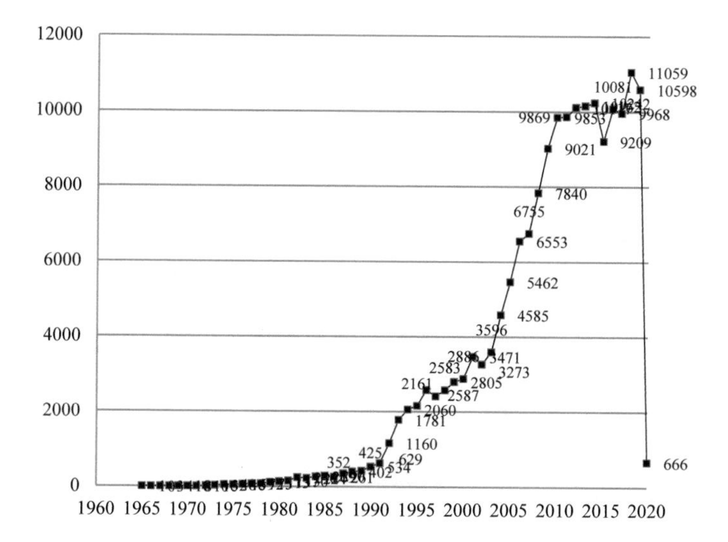

Fig. 1 Frequencies of fuzzy papers with respect to years

Table 1 Sources publishing fuzzy papers

Sources	Frequency
Fuzzy Sets and Systems	5881
IEEE International Conference on Fuzzy Systems	5219
Lecture Notes in Computer Science Including Subseries Lecture Notes in Artificial Intelligence and Lecture Notes in Bioinformatics	5008
Journal of Intelligent and Fuzzy Systems	2479
IEEE Transactions on Fuzzy Systems	2200
Advances in Intelligent Systems and Computing	1915
Information Sciences	1755
Expert Systems with Applications	1588
Annual Conference of the North American Fuzzy Information Processing Society NAFIPS	1528
Applied Mechanics and Materials	1483
Advanced Materials Research	1258
Applied Soft Computing Journal	1225
Studies in Fuzziness and Soft Computing	1166
Proceedings of SPIE The International Society for Optical Engineering	1165
Soft Computing	1058
Communications in Computer and Information Science	1012
International Journal of Fuzzy Systems	893
Proceedings of the World Congress on Intelligent Control and Automation WCICA	879
Studies in Computational Intelligence	740

(continued)

Table 1 (continued)

Sources	Frequency
Proceedings of the IEEE International Conference on Systems Man and Cybernetics	724
Neurocomputing	638
Lecture Notes in Electrical Engineering	616
IFAC Proceedings Volumes IFAC PapersOnLine	614
International Journal of Intelligent Systems	586
American Institute of Physics (AIP) Conference Proceedings	580
International Journal of Uncertainty Fuzziness and Knowledge Based Systems	573
Kongzhi Yu Juece Control and Decision	566
Neural Computing and Applications	517
Mathematical Problems in Engineering	513
IECON Proceedings Industrial Electronics Conference	512
Iranian Journal of Fuzzy Systems	506

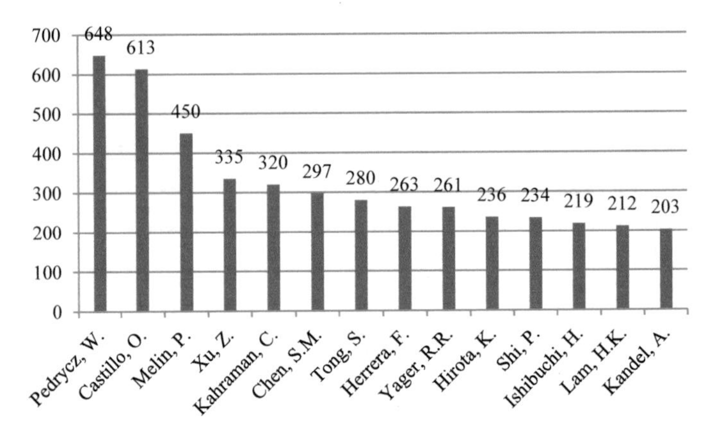

Fig. 2 Authors publishing fuzzy papers over 200

Figure 2 shows the authors publishing over 200 papers up to now, based on the fuzzy set theory. W. Pedrycz, O. Castillo, and P. Melin are the most productive authors in this field.

Figure 3 indicates the affiliations publishing over 750 fuzzy papers up to now. Northeastern University in China takes the first rank publishing papers using the fuzzy set theory.

Figure 4 illustrates the countries publishing fuzzy papers more than 2000 up to now. China, India, USA, Taiwan, and Iran are the first five leading countries.

Table 2 shows the classification of fuzzy publications. Most of the documents are articles and conference papers.

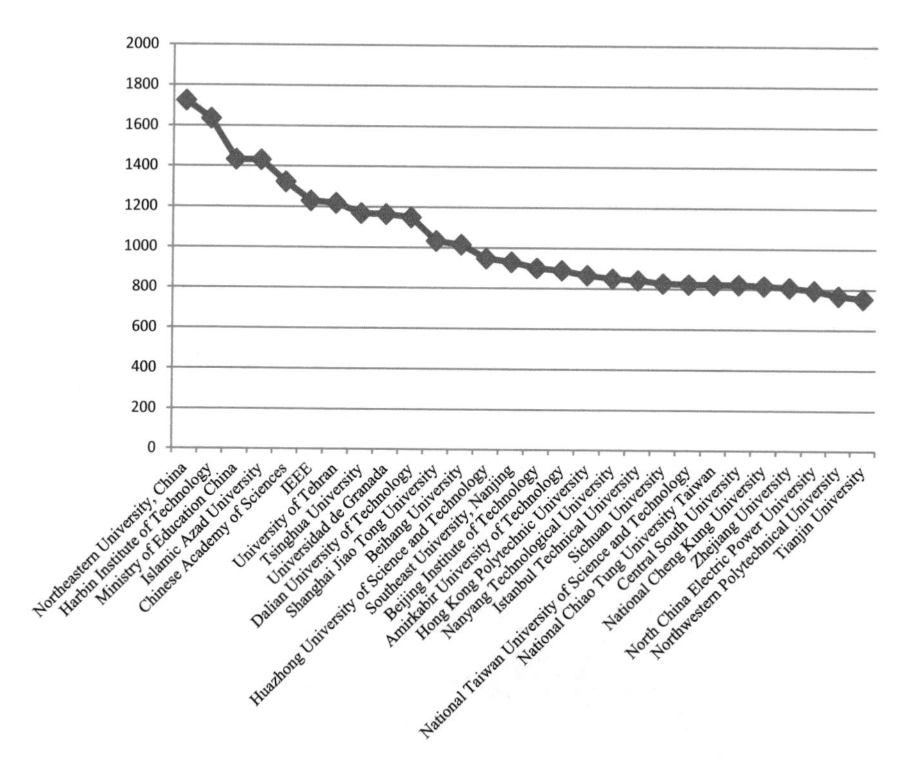

Fig. 3 Affiliations publishing more than 750 fuzzy papers

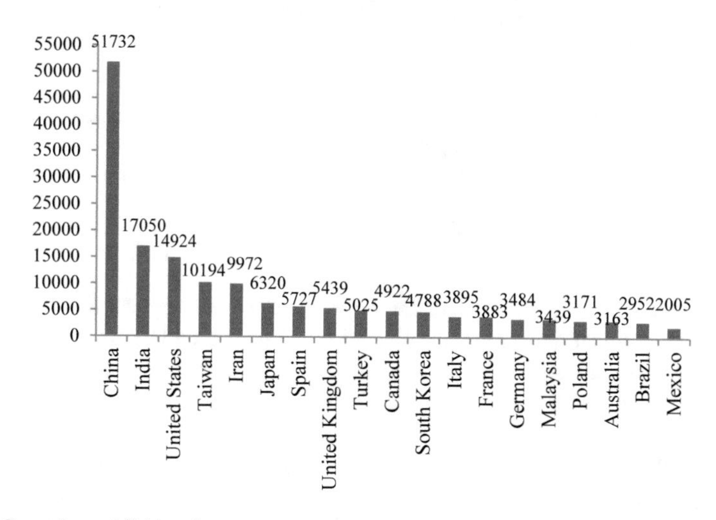

Fig. 4 Countries publishing fuzzy papers more than 2000

Table 2 Classification of fuzzy publications

Document Type	Frequency
Article	95536
Conference Paper	76210
Book Chapter	2585
Review	1067
Erratum	376
Editorial	296
Conference Review	249
Note	247
Book	179
Letter	136
Short Survey	31
Report	11

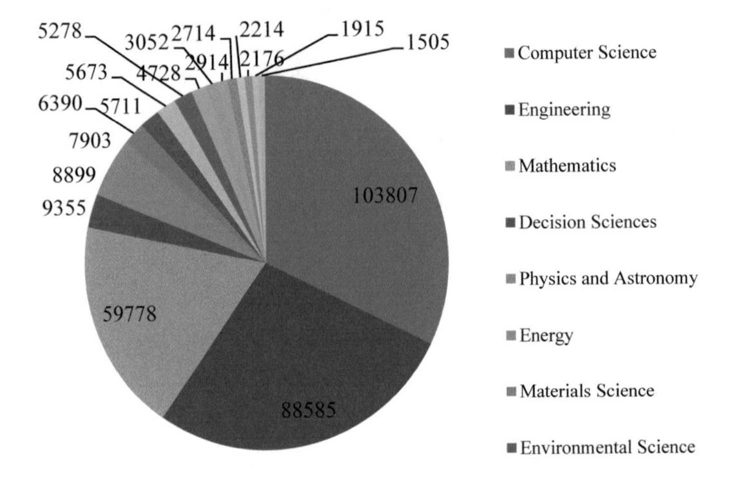

Fig. 5 Subject areas of the published fuzzy papers and the frequencies

We analyze the subject areas of the papers using the fuzzy set theory. Figure 5 illustrates the subject areas of the published fuzzy papers and their frequencies. The fuzzy papers are mostly used in engineering and computer science areas. The number of papers using the fuzzy sets on computer science is 103,807. A significant trend in fuzzy computer science publications begins in 1992 even some fluctuations are observed. The number of papers using the fuzzy sets on engineering is 88,585. The number of papers using the fuzzy sets on mathematics is 59,778, while the number of papers using the fuzzy sets on decision sciences is 9355.

Advanced Fuzzy Extensions

Fuzzy sets have been very popular in almost all branches of science since it has emerged in 1965 (Zadeh 1965). The ordinary fuzzy sets have been extended to many new types: Type 2 fuzzy sets (Zadeh 1975), interval-valued fuzzy sets (Sambuc 1975; Zadeh 1975; Jahn 1975; Grattan-Guinness 1976), intuitionistic fuzzy sets (Atanassov 1986), fuzzy multi-sets (Yager 1986), neutrosophic fuzzy sets (Smarandache 1998), non-stationary fuzzy sets (Garibaldi and Ozen 2007), hesitant fuzzy sets (Torra and Narukawa 2009), Pythagorean fuzzy sets (Yager and Abbasov 2013; Yager 2013), picture fuzzy sets (Cuong 2014), orthopair fuzzy sets (Yager 2016), and spherical fuzzy sets (Kahraman and Gündoğdu 2018). It starts from ordinary fuzzy sets and extends to recently developed types of fuzzy sets as shown in Fig. 6.

Ordinary Fuzzy Sets

Let a set U be a universe of discourse. An ordinary fuzzy set $\tilde{A}$ is an object having the form $\tilde{A} = \{\langle u, \mu_{\tilde{A}}(u) | u \in U\}$ where the function $0 \le \mu_{\tilde{A}}(u) \le 1$ is the degree of membership of u to $\tilde{A}$. Its range is the subset of nonnegative real numbers whose supremum is finite. Zadeh (1965) introduced fuzzy sets as a class of objects with a continuum of grades of membership. He extended the notions of inclusion, union, intersection, complement, relation, convexity, linguistic hedges, etc… to such sets and established various properties of these notions in the context of fuzzy sets.

Type 2 Fuzzy Sets

Zadeh (1975) introduced the concept of type 2 fuzzy sets as an extension of the ordinary fuzzy sets. Such sets are fuzzy sets whose membership grades themselves are type 1 fuzzy sets. They are very useful incircumstances where it is difficult to determine an exact membership function for a fuzzy set. This type of fuzzy sets requires too many parameters to be used in the problem modeling. Besides, the third

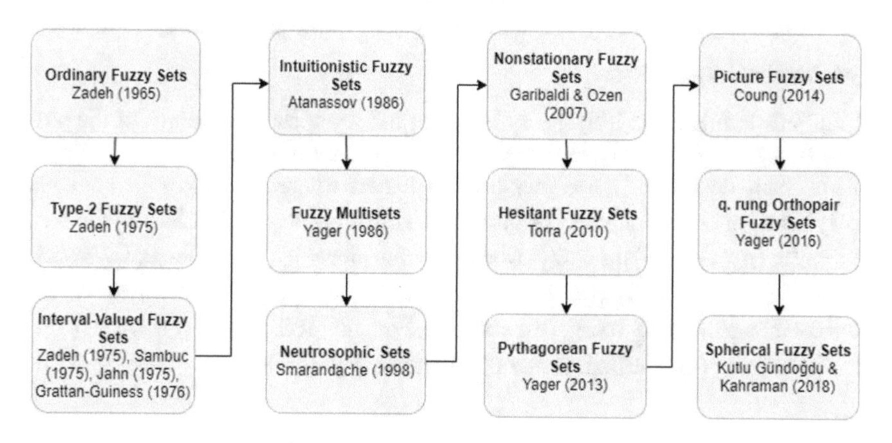

Fig. 6 Extensions of fuzzy sets

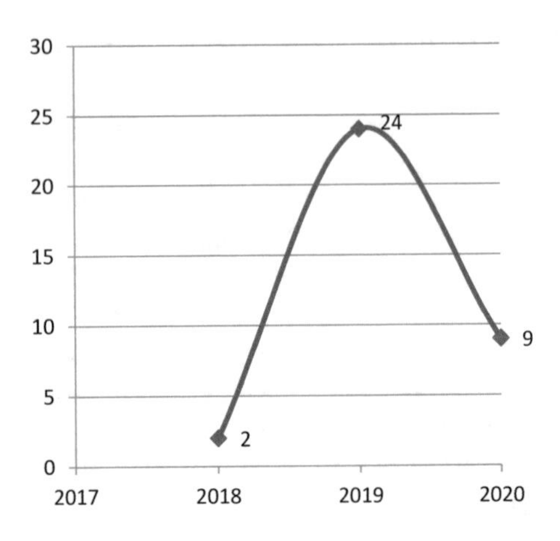

Fig. 7 Frequencies of spherical fuzzy papers with respect to years

dimension is ignored by many researchers for simplification purpose, and these sets are called interval-valued type 2 fuzzy sets. A type 2 fuzzy set in the universe of discourse U can be represented by a type 2 membership function $\mu_{\tilde{\tilde{A}}}(u)$ shown as follows:

$$\tilde{\tilde{A}} = \left\{\left\langle (u,x), \mu_{\tilde{A}}(u,x) \middle| \forall u \in U, \forall x \in I_u \subseteq [0,1], 0 \leq \mu_{\tilde{A}}(u,x) \leq 1\right\}\right. \quad (1)$$

where I_u denotes an interval $[0, 1]$.

Interval-Valued Fuzzy Sets

Interval-valued fuzzy sets were introduced independently by Zadeh (1975), Grattan-Guinness (1976), Jahn (1975), and Sambuc (1975). An interval-valued fuzzy set is a special case of type 2 fuzzy sets. An interval-valued fuzzy set (IVFS) is defined by a mapping F from the universe U to the set of closed intervals in [0, 1]. Let $F(u) = [F_L(u), F^U(u)]$. The union, intersection, and complementation of IVFSs are obtained by canonically extending fuzzy set-theoretical operations to intervals.

Intuitionistic Fuzzy Sets

Intuitionistic fuzzy set (IFS) theory is one of the significant extensions of the ordinary fuzzy set (FS) theory to deal with the vagueness in the data, which utilizes a membership degree and a non-membership degree, whose sum is less than or equal to 1. Let a set U be a universe of discourse. An IFS $\tilde{A}$ is an object having the form $\tilde{A} = \left\{\langle u, (\mu_{\tilde{A}}(u), v_{\tilde{A}}(u)) \middle| u \in U\right\}$ where the functions $\mu_{\tilde{A}}(u) : U \to [0,1]$, $v_{\tilde{A}}(u) : U \to [0,1]$ and $0 \leq \mu_{\tilde{A}}(u) + v_{\tilde{A}}(u) \leq 1$ are the degree of membership, non-membership of u to $\tilde{A}$, respectively. For any IFS $\tilde{A}$ and $u \in U$, $\pi_{\tilde{A}}(u) = 1 - \mu_{\tilde{A}}(u) - v_{\tilde{A}}(u)$ is called degree of indeterminacy of u to $\tilde{A}$.

Fuzzy Multi-sets

Yager (1986) introduced the bag structure as a set-like object in which repeated elements were significant. Basic operations on bags such as intersection, union, and addition were discussed and introduced the operation of selecting elements from a bag based upon their membership in a set and showed the usefulness of the bag structure in relational databases by Yager (1986).

Let U be a non-empty set. A fuzzy multi-set $\tilde{A}$ drawn from U is characterized by a function, "count membership" of $\tilde{A}$ denoted by $CrM_{\tilde{A}}$ such that $CrM_{\tilde{A}} : U \rightarrow X$ where X is the set of all crisp multi-sets drawn from the unit interval [0, 1]. Then, for any $u \in U$, $CrM_{\tilde{A}}$ value is a crisp multi-set drawn from [0, 1]. For each $u \in U$, the membership sequence is defined as the decreasingly ordered sequence of elements in $CrM_{\tilde{A}}$. It is denoted by $\left(\mu^1_{\tilde{A}}(u), \mu^2_{\tilde{A}}(u), \ldots, \mu^n_{\tilde{A}}(u)\right)$ where $\mu^1_{\tilde{A}}(u) \geq \mu^2_{\tilde{A}}(u) \geq \ldots \geq \mu^n_{\tilde{A}}(u)$.

Neutrosophic Sets

Neutrosophic sets (NS) are represented by the three dimensions: a truthiness degree, an indeterminacy degree, and a falsity degree (Smarandache 1998). NS not only deal with the hesitancy of the system but also decrease indecisiveness of inconsistent information. Thus, the truthiness, falsity, and indeterminacy values can be independently assigned (Smarandache 1998).

Let U be a universe of discourse. Neutrosophic set $\tilde{A}$ in U is an object having the form $\tilde{A} = \left\{\langle u, (T_{\tilde{A}}(u), I_{\tilde{A}}(u), F_{\tilde{A}}(u)) \middle| u \in U\right\}$ where $T_{\tilde{A}}$ is the truth-membership function, $I_{\tilde{A}}$ is the indeterminacy-membership function, and $F_{\tilde{A}}$ is the falsity membership function. There is no restriction on their sum and so $0 \leq T_{\tilde{A}}(u) + I_{\tilde{A}}(u) + F_{\tilde{A}}(u) \leq 3$.

Non-stationary Fuzzy Sets

Garibaldi and Ozen (2007) presented a case study in which the introduction of vagueness into the membership functions of a fuzzy system was investigated in order to model the variation exhibited by decision-makers in a medical decision-making context through non-stationary fuzzy reasoning.

Let $\tilde{A}$ denote a fuzzy set of a universe of discourse U characterized by a membership function $\mu_{\tilde{A}}(u)$. Let T be a set of time points t_i, possibly infinite, and $f : T \rightarrow \Re$ denote a perturbation function. A non-stationary fuzzy set $\tilde{A}$of the universe of discourse U is characterized by a non-stationary membership function $\mu_{\tilde{A}}(u)$: $T \times U \rightarrow [0, 1]$ that associates each element (t, x) of with a time-specific variation of $T \times U$. The non-stationary fuzzy set $\tilde{A}$ is denoted by:

$$\tilde{A} = \int_{t \in T} \int_{u \in U} \mu_{\tilde{A}}(t, u)/u/t \tag{2}$$

Hesitant Fuzzy Sets

Torra and Narukawa (2009) defined hesitant fuzzy sets and presented an extension principle, which permits to generalize existing operations on fuzzy sets to this new type of fuzzy sets. Torra and Narukawa (2009) proposed hesitant fuzzy sets (HFSs) and introduced some basic operations for HFS. He proved that the envelope of the hesitant fuzzy sets was an intuitionistic fuzzy set. Hesitant fuzzy sets can be used as a functional tool allowing many potential degrees of membership of an element to a set. These fuzzy sets force the membership degree of an element to be possible values between zero and one. Torra and Narukawa (2009) defined hesitant fuzzy sets (HFSs) as follows: Let $M = \{\mu_1, \ldots, \mu_N\}$ be a set of N membership functions. Then, the hesitant fuzzy set associated with M, that is h_M, is defined as follows:

$$h_M(u) = \bigcup_{\mu \epsilon M} \{\mu(u)\} \tag{3}$$

Pythagorean Fuzzy Sets

In the real life, the decision-makers might express their preferences about membership degrees and non-membership degrees of an alternative with respect to a criterion dissatisfies the condition that the sum of the membership and non-membership degrees should be less than or equal to 1.0. Pythagorean fuzzy sets (PFS) developed by Yager (2013), which had been called as Intuitionistic type 2 fuzzy sets (IFS2) by Atanassov previously (Atanassov 1989), are characterized by a membership degree and a non-membership degree satisfying the condition that their squared sum is at most equal to one, which is a generalization of intuitionistic fuzzy sets (IFS) (Yager and Abbasov 2013). This concept provides a larger preference area for decision-makers. In other words, all the intuitionistic fuzzy degrees are a part of the Pythagorean fuzzy sets, which shows that the PFS is more powerful to handle the uncertain problems.

Definition of PFS is given as follows:

Let a set U be a universe of discourse. A PFS $\tilde{P}$ is an object having the form $\tilde{P} = \{\langle u, (\mu_{\tilde{p}}(u), v_{\tilde{p}}(u)) | u \in U\}$ where the functions $\mu_{\tilde{p}}(u) : U \to [0, 1]$, $v_{\tilde{p}}(u) : U \to [0, 1]$ and $0 \leq \mu^2_{\tilde{p}}(u) + v^2_{\tilde{p}}(u) \leq 1$ are the degree of membership, non-membership of u to P, respectively. For any PFS $\tilde{A}$ and $u \in U$, $\pi_{\tilde{p}}(u) = \sqrt{1 - \mu^2_{\tilde{p}}(u) - v^2_{\tilde{p}}(u)}$ is called degree of hesitancy of u to $\tilde{P}$.

As seen from the above formula, hesitancy degree is based on membership and non-membership degrees in the PFS. However, DM wants to define their hesitancy degree like a third dimension as irrelevant from membership and non-membership degrees. It is not allowed to define hesitancy degree by the decision-maker in the PFS.

Picture Fuzzy Sets

Fundamentally, picture fuzzy set-based models may be adequate in situations when we face human opinions involving more answers of type: yes, abstain, no, and refusal. Voting can be a good example of such a situation as the human voters may be divided into four groups of those who vote for, abstain, vote against, and refusal of the voting. A picture fuzzy set on a $\tilde{A}_S$ of the universe of discourse U is given by; $\tilde{A}_S = \left\{ \left\langle u, (\mu_{\tilde{A}_S}(u), v_{\tilde{A}_S}(u), \pi_{\tilde{A}_S}(u)) \middle| u \in U \right\rangle \right\}$ where $\mu_{\tilde{A}_S}(u) : U \to [0,1]$, $v_{\tilde{A}_S}(u) : U \to [0,1]$, $\pi_{\tilde{A}_S}(u) : U \to [0,1]$ and $0 \leq \mu_{\tilde{A}}(u) + v_{\tilde{A}}(u) + \pi_{\tilde{A}}(u) \leq 1 \quad \forall u \in U$. Then, $\chi = 1 - (\mu_{\tilde{A}}(u) + v_{\tilde{A}}(u) + \pi_{\tilde{A}}(u))$ could be called the degree of refusal membership of u in U (Cuong and Kreinovich 2013).

q-Rung Orthopair Fuzzy Sets (q-ROFS)

Yager (2016) proposed a general class of these sets called q-rung orthopair fuzzy sets in which the sum of the qth power of the support against is bonded by one. They note that as q increases, the space of acceptable orthopairs increases and thus gives the user more freedom in expressing their belief about membership grade. The characteristic of q-ROFS that it allows the sum to be greater than one but qth sum of membership degree and non-membership degrees to be less than one, providing more freedom for DMs, results in less information loss.

Let a set U be a universe of discourse. A q-rung orthopair fuzzy set $\tilde{O}$ is an object having the form $\tilde{O} = \left\{ \langle u, (\mu_{\tilde{O}}(u), v_{\tilde{O}}(u)) | u \in U \right\}$ where the function $\mu_{\tilde{O}}(u) : U \to [0,1]$, $v_{\tilde{O}}(u) : U \to [0,1]$ and $0 \leq \mu_{\tilde{O}}^q(u) + v_{\tilde{O}}^q(u) \leq 1$ are the degree of membership, non-membership of u to P, respectively. For any PFS $\tilde{A}$ and $u \in U$, $\pi_{\tilde{O}}(u) = \sqrt{1 - \mu_{\tilde{O}}^q(u) - v_{\tilde{O}}^q(u)}$ is called degree of hesitancy of u to $\tilde{O}$.

In the following section, we introduce a last extension of fuzzy sets that are spherical fuzzy sets (SFSs), which provides a larger preference domain for decision-makers.

Spherical Fuzzy Sets

In this section, the whole publications including the word *spherical fuzzy* in their titles and abstract are classified with respect to their publication years, authors, the countries of the authors, and subject areas. Each classification type is enhanced by the graphical illustrations.

Figure 7 demonstrates the frequencies of the papers using the spherical fuzzy set theory with respect to years. It shows that there is an increasing trend starting at the beginning of 2018.

Figure 8 shows the authors publishing spherical fuzzy papers up to now. T. Mahmood and C. Kahraman are the pioneer authors in this field. Figure 9 illustrates the subject areas of the published fuzzy papers and their frequencies. The fuzzy papers are mostly used in computer science and mathematics.

Kahraman and Gündoğdu (2018) introduced spherical fuzzy sets as an extension of picture fuzzy sets. The idea behind SFS is to let decision-makers to generalize

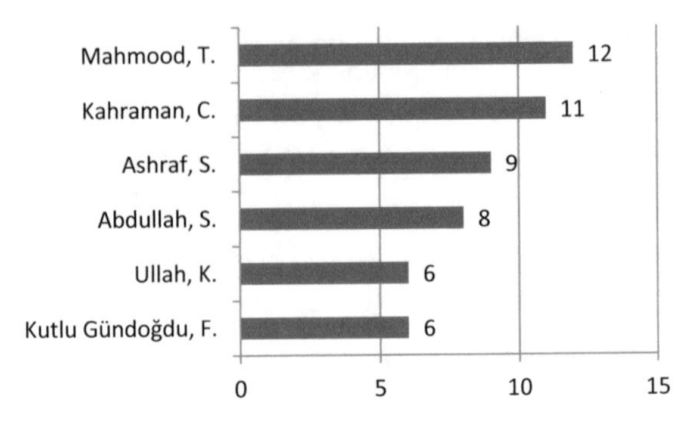

Fig. 8 Authors publishing spherical fuzzy papers

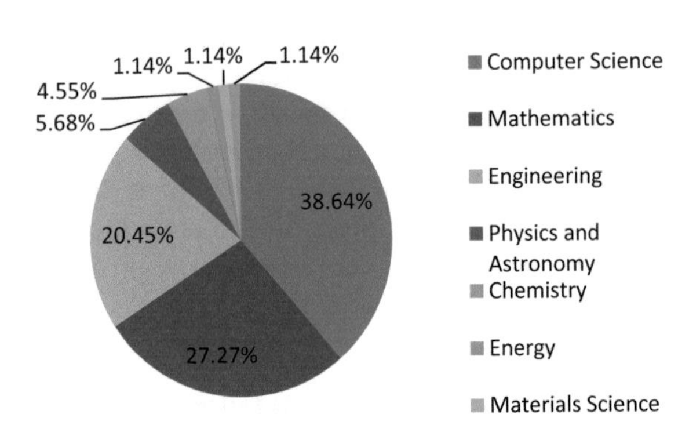

Fig. 9 Subject areas of spherical fuzzy sets

other extensions of fuzzy sets by defining a membership function on a spherical surface and independently assign the parameters of that membership function with a larger domain. A SFS must satisfy the following condition:

A spherical fuzzy sets (SFS) $\tilde{A}_S$ of the universe of discourse U is given by;

$$\tilde{A}_S = \left\{ \left\langle u, (\mu_{\tilde{A}_S}(u), v_{\tilde{A}_S}(u), \pi_{\tilde{A}_S}(u)) \right| u \in U \right\} \tag{4}$$

where

$$\mu_{\tilde{A}_S}(u) : U \to [0,1], \quad v_{\tilde{A}_S}(u) : U \to [0,1], \quad \pi_{\tilde{A}_S}(u) : U \to [0,1]$$

and

$$0 \le \mu^2_{\tilde{A}_S}(u) + v^2_{\tilde{A}_S}(u) + \pi^2_{\tilde{A}_S}(u) \le 1 \quad \forall u \in U \tag{5}$$

For each u, the numbers $\mu_{\tilde{A}_S}(u)$, $v_{\tilde{A}_S}(u)$ and $\pi_{\tilde{A}_S}(u)$ are the degree of *membership, non-membership* and *hesitancy* of u to $\tilde{A}_S$, respectively.

$\chi_{\tilde{A}_S}(u) = \sqrt{1 - \mu^2_{\tilde{A}_S}(u) - v^2_{\tilde{A}_S}(u) - \pi^2_{\tilde{A}_S}(u)}$ is called as a refusal degree (Gündoğdu and Kahraman 2019a; Ashraf and Abdullah 2019).

At the same time, Ashraf and Abdullah (2019) defined spherical fuzzy sets as an extension of picture fuzzy sets and they defined some operational rules and aggregation operations based on Archimedean t-norm and t-conorm. This study was extended with some useful operation such as spherical fuzzy t'-norms and spherical fuzzy t'-conorms by Ashraf et al. (2019a).

Ashraf et al. (2019b) developed spherical fuzzy Dombi weighted averaging, spherical fuzzy Dombi ordered weighted averaging, spherical fuzzy Dombi hybrid weighted averaging, spherical fuzzy Dombi weighted geometric (SFDWG), spherical fuzzy Dombi ordered weighted geometric, and spherical fuzzy Dombi hybrid weighted geometric aggregation operators and discussed several properties of these aggregation operators. These operators were used to get a successful solution of the decision problems.

Ashraf et al. (2019c) described spherical fuzzy distance-weighted averaging, spherical fuzzy distance order-weighted averaging, and spherical fuzzy distance order-weighted average weighted averaging operators and they designed an algorithm to help decision analysis.

Rafiq et al. (2019) investigated the novel similarity measures between spherical fuzzy sets based on cosine function by considering the membership, hesitancy, non-membership and refusal grades in SFS.

Gündoğdu (2019a) summarized the previously introduced spherical fuzzy sets and as an application spherical fuzzy TOPSIS method applied to the site selection of photovoltaic power station in this study. They also presented novel interval-valued spherical fuzzy sets, employed it to develop the extension of TOPSIS under fuzziness, and used in solving a multiple criteria selection problem for 3D printers (Gündoğdu and Kahraman 2019b). Gündoğdu and Kahraman (2019c, d) introduced the spherical fuzzy analytic hierarchy process (SF-AHP), and they applied this method to industrial robot selection and renewable energy selection. Gündoğdu and Kahraman (2019e) extended classical (VIKOR) method to spherical fuzzy VIKOR (SF-VIKOR) method and to show its applicability, this method was applied to a warehouse location selection problem. They were also extended traditional WASPAS method to spherical fuzzy WASPAS (SF-WASPAS) method and showed its application with an industrial robot selection problem (Gündoğdu and Kahraman 2019f).

Gündoğdu (2019b) also proposed some decision-making methods under spherical fuzzy environment like MULTIMOORA, and they applied this method to personnel selection problem.

Boltürk (2019) applied spherical fuzzy TOPSIS and neutrosophic TOPSIS methods and compared the results of each other. The methods are applied to an Automated Storage and Retrieval Systems technology selection problem. Gündoğdu and Kahraman (2020a, b) proposed spherical fuzzy QFD (SF-QFD) method under certainty and uncertainty including linguistic assessment. Gündoğdu and Kahraman (2020a, b) summarized the spherical fuzzy sets and used the spherical fuzzy CODAS method in the hospital location selection problem.

Jin et al. (2019a) proposed a linguistic spherical fuzzy set (LSFS), combining the notion of linguistic fuzzy set and spherical fuzzy sets. They also developed linguistic spherical fuzzy weighted averaging and geometric operators. For the validity, proposed aggregation operators of linguistic spherical fuzzy number were applied to multi-attribute group decision-making problems.

Zeng at al. (2019) adopted a new approach of covering-based spherical fuzzy rough set (CSFRS) models to hybrid spherical fuzzy sets with notions of covering the rough set and presented TOPSIS approach through CSFRS models.

Jin et al. (2019b) introduced some novel logarithmic operations of spherical fuzzy sets and proposed the spherical fuzzy entropy to find the unknown weights information of the criteria.

Ullah et al. (2018) proposed some new similarity measures in the framework of spherical fuzzy sets and T-spherical fuzzy sets including cosine similarity measures, gray similarity measures, and set-theoretical similarity measures. The new similarity measures were applied to a well-known problem of building material recognition.

Garg et al. (2018) improved interactive aggregation operators for the T-spherical fuzzy sets and applied to multi-attribute decision-making problems.

Mahmood et al. (2019) introduced the concept of spherical fuzzy set (SFS) and T-spherical fuzzy set (T-SFS) as a generalization of picture fuzzy sets. In this study, the novelty of SFS and T-SFS is shown by examples and graphical comparison with earlier established concepts. Some operations of SFSs and T-SFSs along with spherical fuzzy relations were defined, and medical diagnostics and decision-making problem were discussed in the environment of SFS and T-SFS as practical applications.

Some basic notions over a universal set U are defined as follows: Let q be a positive real number, a T-spherical fuzzy sets (SFS) $\tilde{A}_S$ of the universe of discourse U, is given by;

$$\tilde{A}_S = \left\{ \left\langle u, (\mu_{\tilde{A}_S}(u), v_{\tilde{A}_S}(u), \pi_{\tilde{A}_S}(u)) \middle| u \in U \right\} \right. \tag{6}$$

where

$$\mu_{\tilde{A}_S}(u) : U \to [0, 1], v_{\tilde{A}_S}(u) : U \to [0, 1], \pi_{\tilde{A}_S}(u) : U \to [0, 1]$$

and

$$0 \le \mu^q_{\tilde{A}_S}(u) + v^q_{\tilde{A}_S}(u) + \pi^q_{\tilde{A}_S}(u) \le 1 \quad \forall u \in U \tag{7}$$

For each u, the numbers $\mu_{\tilde{A}_S}(u), v_{\tilde{A}_S}(u)$ and $\pi_{\tilde{A}_S}(u)$ are the degree of *membership*, *non-membership* and *hesitancy* of u to $\tilde{A}_S$, respectively.

$\chi_{\tilde{A}_S}(u) = \sqrt{1 - \mu^q_{\tilde{A}_S}(u) - v^q_{\tilde{A}_S}(u) - \pi^q_{\tilde{A}_S}(u)}$is called as a refusal degree.

Liu et al. (2019b) extended the generalized Maclaurin symmetric mean (GMSM) operator to T-spherical fuzzy environment and proposed the T-spherical fuzzy

GMSM operator (T-SFGMSM) and the T-spherical fuzzy weighted GMSM operator (T-SFWGMSM). They solved a R&D project selection problem for Yunnan Baiyao Co., Ltd. by the proposed method, successfully.

Quek at al. (2019) developed some new operational laws for T-spherical fuzzy sets, and based on these new operations, proposed two types of Einstein aggregation operators, namely the Einstein interactive averaging aggregation operators and the Einstein interactive geometric aggregation operators under T-spherical fuzzy environment. The T-spherical fuzzy aggregation operators were then applied to a multi-attribute decision-making (MADM) problem related to the degree of pollution of five major cities in China.

Liu et al. (2020) proposed the linguistic spherical fuzzy numbers (Lt-SFNs) to suggest the public's knowledge of language valuation, then they proposed the linguistic spherical fuzzy weighted averaging (Lt-SFSWA) operator for integrating the language assessment knowledge. They also improved TODIM method and a MABAC method based on Lt-SFNs.

Ullah et al. (2019a) developed some correlation coefficients for T-spherical fuzzy sets and used these sets for clustering and multi-attribute decision-making algorithms.

Ullah et al. (2019b) enhanced T-spherical fuzzy sets to interval-valued T-spherical fuzzy sets with their aggregation operators. In this study, the advantages of using the framework of interval-valued T-spherical fuzzy were described theoretically and numerically.

Liu et al. (2019a) proposed Muirhead mean (MM) operator and power average operator, the spherical fuzzy power Muirhead mean (SFPMM) operator, weighted SFPMM operator, spherical fuzzy power dual Muirhead mean (SFPDMM) operator, weighted SFPDMM operator and discussed their anticipated properties under T-spherical fuzzy environment.

Guleria and Bajaj (2019) introduced the concept of T-spherical fuzzy graph along with the operations of product, composition, union, join, and complement. They applied T-spherical fuzzy graphs to solve the decision-making problems in the field of supply chain management and evaluation problem of service centers.

Conclusions

Ordinary fuzzy sets have been introduced by its father, L. A. Zadeh, about 60 years ago. Its aim was to capture the vagueness and impreciseness in the human thoughts. The place where the fuzzy set theory has reached in its 60th year is quite different from its origin, but it seems to be that it needs much time to arrive at the point it deserves. New extensions of fuzzy sets generally focus on the detailed and better definition of discrete or continuous membership functions for the elements belonging to a set. All these efforts try to yield a better definition of uncertainty, imprecision, and vagueness, thus a better control of uncertainty to create a similar system having human thinking style. It is clear that all these extensions will be utilized to improve the quality of the outputs of the fuzzy technologies in the future. Almost every branch of science, from engineering to social sciences, from decision sciences to computer sciences, and from physical sciences to life sciences, utilized the new theory with a great success and will in the future.

We expect new contributions and new applications in different areas of sciences using the theory of spherical fuzzy sets for further research.

Fatma Kutlu Gündoğdu
fatmakutlugundogdu@gmail.com
Cengiz Kahraman
kahramanc@itu.edu.tr

References

Ashraf S, Abdullah S (2019) Spherical aggregation operators and their application in multiattribute group decision-making. Int J Intell Syst 34(3):493–523

Ashraf S, Abdullah S, Aslam M, Qiyas M, Kutbi MA (2019a) Spherical fuzzy sets and its representation of spherical fuzzy t-norms and t-conorms. J Intell Fuzzy Sys 36(6):6089–6102

Ashraf S, Abdullah S, Mahmood T (2019b) Spherical fuzzy Dombi aggregation operators and their application in group decision making problems. J Ambient Intell Humanized Comput 1–19

Ashraf S, Abdullah S, Abdullah L (2019c) Child development influence environmental factors determined using spherical fuzzy distance measures. Math 7(8):661

Atanassov KT (1986) Intuitionistic fuzzy sets. Fuzzy Sets Syst 20(1):87–96

Atanassov KT (2016) Geometrical interpretation of the elements of the intuitionistic fuzzy objects Preprint IM-MFAIS (1989) 1–89, Sofia. Reprinted: Int J Bioautomation 20(S1):27–42

Boltürk E (2019, July) AS/RS Technology selection using spherical fuzzy TOPSIS and neutrosophic TOPSIS. In: International conference on intelligent and fuzzy systems. Springer, Cham, pp 969–976

Cuong BC, Kreinovich V (2013, December) Picture fuzzy sets-a new concept for computational intelligence problems. In: 2013 Third world congress on information and communication technologies (WICT 2013), pp 1–6. IEEE

Cuong BC (2014) Picture fuzzy sets. J Comput Sci Cybern 30(4):409–420

Garg H, Munir M, Ullah K, Mahmood T, Jan N (2018) Algorithm for T-spherical fuzzy multi-attribute decision making based on improved interactive aggregation operators. Symmetry 10(12):670

Garibaldi JM, Ozen T (2007) Uncertain fuzzy reasoning: a case study in modelling expert decision making. IEEE Trans Fuzzy Syst 15(1):16–30

Grattan-Guinness I (1976) Fuzzy membership mapped onto intervals and many-valued quantities. Zeitsehr Math. Logik und Grundlagen Math 22:149–160

Guleria A, Bajaj RK (2019) T-Spherical fuzzy graphs: operations and applications in various selection processes. Arabian J Sci Engg 1–17

Gündoğdu FK (2019a) Principals of spherical fuzzy sets. In: International conference on intelligent and fuzzy systems. Springer, Cham, pp 15–23

Gundoğdu FK (2019b) Extension of MULTIMOORA with spherical fuzzy sets. J Intell Fuzzy Syst, Preprint, 1–16. https://doi.org//10.3233/JIFS-179462

Gündoğdu FK, Kahraman C (2019a) Spherical fuzzy sets and spherical fuzzy TOPSIS method. J Intell Fuzzy Syst 36(1):337–352

Gundogdu FK, Kahraman C (2019b) A novel fuzzy TOPSIS method using emerging interval-valued spherical fuzzy sets. Eng Appl Artif Intell 85:307–323

Gündoğdu FK, Kahraman C (2019c) A novel spherical fuzzy analytic hierarchy process and its renewable energy application. Soft Comput 1–15. https://doi.org/10.1007/s00500-019-04222-w

Gündoğdu FK, Kahraman C (2019d) Spherical fuzzy analytic hierarchy process (AHP) and its application to industrial robot selection. In: International conference on intelligent and fuzzy systems. Springer, Cham, pp 988–996

Gündoğdu FK, Kahraman C (2019e) A novel VIKOR method using spherical fuzzy sets and its application to warehouse site selection. J Intell Fuzzy Syst 37(1):1197–1211

Gundoğdu FK, Kahraman C (2019f) Extension of WASPAS with spherical fuzzy sets. Informatica 30(2):269–292

Gündoğdu FK, Kahraman C (2020a) Extension of CODAS with spherical fuzzy sets. J. of Mult.-Valued Logic & Soft Comput 33:481–505

Gündoğdu FK, Kahraman C (2020b) A novel spherical fuzzy QFD method and its application to the linear delta robot technology development. Eng Appl Artif Intell 87:103348

Jahn KU (1975) Intervall wertige Mengen. Mathematische Nachrichten 68(1):115–132

Jin H, Ashraf S, Abdullah S, Qiyas M, Bano M, Zeng S (2019a) Linguistic spherical fuzzy aggregation operators and their applications in multi-attribute decision making problems. Mathematics 7(5):413

Jin Y, Ashraf S, Abdullah S (2019b) Spherical fuzzy logarithmic aggregation operators based on entropy and their application in decision support systems. Entropy 21(7):628

Kahraman C, Gündoğdu FK (2018) From 1D to 3D membership: spherical fuzzy sets. BOS/SOR2018 Conference, Warsaw, Poland

Kahraman C, Öztayşi B, Çevik Onar S (2016) A comprehensive literature review of 50 years of fuzzy set theory. Int J Comput Intell Syst 9(sup1) 3–24

Liu P, Khan Q, Mahmood T, Hassan N (2019a) T-spherical fuzzy power Muirhead mean operator based on novel operational laws and their application in multi-attribute group decision making. IEEE Access 7:22613–22632

Liu P, Zhu B, Wang P (2019b) A multi-attribute decision-making approach based on spherical fuzzy sets for yunnan baiyao's R&D project selection problem. Int J Fuzzy Syst 21(7):2168–2191

Liu P, Zhu B, Wang P, Shen M (2020) An approach based on linguistic spherical fuzzy sets for public evaluation of shared bicycles in China. Eng Appl Artif Intell 87:103295

Mahmood T, Ullah K, Khan Q, Jan N (2019) An approach toward decision-making and medical diagnosis problems using the concept of spherical fuzzy sets. Neural Comput Appl 31 (11):7041–7053

Quek SG, Selvachandran G, Munir M, Mahmood T, Ullah K, Son LH, Priyadarshini I et al (2019) Multi-attribute multi-perception decision-making based on generalized T-spherical fuzzy weighted aggregation operators on neutrosophic sets. Mathematics 7(9):780

Rafiq M, Ashraf S, Abdullah S, Mahmood T, Muhammad S (2019) The cosine similarity measures of spherical fuzzy sets and their applications in decision making. J Intell Fuzzy Syst 36 (6):6059–6073

Sambuc R (1975) Function Φ-Flous, Application a l'aide au Diagnostic en Pathologie Thyroidienne. University of Marseille

Smarandache F (1998) Neutrosophy: neutrosophic probability, set, and logic: analytic synthesis & synthetic analysis

Torra V, Narukawa Y (2009) On hesitant fuzzy sets and decision. In: 2009 International Conference on Fuzzy Systems, pp 1378–1382. IEEE

Ullah K, Mahmood T, Jan N (2018) Similarity measures for T-spherical fuzzy sets with applications in pattern recognition. Symmetry 10(6):193

Ullah K, Garg H, Mahmood T, Jan N, Ali Z (2019a) Correlation coefficients for T-spherical fuzzy sets and their applications in clustering and multi-attribute decision making. Soft Comput 1–13

Ullah K, Hassan N, Mahmood T, Jan N, Hassan M (2019b) Evaluation of investment policy based on multi-attribute decision-making using interval valued T-spherical fuzzy aggregation operators. Symmetry 11(3):357

Yager R (1986) On the theory of bags. Int Jou Gen Syst 13(1):23–37

Yager RR (2013) Pythagorean fuzzy subsets. Joint IFSA World Congress and NAFIPS Annual Meeting, pp 57–61, Edmonton, Canada

Yager RR (2016) Generalized orthopair fuzzy sets. IEEE Trans. Fuzzy Syst 25(5):1222–1230

Yager RR, Abbasov AM (2013) Membership grades, complex numbers, and decision making. Int J Intell Syst 28(5):436–452

Zadeh LA (1965) Fuzzy Sets. Inf Control 8:338–353

Zadeh LA (1975) The concept of a linguistic variable and its application to approximate reasoning. Inf Sci 8:199–249

Zeng S, Hussain A, Mahmood T, Irfan Ali M, Ashraf S, Munir M (2019) Covering-based spherical fuzzy rough set model hybrid with TOPSIS for multi-attribute decision-making. Symmetry 11(4):547

Contents

Preliminaries of Spherical Fuzzy Set Theory

Properties and Arithmetic Operations of Spherical Fuzzy Sets

Fatma Kutlu Gündoğdu and Cengiz Kahraman

Abstract After the introduction of ordinary fuzzy sets, new extensions have appeared one by one in the literature. Among these extensions, hesitant fuzzy sets are a different extension from the others with more than one membership degree for an element. Intuitionistic fuzzy sets, Pythagorean fuzzy sets, Fermatean fuzzy sets, q-rung Orthopair fuzzy sets are the members of the same class since any element in these sets is represented by a membership degree and a non-membership degree and the hesitancy depends on these degrees. Picture fuzzy sets, neutrosophic sets, and spherical fuzzy sets are the members of the same class since any element in these sets is represented by a membership degree, a non-membership degree, and a hesitancy degree assigned by independently. Spherical fuzzy sets (SFS) have been proposed by Gündoğdu and Kahraman (J Intell Fuzzy Syst 36(1):337–352, 2019a). SFS should satisfy the condition that the squared sum of membership degree and non-membership degree and hesitancy degree should be equal to or less than one. In this chapter, single-valued spherical fuzzy sets and interval-valued spherical fuzzy sets are introduced with their score and accuracy functions; arithmetic and aggregation operations such as spherical fuzzy weighted arithmetic mean operator and interval-valued spherical fuzzy geometric mean operator.

Keywords Intuitionistic fuzzy sets · Pythagorean fuzzy sets · Picture fuzzy sets · Neutrosophic sets · Spherical fuzzy sets

---

F. Kutlu Gündoğdu
Industrial Engineering Department, National Defence University, Turkish Air Force Academy, Istanbul 34149, Turkey
e-mail: fatmakutlugundogdu@gmail.com

C. Kahraman (✉)
Industrial Engineering Department, Istanbul Technical University, Besiktas, Istanbul 34367, Turkey
e-mail: kahramanc@itu.edu.tr

© Springer Nature Switzerland AG 2021

C. Kahraman and F. Kutlu Gündoğdu (eds.), *Decision Making with Spherical Fuzzy Sets*, Studies in Fuzziness and Soft Computing 392,
https://doi.org/10.1007/978-3-030-45461-6_1

1 Introduction

After the presentation of ordinary fuzzy sets by Zadeh (1965) they have been very popular in almost all branches of science. Various researchers (Zadeh 1965; Smarandache 1998; Grattan-Guinness 1976; Sambuc 1975; Zadeh 1975; Atanassov 1986; Torra 2010; Yager 1986, 2013; Garibaldi and Ozen 2007) have introduced several extensions of ordinary fuzzy sets as given in Fig. 1.

A classification of these extensions with their advantage and disadvantages is as follows:

Type-2 fuzzy sets (T2FS): Zadeh (1975) introduced the concept of type-2 fuzzy sets as an extension of the ordinary fuzzy sets. Such sets are fuzzy sets whose membership grades themselves are type-1 fuzzy sets. They are very useful in circumstances where it is difficult to determine an exact membership function for a fuzzy set. This type of

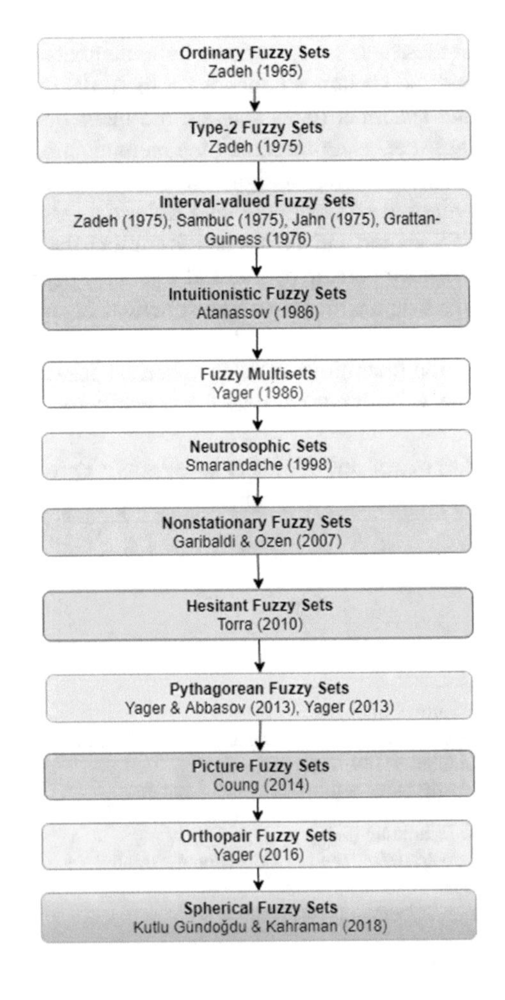

Fig. 1 Extensions of fuzzy sets (Zadeh 1965, 1975; Grattan-Guinness 1976; Sambuc 1975; Garibaldi and Ozen 2007; Jahn 1975; Atanassov 1999; Smarandache 1995; Torra 2010; Yager 1986, 2013, 2016; Cuong 2015; Kahraman and Gündoğdu 2018)

fuzzy sets requires too many parameters to be used in the problem modeling. Besides, the third dimension is ignored by many researchers for simplification purpose and these sets are called interval-valued type-2 fuzzy sets.
Intuitionistic fuzzy sets (IFS): Atanassov (1999) introduced intuitionistic fuzzy sets, which include both membership and non-membership degrees of an element to a fuzzy set as independent elements but with the constraint that their sum can be at most 1.0.
Hesitant fuzzy sets (HFS): Hesitant fuzzy sets can be used as a functional tool allowing many potential membership degrees of an element to a set. These fuzzy sets let several membership degrees of an element to be possible between zero and (Torra 2010).
Pythagorean fuzzy sets: Atanassov's intuitionistic fuzzy sets of second type or Yager's Pythagorean fuzzy sets are characterized by a membership degree and a non-membership degree satisfying the condition that the square sum of membership and non-membership degrees is at most equal to one, which is a generalization of Intuitionistic Fuzzy Sets (IFS). These sets let decision makers use a larger area for assigning membership and non-membership degrees (Yager 2013)
Neutrosophic sets (NS): Neutrosophic logic and neutrosophic sets were developed by Smarandache (1995) as an extension of intuitionistic fuzzy sets. A neutrosophic set is defined as the set where each element of the universe has a degree of truthiness, indeterminacy and falsity. They can be determined independently and their sum may be between 0 and 3.

Picture fuzzy sets were developed by Cuong (2015) Picture fuzzy sets based models may be adequate in situations when we face human opinions involving more answers of types: yes, abstain, no, and refusal. Voting can be a good example of such a situation as the human voters may be divided into four groups of those who: vote for, abstain, vote against, refusal of the voting. These sets let decision makers use a larger area for assigning membership, non-membership, and hesitancy degrees.

Spherical fuzzy sets (SFS) were introduced by Kahraman and Gündoğdu (2018) as an extension of Pythagorean, neutrosophic and picture fuzzy sets. The idea behind SFS is to let decision makers to generalize other extensions of fuzzy sets by defining a membership function on a spherical surface and independently assign the parameters of that membership function with a larger domain. A SFS must satisfy the following condition:

$$0 \le \mu_{\tilde{A}}^2(u) + \nu_{\tilde{A}}^2(u) + \pi_{\tilde{A}}^2(u) \le 1 \quad \forall u \in U \tag{1}$$

On the surface of the sphere, Eq. (1) becomes

$$\mu_{\tilde{A}}^2(u) + \nu_{\tilde{A}}^2(u) + \pi_{\tilde{A}}^2(u) = 1 \quad \forall u \in U \tag{2}$$

where U is universe of discourse; $\mu_{\tilde{A}}(u)$, $\nu_{\tilde{A}}(u)$ and $\pi_{\tilde{A}}(u)$ are the degree of membership, non-membership and hesitancy of u to the fuzzy set $\tilde{A}$, respectively. The advantage of the proposed method is that SFS theory brings together scientifically accepted aspects of Pythagorean fuzzy sets (PFS) and neutrosophic sets (NS) by excluding the criticized aspect of neutrosophic theory, i.e. a sum of $\mu,\ v,\ and\ \pi$ larger than 1 and the criticized aspect of PFS theory, i.e. disregarding an independent hesitancy (Gündoğdu and Kahraman 2019a; Gündoğdu 2019). Moreover, SFS let decision makers use a larger domain than picture fuzzy sets for assigning membership, non-membership, and hesitancy degrees.

Interval-valued spherical fuzzy sets were introduced by Gündoğdu and Kahraman (2019d) as a first time. Interval-valued fuzzy sets are employed for incorporating decision makers' opinions about the parameters of a fuzzy set into the model with an interval instead of a single point.

In this chapter, we summarize single-valued spherical fuzzy sets and interval-valued spherical fuzzy sets. Then some operations on single-valued SFS and interval-valued SFS with some properties are considered. Arithmetic operations involving addition, subtraction and multiplication are presented together with their proofs. Aggregation operators, score and accuracy functions are given with some examples.

The rest of this chapter is organized as follows. Section 2 gives the preliminaries of single-valued spherical fuzzy sets. Section 3 includes the introductory definitions and the preliminaries on interval-valued spherical fuzzy sets. Finally, the study is concluded in the last section.

2 Spherical Fuzzy Sets: Preliminaries

In spherical fuzzy sets, while the squared sum of *membership*, non-membership and hesitancy parameters can be between 0 and 1, each of them can be defined between 0 and 1 independently. Figure 2 illustrates the differences among IFS, PFS, NS and

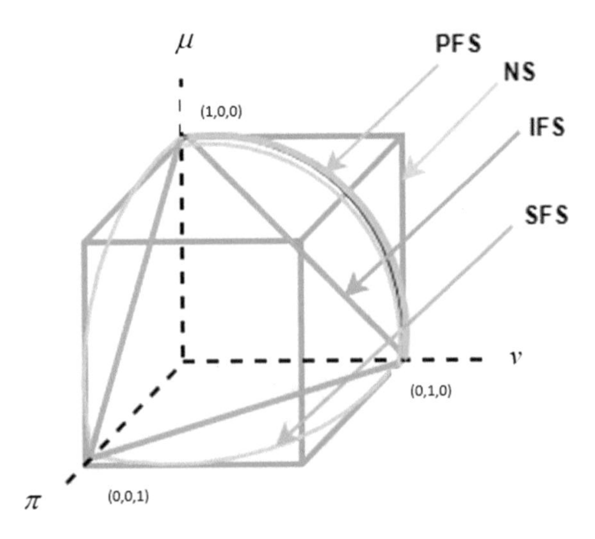

Fig. 2 Geometric representations of IFS, PFS, NS, and SFS

SFS (Gündoğdu and Kahraman 2019a, d; Gündoğdu 2019).

In this section, we give the definition of SFS and summarize spherical distance measurement, arithmetic operations, aggregation and defuzzification operations.

Definition 2.1 Spherical Fuzzy Sets (SFS) $\tilde{A}_S$ spherical fuzzy set $\tilde{A}_S$ of the universe of discourse U is given by

$$\tilde{A}_S = \left\{ \langle u, (\mu_{\tilde{A}_S}(u), \nu_{\tilde{A}_S}(u), \pi_{\tilde{A}_S}(u)) \middle| u \in U \right\} \tag{3}$$

where

$$\mu_{\tilde{A}_S}(u) : U \to [0, 1], \quad \nu_{\tilde{A}_S}(u) : U \to [0, 1], \quad \pi_{\tilde{A}_S}(u) : U \to [0, 1]$$

and

$$0 \le \mu^2_{\tilde{A}_S}(u) + \nu^2_{\tilde{A}_S}(u) + \pi^2_{\tilde{A}_S}(u) \le 1 \quad \forall u \in U \tag{4}$$

For each u, the numbers $\mu_{\tilde{A}_S}(u)$, $\nu_{\tilde{A}_S}(u)$ and $\pi_{\tilde{A}_S}(u)$ are the degree of membership, non-membership and hesitancy of u to $\tilde{A}_S$, respectively.

Definition 2.2 Geometrical representation of SFS and distances between them is illustrated in Fig. 3 (Antonov 1995; Yang and Chiclana 2009).

Spherical distance between $\tilde{A}_S$ and $\tilde{B}_S$ on the surface of a sphere;

$$dis\left(\tilde{A}_S, \tilde{B}_S\right) = \arccos\left\{1 - \frac{1}{2}\left(\left(\mu_{\tilde{A}_s} - \mu_{\tilde{B}_s}\right)^2 + \left(v_{\tilde{A}_s} - v_{\tilde{B}_s}\right)^2 + \left(\pi_{\tilde{A}_s} - \pi_{\tilde{B}_s}\right)^2\right)\right\} \tag{5}$$

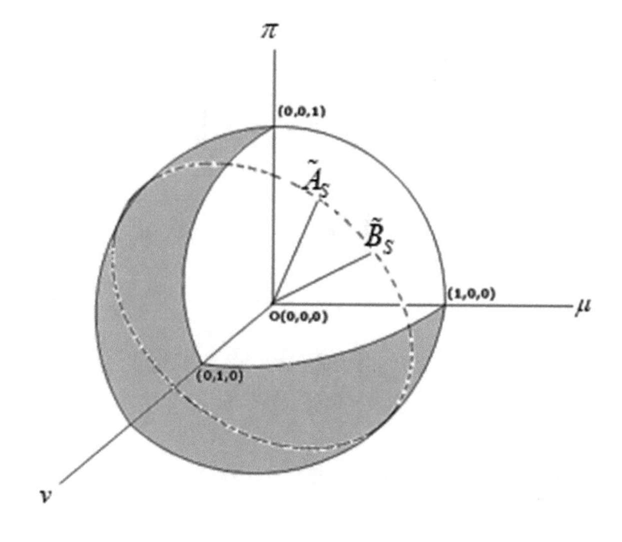

Fig. 3 Geometrical representation of spherical fuzzy sets (Gündoğdu and Kahraman 2019e)

This expression can be used to obtain the spherical distance between two spherical fuzzy sets, as follows:

$$dis\left(\tilde{A}_S, \tilde{B}_S\right) = \frac{2}{\pi}\sum_{i=1}^{n} \arccos\left\{1 - \frac{1}{2}\left(\left(\mu_{\tilde{A}_s} - \mu_{\tilde{B}_s}\right)^2 + \left(v_{\tilde{A}_s} - v_{\tilde{B}_s}\right)^2 + \left(\pi_{\tilde{A}_s} - \pi_{\tilde{B}_s}\right)^2\right)\right\} \quad (6)$$

where the factor $\frac{2}{\pi}$ is defined to get distance values in the range [0, 1] instead of $\left[0, \frac{\pi}{2}\right]$. Because $\mu^2_{\tilde{A}}(u) + \nu^2_{\tilde{A}}(u) + \pi^2_{\tilde{A}}(u) = 1$, we have that,

$$dis\left(\tilde{A}_S, \tilde{B}_S\right) = \frac{2}{\pi}\sum_{i=1}^{n} \arccos\begin{pmatrix} \mu_{\tilde{A}_S}(u_i) \cdot \mu_{\tilde{B}_S}(u_i) \\ +\nu_{\tilde{A}_S}(u_i) \cdot \nu_{\tilde{B}_S}(u_i) \\ +\pi_{\tilde{A}_S}(u_i) \cdot \pi_{\tilde{B}_S}(u_i) \end{pmatrix} \quad (7)$$

Normalized spherical distance between $\tilde{A}_S$ and $\tilde{B}_S$ on the surface of a sphere;

$$dis_n(\tilde{A}_S, \tilde{B}_S) = \frac{2}{n\pi}\sum_{i=1}^{n} \arccos\begin{pmatrix} \mu_{\tilde{A}_S}(u_i) \cdot \mu_{\tilde{B}_S}(u_i) \\ +\nu_{\tilde{A}_S}(u_i) \cdot \nu_{\tilde{B}_S}(u_i) \\ +\pi_{\tilde{A}_S}(u_i) \cdot \pi_{\tilde{B}_S}(u_i) \end{pmatrix} \quad (8)$$

Clearly, we have that $0 \leq dis(\tilde{A}_S, \tilde{B}_S) \leq n$ and $0 \leq dis_n(\tilde{A}_S, \tilde{B}_S) \leq 1$.
Some operations are defined over the Spherical Fuzzy Sets (SFS) as below.

Definition 2.3 Basic Operators

$$\tilde{A}_S \cup \tilde{B}_S = \left\{ \begin{array}{l} \max\{\mu_{\tilde{A}_S}, \mu_{\tilde{B}_S}\}, \min\{\nu_{\tilde{A}_S}, v_{\tilde{B}_S}\}, \\ \min\left\{\left(1 - \left(\left(\max\left\{\mu_{\tilde{A}_S}, \mu_{\tilde{B}_S}\right\}\right)^2 + \left(\min\left\{\nu_{\tilde{A}_S}, v_{\tilde{B}_S}\right\}\right)^2\right)\right)^{1/2}, \max\left\{\pi_{\tilde{A}_S}, \pi_{\tilde{B}_S}\right\}\right\} \end{array} \right\} \quad (9)$$

$$\tilde{A}_S \cap \tilde{B}_S = \left\{ \begin{array}{l} \min\{\mu_{\tilde{A}_S}, \mu_{\tilde{B}_S}\}, \max\{\nu_{\tilde{A}_S}, v_{\tilde{B}_S}\}, \\ \max\left\{\left(1 - \left(\left(\min\left\{\mu_{\tilde{A}_S}, \mu_{\tilde{B}_S}\right\}\right)^2 + \left(\max\left\{\nu_{\tilde{A}_S}, v_{\tilde{B}_S}\right\}\right)^2\right)\right)^{1/2}, \min\left\{\pi_{\tilde{A}_S}, \pi_{\tilde{B}_S}\right\}\right\} \end{array} \right\} \quad (10)$$

$$\tilde{A}_S \oplus \tilde{B}_S = \left\{ \begin{array}{l} \left(\mu^2_{\tilde{A}_S} + \mu^2_{\tilde{B}_S} - \mu^2_{\tilde{A}_S}\mu^2_{\tilde{B}_S}\right)^{1/2}, v_{\tilde{A}_S}v_{\tilde{B}_S}, \\ \left(\left(1 - \mu^2_{\tilde{B}_S}\right)\pi^2_{\tilde{A}_S} + \left(1 - \mu^2_{\tilde{A}_S}\right)\pi^2_{\tilde{B}_S} - \pi^2_{\tilde{A}_S}\pi^2_{\tilde{B}_S}\right)^{1/2} \end{array} \right\} \quad (11)$$

$$\tilde{A}_S \otimes \tilde{B}_S = \left\{ \begin{array}{l} \mu_{\tilde{A}_S}\mu_{\tilde{B}_S}, \left(v^2_{\tilde{A}_S} + v^2_{\tilde{B}_S} - v^2_{\tilde{A}_S}v^2_{\tilde{B}_S}\right)^{1/2}, \\ \left(\left(1 - v^2_{\tilde{B}_S}\right)\pi^2_{\tilde{A}_S} + \left(1 - v^2_{\tilde{A}_S}\right)\pi^2_{\tilde{B}_S} - \pi^2_{\tilde{A}_S}\pi^2_{\tilde{B}_S}\right)^{1/2} \end{array} \right\} \quad (12)$$

Multiplication by a scalar; $\lambda \geq 0$

$$\lambda \cdot \tilde{A}_S = \left\{ \begin{array}{l} \left(1 - \left(1 - \mu_{\tilde{A}_S}^2\right)^{\lambda}\right)^{1/2}, v_{\tilde{A}_S}^{\lambda}, \\ \left(\left(1 - \mu_{\tilde{A}_S}^2\right)^{\lambda} - \left(1 - \mu_{\tilde{A}_S}^2 - \pi_{\tilde{A}_S}^2\right)^{\lambda}\right)^{1/2} \end{array} \right\} \tag{13}$$

λ. Power of $\tilde{A}_S$; $\lambda \geq 0$

$$\tilde{A}_S^{\lambda} = \left\{ \begin{array}{l} \mu_{\tilde{A}_S}^{\lambda}, \left(1 - \left(1 - v_{\tilde{A}_S}^2\right)^{\lambda}\right)^{1/2}, \\ \left(\left(1 - v_{\tilde{A}_S}^2\right)^{\lambda} - \left(1 - v_{\tilde{A}_S}^2 - \pi_{\tilde{A}_S}^2\right)^{\lambda}\right)^{1/2} \end{array} \right\} \tag{14}$$

Definition 2.4 For these SFS $\tilde{A}_S = (\mu_{\tilde{A}_S}, v_{\tilde{A}_S}, \pi_{\tilde{A}_S})$ and $\tilde{B}_S = (\mu_{\tilde{B}_S}, v_{\tilde{B}_S}, \pi_{\tilde{B}_S})$, the followings are valid under the condition $\lambda, \lambda_1, \lambda_2 \geq 0$.

i. $\tilde{A}_S \oplus \tilde{B}_S = \tilde{B}_S \oplus \tilde{A}_S$ (15)

ii. $\tilde{A}_S \otimes \tilde{B}_S = \tilde{B}_S \otimes \tilde{A}_S$ (16)

iii. $\lambda(\tilde{A}_S \oplus \tilde{B}_S) = \lambda\tilde{A}_S \oplus \lambda\tilde{B}_S$ (17)

iv. $\lambda_1\tilde{A}_S \oplus \lambda_2\tilde{A}_S = (\lambda_1 + \lambda_2)\tilde{A}_S$ (18)

v. $(\tilde{A}_S \otimes \tilde{B}_S)^{\lambda} = \tilde{A}_S^{\lambda} \otimes \tilde{B}_S^{\lambda}$ (19)

vi. $\tilde{A}_S^{\lambda_1} \otimes \tilde{A}_S^{\lambda_2} = \tilde{A}_S^{\lambda_1+\lambda_2}$. (20)

Proof According to Definition 2.4, we will prove Eqs. (15), (16), (17), (18), (19), and (20), respectively.

i. $$\tilde{A}_S \oplus \tilde{B}_S = \left\{ \begin{array}{l} \left(\mu_{\tilde{A}_S}^2 + \mu_{\tilde{B}_S}^2 - \mu_{\tilde{A}_S}^2\mu_{\tilde{B}_S}^2\right)^{1/2}, v_{\tilde{A}_S}v_{\tilde{B}_S}, \\ \left(\left(1 - \mu_{\tilde{B}_S}^2\right)\pi_{\tilde{A}_S}^2 + \left(1 - \mu_{\tilde{A}_S}^2\right)\pi_{\tilde{B}_S}^2 - \pi_{\tilde{A}_S}^2\pi_{\tilde{B}_S}^2\right)^{1/2} \end{array} \right\}$$

$$\tilde{B}_S \oplus \tilde{A}_S = \left\{ \begin{array}{l} \left(\mu_{\tilde{B}_S}^2 + \mu_{\tilde{A}_S}^2 - \mu_{\tilde{B}_S}^2\mu_{\tilde{A}_S}^2\right)^{1/2}, v_{\tilde{B}_S}v_{\tilde{A}_S}, \\ \left(\left(1 - \mu_{\tilde{A}_S}^2\right)\pi_{\tilde{B}_S}^2 + \left(1 - \mu_{\tilde{B}_S}^2\right)\pi_{\tilde{A}_S}^2 - \pi_{\tilde{B}_S}^2\pi_{\tilde{A}_S}^2\right)^{1/2} \end{array} \right\}$$

and so, $\tilde{A}_S \oplus \tilde{B}_S = \tilde{B}_S \oplus \tilde{A}_S$.

ii. $$\tilde{A}_S \otimes \tilde{B}_S = \left\{ \begin{array}{l} \mu_{\tilde{A}_S}\mu_{\tilde{B}_S}, \left(v^2_{\tilde{A}_S} + v^2_{\tilde{B}_S} - v^2_{\tilde{A}_S}v^2_{\tilde{B}_S}\right)^{1/2}, \\ \left(\left(1 - v^2_{\tilde{B}_S}\right)\pi^2_{\tilde{A}_S} + \left(1 - v^2_{\tilde{A}_S}\right)\pi^2_{\tilde{B}_S} - \pi^2_{\tilde{A}_S}\pi^2_{\tilde{B}_S}\right)^{1/2} \end{array} \right\}$$

$$\tilde{B}_S \otimes \tilde{A}_S = \left\{ \begin{array}{l} \mu_{\tilde{B}_S}\mu_{\tilde{A}_S}, \left(v^2_{\tilde{B}_S} + v^2_{\tilde{A}_S} - v^2_{\tilde{B}_S}v^2_{\tilde{A}_S}\right)^{1/2}, \\ \left(\left(1 - v^2_{\tilde{A}_S}\right)\pi^2_{\tilde{B}_S} + \left(1 - v^2_{\tilde{B}_S}\right)\pi^2_{\tilde{A}_S} - \pi^2_{\tilde{B}_S}\pi^2_{\tilde{A}_S}\right)^{1/2} \end{array} \right\}$$

and so, $\tilde{A}_S \otimes \tilde{B}_S = \tilde{B}_S \otimes \tilde{A}_S$.

iii. $\lambda(\tilde{A}_S \oplus \tilde{B}_S)$

$$= \lambda \left(\begin{array}{l} \left(\mu^2_{\tilde{A}_S} + \mu^2_{\tilde{B}_S} - \mu^2_{\tilde{A}_S}\mu^2_{\tilde{B}_S}\right)^{1/2}, v_{\tilde{A}_S}v_{\tilde{B}_S}, \\ \left(\left(1 - \mu^2_{\tilde{B}_S}\right)\pi^2_{\tilde{A}_S} + \left(1 - \mu^2_{\tilde{A}_S}\right)\pi^2_{\tilde{B}_S} - \pi^2_{\tilde{A}_S}\pi^2_{\tilde{B}_S}\right)^{1/2} \end{array} \right)$$

$$= \left\{ \begin{array}{l} \left(1 - \left(1 - \left(\mu^2_{\tilde{A}_S} + \mu^2_{\tilde{B}_S} - \mu^2_{\tilde{A}_S}\mu^2_{\tilde{B}_S}\right)\right)^{\lambda}\right)^{1/2}, v^{\lambda}_{\tilde{A}_S}v^{\lambda}_{\tilde{B}_S}, \\ \left(\begin{array}{l} \left(1 - \left(\mu^2_{\tilde{A}_S} + \mu^2_{\tilde{B}_S} - \mu^2_{\tilde{A}_S}\mu^2_{\tilde{B}_S}\right)\right)^{\lambda} \\ -\left(\left(1 - \left(\mu^2_{\tilde{A}_S} + \mu^2_{\tilde{B}_S} - \mu^2_{\tilde{A}_S}\mu^2_{\tilde{B}_S}\right) - \left(1 - \mu^2_{\tilde{B}_S}\right)\pi^2_{\tilde{A}_S} - \left(1 - \mu^2_{\tilde{A}_S}\right)\pi^2_{\tilde{B}_S} + \pi^2_{\tilde{A}_S}\pi^2_{\tilde{B}_S}\right)\right)^{\lambda} \end{array} \right)^{1/2} \end{array} \right\}$$

$\lambda\tilde{A}_S \oplus \lambda\tilde{B}_S$

$$= \left\{ \left(1 - \left(1 - \mu^2_{\tilde{A}_S}\right)^{\lambda}\right)^{1/2}, v^{\lambda}_{\tilde{A}_S}, \left(\left(1 - \mu^2_{\tilde{A}_S}\right)^{\lambda} - \left(1 - \mu^2_{\tilde{A}_S} - \pi^2_{\tilde{A}_S}\right)^{\lambda}\right)^{1/2} \right\}$$

$$\oplus \left\{ \left(1 - \left(1 - \mu^2_{\tilde{B}_S}\right)^{\lambda}\right)^{1/2}, v^{\lambda}_{\tilde{B}_S}, \left(\left(1 - \mu^2_{\tilde{B}_S}\right)^{\lambda} - \left(1 - \mu^2_{\tilde{B}_S} - \pi^2_{\tilde{B}_S}\right)^{\lambda}\right)^{1/2} \right\}$$

$$= \left\{ \begin{array}{l} \left(1 - \left(1 - \mu^2_{\tilde{A}_S}\right)^{\lambda} + 1 - \left(1 - \mu^2_{\tilde{B}_S}\right)^{\lambda} - \left(1 - \left(1 - \mu^2_{\tilde{A}_S}\right)^{\lambda}\right)\left(1 - \left(1 - \mu^2_{\tilde{B}_S}\right)^{\lambda}\right)\right)^{1/2}, v^{\lambda}_{\tilde{A}_S}v^{\lambda}_{\tilde{B}_S}, \\ \left(\begin{array}{l} \left(1 - \left(1 - \left(1 - \mu^2_{\tilde{B}_S}\right)^{\lambda}\right)\left(\left(1 - \mu^2_{\tilde{A}_S}\right)^{\lambda} - \left(1 - \mu^2_{\tilde{A}_S} - \pi^2_{\tilde{A}_S}\right)^{\lambda}\right)\right) \\ + \left(\begin{array}{l} 1 - \left(1 - \left(1 - \mu^2_{\tilde{A}_S}\right)^{\lambda}\right)\left(\left(1 - \mu^2_{\tilde{B}_S}\right)^{\lambda} - \left(1 - \mu^2_{\tilde{B}_S} - \pi^2_{\tilde{B}_S}\right)^{\lambda}\right) \\ -\left(\left(1 - \mu^2_{\tilde{A}_S}\right)^{\lambda} - \left(1 - \mu^2_{\tilde{A}_S} - \pi^2_{\tilde{A}_S}\right)^{\lambda}\right)\left(\left(1 - \mu^2_{\tilde{B}_S}\right)^{\lambda} - \left(1 - \mu^2_{\tilde{B}_S} - \pi^2_{\tilde{B}_S}\right)^{\lambda}\right) \end{array} \right) \end{array} \right)^{1/2} \end{array} \right\}$$

$$= \left\{ \begin{array}{l} \left(1 - \left(1 - \mu^2_{\tilde{A}_S}\right)^{\lambda}\left(1 - \mu^2_{\tilde{B}_S}\right)^{\lambda}\right)^{1/2}, v^{\lambda}_{\tilde{A}_S}v^{\lambda}_{\tilde{B}_S}, \\ \left(\left(1 - \mu^2_{\tilde{A}_S}\right)^{\lambda}\left(1 - \mu^2_{\tilde{B}_S}\right)^{\lambda} - \left(\left(1 - \mu^2_{\tilde{A}_S}\right) - \pi^2_{\tilde{A}_S}\right)^{\lambda}\left(\left(1 - \mu^2_{\tilde{B}_S}\right) - \pi^2_{\tilde{B}_S}\right)^{\lambda}\right)^{1/2} \end{array} \right\}$$

$$= \left\{ \begin{array}{l} \left(1 - \left(1 - \left(\mu^2_{\tilde{A}_S} + \mu^2_{\tilde{B}_S} - \mu^2_{\tilde{A}_S}\mu^2_{\tilde{B}_S}\right)\right)^{\lambda}\right)^{1/2}, v^{\lambda}_{\tilde{A}_S}v^{\lambda}_{\tilde{B}_S}, \\ \left(\begin{array}{l} \left(1 - \left(\mu^2_{\tilde{A}_S} + \mu^2_{\tilde{B}_S} - \mu^2_{\tilde{A}_S}\mu^2_{\tilde{B}_S}\right)\right)^{\lambda} \\ -\left(\left(1 - \left(\mu^2_{\tilde{A}_S} + \mu^2_{\tilde{B}_S} - \mu^2_{\tilde{A}_S}\mu^2_{\tilde{B}_S}\right) - \left(1 - \mu^2_{\tilde{B}_S}\right)\pi^2_{\tilde{A}_S} - \left(1 - \mu^2_{\tilde{A}_S}\right)\pi^2_{\tilde{B}_S} + \pi^2_{\tilde{A}_S}\pi^2_{\tilde{B}_S}\right)\right)^{\lambda} \end{array} \right)^{1/2} \end{array} \right\}$$

and so, $\lambda(\tilde{A}_S \oplus \tilde{B}_S) = \lambda\tilde{A}_S \oplus \lambda\tilde{B}_S$.

iv.

$$\lambda_1\tilde{A}_S \oplus \lambda_2\tilde{A}_S = \left\langle \left(1-\left(1-\mu_{\tilde{A}_S}^2\right)^{\lambda_1}\right)^{1/2}, v_{\tilde{A}_S}^{\lambda_1}, \left(\left(1-\mu_{\tilde{A}_S}^2\right)^{\lambda_1} - \left(1-\mu_{\tilde{A}_S}^2 - \pi_{\tilde{A}_S}^2\right)^{\lambda_1}\right)^{1/2} \right\rangle$$

$$\oplus \left\langle \left(1-\left(1-\mu_{\tilde{A}_S}^2\right)^{\lambda_2}\right)^{1/2}, v_{\tilde{A}_S}^{\lambda_2}, \left(\left(1-\mu_{\tilde{A}_S}^2\right)^{\lambda_2} - \left(1-\mu_{\tilde{A}_S}^2 - \pi_{\tilde{A}_S}^2\right)^{\lambda_2}\right)^{1/2} \right\rangle$$

$$= \left\langle \begin{array}{l} \left(1-\left(1-\mu_{\tilde{A}_S}^2\right)^{\lambda_1} + 1 - \left(1-\mu_{\tilde{A}_S}^2\right)^{\lambda_2} - \left(1-\left(1-\mu_{\tilde{A}_S}^2\right)^{\lambda_1}\right)\left(1-\left(1-\mu_{\tilde{A}_S}^2\right)^{\lambda_2}\right)\right)^{1/2}, v_{\tilde{A}_S}^{\lambda_1+\lambda_2}, \\ \left(\begin{array}{l} \left(1-\mu_{\tilde{A}_S}^2\right)^{\lambda_2}\left(\left(1-\mu_{\tilde{A}_S}^2\right)^{\lambda_1} - \left(1-\mu_{\tilde{A}_S}^2 - \pi_{\tilde{A}_S}^2\right)^{\lambda_1}\right) \\ + \left(1-\mu_{\tilde{A}_S}^2\right)^{\lambda_1}\left(\left(1-\mu_{\tilde{A}_S}^2\right)^{\lambda_2} - \left(1-\mu_{\tilde{A}_S}^2 - \pi_{\tilde{A}_S}^2\right)^{\lambda_2}\right) \\ - \left(\left(1-\mu_{\tilde{A}_S}^2\right)^{\lambda_1} - \left(1-\mu_{\tilde{A}_S}^2 - \pi_{\tilde{A}_S}^2\right)^{\lambda_1}\right)\left(\left(1-\mu_{\tilde{A}_S}^2\right)^{\lambda_2} - \left(1-\mu_{\tilde{A}_S}^2 - \pi_{\tilde{A}_S}^2\right)^{\lambda_2}\right) \end{array} \right)^{1/2} \end{array} \right\rangle$$

$$\lambda_1\tilde{A}_S \oplus \lambda_2\tilde{A}_S$$
$$= \left\langle \left(1-\left(1-\mu_{\tilde{A}_S}^2\right)^{\lambda_1+\lambda_2}\right)^{1/2}, v_{\tilde{A}_S}^{\lambda_1+\lambda_2}, \left(\left(1-\mu_{\tilde{A}_S}^2\right)^{\lambda_1+\lambda_2} - \left(1-\mu_{\tilde{A}_S}^2 - \pi_{\tilde{A}_S}^2\right)^{\lambda_1+\lambda_2}\right)^{1/2} \right\rangle$$

$$(\lambda_1+\lambda_2)\tilde{A}_S$$
$$= \left\langle \left(1-\left(1-\mu_{\tilde{A}_S}^2\right)^{\lambda_1+\lambda_2}\right)^{1/2}, v_{\tilde{A}_S}^{\lambda_1+\lambda_2}, \left(\left(1-\mu_{\tilde{A}_S}^2\right)^{\lambda_1+\lambda_2} - \left(1-\mu_{\tilde{A}_S}^2 - \pi_{\tilde{A}_S}^2\right)^{\lambda_1+\lambda_2}\right)^{1/2} \right\rangle$$

and so, $(\lambda_1+\lambda_2)\tilde{A}_S = \lambda_1\tilde{A}_S \oplus \lambda_2\tilde{A}_S$.

v. $(\tilde{A}_S \otimes \tilde{B}_S)^\lambda$

$$= \left(\begin{array}{l} \mu_{\tilde{A}_S}\mu_{\tilde{B}_S}, \left(v_{\tilde{A}_S}^2 + v_{\tilde{B}_S}^2 - v_{\tilde{A}_S}^2 v_{\tilde{B}_S}^2\right)^{1/2}, \\ \left(\left(1-v_{\tilde{B}_S}^2\right)\pi_{\tilde{A}_S}^2 + \left(1-v_{\tilde{A}_S}^2\right)\pi_{\tilde{B}_S}^2 - \pi_{\tilde{A}_S}^2\pi_{\tilde{B}_S}^2\right)^{1/2} \end{array} \right)^\lambda$$

$$= \left\{ \begin{array}{l} \mu_{\tilde{A}_S}^\lambda \mu_{\tilde{B}_S}^\lambda, \left(1-\left(1-\left(v_{\tilde{A}_S}^2 + v_{\tilde{B}_S}^2 - v_{\tilde{A}_S}^2 v_{\tilde{B}_S}^2\right)\right)^\lambda\right)^{1/2}, \\ \left(\begin{array}{l} \left(1-\left(v_{\tilde{A}_S}^2 + v_{\tilde{B}_S}^2 - v_{\tilde{A}_S}^2 v_{\tilde{B}_S}^2\right)\right)^\lambda \\ -\left(\left(1-\left(v_{\tilde{A}_S}^2 + v_{\tilde{B}_S}^2 - v_{\tilde{A}_S}^2 v_{\tilde{B}_S}^2\right) - \left(1-v_{\tilde{B}_S}^2\right)\pi_{\tilde{A}_S}^2 - \left(1-v_{\tilde{A}_S}^2\right)\pi_{\tilde{B}_S}^2 + \pi_{\tilde{A}_S}^2\pi_{\tilde{B}_S}^2\right)\right)^\lambda \end{array} \right)^{1/2} \end{array} \right\}$$

$$\tilde{A}_S^\lambda \otimes \tilde{B}_S^\lambda$$
$$= \left\{ \mu_{\tilde{A}_S}^\lambda, \left(1-\left(1-v_{\tilde{A}_S}^2\right)^\lambda\right)^{1/2}, \left(\left(1-v_{\tilde{A}_S}^2\right)^\lambda - \left(1-v_{\tilde{A}_S}^2 - \pi_{\tilde{A}_S}^2\right)^\lambda\right)^{1/2} \right\}$$
$$\otimes \left\{ \mu_{\tilde{B}_S}^\lambda, \left(1-\left(1-v_{\tilde{B}_S}^2\right)^\lambda\right)^{1/2}, \left(\left(1-v_{\tilde{B}_S}^2\right)^\lambda - \left(1-v_{\tilde{B}_S}^2 - \pi_{\tilde{B}_S}^2\right)^\lambda\right)^{1/2} \right\}$$

$$= \left\{ \begin{array}{l} \mu_{\tilde{A}_S}^\lambda \mu_{\tilde{B}_S}^\lambda, \left(1-\left(1-v_{\tilde{A}_S}^2\right)^\lambda + 1 - \left(1-v_{\tilde{B}_S}^2\right)^\lambda - \left(1-\left(1-v_{\tilde{A}_S}^2\right)^\lambda\right)\left(1-\left(1-v_{\tilde{B}_S}^2\right)^\lambda\right)\right)^{1/2}, \\ \left(\begin{array}{l} \left(1-\left(1-\left(1-v_{\tilde{B}_S}^2\right)^\lambda\right)\right)\left(\left(1-v_{\tilde{A}_S}^2\right)^\lambda - \left(1-v_{\tilde{A}_S}^2 - \pi_{\tilde{A}_S}^2\right)^\lambda\right) \\ + \left(\begin{array}{l} 1-\left(1-\left(1-v_{\tilde{A}_S}^2\right)^\lambda\right)\left(\left(1-v_{\tilde{B}_S}^2\right)^\lambda - \left(1-v_{\tilde{B}_S}^2 - \pi_{\tilde{B}_S}^2\right)^\lambda\right) \\ -\left(\left(1-v_{\tilde{A}_S}^2\right)^\lambda - \left(1-v_{\tilde{A}_S}^2 - \pi_{\tilde{A}_S}^2\right)^\lambda\right)\left(\left(1-v_{\tilde{B}_S}^2\right)^\lambda - \left(1-v_{\tilde{B}_S}^2 - \pi_{\tilde{B}_S}^2\right)^\lambda\right) \end{array} \right) \end{array} \right)^{1/2} \end{array} \right\}$$

$$= \left\{ \begin{array}{l} \mu_{\tilde{A}_S}^{\lambda}\mu_{\tilde{B}_S}^{\lambda}, \left(1-\left(1-v_{\tilde{A}_S}^2\right)^{\lambda}\left(1-v_{\tilde{B}_S}^2\right)^{\lambda}\right)^{1/2}, \\ \left(\left(1-v_{\tilde{A}_S}^2\right)^{\lambda}\left(1-v_{\tilde{B}_S}^2\right)^{\lambda}-\left(\left(1-v_{\tilde{A}_S}^2\right)-\pi_{\tilde{A}_S}^2\right)^{\lambda}\left(\left(1-v_{\tilde{B}_S}^2\right)-\pi_{\tilde{B}_S}^2\right)^{\lambda}\right)^{1/2} \end{array} \right\}$$

$$= \left\{ \begin{array}{l} \mu_{\tilde{A}_S}^{\lambda}\mu_{\tilde{B}_S}^{\lambda}, \left(1-\left(1-\left(v_{\tilde{A}_S}^2+v_{\tilde{B}_S}^2-v_{\tilde{A}_S}^2v_{\tilde{B}_S}^2\right)\right)^{\lambda}\right)^{1/2}, \\ \left(\begin{array}{l}\left(1-\left(v_{\tilde{A}_S}^2+v_{\tilde{B}_S}^2-v_{\tilde{A}_S}^2v_{\tilde{B}_S}^2\right)\right)^{\lambda} \\ -\left(\left(1-\left(v_{\tilde{A}_S}^2+v_{\tilde{B}_S}^2-v_{\tilde{A}_S}^2v_{\tilde{B}_S}^2\right)-\left(1-v_{\tilde{B}_S}^2\right)\pi_{\tilde{A}_S}^2-\left(1-v_{\tilde{A}_S}^2\right)\pi_{\tilde{B}_S}^2+\pi_{\tilde{A}_S}^2\pi_{\tilde{B}_S}^2\right)\right)^{\lambda}\end{array}\right)^{1/2} \end{array} \right\}$$

and so, $(\tilde{A}_S \otimes \tilde{B}_S)^{\lambda} = \tilde{A}_S^{\lambda} \otimes \tilde{B}_S^{\lambda}$.

vi.

$$\tilde{A}_S^{\lambda_1} \otimes \tilde{A}_S^{\lambda_2} = \left\langle \mu_{\tilde{A}_S}^{\lambda_1}, \left(1-\left(1-v_{\tilde{A}_S}^2\right)^{\lambda_1}\right)^{1/2}, \left(\left(1-v_{\tilde{A}_S}^2\right)^{\lambda_1}-\left(1-v_{\tilde{A}_S}^2-\pi_{\tilde{A}_S}^2\right)^{\lambda_1}\right)^{1/2} \right\rangle$$
$$\otimes \left\langle \mu_{\tilde{A}_S}^{\lambda_2}, \left(1-\left(1-v_{\tilde{A}_S}^2\right)^{\lambda_2}\right)^{1/2}, \left(\left(1-v_{\tilde{A}_S}^2\right)^{\lambda_2}-\left(1-v_{\tilde{A}_S}^2-\pi_{\tilde{A}_S}^2\right)^{\lambda_2}\right)^{1/2} \right\rangle$$
$$= \left\langle \begin{array}{l} \mu_{\tilde{A}_S}^{\lambda_1+\lambda_2}, \left(1-\left(1-v_{\tilde{A}_S}^2\right)^{\lambda_1}+1-\left(1-v_{\tilde{A}_S}^2\right)^{\lambda_2}-\left(1-\left(1-v_{\tilde{A}_S}^2\right)^{\lambda_1}\right)\left(1-\left(1-v_{\tilde{A}_S}^2\right)^{\lambda_2}\right)\right)^{1/2} \\ \left(\begin{array}{l} \left(1-v_{\tilde{A}_S}^2\right)^{\lambda_2}\left(\left(1-v_{\tilde{A}_S}^2\right)^{\lambda_1}-\left(1-v_{\tilde{A}_S}^2-\pi_{\tilde{A}_S}^2\right)^{\lambda_1}\right) \\ +\left(1-v_{\tilde{A}_S}^2\right)^{\lambda_1}\left(\left(1-v_{\tilde{A}_S}^2\right)^{\lambda_2}-\left(1-v_{\tilde{A}_S}^2-\pi_{\tilde{A}_S}^2\right)^{\lambda_2}\right) \\ -\left(\left(1-v_{\tilde{A}_S}^2\right)^{\lambda_1}-\left(1-v_{\tilde{A}_S}^2-\pi_{\tilde{A}_S}^2\right)^{\lambda_1}\right)\left(\left(1-v_{\tilde{A}_S}^2\right)^{\lambda_2}-\left(1-v_{\tilde{A}_S}^2-\pi_{\tilde{A}_S}^2\right)^{\lambda_2}\right) \end{array}\right)^{1/2} \end{array} \right\rangle$$

$$\tilde{A}_S^{\lambda_1} \otimes \tilde{A}_S^{\lambda_2} = \left\langle \mu_{\tilde{A}_S}^{\lambda_1+\lambda_2}, \left(1-\left(1-v_{\tilde{A}_S}^2\right)^{\lambda_1+\lambda_2}\right)^{1/2}, \left(\left(1-v_{\tilde{A}_S}^2\right)^{\lambda_1+\lambda_2}-\left(1-v_{\tilde{A}_S}^2-\pi_{\tilde{A}_S}^2\right)^{\lambda_1+\lambda_2}\right)^{1/2} \right\rangle$$

$$\tilde{A}_S^{\lambda_1+\lambda_2} = \left\langle \mu_{\tilde{A}_S}^{\lambda_1+\lambda_2}, \left(1-\left(1-v_{\tilde{A}_S}^2\right)^{\lambda_1+\lambda_2}\right)^{1/2}, \left(\left(1-v_{\tilde{A}_S}^2\right)^{\lambda_1+\lambda_2}-\left(1-v_{\tilde{A}_S}^2-\pi_{\tilde{A}_S}^2\right)^{\lambda_1+\lambda_2}\right)^{1/2} \right\rangle$$

and so, $\tilde{A}_S^{\lambda_1} \otimes \tilde{A}_S^{\lambda_2} = \tilde{A}_S^{\lambda_1+\lambda_2}$.

Definition 2.5 Spherical Weighted Arithmetic Mean (SWAM) with respect to, $w = (w_1, w_2, \ldots, w_n)$; $w_i \in [0, 1]$; $\sum_{i=1}^{n} w_i = 1$, SWAM is defined as (Gündoğdu and Kahraman 2019f, g);

$$\begin{aligned} & SWAM_w(A_{S1}, \ldots, A_{Sn}) \\ & = w_1 A_{S1} + w_2 A_{S2} + \cdots + w_n A_{Sn} \\ & = \left\{ \begin{array}{l} \left[1-\prod_{i=1}^{n}(1-\mu_{A_{Si}}^2)^{w_i}\right]^{1/2}, \\ \prod_{i=1}^{n} v_{A_{Si}}^{w_i}, \left[\prod_{i=1}^{n}(1-\mu_{A_{Si}}^2)^{w_i}-\prod_{i=1}^{n}(1-\mu_{A_{Si}}^2-\pi_{A_{Si}}^2)^{w_i}\right]^{1/2} \end{array} \right\} \end{aligned} \quad (21)$$

Definition 2.6 Spherical Weighted Geometric Mean (SWGM) with respect to, $w = (w_1, w_2, \ldots, w_n)$; $w_i \in [0, 1]$; $\sum_{i=1}^{n} w_i = 1$, SWGM is defined as;

$$SWGM_w(A_1, \ldots, A_n) = A_{S1}^{w_1} + A_{S2}^{w_2} + \cdots + A_{Sn}^{w_n}$$
$$= \left\{ \begin{array}{l} \prod_{i=1}^{n} \mu_{A_{Si}}^{w_i}, \left[1 - \prod_{i=1}^{n} (1 - v_{A_{Si}}^2)^{w_i}\right]^{1/2}, \\ \left[\prod_{i=1}^{n} (1 - v_{A_{Si}}^2)^{w_i} - \prod_{i=1}^{n} (1 - v_{A_{Si}}^2 - \pi_{A_{Si}}^2)^{w_i}\right]^{1/2} \end{array} \right\} \tag{22}$$

Definition 2.7 Score functions and Accuracy function of sorting SFS are defined by Gündoğdu and Kahraman (2019d, h);

$$Score\left(\tilde{A}_S\right) = \left(\mu_{\tilde{A}_S} - \pi_{\tilde{A}_S}\right)^2 - \left(v_{\tilde{A}_S} - \pi_{\tilde{A}_S}\right)^2 \tag{23}$$

$$Accuracy\left(\tilde{A}_S\right) = \mu_{\tilde{A}_S}^2 + v_{\tilde{A}_S}^2 + \pi_{\tilde{A}_S}^2 \tag{24}$$

Note that: $\tilde{A}_S < \tilde{B}_S$ if and only if

i. $Score(\tilde{A}_S) < Score(\tilde{B}_S)$ or

ii. $Score(\tilde{A}_S) = Score(\tilde{B}_S)$ and $Accuracy(\tilde{A}_S) < Accuracy(\tilde{B}_S)$.

3 Interval-Valued Spherical Fuzzy Sets: Preliminaries

In this section, we give the definition of Interval-valued spherical fuzzy sets (IVSFS) and summarize distance measurement, arithmetic operations, and aggregation and defuzzfication operations with their proofs (Gündoğdu and Kahraman 2019d).

Definition 3.1 An Interval-Valued Spherical Fuzzy Set $\tilde{A}_S$ of the universe of discourse U is defined as in Eq. (25).

$$\tilde{A}_S = \left\{\left\langle u, \left(\left[\mu_{\tilde{A}_S}^L(u), \mu_{\tilde{A}_S}^U(u)\right], \left[v_{\tilde{A}_S}^L(u), v_{\tilde{A}_S}^U(u)\right], \left[\pi_{\tilde{A}_S}^L(u), \pi_{\tilde{A}_S}^U(u)\right]\right) \right| u \in U\right\} \tag{25}$$

where $0 \le \mu_{\tilde{A}_S}^L(u) \le \mu_{\tilde{A}_S}^U(u) \le 1$, $0 \le v_{\tilde{A}_S}^L(u) \le v_{\tilde{A}_S}^U(u) \le 1$ and $0 \le \left(\mu_{\tilde{A}_S}^U(u)\right)^2 + \left(v_{\tilde{A}_S}^U(u)\right)^2 + \left(\pi_{\tilde{A}_S}^U(u)\right)^2 \le 1$. For each $u \in U$, $\mu_{\tilde{A}_S}^U(u)$, $v_{\tilde{A}_S}^U(u)\, and\, \pi_{\tilde{A}_S}^U(u)$ are the upper degrees of membership, non-membership and hesitancy of u to $\tilde{A}_S$, respectively. For each $u \in U$, if $\mu^L_{\tilde{A}_S}(u) = \mu^U_{\tilde{A}_S}(u)$, $v^L_{\tilde{A}_S}(u) = v^U_{\tilde{A}_S}(u)$, and $\pi^L_{\tilde{A}_S}(u) = \pi^U_{\tilde{A}_S}(u)$ then, IVSFS $\tilde{A}_S$ reduces to a single valued SFS.

For an IVSF $\tilde{A}_S$, the pair $\left\langle \left[\mu^L_{\tilde{A}_S}(u), \mu^U_{\tilde{A}_S}(u)\right], \left[v^L_{\tilde{A}_S}(u), v^U_{\tilde{A}_S}(u)\right], \left[\pi^L_{\tilde{A}_S}(u), \pi^U_{\tilde{A}_S}(u)\right]\right\rangle$ is called an interval-valued spherical fuzzy number. For convenience, the pair $\left\langle \left[\mu^L_{\tilde{A}_S}(u), \mu^U_{\tilde{A}_S}(u)\right], \left[v^L_{\tilde{A}_S}(u), v^U_{\tilde{A}_S}(u)\right], \left[\pi^L_{\tilde{A}_S}(u), \pi^U_{\tilde{A}_S}(u)\right]\right\rangle$ is denoted by $\tilde{\alpha} = \langle [a, b], [c, d], [e, f]\rangle$ where $[a, b] \subset [0, 1]$, $[c, d] \subset [0, 1]$, $[e, f] \subset [0, 1]$ and $b^2 + d^2 + f^2 \leq 1$.

Obviously, $\tilde{\alpha}^* = \langle [1, 1], [0, 0], [0, 0]\rangle$ is the largest IVSFS, $\alpha^- = \langle [0, 0], [1, 1], [0, 0]\rangle$ is the smallest IVSFS, and $\tilde{\alpha}^{*/-} = \langle [0, 0], [0, 0], [1, 1]\rangle$ is between largest and smallest IVSFS number.

Some operations are defined over the Interval-valued spherical fuzzy sets as below.

Definition 3.2 Let $\tilde{\alpha} = \langle [a, b], [c, d], [e, f]\rangle$, $\tilde{\alpha}_1 = \langle [a_1, b_1], [c_1, d_1], [e_1, f_1]\rangle$, and $\tilde{\alpha}_2 = \langle [a_2, b_2], [c_2, d_2], [e_2, f_2]\rangle$ be IVSFS then

$$\tilde{\alpha}_1 \cup \tilde{\alpha}_2 = \{[\max\{a_1, a_2\}, \max\{b_1, b_2\}], [\min\{c_1, c_2\}, \min\{d_1, d_2\}], [\min\{e_1, e_2\}, \min\{f_1, f_2\}]\} \tag{26}$$

$$\tilde{\alpha}_1 \cap \tilde{\alpha}_2 = \{[\min\{a_1, a_2\}, \min\{b_1, b_2\}], [\max\{c_1, c_2\}, \max\{d_1, d_2\}], [\min\{e_1, e_2\}, \min\{f_1, f_2\}]\} \tag{27}$$

$$\tilde{\alpha}_1 \oplus \tilde{\alpha}_2 = \left\{ \begin{array}{l} \left[\left((a_1)^2 + (a_2)^2 - (a_1)^2(a_2)^2\right)^{1/2}, \left((b_1)^2 + (b_2)^2 - (b_1)^2(b_2)^2\right)^{1/2}\right], [c_1c_2, d_1d_2], \\ \left[\begin{array}{l} \left(\left(1 - (a_2)^2\right)(e_1)^2 + \left(1 - (a_1)^2\right)(e_2)^2 - (e_1)^2(e_2)^2\right)^{1/2}, \\ \left(\left(1 - (b_2)^2\right)(f_1)^2 + \left(1 - (b_1)^2\right)(f_2)^2 - (f_1)^2(f_2)^2\right)^{1/2} \end{array}\right] \end{array} \right\} \tag{28}$$

$$\tilde{\alpha}_1 \otimes \tilde{\alpha}_2 = \left\{ \begin{array}{l} [a_1a_2, b_1b_2], \left[\left((c_1)^2 + (c_2)^2 - (c_1)^2(c_2)^2\right)^{1/2}, \left((d_1)^2 + (d_2)^2 - (d_1)^2(d_2)^2\right)^{1/2}\right], \\ \left[\begin{array}{l} \left(\left(1 - (c_2)^2\right)(e_1)^2 + \left(1 - (c_1)^2\right)(e_2)^2 - (e_1)^2(e_2)^2\right)^{1/2}, \\ \left(\left(1 - (d_2)^2\right)(f_1)^2 + \left(1 - (d_1)^2\right)(f_2)^2 - (f_1)^2(f_2)^2\right)^{1/2} \end{array}\right] \end{array} \right\} \tag{29}$$

Multiplication by a scalar; $\lambda \geq 0$

$$\lambda \cdot \tilde{\alpha} = \left\{ \begin{array}{l} \left[\left(1 - \left(1 - a^2\right)^\lambda\right)^{1/2}, \left(1 - \left(1 - b^2\right)^\lambda\right)^{1/2}\right], \left[c^\lambda, d^\lambda\right], \\ \left[\left(\left(1 - a^2\right)^\lambda - \left(1 - a^2 - e^2\right)^\lambda\right)^{1/2}, \left(\left(1 - b^2\right)^\lambda - \left(1 - b^2 - f^2\right)^\lambda\right)^{1/2}\right] \end{array} \right\} \tag{30}$$

λth Power of $\tilde{\alpha}$; $\lambda \geq 0$

$$\tilde{\alpha}^{\lambda} = \left\{ \begin{array}{l} [a^{\lambda}, b^{\lambda}], \left[\left(1 - (1 - c^2)^{\lambda}\right)^{1/2}, \left(1 - (1 - d^2)^{\lambda}\right)^{1/2} \right], \\ \left[\left((1 - c^2)^{\lambda} - (1 - c^2 - e^2)^{\lambda}\right)^{1/2}, \left((1 - d^2)^{\lambda} - (1 - d^2 - f^2)^{\lambda}\right)^{1/2} \right] \end{array} \right\} \quad (31)$$

Remark 3.2 In the following, let us look at $\lambda \cdot \tilde{\alpha}$ and $\tilde{\alpha}^{\lambda}$ for some special cases of λ and $\tilde{\alpha}$.

(a) If $\tilde{\alpha} = \langle [a, b], [c, d], [e, f] \rangle = \langle [1, 1], [0, 0], [0, 0] \rangle$ then

$$\lambda \cdot \tilde{\alpha} = \left\{ \begin{array}{l} \left[\left(1 - \left(1 - a^2\right)^{\lambda}\right)^{1/2}, \left(1 - \left(1 - b^2\right)^{\lambda}\right)^{1/2} \right], \left[c^{\lambda}, d^{\lambda}\right], \\ \left[\left(\left(1 - a^2\right)^{\lambda} - \left(1 - a^2 - e^2\right)^{\lambda}\right)^{1/2}, \left(\left(1 - b^2\right)^{\lambda} - \left(1 - b^2 - f^2\right)^{\lambda}\right)^{1/2} \right] \end{array} \right\}$$
$$= \langle [1, 1], [0, 0], [0, 0] \rangle$$

$$\tilde{\alpha}^{\lambda} = \left\{ \begin{array}{l} \left[a^{\lambda}, b^{\lambda}\right], \left[\left(1 - \left(1 - c^2\right)^{\lambda}\right)^{1/2}, \left(1 - \left(1 - d^2\right)^{\lambda}\right)^{1/2} \right], \\ \left[\left(\left(1 - c^2\right)^{\lambda} - \left(1 - c^2 - e^2\right)^{\lambda}\right)^{1/2}, \left(\left(1 - d^2\right)^{\lambda} - \left(1 - d^2 - f^2\right)^{\lambda}\right)^{1/2} \right] \end{array} \right\}$$
$$= \langle [0, 0], [1, 1], [0, 0] \rangle$$

i.e.

$$\lambda \cdot \langle [1, 1], [0, 0], [0, 0] \rangle = \langle [1, 1], [0, 0], [0, 0] \rangle$$
$$(\langle [1, 1], [0, 0], [0, 0] \rangle)^{\lambda} = \langle [1, 1], [0, 0], [0, 0] \rangle$$

(b) If $\tilde{\alpha} = \langle [a, b], [c, d], [e, f] \rangle = \langle [0, 0], [1, 1], [0, 0] \rangle$ then

$$\lambda \cdot \tilde{\alpha} = \left\{ \begin{array}{l} \left[\left(1 - \left(1 - a^2\right)^{\lambda}\right)^{1/2}, \left(1 - \left(1 - b^2\right)^{\lambda}\right)^{1/2} \right], \left[c^{\lambda}, d^{\lambda}\right], \\ \left[\left(\left(1 - a^2\right)^{\lambda} - \left(1 - a^2 - e^2\right)^{\lambda}\right)^{1/2}, \left(\left(1 - b^2\right)^{\lambda} - \left(1 - b^2 - f^2\right)^{\lambda}\right)^{1/2} \right] \end{array} \right\}$$
$$= \langle [0, 0], [1, 1], [0, 0] \rangle$$

$$\tilde{\alpha}^{\lambda} = \left\{ \begin{array}{l} \left[a^{\lambda}, b^{\lambda}\right], \left[\left(1 - \left(1 - c^2\right)^{\lambda}\right)^{1/2}, \left(1 - \left(1 - d^2\right)^{\lambda}\right)^{1/2} \right], \\ \left[\left(\left(1 - c^2\right)^{\lambda} - \left(1 - c^2 - e^2\right)^{\lambda}\right)^{1/2}, \left(\left(1 - d^2\right)^{\lambda} - \left(1 - d^2 - f^2\right)^{\lambda}\right)^{1/2} \right] \end{array} \right\}$$
$$= \langle [0, 0], [1, 1], [0, 0] \rangle$$

i.e.

$$\lambda \cdot \langle [0,0],[1,1],[0,0]\rangle = \langle [0,0],[1,1],[0,0]\rangle$$
$$(\langle [0,0],[1,1],[0,0]\rangle)^{\lambda} = \langle [0,0],[1,1],[0,0]\rangle$$

(c) If $\tilde{\alpha} = \langle [a,b],[c,d],[e,f]\rangle = \langle [0,0],[0,0],[1,1]\rangle$ then

$$\lambda \cdot \tilde{\alpha} = \left\{ \begin{array}{l} \left[\left(1-\left(1-a^2\right)^{\lambda}\right)^{1/2}, \left(1-\left(1-b^2\right)^{\lambda}\right)^{1/2} \right], \left[c^{\lambda}, d^{\lambda}\right], \\ \left[\left(\left(1-a^2\right)^{\lambda} - \left(1-a^2-e^2\right)^{\lambda}\right)^{1/2}, \left(\left(1-b^2\right)^{\lambda} - \left(1-b^2-f^2\right)^{\lambda}\right)^{1/2} \right] \end{array} \right\}$$
$$= \langle [0,0],[0,0],[1,1]\rangle$$

$$\tilde{\alpha}^{\lambda} = \left\{ \begin{array}{l} \left[a^{\lambda}, b^{\lambda}\right], \left[\left(1-\left(1-c^2\right)^{\lambda}\right)^{1/2}, \left(1-\left(1-d^2\right)^{\lambda}\right)^{1/2} \right], \\ \left[\left(\left(1-c^2\right)^{\lambda} - \left(1-c^2-e^2\right)^{\lambda}\right)^{1/2}, \left(\left(1-d^2\right)^{\lambda} - \left(1-d^2-f^2\right)^{\lambda}\right)^{1/2} \right] \end{array} \right\}$$
$$= \langle [0,0],[0,0],[1,1]\rangle$$

i.e.

$$\lambda \cdot \langle [0,0],[0,0],[1,1]\rangle = \langle [0,0],[0,0],[1,1]\rangle$$
$$(\langle [0,0],[0,0],[1,1]\rangle)^{\lambda} = \langle [0,0],[0,0],[1,1]\rangle$$

(d) If $\lambda \to 0$ and $0 < a,b,c,d,e,f < 1$ then

$$\lambda \cdot \tilde{\alpha} = \left\{ \begin{array}{l} \left[\left(1-\left(1-a^2\right)^{\lambda}\right)^{1/2}, \left(1-\left(1-b^2\right)^{\lambda}\right)^{1/2} \right], \left[c^{\lambda}, d^{\lambda}\right], \\ \left[\left(\left(1-a^2\right)^{\lambda} - \left(1-a^2-e^2\right)^{\lambda}\right)^{1/2}, \left(\left(1-b^2\right)^{\lambda} - \left(1-b^2-f^2\right)^{\lambda}\right)^{1/2} \right] \end{array} \right\}$$
$$\to \langle [0,0],[1,1],[0,0]\rangle$$

$$\tilde{\alpha}^{\lambda} = \left\{ \begin{array}{l} \left[a^{\lambda}, b^{\lambda}\right], \left[\left(1-\left(1-c^2\right)^{\lambda}\right)^{1/2}, \left(1-\left(1-d^2\right)^{\lambda}\right)^{1/2} \right], \\ \left[\left(\left(1-c^2\right)^{\lambda} - \left(1-c^2-e^2\right)^{\lambda}\right)^{1/2}, \left(\left(1-d^2\right)^{\lambda} - \left(1-d^2-f^2\right)^{\lambda}\right)^{1/2} \right] \end{array} \right\}$$
$$\to \langle [1,1],[0,0],[0,0]\rangle$$

i.e.

$$\lambda \cdot \tilde{\alpha} \to \langle [0,0],[1,1],[0,0]\rangle,\ as\ \lambda \to 0$$
$$\tilde{\alpha}^{\lambda} \to \langle [1,1],[0,0],[0,0]\rangle,\ as\ \lambda \to 0$$

(e) If $\lambda \to +\propto$ and $0 < a,b,c,d,e,f < 1$ then

$$\lambda \cdot \tilde{\alpha} = \left\{ \begin{bmatrix} \left[\left(1-\left(1-a^2\right)^\lambda\right)^{1/2}, \left(1-\left(1-b^2\right)^\lambda\right)^{1/2}\right], \left[c^\lambda, d^\lambda\right], \\ \left[\left(\left(1-a^2\right)^\lambda - \left(1-a^2-e^2\right)^\lambda\right)^{1/2}, \left(\left(1-b^2\right)^\lambda - \left(1-b^2-f^2\right)^\lambda\right)^{1/2}\right] \end{bmatrix} \right\}$$

$$\to \langle[1, 1], [0, 0], [0, 0]\rangle$$

$$\tilde{\alpha}^\lambda = \left\{ \begin{bmatrix} \left[a^\lambda, b^\lambda\right], \left[\left(1-\left(1-c^2\right)^\lambda\right)^{1/2}, \left(1-\left(1-d^2\right)^\lambda\right)^{1/2}\right], \\ \left[\left(\left(1-c^2\right)^\lambda - \left(1-c^2-e^2\right)^\lambda\right)^{1/2}, \left(\left(1-d^2\right)^\lambda - \left(1-d^2-f^2\right)^\lambda\right)^{1/2}\right] \end{bmatrix} \right\}$$

$$\to \langle[0, 0], [1, 1], [0, 0]\rangle$$

i.e.

$$\lambda \cdot \tilde{\alpha} \to \langle[1, 1], [0, 0], [0, 0]\rangle, \quad as\, \lambda \to +\propto$$
$$\tilde{\alpha}^\lambda \to \langle[0, 0], [1, 1], [0, 0]\rangle, \quad as\, \lambda \to +\propto$$

(f) If $\lambda = 1$ then

$$\lambda \cdot \tilde{\alpha} = \left\{ \begin{bmatrix} \left[\left(1-\left(1-a^2\right)^\lambda\right)^{1/2}, \left(1-\left(1-b^2\right)^\lambda\right)^{1/2}\right], \left[c^\lambda, d^\lambda\right], \\ \left[\left(\left(1-a^2\right)^\lambda - \left(1-a^2-e^2\right)^\lambda\right)^{1/2}, \left(\left(1-b^2\right)^\lambda - \left(1-b^2-f^2\right)^\lambda\right)^{1/2}\right] \end{bmatrix} \right\}$$

$$= \langle[a, b], [c, d], [e, f]\rangle = \alpha$$

$$\tilde{\alpha}^\lambda = \left\{ \begin{bmatrix} \left[a^\lambda, b^\lambda\right], \left[\left(1-\left(1-c^2\right)^\lambda\right)^{1/2}, \left(1-\left(1-d^2\right)^\lambda\right)^{1/2}\right], \\ \left[\left(\left(1-c^2\right)^\lambda - \left(1-c^2-e^2\right)^\lambda\right)^{1/2}, \left(\left(1-d^2\right)^\lambda - \left(1-d^2-f^2\right)^\lambda\right)^{1/2}\right] \end{bmatrix} \right\}$$

$$= \langle[a, b], [c, d], [e, f]\rangle = \alpha$$

i.e.

$$\lambda \cdot \tilde{\alpha} = \tilde{\alpha}$$
$$\tilde{\alpha}^\lambda = \tilde{\alpha}$$

Definition 3.3 Let $\lambda, \lambda_1, \lambda_2 \geq 0$, then

i. $\tilde{\alpha}_1 \oplus \tilde{\alpha}_2 = \tilde{\alpha}_2 \oplus \tilde{\alpha}_1$ (32)

ii. $\tilde{\alpha}_1 \otimes \tilde{\alpha}_2 = \tilde{\alpha}_2 \otimes \tilde{\alpha}_1$ (33)

iii. $\lambda(\tilde{\alpha}_1 \oplus \tilde{\alpha}_2) = \lambda \cdot \tilde{\alpha}_1 \oplus \lambda \cdot \tilde{\alpha}_2$ (34)

iv. $(\tilde{\alpha}_1 \otimes \tilde{\alpha}_2)^{\lambda} = \tilde{\alpha}_1^{\lambda} \otimes \tilde{\alpha}_2^{\lambda}$ (35)

v. $\lambda_1 \cdot \tilde{\alpha} \oplus \lambda_2 \cdot \tilde{\alpha} = (\lambda_1 + \lambda_2) \cdot \tilde{\alpha}$ (36)

vi. $\tilde{\alpha}^{\lambda_1} \otimes \tilde{\alpha}^{\lambda_2} = \tilde{\alpha}^{\lambda_1 + \lambda_2}$ (37)

Proof

(i) By Eq. (28) of Definition 3.2, we have

$$\tilde{\alpha}_1 \oplus \tilde{\alpha}_2 = \left\{ \begin{array}{l} \left[\left((a_1)^2 + (a_2)^2 - (a_1)^2 (a_2)^2 \right)^{1/2}, \left((b_1)^2 + (b_2)^2 - (b_1)^2 (b_2)^2 \right)^{1/2} \right], [c_1 c_2, d_1 d_2], \\ \left[\begin{array}{l} \left((1 - (a_2)^2)(e_1)^2 + \left(1 - (a_1)^2\right)(e_2)^2 - (e_1)^2 (e_2)^2 \right)^{1/2}, \\ \left((1 - (b_2)^2)(f_1)^2 + \left(1 - (b_1)^2\right)(f_2)^2 - (f_1)^2 (f_2)^2 \right)^{1/2} \end{array} \right] \end{array} \right\},$$

$$= \left\{ \begin{array}{l} \left[\left((a_2)^2 + (a_1)^2 - (a_2)^2 (a_1)^2 \right)^{1/2}, \left((b_2)^2 + (b_1)^2 - (b_2)^2 (b_1)^2 \right)^{1/2} \right], [c_2 c_1, d_2 d_1], \\ \left[\begin{array}{l} \left((1 - (a_1)^2)(e_2)^2 + \left(1 - (a_2)^2\right)(e_1)^2 - (e_2)^2 (e_1)^2 \right)^{1/2}, \\ \left((1 - (b_1)^2)(f_2)^2 + \left(1 - (b_2)^2\right)(f_1)^2 - (f_2)^2 (f_1)^2 \right)^{1/2} \end{array} \right] \end{array} \right\},$$

$$= \tilde{\alpha}_2 \oplus \tilde{\alpha}_1$$

(ii) By Eqs. (28) and (30) of Definition 3.2, we have

$$\lambda(\tilde{\alpha}_1 \oplus \tilde{\alpha}_2)$$

$$= \lambda \left\{ \begin{array}{l} \left[\left(a_1^2 + a_2^2 - a_1^2 a_2^2 \right)^{1/2}, \left(b_1^2 + b_2^2 - b_1^2 b_2^2 \right)^{1/2} \right], [c_1 c_2, d_1 d_2], \\ \left[\begin{array}{l} \left(\left(1 - a_2^2\right) e_1^2 + \left(1 - a_1^2\right) e_2^2 - e_1^2 e_2^2 \right)^{1/2}, \\ \left(\left(1 - b_2^2\right) f_1^2 + \left(1 - b_1^2\right) f_2^2 - f_1^2 f_2^2 \right)^{1/2} \end{array} \right] \end{array} \right\}$$

$$= \left\{ \begin{array}{l} \left[\left(1 - \left(1 - \left(a_1^2 + a_2^2 - a_1^2 a_2^2\right)\right)^{\lambda} \right)^{1/2}, \left(1 - \left(1 - \left(b_1^2 + b_2^2 - b_1^2 b_2^2\right)\right)^{\lambda} \right)^{1/2} \right], \left[c_1^{\lambda} c_2^{\lambda}, d_1^{\lambda} d_2^{\lambda} \right], \\ \left[\begin{array}{l} \left(\left(1 - \left(a_1^2 + a_2^2 - a_1^2 a_2^2\right)\right)^{\lambda} - \left(\left(1 - \left(a_1^2 + a_2^2 - a_1^2 a_2^2\right) - \left(1 - a_2^2\right) e_1^2 - \left(1 - a_1^2\right) e_2^2 + e_1^2 e_2^2 \right) \right)^{\lambda} \right)^{1/2}, \\ \left(\left(1 - \left(a_1^2 + a_2^2 - a_1^2 a_2^2\right)\right)^{\lambda} - \left(\left(1 - \left(a_1^2 + a_2^2 - a_1^2 a_2^2\right) - \left(1 - a_2^2\right) e_1^2 - \left(1 - a_1^2\right) e_2^2 + e_1^2 e_2^2 \right) \right)^{\lambda} \right)^{1/2} \end{array} \right] \end{array} \right\}$$

$$\lambda \tilde{\alpha}_1 \oplus \lambda \tilde{\alpha}_2$$

$$= \left\{ \begin{array}{l} \left[\left(1 - \left(1 - a_1^2\right)^{\lambda} \right)^{1/2}, \left(1 - \left(1 - b_1^2\right)^{\lambda} \right)^{1/2} \right], \left[c_1^{\lambda} d_1^{\lambda} \right], \\ \left[\left(\left(1 - a_1^2\right)^{\lambda} - \left(1 - a_1^2 - e_1^2\right)^{\lambda} \right)^{1/2}, \left(\left(1 - b_1^2\right)^{\lambda} - \left(1 - b_1^2 - f_1^2\right)^{\lambda} \right)^{1/2} \right] \end{array} \right\}$$

$$\oplus \left\{ \begin{array}{l} \left[\left(1 - \left(1 - a_2^2\right)^{\lambda} \right)^{1/2}, \left(1 - \left(1 - b_2^2\right)^{\lambda} \right)^{1/2} \right], \left[c_2^{\lambda} d_2^{\lambda} \right], \\ \left[\left(\left(1 - a_2^2\right)^{\lambda} - \left(1 - a_2^2 - e_2^2\right)^{\lambda} \right)^{1/2}, \left(\left(1 - b_2^2\right)^{\lambda} - \left(1 - b_2^2 - f_2^2\right)^{\lambda} \right)^{1/2} \right] \end{array} \right\}$$

$$
= \left\{ \begin{array}{l} \left[\begin{array}{l} \left(1-\left(1-a_1^2\right)^\lambda+1-\left(1-a_2^2\right)^\lambda-\left(1-\left(1-a_1^2\right)^\lambda\right)\left(1-\left(1-a_2^2\right)^\lambda\right)\right)^{1/2}, \\ \left(1-\left(1-b_1^2\right)^\lambda+1-\left(1-b_2^2\right)^\lambda-\left(1-\left(1-b_1^2\right)^\lambda\right)\left(1-\left(1-b_2^2\right)^\lambda\right)\right)^{1/2} \end{array} \right], \left[c_1^\lambda c_2^\lambda, d_1^\lambda d_2^\lambda\right], \\ \left[\begin{array}{l} \left(\begin{array}{l} 1-\left(1-\left(1-a_2^2\right)^\lambda\right)\left(\left(1-a_1^2\right)^\lambda-\left(1-a_1^2-e_1^2\right)^\lambda\right) \\ +\left(\begin{array}{l} 1-\left(1-\left(1-a_1^2\right)^\lambda\right)\left(\left(1-a_2^2\right)^\lambda-\left(1-a_2^2-e_2^2\right)^\lambda\right) \\ -\left(\left(1-a_1^2\right)^\lambda-\left(1-a_1^2-e_1^2\right)^\lambda\right)\left(\left(1-a_2^2\right)^\lambda-\left(1-a_2^2-e_2^2\right)^\lambda\right) \end{array} \right) \end{array} \right)^{1/2}, \\ \left(\begin{array}{l} 1-\left(1-\left(1-b_2^2\right)^\lambda\right)\left(\left(1-b_1^2\right)^\lambda-\left(1-b_1^2-f_1^2\right)^\lambda\right) \\ +\left(\begin{array}{l} 1-\left(1-\left(1-b_1^2\right)^\lambda\right)\left(\left(1-b_2^2\right)^\lambda-\left(1-b_2^2-f_2^2\right)^\lambda\right) \\ -\left(\left(1-b_1^2\right)^\lambda-\left(1-b_1^2-f_1^2\right)^\lambda\right)\left(\left(1-b_2^2\right)^\lambda-\left(1-b_2^2-f_2^2\right)^\lambda\right) \end{array} \right) \end{array} \right)^{1/2} \end{array} \right] \end{array} \right\}
$$

$$
= \left\{ \begin{array}{l} \left[\left(1-\left(1-a_1^2\right)^\lambda\left(1-a_2^2\right)^\lambda\right)^{1/2}, \left(1-\left(1-a_1^2\right)^\lambda\left(1-a_2^2\right)^\lambda\right)^{1/2}\right], \left[c_1^\lambda c_2^\lambda, d_1^\lambda d_2^\lambda\right], \\ \left[\begin{array}{l} \left(\left(1-a_1^2\right)^\lambda\left(1-a_2^2\right)^\lambda-\left(\left(1-a_1^2\right)-e_1^2\right)^\lambda\left(\left(1-a_2^2\right)-e_2^2\right)^\lambda\right)^{1/2}, \\ \left(\left(1-a_1^2\right)^\lambda\left(1-a_2^2\right)^\lambda-\left(\left(1-a_1^2\right)-e_1^2\right)^\lambda\left(\left(1-a_2^2\right)-e_2^2\right)^\lambda\right)^{1/2} \end{array} \right] \end{array} \right\}
$$

$$
= \left\{ \begin{array}{l} \left[\left(1-\left(1-\left(a_1^2+a_2^2-a_1^2a_2^2\right)\right)^\lambda\right)^{1/2}, \left(1-\left(1-\left(b_1^2+b_2^2-b_1^2b_2^2\right)\right)^\lambda\right)^{1/2}\right], \left[c_1^\lambda c_2^\lambda, d_1^\lambda d_2^\lambda\right], \\ \left[\begin{array}{l} \left(\left(1-\left(a_1^2+a_2^2-a_1^2a_2^2\right)\right)^\lambda-\left(\left(1-\left(a_1^2+a_2^2-a_1^2a_2^2\right)-\left(1-a_2^2\right)e_1^2-\left(1-a_1^2\right)e_2^2+e_1^2e_2^2\right)\right)^\lambda\right)^{1/2}, \\ \left(\left(1-\left(b_1^2+b_2^2-b_1^2b_2^2\right)\right)^\lambda-\left(\left(1-\left(b_1^2+b_2^2-b_1^2b_2^2\right)-\left(1-b_2^2\right)f_1^2-\left(1-b_1^2\right)f_2^2+f_1^2f_2^2\right)\right)^\lambda\right)^{1/2} \end{array} \right] \end{array} \right\}
$$

and so, $\lambda(\tilde{\alpha}_1 \oplus \tilde{\alpha}_2) = \lambda \cdot \tilde{\alpha}_1 \oplus \lambda \cdot \tilde{\alpha}_2$

(iii) By Eqs. (32), and (34) of Definition 3.2, we have

$$
(\tilde{\alpha}_1 \otimes \tilde{\alpha}_2)^\lambda
$$

$$
= \left(\begin{array}{l} [a_1a_2, b_1b_2], \left[\left(c_1^2+c_2^2-c_1^2c_2^2\right)^{1/2}, \left(d_1^2+d_2^2-d_1^2d_2^2\right)^{1/2}\right], \\ \left[\begin{array}{l} \left((1-c_2^2)e_1^2+(1-c_1^2)e_2^2-e_1^2e_2^2\right)^{1/2}, \\ \left((1-d_2^2)f_1^2+(1-d_1^2)f_2^2-f_1^2f_2^2\right)^{1/2} \end{array} \right] \end{array} \right)^\lambda
$$

$$
= \left\{ \begin{array}{l} [a_1^\lambda a_2^\lambda, b_1^\lambda b_2^\lambda], \left[\left(1-(1-(c_1^2+c_2^2-c_1^2c_2^2))^\lambda\right)^{1/2}, \left(1-(1-(d_1^2+d_2^2-d_1^2d_2^2))^\lambda\right)^{1/2}\right], \\ \left[\begin{array}{l} \left((1-(c_1^2+c_2^2-c_1^2c_2^2))^\lambda-((1-(c_1^2+c_2^2-c_1^2c_2^2)-(1-c_2^2)e_1^2-(1-c_1^2)e_2^2+e_1^2e_2^2))^\lambda\right)^{1/2}, \\ \left((1-(d_1^2+d_2^2-d_1^2d_2^2))^\lambda-((1-(d_1^2+d_2^2-d_1^2d_2^2)-(1-d_2^2)f_1^2-(1-d_1^2)f_2^2+f_1^2f_2^2))^\lambda\right)^{1/2} \end{array} \right] \end{array} \right\}
$$

$$
\tilde{\alpha}_1^\lambda \otimes \tilde{\alpha}_2^\lambda
$$

$$
= \left\{ \begin{array}{l} [a_1^\lambda, b_1^\lambda], \left[\left(1-(1-c_1^2)^\lambda\right)^{1/2}, \left(1-(1-d_1^2)^\lambda\right)^{1/2}\right], \\ \left[\left((1-c_1^2)^\lambda-(1-c_1^2-e_1^2)^\lambda\right)^{1/2}, \left((1-d_1^2)^\lambda-(1-d_1^2-f_1^2)^\lambda\right)^{1/2}\right] \end{array} \right\}
$$

$$
\otimes \left\{ \begin{array}{l} [a_2^\lambda, b_2^\lambda], \left[\left(1-(1-c_2^2)^\lambda\right)^{1/2}, \left(1-(1-d_2^2)^\lambda\right)^{1/2}\right], \\ \left[\left((1-c_2^2)^\lambda-(1-c_2^2-e_2^2)^\lambda\right)^{1/2}, \left((1-d_2^2)^\lambda-(1-d_2^2-f_2^2)^\lambda\right)^{1/2}\right] \end{array} \right\}
$$

$$= \left\{ \begin{array}{l} [a_1^\lambda a_2^\lambda, b_1^\lambda b_2^\lambda], \left[\begin{array}{l} \left(1-(1-c_1^2)^\lambda + 1-(1-c_2^2)^\lambda - \left(1-(1-c_1^2)^\lambda\right)\left(1-(1-c_2^2)^\lambda\right)\right)^{1/2}, \\ \left(1-(1-d_1^2)^\lambda + 1-(1-d_2^2)^\lambda - \left(1-(1-d_1^2)^\lambda\right)\left(1-(1-d_2^2)^\lambda\right)\right)^{1/2} \end{array} \right], \\ \left[\begin{array}{l} \left(\begin{array}{l} 1-\left(1-(1-c_2^2)^\lambda\right)\left((1-c_1^2)^\lambda-(1-c_1^2-e_1^2)^\lambda\right) \\ +\left(\begin{array}{l} 1-\left(1-(1-c_1^2)^\lambda\right)\left((1-c_2^2)^\lambda-(1-c_2^2-e_2^2)^\lambda\right) \\ -\left((1-c_1^2)^\lambda-(1-c_1^2-e_1^2)^\lambda\right)\left((1-c_2^2)^\lambda-(1-c_2^2-e_2^2)^\lambda\right) \end{array} \right) \end{array} \right)^{1/2}, \\ \left(\begin{array}{l} 1-\left(1-(1-d_2^2)^\lambda\right)\left((1-d_1^2)^\lambda-(1-d_1^2-f_1^2)^\lambda\right) \\ +\left(\begin{array}{l} 1-\left(1-(1-d_1^2)^\lambda\right)\left((1-d_2^2)^\lambda-(1-d_2^2-f_2^2)^\lambda\right) \\ -\left((1-d_1^2)^\lambda-(1-d_1^2-f_1^2)^\lambda\right)\left((1-d_2^2)^\lambda-(1-d_2^2-f_2^2)^\lambda\right) \end{array} \right) \end{array} \right)^{1/2} \end{array} \right] \end{array} \right\}$$

$$= \left\{ \begin{array}{l} [a_1^\lambda a_2^\lambda, b_1^\lambda b_2^\lambda], \left[\left(1-(1-c_1^2)^\lambda(1-c_2^2)^\lambda\right)^{1/2}, \left(1-(1-d_1^2)^\lambda(1-d_2^2)^\lambda\right)^{1/2}\right], \\ \left[\begin{array}{l} \left((1-c_1^2)^\lambda(1-c_2^2)^\lambda - ((1-c_1^2)-e_1^2)^\lambda((1-c_2^2)-e_2^2)^\lambda\right)^{1/2}, \\ \left((1-d_1^2)^\lambda(1-d_2^2)^\lambda - ((1-d_1^2)-f_1^2)^\lambda((1-d_2^2)-f_2^2)^\lambda\right)^{1/2} \end{array} \right] \end{array} \right\}$$

$$= \left\{ \begin{array}{l} [a_1^\lambda a_2^\lambda, b_1^\lambda b_2^\lambda], \left[\left(1-(1-(c_1^2+c_2^2-c_1^2c_2^2))^\lambda\right)^{1/2}, \left(1-(1-(d_1^2+d_2^2-d_1^2d_2^2))^\lambda\right)^{1/2}\right], \\ \left[\begin{array}{l} \left((1-(c_1^2+c_2^2-c_1^2c_2^2))^\lambda - ((1-(c_1^2+c_2^2-c_1^2c_2^2)-(1-c_2^2)e_1^2-(1-c_1^2)e_2^2+e_1^2e_2^2))^\lambda\right)^{1/2}, \\ \left((1-(d_1^2+d_2^2-d_1^2d_2^2))^\lambda - ((1-(d_1^2+d_2^2-d_1^2d_2^2)-(1-d_2^2)f_1^2-(1-d_1^2)f_2^2+f_1^2f_2^2))^\lambda\right)^{1/2} \end{array} \right] \end{array} \right\}$$

and so, $(\tilde{\alpha}_1 \otimes \tilde{\alpha}_2)^\lambda = \tilde{\alpha}_1^\lambda \otimes \tilde{\alpha}_2^\lambda$.

The other proofs of equations are omitted here since their proofs are trivial.

Definition 3.4 Let $\tilde{\alpha}_j = \langle [a_j, b_j], [c_j, d_j], [e_j, f_j] \rangle$ be a collection of Interval-valued Spherical Weighted Arithmetic Mean (IVSWAM) with respect to,$w_j = (w_1, w_2, \ldots, w_n);\ w_j \in [0, 1]\, and\ \sum_{j=1}^n w_j = 1$, IVSWAM is defined as;

$$\begin{array}{l} IVSWAM_w(\tilde{\alpha}_1, \tilde{\alpha}_2, \ldots, \tilde{\alpha}_n) \\ = w_1 \cdot \tilde{\alpha}_1 \oplus w_2 \cdot \tilde{\alpha}_2 \oplus \ldots \oplus w_n \cdot \tilde{\alpha}_n \\ = \left\{ \begin{array}{l} \left[\left(1-\prod_{j=1}^n (1-a_j^2)^{w_j}\right)^{1/2}, \left(1-\prod_{j=1}^n (1-b_j^2)^{w_j}\right)^{1/2}\right], \left[\prod_{j=1}^n c_j^{w_j}, \prod_{j=1}^n d_j^{w_j}\right], \\ \left[\left(\prod_{j=1}^n (1-a_j^2)^{w_j} - \prod_{j=1}^n (1-a_j^2-e_j^2)^{w_j}\right)^{1/2}, \left(\prod_{j=1}^n (1-b_j^2)^{w_j} - \prod_{j=1}^n (1-b_j^2-f_j^2)^{w_j}\right)^{1/2}\right] \end{array} \right\} \end{array} \quad (38)$$

Proof 3.4 We can prove Eq. (38) based on mathematical induction on n: when $n = 2$,

$$IVSWAM_w(\tilde{\alpha}_1, \tilde{\alpha}_2) = w_1 \cdot \tilde{\alpha}_1 \oplus w_2 \cdot \tilde{\alpha}_2$$

According to Definition 3.2, we can see that both $w_1 \cdot \alpha_1$ and $w_2 \cdot \alpha_2$ are IVSFS, and the value of $w_1 \cdot \alpha_1 + w_2 \cdot \alpha_2$ is an IVSFS. By the operational law (v) in Definition 3.3, we have

$$w_1 \cdot \tilde{\alpha}_1 = \left\{ \begin{array}{l} \left[(1-(1-a_1^2)^{w_1})^{1/2}, (1-(1-b_1^2)^{w_1})^{1/2}\right], [c_1^{w_1}, d_1^{w_1}], \\ \left[((1-a_1^2)^{w_1}-(1-a_1^2-e_1^2)^{w_1})^{1/2}, ((1-b_1^2)^{w_1}-(1-b_1^2-f_1^2)^{w_1})^{1/2}\right] \end{array} \right\}$$

$$w_2 \cdot \tilde{\alpha}_2 = \left\{ \begin{array}{l} \left[(1-(1-a_2^2)^{w_2})^{1/2}, (1-(1-b_2^2)^{w_2})^{1/2}\right], [c_2^{w_2}, d_2^{w_2}], \\ \left[((1-a_2^2)^{w_2}-(1-a_2^2-e_2^2)^{w_2})^{1/2}, ((1-b_2^2)^{w_2}-(1-b_2^2-f_2^2)^{w_2})^{1/2}\right] \end{array} \right\}$$

$$\begin{aligned}
&w_1\tilde{\alpha}_1 \oplus w_2\tilde{\alpha}_2 \\
&= \left\{\left[\left(1-(1-a_1^2)^{w_1}\right)^{1/2}, \left(1-(1-b_1^2)^{w_1}\right)^{1/2}\right], \left[c_1^{w_1} d_1^{w_1}\right], \left[\begin{array}{l}((1-a_1^2)^{w_1}-(1-a_1^2-e_1^2)^{w_1})^{1/2}, \\ ((1-b_1^2)^{w_1}-(1-b_1^2-f_1^2)^{w_1})^{1/2}\end{array}\right]\right\} \\
&\oplus \left\{\left[\left(1-(1-a_2^2)^{w_2}\right)^{1/2}, \left(1-(1-b_2^2)^{w_2}\right)^{1/2}\right], \left[c_2^{w_2} d_2^{w_2}\right], \left[\begin{array}{l}((1-a_2^2)^{w_2}-(1-a_2^2-e_2^2)^{w_2})^{1/2}, \\ ((1-b_2^2)^{w_2}-(1-b_2^2-f_2^2)^{w_2})^{1/2}\end{array}\right]\right\} \\
&= \left\{\begin{array}{l}
\left[\begin{array}{l}(1-(1-a_1^2)^{w_1}+1-(1-a_2^2)^{w_2}-(1-(1-a_1^2)^{w_1})(1-(1-a_2^2)^{w_2}))^{1/2}, \\ (1-(1-b_1^2)^{w_1}+1-(1-b_2^2)^{w_2}-(1-(1-b_1^2)^{w_1})(1-(1-b_2^2)^{w_2}))^{1/2}\end{array}\right], \left[c_1^{w_1}c_2^{w_2}, d_1^{w_1}d_2^{w_2}\right], \\
\left[\begin{array}{l}
\left(\begin{array}{l}1-(1-(1-a_2^2)^{w_2})((1-a_1^2)^{w_1}-(1-a_1^2-e_1^2)^{w_1}) \\ +\left(\begin{array}{l}1-(1-(1-a_1^2)^{w_1})((1-a_2^2)^{w_2}-(1-a_2^2-e_2^2)^{w_2}) \\ -((1-a_1^2)^{w_1}-(1-a_1^2-e_1^2)^{w_1})((1-a_2^2)^{w_2}-(1-a_2^2-e_2^2)^{w_2})\end{array}\right)\end{array}\right)^{1/2}, \\
\left(\begin{array}{l}1-(1-(1-b_2^2)^{w_2})((1-b_1^2)^{w_1}-(1-b_1^2-f_1^2)^{w_1}) \\ +\left(\begin{array}{l}1-(1-(1-b_1^2)^{w_1})((1-b_2^2)^{w_2}-(1-b_2^2-f_2^2)^{w_2}) \\ -((1-b_1^2)^{w_1}-(1-b_1^2-f_1^2)^{w_1})((1-b_2^2)^{w_2}-(1-b_2^2-f_2^2)^{w_2})\end{array}\right)\end{array}\right)^{1/2}
\end{array}\right]
\end{array}\right\} \\
&= \left\{\begin{array}{l}\left[(1-(1-a_1^2)^{w_1}(1-a_2^2)^{w_2})^{1/2}, (1-(1-b_1^2)^{w_1}(1-b_2^2)^{w_2})^{1/2}\right], \left[c_1^{w_1}c_2^{w_2}, d_1^{w_1}d_2^{w_2}\right], \\ \left[\begin{array}{l}((1-a_1^2)^{w_1}(1-a_2^2)^{w_2}-(1-a_1^2-e_1^2)^{w_1}(1-a_2^2-e_2^2)^{w_2})^{1/2}, \\ ((1-b_1^2)^{w_1}(1-b_2^2)^{w_2}-(1-b_1^2-f_1^2)^{w_1}(1-b_2^2-f_2^2)^{w_2})^{1/2}\end{array}\right]\end{array}\right\}
\end{aligned}$$

Thus, result is true for $n = 2$. Assume that result is true for $n = k$, Eq. (37) holds, i.e.

$$\begin{aligned}
&IVSWAM_w(\tilde{\alpha}_1, \tilde{\alpha}_2, \ldots, \tilde{\alpha}_k) \\
&= w_1 \cdot \tilde{\alpha}_1 \oplus w_2 \cdot \tilde{\alpha}_2 \oplus \ldots \oplus w_k \cdot \tilde{\alpha}_k \\
&= \left\{\begin{array}{l}\left[\left(1-\prod_{j=1}^{k}(1-a_j^2)^{w_j}\right)^{1/2}, \left(1-\prod_{j=1}^{k}(1-b_j^2)^{w_j}\right)^{1/2}\right], \left[\prod_{j=1}^{k} c_j^{w_j}, \prod_{j=1}^{k} d_j^{w_j}\right], \\ \left\{\left[\left(\prod_{j=1}^{k}(1-a_j^2)^{w_j}-\prod_{j=1}^{k}(1-a_j^2-e_j^2)^{w_j}\right)^{1/2}, \left(\prod_{j=1}^{k}(1-b_j^2)^{w_j}-\prod_{j=1}^{k}(1-b_j^2-f_j^2)^{w_j}\right)^{1/2}\right]\right\}\end{array}\right\}
\end{aligned}$$

Then, when $n = k + 1$,

$$\begin{aligned}
&IVSWAM_w(\tilde{\alpha}_1, \tilde{\alpha}_2, \ldots, \tilde{\alpha}_k) \\
&= IVSWAM_w(\tilde{\alpha}_1, \tilde{\alpha}_2, \ldots, \tilde{\alpha}_k) \oplus w_{k+1} \cdot \tilde{\alpha}_{k+1} \\
&= \left\{\begin{array}{l}\left[\left(1-\prod_{j=1}^{k}(1-a_j^2)^{w_j}\right)^{1/2}, \left(1-\prod_{j=1}^{k}(1-b_j^2)^{w_j}\right)^{1/2}\right], \left[\prod_{j=1}^{k} c_j^{w_j}, \prod_{j=1}^{k} d_j^{w_j}\right], \\ \left[\left(\prod_{j=1}^{k}(1-a_j^2)^{w_j}-\prod_{j=1}^{k}(1-a_j^2-e_j^2)^{w_j}\right)^{1/2}, \left(\prod_{j=1}^{k}(1-b_j^2)^{w_j}-\prod_{j=1}^{k}(1-b_j^2-f_j^2)^{w_j}\right)^{1/2}\right]\end{array}\right\} \\
&\oplus \left\{\begin{array}{l}\left[(1-(1-a_{k+1}^2)^{w_{k+1}})^{1/2}, (1-(1-b_{k+1}^2)^{w_{k+1}})^{1/2}\right], \left[c_{k+1}^{w_{k+1}} d_{k+1}^{w_{k+1}}\right], \\ \left[((1-a_{k+1}^2)^{w_{k+1}}-(1-a_{k+1}^2-e_{k+1}^2)^{w_{k+1}})^{1/2}, ((1-b_{k+1}^2)^{w_{k+1}}-(1-b_{k+1}^2-f_{k+1}^2)^{w_{k+1}})^{1/2}\right]\end{array}\right\} \\
&= \left\{\begin{array}{l}\left[\left(1-\prod_{j=1}^{k+1}(1-a_j^2)^{w_j}\right)^{1/2}, \left(1-\prod_{j=1}^{k+1}(1-b_j^2)^{w_j}\right)^{1/2}\right], \left[\prod_{j=1}^{k+1} c_j^{w_j}, \prod_{j=1}^{k+1} d_j^{w_j}\right], \\ \left[\left(\prod_{j=1}^{k+1}(1-a_j^2)^{w_j}-\prod_{j=1}^{k+1}(1-a_j^2-e_j^2)^{w_j}\right)^{1/2}, \left(\prod_{j=1}^{k+1}(1-b_j^2)^{w_j}-\prod_{j=1}^{k+1}(1-b_j^2-f_j^2)^{w_j}\right)^{1/2}\right]\end{array}\right\}
\end{aligned}$$

i.e. when $n = k + 1$, Eq. (37) also holds. Next, in order to show $IVSWGM_w$ is an IVSFS number. As $\tilde{\alpha}_j = \langle [a_j, b_j], [c_j, d_j], [e_j, f_j] \rangle$ for all j is an IVSFS number, thus $0 \leq a_j, b_j, c_j, d_j, e_j, f_j \leq 1$ and $b_j^2 + d_j^2 + f_j^2 \leq 1$. Thus

$0 \leq \prod_{j=1}^{n}(1-a_j^2)^{w_j} \leq 1$ and hence $0 \leq \left(1-\prod_{j=1}^{k}(1-a_j^2)^{w_j}\right)^{1/2} \leq 1$, $0 \leq \prod_{j=1}^{k} c_j^{w_j} \leq 1$ and $0 \leq \left(\prod_{j=1}^{k}(1-a_j^2)^{w_j} - \prod_{j=1}^{k}(1-a_j^2-e_j^2)^{w_j}\right)^{1/2} \leq 1$. Similarly, $0 \leq \left(1-\prod_{j=1}^{k}(1-b_j^2)^{w_j}\right)^{1/2} \leq 1$, $0 \leq \prod_{j=1}^{k} d_j^{w_j} \leq 1$ and $0 \leq \left(\prod_{j=1}^{k}(1-b_j^2)^{w_j} - \prod_{j=1}^{k}(1-b_j^2-f_j^2)^{w_j}\right)^{1/2} \leq 1$. Thus,

$$\begin{aligned}
&\left(\left(1-\prod_{j=1}^{k}(1-b_j^2)^{w_j}\right)^{1/2}\right)^2 + \left(\prod_{j=1}^{k} d_j^{w_j}\right)^2 \\
&\quad + \left(\left(\prod_{j=1}^{k}(1-b_j^2)^{w_j} - \prod_{j=1}^{k}(1-b_j^2-f_j^2)^{w_j}\right)^{1/2}\right)^2 \\
&= \left(1-\prod_{j=1}^{k}(1-b_j^2)^{w_j}\right) + \prod_{j=1}^{k} d_j^{2w_j} + \left(\prod_{j=1}^{k}(1-b_j^2)^{w_j} - \prod_{j=1}^{k}(1-b_j^2-f_j^2)^{w_j}\right) \\
&\leq 1 - \left(1-\prod_{j=1}^{k}(1-b_j^2)^{w_j}\right) + \prod_{j=1}^{k}(1-b_j^2-f_j^2)^{w_j} \\
&\quad + \left(\prod_{j=1}^{k}(1-b_j^2)^{w_j} - \prod_{j=1}^{k}(1-b_j^2-f_j^2)^{w_j}\right) = 1
\end{aligned}$$

Hence, $IVSWGM_w$ is an $IVSFS$ number and therefore proof is completed.

Definition 3.5 Let $\tilde{\alpha}_j = \langle [a_j, b_j], [c_j, d_j], [e_j, f_j] \rangle$ be a collection of Interval-valued Spherical Geometric Mean (IVSWGM) with respect to, $w_j = (w_1, w_2, \ldots, w_n)$; $w_j \in [0,1]\, and\, \sum_{j=1}^{n} w_j = 1$, IVSWGM is defined as;

$$\begin{aligned}
&IVSWGM_w(\tilde{\alpha}_1, \tilde{\alpha}_2, \ldots, \tilde{\alpha}_n) \\
&= \tilde{\alpha}_1^{w_1} \otimes \tilde{\alpha}_2^{w_2} \otimes \cdots \otimes \tilde{\alpha}_n^{w_n} \\
&= \left\{ \begin{array}{l} \left[\prod_{j=1}^{n} a_j^{w_j}, \prod_{j=1}^{n} b_j^{w_j}\right], \left[\left(1-\prod_{j=1}^{n}(1-c_j^2)^{w_j}\right)^{1/2}, \left(1-\prod_{j=1}^{n}(1-d_j^2)^{w_j}\right)^{1/2}\right], \\ \left[\left(\prod_{j=1}^{n}(1-c_j^2)^{w_j} - \prod_{j=1}^{n}(1-c_j^2-e_j^2)^{w_j}\right)^{1/2}, \left(\prod_{j=1}^{n}(1-d_j^2)^{w_j} - \prod_{j=1}^{n}(1-d_j^2-f_j^2)^{w_j}\right)^{1/2}\right] \end{array} \right\}
\end{aligned} \quad (39)$$

Proof 3.5 The proof of this definition is similar to Definition 3.4, so we omit here.

Example 3.1 Suppose that $\tilde{\alpha}_1 = \langle [0.85, 0.95], [0.1, 0.15], [0.05, 0.15] \rangle$, $\tilde{\alpha}_2 = \langle [0.20, 0.25], [0.65, 0.75], [0.20, 0.25] \rangle$, and $\tilde{\alpha}_3 = \langle [0.55, 0.65], [0.25, 0.30], [0.25, 0.30] \rangle$ and $w = (0.5, 0.4, 0.1)^T$ is the weight vector of $\tilde{a}_j (j = 1, 2, 3)$. Then

$$IVSWAM_w(\tilde{\alpha}_1, \tilde{\alpha}_2, \tilde{\alpha}_3, \tilde{\alpha}_4) = \left\{ \begin{array}{l} \left[\begin{array}{l} \left(1-\left(1-0.85^2\right)^{0.5}\left(1-0.20^2\right)^{0.4}\left(1-0.55^2\right)^{0.1}\right)^{1/2}, \\ \left(1-\left(1-0.95^2\right)^{0.5}\left(1-0.25^2\right)^{0.4}\left(1-0.65^2\right)^{0.1}\right)^{1/2} \end{array} \right], \\ \left[0.1^{0.5}0.65^{0.4}0.25^{0.1},\ 0.15^{0.5}0.75^{0.4}0.30^{0.1}\right], \\ \left[\begin{array}{l} \left(\begin{array}{l} \left(1-0.85^2\right)^{0.5}\left(1-0.20^2\right)^{0.4}\left(1-0.55^2\right)^{0.1} \\ -\left(1-0.85^2-0.05^2\right)^{0.5}\left(1-0.20^2-0.20^2\right)^{0.4}\left(1-0.55^2-0.25^2\right)^{0.1} \end{array} \right)^{1/2}, \\ \left(\begin{array}{l} \left(1-0.95^2\right)^{0.5}\left(1-0.25^2\right)^{0.4}\left(1-0.65^2\right)^{0.1} \\ -\left(1-0.95^2-0.15^2\right)^{0.5}\left(1-0.25^2-0.25^2\right)^{0.4}\left(1-0.65^2-0.30^2\right)^{0.1} \end{array} \right)^{1/2} \end{array} \right] \end{array} \right\}$$

$$= \{[0.71, 0.84], [0.23, 0.31], [0.12, 0.22]\}$$

$$IVSWGM_w(\tilde{\alpha}_1, \tilde{\alpha}_2, \tilde{\alpha}_3, \tilde{\alpha}_4) = \left\{ \begin{array}{l} \left[0.85^{0.5}0.20^{0.4}0.55^{0.1},\ 0.95^{0.5}0.25^{0.4}0.65^{0.1}\right], \\ \left[\begin{array}{l} \left(1-(1-0.10^2)^{0.5}(1-0.65^2)^{0.4}(1-0.25^2)^{0.1}\right)^{1/2}, \\ \left(1-(1-0.15^2)^{0.5}(1-0.75^2)^{0.4}(1-0.30^2)^{0.1}\right)^{1/2} \end{array} \right], \\ \left[\begin{array}{l} \left(\begin{array}{l} (1-0.10^2)^{0.5}(1-0.65^2)^{0.4}(1-0.25^2)^{0.1} \\ -(1-0.10^2-0.05^2)^{0.5}(1-0.65^2-0.20^2)^{0.4}(1-0.25^2-0.25^2)^{0.1} \end{array} \right)^{1/2}, \\ \left(\begin{array}{l} (1-0.15^2)^{0.5}(1-0.75^2)^{0.4}(1-0.30^2)^{0.1} \\ -(1-0.15^2-0.15^2)^{0.5}(1-0.75^2-0.25^2)^{0.4}(1-0.30^2-0.30^2)^{0.1} \end{array} \right)^{1/2} \end{array} \right] \end{array} \right\}$$

$$= \{[0.46, 0.54], [0.45, 0.54], [0.17, 0.24]\}$$

Definition 3.6 The score function of IVSFS number α is defined as

$$Score(\tilde{\alpha}) = S(\tilde{\alpha}) = \frac{a^2 + b^2 - c^2 - d^2 - (e/2)^2 - (f/2)^2}{2} \tag{40}$$

where $Score(\tilde{\alpha}) = S(\tilde{\alpha}) \in [-1, +1]$. Clearly, the greater the $S(\tilde{\alpha})$, the larger the α. In particular, when $S(\tilde{\alpha}) = 1$ then $\tilde{\alpha} = \langle[1, 1], [0, 0], [0, 0]\rangle$; when $S(\tilde{\alpha}) = -1$ then α is the smallest IVSFS number $\tilde{\alpha} = \langle[0, 0], [1, 1], [0, 0]\rangle$.

Definition 3.7 The accuracy function of IVSFS number α is defined as;

$$Accuracy(\tilde{\alpha}) = H(\tilde{\alpha}) = \frac{a^2 + b^2 + c^2 + d^2 + e^2 + f^2}{2} \tag{41}$$

where $H(\tilde{\alpha}) \in [0, 1]$.

Note that: $\tilde{\alpha}_1 < \tilde{\alpha}_2$ if and only if $S(\tilde{\alpha}_1) < S(\tilde{\alpha}_2)$ or $S(\tilde{\alpha}_1) = S(\tilde{\alpha}_2)$ and $H(\tilde{\alpha}_1) < H(\tilde{\alpha}_2)$.

Example 3.2 Let $\tilde{\alpha}_1 = \langle[0.85, 0.95], [0.1, 0.15], [0.05, 0.15]\rangle$, $\tilde{\alpha}_2 = \langle[0.20, 0.25], [0.65, 0.75], [0.20, 0.25]\rangle$ and $\tilde{\alpha}_3 = \langle[0.55, 0.65], [0.25, 0.30], [0.25, 0.30]\rangle$ be three interval-valued spherical fuzzy numbers. According to Eq. (40) we have $S(\tilde{\alpha}_1) = 0.790, S(\tilde{\alpha}_2) = -0.467$, and $S(\tilde{\alpha}_3) = 0.248$. Thus, the alternative $\tilde{\alpha}_1$ is better than others. This example shows that the proposed score function is reasonable.

Definition 3.8 Let $\tilde{\alpha}_1 = \langle [a_1, b_1], [c_1, d_1], [e_1, f_1] \rangle$ and $\tilde{\alpha}_2 = \langle [a_2, b_2], [c_2, d_2], [e_2, f_2] \rangle$ be two IVSFS numbers, then we define the distance between $\tilde{\alpha}_1$ and $\tilde{\alpha}_2$ as follows (Ejegwa 2019);

$$d(\tilde{\alpha}_1, \tilde{\alpha}_2) = \frac{1}{4}\left(\left|a_1^2 - a_2^2\right| + \left|b_1^2 - b_2^2\right| + \left|c_1^2 - c_2^2\right| + \left|d_1^2 - d_2^2\right| + \left|e_1^2 - e_2^2\right| + \left|f_1^2 - f_2^2\right|\right) \quad (42)$$

4 Conclusions

Fuzzy set theory is expanding with a huge acceleration in all branches of science. Especially, control theory and decision making areas are utilizing the fuzzy set theory much more than the other branches. Every new extension of fuzzy sets present new opportunities to the researchers to extend their interest areas by using them. Intuitionistic fuzzy sets have become an important extension which almost all other extensions are based on with some differences. Spherical fuzzy sets are the latest extension of intuitionistic fuzzy sets, proposing an independent hesitancy degree from the other parameters (Gündoğdu 2019a; Gündoğdu and Kahraman 2019b, c, 2020a, b). Spherical fuzzy sets are expected to be preferred by most of the researchers in a short time period since their principals are strong enough to develop further.

For further research, we suggest operations to be developed for new spherical fuzzy numbers such as LR SF numbers, Gaussian SF numbers, triangular SF numbers, and trapezoidal SF numbers. These numbers will require new arithmetic operations, defuzzification and aggregation operators to be defined. Later, they will let classical multi-criteria decision making methods be extended by these new types of fuzzy numbers.

References

Antonov I (1995) On a new geometrical interpretation of the intuitionistic fuzzy sets. Notes Intuitionistic Fuzzy Sets 1(1):29–31

Atanassov KT (1986) Intuitionistic fuzzy sets. Fuzzy sets and Systems 20(1):87–96

Atanassov KT (1999) Intuitionistic fuzzy sets. In: Intuitionistic fuzzy sets. Physica, Heidelberg, pp 1–137

Cuong B (2015) Picture fuzzy sets. J Comput Sci Cybern 30. https://doi.org/10.15625/1813-9663/30/4/5032

Ejegwa PA (2019) Modified Zhang and Xu's distance measure for pythagorean fuzzy sets and its application to pattern recognition problems. Neural Computing and Applications

Garibaldi JM, Ozen T (2007) Uncertain fuzzy reasoning: a case study in modelling expert decision making. IEEE Trans Fuzzy Syst 15(1):16–30

Grattan-Guinness I (1976) Fuzzy membership mapped onto interval and many-valued quantities. Z Math Logik Grundlag Math 22(1):149–160

Gündoğdu FK (2019a) Extension of MULTIMOORA with spherical fuzzy sets. J Intell Fuzzy Syst, 1–16. Preprint, https://doi.org/10.3233/JIFS-179462
Gündoğdu FK (2019b) Principals of spherical fuzzy sets. In: International conference on intelligent and fuzzy systems. Springer, Cham, pp 15–23
Gündoğdu FK, Kahraman C (2019a) Spherical fuzzy sets and spherical fuzzy TOPSIS method. J Intell Fuzzy Syst 36(1):337–352
Gündoğdu FK, Kahraman C (2019b) Spherical fuzzy sets and decision making applications. In: International conference on intelligent and fuzzy systems. Springer, Cham, pp 979–987
Gündoğdu FK, Kahraman C (2019c) Spherical fuzzy analytic hierarchy process (AHP) and its application to industrial robot selection. In: International conference on intelligent and fuzzy systems. Springer, Cham, pp 988–996
Gündoğdu FK, Kahraman C (2019d) A novel fuzzy TOPSIS method using emerging interval-valued spherical fuzzy sets. Eng Appl Artif Intell 85:307–323
Gündoğdu FK, Kahraman C (2019e) A novel spherical fuzzy analytic hierarchy process and its renewable energy application. Soft Comput, 1–15. https://doi.org/10.1007/s00500-019-04222-w
Gündoğdu FK, Kahraman C (2019f) Spherical fuzzy analytic hierarchy process (AHP) and its application to industrial robot selection. In: International conference on intelligent and fuzzy systems. Springer, Cham, pp 988–996
Gündoğdu FK, Kahraman C (2019g) A novel VIKOR method using spherical fuzzy sets and its application to warehouse site selection. J Intell Fuzzy Syst 37(1):1197–1211
Gündoğdu FK, Kahraman C (2019h) Extension of WASPAS with spherical fuzzy sets. Informatica 30(2):269–292
Gündoğdu FK, Kahraman C (2020a) Extension of CODAS with spherical fuzzy sets. J Multi-Valued Log Soft Comput, 1–25
Gündoğdu FK, Kahraman C (2020b) A novel spherical fuzzy QFD method and its application to the linear delta robot technology development. Eng Appl Artif Intell 87:103348
Jahn KU (1975) Intervall wertige Mengen. Math Nachr 68(1):115–132
Kahraman C, Gündoğdu FK (2018) From 1D to 3D membership: spherical fuzzy sets. In: BOS/SOR2018 conference. Warsaw, Poland
Sambuc R (1975) Function Φ-flous. In: Application a l'aide au Diagnostic en Pathologie Thyroidienne. University of Marseille
Smarandache F (1995) Neutrosophic logic and set, mss
Smarandache F (1998) Neutrosophy: neutrosophic probability, set, and logic. American Research Press, Rehoboth, USA, 105p
Torra V (2010) Hesitant fuzzy sets. Int J Intell Syst 25(6):529–539
Yager R (1986) On the theory of bags. Int J Gen Syst 13(1):23–37
Yager RR (2013) Pythagorean fuzzy subsets. In: Joint IFSA world congress and NAFIPS annual meeting. Edmonton, Canada, pp 57–61
Yager RR (2016) Generalized orthopair fuzzy sets. IEEE Trans Fuzzy Syst 25(5):1222–1230
Yang Y, Chiclana F (2009) Intuitionistic fuzzy sets: spherical representation and distances. Int J Intell Syst 24(4):399–420
Zadeh LA (1965) Fuzzy sets. Inf Control 8:338–353
Zadeh LA (1975) The concept of a linguistic variable and its application to approximate reasoning. Inf Sci 8:199–249

Interval Valued Spherical Fuzzy Aggregation Operators and Their Application in Decision Making Problem

M. Lathamaheswari, D. Nagarajan, Harish Garg, and J. Kavikumar

Abstract The collected data may be an interval number rather than the exact number in the process of information fusion for any real-world problem. By taking this into consideration interval numbers under interval spherical fuzzy environment have been used in this present work. To deal with vagueness, impreciseness, and uncertainty we have different environments namely fuzzy, intuitionistic fuzzy and Pythagorean fuzzy sets. All these environments are strictly following the restriction on the characteristic values and their sum. This gives the stress to the decision maker in giving preference values. Hence spherical fuzzy sets are used where the sum of the square of membership, non-membership and refusal degrees is less than or equal to 1. This condition helps the decision maker to give the preference values according to their knowledge without any limitations. Therefore, the interval valued spherical fuzzy sets, operational laws for interval valued spherical fuzzy numbers and aggregation operators with their properties are proposed. Also, the proposed aggregation operators are applied in a decision making problem to choose the best station which scrutinizes the quality of air. A further comparative study is done with the existing method to show the novelty and effectiveness of the proposed method.

1 Introduction

Problems involving multi-criteria the universal one in various fields of modern technology which includes not only the decision-making problem, they meant for retrieval of information pattern recognition, query based on database and reasoning based

M. Lathamaheswari (✉) · D. Nagarajan
Department of Mathematics, Hindustan Institute of Technology & Science, Chennai 603103, India
e-mail: lathamax@gmail.com

H. Garg
School of Mathematics, Thapar University Patiala, Patiala, India

J. Kavikumar
Department of Mathematics and Statistics, Faculty of Applied Science and Technology, Universiti Tun Hussein Onn, Johor, Malaysia

© Springer Nature Switzerland AG 2021

C. Kahraman and F. Kutlu Gündoğdu (eds.), *Decision Making with Spherical Fuzzy Sets*, Studies in Fuzziness and Soft Computing 392,
https://doi.org/10.1007/978-3-030-45461-6_2

on the case. Interrelationship of all the individual criteria is the measurement of satisfaction to the overall assortment of criteria and this process is called aggregation.

Though there are many aggregation operations that exist to solve the real-world problems, it is necessary to derive different types of explicit aggregation operations to empower the modeling of these various types of relationships. Hence, aggregation theory which is a mathematical discipline is developed. According to the knowledge about the connection between the criteria, one can select the function of aggregation (Yager 2009).

Since aggregating various inputs into a single output as the representative is an important task, it emerges commonly in different practical applications. Therefore the research effort has been taken on aggregation functions, their properties and behavior throughout many of the areas including knowledge based systems, image processing, artificial intelligence and decision making process (Beliakov et al. 2010).

Hence Zadeh introduced the theory of fuzzy set in 1965 to deal with impreciseness and uncertainty by considering membership function where the degree of membership lies between 0 and 1 since the data may contain impreciseness and uncertainty in nature, using classing set would not be suitable as it fails to address uncertainty. The concept fuzzy logic has been applied widely in the control system, image processing, and decision making problems.

Since the theory of fuzzy set deals with only membership function, it unable to handle non-membership of function. Also, there is no possibility of having the value of non-membership of the element is always one minus the membership value due to hesitancy. Hence Atanassov introduced intuitionistic fuzzy set theory which coordinates non-membership function with the membership function and characterizes the fuzzy object more admirably (Xu et al. 2011; Ejegwa et al. 2014).

To mitigate some of the difficulties and disadvantages of the fuzzy set, intuitionistic fuzzy set and interval valued intuitionistic fuzzy sets were introduced as the straight forward expansion (Wang et al. 2012). Meta set has a similar opportunity of practical applications as intuitionistic fuzzy set (Starosta and Nski 2013). IFS have three correlated defining functions called membership function, non-membership function and hesitant function.

As hesitancy is also one of the common problems in the process of decision making, it can be dealt with the tool called hesitant fuzzy set which permits many possible degrees of membership of the element of the set. Atanassov and Gargov introduced interval valued intuitionistic fuzzy sets as the combination of interval valued fuzzy sets and intuitionistic fuzzy sets (Broumi and Smarandache 2014).

The fuzzy decision making process is the emerging area of research in the theory of fuzzy decision analysis. It contains the process of choosing the best alternative among the possible alternatives in which the information provided by the decision makers naturally with fuzziness or uncertain. An effective extension of a fuzzy set or type-1 fuzzy is called type-2 fuzzy set (T2FS) where the membership function is characterized by primary and secondary membership functions and footprint of uncertainty (FOU).

FOU is the third dimension of T2FS which emulates more degrees of freedom and makes T2FS more suitable for handling uncertain and imperfect information.

Hence it gets more attention from the researchers and both theoretical and practical aspects have been developed. Further, it has been applied in many real world problems successfully. Though it has an efficiency of handling uncertainty, it has computational complexity also and hence solving real world problems using T2FS is difficult. To sort out this issue and to overcome the limitations of T2FS, interval type-2 fuzzy sets or interval valued type-2 fuzzy sets have been introduced which contains the membership values that are crisp intervals lies between 0 and 1 (Qin and Liu 2014).

A special case of fuzzy sets called Pythagorean fuzzy sets (PFSs) introduced by Yager in 2013 as an advanced tool for handling vagueness where the membership grades are pairs and satisfy the condition. PFSs have a very close link with intuitionistic fuzzy sets (Peng and Yang 2015). The degree of difference between the two sets is described by distance measure. It reflects only the relative differences whereas spherical distance between any two sets calculates the relative difference and their absolute values as well. These kinds of measures are very important in scientific and social economic areas. The distance measure is a dual of the similarity measure. Hamming distance and Euclidean distance are based on two dimensional spaces (Gonga et al. 2016).

A generalization of FS, IFS and picture fuzzy set is called T-spherical fuzzy set which is able to deal with uncertainty without any limitations. It permits the linguistic terms namely in favor or against and that could be some sort of restraint from desires or refusal degree. PFS is the direct generalization of IFS which increases the range of spaces for IFS by adjusting the condition of IFS. Cuong introduced PFS with a better phenomenon by considering four characteristic functions called favor, abstinence, opposition, refusal with the condition that the sum of four membership grades must be 1 or the sum of three membership grades must not exceed the unit interval. By considering these limitations, spherical fuzzy sets and T-Spherical fuzzy sets have been introduced (Ullah et al. 2019).

IFS and IVIFS are getting more attention over the last few decades as they have the capacity of handling with vagueness, impreciseness and uncertainty in the data, due to their restriction on the sum of membership degree and non-membership degree should be less than 1 and hence this concept could not be applied for many situations where this restriction is ruled out. For example, if the decision maker gives his/her preference value for membership and non-membership for a particular alternative is 0.8 and 0.5 then obviously the situation goes out of control with IFS.

Hence Yager introduced Pythagorean fuzzy sets in 2013 with the restriction on the grades of membership that the square sum of the grades of memberships should be less than or equal to one. This condition will be very helpful to reduce the stress of decision makers on the sum of membership values for the particular alternative. Hence it gets more attention and interest among the researchers as it can be applied to any real world problems without any stress on the limitation (Garg 2018). Picture fuzzy set was introduced by Cuong where the elements of the set represented by the triplets namely satisfaction, abstain and dissatisfaction. Also, the degree of refusal is defined by 1-sum of the three membership values (Ashraf and Abdullah 2018).

The rest of the paper is organized as follows. In Sect. 2, the literature review is collected to find the research gap in the spherical field. In Sect. 3, basic definitions are given for a better understanding of the present work. In Sect. 4, the definition of interval valued spherical fuzzy set, score function of interval valued spherical fuzzy number and operational laws for IVSFNs are proposed. In Sect. 5, two new aggregation operators are proposed namely IVSFWA and IVSFWG operators with their properties in detail. In Sect. 6, applied the proposed aggregation operators in decision making problem. In Sect. 7, a comparative analysis is done with the existing method. In Sect. 8, a conclusion is given along with the future direction.

2 Review of Literature

Yager (2009) proposed generalized Bonferroni mean operators. Beliakov et al. (2010) contributed composed aggregation functions. Xu et al. (2011) introduced star shaped intuitionistic fuzzy sets. Wang et al. (2012) proposed interval-valued intuitionistic fuzzy aggregation operators and applied them in a decision making problem. Starosta and Nski (2013) introduced Meta sets. Li and Feng (2013) proposed (λ, μ)-fuzzy ideals and (λ, μ)-fuzzy interior ideals of an ordered Γ—semigroup.

Ejegwa et al. (2014) made an overview of intuitionistic fuzzy sets. Broumi and Smarandache (2014) proposed new operations over interval valued intuitionistic hesitant fuzzy Set. Qin and Liu (2014) proposed Frank aggregation operators under triangular interval type-2 fuzzy environment and applied in multiple attribute group decision making. Peng and Yang (2015) introduced a few results for Pythagorean fuzzy sets. Also proposed aggregation operators and proved their prescribed properties in detail. Further, they applied the proposed aggregation operators in decision making problem.

Rizwan and Iqbal (2016) proposed a novel interval valued intuitionistic fuzzy set of cube toot type. Jamkhaneh (2016) derived new operations under generalized interval valued intuitionistic fuzzy sets. Gonga et al. (2016) proposed the spherical distance for intuitionistic fuzzy sets and applied in decision analysis. Rajesh and Srinivasan (2017a, b) examined some of the operators under interval valued intuitionistic fuzzy set of type-2. Rajesh and Srinivasan introduced interval valued intuitionistic fuzzy sets of the second type.

Rahman et al. (2017) introduced some elemental operations on Pythagorean fuzzy sets. Atanassov (2017) analyzed type-1 fuzzy sets and intuitionistic fuzzy sets. Garg (2018) proposed hesitant Pythagorean fuzzy sets and their aggregation operators. Also, applied those in a multi attribute decision making problem. Dominguez et al. (2018) applied MOORA method in a decision making problem under Pythagorean fuzzy environment. Peng (2018a, b) introduced a new similarity measure and the distance measure for Pythagorean fuzzy set. Ejegwa (2018) proposed distance and similarity measures for Pythagorean fuzzy sets. Peng (2018a, b) proposed new operations for interval-valued Pythagorean fuzzy set. Osawaru et al. (2018) introduced a relative fuzzy set. Ullah et al. (2018) proposed similarity measures for T-Spherical fuzzy sets and applied in pattern recognition.

Garg et al. (2018) Algorithm for T-Spherical Fuzzy Multi-Attribute Decision Making Based on Improved Interactive Aggregation Operators. Li and Zeng (2018) proposed distance measure of Pythagorean fuzzy sets. Ashraf and Abdullah (2018) proposed spherical aggregation operators and applied in multi attribute group decision making. Gündoğdu and Kahraman (2019) introduced spherical fuzzy sets, their operational laws, and spherical fuzzy TOPSIS method.

Ullah et al. (2019) proposed correlation coefficients for T-spherical fuzzy sets and applied them in clustering and multi-attribute decision making. Wu (2018) proposed a dual of fuzzy sets and a dual arithmetics of fuzzy sets. Thao and Smarandache (2019) presented new fuzzy entropy on Pythagorean fuzzy sets. Sunny and Jose (2019) derived Boxdot and star products on interval-valued intuitionistic fuzzy graphs. Zeng et al. (2019) proposed spherical fuzzy rough set model based on covering and collaborated with TOPSIS for decision making problem.

Jin et al. (2019) proposed linguistic spherical fuzzy aggregation operators and applied in multi-attribute decision making problems. Augustine (2019) introduced improved composite relations under the Pythagorean fuzzy environment and applied to medical diagnosis. Ashraf et al. (2019a, b) proposed spherical fuzzy Dombi aggregation operators and applied in group decision making problem. Ashraf et al. (2019a, b) introduced the way of representing spherical fuzzy t-norms and t-conorms. Rafiq et al. (2019) introduced the cosine similarity measures of spherical fuzzy sets and applied in decision making problem. Gundogdu and Kahraman (2018) proposed a novel fuzzy TOPSIS method using prominent interval valued spherical fuzzy sets.

From this literature review, it is found that operational laws and aggregation operators are not proposed yet under interval spherical fuzzy environment. Hence in this present study, interval valued spherical fuzzy set and its operational laws have been introduced. Also proposed a new score function for interval valued spherical fuzzy numbers. Further interval valued spherical weighted arithmetic operator (IVSFWA) and interval valued spherical fuzzy geometric operator (IVSFWG) are also proposed and applied in a decision making problem for checking the quality of air with different alternatives.

3 Basic Concepts

In this section, some of the basic concepts are given for a better understanding of the present work.

3.1 Fuzzy Set (Qin and Liu 2014)

Let U be a nonempty set. A fuzzy set F drawn from U is defined as,

$$F = \{e, \mu_F(e) | e \in U\} \tag{1}$$

where $\mu_F : U \rightarrow [0, 1]$ called memberships function of F defined over a universe of discourse U.

3.2 Type-2 Fuzzy Set (Qin and Liu 2014)

A type-2 fuzzy set, denoted by $\overline{F}$ is characterized by a type-2 membership function $\mu_{\overline{F}}(e, j)$, where $e \in U, j \in J_e \subseteq [0, 1]$, i.e.,

$$\overline{F} = \{((e, j), \mu_{\overline{F}}(e, j))|e \in U, \forall j \in J_e \subseteq [0, 1]\} \tag{2}$$

3.3 Interval Valued Type-2 Fuzzy Set (Qin and Liu 2014)

It is a special case of type-2 fuzzy sets by representing the membership function $\mu_{\overline{F}} = \left[\underline{\mu_{\overline{F}}}, \overline{\mu_{\overline{F}}}\right]$, where $\underline{\mu_{\overline{F}}}$ is a lower membership function and $\overline{\mu_{\overline{F}}}$ is an upper membership function. The area between these two membership functions is called a footprint of uncertainty (FOU), which represents the level of uncertainty of the set.

3.4 Intuitionistic Fuzzy Set (Garg 2018)

Let U be a nonempty set. An intuitionistic fuzzy set (IFS) F in U is an object having the form

$$F = \{\langle e, \mu_F(e), \nu_F(e)\rangle|e \in U\} \tag{3}$$

where the functions $\mu_F(e), \nu_F(e) : U \rightarrow [0, 1]$ define the degree of membership and non-membership respectively, of the element $e \in U$ to F, for the all the elements $e \in U 0 \leq \mu_F(e) + \nu_F(e) \leq 1$.

Also, $\pi_F(e) = 1 - \mu_F(e) - \nu_F(e)$ called the index of IFS, and is the degree of indeterminacy of $e \in U$ of the IFS A, which express the lack of knowledge of whether e belongs to IFS or not. Also $\pi_F(e) \in [0, 1]$, i.e., $\pi_F(e) : U \rightarrow [0, 1]$ and $0 \leq \pi_F(e) \leq 1, \forall e \in U$.

3.5 Interval Valued Intuitionistic Fuzzy Set (Garg 2018)

An Interval valued intuitionistic fuzzy set (IVIFS) F in U is defined as an object of the form

$$F = \{\langle e, P_F(e), Q_F(e)\rangle | e \in U\} \tag{4}$$

where the functions $\mu_F(e) : U \to [0, 1]$, $\nu_F(e) : U \to [0, 1]$ denote the degree of membership and non-membership of F respectively. Also, $\mu_F(e) = \left[\mu_F^L(e), \mu_F^U(e)\right]$ and $\nu_F(e) = \left[\nu_F^L(e), \nu_F^U(e)\right]$ with the condition $0 \le \mu_F^U(e) + \nu_F^U(e) \le 1, \forall e \in U$

3.6 Pythagorean Fuzzy Set (Ashraf et al. 2019a, b)

Let U be the arbitrary non empty set. A Pythagorean fuzzy set P is a mathematical object of the form

$$P = \left\{\langle e, \mu_p(e), \mu_p(e)\rangle / e \in U\right\} \tag{5}$$

where $\mu_p(e) : U \to [0, 1]$ and $\nu_p(e) : U \to [0, 1]$ are the membership degree and non-membership degree of the element $e \in U$ respectively satisfying the condition that $0 \le \mu_p(e) + \nu_p(e) \le 1$. Also the number $\pi_P(e)$ is the Pythagorean index degree of the hesitancy of the element $e \in U$ and is defined by $\pi_P(e) = \sqrt{1 - \left(\mu_P^2(e) + \nu_P^2(e)\right)}$, $\mu_P^2(e) + \nu_P^2(e) \le 1\, \forall e \in U$.

3.7 Interval Valued Pythagorean Fuzzy Set (Peng 2018a, b)

Consider the set of all closed sub intervals Int ([0, 1]) of [0, 1] and U be the universe of discourse. An interval valued Pythagorean fuzzy set $P = \left\{\langle e, \mu_p(e), \nu_p(e)\rangle / e \in U\right\}$, where the membership degree is defined by $\mu_p : U \to Int([0, 1])$, $e \in U \to \mu_p(e) \subseteq [0, 1]$ and the non-membership degree is defined by $\nu_p : U \to Int([0, 1])$, $e \in U \to \nu_p(e) \subseteq [0, 1]$.

Also $\mu_P(e) = \left[\mu_P^L(e), \mu_P^U(e)\right]$ and $\nu_P(e) = \left[\nu_P^L(e), \nu_P^U(e)\right]$ are the closed intervals and $0 \le s\left\{\left(\mu_P^U(e)\right)^2\right\} + \left\{\left(\nu_P^U(e)\right)^2\right\} \le 1$, $\forall e \in U$. The degree of indeterminacy is defined by

$$\begin{aligned} \pi_F(e) &= \left[\pi_P^L(e), \pi_P^U(e)\right] \\ &= \left[\sqrt{1 - \left(\mu_P^U(e)\right)^2 - \left(\nu_P^U(e)\right)^2}, \sqrt{1 - \left(\mu_P^L(e)\right)^2 - \left(\nu_P^L(e)\right)^2}\right]. \end{aligned}$$

3.8 Picture Fuzzy Set (Ashraf et al. 2019a, b)

Let $U \neq \varphi$ be a universal set. Then the picture fuzzy set is defined by

$$R = \{\langle P_R(e), I_R(e), N_R(e)\rangle / e \in U\} \tag{6}$$

where $P_R : U \to [0, 1],\ I_R : U \to [0, 1], N_R : U \to [0, 1]$ are positive, neutral and negative membership degrees for all the values of $e \in U$ and satisfying the condition that $0 \leq P_R(e) + I_R(e) + N_R(e) \leq 1$.

4 Proposed Concepts

In this section, we introduce interval valued spherical fuzzy sets, their operational laws of interval valued spherical fuzzy numbers, aggregation operators and their properties are derived in detail.

4.1 Interval Valued Spherical Fuzzy Set

Let $U \neq \varphi$ be a universal set. Then the interval valued spherical fuzzy set is defined by

$$S = \{\langle P_S(e), N_S(e), I_S(e)\rangle / e \in U\} \tag{7}$$

where, $P_S = \left[P_S^L(e), P_S^U(e)\right]$, $N_S = \left[N_S^L(e), N_S^U(e)\right]$, $I_S = \left[I_S^L(e), I_S^U(e)\right]$ and $P_S : U \to [0, 1]$, $N_S : U \to [0, 1]$, $I_S : U \to [0, 1]$ are membership, non-membership and refusal degrees of membership for all the values of $e \in U$ and satisfying the condition $0 \leq \left(P_S^U(e)\right)^2 + \left(N_S^U(e)\right)^2 + \left(I_S^U(e)\right)^2 \leq 1$.

4.2 Operational Laws

Here we establish the following operational laws are proposed for interval valued spherical fuzzy numbers under interval valued spherical fuzzy environment. Let S, S_1, S_2 be three interval valued spherical fuzzy numbers and let $\lambda > 1$.

1. ***Addition***

$$S_1 \oplus S_2 = \left\langle \left[\sqrt{\left(P_1^L\right)^2 + \left(P_2^L\right)^2 - \left(P_1^L\right)^2 . \left(P_2^L\right)^2}, \right. \right.$$

$$\sqrt{\left(P_1^U\right)^2 + \left(P_2^U\right)^2 - \left(P_1^U\right)^2.\left(P_2^U\right)^2}\Big],$$
$$\left[N_1^L.N_2^L, N_1^U.N_2^U\right], \left[I_1^L.I_2^L, I_1^U.I_2^U\right]\rangle \tag{8}$$

2. ***Multiplication***

$$S_1 \otimes S_2 = \langle\left[P_1^L.P_2^L, P_1^U.P_2^U\right], \left[N_1^L.N_2^L, N_1^U.N_2^U\right],$$
$$\Big[\sqrt{\left(I_1^L\right)^2 + \left(I_2^L\right)^2 - \left(I_1^L\right)^2.\left(I_2^L\right)^2},$$
$$\sqrt{\left(I_1^U\right)^2 + \left(I_2^U\right)^2 - \left(I_1^U\right)^2.\left(I_2^U\right)^2}\Big]\Big\rangle \tag{9}$$

3. ***Multiplication by an Ordinary Number;*** $\lambda > 0$

$$\lambda S = \left\langle\left[\sqrt{1-\left(1-\left(P^L\right)^2\right)^\lambda}, \sqrt{1-\left(1-\left(P^U\right)^2\right)^\lambda}\right],\right.$$
$$\left.\left[\left(N^L\right)^\lambda, \left(N^U\right)^\lambda\right], \left[\left(I^L\right)^\lambda, \left(I^U\right)^\lambda\right]\right\rangle \tag{10}$$

4. ***Power;*** $\lambda > 0$

$$S^\lambda = \left\langle\left[\left(P^L\right)^\lambda, \left(P^U\right)^\lambda\right], \left[\left(N^L\right)^\lambda, \left(N^U\right)^\lambda\right],\right.$$
$$\left.\left[\sqrt{1-\left(1-\left(I^L\right)^2\right)^\lambda}, \sqrt{1-\left(1-\left(I^U\right)^2\right)^\lambda}\right]\right\rangle \tag{11}$$

4.3 Proposed Score Function for Interval Valued Spherical Fuzzy Numbers

Let $S_k = \langle\left[P_{S_k}^L, P_{S_k}^U\right], \left[N_{S_k}^L, N_{S_k}^U\right], \left[I_{S_k}^L, I_{S_k}^U\right]\rangle, k = 1, 2, 3, \ldots, n$ be the set all interval valued spherical fuzzy numbers then the score function is defined by

$$s(S) = \frac{1}{3}\left[\left(P_S^L\right)^2 + \left(P_S^U\right)^2 - \left(N_S^L\right)^2 - \left(N_S^U\right)^2 - \left(I_S^L\right)^2 - \left(I_S^U\right)^2\right]$$
$$\in [-1, 1] \tag{12}$$

5 Aggregation Operators for Interval Valued Spherical Fuzzy Numbers

In this section, we establish the aggregation operators under interval valued spherical fuzzy environment and propose interval valued spherical fuzzy weighted arithmetic operator (IVSFWA) and interval valued spherical fuzzy geometric operator (IVSFWG) based on interval valued spherical fuzzy numbers (IVSFNs).

5.1 *Interval Valued Spherical Fuzzy Weighted Arithmetic Operator*

Let $S_k = \langle [P^L_{S_k}, P^U_{S_k}], [N^L_{S_k}, N^U_{S_k}], [I^L_{S_k}, I^U_{S_k}] \rangle$, $k = 1, 2, 3, \ldots, n$ be a collection of IVSpFNs and let IVSFWA: $\Phi^n \rightarrow \Phi$; if

$$\text{IVSFWA}_\omega\big(S_1, S_{2,} \ldots, S_n\big) = \omega_1 \cdot S_1 \oplus \omega_2 \cdot S_2 \oplus \cdots \oplus \omega_n \cdot S_n \tag{13}$$

then the function IVSFWA is called a interval valued spherical fuzzy weighted arithmetic operator, where $\omega = (\omega_1, \omega_2, \ldots, \omega_n)^T$ is the weight vector of S_k $(k = 1, 2, 3, \ldots, n)$, $\omega_k \geq 0$, $\sum_{k=1}^n \omega_k = 1$. As a particular case, if $\omega = \left(\frac{1}{n}, \frac{1}{n}, \ldots, \frac{1}{n}\right)^T$ then the IVSFWA is reduced into interval valued spherical arithmetic averaging operator (IVSFAA) of dimension n which is defined as follows:

$$\begin{aligned} &\text{IVSFAA}_\omega\big(S_1, S_{2,} \ldots, S_n\big) \\ &= \frac{1}{n} \cdot (S_1 \oplus S_2 \oplus \cdots \oplus S_n) \end{aligned} \tag{14}$$

Based on the proposed operational laws established in Sect. 4, we can derive the following theorem, which shows that the SFWA of IVSFSs is also an IVSFS.

5.1.1 Theorem

Let $S_k = \langle [P^L_{S_k}, P^U_{S_k}], [N^L_{S_k}, N^U_{S_k}], [I^L_{S_k}, I^U_{S_k}] \rangle$, $k = 1, 2, 3, \ldots, n$ be a collection of IVSFNs; then their aggregated value by IVSFWA operator is still an IVSFN and

$$\begin{aligned} &\text{IVSFWA}_\omega\big(S_1, S_{2,} \ldots, S_n\big) \\ &= \left\langle \left[\sqrt{1 - \prod_{k=1}^n \left(1 - (P^L_k)^2\right)^{\omega_k}}, \right.\right. \end{aligned}$$

$$\sqrt{1-\prod_{k=1}^{n}\left(1-(P_k^U)^2\right)^{\omega_k}}\Bigg],$$
$$\left[\prod_{k=1}^{n}(N_k^L)^{\omega_k},\prod_{k=1}^{n}(N_k^U)^{\omega_k},\right.$$
$$\left.\prod_{k=1}^{n}(I_k^L)^{\omega_k},\prod_{k=1}^{n}(N_k^U)^{\omega_k}\right]\Bigg\rangle \tag{15}$$

where $\omega_k > 0,$ is the weight vector of S_k and $\sum_{k=1}^{n}\omega_k = 1$.

Proof We prove this theorem using the law of mathematical induction.

For $n = 2$, based on the operational laws of IVSFNs,

$$\omega_1 S_1 = \left\langle\left[\sqrt{1-\left(1-(P_1^L)^2\right)^{\omega_1}},\sqrt{1-\left(1-(P_1^U)^2\right)^{\omega_1}}\right],\right.$$
$$\left.\left[(N_1^L)^{\omega_1},(N_1^L)^{\omega_1}\right],\left[(I_1^L)^{\omega_1},(I_1^U)^{\omega_1}\right]\right\rangle$$
$$\omega_2 S_2 = \left\langle\left[\sqrt{1-\left(1-(P_2^L)^2\right)^{\omega_2}},\sqrt{1-\left(1-(P_2^U)^2\right)^{\omega_2}}\right],\right.$$
$$\left.\left[(N_2^L)^{\omega_2},(N_2^L)^{\omega_2}\right],\left[(I_2^L)^{\omega_2},(I_2^U)^{\omega_2}\right]\right\rangle$$

$$\text{IVSFWA}_\omega(S_1, S_2)$$
$$=\left\langle\left[\sqrt{1-\prod_{k=1}^{2}\left(1-(P_k^L)^2\right)^{\omega_k}},\right.\right.$$
$$\left.\sqrt{1-\prod_{k=1}^{2}\left(1-(P_k^U)^2\right)^{\omega_k}}\right],$$
$$\left[\prod_{k=1}^{2}(N_k^L)^{\omega_k},\prod_{k=1}^{2}(N_k^U)^{\omega_k}\right],$$
$$\left.\left[\prod_{k=1}^{2}(I_k^L)^{\omega_k},\prod_{k=1}^{2}(I_k^U)^{\omega_k}\right]\right\rangle$$

For $n = t$,

$$\text{IVSFWA}_\omega(S_1, S_2, \ldots, S_t)$$

$$= \left\langle \left[\sqrt{1 - \prod_{k=1}^{t} \left(1 - (P_k^L)^2\right)^{\omega_k}}, \sqrt{1 - \prod_{k=1}^{t} \left(1 - (P_k^U)^2\right)^{\omega_k}} \right], \left[\prod_{k=1}^{t} (N_k^L)^{\omega_k}, \prod_{k=1}^{t} (N_k^U)^{\omega_k} \right], \left[\prod_{k=1}^{t} (I_k^L)^{\omega_k}, \prod_{k=1}^{t} (I_k^U)^{\omega_k} \right] \right\rangle$$

Similarly for $n = t + 1$,

$$\text{IVSFWA}_{\omega}(S_1, S_2, \ldots, S_t, S_{t+1})$$

$$\text{IVSFWA}_{\omega}(S_1, S_2, \ldots, S_t) \oplus \omega_{t+1} \cdot S_{t+1}$$

$$= \left\langle \left[\sqrt{1 - \prod_{k=1}^{t} \left(1 - (P_k^L)^2\right)^{\omega_t}}, \sqrt{1 - \prod_{k=1}^{t} \left(1 - (P_k^U)^2\right)^{\omega_t}} \right], \left[\prod_{k=1}^{t} (N^L)^{\omega_k}, \prod_{k=1}^{t} (N^U)^{\omega_k} \right], \left[\prod_{k=1}^{t} (I^L)^{\omega_k}, \prod_{k=1}^{t} (I^U)^{\omega_k} \right] \right\rangle$$

$$\oplus \left\langle \left[\sqrt{1 - \left(1 - (P_k^L)^2\right)^{\omega_{t+1}}}, \sqrt{1 - \left(1 - (P_k^U)^2\right)^{\omega_{t+1}}} \right], \left[(N_{t+1}^L)^{\omega_{t+1}}, (N_{t+1}^L)^{\omega_{t+1}} \right], \left[(I_{t+1}^L)^{\omega_{t+1}}, (I_{t+1}^L)^{\omega_{t+1}} \right] \right\rangle$$

$$= \left\langle \left[\sqrt{1 - \prod_{k=1}^{t+1} \left(1 - (P_k^L)^2\right)^{\omega_{t+1}}}, \sqrt{1 - \prod_{k=1}^{t+1} \left(1 - (P_k^U)^2\right)^{\omega_{t+1}}} \right], \left[\prod_{k=1}^{t+1} (N_k^L)^{\omega_k}, \prod_{k=1}^{t+1} (N_k^U)^{\omega_k} \right], \right.$$

$$\prod_{k=1}^{t+1}(I_k^L)^{\omega_k}, \prod_{k=1}^{t+1}(I_k^U)^{\omega_k}\Big]\Big\rangle$$

That is IVSFWA (Equation nu.) is holds for $n = t + 1$, therefore (Equation nu.) holds for all n, which completes the proof.

5.1.2 Theorem (Idempotency)

If all S_k $(k = 1, 2, 3, \ldots, n)$ are equal, that is, $S_k = S$ for all k, then

$$\text{IVSFWA}_\omega(S_1, S_{2,} \ldots, S_n) = S. \tag{16}$$

Proof Using Theorem 5.1.1,

$$\begin{aligned}
&\text{IVSpFWA}_\omega(S_1, S_{2,} \ldots, S_n) \\
&= \left\langle \left[\sqrt{1 - \prod_{k=1}^{n}\left(1 - (P_k^L)^2\right)^{\omega_k}}, \sqrt{1 - \prod_{k=1}^{n}\left(1 - (P_k^U)^2\right)^{\omega_k}}\right],\right. \\
&\left.\left[\prod_{k=1}^{n}(N_k^L)^{\omega_k}, \prod_{k=1}^{n}(N_k^U)^{\omega_k}\right], \left[\prod_{k=1}^{n}(I_k^L)^{\omega_k}, \prod_{k=1}^{n}(N_k^U)^{\omega_k}\right]\right\rangle \\
&= \left\langle \left[\sqrt{1 - \left(1 - (P^L)^2\right)^{\sum_{k=1}^{n}\omega_k}}, \sqrt{1 - \left(1 - (P^U)^2\right)^{\sum_{k=1}^{n}\omega_k}}\right],\right. \\
&\left[(N^L)^{\sum_{k=1}^{n}\omega_k}, (N^U)^{\sum_{k=1}^{n}\omega_k}\right], \\
&\left.\left[(I^L)^{\sum_{k=1}^{n}\omega_k}, (I^U)^{\sum_{k=1}^{n}\omega_k}\right]\right\rangle \\
&= \left\langle \left[\sqrt{1 - \left(1 - (P^L)^2\right)}, \sqrt{1 - \left(1 - (P^U)^2\right)}\right],\right. \\
&\left.[(N^L), (N^U)], [(I^L), (I^U)]\right\rangle \\
&= \left\langle \left[\sqrt{(P^L)^2}, \sqrt{(P^U)^2}\right], [(N^L), (N^U)], [(I^L), (I^U)]\right\rangle \\
&= \langle [P^L, P^U], [(N^L), (N^U)], [(I^L), (I^U)]\rangle \\
&= S
\end{aligned}$$

5.1.3 Theorem (Boundedness)

For a collection of IVSpFNs $S_k = \langle [P^L_{S_k}, P^U_{S_k}], [N^L_{S_k}, N^U_{S_k}], [I^L_{S_k}, I^U_{S_k}] \rangle$ $(k = 1, 2, 3, \ldots, n)$ and let

$$S^+ = \left\langle \left[\max_k P_k^-, \max_k P_k^+ \right], \left[\min_k N_k^-, \min_k N_k^+ \right], \left[\min_k I_k^-, \min_k I_k^+ \right] \right\rangle,$$

$$S^- = \left\langle \left[\min_k P_k^-, \min_k P_k^+ \right], \left[\max_k N_k^-, \max_k N_k^+ \right], \left[\max_k I_k^-, \max_k I_k^+ \right] \right\rangle.$$

Then

$$\mathrm{S}^- \leq \mathrm{IVSpFWA}_\omega(S_1, S_2, \ldots, S_n) \leq S^+. \tag{17}$$

Proof Since

$$\min_k P_k^- \leq P_k^- \leq \max P_k^-, \min_k P_k^+ \leq P_k^+ \leq \max P_k^+$$

for all k, it follows that, $1 - \max_k P_k^- \leq 1 - P_k^- \leq 1 - \min_k P_k^-$ and then,

$$\sqrt{1 - \prod_{k=1}^{n} \left(1 - \max\left(P_k^L\right)^2\right)^{\omega_k}} \leq \sqrt{1 - \left(1 - \max\left(P_k^L\right)^2\right)^{\sum_{k=1}^{n} \omega_k}}$$

$$\leq 1 - \left(1 - \max\left(P_k^L\right)^2\right)^{\sum_{k=1}^{n} \omega_k} \leq \max\left(P_k^L\right)^2 \leq \max\left(P_k^L\right),$$

$$\sqrt{1 - \prod_{k=1}^{n} \left(1 - \max\left(P_k^U\right)^2\right)^{\omega_k}} \leq \sqrt{1 - \left(1 - \max\left(P_k^U\right)^2\right)^{\sum_{k=1}^{n} \omega_k}}$$

$$\leq 1 - \left(1 - \max\left(P_k^U\right)^2\right)^{\sum_{k=1}^{n} \omega_k} \leq \max\left(P_k^U\right)^2 \leq \max\left(P_k^U\right)$$

Similarly,

$$\min_k N_k^- \leq N_k^- \leq \max_k N_k^- \Rightarrow \min_k N_k^- \leq \prod_{k=1}^{n} N_k^- \leq \max_k N_k^-,$$

$$\min_k I_k^- \leq I_k^- \leq \max_k I_k^- \Rightarrow \min_k I_k^- \leq \prod_{k=1}^{n} I_k^- \leq \max_k I_k^-$$

Let $\text{IVSFWA}_\omega(S_1, S_{2,} \ldots, S_n) = S$ then by using the score value formula of IVSFN, we have

$$\begin{aligned} s(S) &= \frac{1}{3}\Big[(P_S^L)^2 + (P_S^U)^2 - (N_S^L)^2 - (N_S^U)^2 - (I_S^L)^2 - (I_S^U)^2\Big] \\ &\leq \frac{1}{3}\Big[\max(P_S^L)^2 + \max(P_S^U)^2 - \max(N_S^L)^2 \\ &-\max(N_S^U)^2 - \max(I_S^L)^2 - \max(I_S^U)^2\Big] \\ &= s(S^+) \\ s(S) &= \frac{1}{3}\Big[(P_S^L)^2 + (P_S^U)^2 - (N_S^L)^2 - (N_S^U)^2 - (I_S^L)^2 - (I_S^U)^2\Big] \\ &\geq \frac{1}{3}\Big[\min(P_S^L)^2 + \min(P_S^U)^2 - \min(N_S^L)^2 \\ &-\min(N_S^U)^2 - \min(I_S^L)^2 - \min(I_S^U)^2\Big] \\ &= s(S^-) \end{aligned}$$

Therefore, $S^- \leq \text{IVSFWA}_\omega(S_1, S_{2,} \ldots, S_n) \leq S^+$.

5.1.4 Theorem (Monotonicity)

Let S_k and $S_k'(k = 1, 2, 3, \ldots, n)$ be two sets of interval valued Spherical Fuzzy environments, if $S_k \leq S_k'$ for all k, then $IVSFWA_\omega(S_1, S_2, \ldots, S_n) \leq IVSFWA_\omega(S_1', S_2', \ldots, S_n')$.

Proof Proof follows from the above property and therefore it is omitted here.

5.1.5 Theorem (Invariance)

If $S_{n+1} = \langle [P_{n+1}^L, P_{n+1}^U], [N_{n+1}^L, N_{n+1}^U], [I_{n+1}^L, I_{n+1}^U] \rangle$ is also an IVSFN then for $t > 0$

$$\begin{aligned} &\text{IVSFWA}_\omega(t \cdot S_1 \oplus S_{n+1}, t \cdot S_2 \oplus S_{n+1}, \ldots, t \cdot S_n \oplus S_{n+1}) \\ &= t \cdot \text{IVSFWA}_\omega(S_1, S_2, \ldots, S_n) \oplus S_{n+1} \end{aligned} \tag{18}$$

Proof To prove the theorem it is enough to prove the following results (1) and (2).

1. $\text{IVSFWA}_\omega(S_1 \oplus S_{n+1}, S_2 \oplus S_{n+1}, \ldots, S_n \oplus S_{n+1})$

$$= \text{IVSFWA}_\omega(S_1, S_2, \ldots, S_n) \oplus S_{n+1} \tag{19}$$

2. $\text{IVSFWA}_\omega(t \cdot S_1, t \cdot S_2, \ldots, t \cdot S_n)$
$$= t \cdot \text{IVSFWA}_\omega(S_1, S_2, \ldots, S_n) \tag{20}$$

From (1) and (2), the theorem will be obviously true.
Since,

$$(S_k \oplus S_{n+1}) = \left\langle \left[\sqrt{1 - \prod_{q=\{k,n+1\}} \left(1 - \left(P_q^L\right)^2\right)}, \sqrt{1 - \prod_{q=\{k,n+1\}} \left(1 - \left(P_q^U\right)^2\right)} \right], \left[\prod_{q=\{k,n+1\}} \left(N_q^L\right), \prod_{q=\{k,n+1\}} \left(N_q^U\right) \right], \left[\prod_{q=\{k,n+1\}} \left(I_q^L\right), \prod_{q=\{k,n+1\}} \left(I_q^U\right) \right] \right\rangle$$

LHS of (i) $= \text{IVSpFWA}_\omega\left(S_1 \oplus S_{n+1}, S_2 \oplus S_{n+1}, \ldots, S_n \oplus S_{n+1}\right)$

$$= \left\langle \left[\sqrt{1 - \prod_{q=\{k,n+1\}} \left(1 - \left(P_q^L\right)^2\right)^{\omega_k}}, \sqrt{1 - \prod_{q=\{k,n+1\}} \left(1 - \left(P_q^U\right)^2\right)^{\omega_k}} \right], \left[\prod_{q=\{k,n+1\}} \left(N_q^L\right)^{\omega_k}, \prod_{q=\{k,n+1\}} \left(N_q^U\right)^{\omega_k} \right], \left[\prod_{q=\{k,n+1\}} \left(I_q^L\right)^{\omega_k}, \prod_{q=\{k,n+1\}} \left(I_q^U\right)^{\omega_k} \right] \right\rangle$$

$$= \left\langle \left[\sqrt{1 - \prod_{q=\{k,n+1\}} \left(1 - \left(P_q^L\right)^2\right)^{\omega_k}}, \sqrt{1 - \prod_{q=\{k,n+1\}} \left(1 - \left(P_q^U\right)^2\right)^{\omega_k}} \right], \left[\prod_{q=\{k,n+1\}} \left(N_q^L\right)^{\omega_k}, \prod_{q=\{k,n+1\}} \left(N_q^U\right)^{\omega_k} \right], \left[\prod_{q=\{k,n+1\}} \left(I_q^L\right)^{\omega_k}, \prod_{q=\{k,n+1\}} \left(I_q^U\right)^{\omega_k} \right] \right\rangle$$

$$= \left\langle \left[\left[\sqrt{1 - \prod_{k=1}^{n} \left(1 - \left(P_k^L\right)^2\right)^{\omega_k} \cdot \left(1 - \left(P_{n+1}^L\right)^2\right)^{\sum_{k=1}^{n} \omega_k}}, \right], \sqrt{1 - \prod_{k=1}^{n} \left(1 - \left(P_k^U\right)^2\right)^{\omega_k} \cdot \left(1 - \left(P_{n+1}^U\right)^2\right)^{\sum_{k=1}^{n} \omega_k}} \right. \right.$$

$$\left[\prod_{k=1}^{n} \left(N_k^L\right)^{\omega_k} \cdot \left(N_{n+1}^L\right)^{\sum_{k=1}^{n} \omega_k}, \prod_{k=1}^{n} \left(N_k^U\right)^{\omega_k} \cdot \left(N_{n+1}^U\right)^{\sum_{k=1}^{n} \omega_k} \right],$$

$$\left[\prod_{k=1}^{n}\left(I_k^L\right)^{\omega_k}\cdot\left(I_{n+1}^L\right)^{\sum_{k=1}^{n}\omega_k},\prod_{k=1}^{n}\left(I_k^U\right)^{\omega_k}\cdot\left(I_{n+1}^U\right)^{\sum_{k=1}^{n}\omega_k}\right]\Bigg\rangle$$

$$=\Bigg\langle\left[\sqrt{1-\prod_{k=1}^{n}\left(1-\left(P_k^L\right)^2\right)^{\omega_k}\cdot\left(1-\left(P_{n+1}^L\right)^2\right)},\right.$$

$$\left.\sqrt{1-\prod_{k=1}^{n}\left(1-\left(P_k^U\right)^2\right)^{\omega_k}\cdot\left(1-\left(P_{n+1}^U\right)^2\right)}\right]$$

$$\left[\prod_{k=1}^{n}\left(N_k^L\right)^{\omega_k}\cdot\left(N_{n+1}^L\right),\prod_{k=1}^{n}\left(N_k^U\right)^{\omega_k}\cdot\left(N_{n+1}^U\right)\right],$$

$$\left[\prod_{k=1}^{n}\left(I_k^L\right)^{\omega_k}\cdot\left(I_{n+1}^L\right),\prod_{k=1}^{n}\left(I_k^U\right)^{\omega_k}\cdot\left(I_{n+1}^U\right)\right]\Bigg\rangle \tag{21}$$

Consider RHS of (i) $=\text{IVSFWA}_{\omega}(S_1,S_2,\ldots,S_n)\oplus S_{n+1}$

$$=\Bigg\langle\left[\sqrt{1-\prod_{k=1}^{n}\left(1-(P_k^L)^2\right)^{\omega_k}},\sqrt{1-\prod_{k=1}^{n}\left(1-(P_k^U)^2\right)^{\omega_k}}\right],$$

$$\left[\prod_{k=1}^{n}(N_k^L)^{\omega_k},\prod_{k=1}^{n}(N_k^U)^{\omega_k}\right],\left[\prod_{k=1}^{n}(I_k^L)^{\omega_k},\prod_{k=1}^{n}(I_k^U)^{\omega_k}\right]\Bigg\rangle$$

$$\oplus\left\langle\left[P_{n+1}^L,P_{n+1}^U\right],\left[N_{n+1}^L,N_{n+1}^U\right],\left[I_{n+1}^L,I_{n+1}^U\right]\right\rangle$$

$$=\Bigg\langle\left[\sqrt{1-\prod_{k=1}^{n}\left(1-(P_k^L)^2\right)^{\omega_k}\cdot\left(1-(P_{n+1}^L)^2\right)},\right.$$

$$\left.\sqrt{1-\prod_{k=1}^{n}\left(1-(P_k^U)^2\right)^{\omega_k}\cdot\left(1-(P_{n+1}^U)^2\right)}\right]$$

$$\left[\prod_{k=1}^{n}(N_k^L)^{\omega_k}\cdot(N_{n+1}^L),\prod_{k=1}^{n}(N_k^U)^{\omega_k}\cdot(N_{n+1}^U)\right],$$

$$\left[\prod_{k=1}^{n}(I_k^L)^{\omega_k}\cdot(I_{n+1}^L),\prod_{k=1}^{n}(I_k^U)^{\omega_k}\cdot(I_{n+1}^U)\right]\Bigg\rangle \tag{22}$$

Therefore from Eqs. (21) and (22)

$$\begin{aligned}&\text{IVSFWA}_{\omega}(S_1\oplus S_{n+1},S_2\oplus S_{n+1},\ldots,S_n\oplus S_{n+1})\\&\quad=\text{IVSFWA}_{\omega}(S_1,S_2,\ldots,S_n)\oplus S_{n+1}\end{aligned}$$

Using Eq. (10),

$$t \cdot S_k = \left\langle \left[\sqrt{1 - \prod_{k=1}^{n} \left(1 - (P_k^L)^2\right)^t}, \sqrt{1 - \prod_{k=1}^{n} \left(1 - (P_k^U)^2\right)^t} \right], \left[\prod_{k=1}^{n} (N_k^L)^t, \prod_{k=1}^{n} (N_k^U)^t \right], \left[\prod_{k=1}^{n} (I_k^L)^t, \prod_{k=1}^{n} (I_k^L)^t \right] \right\rangle$$

$$\text{IVSFWA}_\omega(t \cdot S_1, t \cdot S_2, \ldots, t \cdot S_n) = \left\langle \left[\sqrt{1 - \prod_{k=1}^{n} \left(1 - (P_k^L)^2\right)^{t\omega_k}}, \sqrt{1 - \prod_{k=1}^{n} \left(1 - (P_k^U)^2\right)^{t\omega_k}} \right], \left[\prod_{k=1}^{n} (N_k^L)^{t\omega_k}, \prod_{k=1}^{n} (N_k^U)^{t\omega_k} \right], \left[\prod_{k=1}^{n} (I_k^L)^{t\omega_k}, \prod_{k=1}^{n} (I_k^L)^{t\omega_k} \right] \right\rangle \tag{23}$$

$$t \cdot \text{IVSpFWA}_\omega(S_1, S_2, \ldots, S_n) = t \cdot \left(\left\langle \left[\sqrt{1 - \prod_{k=1}^{n} \left(1 - (P_k^L)^2\right)^{\omega_k}}, \sqrt{1 - \prod_{k=1}^{n} \left(1 - (P_k^U)^2\right)^{\omega_k}} \right], \left[\prod_{k=1}^{n} (N_k^L)^{\omega_k}, \prod_{k=1}^{n} (N_k^U)^{\omega_k} \right], \left[\prod_{k=1}^{n} (I_k^L)^{\omega_k}, \prod_{k=1}^{n} (I_k^U)^{\omega_k} \right] \right\rangle \right)$$
$$= \left\langle \left[\sqrt{1 - \prod_{k=1}^{n} \left(1 - (P_k^L)^2\right)^{t\omega_k}}, \sqrt{1 - \prod_{k=1}^{n} \left(1 - (P_k^U)^2\right)^{t\omega_k}} \right], \left[\prod_{k=1}^{n} (N_k^L)^{t\omega_k}, \prod_{k=1}^{n} (N_k^U)^{t\omega_k} \right], \left[\prod_{k=1}^{n} (I_k^L)^{t\omega_k}, \prod_{k=1}^{n} (I_k^L)^{t\omega_k} \right] \right\rangle \tag{24}$$

From Eqs. (23) and (24),

$$\text{IVSFWA}_\omega(t \cdot S_1, t \cdot S_2, \ldots, t \cdot S_n) = t \cdot \text{IVSFWA}_\omega(S_1, S_2, \ldots, S_n), t > 0$$

From the Eqs. (19) and (20), it is obvious that,

$$\text{IVSFWA}_\omega(t \cdot S_1 \oplus S_{n+1}, t \cdot S_2 \oplus S_{n+1}, \ldots, t \cdot S_n \oplus S_{n+1}) = t \cdot \text{IVSFWA}_\omega(S_1, S_2, \ldots, S_n) \oplus S_{n+1}$$

Hence the theorem.

5.2 Interval Valued Spherical Fuzzy Weighted Geometric Operator

Let $S_k = \langle [P^L_{S_k}, P^U_{S_k}], [N^L_{S_k}, N^U_{S_k}], [I^L_{S_k}, I^U_{S_k}] \rangle$, $k = 1, 2, 3, \ldots, n$ be a collection of IVSpFNs and let IVSFWG: $\Phi^n \to \Phi$; if

$$\text{IVSFWGA}_\omega(S_1, S_2, \ldots, S_n) = S_1^{\omega_1} \otimes S_2^{\omega_2} \otimes \cdots \otimes S_n^{\omega_n} \tag{25}$$

then the function IVSFWG is called a interval valued spherical fuzzy weighted geometric operator, where $\omega = (\omega_1, \omega_2, \ldots, \omega_n)^T$ is the weight vector of S_k $(k = 1, 2, 3, \ldots, n)$, $\omega_k \geq 0$, $\sum_{k=1}^n \omega_k = 1$. As a particular case, if $\omega = \left(\frac{1}{n}, \frac{1}{n}, \ldots, \frac{1}{n}\right)^T$ then the IVSFWG is reduced into interval valued spherical arithmetic averaging operator (IVSFGA) of dimension n which is defined as follows:

$$\text{IVSFGA}_\omega(S_1, S_2, \ldots, S_n) = (S_1 \otimes S_2 \otimes \cdots \otimes S_n)^{\frac{1}{n}} \tag{26}$$

and shows that the SFWA of IVSFSs is also an IVSFS.

5.2.1 Theorem

Let $S_k = \langle [P^L_{S_k}, P^U_{S_k}], [N^L_{S_k}, N^U_{S_k}], [I^L_{S_k}, I^U_{S_k}] \rangle$, $k = 1, 2, 3, \ldots, n$ be a collection of IVSFNs; then their aggregated value by IVSFWG operator is still an IVSFN and

$$\begin{aligned} &\text{IVSFWG}_\omega(S_1, S_2, \ldots, S_n) \\ &= \left\langle \left[\prod_{k=1}^n \left(P^L_k\right)^{\omega_k}, \prod_{k=1}^n \left(P^U_k\right)^{\omega_k}\right], \left[\prod_{k=1}^n \left(N^L_k\right)^{\omega_k}, \prod_{k=1}^n \left(N^U_k\right)^{\omega_k}\right], \right. \\ &\left. \left[\sqrt{1 - \prod_{k=1}^n \left(1 - \left(I^L_k\right)^2\right)^{\omega_k}}, \sqrt{1 - \prod_{k=1}^n \left(1 - \left(I^U_k\right)^2\right)^{\omega_k}}\right] \right\rangle \end{aligned} \tag{27}$$

where $\omega_k > 0$, is the weight vector of S_k and $\sum_{k=1}^n \omega_k = 1$.

The proof can be performed in a similar way for Sect. 5.1.1.

5.2.2 Theorem (Idempotency)

If all S_k $(k = 1, 2, 3, \ldots, n)$ are equal, that is, $S_k = S$ for all

$$\text{IVSFWG}_\omega(S_1, S_2, \ldots, S_n) = S. \tag{28}$$

k, then

The proof can be performed in a similar way for Sect. 5.1.2.

5.2.3 Theorem (Boundedness)

For a collection of IVSFNs $S_k = \langle [P^L_{S_k}, P^U_{S_k}], [N^L_{S_k}, N^U_{S_k}], [I^L_{S_k}, I^U_{S_k}] \rangle$ $(k = 1, 2, 3, \ldots, n)$ and let

$$\begin{aligned} S^+ &= \left\langle \left[\max_k P_k^-, \max_k P_k^+ \right], \right. \\ &\quad \left. \left[\min_k N_k^-, \min_k N_k^+ \right], \left[\min_k I_k^-, \min_k I_k^+ \right] \right\rangle, \\ S^- &= \left\langle \left[\min_k P_k^-, \min_k P_k^+ \right], \right. \\ &\quad \left. \left[\max_k N_k^-, \max_k N_k^+ \right], \left[\max_k I_k^-, \max_k I_k^+ \right] \right\rangle. \end{aligned}$$

Then

$$\mathrm{S}^- \leq \mathrm{IVSFWG}_\omega(S_1, S_{2,} \ldots, S_n) \leq S^+. \tag{29}$$

The proof can be performed in a similar way for Sect. 5.1.3.

5.2.4 Theorem (Monotonicity)

Let S_k and $S_k^{'}(k = 1, 2, 3, \ldots, n)$ be two sets of interval valued Spherical Fuzzy environments, if $S_k \leq S_k^{'}$ for all *k*, then

$$IVSFWG_\omega(S_1, S_2, \ldots, S_n) \leq IVSFWG_\omega(S_1', S_2', \ldots, S_n') \tag{30}$$

Proof Proof follows from the above property and therefore it is omitted here.

5.2.5 Theorem (Invariance)

If $S_{n+1} = \langle [P^L_{n+1}, P^U_{n+1}], [N^L_{n+1}, N^U_{n+1}], [I^L_{n+1}, I^U_{n+1}] \rangle$ is also an IVSFN then for $t > 0$,

$$\begin{aligned} &\mathrm{IVSFWG}_\omega(S_1^t \otimes S_{n+1}, S_2^t \otimes S_{n+1}, \ldots, S_n^t \otimes S_{n+1}) \\ &\quad = (\mathrm{IVSFWG}_\omega(S_1, S_2, \ldots, S_n))^t \otimes S_{n+1}. \end{aligned} \tag{31}$$

The proof can be performed in a similar way for Sect. 5.1.5.

6 Application of Proposed Operators

In this section, we proposed a new algorithm to solve a decision making problem to choose the best alternative from all the possible alternatives using the proposed aggregation operators.

6.1 Proposed Methodology

Step 1. Gather the information associated with all the alternatives under various criterions/parameters in the form of interval valued spherical fuzzy matrix (IVSFM) is SD and defined by

$$SD = \langle [P_{ij}^L, P_{ij}^U], [N_{ij}^L, N_{ij}^U], [I_{ij}^L, I_{ij}^U] \rangle_{n \times m}$$

Step 2. Normalize the matrix SD by converting the grade values into benefit type of the cost parameters by applying the following formula.

$$R_{ij} = \begin{cases} S_{ij}, \; for\, benefit\, type\, parameters \\ S_{ij}^c, \; for\, cost\, type\, parameters \end{cases} \tag{32}$$

where, S_{ij}^c is the complement of S_{ij}.

If all the parameters are of the same type then normalization is not necessary.

Step 3. Aggregate the IVSFNs S_{ij} for all the alternatives $A_m (m = 1, 2, \ldots, u)$ into collective decision matrix χ_m using proposed IVSFWA (or IVSFWG) operator.

Step 4. Calculate the score value of all the alternatives $A_m (m = 1, 2, \ldots, u)$

Step 5. Using score value of the alternatives, select the best one.

Step 6. End.

6.2 Application of Proposed Aggregation Operator in Decision Making

Here the application is taken from (Ashraf et al. 2019a, b). The congruity and capability of the proposed aggregation operator in decision-making problems are presented in this part. The quality of air is signified for the three stations is (λ, μ, ρ). Weight vector for the alternatives provided by the decision makers $\omega = (0.3, 0.4, 0.2, 0.1)$. Further, the three calculated indexes are defined by $SO_2(A_1)$, $NO_2(A_2)$ and $PM_{10}(A_3)$. The values obtained for the stations which examined quality of air under these indexes is described Eqs. 28–30, and they are shown in Step 1. Here we consider the same type of data and hence normalization is not necessary for this problem.

Step 1:

Table 1 Air quality data from λ

$$SD1 = \begin{bmatrix} \langle[0.6, 0.7], [0.1, 0.2], [0.8, 0.9]\rangle & \langle[0.3, 0.6], [0.2, 0.3], [0.3, 0.4]\rangle & \langle[0.8, 0.9], [0.3, 0.4], [0.4, 0.5]\rangle \\ \langle[0.4, 0.5], [0.2, 0.3], [0.6, 0.7]\rangle & \langle[0.5, 0.6], [0.3, 0.4], [0.4, 0.5]\rangle & \langle[0.7, 0.8], [0.2, 0.3], [0.3, 0.4]\rangle \\ \langle[0.7, 0.8], [0.2, 0.4], [0.5, 0.6]\rangle & \langle[0.6, 0.7], [0.2, 0.4], [0.3, 0.5]\rangle & \langle[0.8, 0.9], [0.3, 0.4], [0.2, 0.3]\rangle \\ \langle[0.5, 0.7], [0.1, 0.2], [0.4, 0.5]\rangle & \langle[0.7, 0.8], [0.3, 0.4], [0.5, 0.6]\rangle & \langle[0.6, 0.7], [0.1, 0.2], [0.3, 0.4]\rangle \end{bmatrix}$$

Table 2 Air quality data from μ

$$SD2 = \begin{bmatrix} \langle[0.6, 0.8], [0.3, 0.4], [0.7, 0.8]\rangle & \langle[0.3, 0.7], [0.1, 0.2], [0.4, 0.5]\rangle & \langle[0.5, 0.6], [0.1, 0.2], [0.3, 0.4]\rangle \\ \langle[0.8, 0.9], [0.1, 0.2], [0.6, 0.7]\rangle & \langle[0.7, 0.9], [0.2, 0.3], [0.3, 0.6]\rangle & \langle[0.4, 0.6], [0.2, 0.3], [0.2, 0.3]\rangle \\ \langle[0.7, 0.8], [0.2, 0.3], [0.3, 0.4]\rangle & \langle[0.6, 0.8], [0.3, 0.4], [0.2, 0.4]\rangle & \langle[0.7, 0.9], [0.2, 0.4], [0.4, 0.5]\rangle \\ \langle[0.5, 0.7], [0.1, 0.2], [0.3, 0.5]\rangle & \langle[0.7, 0.8], [0.2, 0.3], [0.4, 0.5]\rangle & \langle[0.3, 0.6], [0.3, 0.5], [0.6, 0.8]\rangle \end{bmatrix}$$

Table 3 Air quality data from ρ

$$SD3 = \begin{bmatrix} \langle[0.6, 0.7], [0.2, 0.3], [0.4, 0.5]\rangle & \langle[0.7, 0.8], [0.3, 0.4], [0.3, 0.5]\rangle & \langle[0.8, 0.9], [0.1, 0.2], [0.2, 0.3]\rangle \\ \langle[0.8, 0.9], [0.3, 0.4], [0.4, 0.5]\rangle & \langle[0.6, 0.8], [0.1, 0.2], [0.6, 0.7]\rangle & \langle[0.6, 0.7], [0.3, 0.4], [0.4, 0.5]\rangle \\ \langle[0.7, 0.8], [0.1, 0.2], [0.2, 0.4]\rangle & \langle[0.5, 0.6], [0.3, 0.4], [0.4, 0.5]\rangle & \langle[0.7, 0.8], [0.1, 0.2], [0.4, 0.6]\rangle \\ \langle[0.8, 0.9], [0.2, 0.3], [0.3, 0.5]\rangle & \langle[0.6, 0.9], [0.2, 0.3], [0.8, 0.9]\rangle & \langle[0.6, 0.8], [0.2, 0.3], [0.5, 0.6]\rangle \end{bmatrix}$$

Step 2: Normalization is not necessary
Step 3: The aggregated value is obtained by using Eq. 15 and the obtained values are

$$\chi_1 = \langle[0.88, 0.95], [0.008, 0.032], [0.064, 0.131]\rangle$$
$$\chi_2 = \langle[0.876, 0.977], [0.004, 0.022], [0.045, 0.129]\rangle$$
$$\chi_3 = \langle[0.92, 0.98], [0.006, 0.025], [0.05, 0.131]\rangle$$

Step 4: Using the definition of score function of IVSFN (Eq. 12), the obtained score values are

$$s_1 = 0.551, s_2 = 0.567, s_3 = 0.595$$

Step 5: Since $s_3 > s_2 > s_1$, the best alternative is PM_{10}.
Hence the third station scrutinizes quality air.

7 Comparative Analysis

In Ashraf et al. (2019a, b), operational laws for spherical fuzzy numbers, aggregation operators and their properties are discussed in detail using various triangular norms under spherical fuzzy environment and applied the proposed operators in decision making problem to find the best station which scrutinize the quality of air.

In our proposed work, interval valued spherical fuzzy set is introduced. Also operational laws for interval valued spherical fuzzy numbers, aggregation operators (IVSFWA and IVSFWG) and their properties are derived in detail. Further applied the proposed aggregation operators in decision making problem and selected the best station which scrutinizes the quality of air. Since, uncertainty can be measured effectively in terms of an interval number and provides improved result in fuzzy mathematics, in our proposed work, the preference values are considered as interval numbers rather than an exact number. Hence the novelty and effectiveness of the proposed method.

8 Conclusion

Decision making includes planning, directing and organizing the information and getting the desired output. There are many methodologies and techniques that have been introduced for solving decision making problem. This present work introduces interval valued spherical fuzzy set and operational laws for interval valued spherical fuzzy numbers. Also, a new score function has been introduced. Further two aggregation operators namely interval valued spherical fuzzy weighted arithmetic (IVSFWA) and interval valued spherical fuzzy weighted geometric (IVSFWG) operators with their desired mathematical properties. Furthermore proposed aggregation operators are used in a decision making problem to find the best station which scrutinizes the quality of air. In future this concept may be extended with different triangular norms.

References

Ashraf S, Abdullah S (2018) Spherical aggregation operators and their application in multi attribute group decision making. Int J Intell Syst 1–31

Ashraf S, Abdullah S, Mahmood T (2019a) Spherical fuzzy Dombi aggregation operators and their application in group decision making problems. J Amb Intell Hum Comput. https://doi.org/10.1007/s12652-019-01333-y

Ashraf S, Abdullah S, Aslam M, Qiyasa M, Kutbi MA (2019b) Spherical fuzzy sets and its representation of spherical fuzzy t-norms and t-conorms. J Intell Fuzzy Syst. https://doi.org/10.3233/JIFS-81941

Atanassov KT (2017) Type-1 fuzzy sets and intuitionistic fuzzy sets. Algorithms 106:1–12

Augustine EP (2019) Improved composite relation for Pythagorean fuzzy sets and its application to medical diagnosis. Granular Comput. https://doi.org/10.1007/s41066-019-00156-8

Beliakov Gleb, James Simon, Mordelova Juliana, Ruckschlossová Tatiana, Yager RR (2010) Generalized Bonferroni mean operators in multicriteria aggregation. Fuzzy Sets Syst 161(17):2227–2242

Broumi S, Smarandache F (2014) New operations over interval valued intuitionistic hesitant fuzzy set. Math Stat 2(2):62–71

Dominguez LP, Picon LAR, Iniesta AA, Cruz DL, Xu Z (2018) MOORA under Pythagorean fuzzy set for multiple criteria decision making. Complexity 2018:1–10

Ejegwa PA (2018) Distance and similarity measures for Pythagorean fuzzy sets. Granular Comput. https://doi.org/10.1007/s41066-018-00149-z

Ejegwa PA, Akowe SO, Otene PM, Ikyule JM (2014) An overview on intuitionistic fuzzy sets. Int J Sci Technol 3(3):142–144

Garg H (2018) Hesitant Pythagorean fuzzy sets and their aggregation operators in multiple attribute decision-making. Int J Uncert Quant 8(3):267–289

Garg H, Munir M, Ullah K, Mahmood T, Jan N (2018) Algorithm for T-spherical fuzzy multi-attribute decision making based on improved interactive aggregation operators. Symmetry 10(670):1–24

Gonga Z, Xu X, Yang Y, Zhoud Y, Zhang H (2016) The spherical distance for intuitionistic fuzzy sets and its application in decision analysis. Technol Econ Dev Econ 22(3):393–415

Gundogdu FK, Kahraman C (2018) Spherical fuzzy sets and spherical fuzzy TOPSIS method. J Intell Fuzzy Syst 36:1–16. https://doi.org/10.3233/jifs-181401

Gündoğdu FK, Kahraman C (2019) A novel fuzzy TOPSIS method using emerging interval-valued spherical fuzzy sets. Eng Appl Artif Intell 85:307–323

Jamkhaneh EB (2016) New operations over generalized interval valued intuitionistic fuzzy sets. Gazi Univ J Sci 29(3):667–674

Jin H, Ashraf S, Abdullah S, Qiyas M, Bano M, Zeng S (2019) Linguistic spherical fuzzy aggregation operators and their applications in multi-attribute decision making problems. Mathematics 7(413):1–22

Li and Feng (2013) Intuitionistic (λ, μ)-fuzzy sets in Γ—semigroups. J Inequal Appl 2013(107):1–9

Li D, Zeng W (2018) Distance measure of Pythagorean fuzzy sets. Int J Intell Syst 33:348–361

Osawaru KE, Olaleru JO, Olaoluwa HO (2018) Relative fuzzy set. J Fuzzy Set Valued Anal 2(2018):86–110

Peng X (2018a) New similarity measure and distance measure for Pythagorean fuzzy set. Complex Intell Syst. https://doi.org/10.1007/s40747-018-0084-x

Peng X (2018b) New operations for interval-valued pythagorean fuzzy set. Sci Iran 26(2):1049–1076

Peng X, Yang Y (2015) Some results for Pythagorean fuzzy sets. Int J Intell Syst 30(11):1133–1160

Qin J, Liu X (2014) Frank aggregation operators for triangular interval type-2 fuzzy set and its application in multiple attribute group decision making. J Appl Math 2014(3):1–24

Rafiq M, Ashraf S, Abdullah S, Mahmood T, Muhammad S (2019) The cosine similarity measures of spherical fuzzy sets and their applications in decision making. J Intell Fuzzy Syst. https://doi.org/10.3233/JIFS-181922

Rahman K, Abdullah S, Khan MSA, Ibrar M, Husain E (2017) Some basic operations on Pythagorean fuzzy sets. J Appl Environ Biol Sci 7(1):111–119
Rajesh K, Srinivasan R (2017a) A study on some operators over interval valued intuitionistic fuzzy sets of second type. Int J Innov Res Sci Eng Technol 6(11):21051–21057
Rajesh K, Srinivasan R (2017b) Interval valued intuitionistic fuzzy sets of second type. Adv Fuzzy Math 12(4):845–853
Rizwan U, Iqbal MN (2016) New interval valued intuitionistic fuzzy set of cube root type. Int J Pure Appl Math 111(2):199–208
Starosta BL, Nski WK (2013) Metasets, intuitionistic fuzzy sets and uncertainty. Lect Notes Comput Sci 7894:388–399
Sunny T, Jose SM (2019) Boxdot and star products on interval-valued intuitionistic fuzzy graphs. Int J Innov Technol Explor Eng (IJITEE). 8(7):1238–1242
Thao NX, Smarandache F (2019) A new fuzzy entropy on Pythagorean fuzzy sets. J Intell Fuzzy Syst 37(1):1065–1074
Ullah K, Mahmood T, Jan N (2018) Similarity measures for T-spherical fuzzy sets with applications in pattern recognition. Symmetry 10(193):1–14
Ullah K, Garg H, Mahmood T, Jan N, Ali Z (2019) Correlation coefficients for T-spherical fuzzy sets and their applications in clustering and multi-attribute decision making. Soft Comput. https://doi.org/10.1007/s00500-019-03993-6
Wang W, Liu X, Qin Y (2012) Interval-valued intuitionistic fuzzy aggregation operators. J Syst Eng Electron 23(4):574–580
Wu HC (2018) Duality in fuzzy sets and dual arithmetics of fuzzy sets. Mathematics 7(1):1–24
Xu W, Liu Y, Sun W (2011) On starshaped intuitionistic fuzzy sets. Appl Math 2:1051–1058
Yager RR (2009) On generalized Bonferroni mean operators for multi-criteria aggregation. Int J Approx Reason 50:1279–1286
Zeng S, Hussain A, Mahmood T, Ali MI, Ashraf S, Munir M (2019) Covering-based spherical fuzzy rough set model hybrid with TOPSIS for multi-attribute decision-making. Symmetry 11(547):1–19

Spherical Fuzzy Numbers and Multi-criteria Decision-Making

Irfan Deli and Naim Çağman

Abstract In this work, spherical fuzzy numbers that are generalization of the fuzzy numbers and intuitionistic fuzzy numbers are defined. Then, two special forms of spherical fuzzy numbers called spherical trapezoidal fuzzy numbers (STF-numbers) and spherical triangular fuzzy numbers (STrF-numbers) are defined with laws of their operations and their properties. Also, a distance measure and some arithmetic and geometric operators on both the STF-numbers and STrF-numbers are developed. Finally, by using the STF-numbers and STrF-numbers, two multi-criteria decision making methods are developed and two examples are given to illustrate the proposed methods.

1 Introduction

In our day-to-day life, people often face with many multi-criteria decision-making (MCDM) problems and the MCDM problems contain uncertain information. The information of the MCDM problems cannot always express by crisp numbers and some are more suitable to be denoted by fuzzy sets introduced by Zadeh (1965). The fuzzy sets, especially fuzzy numbers applied to different fields, such as decision-making, pattern recognition, game theory, economics, management, military, engineering technology and so on. However, the fuzzy set just uses single value to character the membership degree of the elements in a universe and ignored the non-membership degree. Therefore, Atanassov (1986) proposed the theory of intuitionistic fuzzy sets, which uses a membership function μ and a non-membership function ν

I. Deli (✉)
Kilis Aralık University, 79000 Kilis, Turkey
e-mail: irfandeli@kilis.edu.tr

N. Çağman
Department of Mathematics, Gaziosmanpaşa University, 60250 Tokat, Turkey
e-mail: naim.cagman@gop.edu.tr

© Springer Nature Switzerland AG 2021

C. Kahraman and F. Kutlu Gündoğdu (eds.), *Decision Making with Spherical Fuzzy Sets*, Studies in Fuzziness and Soft Computing 392,
https://doi.org/10.1007/978-3-030-45461-6_3

for the membership degree of elements in a universe such that $0 \leq \mu(x) + \nu(x) \leq 1$. Also, $\pi(x) = 1 - \mu(x) - \nu(x) \leq 1$ is called hesitancy degree of element x in the universe.

Aggregation operators are the important a field to process the decision making problems in intuitionistic fuzzy theory. Therefore; after Atanassov, aggregation operators are developed in Wei (2010), Xu (2007), Xu and Wang (2012), Xu and Yager (2006), Yu (2013), Zhou and Wei (2013) and geometric operators are developed in He et al. (2014a), He et al. (2014b), Ullah et al. (2018) based on intuitionistic fuzzy sets. Mahapatra and Roy (2009) presented triangular intuitionistic fuzzy numbers which are generalization of triangular fuzzy numbers. Then, Liang et al. (2014) constructed some aggregation operators, are called intuitionistic fuzzy weighted averaging operator, intuitionistic fuzzy order weighted averaging operator, intuitionistic fuzzy hybrid weighted averaging operator with applications to multiatribute decision-making by using the operators. Liang et al. (2014) introduced some aggregation operators on triangular intuitionistic fuzzy numbers and examined their desirable properties. Then, Das and Guha (2013) presented trapezoidal intuitionistic fuzzy numbers which are generalization of trapezoidal fuzzy numbers. Wu and Cao (2013) further introduced some cases of aggregation operators with intuitionistic trapezoidal fuzzy numbers such as; intuitionistic trapezoidal fuzzy weighted geometric operator, intuitionistic trapezoidal fuzzy ordered weighted geometric operator, induced intuitionistic trapezoidal fuzzy ordered weighted geometric operator and the intuitionistic trapezoidal fuzzy hybrid geometric operator. The triangular intuitionistic fuzzy numbers and trapezoidal intuitionistic fuzzy numbers were studied by many researchers such as; on operators [e.g.Wang et al. (2013)], on operations [e.g. Liu (2013), Wan (2013), Wan and Dong (2015)], on application [e.g. Chen and Li (2011), Li (2010), Wang and Zhong (2009), Wan and Dong (2014), Zhang and Liu (2010), Zhang et al. (2013)], and so on.

In recent years, intuitionistic fuzzy sets were extended by Gündoğdu and Kahraman (2019a) to spherical fuzzy sets such that $0 \leq \mu^2(x) + \nu^2(x) + \pi^2(x) \leq 1$ which define hesitancy degree of element x in the universe independently from membership and nonmembership degrees. Also, Gündoğdu and Kahraman (2019a) define spherical fuzzy distances, arithmetic and geometric operations. Then, Ullah et al. (2018) presented T-spherical fuzzy sets includig some similarity measures such as cosine similarity measures, grey similarity measures, and set theoretic similarity measures. Garg et al. (2018) improved aggregation operators for the T-spherical fuzzy sets, which is an extension of the several existing sets, such as intuitionistic fuzzy sets, picture fuzzy sets, neutrosophic sets, and Pythagorean fuzzy sets. Liu et al. (2019) shown some limitations in the operational laws for spherical fuzzy sets and presented novel operational laws on the sets. Moreover, they gave new aggregation operators and applied to multiple-attribute group decision-making (MAGDM). Also, Gündoğdu and Kahraman (2019a) gave definition of interval valued spherical fuzzy sets and developed a new TOPSIS method on the sets.

Neutrosophic sets are presented by Smarandache (1998) that are characterized by a truthiness degree $\mu(x)$, an indeterminacy degree $\nu(x)$ and a falsity degree $\pi(x)$. Then, Ye (2015), Subas (2015) and Deli and Subas (2017) defined the neutrosophic numbers. Since, the neutrosophic numbers contain truthiness degree $\mu(x) \in [0, w]$, an indeterminacy degree $\nu(x) \in [u, 1]$ and a falsity degree $\pi(x) \in [y, 1]$, they do not modelling a information such that $0 \leq \mu^2(x) + \nu^2(x) + \pi^2(x) \leq 1$.

The purpose of this chapter is to develop spherical fuzzy numbers theory that are the generalization of fuzzy numbers and intuitionistic fuzzy numbers. To do this, we first define two special forms of spherical fuzzy numbers are called spherical trapezoidal fuzzy numbers (STF-numbers) and spherical triangular fuzzy numbers(STrF-numbers) with their laws of the operations and their properties. We then develop a distance measure and some arithmetic and geometric operators on both the STF-numbers and STrF-numbers. By using the STF-numbers and STrF-numbers, we also present two multi-criteria decision making approaches. We finally give two practical examples to illustrate the proposed approaches.

2 Preliminary

In this section, we recall some basic notions of the fuzzy sets, intuitionistic fuzzy sets and spherical fuzzy sets.

Definition 1 (Zadeh 1965) Let E be a universe. Then a fuzzy set X over E is defined by

$$X = \{(\mu_X(x)/x) : x \in E\}$$

where μ_X is called membership function of X and defined by $\mu_X : E \to [0.1]$. For each $x \in E$, the value $\mu_X(x)$ represents the degree of x belonging to the fuzzy set X.

Definition 2 (Atanassov 1986) Let E be a universe. An intuitionistic fuzzy set K over E is defined by

$$K = \{< x, \mu_K(x), \upsilon_K(x) >: \ x \in E\}$$

where $\mu_K : E \to [0, 1]$ and $\upsilon_K : E \to [0, 1]$ such that $0 \leq \mu_K(x) + \upsilon_K(x) \leq 1$ for any $x \in E$. For each $x \in E$, the values $\mu_K(x)$ and $\upsilon_K(x)$ are the degree of membership and degree of non-membership of x, respectively.

Definition 3 (Gündoğdu and Kahraman 2019b) Let E be a universe. A spherical fuzzy set A over E is defined by

$$A = \{< x, (\mu_A(x), \upsilon_A(x), \pi_A(x)) >: x \in E\}.$$

where μ_A, υ_A and π_A are called membership function, nonmembership function and hesitancy function, respectively. They are respectively defined by

$$\mu_A : E \to [0, 1], \quad \nu_A : E \to [0, 1], \quad \pi_A : E \to [0, 1]$$

such that $0 \leq \mu_A^2(x) + \upsilon_A^2(x) + \pi_A^2(x) \leq 1$.

Definition 4 (Gündoğdu and Kahraman 2019b) Let $A = \{< x, (\mu_A(x), \upsilon_A(x), \pi_A(x)) >: x \in E\}$ and $B = \{< x, (\mu_B(x), \upsilon_B(x), \pi_B(x)) >: x \in E\}$ be two spherical fuzzy sets and $\gamma > 0$ be any real number. Then,

1. $A \oplus B = \{< x, ((\mu_A^2(x) + \mu_B^2(x) - \mu_A^2(x).\mu_B^2(x))^{\frac{1}{2}}, \upsilon_A(x).\upsilon_B(x), ((1 - \mu_B^2(x)).\pi_A^2(x) + (1 - \mu_A^2(x)).\pi_B^2(x) - \pi_A^2(x).\pi_B^2(x))^{\frac{1}{2}}) >: x \in E\}$

2. $A \otimes B = \{< x, (\mu_A(x).\mu_B(x), (\upsilon_A^2(x) + \upsilon_B^2(x) - \upsilon_A^2(x).\upsilon_B^2(x))^{\frac{1}{2}}, ((1 - \upsilon_B^2(x)).\pi_A^2(x) + (1 - \upsilon_A^2(x)).\pi_B^2(x) - \pi_A^2(x).\pi_B^2(x))^{\frac{1}{2}}) : x \in E\}$
3. $\gamma \odot A = \{< x, ((1 - (1 - \mu_A^2(x))^\gamma)^{\frac{1}{2}}, \upsilon_A^\gamma(x), ((1 - \mu_A^2(x))^\gamma - (1 - \mu_A^2(x) - \pi_A^2(x))^\gamma)^{\frac{1}{2}}) >: x \in E\}$
4. $A^\gamma = \{< x, (\mu_A^\gamma(x), (1 - (1 - \upsilon_A^2(x))^\gamma)^{\frac{1}{2}}, ((1 - \upsilon_A^2(x))^\gamma - (1 - \upsilon_A^2(x) - \pi_A^2(x))^\gamma)^{\frac{1}{2}}) >: x \in E\}$

Definition 5 (Gündoğdu and Kahraman 2019b) Let $A = \{< x, (\mu_A(x), \upsilon_A(x), \pi_A(x)) >: x \in E\}$ and $B = \{< x, (\mu_B(x), \upsilon_B(x), \pi_B(x)) >: x \in E\}$ be two spherical fuzzy sets and $\gamma, \gamma_1, \gamma_2 > 0$ be any real number. Then, the following are valid;

1. $A \oplus B = B \oplus A$
2. $A \otimes B = B \otimes A$
3. $\gamma(A \oplus B) = \gamma B \oplus \gamma A$
4. $\gamma_1 A \oplus \gamma_2 A = (\gamma_1 + \gamma_2)A$
5. $(A \otimes B)^\gamma = A^\gamma \otimes B^\gamma$
6. $A^{\gamma_1} \oplus A^{\gamma_2} = A^{\gamma_1+\gamma_2}$.

Definition 6 (Gündoğdu and Kahraman 2019b) Let $A = \{< x, (\mu_A(x), \upsilon_A(x), \pi_A(x)) >: x \in E\}$ and $B = \{< x, (\mu_B(x), \upsilon_B(x), \pi_B(x)) >: x \in E\}$ be two spherical fuzzy sets. Then,

1. spherical distance between A and B, is denoted by $dis(A, B)$, is defined as;

$$dis(A, B) = \frac{2}{\pi} \sum_{j=1}^{n} \arccos(\mu_A(x_j).\mu_B(x_j)) + \upsilon_A(x_j).\upsilon_B(x_j) + \pi_A(x_j).\pi_B(x_j))$$

2. spherical normalized distance between A and B, is denoted by $dis_n(A, B)$, is defined as;

$$dis_n(A, B) = \frac{2}{n\pi} \sum_{j=1}^{n} \arccos(\mu_A(x_j).\mu_B(x_j)) + \upsilon_A(x_j).\upsilon_B(x_j) + \pi_A(x_j).\pi_B(x_j))$$

Definition 7 (Gündoğdu and Kahraman 2019b) Let $a_j =< \mu_{a_j}, \upsilon_{a_j}, \pi_{a_j} >, j = 1, 2, \ldots, n$ are collection of some spherical fuzzy elements. Then spherical fuzzy weighted aggregation mean operator, denoted by SW_{ao}, is defined as;

$$\begin{aligned} SW_{ao}(a_1, a_2, \ldots, a_n) &= \bigoplus_{j=1}^{n} w_j a_j \\ &= w_1 a_1 \oplus w_2 a_2 \oplus \ldots \oplus w_n a_n \\ &= \Big\langle (1 - \prod_{j=1}^{n} (1 - \mu_{a_j}^2)^{w_j})^{\frac{1}{2}}, \prod_{j=1}^{n} \upsilon_{a_j}^{w_j}, \\ &\quad (\prod_{j=1}^{n} (1 - \mu_{a_j}^2)^{w_j} - \prod_{j=1}^{n} (1 - \mu_{a_j}^2 - \pi_{a_j}^2)^{w_j})^{\frac{1}{2}} \Big\rangle \end{aligned}$$

where, $w = (w_1, w_2, \ldots, w_n)^T$ is a weight vector associated with the SW_{ao} operator, for every $j \in I_n$ such that, $w_j \in [0, 1]$ and $\sum_{j=1}^{n} w_j = 1$.

Definition 8 (Gündoğdu and Kahraman 2019b) Let $a_j =< \mu_{a_j}, \upsilon_{a_j}, \pi_{a_j} >, j = 1, 2, \ldots, n$ are collection of some spherical fuzzy elements. Then spherical fuzzy weighted geometric mean operator, denoted by SW_{go}, is defined as;

$$\begin{aligned} SW_{go}(a_1, a_2, \ldots, a_n) &= \bigotimes_{j=1}^{n} a_j^{w_j} \\ &= a_1^{w_1} \otimes a_2^{w_2} \otimes \ldots \otimes a_n^{w_n} \\ &= \Big\langle \prod_{j=1}^{n} \mu_{a_j}^{w_j}, (1 - \prod_{j=1}^{n} (1 - \upsilon_{a_j}^2)^{w_j})^{\frac{1}{2}}, \\ &\quad (\prod_{j=1}^{n} (1 - \upsilon_{a_j}^2)^{w_j} - \prod_{j=1}^{n} (1 - \upsilon_{a_j}^2 - \pi_{a_j}^2)^{w_j})^{\frac{1}{2}} \Big\rangle \end{aligned}$$

where, $w = (w_1, w_2, \ldots, w_n)^T$ is a weight vector associated with the SW_{go} operator, for every $j \in I_n$ such that, $w_j \in [0, 1]$ and $\sum_{j=1}^{n} w_j = 1$.

3 Spherical Fuzzy Numbers

In this section, we define spherical fuzzy numbers with their properties. Some of the definitions are quoted or inspired by Gündoğdu and Kahraman (2019b), Ye (2015), Subas (2015), Deli and Subas (2017), Li (2014).

Definition 9 Let $a_i, b_i, c_i, d_i \in \mathbb{R}$ and $\alpha_{\tilde{a}}, \beta_{\tilde{a}}, \gamma_{\tilde{a}} \in [0, 1]$ be any real numbers such that $a_i \leq b_i \leq c_i \leq d_i$ for $i = 1, 2, 3$ and $0 \leq \alpha_{\tilde{a}}^2 + \beta_{\tilde{a}}^2 + \gamma_{\tilde{a}}^2 \leq 1$. Then a spherical fuzzy number (SF-number)

$$\tilde{a} = \langle((a_1, b_1, c_1, d_1), \alpha_{\tilde{a}}), ((a_2, b_2, c_2, d_2), \beta_{\tilde{a}},)((a_3, b_3, c_3, d_3), \gamma_{\tilde{a}})\rangle$$

is a special neutrosophic set on the set of real numbers $\mathbb{R}$, whose membership function $\mu_{\tilde{a}}$, hesitancy function $\upsilon_{\tilde{a}}$ and non-membership function $\pi_{\tilde{a}}$ are respectively defined by

$$\mu_{\tilde{a}}(x) = \begin{cases} (x - a_1)\alpha_{\tilde{a}}/(b_1 - a_1), & (a_1 \leq x < b_1) \\ \alpha_{\tilde{a}}, & (b_1 \leq x \leq c_1) \\ (d_1 - x)\alpha_{\tilde{a}}/(d_1 - c_1), & (c_1 < x \leq d_1) \\ 0, & otherwise, \end{cases}$$

$$\upsilon_{\tilde{a}}(x) = \begin{cases} (x - a_2)\alpha_{\tilde{a}}/(b_2 - a_2), & (a_2 \leq x < b_2) \\ \alpha_{\tilde{a}}, & (b_2 \leq x \leq c_2) \\ (d_2 - x)\alpha_{\tilde{a}}/(d_2 - c_2), & (c_2 < x \leq d_2) \\ 0, & otherwise, \end{cases}$$

and

$$\pi_{\tilde{a}}(x) = \begin{cases} (x - a_3)\alpha_{\tilde{a}}/(b_3 - a_3), & (a_3 \leq x < b_3) \\ \alpha_{\tilde{a}}, & (b_3 \leq x \leq c_3) \\ (d_3 - x)\alpha_{\tilde{a}}/(d_3 - c_3), & (c_3 < x \leq d_3) \\ 0, & otherwise, \end{cases}$$

respectively.

Example 1 Assume that $\tilde{a} = \langle((1, 4, 8, 10), 0.4), ((2, 4, 9, 10), 0.5), ((4, 5, 11, 13), 0.6)\rangle$. The meanings of $\tilde{a}$ is interpreted as follows: For example; the membership degree of the element $8 \in R$ belonging to $\tilde{a}$ is 0.4 whereas the hesitancy degree is 0.6 and non-membership degree is 0.5 i.e., $\mu_{\tilde{a}}(4) = 0.4$, $\upsilon_{\tilde{a}}(4) = 0.5$, $\pi_{\tilde{a}}(4) = 0.8$.

3.1 Spherical Trapezoidal Fuzzy Numbers

In the subsection, we define spherical trapezoidal fuzzy numbers with laws of their operations and their properties. Some of the definitions are quoted or inspired by Gündoğdu and Kahraman (2019b), Ye (2015), Subas (2015), Deli and Subas (2017), Li (2014).

Definition 10 Let $a_i, b_i, c_i, d_i \in \mathbb{R}$ and $\alpha_{\tilde{a}}, \beta_{\tilde{a}}, \gamma_{\tilde{a}} \in [0, 1]$ be any real numbers such that $a_i \leq b_i \leq c_i \leq d_i$ for $i = 1, 2, 3$ and $0 \leq \alpha_{\tilde{a}}^2 + \beta_{\tilde{a}}^2 + \gamma_{\tilde{a}}^2 \leq 1$. Then a spherical trapezoidal fuzzy number (STF-number)

$$\tilde{a} = \langle((a, b, c, d), \alpha_{\tilde{a}}, \beta_{\tilde{a}}, \gamma_{\tilde{a}})\rangle$$

is a special neutrosophic set on the set of real numbers $\mathbb{R}$, whose membership functions $\mu_{\tilde{a}}$, hesitancy function $\upsilon_{\tilde{a}}$ and non-membership function $\pi_{\tilde{a}}$ are defined by

$$(\mu_{\tilde{a}}(x), \upsilon_{\tilde{a}}(x), \pi_{\tilde{a}}(x)) = \begin{cases} (\frac{(x-a)\alpha_{\tilde{a}}}{b-a}, \frac{(x-a)\beta_{\tilde{a}}}{b-a}, \frac{(x-a)\gamma_{\tilde{a}}}{b-a}), & (a \leq x < b) \\ (\alpha_{\tilde{a}}, \beta_{\tilde{a}}, \gamma_{\tilde{a}}), & (b \leq x \leq c) \\ (\frac{(d-x)\alpha_{\tilde{a}}}{d-c}, \frac{(d-x)\beta_{\tilde{a}}}{d-c}, \frac{(d-x)\gamma_{\tilde{a}}}{d-c}), & (c < x \leq d) \\ (0, 0, 0), & otherwise, \end{cases}$$

A STF-number $\tilde{a} = \langle(a_1, b_1, c_1); w_{\tilde{a}}, u_{\tilde{a}}, y_{\tilde{a}}\rangle$ may express an ill-known quantity about a , which is approximately equal to a.

Note that the set of all STF-number on R will be denoted by Θ.

Definition 11 Let $\tilde{a} = \langle((a_1, b_1, c_1, d_1), \alpha_{\tilde{a}}, \beta_{\tilde{a}}, \gamma_{\tilde{a}})\rangle$,
$\tilde{b} = \langle((a_2, b_2, c_2, d_2), \alpha_{\tilde{b}}, \beta_{\tilde{b}}, \gamma_{\tilde{b}})\rangle \in \Theta$ and $\lambda \neq 0$ be any real number. Then,

1.

$$\tilde{a}+\tilde{b}=\Big\langle(a_1+a_2, b_1+b_2, c_1+c_2, d_1+d_2);\\ (\alpha_A^2(x)+\alpha_B^2(x)-\alpha_A^2(x).\alpha_B^2(x))^{\frac{1}{2}}, \beta_A(x).\beta_B(x),\\ ((1-\alpha_B^2(x)).\pi_A^2(x)+(1-\alpha_A^2(x)).\gamma_B^2(x)-\gamma_A^2(x).\gamma_B^2(x))^{\frac{1}{2}}\Big\rangle$$

2.

$$\tilde{a}\tilde{b}=\begin{cases}\Big\langle(a_1a_2, b_1b_2, c_1c_2, d_1d_2); (\alpha_A(x)\cdot\alpha_B(x),\\ (\beta_A^2(x)+\beta_B^2(x)-\beta_A^2(x)\cdot\beta_B^2(x))^{\frac{1}{2}},\\ ((1-\beta_B^2(x))\cdot\gamma_A^2(x)+(1-\beta_A^2(x))\cdot\gamma_B^2(x)-\gamma_A^2(x)\cdot\gamma_B^2(x))^{\frac{1}{2}}\Big\rangle\\ \text{if } c_1>0 \text{ and } c_2>0,\\ \Big\langle(a_1d_2, b_1c_2, c_1b_2, d_1a_2); (\alpha_A(x).\alpha_B(x),\\ (\beta_A^2(x)+\beta_B^2(x)-\beta_A^2(x)\cdot\beta_B^2(x))^{\frac{1}{2}},\\ ((1-\beta_B^2(x))\cdot\gamma_A^2(x)+(1-\beta_A^2(x))\cdot\gamma_B^2(x)-\gamma_A^2(x)\cdot\gamma_B^2(x))^{\frac{1}{2}}\Big\rangle\\ \text{if } c_1<0 \text{ and } c_2>0,\\ \Big\langle(d_1d_2, c_1c_2, b_1b_2, a_1a_2); (\alpha_A(x)\cdot\alpha_B(x),\\ (\beta_A^2(x)+\beta_B^2(x)-\beta_A^2(x)\cdot\beta_B^2(x))^{\frac{1}{2}},\\ ((1-\beta_B^2(x))\cdot\gamma_A^2(x)+(1-\beta_A^2(x))\cdot\gamma_B^2(x)-\gamma_A^2(x)\cdot\gamma_B^2(x))^{\frac{1}{2}}\Big\rangle\\ \text{if } c_1<0 \text{ and } c_2<0.\end{cases}$$

3.

$$\lambda\tilde{a}=\begin{cases}\Big\langle(\lambda a_1, \lambda b_1, \lambda c_1, \lambda d_1); ((1-(1-\alpha_A^2(x))^{\lambda})^{\frac{1}{2}}, \beta_A^{\lambda}(x),\\ ((1-\alpha_A^2(x))^{\lambda}-(1-\alpha_A^2(x)-\gamma_A^2(x))^{\lambda})^{\frac{1}{2}})\Big\rangle \quad \text{if } \lambda>0.\\ \\ \Big\langle(\lambda d_1, \lambda c_1, \lambda b_1, \lambda a_1); ((1-(1-\alpha_A^2(x))^{\lambda})^{\frac{1}{2}}, \beta_A^{\lambda}(x),\\ ((1-\alpha_A^2(x))^{\lambda}-(1-\alpha_A^2(x)-\gamma_A^2(x))^{\lambda})^{\frac{1}{2}})\Big\rangle \quad \text{if } \lambda<0.\end{cases}$$

4.

$$\tilde{a}^{\lambda}=\begin{cases}\Big\langle(a_1^{\lambda}, b_1^{\lambda}, c_1^{\lambda}, d_1^{\lambda}); (\alpha_A^{\lambda}(x), (1-(1-\beta_A^2(x))^{\lambda})^{\frac{1}{2}},\\ ((1-\beta_A^2(x))^{\lambda}-(1-\beta_A^2(x)-\gamma_A^2(x))^{\lambda})^{\frac{1}{2}})\Big\rangle \quad \text{if } \lambda>0,\\ \\ \Big\langle(d_1^{\lambda}, c_1^{\lambda}, b_1^{\lambda}, a_1^{\lambda}); (\alpha_A^{\lambda}(x), (1-(1-\beta_A^2(x))^{\lambda})^{\frac{1}{2}},\\ ((1-\beta_A^2(x))^{\lambda}-(1-\beta_A^2(x)-\gamma_A^2(x))^{\lambda})^{\frac{1}{2}})\Big\rangle \quad \text{if } \lambda<0.\end{cases}$$

Proposition 1 *Let* $\tilde{a}=\langle((a_1, b_1, c_1, d_1), \alpha_{\tilde{a}}, \beta_{\tilde{a}}, \gamma_{\tilde{a}})\rangle$, $\tilde{b}=\langle((a_2, b_2, c_2, d_2), \alpha_{\tilde{b}}, \beta_{\tilde{b}}, \gamma_{\tilde{b}})\rangle\in\Theta$ *and* $\lambda, \lambda_1, \lambda_2>0$ *be any real number. Then, the following are valid;*

1. $\tilde{a}+\tilde{b}=\tilde{b}+\tilde{a}$
2. $\tilde{a}\tilde{b}=\tilde{b}\tilde{a}$
3. $\lambda(\tilde{a}+\tilde{b})=\lambda\tilde{a}+\lambda\tilde{b}$
4. $\lambda_1\tilde{a}+\lambda_2\tilde{a}=(\lambda_1+\lambda_2)\tilde{a}$
5. $(\tilde{a}\tilde{b})^\gamma=\tilde{a}^\gamma\tilde{b}^\gamma$
6. $\tilde{a}^{\lambda_1}+\tilde{a}^{\lambda_2}=\tilde{a}^{\lambda_1+\lambda_2}$.

Definition 12 Let $\tilde{a}=\langle((a_1,b_1,c_1,d_1),\alpha_{\tilde{a}},\beta_{\tilde{a}},\gamma_{\tilde{a}})\rangle$, $\tilde{b}=\langle((a_2,b_2,c_2,d_2),\alpha_{\tilde{b}},\beta_{\tilde{b}},\gamma_{\tilde{b}})\rangle\in\Theta$. Then, STF-distance between $\tilde{a}$ and $\tilde{b}$, is denoted by $d(\tilde{a},\tilde{b})$, is defined as;;

$$d(\tilde{a},\tilde{b})=\frac{2}{\pi}\text{arccos}\left(\left(\frac{C(a).C(b)}{C(a)^2+C(b)^2}\right)(\alpha_{\tilde{a}}.\alpha_{\tilde{b}}+\beta_{\tilde{a}}.\beta_{\tilde{b}}+\gamma_{\tilde{a}}.\gamma_{\tilde{b}})\right)$$

where $C(a)=\frac{2a_1+7b_1+7c_1+2d_1}{3}$ and $C(b)=\frac{2a_2+7b_2+7c_2+2d_2}{3}$.

Definition 13 Let $\tilde{a_j}=\langle((a_j,b_j,c_j,d_j);\alpha_{\tilde{a}_j},\beta_{\tilde{a}_j},\gamma_{\tilde{a}_j})\rangle$ be collection of some STF elements for $j\in I_n$. Then, for $s,t\in I_n$,

1. If $d(\tilde{a}_s,\tilde{a}^+)<d(\tilde{a}_t,\tilde{a}^+)$, then $\tilde{a}_t$ is smaller than $\tilde{a}_s$, denoted by $\tilde{a}_t<\tilde{a}_1$
2. If $d(\tilde{a}_s,\tilde{a}^+)=d(\tilde{a}_t,\tilde{a}^+)$
   a. If $d(\tilde{a}_s,\tilde{a}^-)<d(\tilde{a}_t,\tilde{a}^-)$, then $\tilde{a}_s$ is smaller than $\tilde{a}_t$, denoted by $\tilde{a}_s<\tilde{a}_t$
   b. If $d(\tilde{a}_s,\tilde{a}^-)=d(\tilde{a}_t,\tilde{a}^-)$, then $\tilde{a}_s$ and $\tilde{a}_t$ are the same, denoted by $\tilde{a}_s=\tilde{a}_t$

where

$$\begin{aligned}\tilde{a}^+=&\langle((\min_{j\in I_n}\{a_j\},\min_{j\in I_n}\{b_j\},\max_{j\in I_n}\{c_j\},\max_{j\in I_n}\{d_j\});\\&\max_{j\in I_n}\{\alpha_{\tilde{a}_j}\},\min_{j\in I_n}\{\beta_{\tilde{a}_j}\},\min_{j\in I_n}\{\gamma_{\tilde{a}_j}\})\rangle\end{aligned}$$

and

$$\begin{aligned}\tilde{a}^-=&\langle((\min_{j\in I_n}\{a_j\},\min_{j\in I_n}\{b_j\},\max_{j\in I_n}\{c_j\},\max_{j\in I_n}\{d_j\});\\&\min_{j\in I_n}\{\alpha_{\tilde{a}_j}\},\max_{j\in I_n}\{\beta_{\tilde{a}_j}\},\max_{j\in I_n}\{\gamma_{\tilde{a}_j}\})\rangle\end{aligned}$$

Example 2 Let $\tilde{a}_1=\langle(1,2,3,5);0.5,0.4,0.7\rangle$, $\tilde{a}_2=\langle(3,5,7,8);0.9,0.2,0.1\rangle\in\Theta$. Then, we gave

$$\begin{aligned}\tilde{a}^+=&\langle((\min\{1,3\},\min\{2,5\},\max\{3,7\},\max\{5,8\});\\&\max\{0.5,0.9\},\min\{0.4,0.2\},\min\{\{0.7,0.1\})\rangle\\=&\langle((1,2,7,8);0.9,0.2,0.1)\rangle\end{aligned}$$

Now, we calculate the STF-distances $d(\tilde{a}_1,\tilde{a}^+)$ and $d(\tilde{a}_2,\tilde{a}^+)$ between $\tilde{a}_1$ and $\tilde{a}^+$ and $\tilde{a}_2$ and $\tilde{a}^+$, respectively, as

$$\begin{aligned} d(\tilde{a}_1, \tilde{a}^+) &= \tfrac{2}{\pi}\arccos\left(\left(\tfrac{C(\tilde{a}_1).C(\tilde{a}^+)}{C(\tilde{a})^2+C(\tilde{a}^+)^2}\right)(\alpha_{\tilde{a}_1}.\alpha_{\tilde{a}^+} + \beta_{\tilde{a}_1}.\beta_{\tilde{a}^+} + \gamma_{\tilde{a}_1}.\gamma_{\tilde{a}^+})\right) \\ &= \tfrac{2}{\pi}\arccos\left(\left(\frac{\frac{2+14+21+10}{3}.\frac{2+14+49+16}{3}}{(\frac{2+14+21+10}{3})^2+(\frac{2+14+49+16}{3})^2}\right)(0.45 + 0.08 + 0.07)\right) \\ &= \tfrac{2}{\pi}\arccos\left(\left(\frac{\frac{47}{3}.\frac{81}{3}}{(\frac{47}{3})^2+(\frac{81}{3})^2}\right)(0.6)\right) \\ &= \tfrac{2}{\pi}\arccos(0.26) \\ &= 0.83 \end{aligned}$$

and

$$\begin{aligned} d(\tilde{a}_2, \tilde{a}^+) &= \tfrac{2}{\pi}\arccos\left(\left(\tfrac{C(\tilde{a}_1).C(\tilde{a}^+)}{C(\tilde{a})^2+C(\tilde{a}^+)^2}\right)(\alpha_{\tilde{a}_1}.\alpha_{\tilde{a}^+} + \beta_{\tilde{a}_1}.\beta_{\tilde{a}^+} + \gamma_{\tilde{a}_1}.\gamma_{\tilde{a}^+})\right) \\ &= \tfrac{2}{\pi}\arccos\left(\left(\frac{\frac{6+35+49+16}{3}.\frac{2+14+49+16}{3}}{(\frac{6+35+49+16}{3})^2+(\frac{2+14+49+16}{3})^2}\right)(0.81 + 0.04 + 0.01)\right) \\ &= \tfrac{2}{\pi}\arccos\left(\left(\frac{\frac{106}{3}.\frac{81}{3}}{(\frac{106}{3})^2+(\frac{81}{3})^2}\right)(0.86)\right) \\ &= \tfrac{2}{\pi}\arccos(0.41) \\ &= 0.73 \end{aligned}$$

Finally, since $d(\tilde{a}_2, \tilde{a}^+) < d(\tilde{a}_1, \tilde{a}^+)$, we have $\tilde{a}_1 < \tilde{a}_2$.

Definition 14 Let $\tilde{a}_j = \langle((a_j, b_j, c_j, d_j); \alpha_{\tilde{a}_j}, \beta_{\tilde{a}_j}, \gamma_{\tilde{a}_j})\rangle$ be collection of some STF elements for $j \in I_n$. Then,

1. STF weighted aggregation mean operator, denoted by STF_{ao}, is defined as;

$$\begin{aligned} STF_{ao}(a_1, a_2, \ldots, a_n) &= \textstyle\sum_{j=1}^n w_j a_j \\ &= w_1 a_1 + w_2 a_2 + \cdots + w_n a_n \\ &= \Big\langle (\textstyle\sum_{j=1}^n w_j a_j, \sum_{j=1}^n w_j b_j, \sum_{j=1}^n w_j c_j, \sum_{j=1}^n w_j d_j); \\ &\quad (1 - \textstyle\prod_{j=1}^n (1 - \alpha_{a_j}^2)^{w_j})^{\frac{1}{2}}, \prod_{j=1}^n \beta_{a_j}^{w_j}, \\ &\quad (\textstyle\prod_{j=1}^n (1 - \alpha_{a_j}^2)^{w_j} - \prod_{j=1}^n (1 - \alpha_{a_j}^2 - \gamma_{a_j}^2)^{w_j})^{\frac{1}{2}} \Big\rangle \end{aligned}$$

where, $w = (w_1, w_2, \ldots, w_n)^T$ is a weight vector associated with the STF_{ao} operator, for every $j \in I_n$ such that, $w_j \in [0, 1]$ and $\sum_{j=1}^n w_j = 1$.

2. STF weighted geometric mean operator, denoted by STF_{go}, is defined as;

$$\begin{aligned} STF_{go}(a_1, a_2, \ldots, a_n) &= \prod_{j=1}^{n} a_j^{w_j} \\ &= a_1^{w_1} a_2^{w_2} \ldots a_n^{w_n} \\ &= \Big\langle (\prod_{j=1}^{n} a_j^{w_1}, \prod_{j=1}^{n} b_j^{w_1}, \prod_{j=1}^{n} c_j^{w_1}, \prod_{j=1}^{n} d_j^{w_1}); \\ &\quad \prod_{j=1}^{n} \alpha_{a_j}^{w_j}, (1 - \prod_{j=1}^{n}(1-\beta_{a_j}^2)^{w_j})^{\frac{1}{2}}, \\ &\quad (\prod_{j=1}^{n}(1-\beta_{a_j}^2)^{w_j} - \prod_{j=1}^{n}(1-\beta_{a_j}^2-\gamma_{a_j}^2)^{w_j})^{\frac{1}{2}} \Big\rangle \end{aligned}$$

where, $w = (w_1, w_2, \ldots, w_n)^T$ is a weight vector associated with the SW_{go} operator, for every $j \in I_n$ such that, $w_j \in [0, 1]$ and $\sum_{j=1}^{n} w_j = 1$.

Example 3 Let $\tilde{a}_1 = \langle(2, 3, 5, 6); 0.5, 0.4, 0.7\rangle$, $\tilde{a}_2 = \langle(1, 2, 6, 8); 0.4, 0.3, 0.4\rangle$, $\tilde{a}_3 = \langle(0, 4, 5, 9); 0.5, 0.5, 0.5\rangle$ be three STF-numbers, and $w = (0.7, 0.2, 0.1)^T$ be the weight vector of $\tilde{a}_j$ for $j = 1, 2, 3$. Then,

1. STF weighted aggregation mean operator STF_{ao} is calculated as;

$$\begin{aligned} STF_{ao}(\tilde{a}_1, \tilde{a}_2, \tilde{a}_3) &= \sum_{j=1}^{n} w_j a_j \\ &= (0.7)\langle(2, 3, 5, 6); 0.5, 0.4, 0.7\rangle + \\ &\quad (0.2)\langle(1, 2, 6, 8); 0.4, 0.3, 0.4\rangle + \\ &\quad (0.1)\langle(0, 4, 5, 9); 0.5, 0.5, 0.5\rangle \\ &= \Big\langle ((0.7)2 + (0.2)1 + (0.1)0, (0.7)3 + (0.2)2 + (0.1)4, \\ &\quad (0.7)5 + (0.2)6 + (0.1)5, (0.7)6 + (0.2)8 + (0.1)9); \\ &\quad (1 - (0.75)^{0.7})(0.84)^{0.2})(0.75)^{0.1})^{\frac{1}{2}}, \\ &\quad (0.4)^{0.7}(0.3)^{0.2}(0.5)^{0.1}, \\ &\quad ((0.75)^{0.7}(0.84)^{0.2}(0.75)^{0.1} \\ &\quad -(0.26)^{0.7}(0.68)^{0.2}(0.50)^{0.1})^{\frac{1}{2}} \Big\rangle \\ &= \langle(1.6, 2.9, 5.2, 6.7); 0.48, 0.39, 0.66\rangle \end{aligned}$$

2. STF weighted geometric mean operator STF_{go} is calculated as;

$$\begin{aligned} STF_{go}(\tilde{a}_1, \tilde{a}_2, \tilde{a}_3) &= \prod_{j=1}^{n} a_j^{w_j} \\ &= \langle(2, 3, 5, 6); 0.5, 0.4, 0.7\rangle^{0.7} \langle(1, 2, 6, 8); 0.4, 0.3, 0.4\rangle^{0.2} \\ &\quad \langle(0, 4, 5, 9); 0.5, 0.5, 0.5\rangle^{0.1} \\ &= \Big\langle (2^{0.7}1^{0.2}0^{0.1}, 3^{0.7}2^{0.2}4^{0.1}, 5^{0.7}6^{0.2}5^{0.1}, 6^{0.7}8^{0.2}9^{0.1}); \\ &\quad 0.5^{0.7}0.4^{0.2}0.5^{0.1}, \\ &\quad (1 - (1-0.4^2)^{0.7}(1-0.3^2)^{0.2}(1-0.5^2)^{0.1})^{\frac{1}{2}}, \\ &\quad ((1-0.4^2)^{0.7}(1-0.3^2)^{0.2}(1-0.5^2)^{0.1} \\ &\quad -(1-0.4_{a_j}^2 - 0.7^2)^{0.7} \\ &\quad (1-0.3^2-0.4^2)^{0.2}(1-0.5^2-0.5^2)^{0.1})^{\frac{1}{2}} \Big\rangle \\ &= \langle(0.00, 2.85, 5.19, 6.62); 0.48, 0.40, 0.65\rangle \end{aligned}$$

Definition 15 Let $\tilde{a}_j = \langle((a_j, b_j, c_j, d_j); \alpha_{\tilde{a}_j}, \beta_{\tilde{a}_j}, \gamma_{\tilde{a}_j})\rangle$ be a collection of some STF elements for $j \in I_n$.

1. Then, STF ordered weighted aggregation mean operator, denoted by STF_{oao}, is defined as;

$$
\begin{aligned}
STF_{oao}(a_1, a_2, \ldots, a_n) &= \textstyle\sum_{k=1}^{n} w_k b_k \\
&= w_1 b_1 + w_2 b_2 + \cdots + w_n b_n \\
&= \Big\langle (\textstyle\sum_{k=1}^{n} w_k a_k, \sum_{k=1}^{n} w_k b_k, \sum_{k=1}^{n} w_k c_k, \sum_{k=1}^{n} w_k d_k); \\
&\quad (1 - \textstyle\prod_{k=1}^{n} (1 - \alpha_{\tilde{b}_k}^2)^{w_k})^{\frac{1}{2}}, \prod_{k=1}^{n} \beta_{\tilde{b}_k}^{w_k}, \\
&\quad (\textstyle\prod_{k=1}^{n} (1 - \alpha_{\tilde{b}_k}^2)^{w_k} - \prod_{k=1}^{n} (1 - \alpha_{\tilde{b}_k}^2 - \gamma_{\tilde{b}_k}^2)^{w_k})^{\frac{1}{2}} \Big\rangle
\end{aligned}
$$

   where $w = (w_1, w_2, \ldots, w_n)^T$ is a weight vector associated with the mapping STF_{oao}, which satisfies the normalized conditions: $w_k \in [0, 1]$ and $\sum_{k=1}^{n} w_k = 1$; $\tilde{b}_k = \langle((a_k, b_k, c_k, d_k); \alpha_{\tilde{b}_k}, \beta_{\tilde{b}_k}, \gamma_{\tilde{b}_k})\rangle$ is the k-th largest of the n STF-numbers $\tilde{a}_j$ for $j \in I_n$ which is determined through using ranking method such as the above ranking method based on STF-distance.

2. STF weighted ordered geometric mean operator, denoted by STF_{ogo}, is defined as;

$$
\begin{aligned}
STF_{ogo}(a_1, a_2, \ldots, a_n) &= \textstyle\prod_{k=1}^{n} \tilde{b}_k^{w_k} = \tilde{b}_1^{w_1} \tilde{b}_2^{w_2} \ldots \tilde{b}_n^{w_n} \\
&= \Big\langle (\textstyle\prod_{k=1}^{n} a_k^{w_1}, \prod_{k=1}^{n} b_k^{w_1}, \prod_{k=1}^{n} c_k^{w_1}, \prod_{k=1}^{n} d_k^{w_1}); \\
&\quad \textstyle\prod_{j=1}^{n} \alpha_{\tilde{b}_k}^{w_k}, (1 - \prod_{k=1}^{n} (1 - \beta_{\tilde{b}_k}^2)^{w_k})^{\frac{1}{2}}, \\
&\quad (\textstyle\prod_{k=1}^{n} (1 - \beta_{\tilde{b}_k}^2)^{w_k} - \prod_{k=1}^{n} (1 - \beta_{\tilde{b}_k}^2 - \gamma_{\tilde{b}_k}^2)^{w_k})^{\frac{1}{2}} \Big\rangle
\end{aligned}
$$

   where $w = (w_1, w_2, \ldots, w_n)^T$ is a weight vector associated with the mapping STF_{ogo}, which satisfies the normalized conditions: $w_k \in [0, 1]$ and $\sum_{k=1}^{n} w_k = 1$; $\tilde{b}_k = \langle((a_k, b_k, c_k, d_k); \alpha_{\tilde{b}_k}, \beta_{\tilde{b}_k}, \gamma_{\tilde{b}_k})\rangle$ is the k-th largest of the n STF-numbers $\tilde{a}_j$ for $j \in I_n$ which is determined through using ranking method such as the above ranking method based on STF-distance.

Example 4 Let consider Example 3. Then, we compute the comprehensive evaluation of the three STF-number using the STF_{oao} and STF_{ogo}.

We first obtain the STF-distance $d(\tilde{a}_j, \tilde{a}^+)$ for $j = 1, 2, 3$ as follows:

$$d(\tilde{a}_1, \tilde{a}^+) = 0.79, \quad d(\tilde{a}_2, \tilde{a}^+) = 0.86, \quad d(\tilde{a}_3, \tilde{a}^+) = 0.80$$

Obviously, since

$$d(\tilde{a}_1, \tilde{a}^+) < d(\tilde{a}_3, \tilde{a}^+) < d(\tilde{a}_2, \tilde{a}^+)$$

we have $\tilde{a}_2 < \tilde{a}_3 < \tilde{a}_1$. Thereby, according to the above STF-distance method, we have

$$\hat{b}_1 = \tilde{a}_1 = \langle (2, 3, 5, 6); 0.5, 0.4, 0.7 \rangle$$
$$\hat{b}_2 = \tilde{a}_3 = \langle (0, 4, 5, 9); 0.5, 0.5, 0.5 \rangle$$
$$\hat{b}_3 = \tilde{a}_2 = \langle (0, 4, 5, 9); 0.5, 0.5, 0.5 \rangle$$

It follows from Definition 15 that

$$STF_{oao}(\tilde{a}_1, \tilde{a}_2, \tilde{a}_3) = \langle (1.50, 3.10, 5.10, 6.80); 0.49, 0.41, 0.66 \rangle$$
$$STF_{ogo}(\tilde{a}_1, \tilde{a}_2, \tilde{a}_3) = \langle (0.00, 3.05, 5.09, 6.70); 0.49, 0.42, 0.65 \rangle$$

Definition 16 Let $\tilde{a}_j = \langle ((a_j, b_j, c_j, d_j); \alpha_{\tilde{a}_j}, \beta_{\tilde{a}_j}, \gamma_{\tilde{a}_j}) \rangle$ be collection of some STF elements for $j \in I_n$.

1. STF hybrid weighted averaging operator denoted by STF_{hao} is defined as; denoted STF_{hao}

$$\begin{aligned} STF_{hao}(a_1, a_2, \ldots, a_n) &= \sum_{k=1}^n w_k \widehat{b_k} \\ &= w_1 \widehat{b_1} + w_2 \widehat{b_2} + \cdots + w_n \widehat{b_n} \\ &= \Big\langle \big(\sum_{k=1}^n w_k \hat{a}_k, \sum_{k=1}^n w_k \hat{b}_k, \sum_{k=1}^n w_k \hat{c}_k, \sum_{k=1}^n w_k \hat{d}_k\big); \\ &\quad (1 - \prod_{k=1}^n (1 - \alpha^2_{\hat{b}_k})^{w_k})^{\frac{1}{2}}, \prod_{k=1}^n \beta^{w_k}_{\hat{b}_k}, \\ &\quad (\prod_{k=1}^n (1 - \alpha^2_{\hat{b}_k})^{w_k} - \prod_{k=1}^n (1 - \alpha^2_{\hat{b}_k} - \gamma^2_{\hat{b}_k})^{w_k})^{\frac{1}{2}} \Big\rangle \end{aligned}$$

where $w = (w_1, w_2, \ldots, w_n)^T$ is a weight vector associated with the mapping STF_{hao} such that $w_k \in [0, 1]$ and $\sum_{k=1}^n w_k = 1$, $\hat{a}_j \in \Theta$ weighted with $n\omega_j (j \in I_n)$ is denoted by $\widehat{a_j}$, i.e., $\widehat{a_j} = n\omega_j \tilde{a}_j$, here n is regarded as a balance factor; $\omega = (\omega_1, \omega_2, \ldots, \omega_n)^T$ is a weight vector of the $\tilde{a}_j \in \Theta$ $(j \in I_n)$n such that $\omega_j \in [0, 1]$ and $\sum_{j=1}^n \omega_j = 1$; $\widehat{b_k}$ is the k-th largest of the n STF-number $\widehat{a_j} \in \Theta$ $(j \in I_n)$ which are determined through using some ranking method such as the above ranking method based on STF-distance.

2. STF hybrid weighted geometric operator denoted by STF_{hgo}, is denoted by STF_{hgo}, is defined as; denoted

$$\begin{aligned} STF_{hgo}(a_1, a_2, \ldots, a_n) &= \prod_{k=1}^n \widehat{b_k}^{w_k} \\ &= \widehat{b_1}^{w_1} \widehat{b_2}^{w_2} \ldots \widehat{b_n}^{w_n} \\ &= \Big\langle \big(\prod_{k=1}^n \hat{a}_k^{w_k}, \prod_{k=1}^n \hat{b}_k^{w_k}, \prod_{k=1}^n \hat{c}_k^{w_k}, \prod_{k=1}^n \hat{d}_k^{w_k}\big); \\ &\quad \prod_{k=1}^n \alpha^{w_k}_{\hat{b}_k}, (1 - \prod_{k=1}^n (1 - \beta^2_{\hat{b}_k})^{w_k})^{\frac{1}{2}}, \\ &\quad (\prod_{k=1}^n (1 - \beta^2_{\hat{b}_k})^{w_k} - \prod_{k=1}^n (1 - \beta^2_{\hat{b}_k} - \gamma^2_{\hat{b}_k})^{w_k})^{\frac{1}{2}} \Big\rangle \end{aligned}$$

where $w = (w_1, w_2, \ldots, w_n)^T$ is a weight vector associated with the mapping STF_{hgo} such that $w_k \in [0, 1]$ and $\sum_{k=1}^n w_k = 1$, $\hat{a}_j \in \Theta$ weighted with $n\omega_j (j \in I_n)$ is denoted by $\widehat{a_j}$, i.e., $\widehat{a_j} = n\omega_j \tilde{a}_j$, here n is regarded as a balance factor; $\omega = (\omega_1, \omega_2, \ldots, \omega_n)^T$ is a weight vector of the $\tilde{a}_j \in \Theta$ $(j \in I_n)$n such that $\omega_j \in [0, 1]$ and $\sum_{j=1}^n \omega_j = 1$; $\widehat{b_k}$ is the k-th largest of the n STF-number $\widehat{a_j} \in \Theta$ $(j \in I_n)$ which are determined through using some ranking method such as the above ranking method based on STF-distance.

Example 5 Let consider the Example 3. Assume that the weight vector of the three experts is $\omega = (0.1, 0.4, 0.5)^T$ and the position weight vector is $w = (0.7, 0.2, 0.1)^T$. Compute the comprehensive evaluation of the three experts on the decision alternative through using the STF_{hao} and STF_{hgo}.

We can obtain $\tilde{A}_j$ for $j = 1, 2, 3$ as

$$\begin{aligned}\tilde{A}_1 &= 3 \times 0.1 \times \tilde{a}_1 = \langle (0.60, 0.90, 1.50, 1.80); 0.29, 0.76, 0.50\rangle \\ \tilde{A}_2 &= 3 \times 0.4 \times \tilde{a}_2 = \langle (1.20, 2.40, 7.20, 9.60); 0.43, 0.24, 0.43\rangle \\ \tilde{A}_3 &= 3 \times 0.5 \times \tilde{a}_3 = \langle (0.00, 6.00, 7.50, 13.50); 0.59, 0.35, 0.54\rangle\end{aligned}$$

We then obtain the STF-distance $d(\tilde{a}_j, \tilde{a}^+)$ for $j = 1, 2, 3$ as follows:

$$d(\tilde{a}_1, \tilde{a}^+) = 0,91, \quad d(\tilde{a}_2, \tilde{a}^+) = 0.84, \quad d(\tilde{a}_3, \tilde{a}^+) = 0.80$$

Obviously, since

$$d(\tilde{a}_3, \tilde{a}^+) < d(\tilde{a}_2, \tilde{a}^+) < d(\tilde{a}_1, \tilde{a}^+)$$

we have $\tilde{a}_2 < \tilde{a}_3 < \tilde{a}_1$, respectively. Thereby, according to the above the STF-distance ranking method, we have

$$\begin{aligned}\widehat{b_1} &= \tilde{A}_3 = \langle (0.00, 6.00, 7.50, 13.50); 0.59, 0.35, 0.54\rangle \\ \widehat{b_2} &= \tilde{A}_2 = \langle (1.20, 2.40, 7.20, 9.60); 0.43, 0.24, 0.43\rangle \\ \widehat{b_3} &= \tilde{A}_1 = \langle (0.60, 0.90, 1.50, 1.80); 0.29, 0.76, 0.50\rangle\end{aligned}$$

It follows from Definition 24 that

$$\begin{aligned}STF_{hao}(\tilde{a}_1, \tilde{a}_2, \tilde{a}_3) &= \langle (0.30, 4.77, 6.84, 11.55); 0.54, 0.35, 0.53\rangle \\ STF_{hgo}(\tilde{a}_1, \tilde{a}_2, \tilde{a}_3) &= \langle (0.00, 4.13, 6.33, 10.31); 0.52, 0.42, 0.53\rangle\end{aligned}$$

3.2 *Spherical Triangular Fuzzy Numbers*

In the subsection, we define spherical triangular fuzzy numbers with laws of their operations and their properties. Some of the definitions are quoted or inspired by Gündoğdu and Kahraman (2019b), Ye (2015), Subas (2015), Deli and Subas (2017), Li (2014).

Definition 17 Let $a_i, b_i, c_i \in \mathbb{R}$ and $\alpha_{\tilde{a}}, \beta_{\tilde{a}}, \gamma_{\tilde{a}} \in [0, 1]$ be any real numbers such that $a_i \leq b_i \leq c_i$(i=1,2,3) and $0 \leq \alpha_{\tilde{a}}^2 + \beta_{\tilde{a}}^2 + \gamma_{\tilde{a}}^2 \leq 1$. Then a spherical triangular fuzzy number (STrF-number)

$$\tilde{a} = \langle ((a, b, c), \alpha_{\tilde{a}}, \beta_{\tilde{a}}, \gamma_{\tilde{a}})\rangle$$

is a special neutrosophic set on the set of real numbers $\mathbb{R}$, whose membership functions $\mu_{\tilde{a}}$, hesitancy function $\upsilon_{\tilde{a}}$ and non-membership function $\pi_{\tilde{a}}$ are defined by

$$(\mu_{\tilde{a}}(x), \upsilon_{\tilde{a}}(x), \pi_{\tilde{a}}(x)) = \begin{cases} (\frac{(x-a)\alpha_{\tilde{a}}}{b-a}, \frac{(x-a)\beta_{\tilde{a}}}{b-a}, \frac{(x-a)\gamma_{\tilde{a}}}{b-a}), & a \leq x < b, \\ (\frac{(c-x)\alpha_{\tilde{a}}}{c-b}, \frac{(c-x)\beta_{\tilde{a}}}{c-b}, \frac{(c-x)\gamma_{\tilde{a}}}{c-b}), & c < x \leq d, \\ (0, 0, 0), & otherwise. \end{cases}$$

A STrF-number $\tilde{a} = \langle (a_1, b_1, c_1); w_{\tilde{a}}, u_{\tilde{a}}, y_{\tilde{a}} \rangle$ may express an ill-known quantity about a , which is approximately equal to a.

Note that the set of all STrF-number on R will be denoted by Ω.

Definition 18 Let $\tilde{a} = \langle ((a_1, b_1, c_1), \alpha_{\tilde{a}}, \beta_{\tilde{a}}, \gamma_{\tilde{a}}) \rangle$, $\tilde{b} = \langle ((a_2, b_2, c_2), \alpha_{\tilde{b}}, \beta_{\tilde{b}}, \gamma_{\tilde{b}}) \rangle \in \Omega$ and $\lambda \neq 0$ be any real number. Then,

1.

$$\begin{aligned} \tilde{a} + \tilde{b} = \Big\langle & (a_1 + a_2, b_1 + b_2, c_1 + c_2); (\alpha_A^2(x) + \alpha_B^2(x) - \alpha_A^2(x).\alpha_B^2(x))^{\frac{1}{2}}, \\ & \beta_A(x).\beta_B(x), \\ & ((1 - \alpha_B^2(x)).\pi_A^2(x) + (1 - \alpha_A^2(x)).\gamma_B^2(x) - \gamma_A^2(x).\gamma_B^2(x))^{\frac{1}{2}} \Big\rangle \end{aligned}$$

2.

$$\tilde{a}\tilde{b} = \begin{cases} \Big\langle (a_1a_2, b_1b_2, c_1c_2); (\alpha_A(x).\mu_B(x), (\beta_A^2(x) + \beta_B^2(x) - \beta_A^2(x).\beta_B^2(x))^{\frac{1}{2}}, \\ ((1 - \beta_B^2(x)).\gamma_A^2(x) + (1 - \beta_A^2(x)).\gamma_B^2(x) - \gamma_A^2(x).\gamma_B^2(x))^{\frac{1}{2}} \Big\rangle \\ \text{if } c_1 > 0 \text{ and } c_2 > 0 \\ \Big\langle (a_1c_2, b_1b_2, c_1a_2); (\alpha_A(x).\mu_B(x), (\beta_A^2(x) + \beta_B^2(x) - \beta_A^2(x).\beta_B^2(x))^{\frac{1}{2}}, \\ ((1 - \beta_B^2(x)).\gamma_A^2(x) + (1 - \beta_A^2(x)).\gamma_B^2(x) - \gamma_A^2(x).\gamma_B^2(x))^{\frac{1}{2}} \Big\rangle \\ \text{if } c_1 < 0 \text{ and } c_2 > 0 \\ \Big\langle (c_1c_2, b_1b_2, a_1a_2); (\alpha_A(x).\mu_B(x), (\beta_A^2(x) + \beta_B^2(x) - \beta_A^2(x).\beta_B^2(x))^{\frac{1}{2}}, \\ ((1 - \beta_B^2(x)).\gamma_A^2(x) + (1 - \beta_A^2(x)).\gamma_B^2(x) - \gamma_A^2(x).\gamma_B^2(x))^{\frac{1}{2}} \Big\rangle \\ \text{if } (c_1 < 0 \text{ and } c_2 < 0) \end{cases}$$

3.

$$\lambda\tilde{a} = \begin{cases} \Big\langle (\lambda a_1, \lambda b_1, \lambda c_1); ((1 - (1 - \alpha_A^2(x))^{\lambda})^{\frac{1}{2}}, \beta_A^{\lambda}(x), \\ ((1 - \alpha_A^2(x))^{\lambda} - (1 - \alpha_A^2(x) - \gamma_A^2(x))^{\lambda})^{\frac{1}{2}}) \Big\rangle & \text{if } \lambda > 0 \\ \Big\langle (\lambda c_1, \lambda b_1, \lambda a_1); ((1 - (1 - \alpha_A^2(x))^{\lambda})^{\frac{1}{2}}, \beta_A^{\lambda}(x), \\ ((1 - \alpha_A^2(x))^{\lambda} - (1 - \alpha_A^2(x) - \gamma_A^2(x))^{\lambda})^{\frac{1}{2}}) \Big\rangle & \text{if } \lambda < 0 \end{cases}$$

4.

$$\tilde{a}^{\lambda} = \begin{cases} \langle (a_1^{\lambda}, b_1^{\lambda}, c_1^{\lambda}); (\alpha_A^{\lambda}(x), (1-(1-\beta_A^2(x))^{\lambda})^{\frac{1}{2}}, \\ ((1-\beta_A^2(x))^{\lambda} - (1-\beta_A^2(x)-\gamma_A^2(x))^{\lambda})^{\frac{1}{2}}) \rangle & \text{if } \lambda > 0 \\ \langle (c_1^{\lambda}, b_1^{\lambda}, a_1^{\lambda}); (\alpha_A^{\lambda}(x), (1-(1-\beta_A^2(x))^{\lambda})^{\frac{1}{2}}, \\ ((1-\beta_A^2(x))^{\lambda} - (1-\beta_A^2(x)-\gamma_A^2(x))^{\lambda})^{\frac{1}{2}}) \rangle & \text{if } \lambda < 0 \end{cases}$$

Definition 19 Let $\tilde{a} = \langle((a_1, b_1, c_1); \alpha_{\tilde{a}}, \beta_{\tilde{a}}, \gamma_{\tilde{a}})\rangle$, $\tilde{b} = \langle((a_2, b_2, c_2); \alpha_{\tilde{b}}, \beta_{\tilde{b}}, \gamma_{\tilde{b}})\rangle \in \Omega$ and $\lambda, \lambda_1, \lambda_2 > 0$ be any real number. Then, the following are valid;

1. $\tilde{a} + \tilde{b} = \tilde{b} + \tilde{a}$
2. $\tilde{a}\tilde{b} = \tilde{b}\tilde{a}$
3. $\lambda(\tilde{a} + \tilde{b}) = \lambda\tilde{a} + \lambda\tilde{b}$
4. $\lambda_1\tilde{a} + \lambda_2\tilde{a} = (\lambda_1 + \lambda_2)\tilde{a}$
5. $(\tilde{a}\tilde{b})^{\gamma} = \tilde{a}^{\gamma}\tilde{b}^{\gamma}$
6. $\tilde{a}^{\lambda_1} + \tilde{a}^{\lambda_2} = \tilde{a}^{\lambda_1+\lambda_2}$.

Definition 20 Let $\tilde{a} = \langle((a_1, b_1, c_1), \alpha_{\tilde{a}}, \beta_{\tilde{a}}, \gamma_{\tilde{a}})\rangle$, $\tilde{b} = \langle((a_2, b_2, c_2), \alpha_{\tilde{b}}, \beta_{\tilde{b}}, \gamma_{\tilde{b}})\rangle \in \Omega$. Then, STrF-distance between $\tilde{a}$ and $\tilde{b}$, is denoted by $d(\tilde{a}, \tilde{b})$, is defined as;

$$d(\tilde{a}, \tilde{b}) = \frac{2}{\pi}\arccos\left(\left(\frac{C(a).C(b)}{C(a)^2+C(b)^2}\right)(\alpha_{\tilde{a}}.\alpha_{\tilde{b}} + \beta_{\tilde{a}}.\beta_{\tilde{b}} + \gamma_{\tilde{a}}.\gamma_{\tilde{b}})\right)$$

where $C(a) = \frac{a_1+4b_1+c_1}{3}$ and $C(b) = \frac{a_2+4b_2+c_2}{3}$.

Definition 21 $\tilde{a}_j = \langle((a_j, b_j, c_j); \alpha_{\tilde{a}_j}, \beta_{\tilde{a}_j}, \gamma_{\tilde{a}_j})\rangle$, $(j \in I_n)$ are collection of some STrF elements. Then, for $s, t \in I_n$,

1. If $d(\tilde{a}_s, \tilde{a}^+) < d(\tilde{a}_t, \tilde{a}^+)$, then $\tilde{a}_t$ is smaller than $\tilde{a}_s$, denoted by $\tilde{a}_t < \tilde{a}_1$
2. If $d(\tilde{a}_s, \tilde{a}^+) = d(\tilde{a}_t, \tilde{a}^+)$
   a. If $d(\tilde{a}_s, \tilde{a}^-) < d(\tilde{a}_t, \tilde{a}^-)$, then $\tilde{a}_s$ is smaller than $\tilde{a}_t$, denoted by $\tilde{a}_s < \tilde{a}_t$
   b. If $d(\tilde{a}_s, \tilde{a}^-) = d(\tilde{a}_t, \tilde{a}^-)$, then $\tilde{a}_s$ and $\tilde{a}_t$ are the same, denoted by $\tilde{a}_s = \tilde{a}_t$

where

$$\begin{aligned} \tilde{a}^+ = \langle(&\min_{j\in I_n}\{a_j\}, \min_{j\in I_n}\{b_j\}, \max_{j\in I_n}\{c_j\}); \\ &\max_{j\in I_n}\{\alpha_{\tilde{a}_j}\}, \min_{j\in I_n}\{\beta_{\tilde{a}_j}\}, \min_{j\in I_n}\{\gamma_{\tilde{a}_j}\}\rangle \end{aligned}$$

and

$$\begin{aligned} \tilde{a}^- = \langle(((&\min_{j\in I_n}\{a_j\}, \max_{j\in I_n}\{b_j\}, \max_{j\in I_n}\{c_j\}\}); \\ &\min_{j\in I_n}\{\alpha_{\tilde{a}_j}\}, \max_{j\in I_n}\{\beta_{\tilde{a}_j}\}, \max_{j\in I_n}\{\gamma_{\tilde{a}_j}\}))\rangle \end{aligned}$$

Example 6 Let $\tilde{a}_1 = \langle(1, 2, 3); 0.5, 0.4, 0.7\rangle$, $\tilde{a}_2 = \langle(3, 5, 7); 0.9, 0.2, 0.1\rangle \in \Omega$. Then, we gave

$$\begin{aligned} \tilde{a}^+ &= \langle(((\min\{1, 3\}, \min\{2, 5\}, \max\{3, 7\}); \\ &\quad \max\{0.5, 0.9\}, \min\{0.4, 0.2\}, \min\{\{0.7, 0.1\}\})\rangle \\ &= \langle((1, 2, 7); 0.9, 0.2, 0.1)\rangle \end{aligned}$$

Now, we calculate the STrF-distances $d(\tilde{a}_1, \tilde{a}^+)$ and $d(\tilde{a}_2, \tilde{a}^+)$ between $\tilde{a}_1$ and $\tilde{a}^+$ and $\tilde{a}_2$ and $\tilde{a}^+$, respectively, as;

$$\begin{aligned} d(\tilde{a}_1, \tilde{a}^+) &= \frac{2}{\pi}\arccos\left(\left(\frac{C(\tilde{a}_1).C(\tilde{a}^+)}{C(\tilde{a})^2+C(\tilde{a}^+)^2}\right)(\alpha_{\tilde{a}_1}.\alpha_{\tilde{a}^+} + \beta_{\tilde{a}_1}.\beta_{\tilde{a}^+} + \gamma_{\tilde{a}_1}.\gamma_{\tilde{a}^+})\right) \\ &= \frac{2}{\pi}\arccos\left(\left(\frac{\frac{1+8+3}{3}.\frac{1+8+7}{3}}{(\frac{1+8+3}{3})^2+(\frac{1+8+7}{3})^2}\right)(0.45+0.08+0.07)\right) \\ &= \frac{2}{\pi}\arccos\left(\left(\frac{\frac{12}{3}.\frac{16}{3}}{(\frac{12}{3})^2+(\frac{16}{3})^2}\right)(0.6)\right) \\ &= \frac{2}{\pi}\arccos(0.0340) \\ &= 0.98 \end{aligned}$$

and

$$\begin{aligned} d(\tilde{a}_2, \tilde{a}^+) &= \frac{2}{\pi}\arccos\left(\left(\frac{C(\tilde{a}_1).C(\tilde{a}^+)}{C(\tilde{a})^2+C(\tilde{a}^+)^2}\right)(\alpha_{\tilde{a}_1}.\alpha_{\tilde{a}^+} + \beta_{\tilde{a}_1}.\beta_{\tilde{a}^+} + \gamma_{\tilde{a}_1}.\gamma_{\tilde{a}^+})\right) \\ &= \frac{2}{\pi}\arccos\left(\frac{(\frac{3+20+7}{3}.\frac{1+8+7}{3})}{\left(\frac{(3+20+7}{3}\right)^2+(\frac{1+8+7}{3})^2}(0.81+0.04+0.01)\right) \\ &= \frac{2}{\pi}\arccos\left(\left(\frac{\frac{106}{3}.\frac{81}{3}}{(\frac{106}{3})^2+(\frac{81}{3})^2}\right)(0.86)\right) \\ &= \frac{2}{\pi}\arccos(0.1000) \\ &= 0.9333 \end{aligned}$$

Finally, since $d(\tilde{a}_2, \tilde{a}^+) < d(\tilde{a}_1, \tilde{a}^+)$, we have $\tilde{a}_1 < \tilde{a}_2$.

Definition 22 Let $\tilde{a}_j = \langle((a_j, b_j, c_j); \alpha_{\tilde{a}_j}, \beta_{\tilde{a}_j}, \gamma_{\tilde{a}_j})\rangle$, $(j \in I_n)$ are collection of some STrF elements. Then

1. STrF weighted aggregation mean operator, denoted by $STrF_{ao}$, is defined as;

$$\begin{aligned} STrF_{ao}(a_1, a_2, \ldots, a_n) &= \textstyle\sum_{j=1}^n w_j a_j \\ &= w_1a_1 + w_2a_2 + \cdots + w_na_n \\ &= \Big\langle(\textstyle\sum_{j=1}^n w_ja_j, \sum_{j=1}^n w_jb_j, \sum_{j=1}^n w_jc_j); \\ &\quad (1 - \textstyle\prod_{j=1}^n(1-\alpha_{a_j}^2)^{w_j})^{\frac{1}{2}}, \prod_{j=1}^n \beta_{a_j}^{w_j}, \\ &\quad (\textstyle\prod_{j=1}^n(1-\alpha_{a_j}^2)^{w_j} - \prod_{j=1}^n(1-\alpha_{a_j}^2-\gamma_{a_j}^2)^{w_j})^{\frac{1}{2}}\Big\rangle \end{aligned}$$

where, $w = (w_1, w_2, \ldots, w_n)^T$ is a weight vector associated with the $STrF_{ao}$ operator, for every $j \in I_n$ such that, $w_j \in [0, 1]$ and $\sum_{j=1}^n w_j = 1$.

2. STrF weighted geometric mean operator, denoted by $STrF_{go}$, is defined as;

$$\begin{aligned} STrF_{go}(a_1, a_2, \ldots, a_n) &= \prod_{j=1}^{n} a_j^{w_j} \\ &= a_1^{w_1} a_2^{w_2} \ldots a_n^{w_n} \\ &= \big\langle (\prod_{j=1}^{n} a_j^{w_1}, \prod_{j=1}^{n} b_j^{w_1}, \prod_{j=1}^{n} c_j^{w_1}); \\ &\quad \prod_{j=1}^{n} \alpha_{a_j}^{w_j}, (1 - \prod_{j=1}^{n}(1-\beta_{a_j}^2)^{w_j})^{\frac{1}{2}}, \\ &\quad (\prod_{j=1}^{n}(1-\beta_{a_j}^2)^{w_j} - \prod_{j=1}^{n}(1-\beta_{a_j}^2-\gamma_{a_j}^2)^{w_j})^{\frac{1}{2}} \big\rangle \end{aligned}$$

where, $w = (w_1, w_2, \ldots, w_n)^T$ is a weight vector associated with the SW_{go} operator, for every $j \in I_n$ such that, $w_j \in [0, 1]$ and $\sum_{j=1}^{n} w_j = 1$.

Example 7 Let $\tilde{a}_1 = \big\langle (2, 3, 5); 0.5, 0.4, 0.7 \big\rangle$, $\tilde{a}_2 = \big\langle (1, 2, 6); 0.4, 0.3, 0.4 \big\rangle$, $\tilde{a}_3 = \big\langle (0, 4, 5); 0.5, 0.5, 0.5 \big\rangle$ be three STrF-numbers, and $w = (0.7, 0.2, 0.1)^T$ be the weight vector of $\tilde{a}_j$ for $j = 1, 2, 3$. Then,

1. STrF weighted aggregation mean operator $STrF_{ao}$ is calculated as;

$$\begin{aligned} STrF_{ao}(\tilde{a}_1, \tilde{a}_2, \tilde{a}_3) &= \sum_{j=1}^{n} w_j a_j \\ &= (0.7)\big\langle (2, 3, 5); 0.5, 0.4, 0.7 \big\rangle \\ &\quad +(0.2)\big\langle (1, 2, 6); 0.4, 0.3, 0.4 \big\rangle \\ &\quad +(0.1)\big\langle (0, 4, 5); 0.5, 0.5, 0.5 \big\rangle \\ &= < ((0.7)2 + (0.2)1 + (0.1)0, \\ &\quad (0.7)3 + (0.2)2 + (0.1)4, \\ &\quad (0.7)5 + (0.2)6 + (0.1)5); \\ &\quad (1 - (0.75)^{0.7})(0.84)^{0.2})(0.75)^{0.1})^{\frac{1}{2}}, \\ &\quad (0.4)^{0.7}(0.3)^{0.2}(0.5)^{0.1}, \\ &\quad ((0.75)^{0.7}(0.84)^{0.2}(0.75)^{0.1} \\ &\quad -(0.26)^{0.7}(0.68)^{0.2}(0.50)^{0.1})^{\frac{1}{2}} \big\rangle \\ &= < (1.6, 2.9, 5.2); 0.48, 0.39, 0.66 \big\rangle \end{aligned}$$

2. STrF weighted geometric mean operator $STrF_{go}$ is calculated as;

$$\begin{aligned} STrF_{go}(\tilde{a}_1, \tilde{a}_2, \tilde{a}_3) &= \prod_{j=1}^{n} a_j^{w_j} \\ &= \big\langle (2, 3, 5, 6); 0.5, 0.4, 0.7 \big\rangle^{0.7} \big\langle (1, 2, 6, 8); 0.4, 0.3, 0.4 \big\rangle^{0.2} \\ &\quad \big\langle (0, 4, 5, 9); 0.5, 0.5, 0.5 \big\rangle^{0.1} \\ &= \big\langle (2^{0.7}1^{0.2}0^{0.1}, 3^{0.7}2^{0.2}4^{0.1}, 5^{0.7}6^{0.2}5^{0.1}); \\ &\quad 0.5^{0.7}0.4^{0.2}0.5^{0.1}, \\ &\quad (1 - (1 - 0.4^2)^{0.7}(1 - 0.3^2)^{0.2}(1 - 0.5^2)^{0.1})^{\frac{1}{2}}, \\ &\quad ((1 - 0.4^2)^{0.7}(1 - 0.3^2)^{0.2}(1 - 0.5^2)^{0.1} \\ &\quad -(1 - 0.4_{a_j}^2 - 0.7^2)^{0.7}(1 - 0.3^2 - 0.4^2)^{0.2} \\ &\quad (1 - 0.5^2 - 0.5^2)^{0.1})^{\frac{1}{2}} \big\rangle \\ &= \big\langle (0.00, 2.85, 5.19); 0.48, 0.40, 0.65 \big\rangle \end{aligned}$$

Definition 23 Let $\tilde{a}_j = \langle((a_j, b_j, c_j); \alpha_{\tilde{a}_j}, \beta_{\tilde{a}_j}, \gamma_{\tilde{a}_j})\rangle$ be collection of some STrF elements for $(j \in I_n)$. Then,

1. Then, STrF ordered weighted aggregation mean operator, denoted by $STrF_{oao}$, is defined as;

$$\begin{aligned} STrF_{oao}(a_1, a_2, \ldots, a_n) &= \sum_{k=1}^{n} w_k b_k \\ &= w_1 b_1 + w_2 b_2 + \cdots + w_n b_n \\ &= \Big\langle(\sum_{k=1}^{n} w_k a_k, \sum_{k=1}^{n} w_k b_k, \sum_{k=1}^{n} w_k c_k); \\ &\quad (1 - \prod_{k=1}^{n}(1 - \alpha_{b_k}^2)^{w_k})^{\frac{1}{2}}, \prod_{k=1}^{n} \beta_{b_k}^{w_k}, \\ &\quad (\prod_{k=1}^{n}(1 - \alpha_{b_k}^2)^{w_k} - \prod_{k=1}^{n}(1 - \alpha_{b_k}^2 - \gamma_{b_k}^2)^{w_k})^{\frac{1}{2}}\Big\rangle \end{aligned}$$

   where $w = (w_1, w_2, \ldots, w_n)^T$ is a weight vector associated with the mapping $STrF_{oao}$, which satisfies the normalized conditions: $w_k \in [0, 1]$ and $\sum_{k=1}^{n} w_k = 1$; $\tilde{b}_k = \langle((a_k, b_k, c_k); \alpha_{\tilde{b}_k}, \beta_{\tilde{b}_k}, \gamma_{\tilde{b}_k})\rangle$ is the k-th largest of the n STrF-numbers $\tilde{a}_j$ for $j \in I_n$ which is determined through using ranking method such as the above ranking method based on STrF-distance.
2. STrF weighted ordered geometric mean operator, denoted by $STrF_{ogo}$, is defined as;

$$\begin{aligned} STrF_{ogo}(a_1, a_2, \ldots, a_n) &= \prod_{k=1}^{n} \tilde{b}_k^{w_k} = \tilde{b}_1^{w_1} \tilde{b}_2^{w_2} \ldots \tilde{b}_n^{w_n} \\ &= \Big\langle(\prod_{k=1}^{n} a_k^{w_1}, \prod_{k=1}^{n} b_k^{w_1}, \prod_{k=1}^{n} c_k^{w_1}); \\ &\quad \prod_{j=1}^{n} \alpha_{b_k}^{w_k}, (1 - \prod_{k=1}^{n}(1 - \beta_{b_k}^2)^{w_k})^{\frac{1}{2}}, \\ &\quad (\prod_{k=1}^{n}(1 - \beta_{b_k}^2)^{w_k} - \prod_{k=1}^{n}(1 - \beta_{b_k}^2 - \gamma_{b_k}^2)^{w_k})^{\frac{1}{2}}\Big\rangle \end{aligned}$$

   where $w = (w_1, w_2, \ldots, w_n)^T$ is a weight vector associated with the mapping $STrF_{ogo}$, which satisfies the normalized conditions: $w_k \in [0, 1]$ and $\sum_{k=1}^{n} w_k = 1$; $\tilde{b}_k = \langle((a_k, b_k, c_k); \alpha_{\tilde{b}_k}, \beta_{\tilde{b}_k}, \gamma_{\tilde{b}_k})\rangle$ is the k-th largest of the n STrF-numbers $\tilde{a}_j$ for $j \in I_n$ which is determined through using ranking method such as the above ranking method based on STrF-distance.

Example 8 Let consider the Example 7. Then, we compute the comprehensive evaluation of the three STrF-number using the $STrF_{oao}$ and $STrF_{ogo}$.

We first obtain the STrF-distance $d(\tilde{a}_j, \tilde{a}^+)$ $(j = 1, 2, 3)$ as follows:

$$d(\tilde{a}_1, \tilde{a}^+) = 0.80, \quad d(\tilde{a}_2, \tilde{a}^+) = 0.86, \quad d(\tilde{a}_3, \tilde{a}^+) = 0.82$$

Obviously, since

$$d(\tilde{a}_1, \tilde{a}^+) < d(\tilde{a}_3, \tilde{a}^+) < d(\tilde{a}_2, \tilde{a}^+)$$

we have $\tilde{a}_2 < \tilde{a}_3 < \tilde{a}_1$. Thereby, according to the above STrF-distance method, we have

$$\begin{aligned} \hat{b}_1 &= \tilde{a}_1 = \big\langle(2, 3, 5); 0.5, 0.4, 0.7\big\rangle \\ \hat{b}_2 &= \tilde{a}_3 = \big\langle(0, 4, 5); 0.5, 0.5, 0.5\big\rangle \\ \hat{b}_3 &= \tilde{a}_2 = \big\langle(0, 4, 5); 0.5, 0.5, 0.5\big\rangle \end{aligned}$$

It follows from Definition 23 that

$$STrF_{oao}(\tilde{a}_1, \tilde{a}_2, \tilde{a}_3) = \big\langle(1.50, 3.10, 5.10); 0.49, 0.41, 0.66\big\rangle$$
$$STrF_{ogo}(\tilde{a}_1, \tilde{a}_2, \tilde{a}_3) = \big\langle(0.00, 3.05, 5.09); 0.49, 0.42, 0.65\big\rangle$$

Definition 24 Let $\tilde{a}_j = \langle((a_j, b_j, c_j, d_j); \alpha_{\tilde{a}_j}, \beta_{\tilde{a}_j}, \gamma_{\tilde{a}_j})\rangle$ be collection of some STrF elements for $j \in I_n$. Then,

1. STrF hybrid weighted averaging operator denoted by $STrF_{hao}$ is defined as; denoted $STrF_{hao}$

$$\begin{aligned} STrF_{hao}(a_1, a_2, \ldots, a_n) &= \textstyle\sum_{k=1}^n w_k \widehat{b_k} \\ &= w_1\widehat{b_1} + w_2\widehat{b_2} + \cdots + w_n\widehat{b_n} \\ &= \big\langle(\textstyle\sum_{k=1}^n w_k \hat{a}_k, \sum_{k=1}^n w_k \hat{b}_k, \sum_{k=1}^n w_k \hat{c}_k); \\ &\quad (1 - \textstyle\prod_{k=1}^n (1 - \alpha_{\hat{b}_k}^2)^{w_k})^{\frac{1}{2}}, \prod_{k=1}^n \beta_{\hat{b}_k}^{w_k}, \\ &\quad (\textstyle\prod_{k=1}^n (1 - \alpha_{\hat{b}_k}^2)^{w_k} - \prod_{k=1}^n (1 - \alpha_{\hat{b}_k}^2 - \gamma_{\hat{b}_k}^2)^{w_k})^{\frac{1}{2}}\big\rangle \end{aligned}$$

where $w = (w_1, w_2, \ldots, w_n)^T$ is a weight vector associated with the mapping $STrF_{hao}$ such that $w_k \in [0, 1]$ and $\sum_{k=1}^n w_k = 1, \hat{a}_j \in \Theta$ weighted with $n\omega_j (j \in I_n)$ is denoted by $\widehat{a}_j$, i.e., $\widehat{a}_j = n\omega_j\tilde{a}_j$, here n is regarded as a balance factor; $\omega = (\omega_1, \omega_2, \ldots, \omega_n)^T$ is a weight vector of the $\tilde{a}_j \in \Theta$ $(j \in I_n)$n such that $\omega_j \in [0, 1]$ and $\sum_{j=1}^n \omega_j = 1$; $\widehat{b}_k$ is the k-th largest of the n STrF-number $\widehat{a}_j \in \Theta$ for $j \in I_n$ which are determined through using some ranking method such as the above ranking method based on STrF-distance.

2. STrF hybrid weighted geometric operator denoted by $STrF_{hgo}$, is denoted by $STrF_{hgo}$, is defined as; denoted

$$\begin{aligned} STrF_{hgo}(a_1, a_2, \ldots, a_n) &= \textstyle\prod_{k=1}^n \widehat{b}_k^{w_k} \\ &= \widehat{b}_1^{w_1}\widehat{b}_2^{w_2} \ldots \widehat{b}_n^{w_n} \\ &= \big\langle(\textstyle\prod_{k=1}^n \hat{a}_k^{w_k}, \prod_{k=1}^n \hat{b}_k^{w_k}, \prod_{k=1}^n \hat{c}_k^{w_k}); \\ &\quad \textstyle\prod_{k=1}^n \alpha_{\hat{b}_k}^{w_k}, (1 - \prod_{k=1}^n (1 - \beta_{\hat{b}_k}^2)^{w_k})^{\frac{1}{2}}, \\ &\quad (\textstyle\prod_{k=1}^n (1 - \beta_{\hat{b}_k}^2)^{w_k} - \prod_{k=1}^n (1 - \beta_{\hat{b}_k}^2 - \gamma_{\hat{b}_k}^2)^{w_k})^{\frac{1}{2}}\big\rangle \end{aligned}$$

where $w = (w_1, w_2, \ldots, w_n)^T$ is a weight vector associated with the mapping $STrF_{hgo}$ such that $w_k \in [0, 1]$ and $\sum_{k=1}^n w_k = 1$, $\hat{a}_j \in \Theta$ weighted with $n\omega_j (j \in I_n)$ is denoted by $\widehat{a}_j$, i.e., $\widehat{a}_j = n\omega_j\tilde{a}_j$, here n is regarded as a balance factor; $\omega = (\omega_1, \omega_2, \ldots, \omega_n)^T$ is a weight vector of the $\tilde{a}_j \in \Theta$ $(j \in I_n)$n such that $\omega_j \in [0, 1]$ and $\sum_{j=1}^n \omega_j = 1$; $\widehat{b}_k$ is the k-th largest of the n STrF-number $\widehat{a}_j \in \Theta$ for $j \in I_n$ which are determined through using some ranking method such as the above ranking method based on STrF-distance.

Example 9 Let consider the Example 7. Assume that the weight vector of the three experts is $\omega = (0.1, 0.4, 0.5)^T$ and the position weight vector is $w = (0.7, 0.2, 0.1)^T$.

Compute the comprehensive evaluation of the three experts on the decision alternative through using the $STrF_{hao}$ and $STrF_{hgo}$.

We can obtain $\tilde{A}_j$ for $j = 1, 2, 3$ as;

$$\begin{aligned}
\tilde{A}_1 &= 3 \times 0.1 \times \tilde{a}_1 = \big\langle (0.60, 3.60, 7.50); 0.29, 0.76, 0.50 \big\rangle \\
\tilde{A}_2 &= 3 \times 0.4 \times \tilde{a}_2 = \big\langle (0.30, 2.40, 9.00); 0.43, 0.24, 0.43 \big\rangle \\
\tilde{A}_3 &= 3 \times 0.5 \times \tilde{a}_3 = \big\langle (0.00, 4.80, 7.50); 0.59, 0.35, 0.54 \big\rangle
\end{aligned}$$

We then obtain the STrF-distance $d(\tilde{a}_j, \tilde{a}^+)$ $(j = 1, 2, 3)$ as follows:

$$d(\tilde{a}_1, \tilde{a}^+) = 0,82, \quad d(\tilde{a}_2, \tilde{a}^+) = 0.84, \quad d(\tilde{a}_3, \tilde{a}^+) = 0.80$$

Obviously, since

$$d(\tilde{a}_3, \tilde{a}^+) < d(\tilde{a}_1, \tilde{a}^+) < d(\tilde{a}_2, \tilde{a}^+)$$

we have $\tilde{a}_2 < \tilde{a}_1 < \tilde{a}_3$, respectively. Thereby, according to the above the STrF-distance ranking method, we have

$$\begin{aligned}
\widehat{b_1} &= \tilde{A}_3 = \big\langle (0.00, 4.80, 7.50); 0.59, 0.35, 0.54 \big\rangle \\
\widehat{b_2} &= \tilde{A}_2 = \big\langle (0.60, 3.60, 7.50); 0.29, 0.76, 0.50 \big\rangle \\
\widehat{b_3} &= \tilde{A}_1 = \big\langle (0.30, 2.40, 9.00); 0.43, 0.24, 0.43 \big\rangle
\end{aligned}$$

It follows from Definition 24 that

$$\begin{aligned}
STrF_{hao}(\tilde{a}_1, \tilde{a}_2, \tilde{a}_3) &= \big\langle (0.27, 3.00, 8.55); 0.46, 0.29, 0.47 \big\rangle \\
STrF_{hgo}(\tilde{a}_1, \tilde{a}_2, \tilde{a}_3) &= \big\langle (0.00, 2.87, 8.52); 0.44, 0.38, 0.49 \big\rangle
\end{aligned}$$

4 SF-Multi-criteria Group Decision-Making Method

In the section, we developed SF-multi-criteria group decision-making method. Some of the definitions are quoted or inspired by Gündoğdu and Kahraman (2019b), Ye (2015), Subas (2015) Deli and Subas (2017), Li (2014).

4.1 *STF-Multi-criteria Group Decision-Making Method*

In this section, we define a multi-criteria decision making method, so called STF-multi-criteria decision-making method, by using the SFT_{hao} operator. Some of the definitions are quoted or inspired by Gündoğdu and Kahraman (2019b), Ye (2015), Subas (2015), Deli and Subas (2017), Li (2014).

Definition 25 Let $X = (x_1, x_2, \ldots, x_m)$ be a set of alternatives, $U = (u_1, u_2, \ldots, u_n)$ be the set of criterions. If $\tilde{a}_{ij} = \langle (a_{ij}, b_{ij}, c_{ij}, d_{ij}); \alpha_{ij}, \beta_{ij}, \gamma_{ij} \rangle \in \Theta$, then

1.

$$[\tilde{a}_{ij}]_{m\times n} = \begin{array}{c} \\ x_1 \\ x_2 \\ \vdots \\ x_m \end{array} \begin{array}{c} \begin{array}{cccc} u_1 & u_2 & \cdots & u_n \end{array} \\ \begin{pmatrix} \tilde{a}_{11} & \tilde{a}_{12} & \cdots & \tilde{a}_{1n} \\ \tilde{a}_{21} & \tilde{a}_{22} & \cdots & \tilde{a}_{2n} \\ \vdots & \vdots & \vdots & \vdots \\ \tilde{a}_{m1} & \tilde{a}_{m2} & \cdots & \tilde{a}_{mn} \end{pmatrix} \end{array}$$

is called an STF-multi-criteria decision-making matrix of the decision maker.

2.

$$[\tilde{a}_i]_{1\times n} = x_i \begin{array}{c} \begin{array}{cccc} u_1 & u_2 & \cdots & u_n \end{array} \\ \begin{pmatrix} \tilde{a}_{i1} & \tilde{a}_{i2} & \cdots & \tilde{a}_{in} \end{pmatrix} \end{array}$$

is called an STF-multi-criteria decision-making sub-matrix of the decision maker based on the alternatives x_i.

Now, we can give an algorithm of the STF-multi-criteria decision-making method as follows;

Algorithm:
[*Step 1.*] Construct the STF-multi-criteria decision making matrix $[\tilde{a}_{ij}]_{m\times n}$ for $i = 1, 2, \ldots, m$ and $j = 1, 2, \ldots, n$;
[*Step 2.*] Insert the weights of the criterions $\omega_i = (\omega_1, \omega_2, \ldots, \omega_n$;
[*Step 3.*] Insert the (position) weights of the criterions $w_i = (w_1, w_2, \ldots, w_n)$ associated with the operator SFT_{hao};
[*Step 4.*] Compute the STF-numbers $\tilde{A}_{ij} = \langle (\hat{a}_{ij}, \hat{b}_{ij}, \hat{c}_{ij}, \hat{d}_{ij}); \hat{\alpha}_{ij}, \hat{\beta}_{ij}, \hat{\gamma}_{ij} \rangle$ $= n\omega_i \tilde{a}_{ij}$ for $i = 1, 2, \ldots, m$ and $j = 1, 2, \ldots, n$;
[*Step 5.*] Construct the STF-multi-criteria decision making matrix $[\tilde{A}_{ij}]_{m\times n}$, (i=1,2, …,m; j=1,2, …,n);
[*Step 6.*] Obtain the STF-distances $d(\tilde{A}_{ij}, \tilde{a}^+)$ based on the STF-numbers $\tilde{A}_{ij}$ for $i = 1, 2, \ldots, m$ and $j = 1, 2, \ldots, n$, where

$$\tilde{a}^+ = \langle (min_{j\in I_n}\{\hat{a}_{ij}\}, min_{j\in I_n}\{\hat{b}_{ij}\}, max_{j\in I_n}\{\hat{c}_{ij}, max_{j\in I_n}\{\hat{d}_{ij}\}); \\ max_{j\in I_n}\{\hat{\alpha}_{\tilde{a}_j}\}, min_{j\in I_n}\{\hat{\beta}_{\tilde{a}_j}\}, min_{j\in I_n}\{\hat{\gamma}_{\tilde{a}_j}\} \rangle$$

[*Step 7.*] Rank the nonincreasing order of the n STF-numbers $\tilde{A}_{ij}$ for $i = 1, 2, \ldots, m$ by using the ranking method based on STF-distance;
[*Step 8.*] Determine the STF-multi-criteria decision-making sub-matrix $[b_i]_{1\times n} = \tilde{b}_{ik}$ for $k = 1, 2, \ldots, n$ and $i = 1, 2, \ldots, m$, where $\tilde{b}_{ik}$ $(k = 1, 2, \ldots, n)$ is k-th largest of $\tilde{A}_{ij}$ for $i = 1, 2, \ldots, m$;
[*Step 9.*] Compute $r_i = STF_{hao} = (\tilde{b}_{i1}, \tilde{b}_{i2}, \ldots, \tilde{b}_{in}\} = \langle (r_1^i, r_2^i, r_3^i, r_4^i); \hat{\alpha}^i, \hat{\beta}^i, \hat{\gamma}^i \rangle$;
[*Step 10.*] Obtain the STF-distances $d(r_i, \tilde{\hat{a}}^+)$, where

$$\tilde{\hat{a}}^{+} = \langle (min_{i\in I_m}\{r_1^i\}, min_{i\in I_m}\{r_2^i\}, max_{i\in I_m}\{r_3^i\}, max_{i\in I_n}\{r_4^i\}); \\ max_{i\in I_m}\{\hat{\alpha}^i\}, min_{i\in I_m}\{\hat{\beta}^i\}, min_{i\in I_m}\{\hat{\gamma}^i\}\rangle$$

[*Step 11.*] Rank all alternatives x_i based on $d(r_i, \tilde{\hat{a}}^{+})$ by using the by using the ranking method based on STF-distance and determine the best alternative.

Example 10 Let us consider the decision-making problem adapted from Ye (2015). There is an investment company, which wants to invest a sum of money in the best option. There is a panel with the set of the four alternatives is denoted by

$$X = \{x_1 = \text{car company}, x_2 = \text{food company}, x_3 = \text{computer company}, \\ x_4 = \text{arms company}, x_5 = \text{construction company}\}$$

to invest the money. The investment company must take a decision according to the set of the four attributes is denoted by

$$U = \{u_1 = risk, u_2 = growth, u_3 = environmental\ impact, \\ u_4 = performance\}.$$

Then, the weight vector of the attributes is $\omega = (0.4, 0.1, 0.3, 0.2)^T$ and the position weight vector is $w = (0.1, 0.3, 0.4, 0.2)^T$ by using the weight determination based on the normal distribution. For the evaluation of an alternative x_i for $i = 1, 2, 3, 4, 5$ with respect to a criterion u_j for $j = 1, 2, 3, 4$, it is obtained from the questionnaire of a domain expert. Then, Thus, when the four possible alternatives with respect to the above four criteria are evaluated by the expert, we can obtain the following algorithm;

[*Step 1.*] We constructed the STF-multi-criteria decision making matrix $[\tilde{a}_{ij}]_{5\times 4}$ for $i = 1, 2, \ldots, 5$ and $j = 1, 2, \ldots, 4$ as;

$$[\tilde{a}_{ij}]_{5\times 4} = \begin{pmatrix} \langle(1,2,3,4);0.1,0.8,0.5\rangle & \langle(0,2,4,6);0.6,0.5,0.6\rangle & \langle(5,7,7,8);0.9,0.1,0.1\rangle & \langle(3,8,9,9);0.7,0.5,0.5\rangle \\ \langle(0,2,4,6);0.6,0.5,0.6\rangle & \langle(3,8,9,9);0.7,0.5,0.5\rangle & \langle(5,7,7,8);0.9,0.1,0.1\rangle & \langle(1,2,3,4);0.1,0.8,0.5\rangle \\ \langle(0,2,4,6);0.6,0.5,0.6\rangle & \langle(3,8,9,9);0.7,0.5,0.5\rangle & \langle(3,8,9,9);0.7,0.5,0.5\rangle & \langle(1,2,3,4);0.1,0.8,0.5\rangle \\ \langle(5,7,7,8);0.9,0.1,0.1\rangle & \langle(0,2,4,6);0.6,0.5,0.6\rangle & \langle(6,7,8,9);0.1,0.7,0.7\rangle & \langle(5,7,7,8);0.9,0.1,0.1\rangle \\ \langle(1,2,3,4);0.1,0.8,0.5\rangle & \langle(5,5,7,9);0.6,0.5,0.6\rangle & \langle(1,2,3,4);0.1,0.8,0.5\rangle & \langle(5,7,7,8);0.9,0.1,0.1\rangle \end{pmatrix}$$

[*Step 2.*] We inserted the weights of the criterions $\omega = (0.4, 0.1, 0.3, 0.2)^T$ as;
[*Step 3.*] We inserted the (position) weights of the criterions $w = (0.1, 0.3, 0.4, 0.2)^T$ associated with the operator SFT_{hao} as;
[*Step 4.*] We computed the STF-numbers $\tilde{A}_{ij} = \langle(\hat{a}_{ij}, \hat{b}_{ij}, \hat{c}_{ij}, \hat{d}_{ij}); \hat{\alpha}_{ij}, \hat{\beta}_{ij}, \hat{\gamma}_{ij}\rangle = n\omega_i\tilde{a}_{ij}$ for $i = 1, 2, 3, 4, 5$ and $j = 1, 2, 3, 4$) as;

$$\begin{aligned}\tilde{A}_{11} &= 4 \times 0.4 \times \big\langle(1, 2, 3, 4); 0.1, 0.8, 0.5\big\rangle \\ &= \big\langle(1.6000, 3.2000, 4.8000, 6.4000); 0.1263, 0.6998, 0.6053\big\rangle\end{aligned}$$

Likewise, we can obtain other STF-numbers $\tilde{A}_{ij} = n\omega_i\tilde{a}_{ij}$ for $i = 1, 2, 3, 4, 5$ and $j = 1, 2, 3, 4$).
[*Step 5.*] We construct the STF-multi-criteria decision making matrix $[\tilde{A}_{ij}]_{m\times n}$ for $i = 1, 2, 3, 4, 5$ and $j = 1, 2, 3, 4$ as;

$$[\tilde{A}_{ij}]_{5\times 4} = \left(\begin{array}{c}
\big\langle(1.6000, 3.2000, 4.8000, 6.4000); 0.1263, 0.6998, 0.6053\big\rangle \\
\big\langle(0.0000, 3.2000, 6.4000, 9.6000); 0.7144, 0.3299, 0.5993\big\rangle \\
\big\langle(0.0000, 3.2000, 6.4000, 9.6000); 0.7144, 0.3299, 0.5993\big\rangle \\
\big\langle(8.0000, 11.200, 11.200, 12.800); 0.9643, 0.0251, 0.0762\big\rangle \\
\big\langle(1.6000, 3.2000, 4.8000, 6.4000); 0.1263, 0.6998, 0.6053\big\rangle \\
\\
\big\langle(0.0000, 0.8000, 1.6000, 2.4000); 0.4043, 0.7579, 0.4853\big\rangle \\
\big\langle(1.2000, 3.2000, 3.6000, 3.6000); 0.4859, 0.7579, 0.4248\big\rangle \\
\big\langle(1.2000, 3.2000, 3.6000, 3.6000); 0.4859, 0.7579, 0.4248\big\rangle \\
\big\langle(0.0000, 0.8000, 1.6000, 2.4000); 0.4043, 0.7579, 0.4853\big\rangle \\
\big\langle(2.0000, 2.0000, 2.8000, 3.6000); 0.4043, 0.7579, 0.4853\big\rangle \\
\\
\big\langle(6.0000, 8.4000, 8.4000, 9.6000); 0.9294, 0.0631, 0.0925\big\rangle \\
\big\langle(6.0000, 8.4000, 8.4000, 9.6000); 0.9294, 0.0631, 0.0925\big\rangle \\
\big\langle(3.6000, 9.6000, 10.800, 10.800); 0.7445, 0.4353, 0.4971\big\rangle \\
\big\langle(7.2000, 8.4000, 9.6000, 10.800); 0.1095, 0.6518, 0.7435\big\rangle \\
\big\langle(1.2000, 2.4000, 3.6000, 4.8000); 0.1095, 0.7651, 0.5397\big\rangle \\
\\
\big\langle(2.4000, 6.4000, 7.2000, 7.2000); 0.6454, 0.5743, 0.4931\big\rangle \\
\big\langle(0.8000, 1.6000, 2.4000, 3.2000); 0.0895, 0.8365, 0.4539\big\rangle \\
\big\langle(0.8000, 1.6000, 2.4000, 3.2000); 0.0895, 0.8365, 0.4539\big\rangle \\
\big\langle(4.0000, 5.6000, 5.6000, 6.4000); 0.8574, 0.1585, 0.1059\big\rangle \\
\big\langle(4.0000, 5.6000, 5.6000, 6.4000); 0.8574, 0.1585, 0.1059\big\rangle
\end{array}\right)$$

[*Step 6.*] We obtained the STF-distances $d(\tilde{A}_{ij}, \tilde{a}^+)$ based on the STF-numbers $\tilde{A}_{ij}$ for $i = 1, 2, 3, 4, 5$ and $j = 1, 2, 3, 4$ as;

$$\begin{array}{ll}
d(\tilde{A}_{11}, \tilde{a}^{+}) = 0.9457 & d(\tilde{A}_{33}, \tilde{a}^{+}) = 0.7738 \\
d(\tilde{A}_{12}, \tilde{a}^{+}) = 0.9461 & d(\tilde{A}_{34}, \tilde{a}^{+}) = 0.9732 \\
d(\tilde{A}_{13}, \tilde{a}^{+}) = 0.7156 & d(\tilde{A}_{41}, \tilde{a}^{+}) = 0.7406 \\
d(\tilde{A}_{14}, \tilde{a}^{+}) = 0.7813 & d(\tilde{A}_{42}, \tilde{a}^{+}) = 0.9461 \\
d(\tilde{A}_{21}, \tilde{a}^{+}) = 0.7648 & d(\tilde{A}_{43}, \tilde{a}^{+}) = 0.9472 \\
d(\tilde{A}_{22}, \tilde{a}^{+}) = 0.8632 & d(\tilde{A}_{44}, \tilde{a}^{+}) = 0.7259 \\
d(\tilde{A}_{23}, \tilde{a}^{+}) = 0.7156 & d(\tilde{A}_{51}, \tilde{a}^{+}) = 0.9457 \\
d(\tilde{A}_{24}, \tilde{a}^{+}) = 0.9732 & d(\tilde{A}_{52}, \tilde{a}^{+}) = 0.9002 \\
d(\tilde{A}_{31}, \tilde{a}^{+}) = 0.7648 & d(\tilde{A}_{53}, \tilde{a}^{+}) = 0.9581 \\
d(\tilde{A}_{32}, \tilde{a}^{+}) = 0.8632 & d(\tilde{A}_{54}, \tilde{a}^{+}) = 0.7259
\end{array}$$

where

$$\begin{aligned}
\tilde{a}^{+} &= \langle(\min_{j\in I_n}\{\hat{a}_{ij}\}, \min_{j\in I_n}\{\hat{b}_{ij}\}, \max_{j\in I_n}\{\hat{c}_{ij}, \max_{j\in I_n}\{\hat{d}_{ij}\}); \\
&\quad \max_{j\in I_n}\{\hat{\alpha}_{\tilde{a}_j}\}, \min_{j\in I_n}\{\hat{\beta}_{\tilde{a}_j}\}, \min_{j\in I_n}\{\hat{\gamma}_{\tilde{a}_j}\}\rangle \\
&= \big\langle(0.0000, 0.8000, 11.2000, 12.8000); 0.9643, 0.0251, 0.0762\big\rangle
\end{aligned}$$

[*Step 7.*] We ranked the nonincreasing order of the n STF-numbers $\tilde{A}_{ij}$ for $i = 1, 2, 3, 4, 5$ by using the ranking method based on STF-distance as;

$$\begin{array}{c}
\tilde{A}_{13} > \tilde{A}_{14} > \tilde{A}_{11} > \tilde{A}_{12} \\
\tilde{A}_{23} > \tilde{A}_{21} > \tilde{A}_{22} > \tilde{A}_{24} \\
\tilde{A}_{31} > \tilde{A}_{33} > \tilde{A}_{32} > \tilde{A}_{34} \\
\tilde{A}_{44} > \tilde{A}_{41} > \tilde{A}_{42} > \tilde{A}_{43} \\
\tilde{A}_{54} > \tilde{A}_{52} > \tilde{A}_{51} > \tilde{A}_{53}
\end{array}$$

[*Step 8.*] We determined the STF-multi-criteria decision-making sub-matrix $[b_i]_{1\times 4} = \tilde{b}_{ik}$ for $k = 1, 2, \ldots, 4$ and $i = 1, 2, \ldots, 5$ as;

$$\begin{array}{llll}
[b_1]_{1\times 4} = \tilde{b}_{11} = (\tilde{A}_{13}, & \tilde{b}_{12} = \tilde{A}_{14}, & \tilde{b}_{13} = \tilde{A}_{11}, & \tilde{b}_{14} = \tilde{A}_{12}) \\
[b_2]_{1\times 4} = \tilde{b}_{21} = (\tilde{A}_{23}, & \tilde{b}_{22} = \tilde{A}_{21}, & \tilde{b}_{23} = \tilde{A}_{22}, & \tilde{b}_{24} = \tilde{A}_{24}) \\
[b_3]_{1\times 4} = \tilde{b}_{31} = (\tilde{A}_{31}, & \tilde{b}_{32} = \tilde{A}_{33}, & \tilde{b}_{33} = \tilde{A}_{32}, & \tilde{b}_{34} = \tilde{A}_{34}) \\
[b_4]_{1\times 4} = \tilde{b}_{41} = (\tilde{A}_{44}, & \tilde{b}_{42} = \tilde{A}_{41}, & \tilde{b}_{43} = \tilde{A}_{42}, & \tilde{b}_{44} = \tilde{A}_{43}) \\
[b_5]_{1\times 4} = \tilde{b}_{51} = (\tilde{A}_{54}, & \tilde{b}_{52} = \tilde{A}_{52}, & \tilde{b}_{53} = \tilde{A}_{51}, & \tilde{b}_{54} = \tilde{A}_{53})
\end{array}$$

[*Step 9.*] We computed $r_i = STF_{hao} = (\tilde{b}_{i1}, \tilde{b}_{i2}, \ldots, \tilde{b}_{in}\} = \langle(r_1^i, r_2^i, r_3^i, r_4^i); \hat{\alpha}^i, \hat{\beta}^i, \hat{\gamma}^i\rangle$ for $i = 1, 2, 3, 4, 5$ as;

$$\begin{aligned} r_1 &= STF_{hao}(\tilde{b}_{11}, \tilde{b}_{12}, \tilde{b}_{13}, \tilde{b}_{14}) \\ &= \langle (3.4400, 6.0000, 6.6400, 7.5200); 0.7903, 0.2539, 0.3180 \rangle \\ r_2 &= STF_{hao}(\tilde{b}_{21}, \tilde{b}_{22}, \tilde{b}_{23}, \tilde{b}_{24}) \\ &= \langle (2.7200, 5.1200, 6.2400, 7.7600); 0.8097, 0.2206, 0.3681 \rangle \\ r_3 &= STF_{hao}(\tilde{b}_{31}, \tilde{b}_{32}, \tilde{b}_{33}, \tilde{b}_{34}) \\ &= \langle (1.4000, 4.9600, 6.7600, 8.1200); 0.6645, 0.4647, 0.5652 \rangle \\ r_4 &= STF_{hao}(\tilde{b}_{41}, \tilde{b}_{42}, \tilde{b}_{43}, \tilde{b}_{44}) \\ &= \langle (4.7200, 6.6000, 6.8800, 7.9600); 0.8630, 0.1437, 0.2108 \rangle \\ r_5 &= STF_{hao}(\tilde{b}_{51}, \tilde{b}_{52}, \tilde{b}_{53}, \tilde{b}_{54}) \\ &= \langle (2.6400, 3.7200, 4.4000, 5.4000); 0.6673, 0.3992, 0.3469 \rangle \end{aligned}$$

[*Step 10.*] We obtained the STF-distances $d(r_i, \tilde{\hat{a}}^+)$ for $i = 1, 2, 3, 4, 5$ as

$$d(r_1, \tilde{\hat{a}}^+) = 0.7469, \quad (r_2, \tilde{\hat{a}}^+) = 0.7360, \quad d(r_3, \tilde{\hat{a}}^+) = 0.7432,$$
$$d(r_4, \tilde{\hat{a}}^+) = 0.7529, \quad d(r_5, \tilde{\hat{a}}^+) = 0.7773,$$

respectively. Where

$$\begin{aligned} \tilde{\hat{a}}^+ &= \langle (\min_{i\in I_m}\{r_1^i\}, \min_{i\in I_m}\{r_2^i\}, \max_{i\in I_m}\{r_3^i\}, \max_{i\in I_n}\{r_4^i\}); \\ &\quad \max_{i\in I_m}\{\hat{\alpha}^i\}, \min_{i\in I_m}\{\hat{\beta}^i\}, \min_{i\in I_m}\{\hat{\gamma}^i\} \rangle \\ &= \langle (1.4000, 3.7200, 6.8800, 8.1200); 0.8630, 0.1437, 0.2108 \rangle \end{aligned}$$

[*Step 11.*] We ranked all alternatives x_i based on $d(r_i, \tilde{\hat{a}}^+)$ for $i = 1, 2, 3, 4, 5$ by using the ranking method based on STF-distance as;

$$d(r_2, \tilde{\hat{a}}^+) < d(r_3, \tilde{\hat{a}}^+) < d(r_1, \tilde{\hat{a}}^+) < d(r_4, \tilde{\hat{a}}^+) < d(r_5, \tilde{\hat{a}}^+)$$

Therefore, the ranking order of the alternatives x_j $(j = 1, 2, 3, 4)$ is generated as follows:

$$x_5 < x_4 < x_1 < x_3 < x_2$$

The best supplier for the enterprise is x_2.

4.2 STrF-Multi-criteria Group Decision-Making Method

In this section, we define a multi-criteria decision making method, so called STrF-multi-criteria decision-making method, by using the $STrF_{hao}$ operator. Some of the definitions are quoted or inspired by Gündoğdu and Kahraman (2019b), Ye (2015), Subas (2015), Deli and Subas (2017), Li (2014).

Definition 26 Let $X = (x_1, x_2, \ldots, x_m)$ be a set of alternatives, $U = (u_1, u_2, \ldots, u_n)$ be the set of criterions. If $\tilde{a}_{ij} = \langle (a_{ij}, b_{ij}, c_{ij}); \alpha_{ij}, \beta_{ij}, \gamma_{ij} \rangle \in \Theta$, then

1.

$$[\tilde{a}_{ij}]_{m\times n} = \begin{matrix} & u_1 & u_2 & \cdots & u_n \\ x_1 & \tilde{a}_{11} & \tilde{a}_{12} & \cdots & \tilde{a}_{1n} \\ x_2 & \tilde{a}_{21} & \tilde{a}_{22} & \cdots & \tilde{a}_{2n} \\ \vdots & \vdots & \vdots & \vdots & \vdots \\ x_m & \tilde{a}_{m1} & \tilde{a}_{m2} & \cdots & \tilde{a}_{mn} \end{matrix}$$

is called an STrF-multi-criteria decision-making matrix of the decision maker.

2.

$$[\tilde{a}_i]_{1\times n} = \begin{matrix} & u_1 & u_2 & \cdots & u_n \\ x_i & (\tilde{a}_{i1} & \tilde{a}_{i2} & \cdots & \tilde{a}_{in}) \end{matrix}$$

is called an STrF-multi-criteria decision-making sub-matrix of the decision maker based on the alternatives x_i.

Now, we can give an algorithm of the STrF-multi-criteria decision-making method as follows;

Algorithm:

[*Step 1.*] Construct the STrF-multi-criteria decision making matrix $[\tilde{a}_{ij}]_{m\times n}$ for $i = 1, 2, \ldots, m$; $j = 1, 2, \ldots, n$;

[*Step 2.*] Insert the weights of the criterions $\omega_i = (\omega_1, \omega_2, \ldots, \omega_n)$;

[*Step 3.*] Insert the (position) weights of the criterions $w_i = (w_1, w_2, \ldots, w_n)$ associated with the operator $STrF_{hao}$;

[*Step 4.*] Compute the STrF-numbers $\tilde{A}_{ij} = \langle(\hat{a}_{ij}, \hat{b}_{ij}, \hat{c}_{ij}); \hat{\alpha}_{ij}, \hat{\beta}_{ij}, \hat{\gamma}_{ij}\rangle = n\omega_i\tilde{a}_{ij}$ for $i = 1, 2, \ldots, m$; $j = 1, 2, \ldots, n$;

[*Step 5.*] Construct the STrF-multi-criteria decision making matrix $[\tilde{A}_{ij}]_{m\times n}$ for $i = 1, 2, \ldots, m$; $j = 1, 2, \ldots, n$;

[*Step 6.*] Obtain the STrF-distances $d(\tilde{A}_{ij}, \tilde{a}^+)$ based on the STrF-numbers $\tilde{A}_{ij}$ for $i = 1, 2, \ldots, m$; $j = 1, 2, \ldots, n$; where

$$\begin{aligned}\tilde{a}^+ = \langle(&\min_{j\in I_n}\{\hat{a}_{ij}\}, \min_{j\in I_n}\{\hat{b}_{ij}\}, \max_{j\in I_n}\{\hat{c}_{ij}\}); \\ &\max_{j\in I_n}\{\hat{\alpha}_{\tilde{a}_j}\}, \min_{j\in I_n}\{\hat{\beta}_{\tilde{a}_j}\}, \min_{j\in I_n}\{\hat{\gamma}_{\tilde{a}_j}\}\rangle\end{aligned}$$

[*Step 7.*] Rank the nonincreasing order of the n STrF-numbers $\tilde{A}_{ij}$ for $i = 1, 2, \ldots, m$ by using the ranking method based on STrF-distance;

[*Step 8.*] Determine the STrF-multi-criteria decision-making sub-matrix $[b_i]_{1\times n} = \tilde{b}_{ik}$ for $k = 1, 2, \ldots, n$ and $i = 1, 2, \ldots, m$, where $\tilde{b}_{ik}$ (k=1,2, …,n) is k-th largest of $\tilde{A}_{ij}$ for $i = 1, 2, \ldots, m$;

[*Step 9.*] Compute $r_i = STrF_{hao} = (\tilde{b}_{i1}, \tilde{b}_{i2}, \ldots, \tilde{b}_{in}\} = \langle(r_1^i, r_2^i, r_3^i); \hat{\alpha}^i, \hat{\beta}^i, \hat{\gamma}^i\rangle$;

[*Step 10.*] Obtain the STrF-distances $d(r_i, \tilde{\tilde{a}}^+)$ for $i = 1, 2, \ldots, m$, where

$$\begin{aligned}\tilde{\tilde{a}}^+ = \langle(&min_{i\in I_m}\{r_1^i\}, min_{i\in I_m}\{r_2^i\}, max_{i\in I_m}\{r_3^i\}); \\ &max_{i\in I_m}\{\hat{\alpha}^i\}, min_{i\in I_m}\{\hat{\beta}^i\}, min_{i\in I_m}\{\hat{\gamma}^i\}\rangle\end{aligned}$$

[*Step 11.*] Rank all alternatives x_i based on $d(r_i, \tilde{\tilde{a}}^+)$ by using the by using the ranking method based on STrF-distance and determine the best alternative.

Example 11 Let us consider the decision-making problem adapted from Ghorabaee (2016). An auto company desires to select a suitable robot for its production process. For this purpose,they formed a team for making a decision on the problem. The members of this team are the production managers of the factory production lines. The members of the decision-making team must apprise $X = \{x_1, x_2, x_3, x_4, x_5\}$ for selecting the best.Decision makers have access to brochures and data of these alternatives (robots). In this study, four criteria set

$$\begin{aligned} U = \{&u_1 = \textit{Inconsistency with infrastructure}, u_2 = \textit{Manmachine interface}, \\ &u_3 = \textit{Programming flexibility}, u_4 = \textit{Compliance}\} \end{aligned}$$

are considered by decision-making team for the appraisal of alternatives. Then, the weight vector of the attributes is $\omega = (0.3, 0.2, 0.1, 0.4)^T$ and the position weight vector is $w = (0.1, 0.3, 0.4, 0.2)^T$ by using the weight determination based on the normal distribution. For the evaluation of an alternative x_i for $i = 1, 2, 3, 4, 5$ with respect to a criterion u_j for $j = 1, 2, 3, 4$, it is obtained from the questionnaire of a domain members. Then, we use the proposed method to solve this robot selection problem. The calculative procedure is summarized as follows:

[*Step 1.*] We constructed the STrF-multi-criteria decision making matrix $[\tilde{a}_{ij}]_{5\times 4}$ for $i = 1, 2, \ldots, 5$ and $j = 1, 2, \ldots, 4$ as;

$$[\tilde{a}_{ij}]_{5\times 4} = \left(\begin{array}{ll} \langle (2,3,4); 0.9, 0.1, 0.1\rangle & \langle (1,3,5); 0.6, 0.5, 0.6\rangle \\ \langle (2,5,6); 0.1, 0.8, 0.5\rangle & \langle (3,8,9); 0.7, 0.5, 0.5\rangle \\ \langle (1,2,5); 0.6, 0.5, 0.6\rangle & \langle (3,8,9); 0.6, 0.5, 0.6\rangle \\ \langle (5,7,7); 0.9, 0.1, 0.1\rangle & \langle (1,3,5); 0.6, 0.5, 0.6\rangle \\ \langle (1,3,5); 0.1, 0.8, 0.5\rangle & \langle (5,5,7); 0.6, 0.5, 0.6\rangle \end{array} \right.$$

$$\left. \begin{array}{ll} \langle (5,7,8); 0.9, 0.1, 0.1\rangle & \langle (3,8,9); 0.9, 0.1, 0.1\rangle \\ \langle (4,5,3); 0.9, 0.1, 0.1\rangle & \langle (1,2,3); 0.1, 0.8, 0.5\rangle \\ \langle (3,8,9); 0.1, 0.8, 0.5\rangle & \langle (1,2,3); 0.1, 0.8, 0.5\rangle \\ \langle (6,7,8); 0.1, 0.7, 0.7\rangle & \langle (5,7,7); 0.9, 0.1, 0.1\rangle \\ \langle (1,2,3); 0.6, 0.5, 0.6\rangle & \langle (1,3,5); 0.9, 0.1, 0.1\rangle \end{array} \right)$$

[*Step 2.*] We inserted the weights of the criterions $\omega = (0.3, 0.2, 0.1, 0.4)^T$ as;
[*Step 3.*] We inserted the (position) weights of the criterions $w = (0.1, 0.3, 0.4, 0.2)^T$ associated with the operator $STrF_{hao}$ as;
[*Step 4.*] We computed the STrF-numbers $\tilde{A}_{ij} = \langle(\hat{a}_{ij}, \hat{b}_{ij}, \hat{c}_{ij}, \hat{d}_{ij}); \hat{\alpha}_{ij}, \hat{\beta}_{ij}, \hat{\gamma}_{ij}\rangle = n\omega_i \tilde{a}_{ij}$ for $i = 1, 2, 3, 4, 5$ and $j = 1, 2, 3, 4)$ as;

$$\begin{aligned} \tilde{A}_{11} &= 4 \times 0.3 \times \langle (2,3,4); 0.9, 0.1, 0.1\rangle \\ &= \langle (2.40, 3.60, 4.80); 0.9294, 0.0631, 0.0925\rangle \end{aligned}$$

Likewise, we can obtain other STrF-numbers $\tilde{A}_{ij} = n\omega_i\tilde{a}_{ij}$ for $i = 1, 2, 3, 4, 5$ and $j = 1, 2, 3, 4$);

[*Step 5.*] We construct the STrF-multi-criteria decision making matrix $[\tilde{A}_{ij}]_{m\times n}$ for $i = 1, 2, 3, 4, 5$ and $j = 1, 2, 3, 4$ as;

$$[\tilde{A}_{ij}]_{5\times 4} = \begin{pmatrix} \langle(2.40, 3.60, 4.80); 0.9294, 0.0631, 0.0925\rangle & \langle(0.80, 2.40, 4.00); 0.5479, 0.5743, 0.5819\rangle & \langle(2.00, 2.80, 3.20); 0.6967, 0.3981, 0.1049\rangle & \langle(4.80, 12.8, 14.40); 0.9643, 0.0251, 0.0762\rangle \\ \langle(2.40, 6.00, 7.20); 0.1095, 0.7651, 0.5397\rangle & \langle(2.40, 6.40, 7.20); 0.6454, 0.5743, 0.4931\rangle & \langle(1.60, 2.00, 1.20); 0.6967, 0.3981, 0.1049\rangle & \langle(1.60, 3.20, 4.80); 0.1263, 0.6998, 0.6053\rangle \\ \langle(1.20, 2.40, 6.00); 0.6439, 0.4353, 0.6069\rangle & \langle(2.40, 6.40, 7.20); 0.5479, 0.5743, 0.5819\rangle & \langle(1.20, 3.20, 3.60); 0.0633, 0.9146, 0.3308\rangle & \langle(1.60, 3.20, 4.80); 0.1263, 0.6998, 0.6053\rangle \\ \langle(6.00, 8.40, 8.40); 0.9294, 0.0631, 0.0925\rangle & \langle(0.80, 2.40, 4.00); 0.5479, 0.5743, 0.5819\rangle & \langle(2.40, 2.80, 3.20); 0.0633, 0.8670, 0.4880\rangle & \langle(8.00, 11.2, 11.20); 0.9643, 0.0251, 0.0762\rangle \\ \langle(1.20, 3.60, 6.00); 0.1095, 0.7651, 0.5397\rangle & \langle(4.00, 4.00, 5.60); 0.5479, 0.5743, 0.5819\rangle & \langle(0.40, 0.80, 1.20); 0.4043, 0.7579, 0.4853\rangle & \langle(1.60, 4.80, 8.00); 0.9643, 0.0251, 0.0762\rangle \end{pmatrix}$$

[*Step 6.*] We obtained the STrF-distances $d(\tilde{A}_{ij}, \tilde{a}^+)$ based on the STrF-numbers $\tilde{A}_{ij}$ for $i = 1, 2, 3, 4, 5$ and $j = 1, 2, 3, 4$ as;

$$\begin{array}{ll} d(\tilde{A}_{11}, \tilde{a}^+) = 0.7064 & d(\tilde{A}_{33}, \tilde{a}^+) = 0.9652 \\ d(\tilde{A}_{12}, \tilde{a}^+) = 0.8151 & d(\tilde{A}_{34}, \tilde{a}^+) = 0.9410 \\ d(\tilde{A}_{13}, \tilde{a}^+) = 0.7768 & d(\tilde{A}_{41}, \tilde{a}^+) = 0.8077 \\ d(\tilde{A}_{14}, \tilde{a}^+) = 0.8557 & d(\tilde{A}_{42}, \tilde{a}^+) = 0.8151 \\ d(\tilde{A}_{21}, \tilde{a}^+) = 0.9560 & d(\tilde{A}_{43}, \tilde{a}^+) = 0.9619 \\ d(\tilde{A}_{22}, \tilde{a}^+) = 0.8237 & d(\tilde{A}_{44}, \tilde{a}^+) = 0.8431 \\ d(\tilde{A}_{23}, \tilde{a}^+) = 0.8031 & d(\tilde{A}_{51}, \tilde{a}^+) = 0.9480 \\ d(\tilde{A}_{24}, \tilde{a}^+) = 0.9410 & d(\tilde{A}_{52}, \tilde{a}^+) = 0.8218 \\ d(\tilde{A}_{31}, \tilde{a}^+) = 0.7803 & d(\tilde{A}_{53}, \tilde{a}^+) = 0.9292 \\ d(\tilde{A}_{32}, \tilde{a}^+) = 0.8470 & d(\tilde{A}_{54}, \tilde{a}^+) = 0.7235 \end{array}$$

where

$$\begin{aligned}\tilde{a}^+ &= \langle(\min_{j\in I_n}\{\hat{a}_{ij}\}, \min_{j\in I_n}\{\hat{b}_{ij}\}, \max_{j\in I_n}\{\hat{c}_{ij}, \max_{j\in I_n}\{\hat{d}_{ij}\});\\ &\quad \max_{j\in I_n}\{\hat{\alpha}_{\tilde{a}_j}\}, \min_{j\in I_n}\{\hat{\beta}_{\tilde{a}_j}\}, \min_{j\in I_n}\{\hat{\gamma}_{\tilde{a}_j}\}\rangle\\ &= \big\langle(0.4, 0.8, 14.4); 0.9643, 0.0251, 0.0762\big\rangle\end{aligned}$$

[*Step 7.*] We ranked the nonincreasing order of the n STrF-numbers $\tilde{A}_{ij}$ for $i = 1, 2, 3, 4, 5$ by using the ranking method based on STrF-distance as;

$$\begin{array}{c}\tilde{A}_{11} > \tilde{A}_{13} > \tilde{A}_{12} > \tilde{A}_{14}\\ \tilde{A}_{23} > \tilde{A}_{22} > \tilde{A}_{24} > \tilde{A}_{21}\\ \tilde{A}_{31} > \tilde{A}_{32} > \tilde{A}_{34} > \tilde{A}_{33}\\ \tilde{A}_{41} > \tilde{A}_{42} > \tilde{A}_{44} > \tilde{A}_{43}\\ \tilde{A}_{51} > \tilde{A}_{52} > \tilde{A}_{53} > \tilde{A}_{51}\end{array}$$

[*Step 8.*] We determined the STrF-multi-criteria decision-making sub-matrix $[b_i]_{1\times 4} = \tilde{b}_{ik}$ for $k = 1, 2, \ldots, 4$ and $i = 1, 2, \ldots, 5$ as;

$$\begin{aligned}[b_1]_{1\times 4} &= (\tilde{b}_{11} = \tilde{A}_{11},\ \ \tilde{b}_{12} = \tilde{A}_{13},\ \ \tilde{b}_{13} = \tilde{A}_{12},\ \ \tilde{b}_{14} = \tilde{A}_{14})\\ [b_2]_{1\times 4} &= (\tilde{b}_{21} = \tilde{A}_{23},\ \ \tilde{b}_{22} = \tilde{A}_{22},\ \ \tilde{b}_{23} = \tilde{A}_{24},\ \ \tilde{b}_{24} = \tilde{A}_{21})\\ [b_3]_{1\times 4} &= (\tilde{b}_{31} = \tilde{A}_{31},\ \ \tilde{b}_{32} = \tilde{A}_{32},\ \ \tilde{b}_{33} = \tilde{A}_{34},\ \ \tilde{b}_{34} = \tilde{A}_{33})\\ [b_4]_{1\times 4} &= (\tilde{b}_{41} = \tilde{A}_{41},\ \ \tilde{b}_{42} = \tilde{A}_{42},\ \ \tilde{b}_{43} = \tilde{A}_{44},\ \ \tilde{b}_{44} = \tilde{A}_{43})\\ [b_5]_{1\times 4} &= (\tilde{b}_{51} = \tilde{A}_{51},\ \ \tilde{b}_{52} = \tilde{A}_{52},\ \ \tilde{b}_{53} = \tilde{A}_{53},\ \ \tilde{b}_{54} = \tilde{A}_{51})\end{aligned}$$

[*Step 9.*] We computed $r_i = STrF_{hao} = (\tilde{b}_{i1}, \tilde{b}_{i2}, \ldots, \tilde{b}_{in}\} = \langle(r_1^i, r_2^i, r_3^i, r_4^i); \hat{\alpha}^i, \hat{\beta}^i, \hat{\gamma}^i\rangle$ for $i = 1, 2, 3, 4, 5$ as;

$$\begin{aligned}r_1 &= STrF_{hao}(\tilde{b}_{11}, \tilde{b}_{12}, \tilde{b}_{13}, \tilde{b}_{14}) = \big\langle(2.24, 4.76, 5.84); 0.8175, 0.2127, 0.2615\big\rangle\\ r_2 &= STrF_{hao}(\tilde{b}_{21}, \tilde{b}_{22}, \tilde{b}_{23}, \tilde{b}_{24}) = \big\langle(2.00, 4.20, 4.56); 0.5615, 0.5514, 0.4011\big\rangle\\ r_3 &= STrF_{hao}(\tilde{b}_{31}, \tilde{b}_{32}, \tilde{b}_{33}, \tilde{b}_{34}) = \big\langle(1.60, 3.84, 5.76); 0.5259, 0.5754, 0.5928\big\rangle\\ r_4 &= STrF_{hao}(\tilde{b}_{41}, \tilde{b}_{42}, \tilde{b}_{43}, \tilde{b}_{44}) = \big\langle(4.48, 6.60, 7.12); 0.8730, 0.1323, 0.2368\big\rangle\\ r_5 &= STrF_{hao}(\tilde{b}_{51}, \tilde{b}_{52}, \tilde{b}_{53}, \tilde{b}_{54}) = \big\langle(2.36, 3.84, 5.96); 0.7856, 0.2446, 0.3529\big\rangle\end{aligned}$$

[*Step 10.*] We obtained the STrF-distances $d(r_i, \tilde{\hat{a}}^+)$ for $i = 1, 2, 3, 4, 5$ as;
$d(r_1, \tilde{\hat{a}}^+) = 0.7387$, $d(r_2, \tilde{\hat{a}}^+) = 0.7866$, $d(r_3, \tilde{\hat{a}}^+) = 0.7810$,
$d(r_4, \tilde{\hat{a}}^+) = 0.7533$, $d(r_5, \tilde{\hat{a}}^+) = 0.7374$

respectively. Where

$$\tilde{\hat{a}}^{+} = \langle(\min_{i\in I_m}\{r_1^i\}, \min_{i\in I_m}\{r_2^i\}, \max_{i\in I_m}\{r_3^i\}, \max_{i\in I_n}\{r_4^i\}); \\ \max_{i\in I_m}\{\hat{\alpha}^i\}, \min_{i\in I_m}\{\hat{\beta}^i\}, \min_{i\in I_m}\{\hat{\gamma}^i\}\rangle \\ = \big\langle(1.60, 3.84, 218.40); 0.8730, 0.0002, 0.2368\big\rangle$$

[*Step 11.*] We ranked all alternatives x_i based on $d(r_i, \tilde{\hat{a}}^{+})$ for $i = 1, 2, 3, 4, 5$ by using the ranking method based on STrF-distance as;

$$d(r_5, \tilde{\hat{a}}^{+}) < d(r_1, \tilde{\hat{a}}^{+}) < d(r_4, \tilde{\hat{a}}^{+}) < d(r_3, \tilde{\hat{a}}^{+}) < d(r_2, \tilde{\hat{a}}^{+})$$

Therefore, the ranking order of the alternatives x_j $(j = 1, 2, 3, 4)$ is generated as follows:

$$x_2 < x_3 < x_4 < x_1 < x_5$$

Thus the most desirable Robot is x_5.

5 Conclusion

In this chapter, we firstly proposed spherical fuzzy number theory generalization of fuzzy numbers and intuitionistic fuzzy numbers. Then, we defined two special forms of spherical fuzzy numbers is called spherical trapezoidal fuzzy numbers (STF-numbers) and spherical triangular fuzzy numbers (STrF-numbers) with some operations laws and properties. Also, we developed a distance measure and some arithmetic and geometric operators on both STF-numbers and STrF-numbers. Finally, we presented two decision making method approach and gave two practical example, to illustrate the proposed approach. In future research, we will extend and apply the given spherical fuzzy number to other fields, such as supply chain management, risk management, pattern recognition, game theory, society, economics, management, military, and engineering technology and so on.

References

Atanassov K (1986) Intuitionistic fuzzy sets. Fuzzy Sets Syst 20:87–96

Chen Y, Li B (2011) Dynamic multi-attribute decision making model based on triangular intuitionistic fuzzy numbers. Scientia Iranica B 18(2):268–274

Das S, Guha D (2013) Ranking of intuitionistic fuzzy number by centroid point. J Ind Intell Inf 1(2):107–110

Deli I, Subas Y (2017) A ranking method of single valued neutrosophic numbers and its applications to multi-attribute decision making problems. Int J Mach Learn Cybern 8(4):1309–1322

Garg H, Munir M, Ullah K, Mahmood T, Jan N (2018) Algorithm for T-spherical fuzzy multi-attribute decision making based on improved interactive aggregation operators, symmetry. https://doi.org/10.3390/sym10120670

Ghorabaee MK (2016) Developing an MCDM method for robot selection with interval type-2 fuzzy sets. Robot Comput Integr Manuf 37:221–232

Gündoğdu FK, Kahraman C (2019a) A novel fuzzy TOPSIS method using emerging interval-valued spherical fuzzy sets. Eng Appl Artif Intell 85:307–323

Gündoğdu FK, Kahraman C (2019b) Spherical fuzzy sets and spherical fuzzy TOPSIS method. J Intell Fuzzy Syst 36(1):337–352

He Y, Chen H, Zhou L, Han B, Zhao Q, Liu J (2014a) Generalized intuitionistic fuzzy geometric interaction operators and their application to decision making. Expert Syst Appl 41:2484–2495

He Y, Chen H, Zhou L, Liu J, Tao Z (2014b) Intuitionistic fuzzy geometric interaction averaging operators and their application to multi-criteria decision making. Inf Sci 259:142–159

Li DF (2014) Decision and Game theory in management with intuitionistic fuzzy sets. Studies in fuzziness and soft computing, vol 308. Springer

Li DF (2010) A ratio ranking method of triangular intuitionistic fuzzy numbers and its application to MADM problems. Comput Math Appl 60:1557–1570

Liang C, Zhao S, Zhang J (2014) Aggregation operators on triangular intuitionistic fuzzy numbers and its application to multi-criteria decision making problems. Found Comput Decis Sci 39(3):189–208

Liu P (2013) Some generalized dependent aggregation operators with intuitionistic linguistic numbers and their application to group decision making. J Comput Syst Sci 79:131–143

Liu P, Khan Q, Mahmood T, Hassan N (2019) T-spherical fuzzy power muirhead mean operator based on novel operational laws and their application in multi-attribute group decision making. IEEE Access. https://doi.org/10.1109/ACCESS.2019.2896107

Mahapatra GS, Roy TK (2009) Reliability evaluation using triangular intuitionistic fuzzy numbers arithmetic operations. World Acad Sci Eng Technol 3:422–429

Smarandache F (1998) A unifying field in logics. Neutrosophy: neutrosophic probability, set and logic. American Research Press, Rehoboth

Subas Y (2015) Neutrosophic numbers and their application to multi-attribute decision making problems. Master's thesis, Kilis 7 Aralik University, Graduate School of Natural and Applied Science (in Turkish)

Ullah K, Mahmood T, Jan N (2018) Similarity measures for T-spherical fuzzy sets with applications in pattern recognition. symmetry. https://doi.org/10.3390/sym10060193

Wang J, Nie R, Zhang H, Chen X (2013) New operators on triangular intuitionistic fuzzy numbers and their applications in system fault analysis. Inf Sci 251:79–95

Wang JQ, Zhong Z (2009) Aggregation operators on intuitionistic trapezoidal fuzzy number and its application to multi-criteria decision making problems. J Syst Eng Electron 20(2):321–326

Wan SP (2013) Power average operators of trapezoidal intuitionistic fuzzy numbers and application to multi-attribute group decision making. Appl Math Model 37:4112–4126

Wan S, Dong J (2014) Multi-attribute group decision making with trapezoidal intuitionistic fuzzy numbers and application to stock selection. Informatica 25(4):663–697

Wan S, Dong J (2015) Power geometric operators of trapezoidal intuitionistic fuzzy numbers and application to multi-attribute group decision making. Appl Soft Comput 29:153–168

Wei G (2010) Some arithmetic aggregation operators with intuitionistic trapezoidal fuzzy numbers and their application to group decision making. J Comput 5(3):345–351

Wu J, Cao QW (2013) Same families of geometric aggregation operators with intuitionistic trapezoidal fuzzy numbers. Appl Math Model 37:318–327

Xu ZS (2007) Intuitionistic fuzzy aggregation operators. IEEE Trans Fuzzy Syst 15(6):1179–1187

Xu Y, Wang H (2012) The induced generalized aggregation operators for intuitionistic fuzzy sets and their application in group decision making. Appl Soft Comput 12:1168–1179

Xu Z, Yager RR (2006) Some geometric aggregation operators based on intuitionistic fuzzy sets. Int J Gen Syst 35(4):417–433

Ye J (2015) Trapezoidal neutrosophic set and its application to multiple attribute decision-making. Neural Comput Appl 26(5):1157–1166

Yu D (2013) Intuitionistic trapezoidal fuzzy information aggregation methods and their applications to teaching quality evaluation. J Inf Comput Sci 10(6):1861–1869
Zhou L, Wei G (2013) Some intuitionistic fuzzy Einstein hybrid aggregation operators and their application to multiple attribute decision making. Knowl-Based Syst 37:472–479
Zadeh LA (1965) Fuzzy sets. Inf Control 8:338–353
Zhang X, Liu P (2010) Method for aggregating triangular fuzzy intutionistic fuzzy information and its application to decision making. Technol Econ Deve Econ Baltic J Sustain 16(2):280–290
Zhang X, Jin F, Liu P (2013) A grey relational projection method for multi-attribute decision making based on intuitionistic trapezoidal fuzzy number. Appl Math Model 37:3467–3477

The Generalized Dice Similarity Measures for Spherical Fuzzy Sets and Their Applications

Ping Wang, Jie Wang, and Guiwu Wei

Abstract As the extension of fuzzy set, intuitionistic fuzzy set, Pythagorean fuzzy set, and picture fuzzy set, the spherical fuzzy set is characterized by three functions expressing the positive-membership degree, the neutral-membership degree and the negative-membership degree which the sum squares of them is equal or less than 1. In this work, we shall present some novel Dice similarity measures of spherical fuzzy sets and the generalized Dice similarity measures of spherical fuzzy sets and indicates that the Dice similarity measures and asymmetric measures (projection measures) are the special cases of the generalized Dice similarity measures in some parameter values. Then, we propose the generalized Dice similarity measures-based multiple attribute group decision making models with spherical fuzzy information. Then, we apply the generalized Dice similarity measures between spherical fuzzy sets to multiple attribute group decision making. Finally, an illustrative example is given to demonstrate the efficiency of the similarity measures for selecting the desirable ERP system.

1 Introduction

For real multiple attribute decision making (MADM) problems, how to find the most suitable alternative from a given alternative set is an important topic (Tang and Wei 2019; Tang et al. 2019; Wang et al. 2019a, b; Wei 2019). The most common method is fusing the assessment information given by decision makers, and then select the best alternative by compare the fused results. However, due to the fuzziness and uncertainty of practical decision making environment, the assessment information always cannot be expressed by exact real number. To overcome this limitation, Zadeh (1965) defined the fuzzy set which only consider the membership

P. Wang
School of Engineering, Sichuan Normal University, Chengdu, People's Republic of China

J. Wang · G. Wei (✉)
School of Business, Sichuan Normal University, Chengdu 610101, People's Republic of China
e-mail: weiguiwu@163.com

© Springer Nature Switzerland AG 2021

C. Kahraman and F. Kutlu Gündoğdu (eds.), *Decision Making with Spherical Fuzzy Sets*, Studies in Fuzziness and Soft Computing 392,
https://doi.org/10.1007/978-3-030-45461-6_4

degree but ignore the non-membership, on this basis, Atanassov (1989) given the concept of intuitionistic fuzzy set (IFS) which is characterized by the functions of membership degree and non-membership degree, since then, the MADM problems with intuitionistic fuzzy information have been successfully studied by a mass of scholars (Narayanamoorthy et al. 2019; Jiang et al. 2019; Zhou and Xu 2018; Zhang 2018; Zhang et al. 2018; Yu et al. 2018; Yao and Wang 2018; Yang et al. 2018; Xia 2018). But, in the process of the study of IFS, researcher discovered that it still has some limitations for not taking a special case into account, for example, in an election, the voting results are shown as: three votes in favor, four votes in neutral, two votes in against and one vote in abstention. Obviously, we can not only use the membership and non-membership to depict this condition, but use another more convenient tool, which named the picture fuzzy set (PFS) proposed by Cuong (2014), to study the MADM issues. The PFS which contained three measurement index including positive-membership degree, neutral-membership degree and negative-membership degree satisfied the condition of the sum of these index is less or equal to 1. Since that, study literature of PFS has been emerged. Singh (2015) defined the correlation coefficient of PFS. Son (2016) proposed some generalized distance measurement of picture fuzzy number and applied them in picture fuzzy clustering. Wei (2016) built the cross-entropy model under the picture fuzzy environment, which computed the cross-entropy between all the alternatives and the relative ideal solution, to rank all the alternatives. Garg (2017) developed some aggregation operators including picture fuzzy weighted average (PFWA) operator, picture fuzzy ordered weighted average (PFOWA) operator, picture fuzzy hybrid average (PFHA) operator. Similarly, Wei (2017a) proposed the picture fuzzy weighted geometric (PFWG) operator, picture fuzzy ordered weighted geometric (PFOWG) operator, picture fuzzy hybrid geometric (PFHG) operator. In addition, the TODIM model (Wei et al. 2018), projection model (Gao et al. 2018), cosine similarity measures (Wei 2017b) and many other works (Peng 2017; Liu and Zhang 2018; Wang et al. 2018a, b; Wang and Li 2018) about PFS are studied.

However, the application prospect of picture fuzzy set (PFS) is narrow for the limitation of assessment values, that is, the sum of positive-membership, neutral-membership and negative-membership must be less or equal to 1. To overcome this shortcoming, Kutlu Gündoğdu and Kahraman (2019) proposed a generalization form of PFS which called spherical fuzzy set (SFS) to express more widely uncertain information and developed a novel TOPSIS model with spherical fuzzy information. The spherical fuzzy set was depicted by the degree of positive-membership, neutral-membership and negative-membership which the sum squares of them is limited to 1. Since then, some study about SFSs have been worked by scholars (Garg et al. 2018; Ullah et al. 2018; Ashraf et al. 2019a, b; Jin et al. 2019a, b). Ashraf and Abdullah (2018) developed some spherical fuzzy number weighted average aggregation (SFN-WAA) operators and spherical fuzzy number weighted geometric aggregation (SFN-WGA) operators. Wei et al. (2019) developed some cosine and similarity measures between SFS and applied these measures to deal with MADM problems. Gündoğdu and Kahraman (2019) built the spherical fuzzy WASPAS model to handle decision making problems. Based on the interval-valued spherical fuzzy sets, Gündoğdu and

Kahraman (2019) presented a novel fuzzy TOPSIS model. Gündoğdu and Kahraman (2019) extended the VIKOR method to spherical fuzzy environment and studied its application to warehouse site selection. Rafiq et al. (2019) defined some spherical fuzzy cosine similarity measures for MADM. Quek et al. (2019) studied the generalized spherical fuzzy set and proposed some aggregation operators. Mahmood et al. (2018) proposed some operation laws of spherical fuzzy set (SFS) and defined the T-spherical fuzzy set (T-SFS), the applications about medical diagnosis with spherical fuzzy information and T-spherical fuzzy information are also discussed. Ullah et al. (2019) utilized the interval-valued T-spherical fuzzy set (IVT-SFS) to evaluate the investment policy. Based on power operations (Yager 2001; Xu and Yager 2010) and Muirhead Mean (MM) (Muirhead 1902) operators, Liu et al. (2019) developed some T-spherical fuzzy power Muirhead Mean (MM) operators for MADM problems. Although, the spherical fuzzy set (SFS), as the extension of fuzzy set (FS) (Zadeh 1965), intuitionistic fuzzy set (IFS) (Atanassov 1986), Pythagorean fuzzy set (PyFS) (Yager 2013, 2014) and picture fuzzy set (PFS) (Zhang et al. 2018), can be more suitable for handling practical decision making problems. However, up to now, the study about SFS is extremely limited and how to deal with MADM problems with spherical fuzzy information is an interesting topic.

For MADM problems, the way to express the assessment information is only one aspect, another vital aspect is selecting best alternative from a given alternative set. As a powerful tool for handling MADM, the Dice similarity measure has attracted lots of scholar's attention and applied in many different decision making fields. Ye (2012) studied the Dice similarity measures based on the reduct intuitionistic fuzzy sets of interval-valued intuitionistic fuzzy sets. Tang et al. (2017) discussed the Dice similarity measures under intuitionistic fuzzy environment. Shan and Ye (2014) investigated the Dice similarity measures between single valued neutrosophic multisets and applied them in medical diagnosis. Ye (2012) proposed some Dice similarity measures between expected intervals of trapezoidal fuzzy numbers. Ye (2012) defined some Dice similarity measures between generalized trapezoidal fuzzy numbers. Joshi and Kumar (2018) developed some Dice similarity measures under picture fuzzy environment. Wang et al. (2019) extended the Dice similarity measures to the Pythagorean fuzzy environment. From above mentioned literatures, we can easily find that the Dice similarity measure owns the some precious advantages such as: (1) the computing results by Dice similarity measure are stable; (2) the calculating equations are simple; (3) it is available to combine this model with other approaches. However, as far as author knows, the study about Dice similarity measure with spherical fuzzy information doesn't exist. Thus, it's meaningful to work for Dice similarity measure under spherical fuzzy environment. To do that, the main purposes of this paper are: (1) develop two forms of the Dice similarity measures between spherical fuzzy sets (SFSs), (2) develop the generalized Dice similarity measures between T-spherical fuzzy sets (T-SFSs), and (3) develop the generalized Dice measures-based multiple attribute group decision making (MADM) models between spherical fuzzy sets (SFSs). In the multiple attribute group decision making (MADM) process, the main advantage of the proposed methods is more general and more flexible than existing patterns recognition methods with spherical fuzzy information to satisfy the practical requirements.

In order to do so, the remainder of this paper is structured as follows. In the next section, we introduce some basic concepts related to spherical fuzzy set (SFS). In Sect. 3, we shall develop some spherical fuzzy Dice similarity measures and some spherical fuzzy weighted Dice similarity measures. In Sect. 4, the Dice similarity measures for spherical fuzzy set (SFS) are applied to multiple attribute group decision making (MADM) problems. In Sect. 5, an illustrative example is given to demonstrate the efficiency of the similarity measures for selecting the desirable ERP system. Section 6 concludes the paper with some remarks.

2 Preliminaries

In this part, as the extension of fuzzy set (FS), intuitionistic fuzzy set (IFS), Pythagorean fuzzy set (PyFS) and picture fuzzy set (PFS), some fundamental concepts and theories of spherical fuzzy sets (SFSs) which firstly developed by Gundogdu and Kahraman (2019) are briefly introduced as follows.

Definition 1 (Gündoğdu and Kahraman 2019) Assume that X be a fix set. A spherical fuzzy set (SFS) is an object having the form

$$T = \{(x, \mu_T(x), \eta_T(x), v_T(x)) | x \in X\} \tag{1}$$

where $\mu_T(x)$ denotes the positive-membership degree, $\eta_T(x)$ denotes the neutral-membership degree and $v_T(x)$ denotes the negative-membership degree. Absolutely, $\mu_T(x), \eta_T(x)$ *and* $v_T(x) \in [0, 1]$ and the sum squares of $\mu_T(x), \eta_T(x)$ *and* $v_T(x)$ satisfies the condition:

$$0 \leq (\mu_T(x))^2 + (\eta_T(x))^2 + (v_T(x))^2 \leq 1 \tag{2}$$

The degree of indeterminacy is given as:

$$\pi_T(x) = \sqrt{1 - (\mu_T(x))^2 - (\eta_T(x))^2 - (v_T(x))^2} \tag{3}$$

For convenience, we called $T = (\mu, \eta, v)$ a spherical fuzzy number (SFN).

3 Some Dice Similarity Measure for Spherical Fuzzy Sets

The Dice similarity measure can't induce this undefined situation when one vector is zero, which overcomes the disadvantage of the cosine similarity measure (Dice

1945). Therefore, the concept of the Dice similarity measure is introduced in the section (Dice 1945).

Definition 2 (Dice 1945) Let $A = (a_1, a_2, \ldots, a_n)$ and $B = (b_1, b_2, \ldots, b_n)$ be two vectors of length n where all the coordinates are positive real numbers. Then the Dice similarity measure is define as follows:

$$D(A, B) = \frac{2A \cdot B}{\|A\|_2^2 + \|B\|_2^2} = \frac{2\sum_{j=1}^{n} a_j b_j}{\sum_{j=1}^{n} (a_j)^2 \sum_{j=1}^{n} (b_j)^2} \tag{4}$$

where $A \cdot B = \sum_{j=1}^{n} a_j b_j$ is called the inner product of the vector A and B and $\|A\|_2 = \sqrt{\sum_{j=1}^{n} (a_j)^2}$ and $\|B\|_2 = \sqrt{\sum_{j=1}^{n} (b_j)^2}$ are the Euclidean norms of A and B (also called the L_2 norms.

The Dice similarity measure takes value in the interval [0, 1]. However, it is undefined if $a_j = b_j = 0 (j = 1, 2, \ldots, n)$. In this case, let the Dice measure value be zero when $a_j = b_j = 0 (j = 1, 2, \ldots, n)$.

3.1 Dice Similarity Measure for Spherical Fuzzy Sets

In this section, we shall propose some Dice similarity measure and some weighted Dice similarity measure between spherical fuzzy sets (SFSs) based on the concept of the Dice similarity measure (Dice 1945).

Definition 3 Let $T_x = (\mu_{x_j}, \eta_{x_j}, v_{x_j})$ and $T_y = (\mu_{y_j}, \eta_{y_j}, v_{y_j}), j = 1, 2, \ldots, n$, be two groups of spherical fuzzy numbers (SFNs), a Dice similarity measure between SFSs T_x and T_y is proposed as follows:

$$D^1_{SFS}(T_x, T_y) = \frac{1}{n}\sum_{j=1}^{n} \frac{2\left(\mu_{x_j}^2 \mu_{y_j}^2 + \eta_{x_j}^2 \eta_{y_j}^2 + v_{x_j}^2 v_{y_j}^2\right)}{\left(\left(\mu_{x_j}^2\right)^2 + \left(\eta_{x_j}^2\right)^2 + \left(v_{x_j}^2\right)^2\right) + \left(\left(\mu_{y_j}^2\right)^2 + \left(\eta_{y_j}^2\right)^2 + \left(v_{y_j}^2\right)^2\right)} \tag{5}$$

The Dice similarity measure between SFSs T_x and T_y also satisfies the following properties:

1. $0 \leq D^1_{SFS}(T_x, T_y) \leq 1$;
2. $D^1_{SFS}(T_x, T_y) = D^1_{SFS}(T_x, T_y)$;
3. $D^1_{SFS}(T_x, T_y) = 1, \; if \;\; T_x = T_y$,

i.e. $\mu_{x_j} = \mu_{y_j}, \eta_{x_j} = \eta_{y_j}, v_{x_j} = v_{y_j}, j = 1, 2, \ldots, n$.

Proof

1. Let us consider the jth item of the summation in Eq. (5).

$$D^1_{SFS}\left(T_{x_j}, T_{y_j}\right) = \frac{2\left(\mu^2_{x_j}\mu^2_{y_j} + \eta^2_{x_j}\eta^2_{y_j} + v^2_{x_j}v^2_{y_j}\right)}{\left(\left(\mu^2_{x_j}\right)^2 + \left(\eta^2_{x_j}\right)^2 + \left(v^2_{x_j}\right)^2\right) + \left(\left(\mu^2_{y_j}\right)^2 + \left(\eta^2_{y_j}\right)^2 + \left(v^2_{y_j}\right)^2\right)} \tag{6}$$

Obviously, $D^1_{SFS}\left(T_{x_j}, T_{y_j}\right) \geq 0$ and $\left(\left(\mu^2_{x_j}\right)^2 + \left(\eta^2_{x_j}\right)^2 + \left(v^2_{x_j}\right)^2\right) + \left(\left(\mu^2_{y_j}\right)^2 + \left(\eta^2_{y_j}\right)^2 + \left(v^2_{y_j}\right)^2\right) \geq 2\left(\mu^2_{x_j}\mu^2_{y_j} + \eta^2_{x_j}\eta^2_{y_j} + v^2_{x_j}v^2_{y_j}\right)$ according to the inequality $a^2 + b^2 \geq 2ab$. Thus,$0 \leq D^1_{SFS}\left(T_{x_j}, T_{y_j}\right) \leq 1$. From Eq. (5), the summation of n terms is $0 \leq D^1_{SFS}\left(T_x, T_y\right) \leq 1$.

2. It is obvious that the proposition is true.
3. When $T_x = T_y$, there are $\mu_{x_j} = \mu_{y_j}, \eta_{x_j} = \eta_{y_j}, v_{x_j} = v_{y_j}$ for $j = 1, 2, \ldots, n$. So, there is

$$\begin{aligned} & D^1_{SFS}\left(T_x, T_y\right) \\ & = \frac{1}{n}\sum_{j=1}^{n} \frac{2\left(\mu^2_{x_j}\mu^2_{y_j} + \eta^2_{x_j}\eta^2_{y_j} + v^2_{x_j}v^2_{y_j}\right)}{\left(\left(\mu^2_{x_j}\right)^2 + \left(\eta^2_{x_j}\right)^2 + \left(v^2_{x_j}\right)^2\right) + \left(\left(\mu^2_{y_j}\right)^2 + \left(\eta^2_{y_j}\right)^2 + \left(v^2_{y_j}\right)^2\right)} \\ & = \frac{1}{n}\sum_{j=1}^{n} \frac{2\left(\mu^2_{x_j}\mu^2_{x_j} + \eta^2_{x_j}\eta^2_{x_j} + v^2_{x_j}v^2_{x_j}\right)}{\left(\left(\mu^2_{x_j}\right)^2 + \left(\eta^2_{x_j}\right)^2 + \left(v^2_{x_j}\right)^2\right) + \left(\left(\mu^2_{x_j}\right)^2 + \left(\eta^2_{x_j}\right)^2 + \left(v^2_{x_j}\right)^2\right)} \\ & = \frac{1}{n}\sum_{j=1}^{n} \frac{2\left(\left(\mu^2_{x_j}\right)^2 + \left(\eta^2_{x_j}\right)^2 + \left(v^2_{x_j}\right)^2\right)}{2\left(\left(\mu^2_{x_j}\right)^2 + \left(\eta^2_{x_j}\right)^2 + \left(v^2_{x_j}\right)^2\right)} = 1 \end{aligned}$$

Therefore, we have finished the proofs.

If we consider the weights of a_j, a weighted Dice similarity measure between SFSs T_x and T_y is proposed as follows:

$$WD^1_{SFS}\left(T_{x_j}, T_{y_j}\right) = \sum_{j=1}^{n} \omega_j \frac{2\left(\mu^2_{x_j}\mu^2_{y_j} + \eta^2_{x_j}\eta^2_{y_j} + v^2_{x_j}v^2_{y_j}\right)}{\left(\left(\mu^2_{x_j}\right)^2 + \left(\eta^2_{x_j}\right)^2 + \left(v^2_{x_j}\right)^2\right) + \left(\left(\mu^2_{y_j}\right)^2 + \left(\eta^2_{y_j}\right)^2 + \left(v^2_{y_j}\right)^2\right)} \tag{7}$$

where $\omega = (\omega_1, \omega_2, \ldots, \omega_n)^T$ is the weight vector of $a_j (j = 1, 2, \ldots, n)$,with $w_j \in [0, 1], i = 1, 2, \ldots, n, \sum_{j=1}^{n} w_j = 1$. In particular,

if $\omega = (1/n, 1/n, \ldots, 1/n)^T$, then the weighted Dice similarity measure reduces to Dice similarity measure. That's to say, if we take $\omega_j = \frac{1}{n}, j = 1, 2 \ldots, n$, then there is $WD^1_{SFS}(T_x, T_y) = D^1_{SFS}(T_x, T_y)$.

Obviously, the weighted Dice similarity measure of two SFSs T_x and T_y also satisfies the following properties:

1. $0 \leq WD^1_{SFS}(T_{x_j}, T_{y_j}) \leq 1$,
2. $WD^1_{SFS}(T_{x_j}, T_{y_j}) = WD^1_{SFS}(T_{y_j}, T_{x_j})$,
3. $WD^1_{SFS}(T_{x_j}, T_{y_j}) = 1, \ if \ T_x = T_y, j = 1, 2, \ldots, n.$

Similar to the previous proof method, we can prove the above three properties.

When the four terms like the positive-membership degree, the neutral-membership degree, the negative-membership degree and the indeterminacy degree are considered in SFSs, we further propose the Dice similarity measure and weighted Dice similarity measure between SFSs as follows:

$$D^2_{SFS}(T_x, T_y) = \frac{1}{n}\sum_{j=1}^{n} \frac{2\left(\mu^2_{x_j}\mu^2_{y_j} + \eta^2_{x_j}\eta^2_{y_j} + v^2_{x_j}v^2_{y_j} + \pi^2_{x_j}\pi^2_{y_j}\right)}{\left(\left(\mu^2_{x_j}\right)^2 + \left(\eta^2_{x_j}\right)^2 + \left(v^2_{x_j}\right)^2 + \left(\pi^2_{x_j}\right)^2\right) + \left(\left(\mu^2_{y_j}\right)^2 + \left(\eta^2_{y_j}\right)^2 + \left(v^2_{y_j}\right)^2 + \left(\pi^2_{y_j}\right)^2\right)} \tag{8}$$

$$WD^2_{SFS}(T_x, T_y) = \sum_{j=1}^{n} \omega_j \frac{2\left(\mu^2_{x_j}\mu^2_{y_j} + \eta^2_{x_j}\eta^2_{y_j} + v^2_{x_j}v^2_{y_j} + \pi^2_{x_j}\pi^2_{y_j}\right)}{\left(\left(\mu^2_{x_j}\right)^2 + \left(\eta^2_{x_j}\right)^2 + \left(v^2_{x_j}\right)^2 + \left(\pi^2_{x_j}\right)^2\right) + \left(\left(\mu^2_{y_j}\right)^2 + \left(\eta^2_{y_j}\right)^2 + \left(v^2_{y_j}\right)^2 + \left(\pi^2_{y_j}\right)^2\right)} \tag{9}$$

where $\omega = (\omega_1, \omega_2, \ldots, \omega_n)^T$ is the weight vector of $a_j (j = 1, 2, \ldots, n)$,with $\omega_j \in [0, 1], j = 1, 2, \ldots, n, \sum_{j=1}^{n} \omega_j = 1$ and π_{x_j}, π_{y_j} can be obtained by Eq. (3).

3.2 *Another Form of the Dice Similarity Measure for Spherical Fuzzy Sets*

In this section, we shall develop another form of Dice similarity measure for spherical fuzzy sets (SFSs), which is defined as follows;

Definition 4 Let $T_x = (\mu_{x_j}, \eta_{x_j}, v_{x_j})$ and $T_y = (\mu_{y_j}, \eta_{y_j}, v_{y_j})$, $j = 1, 2, \ldots, n$, be two groups of spherical fuzzy numbers (SFNs), then a Dice similarity measure between SFSs T_x and T_y is proposed as follows:

$$D^3_{SFS}(T_x, T_y) = \frac{\sum_{j=1}^{n} 2\left(\mu^2_{x_j}\mu^2_{y_j} + \eta^2_{x_j}\eta^2_{y_j} + v^2_{x_j}v^2_{y_j}\right)}{\sum_{j=1}^{n}\left(\left(\mu^2_{x_j}\right)^2 + \left(\eta^2_{x_j}\right)^2 + \left(v^2_{x_j}\right)^2\right) + \sum_{j=1}^{n}\left(\left(\mu^2_{y_j}\right)^2 + \left(\eta^2_{y_j}\right)^2 + \left(v^2_{y_j}\right)^2\right)} \tag{10}$$

The Dice similarity measure between SFSs T_x and T_y also satisfies the following properties:

1. $0 \le D^3_{SFS}(T_x, T_y) \le 1$;
2. $D^3_{SFS}(T_x, T_y) = D^3_{SFS}(T_y, T_x)$;
3. $D^3_{SFS}(T_x, T_y) = 1,\ if\ T_x = T_y$, there are $\mu_{x_j} = \mu_{y_j}, \eta_{x_j} = \eta_{y_j}, v_{x_j} = v_{y_j}$, for $j = 1, 2, \ldots, n$.

Similar to the previous proof method, we can prove the above three properties.

If we consider the weights of a_j, a weighted Dice similarity measure between SFSs T_x and T_y is proposed as follows:

$$WD^3_{SFS}(T_x, T_y) = \frac{\sum_{j=1}^n 2\omega_j^4\left(\mu_{x_j}^2\mu_{y_j}^2 + \eta_{x_j}^2\eta_{y_j}^2 + v_{x_j}^2 v_{y_j}^2\right)}{\sum_{j=1}^n \omega_j^4\left(\left(\mu_{x_j}^2\right)^2 + \left(\eta_{x_j}^2\right)^2 + \left(v_{x_j}^2\right)^2\right) + \sum_{j=1}^n \omega_j^4\left(\left(\mu_{y_j}^2\right)^2 + \left(\eta_{y_j}^2\right)^2 + \left(v_{y_j}^2\right)^2\right)} \tag{11}$$

where $\omega = (\omega_1, \omega_2, \ldots, \omega_n)^T$ is the weight vector of $a_j (j = 1, 2, \ldots, n)$, with $\omega_j \in [0, 1]$,$i = 1, 2, \ldots, n$,$\sum_{j=1}^n \omega_j = 1$. In particular,

if $\omega = (1/n, 1/n, \ldots, 1/n)^T$, then the weighted Dice similarity measure reduces to Dice similarity measure. That's to say, if we take $\omega_j = \frac{1}{n}, j = 1, 2 \ldots, n$, then there is $WD^3_{SFS}(T_x, T_y) = D^3_{SFS}(T_x, T_y)$.

Obviously, the weighted Dice similarity measure of two SFSs T_x and T_y also satisfies the following properties:

1. $0 \le WD^3_{SFS}(T_x, T_y) \le 1$,
2. $WD^3_{SFS}(T_x, T_y) = WD^3_{SFS}(T_y, T_x)$,
3. $WD^3_{SFS}(T_x, T_y) = 1,\ if\ T_x = T_y, j = 1, 2, \ldots, n$.

When the four terms like the positive-membership degree, the neutral-membership degree, the negative-membership degree and the indeterminacy degree are considered in SFSs, we further propose the Dice similarity measure and weighted Dice similarity measure between SFSs as follows:

$$D^4_{SFS}(T_x, T_y) = \frac{\sum_{j=1}^n 2\left(\mu_{x_j}^2\mu_{y_j}^2 + \eta_{x_j}^2\eta_{y_j}^2 + v_{x_j}^2 v_{y_j}^2 + \pi_{x_j}^2\pi_{y_j}^2\right)}{\sum_{j=1}^n \left(\left(\mu_{x_j}^2\right)^2 + \left(\eta_{x_j}^2\right)^2 + \left(v_{x_j}^2\right)^2 + \left(\pi_{x_j}^2\right)^2\right) + \sum_{j=1}^n \left(\left(\mu_{y_j}^2\right)^2 + \left(\eta_{y_j}^2\right)^2 + \left(v_{y_j}^2\right)^2 + \left(\pi_{y_j}^2\right)^2\right)} \tag{12}$$

$$WD^4_{SFS}(T_x, T_y) = \frac{\sum_{j=1}^n 2\omega_j^4\left(\mu_{x_j}^2\mu_{y_j}^2 + \eta_{x_j}^2\eta_{y_j}^2 + v_{x_j}^2 v_{y_j}^2 + \pi_{x_j}^2\pi_{y_j}^2\right)}{\sum_{j=1}^n \omega_j^4\left(\left(\mu_{x_j}^2\right)^2 + \left(\eta_{x_j}^2\right)^2 + \left(v_{x_j}^2\right)^2 + \left(\pi_{x_j}^2\right)^2\right) + \sum_{j=1}^n \omega_j^4\left(\left(\mu_{y_j}^2\right)^2 + \left(\eta_{y_j}^2\right)^2 + \left(v_{y_j}^2\right)^2 + \left(\pi_{y_j}^2\right)^2\right)} \tag{13}$$

where $\omega = (\omega_1, \omega_2, \ldots, \omega_n)^T$ is the weight vector of $a_j (j = 1, 2, \ldots, n)$,with $\omega_j \in [0, 1]$,$j = 1, 2, \ldots, n$,$\sum_{j=1}^n \omega_j = 1$, and π_{x_j}, π_{y_j} can be obtained by Eq. (3).

3.3 The Generalized Dice Similarity Measure for Spherical Fuzzy Sets

In this section, we develop the generalized Dice similarity measure for spherical fuzzy numbers (SFNs). As the generalization of the Dice similarity measure for SFNs, the generalized Dice similarity measures for SFNs are defined below.

Definition 5 Let $T_x = \left(\mu_{x_j}, \eta_{x_j}, v_{x_j}\right)$ and $T_y = \left(\mu_{y_j}, \eta_{y_j}, v_{y_j}\right), j = 1, 2, \ldots, n$, be two groups of SFNs, then a generalized Dice similarity measure between SFSs T_x and T_y is proposed as follows:

$$GD^1_{SFS}(T_x, T_y) = \frac{1}{n}\sum_{j=1}^{n}\frac{\mu^2_{x_j}\mu^2_{y_j} + \eta^2_{x_j}\eta^2_{y_j} + v^2_{x_j}v^2_{y_j}}{\alpha\left(\left(\mu^2_{x_j}\right)^2 + \left(\eta^2_{x_j}\right)^2 + \left(v^2_{x_j}\right)^2\right) + (1-\alpha)\left(\left(\mu^2_{y_j}\right)^2 + \left(\eta^2_{y_j}\right)^2 + \left(v^2_{y_j}\right)^2\right)} \tag{14}$$

$$GD^2_{SFS}(T_x, T_y) = \frac{\sum_{j=1}^{n}\left(\mu^2_{x_j}\mu^2_{y_j} + \eta^2_{x_j}\eta^2_{y_j} + v^2_{x_j}v^2_{y_j}\right)}{\alpha\sum_{j=1}^{n}\left(\left(\mu^2_{x_j}\right)^2 + \left(\eta^2_{x_j}\right)^2 + \left(v^2_{x_j}\right)^2\right) + (1-\alpha)\sum_{j=1}^{n}\left(\left(\mu^2_{y_j}\right)^2 + \left(\eta^2_{y_j}\right)^2 + \left(v^2_{y_j}\right)^2\right)} \tag{15}$$

where α is a positive parameter for $0 \leq \alpha \leq 1$.

Then, the generalized Dice similarity measure includes some special cases by altering the parameter value α.

If $\alpha = 0.5$, the two generalized Dice similarity measures (14) and (15) reduced to Dice similarity measures (5) and (10):

$$\begin{aligned} GD^1_{SFS}(T_x, T_y) &= \frac{1}{n}\sum_{j=1}^{n}\frac{\mu^2_{x_j}\mu^2_{y_j} + \eta^2_{x_j}\eta^2_{y_j} + v^2_{x_j}v^2_{y_j}}{\alpha\left(\left(\mu^2_{x_j}\right)^2 + \left(\eta^2_{x_j}\right)^2 + \left(v^2_{x_j}\right)^2\right) + (1-\alpha)\left(\left(\mu^2_{y_j}\right)^2 + \left(\eta^2_{y_j}\right)^2 + \left(v^2_{y_j}\right)^2\right)} \\ &= \frac{1}{n}\sum_{j=1}^{n}\frac{\mu^2_{x_j}\mu^2_{y_j} + \eta^2_{x_j}\eta^2_{y_j} + v^2_{x_j}v^2_{y_j}}{0.5\left(\left(\mu^2_{x_j}\right)^2 + \left(\eta^2_{x_j}\right)^2 + \left(v^2_{x_j}\right)^2\right) + 0.5\left(\left(\mu^2_{y_j}\right)^2 + \left(\eta^2_{y_j}\right)^2 + \left(v^2_{y_j}\right)^2\right)} \\ &= \frac{1}{n}\sum_{j=1}^{n}\frac{2\left(\mu^2_{x_j}\mu^2_{y_j} + \eta^2_{x_j}\eta^2_{y_j} + v^2_{x_j}v^2_{y_j}\right)}{\left(\left(\mu^2_{x_j}\right)^2 + \left(\eta^2_{x_j}\right)^2 + \left(v^2_{x_j}\right)^2\right) + \left(\left(\mu^2_{y_j}\right)^2 + \left(\eta^2_{y_j}\right)^2 + \left(v^2_{y_j}\right)^2\right)} \end{aligned} \tag{16}$$

$$\begin{aligned} GD^2_{SFS}(T_x, T_y) &= \frac{\sum_{j=1}^{n}\left(\mu^2_{x_j}\mu^2_{y_j} + \eta^2_{x_j}\eta^2_{y_j} + v^2_{x_j}v^2_{y_j}\right)}{\alpha\sum_{j=1}^{n}\left(\left(\mu^2_{x_j}\right)^2 + \left(\eta^2_{x_j}\right)^2 + \left(v^2_{x_j}\right)^2\right) + (1-\alpha)\sum_{j=1}^{n}\left(\left(\mu^2_{y_j}\right)^2 + \left(\eta^2_{y_j}\right)^2 + \left(v^2_{y_j}\right)^2\right)} \\ &= \frac{\sum_{j=1}^{n}\left(\mu^2_{x_j}\mu^2_{y_j} + \eta^2_{x_j}\eta^2_{y_j} + v^2_{x_j}v^2_{y_j}\right)}{0.5\sum_{j=1}^{n}\left(\left(\mu^2_{x_j}\right)^2 + \left(\eta^2_{x_j}\right)^2 + \left(v^2_{x_j}\right)^2\right) + 0.5\sum_{j=1}^{n}\left(\left(\mu^2_{y_j}\right)^2 + \left(\eta^2_{y_j}\right)^2 + \left(v^2_{y_j}\right)^2\right)} \\ &= \frac{\sum_{j=1}^{n}2\left(\mu^2_{x_j}\mu^2_{y_j} + \eta^2_{x_j}\eta^2_{y_j} + v^2_{x_j}v^2_{y_j}\right)}{\sum_{j=1}^{n}\left(\left(\mu^2_{x_j}\right)^2 + \left(\eta^2_{x_j}\right)^2 + \left(v^2_{x_j}\right)^2\right) + \sum_{j=1}^{n}\left(\left(\mu^2_{y_j}\right)^2 + \left(\eta^2_{y_j}\right)^2 + \left(v^2_{y_j}\right)^2\right)} \end{aligned} \tag{17}$$

If $\alpha = 0, 1$, the two generalized Dice similarity measures reduced to the following asymmetric similarity measures respectively:

$$GD^1_{SFS}(T_x, T_y) = \frac{1}{n}\sum_{j=1}^{n}\frac{\mu^2_{x_j}\mu^2_{y_j} + \eta^2_{x_j}\eta^2_{y_j} + v^2_{x_j}v^2_{y_j}}{\alpha\left(\left(\mu^2_{x_j}\right)^2 + \left(\eta^2_{x_j}\right)^2 + \left(v^2_{x_j}\right)^2\right) + (1-\alpha)\left(\left(\mu^2_{y_j}\right)^2 + \left(\eta^2_{y_j}\right)^2 + \left(v^2_{y_j}\right)^2\right)}$$
$$= \frac{1}{n}\sum_{j=1}^{n}\frac{\mu^2_{x_j}\mu^2_{y_j} + \eta^2_{x_j}\eta^2_{y_j} + v^2_{x_j}v^2_{y_j}}{0\left(\left(\mu^2_{x_j}\right)^2 + \left(\eta^2_{x_j}\right)^2 + \left(v^2_{x_j}\right)^2\right) + (1-0)\left(\left(\mu^2_{y_j}\right)^2 + \left(\eta^2_{y_j}\right)^2 + \left(v^2_{y_j}\right)^2\right)}$$
$$= \frac{1}{n}\sum_{j=1}^{n}\frac{\mu^2_{x_j}\mu^2_{y_j} + \eta^2_{x_j}\eta^2_{y_j} + v^2_{x_j}v^2_{y_j}}{\left(\left(\mu^2_{y_j}\right)^2 + \left(\eta^2_{y_j}\right)^2 + \left(v^2_{y_j}\right)^2\right)} \quad for\ \alpha = 0. \tag{18}$$

$$GD^1_{SFS}(T_x, T_y) = \frac{1}{n}\sum\nolimits_{j=1}^{n}\frac{\mu^2_{x_j}\mu^2_{y_j} + \eta^2_{x_j}\eta^2_{y_j} + v^2_{x_j}v^2_{y_j}}{\alpha\left(\left(\mu^2_{x_j}\right)^2 + \left(\eta^2_{x_j}\right)^2 + \left(v^2_{x_j}\right)^2\right) + (1-\alpha)\left(\left(\mu^2_{y_j}\right)^2 + \left(\eta^2_{y_j}\right)^2 + \left(v^2_{y_j}\right)^2\right)}$$
$$= \frac{1}{n}\sum\nolimits_{j=1}^{n}\frac{\mu^2_{x_j}\mu^2_{y_j} + \eta^2_{x_j}\eta^2_{y_j} + v^2_{x_j}v^2_{y_j}}{1\left(\left(\mu^2_{x_j}\right)^2 + \left(\eta^2_{x_j}\right)^2 + \left(v^2_{x_j}\right)^2\right) + (1-1)\left(\left(\mu^2_{y_j}\right)^2 + \left(\eta^2_{y_j}\right)^2 + \left(v^2_{y_j}\right)^2\right)}$$
$$= \frac{1}{n}\sum\nolimits_{j=1}^{n}\frac{\mu^2_{x_j}\mu^2_{y_j} + \eta^2_{x_j}\eta^2_{y_j} + v^2_{x_j}v^2_{y_j}}{\left(\left(\mu^2_{x_j}\right)^2 + \left(\eta^2_{x_j}\right)^2 + \left(v^2_{x_j}\right)^2\right)} \quad for\ \alpha = 1. \tag{19}$$

$$GD^2_{SFS}(T_x, T_y) = \frac{\sum_{j=1}^{n}\left(\mu^2_{x_j}\mu^2_{y_j} + \eta^2_{x_j}\eta^2_{y_j} + v^2_{x_j}v^2_{y_j}\right)}{\alpha\sum_{j=1}^{n}\left(\left(\mu^2_{x_j}\right)^2 + \left(\eta^2_{x_j}\right)^2 + \left(v^2_{x_j}\right)^2\right) + (1-\alpha)\sum_{j=1}^{n}\left(\left(\mu^2_{y_j}\right)^2 + \left(\eta^2_{y_j}\right)^2 + \left(v^2_{y_j}\right)^2\right)}$$
$$= \frac{\sum_{j=1}^{n}\left(\mu^2_{x_j}\mu^2_{y_j} + \eta^2_{x_j}\eta^2_{y_j} + v^2_{x_j}v^2_{y_j}\right)}{0\sum_{j=1}^{n}\left(\left(\mu^2_{x_j}\right)^2 + \left(\eta^2_{x_j}\right)^2 + \left(v^2_{x_j}\right)^2\right) + (1-0)\sum_{j=1}^{n}\left(\left(\mu^2_{y_j}\right)^2 + \left(\eta^2_{y_j}\right)^2 + \left(v^2_{y_j}\right)^2\right)}$$
$$= \frac{\sum_{j=1}^{n}\left(\mu^2_{x_j}\mu^2_{y_j} + \eta^2_{x_j}\eta^2_{y_j} + v^2_{x_j}v^2_{y_j}\right)}{\sum_{j=1}^{n}\left(\left(\mu^2_{y_j}\right)^2 + \left(\eta^2_{y_j}\right)^2 + \left(v^2_{y_j}\right)^2\right)} \quad for\ \alpha = 0. \tag{20}$$

$$GD^2_{SFS}(T_x, T_y) = \frac{\sum_{j=1}^{n}\left(\mu^2_{x_j}\mu^2_{y_j} + \eta^2_{x_j}\eta^2_{y_j} + v^2_{x_j}v^2_{y_j}\right)}{\alpha\sum_{j=1}^{n}\left(\left(\mu^2_{x_j}\right)^2 + \left(\eta^2_{x_j}\right)^2 + \left(v^2_{x_j}\right)^2\right) + (1-\alpha)\sum_{j=1}^{n}\left(\left(\mu^2_{y_j}\right)^2 + \left(\eta^2_{y_j}\right)^2 + \left(v^2_{y_j}\right)^2\right)}$$
$$= \frac{\sum_{j=1}^{n}\left(\mu^2_{x_j}\mu^2_{y_j} + \eta^2_{x_j}\eta^2_{y_j} + v^2_{x_j}v^2_{y_j}\right)}{1\sum_{j=1}^{n}\left(\left(\mu^2_{x_j}\right)^2 + \left(\eta^2_{x_j}\right)^2 + \left(v^2_{x_j}\right)^2\right) + (1-1)\sum_{j=1}^{n}\left(\left(\mu^2_{y_j}\right)^2 + \left(\eta^2_{y_j}\right)^2 + \left(v^2_{y_j}\right)^2\right)}$$
$$= \frac{\sum_{j=1}^{n}\left(\mu^2_{x_j}\mu^2_{y_j} + \eta^2_{x_j}\eta^2_{y_j} + v^2_{x_j}v^2_{y_j}\right)}{\sum_{j=1}^{n}\left(\left(\mu^2_{y_j}\right)^2 + \left(\eta^2_{y_j}\right)^2 + \left(v^2_{y_j}\right)^2\right)} \quad for\ \alpha = 0. \tag{21}$$

From above analysis, it can be seen that the above four asymmetric similarity measures are the extension of the relative projection measure of the spherical fuzzy numbers (SFNs).

In many situations, the weight of the elements $a_j \in A$ should be taken into account. For example, in multiple attribute decision making, the considered attributes usually have different importance, and thus need to be assigned different weights. Thus, we further propose the following two weighted generalized Dice similarity measures for SFSs, respectively, as follows:

$WGD^1_{SFS}(T_x, T_y)$

$$= \sum_{j=1}^{n} \omega_j \frac{\mu_{x_j}^2 \mu_{y_j}^2 + \eta_{x_j}^2 \eta_{y_j}^2 + v_{x_j}^2 v_{y_j}^2}{\alpha\left(\left(\mu_{x_j}^2\right)^2 + \left(\eta_{x_j}^2\right)^2 + \left(v_{x_j}^2\right)^2\right) + (1-\alpha)\left(\left(\mu_{y_j}^2\right)^2 + \left(\eta_{y_j}^2\right)^2 + \left(v_{y_j}^2\right)^2\right)} \tag{22}$$

$$WGD_{SFS}^2(T_x, T_y) = \frac{\sum_{j=1}^{n} \omega_j^4 \left(\mu_{x_j}^2 \mu_{y_j}^2 + \eta_{x_j}^2 \eta_{y_j}^2 + v_{x_j}^2 v_{y_j}^2\right)}{\alpha \sum_{j=1}^{n} \omega_j^4 \left(\left(\mu_{x_j}^2\right)^2 + \left(\eta_{x_j}^2\right)^2 + \left(v_{x_j}^2\right)^2\right) + (1-\alpha) \sum_{j=1}^{n} \omega_j^4 \left(\left(\mu_{y_j}^2\right)^2 + \left(\eta_{y_j}^2\right)^2 + \left(v_{y_j}^2\right)^2\right)} \tag{23}$$

where $\omega = (\omega_1, \omega_2, \ldots, \omega_n)^T$ is the weight vector of $a_j (j = 1, 2, \ldots, n)$,with $\omega_j \in [0, 1], j = 1, 2, \ldots, n, \sum_{j=1}^{n} \omega_j = 1$. In particular, if $\omega = (1/n, 1/n, \ldots, 1/n)^T$, then the weighted generalized Dice similarity measures reduce to generalized Dice similarity measures. That's to say, if we take $\omega_j = \frac{1}{n}, j = 1, 2 \ldots, n$, then there is $WGD_{SFS}^k(T_x, T_y) = GD_{SFS}^k(T_x, T_y) (k = 1, 2)$.

Then, the weighted generalized Dice similarity measure includes some special cases by altering the parameter value α.

If $\alpha = 0.5$, the two weighted generalized Dice similarity measures (22) and (23) reduced to weighted Dice similarity measures (7) and (11):

$$\begin{aligned} WGD_{SFS}^1(T_x, T_y) &= \sum_{j=1}^{n} \omega_j \frac{\mu_{x_j}^2 \mu_{y_j}^2 + \eta_{x_j}^2 \eta_{y_j}^2 + v_{x_j}^2 v_{y_j}^2}{\alpha\left(\left(\mu_{x_j}^2\right)^2 + \left(\eta_{x_j}^2\right)^2 + \left(v_{x_j}^2\right)^2\right) + (1-\alpha)\left(\left(\mu_{y_j}^2\right)^2 + \left(\eta_{y_j}^2\right)^2 + \left(v_{y_j}^2\right)^2\right)} \\ &= \sum_{j=1}^{n} \omega_j \frac{\mu_{x_j}^2 \mu_{y_j}^2 + \eta_{x_j}^2 \eta_{y_j}^2 + v_{x_j}^2 v_{y_j}^2}{0.5\left(\left(\mu_{x_j}^2\right)^2 + \left(\eta_{x_j}^2\right)^2 + \left(v_{x_j}^2\right)^2\right) + (1-0.5)\left(\left(\mu_{y_j}^2\right)^2 + \left(\eta_{y_j}^2\right)^2 + \left(v_{y_j}^2\right)^2\right)} \\ &= \sum_{j=1}^{n} \omega_j \frac{2\left(\mu_{x_j}^2 \mu_{y_j}^2 + \eta_{x_j}^2 \eta_{y_j}^2 + v_{x_j}^2 v_{y_j}^2\right)}{\left(\left(\mu_{x_j}^2\right)^2 + \left(\eta_{x_j}^2\right)^2 + \left(v_{x_j}^2\right)^2\right) + \left(\left(\mu_{y_j}^2\right)^2 + \left(\eta_{y_j}^2\right)^2 + \left(v_{y_j}^2\right)^2\right)} \end{aligned} \tag{24}$$

$$\begin{aligned} &WGD_{SFS}^2(T_x, T_y) \\ &= \frac{\sum_{j=1}^{n} \omega_j^4 \left(\mu_{x_j}^2 \mu_{y_j}^2 + \eta_{x_j}^2 \eta_{y_j}^2 + v_{x_j}^2 v_{y_j}^2\right)}{\alpha \sum_{j=1}^{n} \omega_j^4 \left(\left(\mu_{x_j}^2\right)^2 + \left(\eta_{x_j}^2\right)^2 + \left(v_{x_j}^2\right)^2\right) + (1-\alpha) \sum_{j=1}^{n} \omega_j^4 \left(\left(\mu_{y_j}^2\right)^2 + \left(\eta_{y_j}^2\right)^2 + \left(v_{y_j}^2\right)^2\right)} \\ &= \frac{\sum_{j=1}^{n} \omega_j^4 \left(\mu_{x_j}^2 \mu_{y_j}^2 + \eta_{x_j}^2 \eta_{y_j}^2 + v_{x_j}^2 v_{y_j}^2\right)}{0.5 \sum_{j=1}^{n} \omega_j^4 \left(\left(\mu_{x_j}^2\right)^2 + \left(\eta_{x_j}^2\right)^2 + \left(v_{x_j}^2\right)^2\right) + (1-0.5) \sum_{j=1}^{n} \omega_j^4 \left(\left(\mu_{y_j}^2\right)^2 + \left(\eta_{y_j}^2\right)^2 + \left(v_{y_j}^2\right)^2\right)} \\ &= \frac{2 \sum_{j=1}^{n} \omega_j^4 \left(\mu_{x_j}^2 \mu_{y_j}^2 + \eta_{x_j}^2 \eta_{y_j}^2 + v_{x_j}^2 v_{y_j}^2\right)}{\sum_{j=1}^{n} \omega_j^4 \left(\left(\mu_{x_j}^2\right)^2 + \left(\eta_{x_j}^2\right)^2 + \left(v_{x_j}^2\right)^2\right) + \sum_{j=1}^{n} \omega_j^4 \left(\left(\mu_{y_j}^2\right)^2 + \left(\eta_{y_j}^2\right)^2 + \left(v_{y_j}^2\right)^2\right)} \end{aligned} \tag{25}$$

If $\alpha = 0, 1$, the two weighted generalized Dice similarity measures reduced to the following asymmetric weighted similarity measures respectively:

$WGD_{SFS}^1(T_x, T_y)$

$$= \sum_{j=1}^{n} \omega_j \frac{\mu_{x_j}^2 \mu_{y_j}^2 + \eta_{x_j}^2 \eta_{y_j}^2 + v_{x_j}^2 v_{y_j}^2}{\alpha\left(\left(\mu_{x_j}^2\right)^2 + \left(\eta_{x_j}^2\right)^2 + \left(v_{x_j}^2\right)^2\right) + (1-\alpha)\left(\left(\mu_{y_j}^2\right)^2 + \left(\eta_{y_j}^2\right)^2 + \left(v_{y_j}^2\right)^2\right)}$$

$$= \sum_{j=1}^{n} \omega_j \frac{\mu_{x_j}^2 \mu_{y_j}^2 + \eta_{x_j}^2 \eta_{y_j}^2 + v_{x_j}^2 v_{y_j}^2}{0\left(\left(\mu_{x_j}^2\right)^2 + \left(\eta_{x_j}^2\right)^2 + \left(v_{x_j}^2\right)^2\right) + (1-0)\left(\left(\mu_{y_j}^2\right)^2 + \left(\eta_{y_j}^2\right)^2 + \left(v_{y_j}^2\right)^2\right)}$$

$$= \sum_{j=1}^{n} \omega_j \frac{\mu_{x_j}^2 \mu_{y_j}^2 + \eta_{x_j}^2 \eta_{y_j}^2 + v_{x_j}^2 v_{y_j}^2}{\left(\left(\mu_{y_j}^2\right)^2 + \left(\eta_{y_j}^2\right)^2 + \left(v_{y_j}^2\right)^2\right)}, \quad for \ \alpha = 0. \tag{26}$$

$$WGD_{SFS}^1(T_x, T_y)$$

$$= \sum_{j=1}^{n} \omega_j \frac{\mu_{x_j}^2 \mu_{y_j}^2 + \eta_{x_j}^2 \eta_{y_j}^2 + v_{x_j}^2 v_{y_j}^2}{\alpha\left(\left(\mu_{x_j}^2\right)^2 + \left(\eta_{x_j}^2\right)^2 + \left(v_{x_j}^2\right)^2\right) + (1-\alpha)\left(\left(\mu_{y_j}^2\right)^2 + \left(\eta_{y_j}^2\right)^2 + \left(v_{y_j}^2\right)^2\right)}$$

$$= \sum_{j=1}^{n} \omega_j \frac{\mu_{x_j}^2 \mu_{y_j}^2 + \eta_{x_j}^2 \eta_{y_j}^2 + v_{x_j}^2 v_{y_j}^2}{1\left(\left(\mu_{x_j}^2\right)^2 + \left(\eta_{x_j}^2\right)^2 + \left(v_{x_j}^2\right)^2\right) + (1-1)\left(\left(\mu_{y_j}^2\right)^2 + \left(\eta_{y_j}^2\right)^2 + \left(v_{y_j}^2\right)^2\right)}$$

$$= \sum_{j=1}^{n} \omega_j \frac{\mu_{x_j}^2 \mu_{y_j}^2 + \eta_{x_j}^2 \eta_{y_j}^2 + v_{x_j}^2 v_{y_j}^2}{\left(\left(\mu_{x_j}^2\right)^2 + \left(\eta_{x_j}^2\right)^2 + \left(v_{x_j}^2\right)^2\right)}, \quad for \ \alpha = 1. \tag{27}$$

$$WGD_{SFS}^2(T_x, T_y)$$

$$= \frac{\sum_{j=1}^{n} \omega_j^4\left(\mu_{x_j}^2 \mu_{y_j}^2 + \eta_{x_j}^2 \eta_{y_j}^2 + v_{x_j}^2 v_{y_j}^2\right)}{\alpha \sum_{j=1}^{n} \omega_j^4\left(\left(\mu_{x_j}^2\right)^2 + \left(\eta_{x_j}^2\right)^2 + \left(v_{x_j}^2\right)^2\right) + (1-\alpha)\sum_{j=1}^{n} \omega_j^4\left(\left(\mu_{y_j}^2\right)^2 + \left(\eta_{y_j}^2\right)^2 + \left(v_{y_j}^2\right)^2\right)}$$

$$= \frac{\sum_{j=1}^{n} \omega_j^4\left(\mu_{x_j}^2 \mu_{y_j}^2 + \eta_{x_j}^2 \eta_{y_j}^2 + v_{x_j}^2 v_{y_j}^2\right)}{0 \sum_{j=1}^{n} \omega_j^4\left(\left(\mu_{x_j}^2\right)^2 + \left(\eta_{x_j}^2\right)^2 + \left(v_{x_j}^2\right)^2\right) + (1-0)\sum_{j=1}^{n} \omega_j^4\left(\left(\mu_{y_j}^2\right)^2 + \left(\eta_{y_j}^2\right)^2 + \left(v_{y_j}^2\right)^2\right)}$$

$$= \frac{\sum_{j=1}^{n} \omega_j^4\left(\mu_{x_j}^2 \mu_{y_j}^2 + \eta_{x_j}^2 \eta_{y_j}^2 + v_{x_j}^2 v_{y_j}^2\right)}{\sum_{j=1}^{n} \omega_j^4\left(\left(\mu_{y_j}^2\right)^2 + \left(\eta_{y_j}^2\right)^2 + \left(v_{y_j}^2\right)^2\right)}, \quad for \ \alpha = 0. \tag{28}$$

$$WGD_{SFS}^2(T_x, T_y)$$

$$= \frac{\sum_{j=1}^{n} \omega_j^4\left(\mu_{x_j}^2 \mu_{y_j}^2 + \eta_{x_j}^2 \eta_{y_j}^2 + v_{x_j}^2 v_{y_j}^2\right)}{\alpha \sum_{j=1}^{n} \omega_j^4\left(\left(\mu_{x_j}^2\right)^2 + \left(\eta_{x_j}^2\right)^2 + \left(v_{x_j}^2\right)^2\right) + (1-\alpha)\sum_{j=1}^{n} \omega_j^4\left(\left(\mu_{y_j}^2\right)^2 + \left(\eta_{y_j}^2\right)^2 + \left(v_{y_j}^2\right)^2\right)}$$

$$= \frac{\sum_{j=1}^{n} \omega_j^4\left(\mu_{x_j}^2 \mu_{y_j}^2 + \eta_{x_j}^2 \eta_{y_j}^2 + v_{x_j}^2 v_{y_j}^2\right)}{1 \sum_{j=1}^{n} \omega_j^4\left(\left(\mu_{x_j}^2\right)^2 + \left(\eta_{x_j}^2\right)^2 + \left(v_{x_j}^2\right)^2\right) + (1-1)\sum_{j=1}^{n} \omega_j^4\left(\left(\mu_{y_j}^2\right)^2 + \left(\eta_{y_j}^2\right)^2 + \left(v_{y_j}^2\right)^2\right)}$$

$$= \frac{\sum_{j=1}^{n} \omega_j^4\left(\mu_{x_j}^2 \mu_{y_j}^2 + \eta_{x_j}^2 \eta_{y_j}^2 + v_{x_j}^2 v_{y_j}^2\right)}{\sum_{j=1}^{n} \omega_j^4\left(\left(\mu_{x_j}^2\right)^2 + \left(\eta_{x_j}^2\right)^2 + \left(v_{x_j}^2\right)^2\right)}, \quad for \ \alpha = 1. \tag{29}$$

From above analysis, it can be seen that the above four asymmetric weighted similarity measures are the extension of the relative weighted projection measure of the T-SFNs.

When the four terms like the positive-membership degree, the neutral-membership degree, the negative-membership degree and the indeterminacy degree are considered in SFSs, we further propose the generalized Dice similarity measure and weighted generalized Dice similarity measure between SFSs as follows:

$$GD_{SFS}^{3}(T_x, T_y) = \frac{1}{n}\sum_{j=1}^{n} \frac{\mu_{x_j}^2\mu_{y_j}^2 + \eta_{x_j}^2\eta_{y_j}^2 + v_{x_j}^2 v_{y_j}^2 + \pi_{x_j}^2\pi_{y_j}^2}{\alpha\left(\left(\mu_{x_j}^2\right)^2 + \left(\eta_{x_j}^2\right)^2 + \left(v_{x_j}^2\right)^2 + \left(\pi_{x_j}^2\right)^2\right) + (1-\alpha)\left(\left(\mu_{y_j}^2\right)^2 + \left(\eta_{y_j}^2\right)^2 + \left(v_{y_j}^2\right)^2 + \left(\pi_{y_j}^2\right)^2\right)} \tag{30}$$

$$GD_{SFS}^{4}(T_x, T_y) = \frac{\sum_{j=1}^{n}\left(\mu_{x_j}^2\mu_{y_j}^2 + \eta_{x_j}^2\eta_{y_j}^2 + v_{x_j}^2 v_{y_j}^2 + \pi_{x_j}^2\pi_{y_j}^2\right)}{\alpha\sum_{j=1}^{n}\left(\left(\mu_{x_j}^2\right)^2 + \left(\eta_{x_j}^2\right)^2 + \left(v_{x_j}^2\right)^2 + \left(\pi_{x_j}^2\right)^2\right) + (1-\alpha)\sum_{j=1}^{n}\left(\left(\mu_{y_j}^2\right)^2 + \left(\eta_{y_j}^2\right)^2 + \left(v_{y_j}^2\right)^2 + \left(\pi_{y_j}^2\right)^2\right)} \tag{31}$$

$$WGD_{SFS}^{3}(T_x, T_y) = \sum_{j=1}^{n}\omega_j \frac{\mu_{x_j}^2\mu_{y_j}^2 + \eta_{x_j}^2\eta_{y_j}^2 + v_{x_j}^2 v_{y_j}^2 + \pi_{x_j}^2\pi_{y_j}^2}{\alpha\left(\left(\mu_{x_j}^2\right)^2 + \left(\eta_{x_j}^2\right)^2 + \left(v_{x_j}^2\right)^2 + \left(\pi_{x_j}^2\right)^2\right) + (1-\alpha)\left(\left(\mu_{y_j}^2\right)^2 + \left(\eta_{y_j}^2\right)^2 + \left(v_{y_j}^2\right)^2 + \left(\pi_{y_j}^2\right)^2\right)} \tag{32}$$

$$WGD_{SFS}^{4}(T_x, T_y) = \frac{\sum_{j=1}^{n}\omega_j^4\left(\mu_{x_j}^2\mu_{y_j}^2 + \eta_{x_j}^2\eta_{y_j}^2 + v_{x_j}^2 v_{y_j}^2 + \pi_{x_j}^2\pi_{y_j}^2\right)}{\alpha\sum_{j=1}^{n}\omega_j^4\left(\left(\mu_{x_j}^2\right)^2 + \left(\eta_{x_j}^2\right)^2 + \left(v_{x_j}^2\right)^2 + \left(\pi_{x_j}^2\right)^2\right) + (1-\alpha)\sum_{j=1}^{n}\omega_j^4\left(\left(\mu_{y_j}^2\right)^2 + \left(\eta_{y_j}^2\right)^2 + \left(v_{y_j}^2\right)^2 + \left(\pi_{y_j}^2\right)^2\right)} \tag{33}$$

where $\omega = (\omega_1, \omega_2, \ldots, \omega_n)^T$ is the weight vector of $a_j (j = 1, 2, \ldots, n)$, with $\omega_j \in [0, 1], j = 1, 2, \ldots, n, \sum_{j=1}^{n}\omega_j = 1$, and π_{x_j}, π_{y_j} can be obtained by Eq. (3), α is a positive parameter for $0 \le \alpha \le 1$.

4 Generalized Dice Similarity Measures for MADM with SFNs

In this section, we shall extend the generalized Dice similarity measures for multiple attribute group decision making with spherical fuzzy information. Let $Y = \{Y_1, Y_2, \ldots, Y_m\}$ be a discrete set of alternatives, and $x = \{x_1, x_2, \ldots, x_n\}$ be the set of attributes, $\omega = (\omega_1, \omega_2, \ldots, \omega_n)$ is the weighting vector of the attribute $x_j (j = 1, 2, \ldots, n)$, where $\omega_j > 0, \sum_{j=1}^{n}\omega_j = 1$. Let $d = \{d_1, d_2, \ldots, d_t\}$ be the set of decision makers, $w = (w_1, w_2, \ldots, w_t)$ be the weighting vector of decision makers, with $w_k > 0, \sum_{k=1}^{t} w_k = 1$. Suppose that $R^{(k)} = \left(r_{ij}^{(k)}\right)_{m\times n} = \left(\mu_{ij}^{(k)}, \eta_{ij}^{(k)}, v_{ij}^{(k)}\right)_{m\times n}$ is the spherical fuzzy decision matrix, where $\mu_{ij}^{(k)}$ indicates the positive-membership degree given by the decision maker d_k, $\eta_{ij}^{(k)}$ indicates the neutral-membership degree given by the decision maker d_k, $v_{ij}^{(k)}$ indicates the negative-membership degree given by the decision maker d_k, $\mu_{ij}^{(k)} \subset [0,1]$, $\eta_{ij}^{(k)} \subset [0,1]$, $v_{ij}^{(k)} \subset [0,1]$, $\left(\mu_{ij}^{(k)}\right)^2 + \left(\eta_{ij}^{(k)}\right)^2 + \left(v_{ij}^{(k)}\right)^2 \leq 1, q \geq 1, i = 1, 2, \ldots, m$, $j = 1, 2, \ldots, n, k = 1, 2, \ldots, t$.

Then, in the following, we shall develop an algorithm to utilize the generalized Dice similarity measures to solve the multiple attribute group decision making with spherical fuzzy information.

Step 1. Utilize the decision information given in matrix $R^{(k)}$ by using the spherical weighted Arithmetic mean (SWAM) operator (Gündoğdu and Kahraman 2019)

$$r_{ij} = \left(\mu_{ij}, \eta_{ij}, v_{ij}\right) = SWAM\left(r_{ij}^{(1)}, r_{ij}^{(2)}, \ldots, r_{ij}^{(t)}\right) = \mathop{\oplus}_{k=1}^{t} w_k r_{ij}^{(k)}$$
$$= \left(\begin{array}{c} \sqrt{1 - \prod_{k=1}^{t}\left(1 - \left(\mu_{ij}^{(k)}\right)^2\right)^{w_k}}, \prod_{k=1}^{t}\left(\eta_{ij}^{(k)}\right)^{w_k}, \\ \sqrt{\prod_{k=1}^{t}\left(1 - \left(\mu_{ij}^{(k)}\right)^2\right)^{w_k} - \prod_{k=1}^{t}\left(1 - \left(\mu_{ij}^{(k)}\right)^2 - \left(v_{ij}^{(k)}\right)^2\right)^{w_k}} \end{array}\right) \quad (34)$$

to aggregate all the decision matrices $R^{(k)} (k = 1, 2, \ldots, t)$ into a collective decision matrix $R = \left(r_{ij}\right)_{m \times n}, i = 1, 2, \ldots, m, j = 1, 2, \ldots, n$, where $w = (w_1, w_2, \ldots, w_n)$ is the weighting vector of decision makers.

Step 2. Defining the spherical fuzzy positive ideal solution (SFPIS)Y^+ as

$$\begin{aligned} \left(\mu^+, \eta^+, v^+\right) &= \left(\left(\mu_1^+, \eta_1^+, v_1^+\right), \left(\mu_2^+, \eta_2^+, v_2^+\right), \ldots, \left(\mu_n^+, \eta_n^+, v_n^+\right)\right) \\ &= ((1, 0, 0), (1, 0, 0), \ldots, (1, 0, 0)) \end{aligned} \quad (35)$$

Step 3. Calculating the weighted generalized Dice similarity measures between $Y_i (i = 1, 2, \ldots, m)$ and Y^+ as follows:

$$\begin{aligned} &WGD_{SFS}^{1}\left(Y_i, Y^+\right) \\ &= \sum_{j=1}^{n} \omega_j \frac{\left(\mu_{ij}\right)^2\left(\mu_j^+\right)^2 + \left(\eta_{ij}\right)^2\left(\eta_j^+\right)^2 + \left(v_{ij}\right)^2\left(v_j^+\right)^2}{\alpha\left(\left(\mu_{ij}^2\right)^2 + \left(\eta_{ij}^2\right)^2 + \left(v_{ij}^2\right)^2\right) + (1-\alpha)\left(\left(\left(\mu_j^+\right)^2\right)^2 + \left(\left(\eta_j^+\right)^2\right)^2 + \left(\left(v_j^+\right)^2\right)^2\right)} \end{aligned} \quad (36)$$

or

$$\begin{aligned} &WGD_{SFS}^{2}\left(Y_i, Y^+\right) \\ &= \frac{\sum_{j=1}^{n} \omega_j^4\left(\left(\mu_{ij}\right)^2\left(\mu_j^+\right)^2 + \left(\eta_{ij}\right)^2\left(\eta_j^+\right)^2 + \left(v_{ij}\right)^2\left(v_j^+\right)^2\right)}{\alpha \sum_{j=1}^{n} \omega_j^4\left(\left(\mu_{ij}^2\right)^2 + \left(\eta_{ij}^2\right)^2 + \left(v_{ij}^2\right)^2\right) + (1-\alpha)\sum_{j=1}^{n} \omega_j^4\left(\left(\left(\mu_j^+\right)^2\right)^2 + \left(\left(\eta_j^+\right)^2\right)^2 + \left(\left(v_j^+\right)^2\right)^2\right)} \end{aligned} \quad (37)$$

Step 4. Rank all the alternatives $Y_i(i = 1, 2, \ldots, m)$ and select the best one(s) in accordance with the weighted generalized Dice similarity measures $WGD^1_{SFS}(Y_i, Y^+)(WGD^2_{SFS}(Y_i, Y^+))$ $(i = 1, 2, \ldots, m)$. If any alternative has the highest $WGD^1_{SFS}(Y_i, Y^+)(WGD^2_{SFS}(Y_i, Y^+))$ value, then, it is the most important alternative.

Step 5. End.

5 Numerical Example and Comparative Analysis

5.1 Numerical Example

The enterprise resource planning (ERP) system selection is a classical MADM issued (He et al. 2019; Lu et al. 2019a, b; Wei et al. 2019a, b). In this section, we utilize a practical multiple attribute group decision making problems to illustrate the application of the developed approaches. Suppose an organization plans to implement enterprise resource planning (ERP) system. The first step is to form a project team that consists of CIO and two senior representatives from user departments. By collecting all possible information about ERP vendors and systems, project term choose five potential ERP systems $Y_i(i = 1, 2, \ldots, 5)$ as candidates. The company employs some external professional organizations (or experts) to aid this decision-making. The project team selects four attributes to evaluate the alternatives: (1) function and technology x_1, (2) strategic fitness x_2, (3) vendor's ability x_3; (4) vendor's reputation x_4. The five possible ERP systems $Y_i(i = 1, 2, \ldots, 5)$ are to be evaluated using the spherical fuzzy numbers (SFNs) by the three decision makers (whose weighting vector $w = (0.35, 0.25, 0.40)^T$) under the above four attributes (whose weighting vector $\omega = (0.35, 0.19, 0.24, 0.22)^T$), and construct, respectively, the decision matrices as listed in the following matrices $R^{(k)} = \left(r_{ij}^{(k)}\right)_{5\times 4}$ $(k = 1, 2, 3)$ as follows:

$$R^1 = \begin{bmatrix} (0.6, 0.6, 0.3) & (0.4, 0.5, 0.4) & (0.2, 0.3, 0.5) & (0.4, 0.6, 0.1) \\ (0.6, 0.4, 0.5) & (0.3, 0.1, 0.2) & (0.5, 0.3, 0.8) & (0.3, 0.7, 0.4) \\ (0.5, 0.2, 0.1) & (0.7, 0.1, 0.3) & (0.8, 0.2, 0.4) & (0.5, 0.1, 0.2) \\ (0.7, 0.3, 0.4) & (0.5, 0.4, 0.7) & (0.2, 0.3, 0.6) & (0.4, 0.4, 0.7) \\ (0.5, 0.1, 0.6) & (0.4, 0.6, 0.2) & (0.7, 0.2, 0.3) & (0.4, 0.1, 0.8) \end{bmatrix}$$

$$R^2 = \begin{bmatrix} (0.3, 0.6, 0.4) & (0.5, 0.4, 0.6) & (0.8, 0.2, 0.5) & (0.6, 0.3, 0.4) \\ (0.6, 0.5, 0.6) & (0.6, 0.3, 0.7) & (0.4, 0.4, 0.8) & (0.3, 0.1, 0.3) \\ (0.7, 0.1, 0.4) & (0.5, 0.2, 0.2) & (0.3, 0.1, 0.1) & (0.6, 0.3, 0.2) \\ (0.4, 0.5, 0.5) & (0.6, 0.4, 0.3) & (0.4, 0.7, 0.5) & (0.3, 0.4, 0.8) \\ (0.2, 0.1, 0.4) & (0.5, 0.4, 0.7) & (0.3, 0.2, 0.4) & (0.6, 0.3, 0.2) \end{bmatrix}$$

$$R^3 = \begin{bmatrix} (0.6, 0.1, 0.2) & (0.4, 0.1, 0.9) & (0.2, 0.6, 0.5) & (0.3, 0.2, 0.6) \\ (0.4, 0.3, 0.2) & (0.5, 0.4, 0.3) & (0.1, 0.5, 0.4) & (0.5, 0.1, 0.5) \\ (0.9, 0.1, 0.3) & (0.4, 0.2, 0.2) & (0.6, 0.3, 0.5) & (0.3, 0.1, 0.1) \\ (0.1, 0.4, 0.5) & (0.3, 0.2, 0.7) & (0.1, 0.3, 0.8) & (0.2, 0.1, 0.3) \\ (0.4, 0.2, 0.1) & (0.6, 0.6, 0.2) & (0.5, 0.3, 0.4) & (0.2, 0.4, 0.1) \end{bmatrix}$$

In the following, we shall utilize the proposed approach in this paper getting the most desirable ERP system(s):

Step 1. Deriving the collective overall spherical fuzzy decision matrix $R = \left(r_{ij}\right)_{m \times n}$ by using Eq. (34):

$$R = \begin{bmatrix} (0.5488, 0.2930, 0.2889) & (0.4283, 0.2484, 0.8042) \\ (0.5352, 0.3770, 0.4848) & (0.4782, 0.2291, 0.5190) \\ (0.7789, 0.1275, 0.3430) & (0.5606, 0.1569, 0.2569) \\ (0.4967, 0.3824, 0.4640) & (0.4703, 0.3031, 0.6316) \\ (0.4065, 0.1320, 0.4510) & (0.5173, 0.5422, 0.4405) \end{bmatrix}$$

$$\begin{matrix} (0.4988, 0.3577, 0.5530) & (0.4355, 0.3251, 0.4431) \\ (0.3712, 0.3955, 0.7586) & (0.3972, 0.1976, 0.4361) \\ (0.6547, 0.1978, 0.4369) & (0.4703, 0.1316, 0.1742) \\ (0.2450, 0.3708, 0.6770) & (0.3098, 0.2297, 0.6572) \\ (0.5588, 0.2352, 0.3627) & (0.4149, 0.2291, 0.5805) \end{matrix} \Bigg]$$

Step 2. Defining the spherical fuzzy positive ideal solution (SF-PIS) Y^+ as

$$\left(\mu^+, \eta^+, \nu^+\right) = ((1, 0, 0), (1, 0, 0), \ldots, (1, 0, 0))$$

Step 3. According to Eqs. (36)and (37) and different values of the parameter α, the weighted generalized Dice measure values between $Y_i (i = 1, 2, 3, 4, 5)$ and Y^+ can be obtained, which are shown in Tables 1 and 2, respectively.

From the Tables 1 and 2, different ranking orders are shown by taking different values of α and different Dice similarity measures. Then the best ERP system should belong to Y_3 or Y_5 according to the principle of the maximum degree of Dice similarity measures between spherical fuzzy numbers (SFNs).

Furthermore, for the special cases of the two generalized Dice measures we obtain the following results:

- When α=0, the two weighted generalized Dice measures will reduce to the weighted projection measures of $Y_i (i = 1, 2, 3, 4, 5)$ on Y^+. Thus, the best ERP system should belong to Y_3 according to the principle of the maximum degree of Dice similarity measures between spherical fuzzy numbers (SFNs).

Table 1 The generalized Dice similarity measures of Eq. (36) and ranking orders

	(Y_1, Y^+)	(Y_2, Y^+)	(Y_3, Y^+)	(Y_4, Y^+)	(Y_5, Y^+)	Ranking orders
$\alpha = 0.0$	0.2417	0.2115	0.4236	0.1639	0.2215	$Y_3 > Y_1 > Y_5 > Y_2 > Y_4$
$\alpha = 0.1$	0.2636	0.2307	0.4573	0.1787	0.2427	$Y_3 > Y_1 > Y_5 > Y_2 > Y_4$
$\alpha = 0.2$	0.2901	0.2537	0.4970	0.1966	0.2683	$Y_3 > Y_1 > Y_5 > Y_2 > Y_4$
$\alpha = 0.3$	0.3227	0.2820	0.5446	0.2184	0.3001	$Y_3 > Y_1 > Y_5 > Y_2 > Y_4$
$\alpha = 0.4$	0.3638	0.3176	0.6030	0.2456	0.3405	$Y_3 > Y_1 > Y_5 > Y_2 > Y_4$
$\alpha = 0.5$	0.4174	0.3637	0.6766	0.2807	0.3934	$Y_3 > Y_1 > Y_5 > Y_2 > Y_4$
$\alpha = 0.6$	0.4905	0.4262	0.7728	0.3276	0.4662	$Y_3 > Y_1 > Y_5 > Y_2 > Y_4$
$\alpha = 0.7$	0.5965	0.5158	0.9056	0.3937	0.5725	$Y_3 > Y_1 > Y_5 > Y_2 > Y_4$
$\alpha = 0.8$	0.7652	0.6564	1.1061	0.4939	0.7435	$Y_3 > Y_1 > Y_5 > Y_2 > Y_4$
$\alpha = 0.9$	1.0802	0.9146	1.4658	0.6651	1.0684	$Y_3 > Y_1 > Y_5 > Y_2 > Y_4$
$\alpha = 1.0$	1.9145	1.6200	2.5656	1.0339	1.9938	$Y_3 > Y_5 > Y_1 > Y_2 > Y_4$

- When α=0.5, the two weighted generalized Dice measures will reduce to the weighted Dice similarity measures of $Y_i (i = 1, 2, 3, 4, 5)$ on Y^+. Thus, the best ERP system should belong to Y_3 according to the principle of the maximum degree of Dice similarity measures between spherical fuzzy numbers (SFNs).
- When α=1, the two weighted generalized Dice measures will reduce the weighted projection measures of $Y_i (i = 1, 2, 3, 4, 5)$ on Y^+. Thus, the best ERP system should belong to Y_3 or Y_5 according to the principle of the maximum degree of Dice similarity measures between spherical fuzzy numbers (SFNs).

Therefore, according to different Dice similarity measures and different values of the parameter α, ranking orders may be also different. Thus the proposed multiple attribute group decision making (MADM) methods can be assigned some value of α and some measure to satisfy the requirements of decision makers' preference and flexible decision making.

Obviously, the multiple attribute group decision making methods based the Dice measures and the projection measures are the special cases of the proposed multiple attribute group decision making (MADM) methods based on generalized Dice measures. Therefore, in the multiple attribute group decision making process, the multiple attribute group decision making models developed in this paper are more

Table 2 The generalized Dice similarity measures of Eq. (37) and ranking orders

	(Y_1, Y^+)	(Y_2, Y^+)	(Y_3, Y^+)	(Y_4, Y^+)	(Y_5, Y^+)	Ranking orders
$\alpha = 0.0$	0.2132	0.2129	0.4055	0.0775	0.1540	$Y_3 > Y_1 > Y_2 > Y_5 > Y_4$
$\alpha = 0.1$	0.2334	0.2326	0.4414	0.0849	0.1698	$Y_3 > Y_1 > Y_2 > Y_5 > Y_4$
$\alpha = 0.2$	0.2578	0.2563	0.4844	0.0939	0.1892	$Y_3 > Y_1 > Y_2 > Y_5 > Y_4$
$\alpha = 0.3$	0.2879	0.2854	0.5366	0.1049	0.2137	$Y_3 > Y_1 > Y_2 > Y_5 > Y_4$
$\alpha = 0.4$	0.3260	0.3220	0.6015	0.1190	0.2453	$Y_3 > Y_1 > Y_2 > Y_5 > Y_4$
$\alpha = 0.5$	0.3757	0.3694	0.6842	0.1373	0.2880	$Y_3 > Y_1 > Y_2 > Y_5 > Y_4$
$\alpha = 0.6$	0.4433	0.4330	0.7933	0.1624	0.3487	$Y_3 > Y_1 > Y_2 > Y_5 > Y_4$
$\alpha = 0.7$	0.5406	0.5231	0.9437	0.1986	0.4418	$Y_3 > Y_1 > Y_2 > Y_5 > Y_4$
$\alpha = 0.8$	0.6927	0.6607	1.1645	0.2556	0.6025	$Y_3 > Y_1 > Y_2 > Y_5 > Y_4$
$\alpha = 0.9$	0.9636	0.8963	1.5203	0.3585	0.9474	$Y_3 > Y_1 > Y_5 > Y_2 > Y_4$
$\alpha = 1.0$	1.5826	1.3931	2.1891	0.6002	2.2149	$Y_5 > Y_3 > Y_1 > Y_2 > Y_4$

general and more flexible than existing multiple attribute group decision making models under spherical fuzzy environment.

5.2 *Comparative Analysis*

Gündoğdu and Kahraman (2019) firstly proposed the concepts of spherical fuzzy set (SFS) and defined the spherical weighted arithmetic mean (SFAM) operator and the spherical weighted geometric mean (SFGM) operator to fuse spherical fuzzy numbers (SFNs), and built the spherical fuzzy TOPSIS model to deal with MADM problems. Based on attribute's weights and spherical fuzzy information in matrix R, the ordering results by the SFAM operator, SFGM operator and the spherical fuzzy TOPSIS model are listed as follows (Table 3).

1. If we utilize the SFAM operator to fuse the spherical fuzzy number, the fused results are shown as:

$$r_1 = SFAM(r_{11}, r_{12}, r_{13}, r_{14})$$

Table 3 The ordering of all alternatives by different methods

Methods	Ordering
SFAM operator	$Y_3 > Y_5 > Y_1 > Y_2 > Y_4$
SFGM operator	$Y_3 > Y_1 > Y_5 > Y_2 > Y_4$
Spherical fuzzy TOPSIS model	$Y_3 > Y_1 > Y_5 > Y_2 > Y_4$
Generalized Dice similarity measures of Eq. (36) ($\alpha = 0.5$)	$Y_3 > Y_1 > Y_5 > Y_2 > Y_4$
Generalized Dice similarity measures of Eq. (37) ($\alpha = 0.5$)	$Y_3 > Y_1 > Y_2 > Y_5 > Y_4$

$$= \left(\sqrt{1 - \prod_{j=1}^{4} \left(1 - \mu_{1j}^2\right)^{w_j}}, \prod_{j=1}^{4} (\eta_{1j})^{w_j}, \sqrt{\prod_{j=1}^{4} \left(1 - \mu_{1j}^2\right)^{w_j} - \prod_{j=1}^{4} \left(1 - \mu_{1j}^2 - v_{1j}^2\right)^{w_j}} \right)$$
$$= (0.4933, 0.3048, 0.5577)$$

$$r_2 = SFAM(r_{21}, r_{22}, r_{23}, r_{24})$$
$$= \left(\sqrt{1 - \prod_{j=1}^{4} \left(1 - \mu_{2j}^2\right)^{w_j}}, \prod_{j=1}^{4} (\eta_{2j})^{w_j}, \sqrt{\prod_{j=1}^{4} \left(1 - \mu_{2j}^2\right)^{w_j} - \prod_{j=1}^{4} \left(1 - \mu_{2j}^2 - v_{2j}^2\right)^{w_j}} \right)$$
$$= (0.4626, 0.3010, 0.5735)$$

$$r_3 = SFAM(r_{31}, r_{32}, r_{33}, r_{34})$$
$$= \left(\sqrt{1 - \prod_{j=1}^{4} \left(1 - \mu_{3j}^2\right)^{w_j}}, \prod_{j=1}^{4} (\eta_{3j})^{w_j}, \sqrt{\prod_{j=1}^{4} \left(1 - \mu_{3j}^2\right)^{w_j} - \prod_{j=1}^{4} \left(1 - \mu_{3j}^2 - v_{3j}^2\right)^{w_j}} \right)$$
$$= (0.6666, 0.1484, 0.3506)$$

$$r_4 = SFAM(r_{41}, r_{42}, r_{43}, r_{44})$$
$$= \left(\sqrt{1 - \prod_{j=1}^{4} \left(1 - \mu_{4j}^2\right)^{w_j}}, \prod_{j=1}^{4} (\eta_{4j})^{w_j}, \sqrt{\prod_{j=1}^{4} \left(1 - \mu_{4j}^2\right)^{w_j} - \prod_{j=1}^{4} \left(1 - \mu_{4j}^2 - v_{4j}^2\right)^{w_j}} \right)$$
$$= (0.4097, 0.3247, 0.5964)$$

$$r_5 = SFAM(r_{51}, r_{52}, r_{53}, r_{54})$$
$$= \left(\sqrt{1 - \prod_{j=1}^{4} \left(1 - \mu_{5j}^2\right)^{w_j}}, \prod_{j=1}^{4} (\eta_{5j})^{w_j}, \sqrt{\prod_{j=1}^{4} \left(1 - \mu_{5j}^2\right)^{w_j} - \prod_{j=1}^{4} \left(1 - \mu_{5j}^2 - v_{5j}^2\right)^{w_j}} \right)$$
$$= (0.4734, 0.2239, 0.4622)$$

Then according to the equation of score function $S(r_i) = \left(\mu_i^2 - v_i^2\right) - \left(\eta_i^2 - v_i^2\right)$ given in (Gündoğdu and Kahraman 2019), we can obtain:

$$S(r_1) = 0.1504,\ S(r_2) = 0.1234,\ S(r_3) = 0.4224,$$
$$S(r_4) = 0.0624,\ S(r_5) = 0.1740.$$

Thus, the rank of alternatives is:$Y_3 > Y_5 > Y_1 > Y_2 > Y_4$.

2. If we utilize the SFGM operator to fuse the spherical fuzzy number, the fused results are shown as:

$$\begin{aligned} r_1 &= SFGM(r_{11}, r_{12}, r_{13}, r_{14}) \\ &= \left(\prod_{j=1}^{4} (\mu_{1j})^{w_j}, \sqrt{1 - \prod_{j=1}^{4} \left(1 - \eta_{1j}^2\right)^{w_j}}, \sqrt{\prod_{j=1}^{4} \left(1 - \eta_{1j}^2\right)^{w_j} - \prod_{j=1}^{4} \left(1 - \eta_{1j}^2 - v_{1j}^2\right)^{w_j}} \right) \\ &= (0.4863, 0.3099, 0.5534) \end{aligned}$$

$$\begin{aligned} r_2 &= SFGM(r_{21}, r_{22}, r_{23}, r_{24}) \\ &= \left(\prod_{j=1}^{4} (\mu_{2j})^{w_j}, \sqrt{1 - \prod_{j=1}^{4} \left(1 - \eta_{2j}^2\right)^{w_j}}, \sqrt{\prod_{j=1}^{4} \left(1 - \eta_{2j}^2\right)^{w_j} - \prod_{j=1}^{4} \left(1 - \eta_{2j}^2 - v_{2j}^2\right)^{w_j}} \right) \\ &= (0.4494, 0.3275, 0.5923) \end{aligned}$$

$$\begin{aligned} r_3 &= SFGM(r_{31}, r_{32}, r_{33}, r_{34}) \\ &= \left(\prod_{j=1}^{4} (\mu_{3j})^{w_j}, \sqrt{1 - \prod_{j=1}^{4} \left(1 - \eta_{3j}^2\right)^{w_j}}, \sqrt{\prod_{j=1}^{4} \left(1 - \eta_{3j}^2\right)^{w_j} - \prod_{j=1}^{4} \left(1 - \eta_{3j}^2 - v_{3j}^2\right)^{w_j}} \right) \\ &= (0.6281, 0.1536, 0.3296) \end{aligned}$$

$$\begin{aligned} r_4 &= SFGM(r_{41}, r_{42}, r_{43}, r_{44}) \\ &= \left(\prod_{j=1}^{4} (\mu_{4j})^{w_j}, \sqrt{1 - \prod_{j=1}^{4} \left(1 - \eta_{4j}^2\right)^{w_j}}, \sqrt{\prod_{j=1}^{4} \left(1 - \eta_{4j}^2\right)^{w_j} - \prod_{j=1}^{4} \left(1 - \eta_{4j}^2 - v_{4j}^2\right)^{w_j}} \right) \\ &= (0.3740, 0.3377, 0.6026) \end{aligned}$$

$$\begin{aligned} r_5 &= SFGM(r_{51}, r_{52}, r_{53}, r_{54}) \\ &= \left(\prod_{j=1}^{4} (\mu_{5j})^{w_j}, \sqrt{1 - \prod_{j=1}^{4} \left(1 - \eta_{5j}^2\right)^{w_j}}, \sqrt{\prod_{j=1}^{4} \left(1 - \eta_{5j}^2\right)^{w_j} - \prod_{j=1}^{4} \left(1 - \eta_{5j}^2 - v_{5j}^2\right)^{w_j}} \right) \\ &= (0.4614, 0.3052, 0.4668) \end{aligned}$$

Then according to the equation of score function $S(r_i) = \left(\mu_i^2 - \upsilon_i^2\right) - \left(\eta_i^2 - \upsilon_i^2\right)$ given in (Gündoğdu and Kahraman 2019), we can obtain:

$$\begin{aligned} &S(r_1) = 0.1405,\ S(r_2) = 0.0947,\ S(r_3) = 0.3709, \\ &S(r_4) = 0.0258,\ S(r_5) = 0.1197. \end{aligned}$$

Thus, the rank of alternatives is: $Y_3 > Y_1 > Y_5 > Y_2 > Y_4$.

3. If we utilize the spherical fuzzy TOPSIS model to deal with this MADM problem, then the decision process can be depicted as follows.

According to the score function $S(r_i)$ and the Eqs. (31) and (35) in literature (Gündoğdu and Kahraman 2019), we can derive the spherical fuzzy positive ideal solution (SF-PIS) Y^* and spherical fuzzy negative ideal solution (SF-NIS) Y^- as:

$$Y^* = \left\langle \begin{matrix} (0.7789, 0.1275, 0.3430), (0.5606, 0.1569, 0.2569) \\ (0.6547, 0.1978, 0.4369), (0.4703, 0.1316, 0.1742) \end{matrix} \right\rangle$$

$$Y^- = \left\langle \begin{matrix} (0.4967, 0.3824, 0.4640), (0.5173, 0.5422, 0.4405) \\ (0.2450, 0.3708, 0.6770), (0.3098, 0.2297, 0.6572) \end{matrix} \right\rangle$$

Then the normalized Euclidean distane between the alternatives and the SF-PIS as well as the SF-NIS can be calculated as:

$$d(Y_1, Y^*) = 0.2556, d(Y_2, Y^*) = 0.2605, d(Y_3, Y^*) = 0.0000,$$
$$d(Y_4, Y^*) = 0.3228, d(Y_5, Y^*) = 0.2555$$

and

$$d(Y_1, Y^-) = 0.2148, d(Y_2, Y^-) = 0.1396, d(Y_3, Y^-) = 0.3249,$$
$$d(Y_4, Y^-) = 0.0954, d(Y_5, Y^-) = 0.2004$$

Next, according to the distance to the SF-NIS and the SF-PIS, we can compute the relative closeness $\xi(Y_i)$ by the equation $\xi(Y_i) = d(Y_i, Y^-) / (d(Y_i, Y^*) + d(Y_i, Y^-))$, the calculated results are shown as:

$$\xi(Y_1) = 0.4566, \xi(Y_2) = 0.3490, \xi(Y_3) = 1.0000,$$
$$\xi(Y_4) = 0.2282, \xi(Y_5) = 0.4396$$

Thus, the rank of alternatives is: $Y_3 > Y_1 > Y_5 > Y_2 > Y_4$.

From above analysis, we can compare our developed method with the SFAM operator, SFGM operator and the spherical fuzzy TOPSIS model as follows.

From above mentioned analysis, we can easily find that different methods can derive different ordering of alternatives, but the best alternatives are same. However, the SFAM and SFGM operators doesn't take the interrelationships of being fused arguments into consideration; our developed Dice similarity measures can overcome this limitation. In addition, the Dice similarity measures with spherical fuzzy information own some advantages: (1) the computing results by our developed methods are stable; (2) the calculating equations are simple; (3) it is available to combine this model with other approaches.

From what has been discussed above, the application prospect of the Dice similarity measures with spherical fuzzy information is extensive, therefore, in the multiple attribute group decision making (MADM) process, the multiple attribute group decision making models developed in this paper are more general and more flexible than existing multiple attribute group decision making models under spherical fuzzy environment.

6 Conclusion

The spherical fuzzy set (SFS) is characterized by three functions expressing the degree of positive-membership, the degree of neutral-membership and the degree of positive-membership. It was proposed as a generalization of FS, IFS, PFS and PyFS in order to deal with indeterminate and inconsistent information. In this work, we shall present some novel Dice similarity measures of spherical fuzzy sets (SFSs) and the generalized Dice similarity measures of spherical fuzzy sets (SFSs) and indicates that the Dice similarity measures and asymmetric measures (projection measures) are the special cases of the generalized Dice similarity measures in some parameter values. Then, we propose the generalized Dice similarity measures-based multiple attribute group decision making (MADM) models with spherical fuzzy information. Moreover, we apply the generalized Dice similarity measures between spherical fuzzy sets (SFSs) to multiple attribute group decision making. Finally, an illustrative example is given to demonstrate the efficiency of the similarity measures for selecting the desirable ERP system. In the future, the application of the proposed Dice similarity measure of SFSs needs to be explored in decision making (Deng and Gao 2019; Gao et al. 2019; Li and Lu 2019; Lu and Wei 2019; Wang et al. 2019; Wang 2019; Wu et al. 2019a, b), risk analysis and many other fields under uncertain environment (Liang et al. 2019; Feng et al. 2019; Zhu and Li 2018; Yu et al. 2018; Xu et al. 2018; Xia 2018).

Acknowledgements The work was supported by the National Natural Science Foundation of China under Grant No. 71571128 and the Humanities and Social Sciences Foundation of Ministry of Education of the People's Republic of China (No. 14YJCZH091, 15XJA630006).

References

Ashraf S, Abdullah S (2018) Spherical aggregation operators and their application in multiattribute group decision-making. Int J Intell Syst. https://doi.org/10.1002/int.22062

Ashraf S, Abdullah S, Aslam M, Qiyas M, Kutbi MA (2019a) Spherical fuzzy sets and its representation of spherical fuzzy t-norms and t-conorms. J Intell Fuzzy Syst 36(6):6089–6102. https://doi.org/10.3233/jifs-181941

Ashraf S, Abdullah S, Mahmood T, Ghani F, Mahmood T (2019b) Spherical fuzzy sets and their applications in multi-attribute decision making problems. J Intell Fuzzy Syst 36(3):2829–2844. https://doi.org/10.3233/jifs-172009

Atanassov K (1986) Intuitionistic fuzzy sets. Fuzzy Sets Syst 20:87–96

Atanassov K (1989) More on intuitionistic fuzzy sets. Fuzzy Sets Syst 33:37–46
Cuong BC (2014) Picture fuzzy sets. J Comput Sci Cybern 30(4):409–420
Deng XM, Gao H (2019) TODIM method for multiple attribute decision making with 2-tuple linguistic Pythagorean fuzzy information. J Intell Fuzzy Syst 37(2):1769–1780
Dice LR (1945) Measures of the amount of ecologic association between species. Ecology 26:297–302
Feng M, Liu PD, Geng YS (2019) A method of multiple attribute group decision making based on 2-tuple linguistic dependent Maclaurin symmetric mean operators. Symmetry-Basel 11(1). https://doi.org/10.3390/sym11010031
Gao H, Wei GW, Huang YH (2018) Dual hesitant bipolar fuzzy Hamacher prioritized aggregation operators in multiple attribute decision making. IEEE Access 6:11508–11522. https://doi.org/10.1109/access.2017.2784963
Gao H, Lu M, Wei Y (2019) Dual hesitant bipolar fuzzy Hamacher aggregation operators and their applications to multiple attribute decision making. J Intell Fuzzy Syst 37(4):5755–5766
Garg H (2017) Some picture fuzzy aggregation operators and their applications to multicriteria decision-making. Arab J Sci Eng 42(12):5275–5290. https://doi.org/10.1007/s13369-017-2625-9
Garg H, Munir M, Ullah K, Mahmood T, Jan N (2018) Algorithm for T-spherical fuzzy multi-attribute decision making based on improved interactive aggregation operators. Symmetry-Basel 10(12). https://doi.org/10.3390/sym10120670
Gundogdu FK, Kahraman C (2019a) Extension of WASPAS with spherical fuzzy sets. Informatica 30(2):269–292. https://doi.org/10.15388/Informatica.2019.206
Gundogdu FK, Kahraman C (2019b) A novel fuzzy TOPSIS method using emerging interval-valued spherical fuzzy sets. Eng Appl Artif Intell 85:307–323. https://doi.org/10.1016/j.engappai.2019.06.003
Gundogdu FK, Kahraman C (2019c) A novel VIKOR method using spherical fuzzy sets and its application to warehouse site selection. J Intell Fuzzy Syst 37(1):1197–1211. https://doi.org/10.3233/jifs-182651
Gündoğdu FK, Kahraman C (2019) Spherical fuzzy sets and spherical fuzzy TOPSIS method. J Intell Fuzzy Syst 36(1):337–352. https://doi.org/10.3233/jifs-181401
He TT, Wei GW, Lu JP, Wei C, Lin R (2019) Pythagorean 2-tuple linguistic VIKOR method for evaluating human factors in construction project management. Mathematics 7(12):1149
Jiang Q, Jin X, Lee SJ, Yao SW (2019) A new similarity/distance measure between intuitionistic fuzzy sets based on the transformed isosceles triangles and its applications to pattern recognition. Expert Syst Appl 116:439–453. https://doi.org/10.1016/j.eswa.2018.08.046
Jin HH, Ashraf S, Abdullah S, Qiyas M, Bano M, Zeng SZ (2019a) Linguistic spherical fuzzy aggregation operators and their applications in multi-attribute decision making problems. Mathematics 7(5). https://doi.org/10.3390/math7050413
Jin Y, Ashraf S, Abdullah S (2019b) Spherical fuzzy logarithmic aggregation operators based on entropy and their application in decision support systems. Entropy 21(7). https://doi.org/10.3390/e21070628
Joshi D, Kumar S (2018) An approach to multi-criteria decision making problems using dice similarity measure for picture fuzzy sets. Commun Comput Inf Sci
Li ZX, Lu M (2019) Some novel similarity and distance and measures of Pythagorean fuzzy sets and their applications. J Int Fuzzy Syst 37(2):1781–1799
Liang W, Zhao G, Wu H, Dai B (2019) Risk assessment of rockburst via an extended MABAC method under fuzzy environment. Tunn Undergr Space Technol 83:533–544. https://doi.org/10.1016/j.tust.2018.09.037
Liu PD, Zhang XH (2018) A novel picture fuzzy linguistic aggregation operator and its application to group decision-making. Cogn Comput 10(2):242–259. https://doi.org/10.1007/s12559-017-9523-z
Liu P, Khan Q, Mahmood T, Hassan N (2019) T-spherical fuzzy power muirhead mean operator based on novel operational laws and their application in multi-attribute group decision making. IEEE Access 7:22613–22632

Lu JP, Wei C (2019) TODIM method for performance appraisal on social-integration-based rural reconstruction with interval-valued intuitionistic fuzzy information. J Intell Fuzzy Syst 37(2):1731–1740
Lu JP, Tang XY, Wei GW, Wei C, Wei Y (2019a) Bidirectional project method for dual hesitant Pythagorean fuzzy multiple attribute decision-making and their application to performance assessment of new rural construction. Int J Intell Syst 34(8):1920–1934. https://doi.org/10.1002/int.22126
Lu JP, Wei C, Wu J, Wei GW (2019b) TOPSIS method for probabilistic linguistic MAGDM with entropy weight and its application to supplier selection of new agricultural machinery products. Entropy 21(10):953
Mahmood T, Ullah K, Khan Q, Jan N (2018) An approach toward decision-making and medical diagnosis problems using the concept of spherical fuzzy sets. Neural Comput Appl. https://doi.org/10.1007/s00521-018-3521-2
Muirhead RF (1902) Some methods applicable to identities and inequalities of symmetric algebraic functions of n letters. Proc Edinburgh Math Soc 21(3):144–162
Narayanamoorthy S, Geetha S, Rakkiyappan R, Joo YH (2019) Interval-valued intuitionistic hesitant fuzzy entropy based VIKOR method for industrial robots selection. Expert Syst Appl 121:28–37. https://doi.org/10.1016/j.eswa.2018.12.015
Peng SM (2017) Study on enterprise risk management assessment based on picture fuzzy multiple attribute decision-making method. J Intell Fuzzy Syst 33(6):3451–3458. https://doi.org/10.3233/jifs-16298
Quek SG, Selvachandran G, Munir M, Mahmood T, Ullah K, Son LH, Thong PH, Kumar R, Priyadarshini I (2019) Multi-attribute multi-perception decision-making based on generalized T-spherical fuzzy weighted aggregation operators on neutrosophic sets. Mathematics 7(9). https://doi.org/10.3390/math7090780
Rafiq M, Ashraf S, Abdullah S, Mahmood T, Muhammad S (2019) The cosine similarity measures of spherical fuzzy sets and their applications in decision making. J Intell Fuzzy Syst 36(6):6059–6073. https://doi.org/10.3233/jifs-181922
Shan Y, Ye J (2014) Dice similarity measure between single valued neutrosophic multisets and its application in medical diagnosis. In: Neutrosophic sets & systems
Singh P (2015) Correlation coefficients for picture fuzzy sets. J Intell Fuzzy Syst 28(2):591–604. https://doi.org/10.3233/ifs-141338
Son LH (2016) Generalized picture distance measure and applications to picture fuzzy clustering. Appl Soft Comput 46:284–295. https://doi.org/10.1016/j.asoc.2016.05.009
Tang XY, Wei GW (2019) Dual hesitant Pythagorean fuzzy Bonferroni mean operators in multi-attribute decision making. Arch Control Sci 29(2):339–386. https://doi.org/10.24425/acs.2019.129386
Tang Y, Wen L-L, Wei G-W (2017) Approaches to multiple attribute group decision making based on the generalized Dice similarity measures with intuitionistic fuzzy information. Int J Knowl-Based Intell Engineering Syst 21(2):85–95. https://doi.org/10.3233/kes-170354
Tang XY, Wei GW, Gao H (2019) Models for multiple attribute decision making with interval-valued Pythagorean fuzzy Muirhead mean operators and their application to green suppliers selection. Informatica 30(1):153–186. https://doi.org/10.15388/Informatica.2019.202
Ullah K, Mahmood T, Jan N (2018) Similarity measures for T-spherical fuzzy sets with applications in pattern recognition. Symmetry-Basel 10(6). https://doi.org/10.3390/sym10060193
Ullah K, Hassan N, Mahmood T, Jan N, Hassan M (2019) Evaluation of investment policy based on multi-attribute decision-making using interval valued T-spherical fuzzy aggregation operators. Symmetry-Basel 11(3). https://doi.org/10.3390/sym11030357
Wang R (2019) Research on the application of the financial investment risk appraisal models with some interval number Muirhead mean operators. J Intell Fuzzy Syst 37(2):1741–1752
Wang R, Li YL (2018) Picture hesitant fuzzy set and its application to multiple criteria decision-making. Symmetry-Basel 10(7). https://doi.org/10.3390/sym10070295

Wang L, Zhang HY, Wang JQ, Li L (2018a) Picture fuzzy normalized projection-based VIKOR method for the risk evaluation of construction project. Appl Soft Comput 64:216–226. https://doi.org/10.1016/j.asoc.2017.12.014
Wang J, Tang XY, Wei GW (2018b) Models for multiple attribute decision-making with dual generalized single-valued neutrosophic Bonferroni mean operators. Algorithms 11(1). https://doi.org/10.3390/a11010002
Wang J, Lu JP, Wei GW, Lin R, Wei C (2019a) Models for MADM with single-valued neutrosophic 2-tuple linguistic Muirhead mean operators. Mathematics 7(5). https://doi.org/10.3390/math7050442
Wang P, Wang J, Wei GW (2019b) EDAS method for multiple criteria group decision making under 2-tuple linguistic neutrosophic environment. J Intell Fuzzy Syst 37(2):1597–1608. https://doi.org/10.3233/jifs-179223
Wang J, Gao H, Wei G (2019c) The generalized Dice similarity measures for Pythagorean fuzzy multiple attribute group decision making. Int J Intell Syst 34(6):1158–1183. https://doi.org/10.1002/int.22090
Wang J, Gao H, Lu M (2019d) Approaches to strategic supplier selection under interval neutrosophic environment. J Intell Fuzzy Syst 37(2):1707–1730
Wei GW (2016) Picture fuzzy cross-entropy for multiple attribute decision making problems. J Bus Econ Manage 17(4):491–502. https://doi.org/10.3846/16111699.2016.1197147
Wei GW (2017a) Picture fuzzy aggregation operators and their application to multiple attribute decision making. J Intell Fuzzy Syst 33(2):713–724. https://doi.org/10.3233/jifs-161798
Wei GW (2017b) Some cosine similarity measures for picture fuzzy sets and their applications to strategic decision making. Informatica 28(3):547–564. https://doi.org/10.15388/Informatica.2017.144
Wei GW (2019) 2-tuple intuitionistic fuzzy linguistic aggregation operators in multiple attribute decision making. Iran J Fuzzy Syst 16(4):159–174
Wei GW, Gao H, Wang J, Huang YH (2018) Research on risk evaluation of enterprise human capital investment with interval-valued bipolar 2-tuple linguistic information. IEEE Access 6:35697–35712. https://doi.org/10.1109/access.2018.2836943
Wei GW, Wang J, Lu M, Wu J, Wei C (2019a) Similarity measures of spherical fuzzy sets based on cosine function and their applications. IEEE Access 7:159069–159080. https://doi.org/10.1109/access.2019.2949296
Wei GW, Wu J, Wei C, Wang J, Lu JP (2019b) Models for MADM with 2-tuple linguistic neutrosophic Dombi Bonferroni mean operators. IEEE Access 7:108878–108905. https://doi.org/10.1109/access.2019.2930324
Wei GW, Zhang SQ, Lu JP, Wu J, Wei C (2019c) An extended bidirectional projection method for picture fuzzy MAGDM and its application to safety assessment of construction project. IEEE Access 7(1):166138–166147
Wu LP, Gao H, Wei C (2019a) VIKOR method for financing risk assessment of rural tourism projects under interval-valued intuitionistic fuzzy environment. J Intell Fuzzy Syst 37(2):2001–2008
Wu LP, Wang J, Gao H (2019b) Models for competiveness evaluation of tourist destination with some interval-valued intuitionistic fuzzy Hamy mean operators. J Intell Fuzzy Syst 36(6):5693–5709
Xia MM (2018a) Interval-valued intuitionistic fuzzy matrix games based on Archimedean t-conorm and t-norm. Int J Gen Syst 47(3):278–293. https://doi.org/10.1080/03081079.2017.1413100
Xia MM (2018b) A hesitant fuzzy linguistic multi-criteria decision-making approach based on regret theory. Int J Fuzzy Syst 20(7):2135–2143. https://doi.org/10.1007/s40815-018-0502-7
Xu Z, Yager RR (2010) Power-geometric operators and their use in group decision making. IEEE Trans Fuzzy Syst 18(1):94–105. https://doi.org/10.1109/tfuzz.2009.2036907
Xu Y, Shang X, Wang J, Wu W, Huang H (2018) Some q-Rung dual hesitant fuzzy Heronian mean operators with their application to multiple attribute group decision-making. Symmetry-Basel 10(10). https://doi.org/10.3390/sym10100472
Yager RR (2001) The power average operator. IEEE Trans Syst Man Cybern Part A: Syst Hum 31(6):724–731. https://doi.org/10.1109/3468.983429

Yager RR (2013) Pythagorean fuzzy subsets. In: Proceedings of the 2013 joint IFSA world congress and NAFIPS annual meeting

Yager RR (2014) Pythagorean membership grades in multicriteria decision making. IEEE Trans Fuzzy Syst 22(4):958–965

Yang W, Pang YF, Shi JR, Wang CJ (2018) Linguistic hesitant intuitionistic fuzzy decision-making method based on VIKOR. Neural Comput Appl 29(7):613–626. https://doi.org/10.1007/s00521-016-2526-y

Yao DB, Wang CC (2018) Hesitant intuitionistic fuzzy entropy/cross-entropy and their applications. Soft Comput 22(9):2809–2824. https://doi.org/10.1007/s00500-017-2753-x

Ye J (2012a) Multicriteria decision-making method using the Dice similarity measure based on the reduct intuitionistic fuzzy sets of interval-valued intuitionistic fuzzy sets. Appl Math Model 36(9):4466–4472. https://doi.org/10.1016/j.apm.2011.11.075

Ye J (2012b) Multicriteria decision-making method using the Dice similarity measure between expected intervals of trapezoidal fuzzy numbers. J Decis Syst 21(4):307–317

Ye J (2012c) The Dice similarity measure between generalized trapezoidal fuzzy numbers based on the expected interval and its multicriteria group decision-making method. J Chin Inst Ind Eng 29(6):375–382

Yu GF, Li DF, Qiu JM, Zheng XX (2018a) Some operators of intuitionistic uncertain 2-tuple linguistic variables and application to multi-attribute group decision making with heterogeneous relationship among attributes. J Intell Fuzzy Syst 34(1):599–611. https://doi.org/10.3233/jifs-17821

Yu SM, Zhang HY, Wang JQ (2018b) Hesitant fuzzy linguistic Maclaurin symmetric mean operators and their applications to multi-criteria decision-making problem. Int J Intell Syst 33(5):953–982. https://doi.org/10.1002/int.21907

Zadeh LA (1965) Fuzzy sets. In: Information and control, vol 8

Zhang Z (2018) Geometric Bonferroni means of interval-valued intuitionistic fuzzy numbers and their application to multiple attribute group decision making. Neural Comput Appl 29(11):1139–1154. https://doi.org/10.1007/s00521-016-2621-0

Zhang GF, Zhang ZM, Kong H (2018) Some normal intuitionistic fuzzy Heronian mean operators using Hamacher operation and their application. Symmetry-Basel 10(6). https://doi.org/10.3390/sym10060199

Zhou W, Xu ZS (2018) Extended intuitionistic fuzzy sets based on the hesitant fuzzy membership and their application in decision making with risk preference. Int J Intell Syst 33(2):417–443. https://doi.org/10.1002/int.21938

Zhu JH, Li YL (2018) Hesitant Fuzzy Linguistic Aggregation Operators Based on the Hamacher t-norm and t-conorm. Symmetry-Basel 10 (6). https://doi.org/10.3390/sym10060189

Spherical Fuzzy Bonferroni Mean Aggregation Operators and Their Applications to Multiple-Attribute Decision Making

Elmira Farrokhizadeh, Seyed Amin Seyfi Shishavan, Yaser Donyatalab, Fatma Kutlu Gündoğdu, and Cengiz Kahraman

Abstract The Bonferroni Mean (BM) which was introduced by Bonferroni has been extensively applied in Multi-Attribute Group Decision-Making (MAGDM) and support system problems because of its usefulness in the aggregation techniques. One of the most important and distinguishing characteristic of the BM is its capability to capture the interrelationship between arguments. Motivated by the applications of Spherical Fuzzy Sets (SFS) in recent studies and in order to consider the interrelationship between arguments, it seems necessary to develop novel aggregation operators to use in this kind of fuzzy sets in MAGDM problems. Therefore, in this paper we tried to adopt BM and spherical fuzzy sets operators in order to propose newfound aggregation operators such as: Spherical Fuzzy Bonferroni mean (SFBM) and Spherical Fuzzy Normalized Weighted Bonferroni mean (SFNWBM). Finally, based on the proposed aggregation operators (SFNWBM), we present an approach to multi-criteria group-decision making problems under the spherical fuzzy environment, and to illustrate the validity of the novel aggregation operator, a practical example is provided.

Keywords Spherical fuzzy sets · Aggregation operator · Bonferroni mean · MAGDM

1 Introduction

Multi-Attribute Group Decision Making (MAGDM) is a process that can give the ranking results for the finite alternatives according to the attribute values (which some

E. Farrokhizadeh · S. A. Seyfi Shishavan (✉) · Y. Donyatalab · C. Kahraman
Department of Industrial Engineering, Istanbul Technical University, Macka, 34367 Besiktas, Istanbul, Turkey
e-mail: aminseyfi@ymail.com

F. Kutlu Gündoğdu
Industrial Engineering Department, National Defence University, Turkish Air Force Academy, 34149 Istanbul, Turkey
e-mail: fatmakutlugundogdu@gmail.com

© Springer Nature Switzerland AG 2021

C. Kahraman and F. Kutlu Gündoğdu (eds.), *Decision Making with Spherical Fuzzy Sets*, Studies in Fuzziness and Soft Computing 392,
https://doi.org/10.1007/978-3-030-45461-6_5

of them are tangible or intangible) of different alternatives, and it is the important aspect of decision sciences (Liu et al. 2019). MAGDM plays an important role in modern decision science, and it has been widely used in economics, management and the other fields in recent years. Thus, how to effectively aggregate attribute values is a core issue of any MAGDM methods. On the other hand, due to the subjective nature of human thinking in real decision-making problems, decision makers' evaluations over alternatives are always imprecise and uncertain (Xing et al. 2019). To deal with uncertainty nature of this type of problems, using fuzzy logic can be an appropriate approach.

Fuzzy sets were proposed by Zadeh (Zadeh 1965) which are used to represent the degree of membership corresponding to the complexity. After the introduction of ordinary fuzzy sets, several extensions of ordinary fuzzy sets have been developed by various researchers. In recent years, several researchers have utilized these extensions in the solution of multi-criteria decision-making problems.

The ordinary fuzzy sets have been extended to many new types (Gündoğdu and Kahraman 2019a): Type-2 fuzzy sets (Zadeh 1975) to deal with the uncertainty of the membership function in the fuzzy set theory; intuitionistic fuzzy sets (Atanassov 1986) to deal with the problem of assigning non-membership degrees; neutrosophic fuzzy sets (Smarandache 1999) where each element of the universe has a degree of truthiness, indeterminacy, and falsity; hesitant fuzzy sets (Torra 2010) where a single element can have a set of values for its membership degree; Pythagorean fuzzy sets (Yager 2013) which facilitate the representation on a larger body of membership and non-standard membership grades; Picture fuzzy sets (Cuong 2014) which is direct extensions of the fuzzy sets and the intuitionistic fuzzy sets and spherical fuzzy sets (SFNs) (Gündoğdu and Kahraman 2019b, c, d) where the hesitancy of a decision maker can be defined independently from membership and non-membership degrees.

To deal with some decision problems, the initial and important issue is to select an appropriate information expression tool to describe the assessment information. In order to fully express people's decision-making awareness and complex conditions, SFNs can fully express people's decision-making consciousness and describe the decision information precisely by a parameter which can flexibly adjust the scope of information expression (Liu et al. 2019).

In the practical applications, it is usually necessary to aggregate several inputs into a single representative output, such as decision making. Decision-making attempts to choose the best-performing option or alternative among the feasible ones in order to satisfy a certain objective represented by an attribute. The aggregation operators are useful models for combining and summarizing a finite set of numerical values into a single numerical value. Such operators are essential for solving multi-criteria and group decision-making (MCGDM) problems (Donyatalab et al. 2019). Up to now, various aggregation operators in spherical fuzzy environment have been developed by researchers in order to aggregate expert's comments. Spherical Weighted Arithmetic Mean (SWAM) and Spherical Weighted Geometric Mean (SWGM) introduced by Gündoğdu and Kahraman (2019a), Harmonic mean aggregation operators introduced by Donyatalab et al. (2019), Spherical fuzzy Dombi aggregation operators introduced by Ashraf et al. (2019), and linguistic spherical fuzzy aggregation operators by Jin et al. (2019). Furthermore, the Bonferroni mean that was first introduced by Bonferroni (1950), is one of the aggregation operators. It is composed of arithmetic means and products. It has the desirable properties which capture the interrelationships among arguments.

The Bonferroni mean which can provide for the aggregation lying between the max and min operators and logical "oring" and "anding" operators, has been identified as a useful averaging aggregation function with the potential for interesting applications in fuzzy systems and multi-criteria decision making and has attracted a lot of attention from researchers (Yager 2009). A significant characteristic of BM is that it cannot only consider the importance of each criterion but also reflect the interrelationship of the individual criterion. As Bonferroni mean (BM) can capture the interrelationships among arguments, it has been applied widely in MCDM under different circumstances. Yager (2009) proposed several generalized Bonferroni mean operators and applied them to multi-criteria aggregation. Xia et al. (2013) developed the geometric Bonferroni means and utilized them to deal with the MCDM problems. The intuitionistic fuzzy Bonferroni means was introduced by Xu and Yager (2011). Under hesitant fuzzy environment, the hesitant fuzzy Bonferroni mean operator and the hesitant fuzzy geometric Bonferroni mean operator were proposed by Zhu et al. (2012). Recently, Pythagorean fuzzy interaction power Bonferroni mean aggregation operators in multiple attribute decision making was proposed by Wang and Li (2019). Pythagorean triangular fuzzy linguistic Bonferroni mean operators and their application for multi-attribute decision making were proposed by Jing et al. (2017).

To the best of our knowledge, there is no study in literature about Bonferroni mean aggregation operator in spherical fuzzy environment. In this paper, we first develop novel aggregation operators named "Bonferroni Mean Aggregation Operator with spherical fuzzy Sets" and "Normalized Weighted Bonferroni Mean Aggregation Operator with spherical fuzzy Sets" and then investigate the application of proposed aggregation operators in Multi-criteria Decision-Making problems.

The reminder of the paper is organized as follows. Some basic concepts are reviewed in Sect. 2. Novel aggregation operators are proposed in Sect. 3. Descriptive analyses and a numerical example is provided and discussed in Sect. 4 and finally, the paper ends with some concluding remarks in Sect. 5.

2 Preliminaries

In this section, we define some basic and primal notions and operations related to spherical fuzzy sets and the Bonferroni mean.

2.1 Spherical Fuzzy Sets

The novel concept of SFS (Spherical Fuzzy Sets) provides a larger preference domain for decision makers to assign membership degrees since the squared sum of the spherical parameters is allowed to be at most 1.0. DMs can define their hesitancy information independently under spherical fuzzy environment. SFS are a generalization of Pythagorean Fuzzy Sets (Gündoğdu and Kahraman 2019b, c, d). In the following, definition of SFS is presented:

Definition 2.1 Single valued Spherical Fuzzy Sets (SFS) $\tilde{A}_S$ of the universe of discourse U is given by

$$\tilde{A}_S = \left\{ \langle u, (\mu_{\tilde{A}_S}(u), \nu_{\tilde{A}_S}(u), \pi_{\tilde{A}_S}(u)) \middle| u \in U \right\} \tag{1}$$

where

$$\mu_{\tilde{A}_S} : U \to [0, 1], \nu_{\tilde{A}_S}(u) : U \to [0, 1], \pi_{\tilde{A}_S} : U \to [0, 1]$$

and

$$0 \le \mu^2_{\tilde{A}_S}(u) + \nu^2_{\tilde{A}_S}(u) + \pi^2_{\tilde{A}_S}(u) \le 1 \qquad \forall u \in U \tag{2}$$

For each u, the numbers $\mu_{\tilde{A}_S}(u)$, $\nu_{\tilde{A}_S}(u)$ and $\pi_{\tilde{A}_S}(u)$ are the degree of membership, non-membership and hesitancy of u to $\tilde{A}_S$, respectively.

Definition 2.2 Basic operators of Single-valued SFS;

i. $$\tilde{A}_S \oplus \tilde{B}_S = \left\{ \begin{array}{c} \left(\mu^2_{\tilde{A}_S} + \mu^2_{\tilde{B}_S} - \mu^2_{\tilde{A}_S}\mu^2_{\tilde{B}_S}\right)^{\frac{1}{2}}, \upsilon_{\tilde{A}_S}\upsilon_{\tilde{B}_S}, \\ \left(\left(1 - \mu^2_{\tilde{B}_S}\right)\pi^2_{\tilde{A}_S} + \left(1 - \mu^2_{\tilde{A}_S}\right)\pi^2_{\tilde{B}_S} - \pi^2_{\tilde{A}_S}\pi^2_{\tilde{B}_S}\right)^{\frac{1}{2}} \end{array} \right\} \tag{3}$$

ii. $$\tilde{A}_S \otimes \tilde{B}_S = \left\{ \begin{array}{c} \mu_{\tilde{A}_S}\mu_{\tilde{B}_S}, \left(\upsilon^2_{\tilde{A}_S} + \upsilon^2_{\tilde{B}_S} - \upsilon^2_{\tilde{A}_S}\upsilon^2_{\tilde{B}_S}\right)^{\frac{1}{2}}, \\ \left(\left(1 - \upsilon^2_{\tilde{B}_S}\right)\pi^2_{\tilde{A}_S} + \left(1 - \upsilon^2_{\tilde{A}_S}\right)\pi^2_{\tilde{B}_S} - \pi^2_{\tilde{A}_S}\pi^2_{\tilde{B}_S}\right)^{\frac{1}{2}} \end{array} \right\} \tag{4}$$

iii. $$\lambda \cdot \tilde{A}_S = \left\{ \begin{array}{c} \left(1 - \left(1 - \mu^2_{\tilde{A}_S}\right)^{\lambda}\right)^{\frac{1}{2}}, \upsilon^{\lambda}_{\tilde{A}_S}, \\ \left(\left(1 - \mu^2_{\tilde{A}_S}\right)^{\lambda} - \left(1 - \mu^2_{\tilde{A}_S} - \pi^2_{\tilde{A}_S}\right)^{\lambda}\right)^{\frac{1}{2}} \end{array} \right\} for \lambda > 0 \tag{5}$$

iv. $$\tilde{A}^{\lambda}_s = \left\{ \mu^{\lambda}_{\tilde{A}_s}, \sqrt{1 - \left(1 - \vartheta^2_{\tilde{A}_s}\right)^{\lambda}}, \sqrt{\left(1 - \vartheta^2_{\tilde{A}_s}\right)^{\lambda} - \left(1 - \vartheta^2_{\tilde{A}_s} - \pi^2_{\tilde{A}_s}\right)^{\lambda}} \right\} \tag{6}$$

Definition 2.3 Single-valued Spherical Weighted Arithmetic Mean (SWAM) with respect to, $w = (w_1, w_2 \ldots, w_n)$; $w_i \in [0, 1]$; $\sum_{i=1}^{n} w_i = 1$, SWAM is defined as;

$$SWAM_w(\tilde{A}_{S1}, \ldots, \tilde{A}_{Sn}) = w_1\tilde{A}_{S1} + w_2\tilde{A}_{S2} + \cdots + w_n\tilde{A}_{Sn}$$
$$= \left\{ \begin{array}{c} \left[1 - \prod_{i=1}^{n}(1 - \mu_{\tilde{A}_{S_i}}^2)^{w_i}\right]^{\frac{1}{2}}, \\ \prod_{i=1}^{n} v_{\tilde{A}_{S_i}}^{w_i}, \left[\prod_{i=1}^{n}(1 - \mu_{\tilde{A}_{S_i}}^2)^{w_i} - \prod_{i=1}^{n}(1 - \mu_{\tilde{A}_{S_i}}^2 - \pi_{\tilde{A}_{S_i}}^2)^{w_i}\right]^{\frac{1}{2}} \end{array} \right\} \tag{7}$$

Definition 2.4 Single-valued Spherical Weighted Geometric Mean (SWGM) with respect to,$w = (w_1, w_2 \ldots, w_n)$; $w_i \in [0, 1]$; $\sum_{i=1}^{n} w_i = 1$, SWGM is defined as;

$$SWGM_w(\tilde{A}_1, \ldots, \tilde{A}_n) = \tilde{A}_{S1}^{w_1} + \tilde{A}_{S2}^{w_2} + \cdots + \tilde{A}_{Sn}^{w_n}$$
$$= \left\{ \begin{array}{c} \prod_{i=1}^{n} \mu_{\tilde{A}_{Si}}^{w_i}, \left[1 - \prod_{i=1}^{n}(1 - v_{\tilde{A}_{Si}}^2)^{w_i}\right]^{\frac{1}{2}}, \\ \left[\prod_{i=1}^{n}(1 - v_{\tilde{A}_{Si}}^2)^{w_i} - \prod_{i=1}^{n}(1 - v_{\tilde{A}_{Si}}^2 - \pi_{\tilde{A}_{Si}}^2)^{w_i}\right]^{\frac{1}{2}} \end{array} \right\} \tag{8}$$

Definition 2.5 Score functions and Accuracy functions of sorting SFS are defined by;

$$Score\left(\tilde{A}_S\right) = (\mu_{\tilde{A}_S} - \pi_{\tilde{A}_S}/2)^2 - (v_{\tilde{A}_S} - \pi_{\tilde{A}_S}/2)^2 \tag{9}$$

$$Accuracy\left(\tilde{A}_S\right) = \mu_{\tilde{A}_S}^2 + v_{\tilde{A}_S}^2 + \pi_{\tilde{A}_S}^2 \tag{10}$$

Note that: $\tilde{A}_S < \tilde{B}_S$ if and only if

i. $Score(\tilde{A}_S) < Score(\tilde{B}_S)$ or
ii. $Score(\tilde{A}_S) = Score(\tilde{B}_S)$ and $Accuracy(\tilde{A}_S) < Accuracy(\tilde{B}_S)$

2.2 Bonferroni Mean

Definition 2.6 The Bonferroni Mean (BM) which is introduced by Bonferroni for crisp number is defined as follow:

$$BM^{p,q}(a_1, a_2, \ldots, a_n) = \left(\frac{1}{n(n-1)}\left(\sum_{i \neq j=1}^{n} a_i^p \otimes a_j^q\right)\right)^{\frac{1}{p+q}}. \tag{11}$$

where a_i $(i = 1, 2, 3, \ldots, n)$ be a collection of nonnegative crisp number and $p, q \geq 0$.

One explanation of the Bonferroni Mean is as a kind of composed "anding" and "averaging" operator (Yager 2009). Here a_i^p and a_j^q indicates the degree to which both criteria a_i and a_j are satisfied under the given conditions and the special case when $p = q = 1$.

Some of the BM's properties can be shown as follows:

I. $BM^{p,q}(0, 0, \ldots, 0) = 0$;
II. $BM^{p,q}(a, a, \ldots, a) = a$ if $a_i = a$ for all i;
III. $BM^{p,q}(a_1, a_2, \ldots, a_n) \geq BM^{p,q}(b_1, b_2, \ldots, b_n)$ if $a_i \geq b_i$ for all i so the BM is monotonic;
IV. $\min\{a_i\} \leq BM^{p,q}(a_1, a_2, \ldots, a_n) \leq \max\{a_i\}$;

Definition 2.7 (Zhou and He 2012) Let $p, q \geq 0$ and a_i $(i = 1, 2, 3, \ldots, n)$ be a collection of nonnegative numbers with corresponding weight vector $w_i = (w_1, w_2, \ldots, w_n)$ where $w_i \geq 0$ and $\sum_{i=1}^{n} w_i = 1$. If:

$$NWBM^{p,q}(a_1, a_2, \ldots, a_n) = \left(\sum_{i \neq j=1}^{n} \frac{w_i . w_j}{1 - w_i} a_i^p \otimes a_j^q \right)^{\frac{1}{p+q}} \tag{12}$$

then $NWBM^{p,q}$ is called the normalized weighted Bonferroni mean (NWBM).

Moreover, the NWBM has the properties as follow:

I. Reducibility: it means that if $p, q \geq 0$ and a_i $(i = 1, 2, 3, \ldots, n)$ be a collection of nonnegative numbers with corresponding weight vector $w_i = \left(\frac{1}{n}, \frac{1}{n}, \ldots, \frac{1}{n}\right)$ then:

$$NWBM^{p,q}(a_1, a_2, \ldots, a_n) = BM^{p,q}(a_1, a_2, \ldots, a_n);$$

II. Idempotency:
$NWBM^{p,q}(a, a, \ldots, a) = a$ if $a_i = a$ for all i;
III. Monotonicity:
$NWBM^{p,q}(a_1, a_2, \ldots, a_n) \geq NWBM^{p,q}(b_1, b_2, \ldots, b_n)$ if $a_i \geq b_i$ for all i;
IV. Boundedness:

$$\min\{a_i\} \leq NWBM^{p,q}(a_1, a_2, \ldots, a_n) \leq \max\{a_i\};$$

3 Spherical Fuzzy Bonferroni Mean Aggregation Operators

In this section, we extend the BM to Spherical fuzzy information and introduce some Spherical fuzzy Bonferroni mean aggregation operators. Some attractive properties of these operators are also presented and discussed.

3.1 Spherical Fuzzy Bonferroni Mean Operator (SFBM)

Let $p, q \geq 0$ and $\tilde{A}_i = \left(\mu_{\tilde{A}_i}, \vartheta_{\tilde{A}_i}, \pi_{\tilde{A}_i}\right)$ with $i = 1, 2, \ldots, n$ be a collection of SFSs. If:

$$SFBM^{p,q}\left(\tilde{A}_1, \tilde{A}_2, \ldots, \tilde{A}_n\right) = \left(\frac{1}{n(n-1)}\left(\oplus_{i\neq j=1}^{n} \tilde{A}_i^p \otimes \tilde{A}_j^q\right)\right)^{\frac{1}{p+q}} \tag{13}$$

is called the Spherical fuzzy Bonferroni mean (SFBM).

By the operations of SFNs and definition of Bonferroni mean, we can obtain the following result.

Theorem 3.1.1 Let $p, q \geq 0$ and $\tilde{A}_i = \left(\mu_{\tilde{A}_i}, \vartheta_{\tilde{A}_i}, \pi_{\tilde{A}_i}\right)$ with $i = 1, 2, \ldots, n$ be a collection of SFSs. Then the aggregated value by the $SFBM$ operator is also be a spherical fuzzy set with the below formulation:

$$\begin{aligned} &SFBM^{p,q}\left(\tilde{A}_1, \tilde{A}_2, \ldots, \tilde{A}_n\right) \\ &= \left\{\left[1 - \left[\prod_{i\neq j=1}^{n}\left(1 - \mu_{\tilde{A}_i}^{2p}\mu_{\tilde{A}_j}^{2q}\right)\right]^{\frac{1}{n(n-1)}}\right]^{\frac{1}{2(p+q)}},\right. \\ &\left[1 - \left[1 - \left[\prod_{i\neq j=1}^{n}\left(1 - \alpha_i^p\alpha_j^q\right)\right]^{\frac{1}{n(n-1)}}\right]^{\frac{1}{p+q}}\right]^{\frac{1}{2}}, \\ &\left[1 - \vartheta_{SFBM^{p,q}}^2 - \left[1 - \left[\prod_{i\neq j=1}^{n}\left(1 - \alpha_i^p\alpha_j^q\right)\right]^{\frac{1}{n(n-1)}} - \left(1 - \mu_{SFBM^{p,q}}^{2(p+q)}\right)\right.\right. \\ &\left.\left.\left. + \left[\prod_{i\neq j=1}^{n}\left(1 - \mu_{\tilde{A}_i}^{2p}\mu_{\tilde{A}_j}^{2q} - \alpha_i^p\alpha_j^q + \beta_i^p\beta_j^q\right)\right]^{\frac{1}{n(n-1)}}\right]^{\frac{1}{p+q}}\right]^{\frac{1}{2}}\right\} \end{aligned} \tag{14}$$

where $1-\vartheta^2_{\tilde{A}_i}=\alpha_i$, $1-\vartheta^2_{\tilde{A}_j}=\alpha_j$, $1-\vartheta^2_{\tilde{A}_i}-\pi^2_{\tilde{A}_i}=\beta_i$ and $1-\vartheta^2_{\tilde{A}_j}-\pi^2_{\tilde{A}_j}=\beta_j$;

Proof To prove the Theorem 3.1.1, we apply the operations of Spherical Fuzzy Sets to Eq. (13) the same as follow:

Step 1: By applying Eq. (6) in Definition 2.2, we have:

$$\tilde{A}_i^p=\left\{\mu^p_{\tilde{A}_i},\sqrt{1-\left(1-\vartheta^2_{\tilde{A}_i}\right)^p},\sqrt{\left(1-\vartheta^2_{\tilde{A}_i}\right)^p-\left(1-\vartheta^2_{\tilde{A}_i}-\pi^2_{\tilde{A}_i}\right)^p}\right\}$$

$$\tilde{A}_j^q=\left\{\mu^q_{\tilde{A}_j},\sqrt{1-\left(1-\vartheta^2_{\tilde{A}_j}\right)^q},\sqrt{\left(1-\vartheta^2_{\tilde{A}_j}\right)^q-\left(1-\vartheta^2_{\tilde{A}_j}-\pi^2_{\tilde{A}_j}\right)^q}\right\}$$

Step 2: By applying Eq. (4) in Definition 2.2 and considering $1-\vartheta^2_{\tilde{A}_i}=\alpha_i$, $1-\vartheta^2_{\tilde{A}_j}=\alpha_j$, $1-\vartheta^2_{\tilde{A}_i}-\pi^2_{\tilde{A}_i}=\beta_i$ and $1-\vartheta^2_{\tilde{A}_j}-\pi^2_{\tilde{A}_j}=\beta_j$, so we have:

$$\tilde{A}_i^p\otimes\tilde{A}_j^q=\left\{\mu^p_{\tilde{A}_i}\mu^q_{\tilde{A}_j},\left[1-\alpha_i^p.\alpha_j^q\right]^{\frac{1}{2}},\left[\alpha_i^p\alpha_j^q-\beta_i^p\beta_j^q\right]^{\frac{1}{2}}\right\}$$

Step 3. By applying Eq. (3) in Definition 2.2, we have:

$$\sum_{i\neq j=1}^{n}\tilde{A}_i^p\otimes\tilde{A}_j^q=\left\{\left[1-\prod_{i\neq j=1}^{n}\left(1-\mu^{2p}_{\tilde{A}_i}\cdot\mu^{2q}_{\tilde{A}_j}\right)\right]^{\frac{1}{2}},\right.$$

$$\left[\prod_{i\neq j=1}^{n}\left(1-\alpha_i^p.\alpha_j^q\right)\right]^{\frac{1}{2}},\left[\prod_{i\neq j=1}^{n}\left(1-\mu^{2p}_{\tilde{A}_i}\cdot\mu^{2q}_{\tilde{A}_j}\right)\right.$$

$$\left.\left.-\prod_{i\neq j=1}^{n}\left(1-\mu^{2p}_{\tilde{A}_i}\cdot\mu^{2q}_{\tilde{A}_j}-\alpha_i^p\alpha_j^q+\beta_i^p\beta_j^q\right)\right]^{\frac{1}{2}}\right\}$$

which is obtained by mathematical induction on n as follow:
for $n=2$

$$\mu_{\sum_{i\neq j=1}^{2}\tilde{A}_i^p\otimes\tilde{A}_j^q}$$
$$=\left[\left(\mu^{2p}_{\tilde{A}_1}\cdot\mu^{2q}_{\tilde{A}_2}\right)+\left(\mu^{2p}_{\tilde{A}_2}\cdot\mu^{2q}_{\tilde{A}_1}\right)-\left(\mu^{2p}_{\tilde{A}_1}\cdot\mu^{2q}_{\tilde{A}_2}\right).\left(\mu^{2p}_{\tilde{A}_2}\cdot\mu^{2q}_{\tilde{A}_1}\right)\right]^{\frac{1}{2}}$$
$$=\left[1-\left(1-\mu^{2p}_{\tilde{A}_1}\cdot\mu^{2q}_{\tilde{A}_2}\right).\left(1-\mu^{2p}_{\tilde{A}_2}\cdot\mu^{2q}_{\tilde{A}_1}\right)\right]^{\frac{1}{2}};$$

$$\vartheta_{\sum_{i\neq j=1}^{2}\tilde{A}_i^p\otimes\tilde{A}_j^q} = \left[\left(1-\alpha_1^p\alpha_2^q\right).\left(1-\alpha_2^p\alpha_1^q\right)\right]^{\frac{1}{2}};$$

$$\begin{aligned}&\pi_{\sum_{i\neq j=1}^{2}\tilde{A}_i^p\otimes\tilde{A}_j^q}\\&=\left[\left(1-\mu_{\tilde{A}_2}^{2p}\cdot\mu_{\tilde{A}_1}^{2q}\right).\left(1-\mu_{\tilde{A}_1}^{2p}\cdot\mu_{\tilde{A}_2}^{2q}-\alpha_1^p\alpha_2^q+\beta_1^p\beta_2^q\right)\right.\\&\left.+\left(1-\mu_{\tilde{A}_1}^{2p}\cdot\mu_{\tilde{A}_2}^{2q}\right).\left(1-\mu_{\tilde{A}_1}^{2p}\cdot\mu_{\tilde{A}_2}^{2q}-\alpha_2^p\alpha_1^q+\beta_2^p\beta_1^q\right)\right]^{\frac{1}{2}};\end{aligned}$$

After applying the multiplication

$$\begin{aligned}&\pi_{\sum_{i\neq j=1}^{2}\tilde{A}_i^p\otimes\tilde{A}_j^q}\\&=\left[\left(1-\mu_{\tilde{A}_1}^{2p}\cdot\mu_{\tilde{A}_2}^{2q}\right).\left(1-\mu_{\tilde{A}_2}^{2p}\cdot\mu_{\tilde{A}_1}^{2q}\right)\right.\\&-\left(1-\mu_{\tilde{A}_1}^{2p}\cdot\mu_{\tilde{A}_2}^{2q}-\alpha_1^p\alpha_2^q+\beta_1^p\beta_2^q\right)\\&\left.\times\left(1-\mu_{\tilde{A}_1}^{2p}\cdot\mu_{\tilde{A}_2}^{2q}-\alpha_2^p\alpha_1^q+\beta_2^p\beta_1^q\right)\right]^{\frac{1}{2}};\end{aligned}$$

So if n = k:

$$\begin{aligned}&\sum_{i\neq j=1}^{k}\tilde{A}_i^p\otimes\tilde{A}_j^q\\&=\left\{\left[1-\prod_{i\neq j=1}^{k}\left(1-\mu_{\tilde{A}_i}^{2p}\cdot\mu_{\tilde{A}_j}^{2q}\right)\right]^{\frac{1}{2}},\right.\\&\left[\prod_{i\neq j=1}^{k}\left(1-\alpha_i^p.\alpha_j^q\right)\right]^{\frac{1}{2}},\left[\prod_{i\neq j=1}^{k}\left(1-\mu_{\tilde{A}_i}^{2p}\cdot\mu_{\tilde{A}_j}^{2q}\right)\right.\\&\left.\left.-\prod_{i\neq j=1}^{k}\left(1-\mu_{\tilde{A}_i}^{2p}\cdot\mu_{\tilde{A}_j}^{2q}-\alpha_i^p\alpha_j^q+\beta_i^p\beta_j^q\right)\right]^{\frac{1}{2}}\right\};\end{aligned}$$

Then when $n = k + 1$, we have:

$$\sum_{i\neq j=1}^{k+1}\tilde{A}_i^p\otimes\tilde{A}_j^q=\left(\sum_{i\neq j=1}^{k}\tilde{A}_i^p\otimes\tilde{A}_j^q\right)$$

$$+\left(\sum_{i=1}^{k} \tilde{A}_i^p \otimes \tilde{A}_{k+1}^q\right) + \left(\sum_{j=1}^{k} \tilde{A}_{k+1}^p \otimes \tilde{A}_j^q\right);$$

So with operations defined in Definition 2.2, we can easily prove that:

$$\left(\sum_{i=1}^{k} \tilde{A}_i^p \otimes \tilde{A}_{k+1}^q\right) = \left\{\left[1 - \prod_{i=1}^{k}\left(1 - \mu_{\tilde{A}_i}^{2p} \cdot \mu_{\tilde{A}_{k+1}}^{2q}\right)\right]^{\frac{1}{2}},\right.$$

$$\left[\prod_{i=1}^{k}(1 - \alpha_i^p.\alpha_{k+1}^q)\right]^{\frac{1}{2}}, \left[\prod_{i=1}^{k}\left(1 - \mu_{\tilde{A}_i}^{2p} \cdot \mu_{\tilde{A}_{k+1}}^{2q}\right)\right.$$

$$\left.\left. - \prod_{i=1}^{k}\left(1 - \mu_{\tilde{A}_i}^{2p} \cdot \mu_{\tilde{A}_{k+1}}^{2q} - \alpha_i^p \alpha_{k+1}^q + \beta_i^p \beta_{k+1}^q\right)\right]\right\};$$

And

$$\sum_{j=1}^{k} \tilde{A}_{k+1}^p \otimes \tilde{A}_j^q = \left\{\left[1 - \prod_{j=1}^{k}\left(1 - \mu_{\tilde{A}_{k+1}}^{2p} \cdot \mu_{\tilde{A}_j}^{2q}\right)\right]^{\frac{1}{2}},\right.$$

$$\left[\prod_{j=1}^{k}\left(1 - \alpha_{k+1}^p.\alpha_j^q\right)\right]^{\frac{1}{2}}, \left[\prod_{j=1}^{k}\left(1 - \mu_{\tilde{A}_{k+1}}^{2p} \cdot \mu_{\tilde{A}_j}^{2q}\right)\right.$$

$$\left.\left. - \prod_{j=1}^{k}\left(1 - \mu_{\tilde{A}_{k+1}}^{2p} \cdot \mu_{\tilde{A}_j}^{2q} - \alpha_{k+1}^p \alpha_j^q + \beta_{k+1}^p \beta_j^q\right)\right]\right\};$$

Then

$$\sum_{i \neq j=1}^{k+1} \tilde{A}_i^p \otimes \tilde{A}_j^q$$

$$= \left\{\left[1 - \prod_{i \neq j=1}^{k+1}\left(1 - \mu_{\tilde{A}_i}^{2p} \cdot \mu_{\tilde{A}_j}^{2q}\right)\right]^{\frac{1}{2}}, \left[\prod_{i \neq j=1}^{k+1}\left(1 - \alpha_i^p.\alpha_j^q\right)\right]^{\frac{1}{2}},\right.$$

$$\left.\left[\prod_{i \neq j=1}^{k+1}\left(1 - \mu_{\tilde{A}_i}^{2p} \cdot \mu_{\tilde{A}_j}^{2q}\right) - \prod_{i \neq j=1}^{k+1}\left(1 - \mu_{\tilde{A}_i}^{2p} \cdot \mu_{\tilde{A}_j}^{2q} - \alpha_i^p \alpha_j^q + \beta_i^p \beta_j^q\right)\right]\right\};$$

It means that $n = k$ formula holds for $n = k + 1$, then it holds for all n.
Step 4:

$$\left(\frac{1}{n(n-1)}\left(\sum_{i\neq j=1}^{n}\tilde{A}_i^p\otimes\tilde{A}_j^q\right)\right)$$

$$=\left\{\left[1-\left(\prod_{i\neq j=1}^{n}\left(1-\mu_{\tilde{A}_i}^{2p}\cdot\mu_{\tilde{A}_j}^{2q}\right)\right)^{\frac{1}{n(n-1)}}\right]^{\frac{1}{2}},\right.$$

$$\left(\prod_{i\neq j=1}^{n}\left(1-\alpha_i^p\alpha_j^q\right)\right)^{\frac{1}{2n(n-1)}},$$

$$\left[\left[\prod_{i\neq j=1}^{n}\left(1-\mu_{\tilde{A}_i}^{2p}\cdot\mu_{\tilde{A}_j}^{2q}\right)\right]^{\frac{1}{n(n-1)}}\right.$$

$$\left.\left.-\left[\prod_{i\neq j=1}^{k}\left(1-\mu_{\tilde{A}_i}^{2p}\cdot\mu_{\tilde{A}_j}^{2q}-\alpha_i^p\alpha_j^q+\beta_i^p\beta_j^q\right)\right]^{\frac{1}{n(n-1)}}\right]^{\frac{1}{2}}\right\}$$

$$\mu_{\frac{1}{n(n-1)}\left(\sum_{i\neq j=1}^{n}\tilde{A}_i^p\otimes\tilde{A}_j^q\right)}$$

$$=\left[1-\left(1-\left(1-\prod_{i\neq j=1}^{n}\left(1-\mu_{\tilde{A}_i}^{2p}\cdot\mu_{\tilde{A}_j}^{2q}\right)\right)\right)^{\frac{1}{n(n-1)}}\right]^{\frac{1}{2}}$$

$$=\left[1-\left(\prod_{i\neq j=1}^{n}\left(1-\mu_{\tilde{A}_i}^{2p}\cdot\mu_{\tilde{A}_j}^{2q}\right)\right)^{\frac{1}{n(n-1)}}\right]^{\frac{1}{2}};$$

$$\vartheta_{\frac{1}{n(n-1)}\left(\sum_{i\neq j=1}^{n}\tilde{A}_i^p\otimes\tilde{A}_j^q\right)}=\left(\prod_{i\neq j=1}^{n}\left(1-\alpha_i^p\alpha_j^q\right)\right)^{\frac{1}{2n(n-1)}};$$

$$\pi_{\frac{1}{n(n-1)}\left(\sum_{i\neq j=1}^{n}\tilde{A}_i^p\otimes\tilde{A}_j^q\right)}$$

$$=\left[\left(1-\left(1-\prod_{i\neq j=1}^{n}\left(1-\mu_{\tilde{A}_i}^{2p}\cdot\mu_{\tilde{A}_j}^{2q}\right)\right)\right)^{\frac{1}{n(n-1)}}\right.$$

$$-\left(1-\left(1-\prod_{i\neq j=1}^{n}\left(1-\mu_{\tilde{A}_i}^{2p}\cdot\mu_{\tilde{A}_j}^{2q}\right)\right)-\prod_{i\neq j=1}^{n}\left(1-\mu_{\tilde{A}_i}^{2p}\cdot\mu_{\tilde{A}_j}^{2q}\right)\right.$$

$$+\prod_{i\neq j=1}^{k}\left(1-\mu_{\tilde{A}_i}^{2p}\cdot\mu_{\tilde{A}_j}^{2q}-\alpha_i^p\alpha_j^q+\beta_i^p\beta_j^q\right)\Bigg)^{\frac{1}{n(n-1)}}\Bigg]^{\frac{1}{2}}$$

$$\pi_{\frac{1}{n(n-1)}\left(\sum_{i\neq j=1}^{n}\tilde{A}_i^p\otimes\tilde{A}_j^q\right)}$$

$$=\left[\left(1-\left(1-\prod_{i\neq j=1}^{n}\left(1-\mu_{\tilde{A}_i}^{2p}\cdot\mu_{\tilde{A}_j}^{2q}\right)\right)\right)^{\frac{1}{n(n-1)}}\right.$$

$$-\left(1-\left(1-\prod_{i\neq j=1}^{n}\left(1-\mu_{\tilde{A}_i}^{2p}\cdot\mu_{\tilde{A}_j}^{2q}\right)\right)-\prod_{i\neq j=1}^{n}\left(1-\mu_{\tilde{A}_i}^{2p}\cdot\mu_{\tilde{A}_j}^{2q}\right)\right.$$

$$\left.\left.+\prod_{i\neq j=1}^{k}\left(1-\mu_{\tilde{A}_i}^{2p}\cdot\mu_{\tilde{A}_j}^{2q}-\alpha_i^p\alpha_j^q+\beta_i^p\beta_j^q\right)\right)^{\frac{1}{n(n-1)}}\right]^{\frac{1}{2}}$$

$$=\left[\left[\prod_{i\neq j=1}^{n}\left(1-\mu_{\tilde{A}_i}^{2p}\cdot\mu_{\tilde{A}_j}^{2q}\right)\right]^{\frac{1}{n(n-1)}}\right.$$

$$\left.-\left[\prod_{i\neq j=1}^{k}\left(1-\mu_{\tilde{A}_i}^{2p}\cdot\mu_{\tilde{A}_j}^{2q}-\alpha_i^p\alpha_j^q+\beta_i^p\beta_j^q\right)\right]^{\frac{1}{n(n-1)}}\right]^{\frac{1}{2}}$$

Step 5:

$$\left(\frac{1}{n(n-1)}\left(\sum_{i\neq j=1}^{n}\tilde{A}_i^p\otimes\tilde{A}_j^q\right)\right)^{\frac{1}{p+q}}$$

$$=\left\{\left(1-\left(\prod_{i\neq j=1}^{n}\left(1-\mu_{\tilde{A}_i}^{2p}\cdot\mu_{\tilde{A}_j}^{2q}\right)\right)^{\frac{1}{n(n-1)}}\right)^{\frac{1}{2(p+q)}},\right.$$

$$\left[1-\left[1-\left(\prod_{i\neq j=1}^{n}\left(1-\alpha_i^p\alpha_j^q\right)\right)^{\frac{1}{n(n-1)}}\right]^{\frac{1}{p+q}}\right]^{\frac{1}{2}},$$

$$\left[\left(1-\left(\prod_{i\neq j=1}^{n}\left(1-\alpha_i^p\alpha_j^q\right)\right)^{\frac{1}{n(n-1)}}\right)^{\frac{1}{p+q}}\right.$$

$$-\left[1-\left(\prod_{i\neq j=1}^{n}\left(1-\alpha_i^p\alpha_j^q\right)\right)^{\frac{1}{n(n-1)}}\right.$$

$$-\left(\prod_{i\neq j=1}^{n}\left(1-\mu_{\tilde{A}_i}^{2p}\cdot\mu_{\tilde{A}_j}^{2q}\right)\right)^{\frac{1}{n(n-1)}}$$

$$\left.\left.\left.+\left(\prod_{i\neq j=1}^{n}\left(1-\mu_{\tilde{A}_i}^{2p}\cdot\mu_{\tilde{A}_j}^{2q}-\alpha_i^p\alpha_j^q+\beta_i^p\beta_j^q\right)\right)^{\frac{1}{n(n-1)}}\right]^{\frac{1}{p+q}}\right]^{\frac{1}{2}}\right\};$$

Thus we can prove that:

$$SFBM^{p,q}\left(\tilde{A}_1,\tilde{A}_2,\ldots,\tilde{A}_n\right)$$

$$=\left\{\left[1-\left[\prod_{i\neq j=1}^{n}\left(1-\mu_{\tilde{A}_i}^{2p}\mu_{\tilde{A}_j}^{2q}\right)\right]^{\frac{1}{n(n-1)}}\right]^{\frac{1}{2(p+q)}},\right.$$

$$\left[1-\left[1-\left[\prod_{i\neq j=1}^{n}\left(1-\alpha_i^p\alpha_j^q\right)\right]^{\frac{1}{n(n-1)}}\right]^{\frac{1}{p+q}}\right]^{\frac{1}{2}},$$

$$\left[1-\vartheta_{SFBM^{p,q}}^{2}-\left[1-\left[\prod_{i\neq j=1}^{n}\left(1-\alpha_i^p\alpha_j^q\right)\right]^{\frac{1}{n(n-1)}}-\left(1-\mu_{SFBM^{p,q}}^{2(p+q)}\right)\right.\right.$$

$$\left.\left.\left.+\left[\prod_{i\neq j=1}^{n}\left(1-\mu_{\tilde{A}_i}^{2p}\mu_{\tilde{A}_j}^{2q}-\alpha_i^p\alpha_j^q+\beta_i^p\beta_j^q\right)\right]^{\frac{1}{n(n-1)}}\right]^{\frac{1}{p+q}}\right]^{\frac{1}{2}}\right\};$$

Theorem 3.1.2 Let $p, q \geq 0$ and $\tilde{A}_i = \left(\mu_{\tilde{A}_i}, \vartheta_{\tilde{A}_i}, \pi_{\tilde{A}_i}\right)$ with $i = 1, 2, \ldots, n$ be a collection of SFSs. So Spherical Fuzzy Bonferroni mean (SFBM) has the following properties:

I. **Monotonicity**: If $\tilde{A}_i \geq \tilde{B}_i$ for all i then $SFBM^{p,q}\left(\tilde{A}_1, \tilde{A}_2, \ldots, \tilde{A}_n\right) \geq SFBM^{p,q}\left(\tilde{B}_1, \tilde{B}_2, \ldots, \tilde{B}_n\right)$.

II. **Commutativity**: $SFBM^{p,q}\left(\tilde{A}_1, \tilde{A}_2, \ldots, \tilde{A}_n\right) = SFBM^{p,q}\left(\overline{\tilde{A}_1}, \overline{\tilde{A}_2}, \ldots, \overline{\tilde{A}_n}\right)$ where $\overline{\tilde{A}_i}$ for all i = n be any permution of $\tilde{A}_i$.

III. **Boundedness**: Let $\tilde{A}_i^- = \left\{\min\left\{\mu_{\tilde{A}_i}\right\}, \max\left\{\vartheta_{\tilde{A}_i}\right\}, \max\left\{\pi_{\tilde{A}_i}\right\}\right\}$ and $\tilde{A}_i^+ = \left\{\max\left\{\mu_{\tilde{A}_i}\right\}, \min\left\{\vartheta_{\tilde{A}_i}\right\}, \min\left\{\pi_{\tilde{A}_i}\right\}\right\}$ So:

$$\tilde{A}_i^- \leq SFBM^{p,q}\left(\tilde{A}_1, \tilde{A}_2, \ldots, \tilde{A}_n\right) \leq \tilde{A}_i^+$$

Parameters p and q play a very significant role in the SFBM operator. The Specific cases can be derived By assigning some special values to the SFBM. For example if p = 1 and q = 0, then:

$$\begin{aligned}
&SFBM^{p,q}\left(\tilde{A}_1, \tilde{A}_2, \ldots, \tilde{A}_n\right) \\
&= \left\{\left(1 - \left(\prod_{i=1}^{n}\left(1 - \mu_{\tilde{A}_i}^2\right)\right)^{\frac{1}{n}}\right)^{\frac{1}{2}}, \left[1 - \left[1 - \left(\prod_{i=1}^{n}(1 - \alpha_i)\right)^{\frac{1}{n}}\right]\right]^{\frac{1}{2}}, \right. \\
&\left[\left(1 - \left(\prod_{i=1}^{n}(1 - \alpha_i)\right)^{\frac{1}{n}}\right) - \left[1 - \left(\prod_{i=1}^{n}(1 - \alpha_i)\right)^{\frac{1}{n}}\right.\right. \\
&\left.\left.\left. - \left(\prod_{i=1}^{n}\left(1 - \mu_{\tilde{A}_i}^2\right)\right)^{\frac{1}{n}} + \left(\prod_{i=1}^{n}\left(1 - \mu_{\tilde{A}_i}^2 - \alpha_i + \beta_i\right)\right)^{\frac{1}{n}}\right]^{\frac{1}{2}}\right]\right\} \\
&= \left\{\left(1 - \left(\prod_{i=1}^{n}\left(1 - \mu_{\tilde{A}_i}^2\right)\right)^{\frac{1}{n}}\right)^{\frac{1}{2}}, \left(\prod_{i=1}^{n}\left(\vartheta_{\tilde{A}_i}\right)\right)^{\frac{1}{n}}, \right. \\
&\left.\left[\left[\left(\prod_{i=1}^{n}\left(1 - \mu_{\tilde{A}_i}^2\right)\right)^{\frac{1}{n}} - \left(\prod_{i=1}^{n}\left(1 - \mu_{\tilde{A}_i}^2 - \pi_{\tilde{A}_i}^2\right)\right)^{\frac{1}{n}}\right]\right]^{\frac{1}{2}}\right\} \\
&= \frac{1}{n}\tilde{A}_1 + \frac{1}{n}\tilde{A}_2 + \cdots + \frac{1}{n}\tilde{A}_n;
\end{aligned}$$

Which is the Spherical Arithmetic mean operator or Spherical Weighted Arithmetic Mean (SWAM) when $w_i = \frac{1}{n}$.

3.2 Spherical Fuzzy Normalized Weighted Bonferroni Mean Operator (SFNWBM)

In the SFBM weights of the SFSs $\tilde{A}_i = \left(\mu_{\tilde{A}_i}, \vartheta_{\tilde{A}_i}, \pi_{\tilde{A}_i}\right)$ with $i = 1, 2, \ldots, n$ are not taken into account. In this sub-section, we suggest the Spherical Fuzzy Weighted Bonferroni Mean which can capture the weight of each SFN.

Let $p, q \geq 0$ and $\tilde{A}_i = \left(\mu_{\tilde{A}_i}, \vartheta_{\tilde{A}_i}, \pi_{\tilde{A}_i}\right)$ for $i = 1, 2, \ldots, n$ 1 be a collection of SFSs with weight vector $w_i = (w_1, w_2, \ldots, w_n)^T$ where $w_i \in [0, 1]$ and $\sum_{i=1}^{n} w_i = 1$ If:

$$SFWBM^{p,q}\left(\tilde{A}_1, \tilde{A}_2, \ldots, \tilde{A}_n\right) = \left(\frac{1}{n(n-1)}\left(\oplus_{i\neq j=1}^{n}\left(w_i\tilde{A}_i\right)^p \otimes w_j\tilde{A}_j^q\right)\right)^{\frac{1}{p+q}} \quad (15)$$

Then $SFWBM^{p,q}$ is called the Spherical fuzzy Weighted Bonferroni mean with parameter p and q.

By the operations of SFNs and definition of Weighted Bonferroni mean, we can obtain the following result.

Theorem 3.2.1 Let $p, q \geq 0$ and $\tilde{A}_i = \left(\mu_{\tilde{A}_i}, \vartheta_{\tilde{A}_i}, \pi_{\tilde{A}_i}\right)$ with $i = 1, 2, \ldots, n$ be a collection of SFSs and weight vector $w_i = (w_1, w_2, \ldots, w_n)^T$ which must satisfy $w_i \in [0, 1]$ and $\sum_{i=1}^{n} w_i = 1$. Then the aggregated value by the $SFWBM$ operator is also be a spherical fuzzy set with the below formulation:

$$\begin{aligned} & SFWBM^{p,q}\left(\tilde{A}_1, \tilde{A}_2, \ldots, \tilde{A}_n\right) \\ & = \left\{\left[1 - \left(\prod_{i\neq j=1}^{n}\left(1 - (1-\gamma_i^{w_i})^p\left(1-\gamma_j^{w_j}\right)^q\right)\right)^{\frac{1}{n(n-1)}}\right]^{\frac{1}{2(p+q)}},\right. \\ & \left[1 - \left[1 - \left(\prod_{i\neq j=1}^{n}\left(1 - \alpha_i^p\alpha_j^q\right)\right)^{\frac{1}{n(n-1)}}\right]^{\frac{1}{p+q}}\right]^{\frac{1}{2}}, \\ & \left[\left(1 - \vartheta_{SFWBM}^2\right) - \left[1 - \left(\prod_{i\neq j=1}^{n}\left(1 - \alpha_i^p\alpha_j^q\right)\right)^{\frac{1}{n(n-1)}} - \left(1 - \mu_{SFWBM}^{2(p+q)}\right)\right.\right. \\ & \left.\left.\left. + \left(\prod_{i\neq j=1}^{n}\left(1 - (1-\gamma_i^{w_i})^p\left(1-\gamma_j^{w_j}\right)^q - \alpha_i^p\alpha_j^q + \beta_i^p\beta_j^q\right)\right)\right]^{\frac{1}{2}}\right]\right\}; \end{aligned} \quad (16)$$

where $1-\vartheta_{\tilde{A}_i}^2=\alpha_i$, $1-\vartheta_{\tilde{A}_j}^2=\alpha_j$, $1-\vartheta_{\tilde{A}_i}^2-\pi_{\tilde{A}_i}^2=\beta_i$, $1-\vartheta_{\tilde{A}_j}^2-\pi_{\tilde{A}_j}^2=\beta_j$, $1-\mu_{\tilde{A}_i}^2=\gamma_i$ and $1-\mu_{\tilde{A}_j}^2=\gamma_j$.

The proof of Theorem 3.2.1 is omitted because of similarity of the proof's steps to the proof of Theorem 3.1.1 and To avoid duplication.

The classical SFBM ignores the weight vector of aggregated arguments, the SFWBM does not have reducibility, boundedness and cannot reflect the interrelationship between the individual criterion and other criteria. To overcome with these issues, in this subsection, we propose the spherical fuzzy normalized weighted bonferroni mean aggregation operator.

Let $p,q\geq 0$ and $\tilde{A}_i=\left(\mu_{\tilde{A}_i},\vartheta_{\tilde{A}_i},\pi_{\tilde{A}_i}\right)$ for $i=1,2,\ldots,n$ 1 be a collection of SFSs with weight vector $w_i=(w_1,w_2,\ldots,w_n)^T$ where $w_i\in[0,1]$ and $\sum_{i=1}^n w_i=1$ If:

$$SFNWBM^{p,q}\left(\tilde{A}_1,\tilde{A}_2,\ldots,\tilde{A}_n\right)=\left(\oplus_{i\neq j=1}^n\left(\frac{w_iw_j}{1-w_i}\tilde{A}_i^p\otimes\tilde{A}_j^q\right)\right)^{\frac{1}{p+q}}\tag{17}$$

Then $SFNWBM^{p,q}$ is called the Spherical fuzzy Normalized Weighted Bonferroni mean with parameter p and q.

Theorem 3.2.2 Let $p,q\geq 0$ and $\tilde{A}_i=\left(\mu_{\tilde{A}_i},\vartheta_{\tilde{A}_i},\pi_{\tilde{A}_i}\right)$ with $i=1,2,\ldots,n$ be a collection of SFSs and weight vector $w_i=(w_1,w_2,\ldots,w_n)^T$ which must satisfy $w_i\in[0,1]$ and $\sum_{i=1}^n w_i=1$. Then the aggregated value by the $SFNWBM$ operator is also a spherical fuzzy set with the below formulation

$$\begin{aligned}
&SFNWBM^{p,q}\left(\tilde{A}_1,\tilde{A}_2,\ldots,\tilde{A}_n\right)\\
&=\left[1-\prod_{i\neq j=1}^n\left(1-\mu_{\tilde{A}_i}^{2p}\mu_{\tilde{A}_j}^{2q}\right)^{\frac{w_iw_j}{1-w_i}}\right]^{\frac{1}{2(p+q)}},\\
&\left[1-\left(1-\prod_{i\neq j=1}^n\left(1-\alpha_i^p\alpha_j^q\right)^{\frac{w_iw_j}{1-w_i}}\right)^{\frac{1}{p+q}}\right]^{\frac{1}{2}},\\
&\left[\left(1-\vartheta_{SFNWBM}^2\right)-\left(1-\prod_{i\neq j=1}^n\left(1-\alpha_i^p\alpha_j^q\right)^{\frac{w_iw_j}{1-w_i}}-\left(1-\mu_{SFNWBM}^{2(p+q)}\right)\right.\right.\\
&\left.\left.+\prod_{i\neq j=1}^n\left(1-\mu_{\tilde{A}_i}^{2p}\mu_{\tilde{A}_j}^{2q}-\alpha_i^p\alpha_j^q+\beta_i^p\beta_j^q\right)^{\frac{w_iw_j}{1-w_i}}\right)^{\frac{1}{p+q}}\right]^{\frac{1}{2}}
\end{aligned}\tag{18}$$

where $1-\vartheta_{\tilde{A}_i}^2=\alpha_i, 1-\vartheta_{\tilde{A}_j}^2=\alpha_j, 1-\vartheta_{\tilde{A}_i}^2-\pi_{\tilde{A}_i}^2=\beta_i$ and $1-\vartheta_{\tilde{A}_j}^2-\pi_{\tilde{A}_j}^2=\beta_j$.

Proof To prove the Theorem 3.2.2, we apply the operations of Spherical Fuzzy Sets to Eq. (17) the same as follow:

Step 1: From second step of proof 3.2.2, we have:

$$\tilde{A}_i^p \otimes \tilde{A}_j^q=\left\{\mu_{\tilde{A}_i}^p \mu_{\tilde{A}_j}^q,\left[1-\alpha_i^p.\alpha_j^q\right]^{\frac{1}{2}},\left[\alpha_i^p \alpha_j^q-\beta_i^p \beta_j^q\right]^{\frac{1}{2}}\right\};$$

where $1-\vartheta_{\tilde{A}_i}^2=\alpha_i, 1-\vartheta_{\tilde{A}_j}^2=\alpha_j, 1-\vartheta_{\tilde{A}_i}^2-\pi_{\tilde{A}_i}^2=\beta_i$ and $1-\vartheta_{\tilde{A}_j}^2-\pi_{\tilde{A}_j}^2=\beta_j$.
Step 2: With apply the Eq. (5) in Definition 2.2, we can obtain:

$$\begin{aligned}
&\frac{w_i w_j}{1-w_i} \tilde{A}_i^p \otimes \tilde{A}_j^q \\
&=\left\{\left[1-\left(1-\mu_{\tilde{A}_i}^{2p} \mu_{\tilde{A}_j}^{2q}\right)^{\frac{w_i w_j}{1-w_i}}\right]^{\frac{1}{2}},\left[1-\alpha_i^p \alpha_j^q\right]^{\frac{w_i w_j}{2(1-w_i)}},\right. \\
&\left.\left[\left(1-\mu_{\tilde{A}_i}^{2p} \mu_{\tilde{A}_j}^{2q}\right)^{\frac{w_i w_j}{1-w_i}}-\left(1-\mu_{\tilde{A}_i}^{2p} \mu_{\tilde{A}_j}^{2q}-\alpha_i^p \alpha_j^q+\beta_i^p \beta_j^q\right)^{\frac{w_i w_j}{1-w_i}}\right]^{\frac{1}{2}}\right\};
\end{aligned}$$

Step 3: With apply the Eq. (3) in Definition 2.2 and mathematical induction on n the same as proof of Theorem 3.1.1 we can obtain:

$$\begin{aligned}
&\oplus_{i \neq j=1}^n\left(\frac{w_i w_j}{1-w_i} \tilde{A}_i^p \otimes \tilde{A}_j^q\right) \\
&=\left\{\left[1-\prod_{i \neq j=1}^n\left(\left(1-\mu_{\tilde{A}_i}^{2p} \mu_{\tilde{A}_j}^{2q}\right)^{\frac{w_i w_j}{1-w_i}}\right)\right]^{\frac{1}{2}},\right. \\
&\prod_{i \neq j=1}^n\left(\left[1-\alpha_i^p \alpha_j^q\right]^{\frac{w_i w_j}{2(1-w_i)}}\right), \\
&\left[\prod_{i \neq j=1}^n\left(\left(1-\mu_{\tilde{A}_i}^{2p} \mu_{\tilde{A}_j}^{2q}\right)^{\frac{w_i w_j}{1-w_i}}\right)\right. \\
&\left.\left.-\prod_{i \neq j=1}^n\left(\left(1-\mu_{\tilde{A}_i}^{2p} \mu_{\tilde{A}_j}^{2q}-\alpha_i^p \alpha_j^q+\beta_i^p \beta_j^q\right)^{\frac{w_i w_j}{1-w_i}}\right)\right]^{\frac{1}{2}}\right\};
\end{aligned}$$

Step 4: Finally, with applying the Eq. (6) in Definition 2.2, we have:

$$SFNWBM^{p,q}\left(\tilde{A}_1, \tilde{A}_2, \ldots, \tilde{A}_n\right)$$

$$= \left[1 - \prod_{i \neq j = 1}^{n} \left(1 - \mu_{\tilde{A}_i}^{2p} \mu_{\tilde{A}_j}^{2q} \right)^{\frac{w_i w_j}{1 - w_i}} \right]^{\frac{1}{2(p+q)}},$$

$$\left[1 - \left(1 - \prod_{i \neq j = 1}^{n} \left(1 - \alpha_i^p \alpha_j^q \right)^{\frac{w_i w_j}{1 - w_i}} \right)^{\frac{1}{p+q}} \right]^{\frac{1}{2}},$$

$$\left[\left(1 - \vartheta_{SFNWBM}^2 \right) - \left(1 - \prod_{i \neq j = 1}^{n} \left(1 - \alpha_i^p \alpha_j^q \right)^{\frac{w_i w_j}{1 - w_i}} - \left(1 - \mu_{SFNWBM}^{2(p+q)} \right) \right. \right.$$

$$\left. \left. + \prod_{i \neq j = 1}^{n} \left(1 - \mu_{\tilde{A}_i}^{2p} \mu_{\tilde{A}_j}^{2q} - \alpha_i^p \alpha_j^q + \beta_i^p \beta_j^q \right)^{\frac{w_i w_j}{1 - w_i}} \right)^{\frac{1}{p+q}} \right]^{\frac{1}{2}} \Bigg\};$$

Theorem 3.2.3 Let $p, q \geq 0$ and $\tilde{A}_i = \left(\mu_{\tilde{A}_i}, \vartheta_{\tilde{A}_i}, \pi_{\tilde{A}_i} \right)$ with $i = 1, 2, \ldots, n$ be a collection of SFSs with weight vector $w_i = (w_1, w_2, \ldots, w_n)^T$ Then $SFNWBM$ aggregation operator has the following properties:

I. Reducibility: Let $p, q \geq 0$ and $\tilde{A}_i = \left(\mu_{\tilde{A}_i}, \vartheta_{\tilde{A}_i}, \pi_{\tilde{A}_i} \right)$ with $i = 1, 2, \ldots, n$ be a collection of SFSs with weight vector $w_i = \left(\frac{1}{n}, \frac{1}{n}, \ldots, \frac{1}{n} \right)^T$ Then:

$$SFNWBM^{p,q} \left(\tilde{A}_1, \tilde{A}_2, \ldots, \tilde{A}_n \right)$$
$$= \left(\oplus_{i \neq j = 1}^{n} \left(\frac{w_i w_j}{1 - w_i} \tilde{A}_i^p \otimes \tilde{A}_j^q \right) \right)^{\frac{1}{p+q}}$$
$$= \left(\oplus_{i \neq j = 1}^{n} \left(\frac{1}{n(n-1)} \tilde{A}_i^p \otimes \tilde{A}_j^q \right) \right)^{\frac{1}{p+q}}$$
$$= SFBM^{p,q} \left(\tilde{A}_1, \tilde{A}_2, \ldots, \tilde{A}_n \right);$$

II. Idempotency:

$$SFNWBM^{p,q} \left(\tilde{A}, \tilde{A}, \ldots, \tilde{A} \right)$$
$$= \left(\oplus_{i \neq j = 1}^{n} \left(\frac{w_i w_j}{1 - w_i} \tilde{A}_i^p \otimes \tilde{A}_j^q \right) \right)^{\frac{1}{p+q}}$$
$$= \left(\tilde{A}^{p+q} \oplus_{i \neq j = 1}^{n} \left(\frac{w_i w_j}{1 - w_i} \right) \right)^{\frac{1}{p+q}} = \tilde{A};$$

III. Monotonicity: If $\tilde{A}_i \geq \tilde{B}_i$ and $p, q \geq 0$, then:
$\tilde{A}_i^p \otimes \tilde{A}_j^q \geq \tilde{B}_i^p \otimes \tilde{B}_j^q$ and

$\left(\oplus_{i\neq j=1}^{n}\left(\frac{w_i w_j}{1-w_i}\tilde{A}_i^p \otimes \tilde{A}_j^q\right)\right)^{\frac{1}{p+q}} \geq \left(\oplus_{i\neq j=1}^{n}\left(\frac{w_i w_j}{1-w_i}\tilde{B}_i^p \otimes \tilde{B}_j^q\right)\right)^{\frac{1}{p+q}}$ which is proved that:

$$SFNWBM^{p,q}\left(\tilde{A}_1, \tilde{A}_2, \ldots, \tilde{A}_n\right) \geq SFNWBM^{p,q}\left(\tilde{B}_1, \tilde{B}_2, \ldots, \tilde{B}_n\right);$$

IV. Boundedness: By idempotency property, we can say that:

$$SFNWBM^{p,q}\left(\min\left\{\tilde{A}_1\right\}, \min\left\{\tilde{A}_2\right\}, \ldots, \min\left\{\tilde{A}_n\right\}\right) = \min\left\{\tilde{A}_i\right\}$$

$$SFNWBM^{p,q}\left(\max\left\{\tilde{A}_1\right\}, \max\left\{\tilde{A}_2\right\}, \ldots, \max\left\{\tilde{A}_n\right\}\right) = \max\left\{\tilde{A}_i\right\}$$

And since $\min\left\{\tilde{A}_i\right\} \leq \tilde{A}_i \leq \max\left\{\tilde{A}_i\right\}$, so:

$$\min\left\{\tilde{A}_i\right\} \leq SFNWBM^{p,q}\left(\tilde{A}_1, \tilde{A}_2, \ldots, \tilde{A}_n\right) \leq \max\left\{\tilde{A}_i\right\};$$

4 Descriptive Analyses

A group of decision-makers try to choice the best alternative among all possible alternatives which are evaluated on multiple attributes in MAGDM methods. Assume that m decision makers $D = \{D_1, D_2, \cdots, D_m\}$ with related weight vector $w = \{w_1, w_2, \ldots, w_n\}$, $\sum_{i=1}^{n} w_i = 1$, $w_i \geq 0$ are going to select the best alternative among n alternatives $A = \{A_1, A_2, \ldots, A_n\}$ based on p criteria $C\left\{C_1, C_2, \ldots, C_p\right\}$ with corresponding weight vector $\omega_i = \left(\omega_1, \omega_2, \ldots, \omega_p\right)$, $\sum_{i=1}^{p} \omega_i = 1, \omega_i \geq 0$. The steps of MAGDM method based on our proposed aggregation operator are defined as follows:

Step 1. Prepare Alternative-Criteria matrix for each decision-maker and determine the weight vectors for decision makers and criteria.

Step 2. Normalize the established decision matrices with considering the positive or negative form of criteria.

Step 3. Aggregate the established decision matrices including spherical fuzzy numbers with respect to corresponding weight vectors by using proposed SFNWBM aggregation operator using Eq (18).

Step 4. Aggregate the obtained decision matrix from Step 3 by considering the related weight vector of each criterion using proposed SFNWBM in Eq (18).

Step 5. Calculate the comparison functions given in Definition (2.5) for the obtained decision matrix from Step 4.

Step 6. Rank the alternatives based on the comparison rules given in Definition (2.5).

Step 7. Select the best alternative.

4.1 Illustrative Example

In this section, we present an example using the proposed MAGDM method to show the effectiveness of new aggregation operator. We choose an example which describes the evaluation of air quality. In this example there are three stations which analyze the air quality by (α, β, γ) as Decision-Makers with corresponding weight vector $w_m = \{0.314_1, 0.355_2, 0.331_3\}$ and consider $SO_2, NO_2\, and\, PM_{10}$ as criteria for each alternative and their weight vectors are $w_p = \{0.4_1, 0.2_2, 0.4_3\}$. The values are shown in Tables 1, 2 and 3.

Because all criteria are positive, there is no need to normalize decision matrices. Apply the proposed aggregation operator to aggregate three stations into one matrix with parameter $p = 2$ and $q = 1$, as given in Table 4.

In the following, SFNWBM is used again to aggregate the above-mentioned matrix (Table 4) with parameter $p = 2$ and $q = 1$. The results are shown in Table 5.

Table 1 Air quality data derived from station α

	SO_2			NO_2			PM_{10}		
A_1	0.5	0.8	0.3	0.8	0.4	0.3	0.7	0.5	0.4
A_2	0.7	0.6	0.2	0.3	0.9	0.1	0.5	0.3	0.7
A_3	0.4	0.8	0.4	0.5	0.8	0.2	0.7	0.3	0.2
A_4	0.8	0.3	0.5	0.6	0.6	0.3	0.3	0.6	0.5

Table 2 Air quality data derived from station β

	SO_2			NO_2			PM_{10}		
A_1	0.7	0.5	0.1	0.6	0.3	0.4	0.3	0.8	0.5
A_2	0.8	0.4	0.4	0.5	0.7	0.1	0.7	0.2	0.4
A_3	0.3	0.9	0.2	0.7	0.1	0.4	0.4	0.6	0.3
A_4	0.8	0.4	0.3	0.4	0.6	0.5	0.5	0.1	0.8

Table 3 Air quality data derived from station γ

	SO_2			NO_2			PM_{10}		
A_1	0.8	0.2	0.4	0.4	0.4	0.3	0.6	0.4	0.5
A_2	0.7	0.5	0.2	0.6	0.5	0.4	0.8	0.5	0.2
A_3	0.5	0.4	0.6	0.9	0.3	0.1	0.3	0.1	0.9
A_4	0.7	0.3	0.5	0.8	0.5	0.2	0.2	0.8	0.4

The score and accuracy values of each alternative are calculated by using Eqs (9) and (10), respectively. The results are presented in Table 6.

Finally, based on comparison rules which are defined in Definition 2.5, ranking of the alternatives are as follow:

$$A_2 > A_1 > A_4 > A_3$$

So the second alternative (A_2) is selected (Table 6).

It is noted that the parameters p and q play a very important role in the final results. With the change of the parameters, the final results may change. Effects of the parameters on the result are the same as in Table 7. Results show the direct relationship between parameter changes and the scores of overall values. Parameters can be described as the degrees of decision makers' pessimism. It means that if decision makers are pessimistic, then higher values can be assigned to the parameters and vice versa. Results of Table 7 present the independence of the ranking results, which means that the results are objective and cannot be influenced by decision makes' preference of pessimism and optimism. Thus, this fact shows the reliability of ranking results.

5 Conclusion

To aggregate the spherical fuzzy information, some aggregation operators have been developed and investigated in the literature but, the Bonferroni mean aggregation operators which reflect the interrelationship of the aggregated arguments have not yet been investigated in the spherical environment. The desirable characteristic of the BM is its capability to capture the interrelationship between the input arguments.

Table 4 Aggregated air quality data derived from stations α, β and γ

	SO_2			NO_2			PM_{10}		
A_1	0.68	0.52	0.28	0.62	0.36	0.34	0.55	0.58	0.45
A_2	0.73	0.5	0.29	0.48	0.7	0.24	0.68	0.34	0.47
A_3	0.4	0.73	0.42	0.72	0.4	0.25	0.48	0.36	0.6
A_4	0.77	0.33	0.44	0.61	0.56	0.35	0.35	0.55	0.6

Table 5 Aggregated air quality data based on criteria

	$\mu_{\tilde{e}_i}$	$\vartheta_{\tilde{e}_i}$	$\pi_{\tilde{e}_i}$
A_1	0.62	0.49	0.35
A_2	0.67	0.48	0.36
A_3	0.52	0.52	0.45
A_4	0.59	0.47	0.48

Table 6 Score and accuracy values of each alternative

	Score	Accuracy
A_1	0.13	0.76
A_2	0.18	0.82
A_3	0.004	0.75
A_4	0.12	0.81

The main contribution of this paper is to propose novel aggregation operators such as SFBM and SFNWBM which are developed for spherical fuzzy information. All three operators have their own specifications. The SFBM ignores the weight vector of aggregated arguments and the SFWBM does not have reducibility, and cannot reflect the interrelationship between the criteria but SFNWBM aggregation operator considers the weight vector and reflects the interrelationship between the criteria. Finally, for the situation that the criteria have connections in spherical fuzzy multi-criteria group decision making, an approach is proposed on the basis of the SFNWBM aggregation operator. In order to illustrate the validity of the developed aggregation operator, a numerical example is employed. Also a sensitivity analysis is performed to investigate the amount of change in score values of alternatives considering different p and q parameters. The sensitivity analysis results present that either decision-maker is optimistic or pessimistic; the aggregation operator's results are objective which means that the results are not affected by decision makers' preference of pessimism and optimism. As the future study, the novel aggregation operators will be extend to interval spherical fuzzy sets, triangular spherical fuzzy sets and other type of spherical fuzzy sets.

Table 7 Score value of each alternative in different p and q

	p = q = 1	p = q = 2	p = q = 4	p = q = 6	p = q = 8
A_1	0.097	0.128	0.180	0.213	0.234
A_2	0.158	0.181	0.221	0.247	0.266
A_3	−0.054	−0.005	0.078	0.138	0.182
A_4	0.074	0.112	0.175	0.213	0.237
Ranking	$A_2 > A_1 > A_4 > A_3$	$A_2 > A_1 > A_4 > A_3$	$A_2 > A_1 > A_4 > A_3$	$A_2 > A_1 > A_4 > A_3$	$A_2 > A_1 > A_4 > A_3$

References

Ashraf S, Abdullah S, Mahmood T (2019) Spherical fuzzy Dombi aggregation operators and their application in group decision making problems. J Ambient Intell HumIzed Comput. https://doi.org/10.1007/s12652-019-01333-y
Atanassov KT (1986) Intuitionistic fuzzy sets. Fuzzy Sets Syst 20(1):87–96. https://doi.org/10.1016/S0165-0114(86)80034-3
Bonferroni C (1950) Sulle medie multiple di potenze. Bolletino Matematica Italiana 5:267–270
Cuong BC (2014) Picture fuzzy sets. J Comput Sci Cyberne Ics 30(4):409–420. https://doi.org/10.15625/1813-9663/30/4/5032
Donyatalab Y, Farrokhizadeh E, Garmroodi SD, Shishavan SA (2019) Harmonic mean aggregation operators in spherical fuzzy environment and their group decision making applications. Mult Valued Log Soft Comput 33(6):565–592
Gündoğdu FK, Kahraman C (2019a) A novel spherical fuzzy analytic hierarchy process and its renewable energy application. Soft Comput. https://doi.org/10.1007/s00500-019-04222-w
Gündoğdu FK, Kahraman C (2019b) Spherical fuzzy sets and spherical fuzzy TOPSIS method. J Intell Fuzzy Syst 36(1):337–352. https://doi.org/10.3233/JIFS-181401
Gündoğdu FK, Kahraman C (2019c) Extension of WASPAS with spherical fuzzy sets. Informatica 30(2):269–292. https://doi.org/10.15388/Informatica.2019.206
Gündoğdu FK, Kahraman C (2019d) A novel VIKOR method using spherical fuzzy sets and its application to warehouse site selection. J Intell Fuzzy Syst
Jin H, Ashraf S, Abdullah S, Qiyas M, Bano M, Zeng S (2019) Linguistic spherical fuzzy aggregation operators and their applications in multi-attribute decision making problems. Mathematics 7(5). https://doi.org/10.3390/math7050413
Jing N, Xian S, Xiao Y (2017) Pythagorean triangular fuzzy linguistic bonferroni mean operators and their application for multi-attribute decision making. https://doi.org/10.1109/ciapp.2017.8167255
Liu P, Zhu B, Wang P (2019) A multi-attribute decision-making approach based on spherical fuzzy sets for Yunnan Baiyao's R&D project selection problem. Int J Fuzzy Syst 21(7):2168–2191. https://doi.org/10.1007/s40815-019-00687-x
Smarandache F (1999) A unifying field in logics neutrosophy: neutrosophic probability, set and logic. American Research Press, pp 1–141. https://doi.org/10.6084/m9.figshare.1014204
Torra V (2010) Hesitant fuzzy sets. Int J Intell Syst 25(6):529–539
Wang L, Li N (2019) Pythagorean fuzzy interaction power Bonferroni mean aggregation operators in multiple attribute decision making. Int J Intell Syst. https://doi.org/10.1002/int.22204
Xia MM, Xu ZS, Zhu B (2013) Geometric Bonferroni means with their application in multi-criteria decision making. Knowl Based Syst 40:88–100
Xing Y, Zhang R, Wang J, Bai K, Xue J (2019) A new multi-criteria group decision-making approach based on q-rung orthopair fuzzy interaction Hamy mean operators. Neural Comput Appl. https://doi.org/10.1007/s00521-019-04269-8
Xu ZS, Yager RR (2011) Intuitionistic fuzzy Bonferroni mean. IEEE Trans Syst Man Cybern Part B 41(2):568–578
Yager RR (2009) On generalized Bonferroni mean operators for multi-criteria aggregation. Int J Approx Reason 50(8):1279–1286
Yager RR (2013) Pythagorean fuzzy subsets. In: 2013 joint IFSA world congress and NAFIPS annual meeting (IFSA/NAFIPS). IEEE, pp 57–61
Zadeh LA (1965) Fuzzy sets. Inf Control 8(338–353):1965
Zadeh LA (1975) The concept of a linguistic variable and its application to approximate reasoning. Inf Sci 8(3):199–249. https://doi.org/10.1016/0020-0255(75)90036-5
Zhou W, He JM (2012) Intuitionistic fuzzy normalized weighted Bonferroni mean and Its application in multicriteria decision making. J Appl Math. https://doi.org/10.1155/2012/136254
Zhu B, Xia MM, Xu ZS (2012) Hesitant fuzzy geometric Bonferroni mean. Inf Sci 205(1):72–85

Spherical Fuzzy Soft Sets

P. A. Fathima Perveen, Sunil Jacob John, and K. V. Babitha

Abstract In this chapter, the definition of spherical fuzzy soft sets (SFSSs) and a few of its properties are introduced. SFSSs are presented as a generalization of soft sets. Notably, we tend to showed DeMorgan's laws that are valid is SFSS theory. Also, we advocate an algorithm to solve the decision-making problem primarily based on adjustable soft discernibility matrix. It offers an order relation among all the items of our universe. Finally, an illustrative example is mentioned to show that they can be successfully used to solve problems with uncertainties.

1 Introduction

The real world is complete of uncertainty, imprecision and vagueness. Traditionally, the information extracted from various resources is crisp, settled and precise, that is a smaller amount adaptable to use in real-life problems because of uncertainties arises all over. To influence such uncertainties, many theories like fuzzy set (Zadeh 1965), rough set (Pawlak 1982), vague set (Gau and Buehrer 1993), etc., came into existence. However these theories have their obstacles and inherent difficulties. For example, fuzzy sets developed by Zadeh (1965) was at first used for several applications. However the non-membership value of an element in the fuzzy sets could not be derived from its membership value. So to triumph over some drawbacks of fuzzy sets, Atanassov (1986) evolved intuitionistic fuzzy sets(IFSs) with the aid of introducing indices, membership degree ($\mu(x)$) and non-membership degree ($\vartheta(x)$) such that $0 \leq \mu(x) + \vartheta(x) \leq 1$. Afterward, the interval-valued intuitionistic fuzzy sets were

P. A. Fathima Perveen (✉) · S. J. John
Department of Mathematics, National Institute of Technology Calicut, Calicut 673 601, India
e-mail: perveenpa@gmail.com

S. J. John
e-mail: sunil@nitc.ac.in

K. V. Babitha
Department of Mathematics, Government College, Kasargod 671 123, India
e-mail: babikvkurup@yahoo.co.in

© Springer Nature Switzerland AG 2021

C. Kahraman and F. Kutlu Gündoğdu (eds.), *Decision Making with Spherical Fuzzy Sets*, Studies in Fuzziness and Soft Computing 392,
https://doi.org/10.1007/978-3-030-45461-6_6

developed by Atanassov and Gargov (1989). It is the extension of fuzzy sets and IFSs. But in some real-life situations, we have to face the situations that the sum of membership grade and non-membership grade exceed 1. To overcome this situation Yager (2013), developed the Pythagorean fuzzy set such that $0 \leq \mu^2(x) + \vartheta^2(x) \leq 1$. It is a generalization of IFSs. But these theories can't display the neutral state (which neither favor nor disfavor). Based on these circumstances, to conquer the situation, the concept of picture fuzzy sets was developed by Cuong and Kreinovich (2013) with the condition $0 \leq \mu(x) + \eta(x) + \vartheta(x) \leq 1$. Surely, it is greater beneficial and it has more applications than IFSs and Pythagorean fuzzy sets. These sets are extensively applied in different fields (Athira et al. 2019a, b; Garg 2017, 2019; Xu 2007). But there are a few sorts of issues which can't be dealt with by these IFSs, Pythagorean fuzzy sets or picture fuzzy sets. That is, in some cases there may be situations that the sum of positive membership degree, neutral membership degree, and negative membership degree exceeds 1. To overcome this situation Ashraf et al. (2019), proposed the idea of spherical fuzzy sets with the condition $0 \leq \mu^2(x) + \eta^2(x) + \vartheta^2(x) \leq 1$. The idea of spherical sets gives us a other way of representation to a few of the problems which see very troublesome to clarify utilizing existing other expansions of fuzzy set theory, such as problems involving human suppositions, including distinctive responses of the sort yes, abstain, no and refusal. At the same time, another version of the spherical fuzzy set was delivered with the aid of Gündogdu and Kahraman (2019). Instead of neutral membership degree, they utilized hesitancy degree $\pi(x)$ as one of the three indices. The other two indices are positive membership degree ($\mu(x)$) and negative membership degree ($\vartheta(x)$). And these three indices should satisfy the condition $0 \leq \mu^2(x) + \pi^2(x) + \vartheta^2(x) \leq 1$. Afterward, an extension of WASPAS with spherical fuzzy sets, VIKOR method using spherical fuzzy sets and correlation coefficients were presented by the researchers in (Gündogdu and Kahraman 2019, ?; Gündogdu et al. 2018; Ullah et al. 2019).

Apart from these sets Molodtsov (1999), proposed soft set theory as a general mathematical tool for dealing with uncertainty. Concretely, a soft set is a parameterized family of subsets of the universe. But Maji et al. (2001a) combined the ideas of fuzzy sets and soft set and introduced the concept of fuzzy soft sets (FSSs). Afterward Maji et al. (2001b), proposed the concept of intuitionistic fuzzy soft sets (IFSSs). Again Yang et al. (2009), combined the concept of soft set with interval-valued fuzzy sets. Since its appearance, several researchers have presented different styles of methods and algorithms for fixing the decision-making problems underneath the soft set environment. For illustration Feng and Zhou (2014), introduced the soft discernibility matrix. Çagman and Enginoglu (2010) constructed a soft max-min decision-making method which can be successfully applied to the problems that contain uncertainiries. Arora and Garg (2018) proposed an algorithm for solving the decision-making problems based on the aggregation operators under the IFSSs. However, a few extension models of soft sets are quickly developed such as soft rough sets (Feng et al. 2010), generalized FSS (Majumdar and Samanta 2010), group generalized FSS (Garg and Arora 2018), Pythagorean fuzzy soft sets (Peng et al. 2015), picture FSS (Yang et al. 2015), interval-valued picture FSS (Khalil et al. 2019).

Decision-making problems are a big part of human society and applied widely to practical fields like economics, management, engineering. However, with the development of science and technology, the uncertainty also plays a dominant role at some point of the decision-making analysis. As an application of soft sets Maji et al. (2002), introduced the decision-making problem with the help of rough mathematics by Pawlak (1982). Roy and Maji (2007) presented some results on an application of FSSs in decision-making problem. They introduced a novel method of object recognition from an imprecise multi observer data. This model involves construction of a comparison table from a FSS in a parametric sense for decision-making. Later, a modified form of the Roy and Maji (2007) algorithm is presented by Kong et al. (2009). Also Jiang et al. (2011), presented an adjustable approach of IFSSs based on decision-making. Peng and Garg (2018) proposed three algorithms to solve interval-valued fuzzy soft decision-making problem by weighted distance based approximation, combinative distance based assessment and similarity measure. Later, based on adjustable soft discernibility matrix Yang et al. (2015), presented a decision-making approach on picture FSSs. Subsequently, a new decision-making algorithm using matrix representation of the inverse soft set is presented by Kamaci et al. (2018). Kamaci (2018) also introduced a bijective soft decision system based on the bijective soft matrix theory. Recently Riaz et al. (2019), presented multi-criteria group decision-making methods by using N-soft set and N-soft topology to deal with uncertainties in the real world problems.

In this chapter, we combined spherical fuzzy sets and soft sets, which results a new soft set model called spherical fuzzy soft set. This new soft set model is more realistic, sensible and accurate than the present alternative soft set models. Through this chapter, the goal is to solve the decision-making problem using adjustable soft discernibility matrix. The remainder of this chapter is organized as follows. In Sect. 2, the definitions of soft sets, discernibility matrix, soft discernibility matrix, fuzzy soft sets, Pythagorean fuzzy soft sets, and spherical fuzzy sets are recalled. In Sect. 3, the definition of spherical fuzzy soft set is proposed and some of its properties included. In Sect. 4, as an application of spherical fuzzy soft set, a decision-making problem is evaluated and an adjustable algorithm is presented. The definitions and results discussed in the Sects. 4 and 5 are taken from (Perveen et al. 2019). The chapter is concluded in Sect. 5.

2 Background

Let U be a universe of objects and E be a set of parameters that are describe the elements of U. Also, $P(U)$ be the power set of U and $A \subseteq E$.

Definition 1 A (Molodtsov 1999) pair $\langle F, A\rangle$ is called a soft set over U, where F is a mapping given by $F : A \to P(U)$. In other words, a soft set over U is a parametrized family of subsets of the universe U. For $e \in A$, $F(e)$ may be considered as the set of e-approximate elements of the soft set $\langle F, A\rangle$.

For two soft sets $\langle F, A\rangle$ and $\langle G, B\rangle$ over U, $\langle F, A\rangle$ is called a soft subset of $\langle G, B\rangle$ if,

1. $A \subseteq B$ and
2. $\forall e \in A, F(e) \subseteq G(e)$.

Definition 2 If (Ali 2011) $F : A \to P(U \times U)$ be a mapping from a subset of parameter A to the power set of $U \times U$, then the soft set $\langle F, A\rangle$ over $U \times U$ is called a soft binary relation over U.

Definition 3 A (Ali 2012) soft binary relation $\langle F, A\rangle$ over a set U is called a soft equivalence relation over U, if $F(e) \neq \phi$ is an equivalence relation on U for all $e \in A$.

Definition 4 Suppose (Ali 2012) $\langle F, A\rangle$ is a soft equivalence relation over U, for each equivalence relation $F(e)$, the notion of equivalence class over it is defined as $[x]_{F(e)} = \{y : (x, y) \in F(e), y \in U\}$.

We also observe that there is an indiscernibility relation induced by the soft set $\langle F, A\rangle$ itself. This indiscernibility relation is obtained by intersection of all equivalence relations defined by parameters. Let us say $IND\langle F, A\rangle = \bigcap_{e_i \in A} F(e_i)$.

Suppose $\langle F, A\rangle$ is a soft set over U, where $U = \{h_1, h_2, \ldots, h_m\}$. The partition of U determined by the indiscernibility relation $IND\langle F, A\rangle$ can be denoted by $U|IND\langle F, A\rangle$, and $U|IND\langle F, A\rangle = \{C_1, C_2, \ldots, C_i\}$ $(i \leq m)$, where $C_i = \{[h_j]_{IND\langle F,A\rangle} : h_j \in U\}$.

Definition 5 Let (Feng and Zhou 2014) $\langle F, A\rangle$ be a soft set over U. Partition $U|IND\langle F, A\rangle = \{C_i : i \leq |U|\}$ is determined by F. $\mathcal{D} = (D(C_i, C_j))_{i,j\leq|U|}$ is called partition class-based discernibility matrix over $\langle F, A\rangle$, where $D(C_i, C_j) = \{e_l \in A : F(h_i, e_l) \neq F(h_j, e_l), \forall h_i \in C_i, \forall h_j \in C_j\}$ is called the set of discernibility parameter between C_i and C_j, and $F(h_i, e_l)$ is the value of objects in C_i associated with parameter e_l.

Definition 6 Let (Feng and Zhou 2014) $\langle F, A\rangle$ be a soft set over U. Partition $U|IND\langle F, A\rangle = \{C_i : i \leq |U|\}$ is determined by F. The soft discernibility matrix is defined as $\mathcal{D} = (D(C_i, C_j))_{i,j\leq|U|}$, where $D(C_i, C_j) = \{E^i \cup E^j : i, j \leq |U|\}$ is called the set of soft discernibility parameters between C_i and C_j in which $E^i = \{e^i_l : F(h_i, e_l) = 1$ and $F(h_j, e_l) = 0, \forall h_i \in C_i, \forall h_j \in C_j, e_l \in A\}$, and $E^j = \{e^j_l : F(h_j, e_l) = 1$ and $F(h_i, e_l) = 0, \forall h_i \in C_i, \forall h_j \in C_j, e_l \in A\}$, the symbol e^i_l (or e^j_l) represents the objects in C_i (or C_j) have the value 1 at parameter e_l, that is, $F(h_i, e_l) = 1, h_i \in C_i$ (or $F(h_j, e_l) = 1, h_j \in C_j$).

Definition 7 Let (Maji et al. 2001a) $\mathcal{F}(U)$ be the set of all fuzzy subsets of U. Let E be a set of parameters and $A \subseteq E$. A pair $\langle F, A\rangle$ is called a fuzzy soft set over U, where F is a mapping given by $F : A \to \mathcal{F}(U)$.

For $\forall e \in A$, $F(e)$ is a fuzzy subset of U and it is called fuzzy value set of parameter E. We can see that every soft set may be considered as a fuzzy soft set. If $\forall e \in A$, $F(e)$ is a crisp subset of U, then $\langle F, A\rangle$ is degenerated to be a standard soft set.

Definition 8 Let (Peng et al. 2015) U be the universe set and E be a set of parameters. $\mathcal{PF}(U)$ denotes the set of all Pythagorean fuzzy sets of U. Let $A \subseteq E$. A pair $\langle F, A\rangle$ is called a Pythagorean fuzzy soft set over U, where F is a mapping is given by $F : A \to \mathcal{PF}(U)$.

For any parameter $e \in A$, $F(e)$ is a Pythagorean fuzzy subsets of U and it is called Pythagorean fuzzy value set of parameter e.

Definition 9 A (Ashraf et al. 2019) Spherical fuzzy set (*SFS*), $\mathcal{S}$ on a universe U is an object of the form,

$$\mathcal{S} = \{(x, \mu_{\mathcal{S}}(x), \eta_{\mathcal{S}}(x), \vartheta_{\mathcal{S}}(x))|x \in U\}$$

where, $\mu_{\mathcal{S}}(x) \in [0, 1]$ is called the "degree of positive membership of x in U", $\eta_{\mathcal{S}}(x) \in [0, 1]$ is called the "degree of neutral membership of x in U" and $\vartheta_{\mathcal{S}}(x) \in [0, 1]$ is called the "degree of negative membership of x in U" and where $\mu_{\mathcal{S}}, \eta_{\mathcal{S}}$ and $\vartheta_{\mathcal{S}}$ satisfy the following condition,

$$\forall x \in U, \mu_{\mathcal{S}}^2(x) + \eta_{\mathcal{S}}^2(x) + \vartheta_{\mathcal{S}}^2(x) \leq 1.$$

Then for $x \in U$, $\pi_{\mathcal{S}}(x) = \sqrt{1 - \mu_{\mathcal{S}}^2(x) - \eta_{\mathcal{S}}^2(x) - \vartheta_{\mathcal{S}}^2(x)}$ is called the degree of refusal-membership of x in U. For simplicity, we call $\mathcal{S}(\mu_{\mathcal{S}}(x), \eta_{\mathcal{S}}(x), \vartheta_{\mathcal{S}}(x))$ a spherical fuzzy number (SFN) denoted by $e = \mathcal{S}(\mu_e, \eta_e, \vartheta_e)$ where $\mu_e, \eta_e, \vartheta_e \in [0, 1]$, $\pi_e = \sqrt{1 - (\mu_e)^2 - (\eta_e)^2 - (\vartheta_e)^2}$, and $(\mu_e)^2 + (\eta_e)^2 + (\vartheta_e)^2 \leq 1$.

Definition 10 For (Ashraf et al. 2019) every two SFSs A and B over a universe U, the union, intersection and complement are defined as follows:

1. $A \subseteq B$ if $\forall x \in U, \mu_A(x) \leq \mu_B(x), \eta_A(x) \leq \eta_B(x)$ and $\vartheta_A(x) \geq \vartheta_B(x)$.
2. $A = B$ if and only if $A \subseteq B$ and $B \subseteq A$.
3. $A \cup B = \{(x, \max\{\mu_A(x), \mu_B(x)\}, \min\{\eta_A(x), \eta_B(x)\}, \min\{\vartheta_A(x), \vartheta_B(x)\})|x \in U\}$.
4. $A \cap B = \{(x, \min\{\mu_A(x), \mu_B(x)\}, \min\{\eta_A(x), \eta_B(x)\}, \max\{\vartheta_A(x), \vartheta_B(x)\})|x \in U\}$.
5. $\text{Co}(A) = A^c = \{(x, \vartheta_A(x), \eta_A(x), \mu_A(x))|x \in U\}$.

Let $SFS(U)$ denote the set of all spherical fuzzy sets of U.

Proposition 1 *Let $A, B, C \in SFS(U)$. Then*

1. *If $A \subseteq B$ and $B \subseteq C$, then $A \subseteq C$.*
2. $(A^c)^c = A$.
3. *operations $\cap$ and $\cup$ are commutative, associative and distributive.*
4. *operations $\cap$ and $\cup$ satisfy DeMorgan's laws.*

Proof Follows directly from the definitions of SFSs. □

3 Spherical Fuzzy Soft Sets

In this chapter, we present a new concept of spherical FSS (SFSS) and investigated their numerous properties over the universal set U.

Definition 11 Let U be an initial universal set, E be a set of parameters and $A \subseteq E$. A pair $\langle F, A \rangle$ is called a spherical fuzzy soft set (SFSS) over U, where F is a mapping given by $F : A \to SFS(U)$.

Here, for any parameter $e \in E$, $F(e)$ can be written as a spherical fuzzy set such that

$$F(e) = \{(x, \mu_{F(e)}(x), \eta_{F(e)}(x), \vartheta_{F(e)}(x))|x \in U\}$$

where $\mu_{F(e)}(x)$ is the degree of positive membership, $\eta_{F(e)}(x)$ is the degree of neutral membership and $\vartheta_{F(e)}(x)$ is the degree of negative membership function respectively with the condition, $\mu^2_{F(e)}(x) + \eta^2_{F(e)}(x) + \vartheta^2_{F(e)}(x) \leq 1$. If for any parameter $e \in A$ and for any $x \in U$, $\eta_{F(e)}(x) = 0$, then $F(e)$ will become a Pythagorean fuzzy set and $\langle F, A \rangle$ will become a Pythagorean fuzzy soft set if it is true for all $e \in A$. We denote the set of all Spherical fuzzy soft sets over U by $SFSS(U)$.

Example 1 Consider a SFSS $\langle F, A \rangle$ where U is the set of different plots, that are suitable to build a house, which is denoted by $U = \{p_1, p_2, p_3\}$ and A is the set of parameters, where $A = \{e_1, e_2, e_3\}$. Let e_1 stands for availability of water, e_2 stands for in the green surroundings and e_3 stand for reasonable price. The SFSS $\langle F, A \rangle$ describe the "suitability of different plots". Suppose that
$F(e_1) = \{(p_1, 0.5, 0.2, 0.6), (p_2, 0.7, 0.3, 0.2), (p_3, 0.9, 0.1, 0.2)\}$
$F(e_2) = \{(p_1, 0.7, 0.1, 0.4), (p_2, 0.3, 0.2, 0.8), (p_3, 0.3, 0.1, 0.6)\}$
$F(e_3) = \{(p_1, 0.8, 0.2, 0.2), (p_2, 0.9, 0.3, 0.1), (p_3, 0.5, 0.1, 0.7)\}$

The Spherical fuzzy soft set $\langle F, A \rangle$ is a parameterized family $\{F(e_i) : i = 1, 2, 3\}$ of SFSs over U. The matrix form of the SFSS $\langle F, A \rangle$, is given as follows:

$$\langle F, A \rangle = \begin{pmatrix} & e_1 & e_2 & e_3 \\ p_1 & (0.5, 0.2, 0.6) & (0.7, 0.1, 0.4) & (0.8, 0.2, 0.2) \\ p_2 & (0.7, 0.3, 0.2) & (0.3, 0.2, 0.8) & (0.9, 0.3, 0.1) \\ p_3 & (0.9, 0.1, 0.2) & (0.3, 0.1, 0.6) & (0.5, 0.1, 0.7) \end{pmatrix}$$

Definition 12 Let $\langle F, A \rangle$ and $\langle G, B \rangle$ be two SFSSs over U. Then $\langle F, A \rangle$ is called a spherical fuzzy soft subset of $\langle G, B \rangle$, if

1. $A \subseteq B$, and
2. $\forall e \in A, F(e) \subseteq G(e)$.

Definition 13 Two SFSSs $\langle F, A \rangle$ and $\langle G, B \rangle$ over U are called to be spherical fuzzy soft equal, if and only if $\langle F, A \rangle$ is a spherical fuzzy soft subset of $\langle G, B \rangle$ and $\langle G, B \rangle$ is a spherical fuzzy soft subset of $\langle F, A \rangle$.

Definition 14 Let $\langle F, A \rangle$ be a $SFSS$ over U. The complement of $\langle F, A \rangle$, denoted by $\langle F, A \rangle^c$, is defined by $\langle F, A \rangle^c = \langle F^c, A \rangle$, where $F^c : A \to SFS(U)$ is a mapping given by $F^c(e) = (F(e))^c$ for every $e \in A$.

Since $(F^c(e))^c$ is equal to the $F(e)$, we get $(\langle F, A\rangle^c)^c = \langle F, A\rangle$.

Definition 15 Let $\langle F, A\rangle$ and $\langle G, B\rangle$ be two *SFSS*s over U, then $\langle F, A\rangle$ AND $\langle G, B\rangle$ is a spherical fuzzy soft set denoted by $\langle F, A\rangle \wedge \langle G, B\rangle$ is defined by $\langle F, A\rangle \wedge \langle G, B\rangle = \langle H, A \times B\rangle$ where $H(\alpha, \beta) = F(\alpha) \cap G(\beta)$ $\forall \alpha, \beta \in A \times B$. That is, $H(\alpha, \beta)(x) = (x, \min\{\mu_{F(\alpha)}(x), \mu_{G(\beta)}(x)\}, \min\{\eta_{F(\alpha)}(x), \eta_{G(\beta)}(x)\}, \max\{\vartheta_{F(\alpha)}(x), \vartheta_{G(\beta)}(x)\})$, $\forall \alpha, \beta \in A \times B$ and $\forall x \in U$.

Definition 16 Let $\langle F, A\rangle$ and $\langle G, B\rangle$ be two *SFSS*s over U, then $\langle F, A\rangle$ OR $\langle G, B\rangle$ is a spherical fuzzy soft set denoted by $\langle F, A\rangle \vee \langle G, B\rangle$ is defined by $\langle F, A\rangle \vee \langle G, B\rangle = \langle H, A \times B\rangle$ where $H(\alpha, \beta) = F(\alpha) \cup G(\beta)$ $\forall \alpha, \beta \in A \times B$. That is, $H(\alpha, \beta)(x) = (x, \max\{\mu_{F(\alpha)}(x), \mu_{G(\beta)}(x)\}, \min\{\eta_{F(\alpha)}(x), \eta_{G(\beta)}(x)\}, \min\{\vartheta_{F(\alpha)}(x), \vartheta_{G(\beta)}(x)\})$, $\forall \alpha, \beta \in A \times B$ and $\forall x \in U$.

Theorem 1 *Let $\langle F, A\rangle$ and $\langle G, B\rangle$ be two SFSSs over U. Then*

(i) $(\langle F, A\rangle \wedge \langle G, B\rangle)^c = \langle F, A\rangle^c \vee \langle G, B\rangle^c$
(ii) $(\langle F, A\rangle \vee \langle G, B\rangle)^c = \langle F, A\rangle^c \wedge \langle G, B\rangle^c$

Proof We shall prove the part (i), while other can be obtained similarly. Suppose that $\langle F, A\rangle \wedge \langle G, B\rangle = \langle H, A \times B\rangle$, where $H(\alpha, \beta) = F(\alpha) \cap G(\beta)$ $\forall (\alpha, \beta) \in A \times B$. That is, $H(\alpha, \beta) = (x, \min\{\mu_{F(\alpha)}(x), \mu_{G(\beta)}(x)\}, \min\{\eta_{F(\alpha)}(x), \eta_{G(\beta)}(x)\}, \max\{\vartheta_{F(\alpha)}(x), \vartheta_{G(\beta)}(x)\})$, $\forall (\alpha, \beta) \in A \times B$ and $\forall x \in U$.

Now $(\langle F, A\rangle \wedge \langle G, B\rangle)^c = \langle H, A \times B\rangle^c = \langle H^c, A \times B\rangle$. That is, $\forall (\alpha, \beta) \in A \times B$ and $\forall x \in U$, we have

$$H^c(\alpha, \beta) = (x, \max\{\vartheta_{F(\alpha)}(x), \vartheta_{G(\beta)}(x)\}, \min\{\eta_{F(\alpha)}(x), \eta_{G(\beta)}(x)\}, \min\{\mu_{F(\alpha)}(x), \mu_{G(\beta)}(x)\}) \quad (1)$$

Next consider, $\langle F, A\rangle^c \vee \langle G, B\rangle^c = \langle F^c, A\rangle \vee \langle G^c, B\rangle = \langle I, A \times B\rangle$ where $I(\alpha, \beta) = F^c(\alpha) \cup G^c(\beta)$
$\forall (\alpha, \beta) \in A \times B$. That is, for $x \in U$, we get

$$\begin{aligned} I(\alpha, \beta) &= (x, \max\{\mu_{F^c(\alpha)}(x), \mu_{G^c(\beta)}(x)\}, \min\{\eta_{F^c(\alpha)}(x), \eta_{G^c(\beta)}(x)\}, \\ &\quad \min\{\vartheta_{F^c(\alpha)}(x), \vartheta_{G^c(\beta)}(x)\}) \\ &= (x, \max\{\vartheta_{F(\alpha)}(x), \vartheta_{G(\beta)}(x)\}, \min\{\eta_{F(\alpha)}(x), \eta_{G(\beta)}(x)\}, \\ &\quad \min\{\mu_{F(\alpha)}(x), \mu_{G(\beta)}(x)\}) \qquad (2) \end{aligned}$$

From (1) and (2) we get $(\langle F, A\rangle \wedge \langle G, B\rangle)^c = \langle F, A\rangle^c \vee \langle G, B\rangle^c$. □

Definition 17 The union of two *SFSS*s $\langle F, A\rangle$ and $\langle G, B\rangle$ over a common universe U is a SFSS $\langle H, C\rangle$ where $C = A \cup B$ and $\forall e \in C$.

$$H(e) = \begin{cases} F(e), & \text{if } e \in A - B \\ G(e), & \text{if } e \in B - A \\ F(e) \cup G(e), & \text{if } e \in A \cap B \end{cases}$$

That is, $\forall e \in A \cap B$, we have $F(e) \cup G(e) = \{x, \max\{\mu_{F(e)}(x), \mu_{G(e)}(x)\}, \min\{\eta_{F(e)}(x), \eta_{G(e)}(x)\}, \min\{\vartheta_{F(e)}(x), \vartheta_{G(e)}(x)\} | x \in U\}$. This relation is denoted by $\langle F, A\rangle \cup \langle G, B\rangle = \langle H, C\rangle$.

Theorem 2 *The union $\langle H, C\rangle$ of two SFSSs $\langle F, A\rangle$ and $\langle G, B\rangle$ is a SFSS.*

Proof From Definition 17, we have $\forall e \in C$ if $e \in A - B$ or $e \in B - A$ then $H(e) = F(e)$ or $H(e) = G(e)$. So in either case, we get $H(e)$ is a spherical fuzzy soft set.

If $e \in A \cap B$, for a fixed $x \in U$, without loss of generality, assume $\mu_{F(e)}(x) \leq \mu_{G(e)}(x)$, then we have,

$$\begin{aligned}
\mu^2_{H(e)}(x) + \eta^2_{H(e)}(x) + \vartheta^2_{H(e)}(x) &= \{\mu^2_{F(e)}(x) \vee \mu^2_{G(e)}(x)\} + \{\eta^2_{F(e)}(x) \wedge \eta^2_{G(e)}(x)\} \\
&\quad + \{\vartheta^2_{F(e)}(x) \wedge \vartheta^2_{G(e)}(x)\} \\
&= \{\mu^2_{G(e)}(x)\} + \{\eta^2_{F(e)}(x) \wedge \eta^2_{G(e)}(x)\} \\
&\quad + \{\vartheta^2_{F(e)}(x) \wedge \vartheta^2_{G(e)}(x)\} \\
&\leq \mu^2_{G(e)}(x) + \eta^2_{G(e)}(x) + \vartheta^2_{G(e)}(x) \\
&\leq 1
\end{aligned}$$

Therefore $\langle H, C\rangle$ is a SFSS. □

Definition 18 The intersection of two spherical fuzzy soft sets $\langle F, A\rangle$ and $\langle G, B\rangle$ over a common universe U is a spherical fuzzy soft set $\langle H, C\rangle$ where $C = A \cup B$ and $\forall e \in C$

$$H(e) = \begin{cases} F(e), & \text{if } e \in A - B \\ G(e), & \text{if } e \in B - A \\ F(e) \cap G(e), & \text{if } e \in A \cap B \end{cases}$$

That is $\forall e \in A \cap B$, we have $F(e) \cap G(e) = \{x, \min\{\mu_{F(e)}(x), \mu_{G(e)}(x)\}, \min\{\eta_{F(e)}(x), \eta_{G(e)}(x)\}, \max\{\vartheta_{F(e)}(x), \vartheta_{G(e)}(x)\} | x \in U\}$. This relation is denoted by $\langle F, A\rangle \cap \langle G, B\rangle = \langle H, C\rangle$.

Theorem 3 *The intersection $\langle H, C\rangle$ of two SFSSs $\langle F, A\rangle$ and $\langle G, B\rangle$ is a SFSS.*

Proof From Definition 18, we have $\forall e \in C$ if $e \in A - B$ or $e \in B - A$ then $H(e) = F(e)$ or $H(e) = G(e)$. So in either case, we get $H(e)$ is a SFSS.

If $e \in A \cap B$, for a fixed $x \in U$, without loss of generality, assume $\vartheta_{F(e)}(x) \leq \vartheta_{G(e)}(x)$, then we have,

$$\begin{aligned}\mu^2_{H(e)}(x) + \eta^2_{H(e)}(x) + \vartheta^2_{H(e)}(x) &= \{\mu^2_{F(e)}(x) \wedge \mu^2_{G(e)}(x)\} + \{\eta^2_{F(e)}(x) \wedge \eta^2_{G(e)}(x)\} \\ &\quad + \{\vartheta^2_{F(e)}(x) \vee \vartheta^2_{G(e)}(x)\} \\ &= \{\mu^2_{F(e)}(x) \vee \mu^2_{G(e)}(x)\} + \{\eta^2_{F(e)}(x) \wedge \eta^2_{G(e)}(x)\} \\ &\quad + \{\vartheta^2_{G(e)}(x)\} \\ &\leq \mu^2_{G(e)}(x) + \eta^2_{G(e)}(x) + \vartheta^2_{G(e)}(x) \\ &\leq 1\end{aligned}$$

Therefore $\langle H, C\rangle$ is a SFSS.

Theorem 4 *Let $\langle F, A\rangle$, $\langle G, B\rangle$ and $\langle H, C\rangle$ be SFSSs over U. Then*

1. $\langle F, A\rangle \cup \langle F, A\rangle = \langle F, A\rangle$.
2. $\langle F, A\rangle \cap \langle F, A\rangle = \langle F, A\rangle$.
3. $\langle F, A\rangle \cup \langle G, B\rangle = \langle G, B\rangle \cup \langle F, A\rangle$.
4. $\langle F, A\rangle \cap \langle G, B\rangle = \langle G, B\rangle \cap \langle F, A\rangle$.
5. $(\langle F, A\rangle \cup \langle G, B\rangle) \cup \langle H, C\rangle = \langle F, A\rangle \cup (\langle G, B\rangle \cup \langle H, C\rangle)$.
6. $(\langle F, A\rangle \cap \langle G, B\rangle) \cap \langle H, C\rangle = \langle F, A\rangle \cap (\langle G, B\rangle \cap \langle H, C\rangle)$.

Proof Follows from the definitions of union and intersection and Proposition 1. □

Theorem 5 *Let $\langle F, A\rangle$, $\langle G, B\rangle$ and $\langle H, C\rangle$ be spherical fuzzy soft sets over U. Then*

1. $\langle F, A\rangle \cup (\langle G, B\rangle \cap \langle H, C\rangle) = (\langle F, A\rangle \cup \langle G, B\rangle) \cap (\langle F, A\rangle \cup \langle H, C\rangle)$.
2. $\langle F, A\rangle \cap (\langle G, B\rangle \cup \langle H, C\rangle) = (\langle F, A\rangle \cap \langle G, B\rangle) \cup (\langle F, A\rangle \cap \langle H, C\rangle)$.

Proof It is obtained directly obtained from the fact that the union and intersection of SFSS are distributive in Proposition 1. □

Theorem 6 *Let $\langle F, A\rangle$ and $\langle G, B\rangle$ be two SFSSs over U, then we have the following properties*

1. $(\langle F, A\rangle \cap \langle G, B\rangle)^c = \langle F, A\rangle^c \cup \langle G, B\rangle^c$.
2. $(\langle F, A\rangle \cup \langle G, B\rangle)^c = \langle F, A\rangle^c \cap \langle G, B\rangle^c$

Proof We shall prove only the part (i), while other can be obtained similarly.

We have $\langle F, A\rangle \cap \langle G, B\rangle = \langle H, C\rangle$ where $C = A \cup B$ and $\forall e \in C$

$$H(e) = \begin{cases} F(e), & \text{if } e \in A - B \\ G(e), & \text{if } e \in B - A \\ F(e) \cap G(e), & \text{if } e \in A \cap B \end{cases}$$

That is, $\forall e \in A \cap B$, we have $F(e) \cap G(e) = \{x, \min\{\mu_{F(e)}(x), \mu_{G(e)}(x)\}, \min\{\eta_{F(e)}(x), \eta_{G(e)}(x)\}, \max\{\vartheta_{F(e)}(x), \vartheta_{G(e)}(x) | x \in U\}$. So that $(\langle F, A\rangle \cap \langle G, B\rangle)^c = \langle H, C\rangle^c = \langle H^c, C\rangle$ and $H^c(e) = (H(e))^c$. Therefore,

$$(H(e))^c = \begin{cases} (F(e))^c, & \text{if } e \in A - B \\ (G(e))^c, & \text{if } e \in B - A \\ (F(e) \cap G(e))^c, & \text{if } e \in A \cap B \end{cases}$$

That is, $\forall e \in A \cap B$, we get

$$\begin{aligned}(F(e) \cap G(e))^c &= \{x, \min\{\mu_{F(e)}(x), \mu_{G(e)}(x)\}, \min\{\eta_{F(e)}(x), \\ &\qquad \eta_{G(e)}(x)\}, \max\{\vartheta_{F(e)}(x), \vartheta_{G(e)}(x)|x \in U\}^c \\ &= \{x, \max\{\vartheta_{F^c(e)}(x), \vartheta_{G^c(e)}(x)\}, \min\{\eta_{F^c(e)}(x), \\ &\qquad \eta_{G^c(e)}(x)\}, \min\{\mu_{F^c(e)}(x), \mu_{G^c(e)}(x)|x \in U\}\end{aligned}$$

Now $\langle F, A\rangle^c = \langle F^c, A\rangle$ and $\langle G, B\rangle^c = \langle G^c, B\rangle$. So that $\langle F, A\rangle^c \cup \langle G, B\rangle^c = \langle F^c, A\rangle \cup \langle G^c, B\rangle = \langle H^c, C\rangle$ where $C = A \cup B$ and

$$H^c(e) = \begin{cases} (F^c(e)), & \text{if } e \in A - B \\ (G^c(e)), & \text{if } e \in B - A \\ (F^c(e) \cup G^c(e)), & \text{if } e \in A \cap B \end{cases}$$

That is, $\forall e \in A \cap B$, $(F^c(e) \cup G^c(e)) = \{x, \max\{\vartheta_{F^c(e)}(x), \vartheta_{G^c(e)}(x)\}, \min\{\eta_{F^c(e)}(x), \eta_{G^c(e)}(x)\}, \min\{\mu_{F^c(e)}(x), \mu_{G^c(e)}(x)|x \in U\}$. Therefore, we get $(\langle F, A\rangle \cap \langle G, B\rangle)^c = \langle F, A\rangle^c \cup \langle G, B\rangle^c$ □

Example 2 Let $U = \{x_1, x_2, x_3\}$ and $E = \{e_1, e_2, e_3, e_4, e_5, e_6\}$, Suppose that $\langle F, A\rangle$, $\langle G, B\rangle$ and $\langle I, D\rangle$ are three SFSSs over U given by $A = \{e_1, e_2, \}$, $B = \{e_1, e_2, e_3\}$, and $D = \{e_1, e_2, e_5, e_6\}$ defined as:

$$\langle F, A\rangle = \begin{pmatrix} & e_1 & e_2 \\ x_1 & (0.2, 0.1, 0.5) & (0.8, 0.2, 0.3) \\ x_2 & (0.8, 0.1, 0.5) & (0.6, 0.3, 0.4) \\ x_3 & (0.1, 0.3, 0.4) & (0.5, 0.1, 0.7) \end{pmatrix},$$

$$\langle G, B\rangle = \begin{pmatrix} & e_1 & e_2 & e_3 \\ x_1 & (0.3, 0.2, 0.4) & (0.9, 0.2, 0.1) & (0.7, 0.3, 0.4) \\ x_2 & (0.8, 0.2, 0.3) & (0.7, 0.3, 0.2) & (0.9, 0.1, 0.1) \\ x_3 & (0.3, 0.4, 0.2) & (0.5, 0.3, 0.6) & (0.4, 0.5, 0.5) \end{pmatrix} \text{ and}$$

$$\langle I, D\rangle = \begin{pmatrix} & e_1 & e_2 & e_5 & e_6 \\ x_1 & (0.2, 0.1, 0.8) & (0.8, 0.2, 0.5) & (0.3, 0.5, 0.2) & (0.8, 0.2, 0.5) \\ x_2 & (0.9, 0.1, 0.2) & (0.7, 0.2, 0.4) & (0.9, 0.2, 0.3) & (0.6, 0.1, 0.6) \\ x_3 & (0.3, 0.2, 0.4) & (0.6, 0.3, 0.2) & (0.1, 0.1, 0.7) & (0.7, 0.5, 0.1) \end{pmatrix}.$$

Then $\langle F, A\rangle$ is a spherical fuzzy soft subset of $\langle G, B\rangle$.
The complement of $\langle F, A\rangle$ is given by,

$$\langle F, A\rangle^c = \langle F^c, A\rangle = \begin{pmatrix} & e_1 & e_2 \\ x_1 & (0.5, 0.1, 0.2) & (0.3, 0.2, 0.8) \\ x_2 & (0.5, 0.1, 0.8) & (0.4, 0.3, 0.6) \\ x_3 & (0.4, 0.3, 0.1) & (0.7, 0.1, 0.5) \end{pmatrix}.$$

The AND and OR operations are given by,

$$\langle F, A\rangle \wedge \langle G, B\rangle = \langle H, A \times B\rangle =$$

$$\begin{pmatrix} & (e_1, e_1) & (e_1, e_2) & (e_1, e_3) & (e_2, e_1) & (e_2, e_2) & (e_2, e_3) \\ x_1 & (0.2, 0.1, 0.5) & (0.2, 0.1, 0.5) & (0.2, 0.1, 0.5) & (0.3, 0.2, 0.4) & (0.8, 0.2, 0.3) & (0.7, 0.2, 0.4) \\ x_2 & (0.8, 0.1, 0.5) & (0.7, 0.1, 0.5) & (0.8, 0.1, 0.5) & (0.6, 0.2, 0.4) & (0.6, 0.3, 0.4) & (0.6, 0.1, 0.4) \\ x_3 & (0.1, 0.3, 0.4) & (0.1, 0.3, 0.6) & (0.1, 0.3, 0.5) & (0.3, 0.1, 0.7) & (0.5, 0.1, 0.7) & (0.4, 0.1, 0.7) \end{pmatrix}$$

and
$\langle F, A\rangle \vee \langle G, B\rangle = \langle I, A\times B\rangle =$

$$\begin{pmatrix} & (e_1,e_1) & (e_1,e_2) & (e_1,e_3) & (e_2,e_1) & (e_2,e_2) & (e_2,e_3) \\ x_1 & (0.3,0.1,0.4) & (0.9,0.1,0.1) & (0.7,0.1,0.4) & (0.8,0.2,0.3) & (0.9,0.2,0.1) & (0.8,0.2,0.3) \\ x_2 & (0.8,0.1,0.3) & (0.8,0.1,0.2) & (0.9,0.1,0.1) & (0.8,0.2,0.3) & (0.7,0.3,0.2) & (0.9,0.1,0.1) \\ x_3 & (0.3,0.3,0.2) & (0.5,0.3,0.4) & (0.4,0.3,0.4) & (0.5,0.1,0.2) & (0.5,0.1,0.6) & (0.5,0.1,0.5) \end{pmatrix}.$$

The union and intersection of SFSSs $\langle F, A\rangle$ and $\langle I, D\rangle$ are given by $\langle F, A\rangle \cup \langle I, D\rangle = \langle H, C\rangle$ where $C = A\cup D = \{e_1, e_2, e_5, e_6\}$ and defined as

$$\langle F, A\rangle \cup \langle I, D\rangle = \langle H, C\rangle = \begin{pmatrix} & e_1 & e_2 & e_5 & e_6 \\ x_1 & (0.2,0.1,0.5) & (0.8,0.2,0.3) & (0.3,0.5,0.2) & (0.8,0.2,0.5) \\ x_2 & (0.9,0.1,0.2) & (0.7,0.2,0.4) & (0.9,0.2,0.3) & (0.6,0.1,0.6) \\ x_3 & (0.3,0.2,0.4) & (0.6,0.1,0.2) & (0.1,0.1,0.7) & (0.7,0.5,0.1) \end{pmatrix}$$

and $\langle F, A\rangle \cap \langle I, D\rangle = \langle H, C\rangle$, where $C = A\cup D = \{e_1, e_2, e_3, e_4\}$ given by

$$\langle F, A\rangle \cap \langle I, D\rangle = \langle H, C\rangle = \begin{pmatrix} & e_1 & e_2 & e_5 & e_6 \\ x_1 & (0.2,0.1,0.8) & (0.8,0.2,0.5) & (0.3,0.5,0.2) & (0.8,0.2,0.5) \\ x_2 & (0.8,0.1,0.5) & (0.6,0.2,0.4) & (0.9,0.2,0.3) & (0.6,0.1,0.6) \\ x_3 & (0.1,0.2,0.4) & (0.5,0.1,0.7) & (0.1,0.1,0.7) & (0.7,0.5,0.1) \end{pmatrix}$$

4 Application of the Spherical Fuzzy Soft Set Model Based on Adjustable Soft Discernibility Matrix

In this section, we present an algorithm for SFSS by combining the algorithm based on soft discernibility matrix (Yang et al. 2015) and decision-making method used in (Feng and Zhou 2014). By using the new algorithm, we are capable of locate now not most effective the optimal object however we get an order relation among the all objects in our universe.

First we present some basic ideas.

Definition 19 Let $\varpi = \langle F, A\rangle$ be a *SFSS* over U. For a triple $(p, q, r) \in [0, 1]^3$, the (p, q, r)-level soft set of ϖ is a crisp soft set $L(\varpi; (p, q, r)) = \langle F_{(p,q,r)}, A\rangle$ is defined by

$$\begin{aligned} F_{(p,q,r)}(e) &= L(F(e); (p, q, r)) \\ &= \{x \in U | \mu_{F(e)}(x) \geq p, \eta_{F(e)}(x) \leq q \text{ and } \vartheta_{F(e)} \leq r\}, \forall e \in A. \end{aligned}$$

Example 3 Consider Example 1. Now we take $(p, q, r) = (0.3, 0.2, 0.5)$, we get
$L(F(e_1); (0.3, 0.2, 0.5)) = \{p_3\}$
$L(F(e_2); (0.3, 0.2, 0.5)) = \{p_1\}$
$L(F(e_3); (0.3, 0.2, 0.5)) = \{p_1\}$

Here, the $(0.3, 0.2, 0.5)$-level soft set of $\varpi = \langle F, A\rangle$ is a crisp soft set $L(F(\varpi; (0.3, 0.2, 0.5))$.

Theorem 7 *Let $\varpi = \langle F, A\rangle$ be a spherical fuzzy soft set over U, where $A \subseteq E$ and E is a set of parameters. $L(\varpi; (p_1, q_1, r_1))$ and $L(\varpi; (p_2, q_2, r_2))$ are (p_1, q_1, r_1)-level soft set and (p_2, q_2, r_2)-level soft set of ϖ, respectively, where $p_1, q_1, r_1, p_2, q_2, r_2 \in$*

$[0, 1]$. *If* $p_2 \leq p_1, q_2 \geq q_1$ *and* $r_2 \geq r_1$, *then we have* $L(\varpi; (p_1, q_1, r_1)) \subseteq L(\varpi; (p_2, q_2, r_2))$.

Proof Let $L(\varpi; (p_1, q_1, r_1)) = \langle F_{(}p_1, q_1, r_1), A\rangle$ where, $F_{(}p_1, q_1, r_1)(e) = L(F(e); (p_1, q_1, r_1)) = \{x \in U | \mu_{F(e)}(x) \geq p_1, \eta_{F(e)}(x) \leq q_1$ and $\vartheta_{F(e)} \leq r_1\}$, for all $e \in A$. And let $L(\varpi; (p_2, q_2, r_2)) = \langle F_{(}p_2, q_2, r_2), A\rangle$ where, $F_{(}p_2, q_2, r_2)(e) = L(F(e); (p_2, q_2, r_2)) = \{x \in U | \mu_{F(e)}(x) \geq p_2, \eta_{F(e)}(x) \leq q_2$ and $\vartheta_{F(e)} \leq r_2\}$, for all $e \in A$. Obviously, $A \subseteq A$. To prove the theorem it is enough to show that $\forall e \in A$, $F_{(}p_1, q_1, r_1)(e) \subseteq F_{(}p_2, q_2, r_2)(e)$. Since $p_2 \leq p_1, q_2 \geq q_1$ and $r_2 \geq r_1$, $\{x \in U | \mu_{F(e)}(x) \geq p_1, \eta_{F(e)}(x) \leq q_1$ and $\vartheta_{F(e)} \leq r_1\} \subseteq \{x \in U | \mu_{F(e)}(x) \geq p_2, \eta_{F(e)}(x) \leq q_2$ and $\vartheta_{F(e)} \leq r_2\}$. Thus we have, $F_{(}p_1, q_1, r_1)(e) \subseteq F_{(}p_2, q_2, r_2)(e)$, $\forall e \in A$. Therefore, $L(\varpi; (p_1, q_1, r_1)) \subseteq L(\varpi; (p_2, q_2, r_2))$.

Theorem 8 *Let* $\varpi = \langle F, A\rangle$ *and* $\varsigma = \langle G, A\rangle$ *be two spherical fuzzy sets over* U, *where* $A \subseteq E$ *and* E *is a set of parameters.* $L(\varpi; (p, q, r))$ *and* $L(\varsigma; (p, q, r))$ *are* (p, q, r)*-level soft sets of* ϖ *and* ς *respectively, where* $p, q, r \in [0, 1]$. *If* $\varpi \subseteq \varsigma$, *then* $L(\varpi; (p, q, r)) \subseteq L(\varsigma; (p, q, r))$.

Proof Let $L(\varpi; (p, q, r)) = \langle F_{(}p, q, r), A\rangle$ where, $F_{(}p, q, r)(e) = L(F(e); (p, q, r)) = \{x \in U | \mu_{F(e)}(x) \geq p, \eta_{F(e)}(x) \leq q$ and $\vartheta_{F(e)} \leq r\}$, for all $e \in A$. And let $L(\varsigma; (p, q, r)) = \langle G_{(}p, q, r), A\rangle$ where, $G_{(}p, q, r)(e) = L(G(e); (p, q, r)) = \{x \in U | \mu_{G(e)}(x) \geq p, \eta_{G(e)}(x) \leq q$ and $\vartheta_{G(e)} \leq r\}$, for all $e \in A$. Obviously, $A \subseteq A$. To prove the theorem it is enough to show that $\forall e \in A$, $F_{(}p, q, r)(e) \subseteq G_{(}p, q, r)(e)$.
Since $\varpi \subseteq \varsigma$, we have the following $\mu_{F(e)}(x) \leq \mu_{G(e)}(x)$, $\eta_{F(e)}(x) \geq \eta_{G(e)}(x)$ and $\vartheta_{F(e)}(x) \geq \vartheta_{G(e)}(x)$ for all $x \in U$ and $e \in A$. Assume, $x \in F_{(}p, q, r)(e)$. Then, we have, $\mu_{F(e)}(x) \geq p$, $\eta_{F(e)}(x) \leq q$ and $\vartheta_{F(e)}(x) \leq r$. Further, $\mu_{F(e)}(x) \leq \mu_{G(e)}(x)$, $\eta_{F(e)}(x) \geq \eta_{G(e)}(x)$ and $\vartheta_{F(e)}(x) \geq \vartheta_{G(e)}(x)$, we get $\mu_{G(e)}(x) \geq p$, $\eta_{G(e)}(x) \leq q$ and $\vartheta_{G(e)}(x) \leq r$. Hence $x \in \{x \in U | \mu_{G(e)}(x) \geq p, \eta_{G(e)}(x) \leq q$ and $\vartheta_{G(e)} \leq r\}$. Therefore, $F_{(}p, q, r)(e) \subseteq G_{(}p, q, r)(e)$, $\forall e \in A$. Consequently, $L(\varpi; (p, q, r)) \subseteq L(\varsigma; (p, q, r))$. □

While making decisions every now and then, we want to impose special thresholds on specific parameters. To triumph over this situation, we use a function to replace a constant value triple as the thresholds on positive membership value, neutral membership value and negative membership value respectively.

Definition 20 Let $\varpi = \langle F, A\rangle$ be a SFSS over U. Let $\lambda : A \to [0, 1]^3$ be a function. That is, forall $e \in A$, $\lambda(e) = (p(e), q(e), r(e))$, where $p(e), q(e), r(e) \in [0, 1]$. The level soft of ϖ with respect to λ is a crisp soft set $L(\varpi; \lambda) = \langle F_\lambda, A\rangle$ defined by $F_\lambda(e) = L(F(e); \lambda(e)) = \{x \in U | \mu_{F(e)}(x) \geq p(e), \eta_{F(e)}(x) \leq q(e)$ and $\vartheta_{F(e)} \leq r(e)\}$, $\forall e \in A$.

Remark 1 Here $\lambda : A \to [0, 1]^3$ is the function, where $p(e)$ is the least threshold with respect to the parameter e on the degree of positive membership, $q(e)$ is the greatest threshold with respect to the parameter e on the degree of neutral membership and $r(e)$ is the greatest threshold with respect to the parameter e on the degree of negative membership.

Let $\varpi = \langle F, A\rangle$ be a SFSS over U. The four threshold functions that all are already familiar with us are shown below,

1. The Mid-level threshold function (mid_ϖ): The mid-threshold function of $\varpi = \langle F, A\rangle$, mid_ϖ defined from A to $[0, 1]^3$ is given by $\text{mid}_\varpi(e) = (p_{\text{mid}_\varpi}(e), q_{\text{mid}_\varpi}(e), r_{\text{mid}_\varpi}(e))$, $\forall e \in A$, where $p_{\text{mid}_\varpi}(e) = \frac{1}{|U|}\sum\limits_{x\in U}\mu_{F(e)}(x)$, $q_{\text{mid}_\varpi}(e) = \frac{1}{|U|}\sum\limits_{x\in U}\eta_{F(e)}(x)$, $r_{\text{mid}_\varpi}(e) = \frac{1}{|U|}\sum\limits_{x\in U}\vartheta_{F(e)}(x)$. The level soft set with respect to mid_ϖ, $L(\varpi; \text{mid}_\varpi)$ is called the mid-level soft set of ϖ.
2. The Top-bottom-bottom-level threshold function (tbb_ϖ): The top-bottom-bottom threshold function of $\varpi = \langle F, A\rangle$, tbb_ϖ defined from A to $[0, 1]^3$ is given by $\text{tbb}_\varpi(e) = (p_{\text{tbb}_\varpi}(e), q_{\text{tbb}_\varpi}(e), r_{\text{tbb}_\varpi}(e))$, $\forall e \in A$, where $p_{\text{tbb}_\varpi}(e) = \max\limits_{x\in U}\mu_{F(e)}(x)$, $q_{\text{tbb}_\varpi}(e) = \min\limits_{x\in U}\eta_{F(e)}(x)$, $r_{\text{tbb}_\varpi}(e) = \min\limits_{x\in U}\vartheta_{F(e)}(x)$. The level soft set with respect to tbb_ϖ, $L(\varpi; \text{tbb}_\varpi)$ is called the top-bottom-bottom-level soft set of ϖ.
3. The Bottom-bottom-bottom-level threshold function (bbb_ϖ): The bottom-bottom-bottom threshold function of $\varpi = \langle F, A\rangle$, bbb_ϖ defined from A to $[0, 1]^3$ is given by $\text{bbb}_\varpi(e) = (p_{\text{bbb}_\varpi}(e), q_{\text{bbb}_\varpi}(e), r_{\text{bbb}_\varpi}(e))$, $\forall e \in A$, where $p_{\text{bbb}_\varpi}(e) = \min\limits_{x\in U}\mu_{F(e)}(x)$, $q_{\text{bbb}_\varpi}(e) = \min\limits_{x\in U}\eta_{F(e)}(x)$, $r_{\text{bbb}_\varpi}(e) = \min\limits_{x\in U}\vartheta_{F(e)}(x)$. The level soft set with respect to bbb_ϖ, $L(\varpi; \text{bbb}_\varpi)$ is called the bottom-bottom-bottom-level soft set of ϖ.
4. The Med-level threshold function (med_ϖ): The med-threshold function of $\varpi = \langle F, A\rangle$, med_ϖ defined from A to $[0, 1]^3$ is given by $\text{med}_\varpi(e) = (p_{\text{med}_\varpi}(e), q_{\text{med}_\varpi}(e), r_{\text{med}_\varpi}(e))$, $\forall e \in A$, where $\forall e \in A$, $p_{\text{med}_\varpi}(e)$, $q_{\text{med}_\varpi}(e)$, $r_{\text{med}_\varpi}(e)$ are the medians by ranking the degree of positive, neutral and negative membership respectively of all alternatives according to order from large to small (or small to large), that is,

$$p_{\text{med}_\varpi}(e) = \begin{cases} \mu_{F(e)}\left(x_{\left(\frac{|U|+1}{2}\right)}\right), & \text{if } |U| \text{ is odd,} \\ \dfrac{\mu_{F(e)}\left(x_{\left(\frac{|U|}{2}\right)}\right)+\mu_{F(e)}\left(x_{\left(\frac{|U|}{2}+1\right)}\right)}{2}, & \text{if } |U| \text{ is even.} \end{cases}$$

$$q_{\text{med}_\varpi}(e) = \begin{cases} \eta_{F(e)}\left(x_{\left(\frac{|U|+1}{2}\right)}\right), & \text{if } |U| \text{ is odd,} \\ \dfrac{\eta_{F(e)}\left(x_{\left(\frac{|U|}{2}\right)}\right)+\eta_{F(e)}\left(x_{\left(\frac{|U|}{2}+1\right)}\right)}{2}, & \text{if } |U| \text{ is even} \end{cases}$$

and

$$r_{\text{med}_\varpi}(e) = \begin{cases} \vartheta_{F(e)}\left(x_{\left(\frac{|U|+1}{2}\right)}\right), & \text{if } |U| \text{ is odd,} \\ \dfrac{\vartheta_{F(e)}\left(x_{\left(\frac{|U|}{2}\right)}\right)+\vartheta_{F(e)}\left(x_{\left(\frac{|U|}{2}+1\right)}\right)}{2}, & \text{if } |U| \text{ is even.} \end{cases}$$

The level soft set with respect to med_ϖ, $L(\varpi; \text{med}_\varpi)$ is called the med-level soft set of ϖ.

Example 4 Let us reconsider the Example 1. From it, the above mentioned thresholds and the level soft sets are given below,

1. $\text{mid}_{\varpi} = \{\langle e_1, (0.70, 0.20, 0.33)\rangle, \langle e_2, (0.43, 0.13, 0.60)\rangle, \langle e_3, (0.73, 0.20, 0.33)\rangle\}$
2. $\text{tbb}_{\varpi} = \{\langle e_1, (0.90, 0.10, 0.20)\rangle, \langle e_2, (0.70, 0.10, 0.40)\rangle, \langle e_3, (0.90, 0.10, 0.10)\rangle\}$
3. $\text{bbb}_{\varpi} = \{\langle e_1, (0.50, 0.10, 0.20)\rangle, \langle e_2, (0.30, 0.10, 0.40)\rangle, \langle e_3, (0.50, 0.10, 0.10)\rangle\}$
4. $\text{med}_{\varpi} = \{\langle e_1, (0.70, 0.20, 0.20)\rangle, \langle e_2, (0.30, 0.10, 0.60)\rangle, \langle e_3, (0.80, 0.20, 0.20)\rangle\}$

From the above discussions, when the level soft set has been proposed using threshold functions, an order relation of the objects will be simply obtained from the soft discernibility matrix as noted in Feng and Zhou (2014). Currently, we are presenting a persuasive algorithm based on adjustable soft discernibility matrix as follows:

Algorithm 1: Decision making based on adjustable soft discernibility matrix.

Step 1: Input the spherical fuzzy soft set $\varpi = \langle F, A\rangle$ over U, where $U = \{x_1, x_2, \ldots, x_m\}$.

Step 2: Input any one of the threshold function $\lambda : A \to [0, 1]^3$ (mid-threshold function; or top-bottom-bottom-threshold function; or bottom-bottom-bottom-threshold function; or med-threshold function) or input a threshold triple $(p, q, r) \in [0, 1]^3$ for decision-making.

Step 3: Compute the corresponding level soft set $L(\varpi, \lambda)$ of the *SFSS* $\varpi = \langle F, A\rangle$.

Step 4: Present the level soft set $L(\varpi, \lambda)$ in tabular form.

Step 5: Compute the partition of U and the discernibility matrix.

Step 6: Compute the soft discernibility matrix $\mathcal{D} = D(C_i, C_j)_{i,j \leq m}$

Step 7: Select the items D_1 and D_2 from the soft discernibility matrix, where
$D_1 = \{D(C_i, C_j) : |D(C_i, C_j)| = 2n, n \in N\}$
$D_2 = \{D(C_i, C_j) : |D(C_i, C_j)| = 2n + 1, n \in N\}$

Step 8: For every element of D_1, make the comparison of $|E^i|$ with $|E^j|$. If $|E^i| = |E^j|$ then the object(s) $x_i \in C_i$ and $x_j \in C_j$ are kept in the same decision class. Otherwise there must exist an order relation between the objects in C_i and the objects in C_j. That is, either x_i is superior to x_j or x_j is superior to x_i.

Step 9: If we get an order relation including all the objects in U, then turn to step 11; otherwise, go to the next step.

Step 10: Combine with the result of step 9, find the corresponding element in D_2 to compare the order relation.

Step 11: Output the order relation among all the objects by combining step 9 and step 10.

Step 12: Choose the optimal object(s) from the order relation by selecting the object(s) in the first place of the order relation which is arranged from large to small. If there exist more than one optimal object take any one of them as the optimal solution.

The following example is used to illustrate the fundamental idea of Algorithm given on top of.

Example 5 Mr.X wishes to buy a car for which he is now shortlisting the available options in the market. Let us consider a SFSS $\varpi = \langle F, A\rangle$ which describes "the best quality cars". Suppose there are 6 cars under consideration, $U = \{x_1, x_2, x_3, x_4, x_5, x_6\}$. Mr.X must take in line with the criteria set $A = \{e_1, e_2, e_3, e_4, e_5, e_6\}$ where e_1 stands for affordability, e_2 stands for fuel efficiency, e_3 stands for safety measurement, e_4 stands for space inside the car, e_5 stands for mileage and e_6 stands for speed. The cars $U = \{x_1, x_2, x_3, x_4, x_5, x_6\}$ are evaluated by Mr.X with respect to the criteria e_j, $(j = 1, 2, 3, 4, 5, 6)$ and construct a SFSS $\varpi = \langle F, A\rangle$ given as Table 1.

Suppose that Mr.X would like to select the best car from the different available cars. From the different rules in decision-making problem, if during this case he (she) deals with med-level decision rule. Clearly, the med threshold of $\varpi = \langle F, A\rangle$ is $\text{med}_{\varpi} = \{\langle e_1, (0.70, 0.15, 0.40)\rangle, \langle e_2, (0.50, 0.15, 0.55)\rangle, \langle e_3, (0.55, 0.10, 0.50)\rangle, \langle e_4, (0.55, 0.20, 0.25)\rangle\ \langle e_5, (0.70, 0.10, 0.20)\rangle, \langle e_3, (0.70, 0.15, 0.35)\rangle\}$.
Using this method we shall obtain the med-level soft set $L(\varpi; \text{med}_{\varpi})$ of ϖ with tabular representation as in Table 2.

From this table, it is obvious that each of the parameter e_i; $i = 1, 2, 3, 4, 5, 6$ induces an equivalence relation on U. So we have a soft equivalence relation say $\langle F, A\rangle$ over U. Hence we get the following equivalence classes (ECs) for each of the equivalence relations as

(i) For $F(e_1)$, ECs are $\{x_1, x_3, x_6\}$, $\{x_2, x_4, x_5\}$.
(ii) For $F(e_2)$, ECs are $\{x_1, x_4, x_5, x_6\}$, $\{x_2, x_3\}$.
(iii) For $F(e_3)$, ECs are $\{x_1, x_2, x_4, x_6\}$, $\{x_3, x_5\}$.

Table 1 Rating values of SFSS $\varpi = \langle F, A\rangle$

	e_1	e_2	e_3	e_4	e_5	e_6
x_1	(0.8,0.2,0.3)	(0.5,0.2,0.6)	(0.2,0.1,0.9)	(0.6,0.2,0.2)	(0.6,0.2,0.1)	(0.8,0.1,0.2)
x_2	(0.7,0.1,0.3)	(0.8,0.1,0.3)	(0.9,0.2,0.1)	(0.6,0.2,0.4)	(0.5,0.1,0.2)	(0.5,0.2,0.4)
x_3	(0.4,0.2,0.6)	(0.9,0.1,0.1)	(0.7,0.1,0.4)	(0.5,0.2,0.3)	(0.7,0.1,0.5)	(0.9,0.1,0.3)
x_4	(0.7,0.1,0.4)	(0.3,0.2,0.7)	(0.4,0.2,0.5)	(0.8,0.1,0.4)	(0.7,0.1,0.3)	(0.2,0.3,0.6)
x_5	(0.8,0.1,0.4)	(0.5,0.2,0.6)	(0.6,0.1,0.5)	(0.5,0.1,0.2)	(0.8,0.1,0.1)	(0.6,0.2,0.4)
s_6	(0.7,0.5,0.4)	(0.4,0.1,0.5)	(0.5,0.1,0.5)	(0.5,0.2,0.2)	(0.7,0.5,0.2)	(0.9,0.1,0.2)

Table 2 Med-level soft set $L(\varpi; \text{med}_{\varpi})$ of ϖ

	e_1	e_2	e_3	e_4	e_5	e_6
x_1	0	0	0	1	0	1
x_2	1	1	0	0	0	0
x_3	0	1	1	0	0	1
x_4	1	0	0	0	0	0
x_5	1	0	1	0	1	0
x_6	0	0	0	1	0	1

(iv) For $F(e_4)$, ECs are $\{x_1, x_6\}$, $\{x_2, x_3, x_4, x_5\}$.
(v) For $F(e_5)$, ECs are $\{x_1, x_2, x_3, x_4, x_6\}$, $\{x_5\}$.
(vi) For $F(e_6)$, ECs are $\{x_1, x_3, x_6\}$, $\{x_2, x_4, x_5\}$.

We also observe that there is an indiscernibility relation defined by the soft set $\langle F, A\rangle$ itself. Which can be obtained as
$IND\langle F, A\rangle = \{(x_1, x_1), (x_1, x_6), (x_6, x_1), (x_2, x_2), (x_3, x_3), (x_4, x_4), (x_5, x_5), (x_6, x_6)\}$
Therefore the partition of U is
$U|IND\langle F, A\rangle = \{\{x_1, x_6\}, \{x_2\}, \{x_3\}, \{x_4\}, \{x_5\}\}$
From this, we can obtain the partition of U is $C_1 = \{x_1, x_6\}$, $C_2 = \{x_2\}$, $C_3 = \{x_3\}$, $C_4 = \{x_5\}$, $C_5 = \{x_6\}$. Now the discernibility matrix is constructed in Table 3.

Further, the soft discernibility matrix is constructed in Table 4
$D_1 = D(C_2, C_1), D(C_4, C_3), D(C_5, C_3), D(C_5, C_4)$
$D_2 = D(C_3, C_1), D(C_3, C_2), D(C_4, C_1), D(C_4, C_2), D(C_5, C_1), D(C_5, C_2)$

In D_1, we have $|E^1| = |E^2|$ in $D(C_2, C_1)$. So x_1, x_2 and x_6 are in the same decision class. Again, we get $|E^4| \leq |E^3|$ in $D(C_4, C_3)$. Thus $x_3 \geq x_4$. Since $|E^3| = |E^5|$ in $D(C_5, C_3)$, x_3 and x_5 are also in the same decision class. In $D(C_5, C_4)$, $|E^5| \geq |E^4|$. So $x_5 \geq x_4$. Overall we get, $\{x_3, x_5\} \geq x_4$ and x_1, x_2 and x_6 are in the same decision class.

Next consider D_2, we get $|E^3| \geq |E^1|$. Thus $x_3 \geq \{x_1, x_6\}$. That is, $\{x_3, x_5\} \geq \{x_1, x_2, x_6\} \geq x_4$. So we obtained an order relation and optimal decision is obtained to select either x_3 or x_5. Therefore, Mr.x should select either car x_3 or car x_5.

Table 3 Discernibility matrix of the considered example

	C_1	C_2	C_3	C_4	C_5
C_1	ϕ				
C_2	$\{e_1, e_2, e_4, e_6\}$	ϕ			
C_3	$\{e_2, e_3, e_4\}$	$\{e_1, e_3, e_6\}$	ϕ		
C_4	$\{e_1, e_4, e_6\}$	$\{e_2\}$	$\{e_1, e_2, e_3, e_6\}$	ϕ	
C_5	$\{e_1, e_3, e_4, e_5, e_6\}$	$\{e_2, e_3, e_5\}$	$\{e_1, e_2, e_5, e_6\}$	$\{e_3, e_5\}$	ϕ

Table 4 Soft discernibility matrix of the med-level soft set $L\langle \varpi, \text{med}_{\varpi}\rangle$

	C_1	C_2	C_3	C_4	C_5
C_1	ϕ				
C_2	$\{e_1^2, e_2^2, e_4^1, e_6^1\}$	ϕ			
C_3	$\{e_2^3, e_3^3, e_4^1\}$	$\{e_1^2, e_3^3, e_6^3\}$	ϕ		
C_4	$\{e_1^4, e_4^1, e_6^1\}$	$\{e_2^2\}$	$\{e_1^4, e_2^3, e_3^3, e_6^3\}$	ϕ	
C_5	$\{e_1^5, e_3^5, e_4^1, e_5^5, e_6^1\}$	$\{e_2^2, e_3^5, e_5^5\}$	$\{e_1^5, e_2^3, e_5^5, e_6^3\}$	$\{e_3^5, e_5^5\}$	ϕ

5 Conclusions

Spherical fuzzy soft sets are a generalization of fuzzy soft sets. This new model is more realistic, practical and accurate in some cases than the existing other soft set models. In this chapter, we tend to introduced the definition of spherical fuzzy soft sets and mentioned various operations on them. Later, we proved some theorems based on proposed definitions included DeMorgan's law. Then a decision-making problem based on adjustable soft discernibility matrix on spherical fuzzy soft set is proposed as an application. This new extension can be successfully applied to many other convenient problems that contain uncertainties.

Acknowledgements The first author is gratefully acknowledge the financial assistance provided by University Grants Commission (UGC) India, throughout the preparation of the chapter.

References

Ali MI (2011) A note on soft sets, rough soft sets and fuzzy soft sets. Appl Soft Comput 11(4):3329–3332

Ali MI (2012) Another view on reduction of parameters in soft sets. Appl Soft Comput 12(6):1814–1821

Arora R, Garg H (2018) Robust aggregation operators for multi-criteria decision making with intuitionistic fuzzy soft set environment. Sci Iran E 25(2):931–942

Ashraf S, Abdullah S, Mahmood T, Ghani F, Mahmood T (2019) Spherical fuzzy sets and their applications in multi-attribute decision making problems. J Intell Fuzzy Syst 36:2829–2844

Atanassov KT (1986) Intuitionistic fuzzy sets. Fuzzy Sets Syst 20:87–96

Atanassov KT, Gargov G (1989) Interval valued intuitionistic fuzzysets. Fuzzy Sets Syst 31:343–349

Athira TM, John SJ, Garg H (2019) Entropy and distance measures of Pythagorean fuzzy soft sets and their applications. J Intell Fuzzy Syst 37(3):4071–4084

Athira TM, John SJ, Garg H (2019) A novel entropy measure of Pythagorean fuzzy soft sets. AIMS Math. https://doi.org/10.3934/math.2019398

Çagman N, Enginoglu S (2010) Soft matrix theory and its decision making. Comput Math Appl 59(10):3308–3314

Cuong BC, Kreinovich V (2013) Picture fuzzy sets-a new concept for computational intelligence problems. In: Proceedings of 2013 3rd world congress on information and communication technologies (WICT 2013) IEEE, pp 1–6

Feng Q, Zhou Y (2014) Soft discernibility matrix and its applications in decision making. Appl Soft Comput 24:749–756

Feng F, Li C, Davvaz B, Ali MI (2010) Soft sets combined with fuzzy sets and rough sets: a tentative approach. Soft Comput 14(9):899–911

Garg H (2017) Some picture fuzzy aggregation operators and their applications to multi criteria decision-making. Arab J Sci Eng 42(12):5275–5290

Garg H (2019) New logarithmic operational laws and their aggregation operators for Pythagorean fuzzy set and their applications. Int J Intell Syst 34(1):82–106

Garg H, Arora R (2018) Generalized and group-based generalized intuitionistic fuzzy soft sets with applications in decision-making. Appl Intell 48(2):343–356

Gau WL, Buehrer DJ (1993) Vague sets. IEEE Trans Syst Man Cybern 23(2):610–614

Gündogdu FK, Kahraman C (2019) Spherical fuzzy sets and spherical fuzzy TOPSIS method. J Intell Fuzzy Syst 36(1):337–352
Gündogdu FK, Kahraman C (2019) Extension of WASPAS with spherical fuzzy sets. Informatica 30(2):269–292
Gündogdu FK, Kahraman C (2019) A novel VIKOR method using spherical fuzzy sets and its application to warehouse site selection. J Intell Fuzzy Syst 37(1):1197–1211
Gündogdu FK, Kahraman C, Civan HN (2018) A novel hesitant fuzzy EDAS method and its application to hospital selection. J Intell Fuzzy Syst 35(6):6353–6365
Jiang Y, Tang Y, Chen Q (2011) An adjustable approach to intuitionistic fuzzy soft sets based decision making. Appl Math Model 35(2):824–836
Kamaci H (2018) Bijective soft matrix theory and multi-bijective linguistic soft decision system. FILOMAT 32(11):3799–3814
Kamaci H, Saltik K, Fulya Akiz H, Osman Atagün A (2018) Cardinality inverse soft matrix theory and its applications in multi criteria group decision making. J Intell Fuzzy Syst 34(3):2031–2049
Khalil AM, Li SG, Garg H, Li HX, Ma SQ (2019) New operations on interval-valued picture fuzzy set, interval-valued picture fuzzy soft set and their applications. IEEE Access 7:51236–51253
Kong Z, Gao L, Wang L (2009) Comment on a fuzzy soft set theoretic approach to decision making problems. J Comput Appl Math 223(2):540–542
Maji PK, Biswas R, Roy AR (2001) Fuzzy soft sets. J Fuzzy Math 9(3):589–602
Maji PK, Biswas R, Roy AR (2001) Intuitionistic fuzzy soft sets. J Fuzzy Math 9(3):677–692
Maji P, Roy AR, Biswas R (2002) An application of soft sets in a decision making problem. Comput Math Appl 44(8–9):1077–1083
Majumdar P, Samanta SK (2010) Generalised fuzzy soft sets. Comput Math Appl 59(4):1425–1432
Molodtsov D (1999) Soft set theory-first results. Comput Math Appl 37(4–5):19–31
Pawlak Z (1982) Rough sets. Int J Comput Inf Sci 11(5):341–356
Peng XD, Garg H (2018) Algorithms for interval-valued fuzzy soft sets in emergency decision making based on WDBA and CODAS with new information measure. Comput Ind Eng 119:439–452
Peng X, Yang Y, Song J, Jiang Y (2015) Pythagoren fuzzy soft set and its application. Comput Eng 41(7):224–229
Perveen FA, John SJ, Bibitha KV, Garg H (2019) Spherical fuzzy soft sets and its applications in decision-making problems. J Intell Fuzzy Syst. https://doi.org/10.3233/JIFS-190728
Riaz M, Çagman N, Zareef I, Aslam M (2019) N-Soft Topology and its Applications to Multi-Criteria Group Decision Making. J Intell Fuzzy Syst 36(6):6521–6536
Roy AR, Maji P (2007) A fuzzy soft set theoretic approach to decision making problems. J Comput Appl Math 203(2):412–418
Ullah K, Garg H, Mahmood T, Jan N, Ali Z (2019) Correlation coefficients for T-spherical fuzzy sets and their applications in clustering and multi-attribute decision making. Soft Comput 1–13: https://doi.org/10.1007/s00500-019-03993-6
Xu ZS (2007) Intuitionistic fuzzy aggregation operators. IEEE Trans Fuzzy Syst 15:1179–1187
Yager RR (2013) Pythagorean fuzzy subsets. In: 2013 joint IFSA world congress and NAFIPS annual meeting (IFSA/NAFIPS) IEEE, pp 57–61
Yang X, Lin TY, Yang J, Li Y, Yu D (2009) Combination of interval valued fuzzy set and soft set. Comput Math Appl 58(3):521–527
Yang Y, Liang C, Ji S, Liu T (2015) Adjustable soft discernibility matrix based on picture fuzzy soft sets and its applications in decision making. J Intell Fuzzy Syst 29(4):1711–1722
Zadeh LA (1965) Fuzzy sets. Inf Control 8(3):338–353

Decision Making Method Based on Spherical Fuzzy Graphs

Muhammad Akram

Abstract In this chapter, we present certain concepts of spherical fuzzy graphs and describe various methods of their construction. We compute degree and total degree of spherical fuzzy graphs. We discuss some of their important properties. Moreover, we consider applications of spherical fuzzy graphs to decision making.

1 Introduction

Zadeh (1965) introduced the concept of fuzzy sets. Atanassov (1986) considered the intuitionistic fuzzy sets as an extension of fuzzy sets by defining membership function $\alpha : X \rightarrow [0, 1]$ and nonmembership function $\beta : X \rightarrow [0, 1]$ satisfying $0 \leq \alpha(a) + \beta(a) \leq 1$. To improve the adeptness of intuitionistic fuzzy sets and to capture evaluation information given by decision-makers, (Cuong and Kreinovich 2013; Cuong 2013a, b) proposed the concept of picture fuzzy set (PFS) as direct extension of intuitonistic fuzzy sets, which may be appropriate in cases when human opinions are of types: yes, abstain, no and refusal. Picture fuzzy set gives three degrees to an element named, degree of positive membership $\alpha : X \rightarrow [0, 1]$, degree of neutral membership $\gamma : X \rightarrow [0, 1]$ and degree of negative membership $\beta : X \rightarrow [0, 1]$ under the constraint $0 \leq \alpha(a) + \gamma(a) + \beta(a) \leq 1$, where $\pi(a) = 1 - (\alpha(a) + \gamma(a) + \beta(a))$ is the degree of refusal membership. Voting can be a good example of such a situation as the human voters may be divided into four groups of those who: vote for, abstain, vote against and refusal of the voting. PFSs have many applications in decision making, fuzzy inference, clustering etc. Garg (2017) presented some picture fuzzy aggregation operators. Zhang et al. (2018) explored picture fuzzy Dombi Heronian mean operators. There have been other attempts (Cuong 2014; Cuong et al. 2015; Singh 2015; Wang et al. 2017) on PFSs. Yager (2013) proposed Pythagorean fuzzy sets as an extension of intuitionistic fuzzy set, which some how broadened the space of participation by replacing the condition with $0 \leq \alpha(a)^2 + \beta(a)^2 \leq 1$. Similarly,

M. Akram (✉)
Department of Mathematics, University of the Punjab, New Campus, Lahore, Pakistan
e-mail: m.akram@pucit.edu.pk

© Springer Nature Switzerland AG 2021

C. Kahraman and F. Kutlu Gündoğdu (eds.), *Decision Making with Spherical Fuzzy Sets*, Studies in Fuzziness and Soft Computing 392,
https://doi.org/10.1007/978-3-030-45461-6_7

the structure of spherical fuzzy set (SFS) proposed by Gündo and Kahraman (2019) along with spherical fuzzy TOPSIS method provides incredible significance to manage human opinions efficiently. This model extends the space of acceptable triplets (α, γ, β) with the constraint $0 \leq \alpha(a)^2 + \gamma(a)^2 + \beta(a)^2 \leq 1$. For instance, when a decision-maker provides the evaluation information with positive membership degree 0.5, abstinence membership degree 0.3 and negative membership degree 0.8. It can be seen that the PFS fails to tackle this issue because $0.5 + 0.3 + 0.8 > 1$. However, $0.5^2 + 0.3^2 + 0.8^2 < 1$ leads to the fact that SFS is capable to show this evaluation information. Figure 1 provides a view on comparison of spaces of PFS and SFS. Mahmood et al. (2018) described SFSs as an approach toward decision-making and medical diagnosis problems. Further, Ashraf et al. (2019) discussed the concept of spherical fuzzy sets and their applications in multi-attribute decision making problems.

Graph theory is one of the powerful tools to represent a wide range of real-life problems. Graphs have a number of applications in different disciplines. Kaufmann (1973) considered the notion of fuzzy graphs in 1973 based on Zadeh's fuzzy relations (Zadeh 1971). Rosenfeld (1975) developed the structure of fuzzy graphs by getting distinctive fuzzy analogues of graph-theoretic thoughts, including cycles, paths, connectedness. Parvathi and Karunambigai (2006) enlarged the notion of fuzzy graphs to intuitionistic fuzzy graphs. Naz et al. (2018) introduced the idea of Pythagorean fuzzy graphs, an expansion of the notion of Akram and Davvaz's intuitionistic fuzzy graphs (Akram and Davvaz 2012) with applications in decision making. Due to their flexibility and applicability, many researchers have explored IFGs and their extensions with a number of interesting results, one can see (Akram and Davvaz 2012; Akram and Naz 2018; Akram et al. 2014, 2018a, b, 2019a, b, c; Akram and Habib 2019; Akram and Ali 2018; Habib et al. 2019; Sarwar and Akram 2016).

Human opinions can never be confined to yes or no. However, they may be yes, abstain, no and refusal. Thus spherical fuzzy model being more versatile than picture fuzzy model, motivates us to define spherical fuzzy graphs. In this chapter, we present certain concepts of spherical fuzzy graphs and describe various methods of their construction. We compute degree and total degree of spherical fuzzy graphs. We discuss some of their important properties. Moreover, we consider applications of spherical fuzzy graphs to decision making.

Definition 1 (Gündo and Kahraman 2019) A spherical fuzzy set X on an underlying set V is defined as

$$X = \{(a, \alpha_X(a), \gamma_X(a), \beta_X(a)) \mid a \in V\}$$

where $\alpha_X(a) \in [0, 1]$ is known as degree of truthness of a in X, $\gamma_X(a) \in [0, 1]$ is known as degree of abstinence of a in X and $\beta_X(a) \in [0, 1]$ is known as degree of falseness of a in X, where α_X, γ_X and β_X fulfil the following condition $0 \leq \alpha_X^2(a) + \gamma_X^2(a) + \beta_X^2(a) \leq 1$. Further, for all $a \in V$, $\delta_X(a) = \sqrt{1 - (\alpha_X^2(a) + \gamma_X^2(a) + \beta_X^2(a))}$ is called degree of refusal membership of a in X.

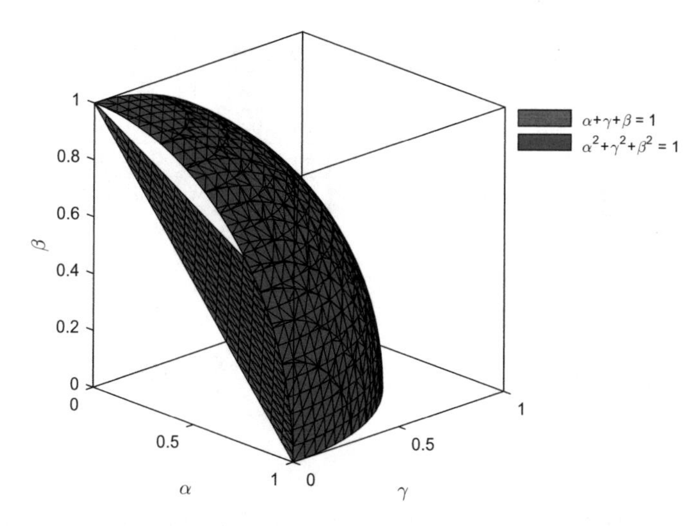

Fig. 1 Comparison of spaces of PFS and SFS

Figure 1 displays the space of spherical fuzzy sets as an extension of the space of picture fuzzy sets.

2 Spherical Fuzzy Graphs

Definition 2 A *spherical fuzzy graph* (SFG) on a non-empty set V is a pair $G = (X, Y)$ with X a spherical fuzzy set on V and Y a spherical fuzzy relation on V such that

$$\alpha_Y(a, b) \leq \min\{\alpha_X(a), \alpha_X(b)\}.$$
$$\gamma_Y(a, b) \leq \min\{\gamma_X(a), \gamma_X(b)\},$$
$$\beta_Y(a, b) \leq \max\{\beta_X(a), \beta_X(b)\}$$

and $0 \leq \alpha_Y^2(a, b) + \gamma_Y^2(a, b) + \beta_Y^2(a, b) \leq 1$ for all $a, b \in V$. Note that $Y(a, b) = 0$ for all $(a, b) \in V \times V - E$. X is called *spherical fuzzy vertex set* of G and Y is a *spherical fuzzy edge set* of G. Y is a spherical fuzzy relation on X. If Y is not symmetric on X, $G = (X, Y)$ is called *spherical fuzzy digraph.*

Example 1 Consider a graph $G^* = (V, E)$, where $V = \{w, x, y, z\}$ and $E = \{wx, xy, yz, wz\}$. Let $X = (\alpha_X, \gamma_X, \beta_X)$ be a spherical fuzzy set in V and let $Y = (\alpha_Y, \gamma_Y, \beta_Y)$ be a spherical fuzzy subset of $E \subseteq V \times V$ defined by

By routine calculations, it is easy to see that $G = (X, Y)$ is a SFG of G^* (Table 1).

Table 1 Spherical fuzzy vertex set X and spherical fuzzy edge set Y

X	w	x	y	z
α_X	0.6	0.7	0.3	0.6
γ_X	0.5	0.3	0.8	0.4
β_X	0.3	0.6	0.4	0.5

Y	wx	xy	yz	wz
α_Y	0.6	0.2	0.3	0.5
γ_Y	0.3	0.3	0.4	0.4
β_Y	0.6	0.6	0.5	0.5

Definition 3 Let $G = (X, Y)$ be a SFG defined on $G^* = (V, E)$. The *degree* of a vertex a of G is denoted by $d_G(a) = (d_\alpha(a), d_\gamma(a), d_\beta(a))$ and defined as

$$d_G(a) = \left(\sum_{a \neq b} \alpha_Y(a, b), \sum_{a \neq b} \gamma_Y(a, b), \sum_{a \neq b} \beta_Y(a, b) \right)$$

for all $(a, b) \in E$.

Example 2 Consider a SFG G displayed in Fig. 2. The degree of vertex x in G can be computed as

$$\begin{aligned} d_G(x) &= (\alpha_Y(x, y) + \alpha_Y(x, w),\ \gamma_Y(x, y) + \gamma_Y(x, w),\ \beta_Y(x, y) + \beta_Y(x, w)) \\ &= (0.2 + 0.6,\ 0.3 + 0.3,\ 0.6 + 0.6) = (0.8, 0.6, 1.2). \end{aligned}$$

Definition 4 Let $G = (X, Y)$ be a SFG defined on $G^* = (V, E)$. The *total degree* of a vertex a of G is denoted by $td_G(a) = (td_\alpha(a), td_\gamma(a), td_\beta(a))$ and defined as

$$td_G(a) = \left(\sum_{a \neq b} \alpha_Y(a, b) + \alpha_X(a), \sum_{a \neq b} \gamma_Y(a, b) + \gamma_X(a), \sum_{a \neq b} \beta_Y(a, b) + \beta_X(a) \right)$$

for all $(a, b) \in E$.

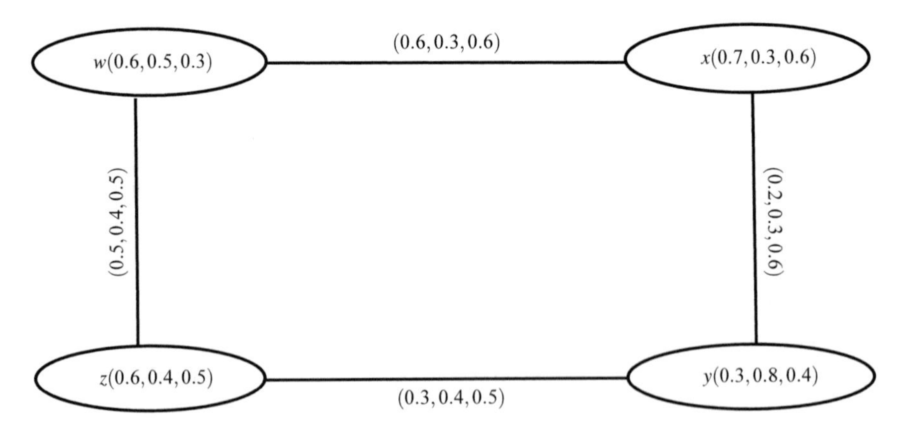

Fig. 2 Spherical fuzzy graph

Example 3 Consider a SFG G displayed in Fig. 2. The total degree of vertex x in G can be computed as

$$\begin{aligned} td_G(x) &= (d_\alpha(x) + \alpha_X(x),\ d_\gamma(x) + \gamma_X(x),\ d_\beta(x) + \beta_X(x)) \\ &= (0.8 + 0.7,\ 0.6 + 0.3,\ 1.2 + 0.6) = (1.5, 0.9, 1.8). \end{aligned}$$

We now describe some basic operations on SFGs.

Definition 5 Let $X_1 = (\alpha_{X_1}, \gamma_{X_1}, \beta_{X_1})$ and $X_2 = (\alpha_{X_2}, \gamma_{X_2}, \beta_{X_2})$ be spherical fuzzy subsets defined on V_1 and V_2, let $Y_1 = (\alpha_{Y_1}, \gamma_{Y_1}, \beta_{Y_1})$ and $Y_2 = (\alpha_{Y_2}, \gamma_{Y_2}, \beta_{Y_2})$ be spherical fuzzy subsets on E_1 and E_2, respectively. Then, the direct product of G_1 and G_2 is a SFG $G_1 \times G_2 = (X_1 \times X_2, Y_1 \times Y_2)$ such that:

$$\begin{aligned} \text{(i) } &\big(\alpha_{X_1} \times \alpha_{X_2}\big)(a_1, a_2) = \min\{\alpha_{X_1}(a_1), \alpha_{X_2}(a_2)\} \\ &\big(\gamma_{X_1} \times \gamma_{X_2}\big)(a_1, a_2) = \min\{\gamma_{X_1}(a_1), \gamma_{X_2}(a_2)\} \\ &\big(\beta_{X_1} \times \beta_{X_2}\big)(a_1, a_2) = \max\{\beta_{X_1}(a_1), \beta_{X_2}(a_2)\} \text{ for all } (a_1, a_2) \in V_1 \times V_2, \end{aligned}$$

$$\begin{aligned} \text{(ii) } &\big(\alpha_{Y_1} \times \alpha_{Y_2}\big)(a_1, a_2)(b_1, b_2) = \min\{\alpha_{Y_1}(a_1b_1), \alpha_{Y_2}(a_2b_2)\} \\ &\big(\gamma_{Y_1} \times \gamma_{Y_2}\big)(a_1, a_1)(b_1, b_2) = \min\{\gamma_{Y_1}(a_1b_1), \gamma_{Y_2}(a_2b_2)\} \\ &\big(\beta_{Y_1} \times \beta_{Y_2}\big)(a_1, a_1)(b_1, b_2) = \max\{\beta_{Y_1}(a_1b_1), \beta_{Y_2}(a_2b_2)\} \text{ for all } a_1b_1 \in E_1, \\ &\qquad\qquad\qquad\qquad \text{for all } a_2b_2 \in E_2. \end{aligned}$$

Proposition 1 *If G_1 and G_2 are SFGs, then $G_1 \times G_2$ is a SFG.*

Proof Let $a_1b_1 \in E_1$ and $a_2b_2 \in E_2$. Then we have

$$\begin{aligned} \Big(\alpha_{Y_1} \times \alpha_{Y_2}\Big)(a_1, a_2)(b_1, b_2) &= \min\Big(\alpha_{Y_1}(a_1b_1), \alpha_{Y_2}(a_2b_2)\Big) \\ &\leq \min\Big(\min\big(\alpha_{X_1}(a_1), \alpha_{X_1}(b_1)\big), \min\big(\alpha_{X_2}(a_2), \alpha_{X_2}(b_2)\big)\Big) \\ &= \min\Big(\min\big(\alpha_{X_1}(a_1), \alpha_{X_2}(a_2)\big), \min\big(\alpha_{X_1}(b_1), \alpha_{X_2}(b_2)\big)\Big) \\ &= \min\Big(\big(\alpha_{X_1} \times \alpha_{X_2}\big)(a_1, a_2), \big(\alpha_{X_1} \times \alpha_{X_2}\big)(b_1, b_2)\Big), \end{aligned}$$

$$\begin{aligned} \Big(\gamma_{Y_1} \times \gamma_{Y_2}\Big)(a_1, a_2)(b_1, b_2) &= \min\Big(\gamma_{Y_1}(a_1b_1), \gamma_{Y_2}(a_2b_2)\Big) \\ &\leq \min\Big(\min\big(\gamma_{X_1}(a_1), \gamma_{X_1}(b_1)\big), \min\big(\gamma_{X_2}(a_2), \gamma_{X_2}(b_2)\big)\Big) \\ &= \min\Big(\min\big(\gamma_{X_1}(a_1), \gamma_{X_2}(a_2)\big), \min\big(\gamma_{X_1}(b_1), \gamma_{X_2}(b_2)\big)\Big) \\ &= \min\Big(\big(\gamma_{X_1} \times \gamma_{X_2}\big)(a_1, a_2), \big(\gamma_{X_1} \times \gamma_{X_2}\big)(b_1, b_2)\Big), \end{aligned}$$

$$\begin{aligned}\Big(\beta_{Y_1}\times\beta_{Y_2}\Big)(a_1,a_2)(b_1,b_2) &= \max\Big(\beta_{Y_1}(a_1b_1),\beta_{Y_2}(a_2b_2)\Big)\\ &\leq \max\Big(\max\big(\beta_{X_1}(a_1),\beta_{X_1}(b_1)\big),\max\big(\beta_{X_2}(a_2),\beta_{X_2}(b_2)\big)\Big)\\ &= \max\Big(\max\big(\beta_{X_1}(a_1),\beta_{X_2}(a_2)\big),\max\big(\beta_{X_1}(b_1),\beta_{X_2}(b_2)\big)\Big)\\ &= \max\Big(\big(\beta_{X_1}\times\beta_{X_2}\big)(a_1,a_2),\big(\beta_{X_1}\times\beta_{X_2}\big)(b_1,b_2)\Big).\end{aligned}$$

This completes the proof.

Definition 6 Let $G_1=(X_1,Y_1)$ and $G_2=(X_2,Y_2)$ be two SFGs. Then for any vertex, $(a_1,a_2)\in V_1\times V_2$,

$$\begin{aligned}(d_\alpha)_{G_1\times G_2}(a_1,a_2) &= \sum_{(a_1,a_2)(b_1,b_2)\in E_1\times E_2}\Big(\alpha_{Y_1}\times\alpha_{Y_2}\Big)(a_1,a_2)(b_1,b_2)\\ &= \sum_{a_1b_1\in E_1,\ a_2b_2\in E_2}\min\{\alpha_{Y_1}(a_1b_1),\alpha_{Y_2}(a_2b_2)\},\end{aligned}$$

$$\begin{aligned}(d_\gamma)_{G_1\times G_2}(a_1,a_2) &= \sum_{(a_1,a_2)(b_1,b_2)\in E_1\times E_2}\Big(\gamma_{Y_1}\times\gamma_{Y_2}\Big)(a_1,a_2)(b_1,b_2)\\ &= \sum_{a_1b_1\in E_1,\ a_2b_2\in E_2}\min\{\gamma_{Y_1}(a_1b_1),\gamma_{Y_2}(a_2b_2)\},\end{aligned}$$

$$\begin{aligned}(d_\beta)_{G_1\times G_2}(a_1,a_2) &= \sum_{(a_1,a_2)(b_1,b_2)\in E_1\times E_2}\Big(\beta_{Y_1}\times\beta_{Y_2}\Big)(a_1,a_2)(b_1,b_2)\\ &= \sum_{a_1b_1\in E_1,\ a_2b_2\in E_2}\max\{\beta_{Y_1}(a_1b_1),\beta_{Y_2}(a_2b_2)\}.\end{aligned}$$

Theorem 1 *Let $G_1=(X_1,Y_1)$ and $G_2=(X_2,Y_2)$ be two SFGs. If $\alpha_{Y_2}\geq\alpha_{Y_1}$, $\gamma_{Y_2}\geq\gamma_{Y_1}$ and $\beta_{Y_2}\leq\beta_{Y_1}$, then $d_{G_1\times G_2}(a_1,a_2)=d_{G_1}(a_1)$ and if $\alpha_{Y_1}\geq\alpha_{Y_2}$, $\gamma_{Y_1}\geq\gamma_{Y_2}$ and $\beta_{Y_1}\leq\beta_{Y_2}$, then $d_{G_1\times G_2}(a_1,a_2)=d_{G_2}(a_2)$ for all $(a_1,a_2)\in V_1\times V_2$.*

Proof By definition of vertex degree of $G_1\times G_2$, we have

$$\begin{aligned}(d_\alpha)_{G_1\times G_2}(a_1,a_2) &= \sum_{(a_1,a_2)(b_1,b_2)\in E_1\times E_2}\Big(\alpha_{Y_1}\times\alpha_{Y_2}\Big)(a_1,a_2)(b_1,b_2)\\ &= \sum_{a_1b_1\in E_1,\ a_2b_2\in E_2}\min\{\alpha_{Y_1}(a_1b_1),\alpha_{Y_2}(a_2b_2)\}\\ &= \sum_{a_1b_1\in E_1}\alpha_{Y_1}(a_1b_1)\quad(\text{since }\alpha_{Y_2}\geq\alpha_{Y_1})\\ &= (d_\alpha)_{G_1}(a_1),\end{aligned}$$

$$
\begin{aligned}
(d_\gamma)_{G_1\times G_2}(a_1,a_2) &= \sum_{(a_1,a_2)(b_1,b_2)\in E_1\times E_2} \Big(\gamma_{Y_1}\times\gamma_{Y_2}\Big)(a_1,a_2)(b_1,b_2)\\
&= \sum_{a_1b_1\in E_1,\ a_2b_2\in E_2} \min\{\gamma_{Y_1}(a_1b_1),\gamma_{Y_2}(a_2b_2)\}\\
&= \sum_{a_1b_1\in E_1} \gamma_{Y_1}(a_1b_1) \quad (\text{since } \gamma_{Y_2}\geq\gamma_{Y_1})\\
&= (d_\gamma)_{G_1}(a_1),
\end{aligned}
$$

$$
\begin{aligned}
(d_\beta)_{G_1\times G_2}(a_1,a_2) &= \sum_{(a_1,a_2)(b_1,b_2)\in E_1\times E_2} \Big(\beta_{Y_1}\times\beta_{Y_2}\Big)(a_1,a_2)(b_1,b_2)\\
&= \sum_{a_1b_1\in E_1,\ a_2b_2\in E_2} \max\{\beta_{Y_1}(a_1b_1),\beta_{Y_2}(a_2b_2)\}\\
&= \sum_{a_1b_1\in E_1} \beta_{Y_1}(a_1b_1) \quad (\text{since } \beta_{Y_2}\leq\beta_{Y_1})\\
&= (d_\beta)_{G_1}(a_1).
\end{aligned}
$$

Hence, $d_{G_1\times G_2}(a_1,a_2)=d_{G_1}(a_1)$. Analogously, we can show that, if $\alpha_{Y_1}\geq\alpha_{Y_2}$, $\gamma_{Y_1}\geq\gamma_{Y_2}$ and $\beta_{Y_1}\leq\beta_{Y_2}$, then $d_{G_1\times G_2}(a_1,a_2)=d_{G_2}(a_2)$.

Definition 7 Let $G_1=(X_1,Y_1)$ and $G_2=(X_2,Y_2)$ be two SFGs. Then for any vertex, $(a_1,a_2)\in V_1\times V_2$,

$$
\begin{aligned}
(td_\alpha)_{G_1\times G_2}(a_1,a_2) &= \sum_{(a_1,a_2)(b_1,b_2)\in E_1\times E_2} \Big(\alpha_{Y_1}\times\alpha_{Y_2}\Big)(a_1,a_2)(b_1,b_2)\\
&\quad + \big(\alpha_{X_1}\times\alpha_{X_2}\big)(a_1,a_2)\\
&= \sum_{a_1b_1\in E_1,\ a_2b_2\in E_2} \min\{\alpha_{Y_1}(a_1b_1),\alpha_{Y_2}(a_2b_2)\}\\
&\quad + \min\{\alpha_{X_1}(a_1),\alpha_{X_2}(a_2)\},
\end{aligned}
$$

$$
\begin{aligned}
(td_\gamma)_{G_1\times G_2}(a_1,a_2) &= \sum_{(a_1,a_2)(b_1,b_2)\in E_1\times E_2} \Big(\gamma_{Y_1}\times\gamma_{Y_2}\Big)(a_1,a_2)(b_1,b_2) + \big(\gamma_{X_1}\times\gamma_{X_2}\big)(a_1,a_2)\\
&= \sum_{a_1b_1\in E_1,\ a_2b_2\in E_2} \min\{\gamma_{Y_1}(a_1b_1),\gamma_{Y_2}(a_2b_2)\}\\
&\quad + \min\{\gamma_{X_1}(a_1),\gamma_{X_2}(a_2)\},
\end{aligned}
$$

$$(td_\beta)_{G_1\times G_2}(a_1,a_2)=\sum_{(a_1,a_2)(b_1,b_2)\in E_1\times E_2}\Big(\beta_{Y_1}\times\beta_{Y_2}\Big)(a_1,a_2)(b_1,b_2)+\big(\beta_{X_1}\times\beta_{X_2}\big)(a_1,a_2)$$
$$=\sum_{a_1b_1\in E_1,\ a_2b_2\in E_2}\max\{\beta_{Y_1}(a_1b_1),\beta_{Y_2}(a_2b_2)\}$$
$$+\max\{\beta_{X_1}(a_1),\beta_{X_2}(a_2)\}.$$

Theorem 2 *Let $G_1=(X_1,Y_1)$ and $G_2=(X_2,Y_2)$ be two SFGs. If*

(i) $\alpha_{Y_2}\geq\alpha_{Y_1}$, then $(td_\alpha)_{G_1\times G_2}(a_1,a_2)=(d_\alpha)_{G_1}(a_1)+\min\{\alpha_{X_1}(a_1),\alpha_{X_2}(a_2)\}$;
(ii) $\gamma_{Y_2}\geq\gamma_{Y_1}$, then $(td_\gamma)_{G_1\times G_2}(a_1,a_2)=(d_\gamma)_{G_1}(a_1)+\min\{\gamma_{X_1}(a_1),\gamma_{X_2}(a_2)\}$;
(iii) $\beta_{Y_2}\leq\beta_{Y_1}$, then $(td_\beta)_{G_1\times G_2}(a_1,a_2)=(d_\beta)_{G_1}(a_1)+\max\{\beta_{X_1}(a_1),\beta_{X_2}(a_2)\}$;
(iv) $\alpha_{Y_1}\geq\alpha_{Y_2}$, then $(td_\alpha)_{G_1\times G_2}(a_1,a_2)=(d_\alpha)_{G_2}(a_2)+\min\{\alpha_{X_1}(a_1),\alpha_{X_2}(a_2)\}$;
(v) $\gamma_{Y_1}\geq\gamma_{Y_2}$, then $(td_\gamma)_{G_1\times G_2}(a_1,a_2)=(d_\gamma)_{G_2}(a_2)+\min\{\gamma_{X_1}(a_1),\gamma_{X_2}(a_2)\}$;
(vi) $\beta_{Y_1}\leq\beta_{Y_2}$, then $(td_\beta)_{G_1\times G_2}(a_1,a_2)=(d_\beta)_{G_2}(a_2)+\max\{\beta_{X_1}(a_1),\beta_{X_2}(a_2)\}$;

for all $(a_1,a_2)\in V_1\times V_2$.

Proof The proof is obvious using Definition 7 and Theorem 1.

Example 4 Consider two SFGs $G_1=(X_1,Y_1)$ and $G_2=(X_2,Y_2)$ on $V_1=\{a,b,c\}$ and $V_2=\{d,e\}$, respectively, as shown in Fig. 3. Their direct product $G_1\times G_2$ is shown in Fig. 4.

As, $\alpha_{Y_2}\geq\alpha_{Y_1}$, $\gamma_{Y_2}\geq\gamma_{Y_1}$ and $\beta_{Y_2}\leq\beta_{Y_1}$. So, by Theorem 1, we have

$$(d_\alpha)_{G_1\times G_2}(b,d)=(d_\alpha)_{G_1}(b)=0.5,\ (d_\gamma)_{G_1\times G_2}(b,d)$$
$$=(d_\gamma)_{G_1}(b)=0.3,\ (d_\beta)_{G_1\times G_2}(b,d)=(d_\beta)_{G_1}(b)=1.2.$$

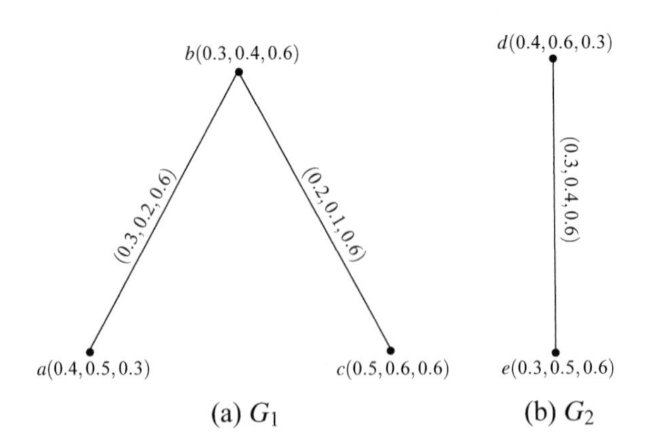

Fig. 3 Spherical fuzzy graphs

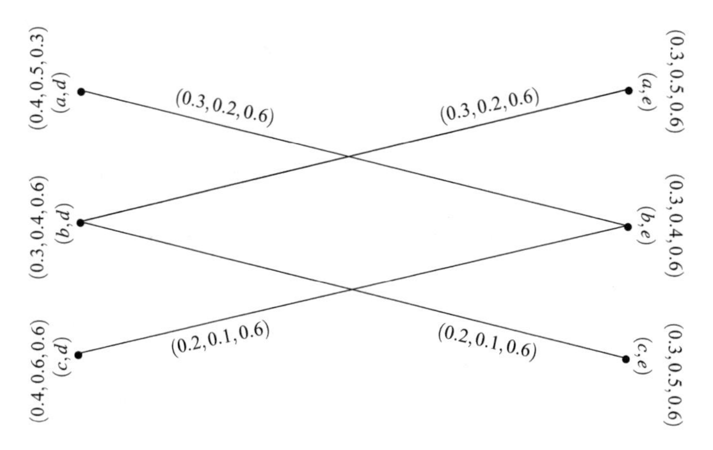

Fig. 4 $G_1 \times G_2$

Hence, $d_{G_1 \times G_2}(b, d) = (0.5, 0.3, 1.2)$.
Furthermore, by Theorem 2, we have

$$(td_\alpha)_{G_1 \times G_2}(b, d) = (d_\alpha)_{G_1}(b) + \min\{\alpha_{X_1}(b), \alpha_{X_2}(d)\} = 0.8,$$

$$(td_\gamma)_{G_1 \times G_2}(b, d) = (d_\gamma)_{G_1}(b) + \min\{\gamma_{X_1}(b), \gamma_{X_2}(d)\} = 0.7,$$

$$(td_\beta)_{G_1 \times G_2}(b, d) = (d_\beta)_{G_1}(b) + \max\{\beta_{X_1}(b), \beta_{X_2}(d)\} = 1.8.$$

Hence, $(td)_{G_1 \times G_2}(b, d) = (0.8, 0.7, 1.8)$.
Analogously, we can calculate the degree and total degree of all vertices in $G_1 \times G_2$.

Definition 8 Let $X_1 = (\alpha_{X_1}, \gamma_{X_1}, \beta_{X_1})$ and $X_2 = (\alpha_{X_2}, \gamma_{X_2}, \beta_{X_2})$ be spherical fuzzy subset defined on V_1 and V_2, and let $Y_1 = (\alpha_{Y_1}, \gamma_{Y_1}, \beta_{Y_1})$ and $Y_2 = (\alpha_{Y_2}, \gamma_{Y_2}, \beta_{Y_2})$ be spherical fuzzy subset on E_1 and E_2, respectively. Then, Cartesian product of G_1 and G_2 is a SFG $G_1 \square G_2 = (X_1 \square X_2, Y_1 \square Y_2)$ such that:

(i) $\left(\alpha_{X_1} \square \alpha_{X_2}\right)(a_1, a_2) = \min\{\alpha_{X_1}(a_1), \alpha_{X_2}(a_2)\}$
$\left(\gamma_{X_1} \square \gamma_{X_2}\right)(a_1, a_2) = \min\{\gamma_{X_1}(a_1), \gamma_{X_2}(a_2)\}$
$\left(\beta_{X_1} \square \beta_{X_2}\right)(a_1, a_2) = \max\{\beta_{X_1}(a_1), \beta_{X_2}(a_2)\}$ for all $(a_1, a_2) \in V_1 \times V_2$,

(ii) $\left(\alpha_{Y_1} \square \alpha_{Y_2}\right)(a, a_2)(a, b_2) = \min\{\alpha_{X_1}(a), \alpha_{Y_2}(a_2 b_2)\}$
$\left(\gamma_{Y_1} \square \gamma_{Y_2}\right)(a, a_2)(a, b_2) = \min\{\gamma_{X_1}(a), \gamma_{Y_2}(a_2 b_2)\}$
$\left(\beta_{Y_1} \square \beta_{Y_2}\right)(a, a_2)(a, b_2) = \max\{\beta_{X_1}(a), \beta_{Y_2}(a_2 b_2)\}$
for all $a \in V_1$, for all $a_2 b_2 \in E_2$,

$$\text{(iii)}\ (\alpha_{Y_1} \Box \alpha_{Y_2})(a_1, c)(b_1, c) = \min\{\alpha_{Y_1}(a_1b_1), \alpha_{X_2}(c)\}$$
$$(\gamma_{Y_1} \Box \gamma_{Y_2})(a_1, c)(b_1, c) = \min\{\gamma_{Y_1}(a_1b_1), \gamma_{X_2}(c)\}$$
$$(\beta_{Y_1} \Box \beta_{Y_2})(a_1, c)(b_1, c) = \max\{\beta_{Y_1}(a_1b_1), \beta_{X_2}(c)\}$$
$$\text{for all } c \in V_2, \quad \text{for all } a_1b_1 \in E_1.$$

Proposition 2 *If G_1 and G_2 are SFGs, then $G_1 \Box G_2$ is a SFG.*

Proof Let $a \in V_1$ and $a_2b_2 \in E_2$. Then we have

$$\begin{aligned}
(\alpha_{Y_1} \Box \alpha_{Y_2})(a, a_2)(a, b_2) &= \min\left(\alpha_{X_1}(a), \alpha_{Y_2}(a_2b_2)\right)\\
&\leq \min\Big(\alpha_{X_1}(a), \min\left(\alpha_{X_2}(a_2), \alpha_{X_2}(b_2)\right)\Big)\\
&= \min\Big(\min\left(\alpha_{X_1}(a), \alpha_{X_2}(a_2)\right), \min\left(\alpha_{X_1}(a), \alpha_{X_2}(b_2)\right)\Big)\\
&= \min\Big(\left(\alpha_{X_1} \Box \alpha_{X_2}\right)(a, a_2), \left(\alpha_{X_1} \Box \alpha_{X_2}\right)(a, b_2)\Big),
\end{aligned}$$

$$\begin{aligned}
(\gamma_{Y_1} \Box \gamma_{Y_2})(a, a_2)(a, b_2) &= \min\left(\gamma_{X_1}(a), \gamma_{Y_2}(a_2b_2)\right)\\
&\leq \min\Big(\gamma_{X_1}(a), \min\left(\gamma_{X_2}(a_2), \gamma_{X_2}(b_2)\right)\Big)\\
&= \min\Big(\min\left(\gamma_{X_1}(a), \gamma_{X_2}(a_2)\right), \min\left(\gamma_{X_1}(a), \gamma_{X_2}(b_2)\right)\Big)\\
&= \min\Big(\left(\gamma_{X_1} \Box \gamma_{X_2}\right)(a, a_2), \left(\gamma_{X_1} \Box \gamma_{X_2}\right)(a, b_2)\Big),
\end{aligned}$$

$$\begin{aligned}
(\beta_{Y_1} \Box \beta_{Y_2})(a, a_2)(a, b_2) &= \max\left(\beta_{X_1}(a), \beta_{Y_2}(a_2b_2)\right)\\
&\leq \max\Big(\beta_{X_1}(a), \max\left(\beta_{X_2}(a_2), \beta_{X_2}(b_2)\right)\Big)\\
&= \max\Big(\max\left(\beta_{X_1}(a), \beta_{X_2}(a_2)\right), \max\left(\beta_{X_1}(a), \beta_{X_2}(b_2)\right)\Big)\\
&= \max\Big(\left(\beta_{X_1} \Box \beta_{X_2}\right)(a, a_2), \left(\beta_{X_1} \Box \beta_{X_2}\right)(a, b_2)\Big).
\end{aligned}$$

Let $c \in V_2$ and $a_1b_1 \in E_1$. Then we have

$$\begin{aligned}
(\alpha_{Y_1} \Box \alpha_{Y_2})(a_1, c)(b_1, c) &= \min\left(\alpha_{Y_1}(a_1b_1), \alpha_{X_2}(c)\right)\\
&\leq \min\Big(\min\left(\alpha_{X_1}(a_1), \alpha_{X_1}(b_1)\right), \alpha_{X_2}(c)\Big)\\
&= \min\Big(\min\left(\alpha_{X_1}(a_1), \alpha_{X_2}(c)\right), \min\left(\alpha_{X_1}(b_1), \alpha_{X_2}(c)\right)\Big)\\
&= \min\Big(\left(\alpha_{X_1} \Box \alpha_{X_2}\right)(a_1, c), \left(\alpha_{X_1} \Box \alpha_{X_2}\right)(b_1, c)\Big),
\end{aligned}$$

$$\begin{aligned}\left(\gamma_{Y_1} \,\square\, \gamma_{Y_2}\right)(a_1, c)(b_1, c) &= \min\left(\gamma_{Y_1}(a_1 b_1), \gamma_{X_2}(c)\right)\\ &\leq \min\Big(\min\left(\gamma_{X_1}(a_1), \gamma_{X_1}(b_1)\right), \gamma_{X_2}(c)\Big)\\ &= \min\Big(\min\left(\gamma_{X_1}(a_1), \gamma_{X_2}(c)\right), \min\left(\gamma_{X_1}(b_1), \gamma_{X_2}(c)\right)\Big)\\ &= \min\Big(\left(\gamma_{X_1} \,\square\, \gamma_{X_2}\right)(a_1, c), \left(\gamma_{X_1} \,\square\, \gamma_{X_2}\right)(b_1, c)\Big),\end{aligned}$$

$$\begin{aligned}\left(\beta_{Y_1} \,\square\, \beta_{Y_2}\right)(a_1, c)(b_1, c) &= \max\left(\beta_{Y_1}(a_1 b_1), \beta_{X_2}(c)\right)\\ &\leq \max\Big(\max\left(\beta_{X_1}(a_1), \beta_{X_1}(b_1)\right), \beta_{X_2}(c)\Big)\\ &= \max\Big(\max\left(\beta_{X_1}(a_1), \beta_{X_2}(c)\right), \max\left(\beta_{X_1}(b_1), \beta_{X_2}(c)\right)\Big)\\ &= \max\Big(\left(\beta_{X_1} \,\square\, \beta_{X_2}\right)(a_1, c), \left(\beta_{X_1} \,\square\, \beta_{X_2}\right)(b_1, c)\Big).\end{aligned}$$

This completes the proof.

Definition 9 Let $G_1 = (X_1, Y_1)$ and $G_2 = (X_2, Y_2)$ be two SFGs. Then for any vertex, $(a_1, a_2) \in V_1 \times V_2$,

$$\begin{aligned}(d_\alpha)_{G_1 \,\square\, G_2}(a_1, a_2) &= \sum_{(a_1,a_2)(b_1,b_2)\in E_1 \,\square\, E_2} \Big(\alpha_{Y_1} \,\square\, \alpha_{Y_2}\Big)(a_1, a_2)(b_1, b_2)\\ &= \sum_{a_1=b_1,\ a_2b_2\in E_2} \min\{\alpha_{X_1}(a_1), \alpha_{Y_2}(a_2 b_2)\}\\ &+ \sum_{a_2=b_2,\ a_1b_1\in E_1} \min\{\alpha_{X_2}(a_2), \alpha_{Y_1}(a_1 b_1)\},\end{aligned}$$

$$\begin{aligned}(d_\gamma)_{G_1 \,\square\, G_2}(a_1, a_2) &= \sum_{(a_1,a_2)(b_1,b_2)\in E_1 \,\square E_2} \Big(\gamma_{Y_1} \,\square\, \gamma_{Y_2}\Big)(a_1, a_2)(b_1, b_2)\\ &= \sum_{a_1=b_1,\ a_2b_2\in E_2} \min\{\gamma_{X_1}(a_1), \gamma_{Y_2}(a_2 b_2)\}\\ &+ \sum_{a_2=b_2,\ a_1b_1\in E_1} \min\{\gamma_{X_2}(a_2), \gamma_{Y_1}(a_1 b_1)\},\end{aligned}$$

$$
\begin{aligned}
(d_\beta)_{G_1 \,\square\, G_2}(a_1, a_2) &= \sum_{(a_1,a_2)(b_1,b_2)\in E_1 \,\square\, E_2} \Big(\beta_{Y_1} \,\square\, \beta_{Y_2}\Big)(a_1, a_2)(b_1, b_2) \\
&= \sum_{a_1=b_1,\ a_2b_2\in E_2} \max\{\beta_{X_1}(a_1), \beta_{Y_2}(a_2b_2)\} \\
&\quad + \sum_{a_2=b_2,\ a_1b_1\in E_1} \max\{\beta_{X_2}(a_2), \beta_{Y_1}(a_1b_1)\}.
\end{aligned}
$$

Theorem 3 *Let $G_1 = (X_1, Y_1)$ and $G_2 = (X_2, Y_2)$ be two SFGs. If $\alpha_{X_1} \geq \alpha_{Y_2}$, $\gamma_{X_1} \geq \gamma_{Y_2}$, $\beta_{X_1} \leq \beta_{Y_2}$ and $\alpha_{X_2} \geq \alpha_{Y_1}$, $\gamma_{X_2} \geq \gamma_{Y_1}$, $\beta_{X_2} \leq \beta_{Y_1}$. Then $d_{G_1 \,\square\, G_2}(a_1, a_2) = d_{G_1}(a_1) + d_{G_2}(a_2)$ for all $(a_1, a_2) \in V_1 \times V_2$.*

Proof By definition of vertex degree of $G_1 \,\square\, G_2$, we have

$$
\begin{aligned}
(d_\alpha)_{G_1 \,\square\, G_2}(a_1, a_2) &= \sum_{(a_1,a_2)(b_1,b_2)\in E_1 \,\square\, E_2} \Big(\alpha_{Y_1} \,\square\, \alpha_{Y_2}\Big)(a_1, a_2)(b_1, b_2) \\
&= \sum_{a_1=b_1,\ a_2b_2\in E_2} \min\{\alpha_{X_1}(a_1), \alpha_{Y_2}(a_2b_2)\} \\
&\quad + \sum_{a_2=b_2,\ a_1b_1\in E_1} \min\{\alpha_{X_2}(a_2), \alpha_{Y_1}(a_1b_1)\} \\
&= \sum_{a_2b_2\in E_2} \alpha_{Y_2}(a_2b_2) \\
&\quad + \sum_{a_1b_1\in E_1} \alpha_{Y_1}(a_1b_1) \ \ \text{(by using } \alpha_{X_1} \geq \alpha_{Y_2} \text{ and } \alpha_{X_2} \geq \alpha_{Y_1}) \\
&= (d_\alpha)_{G_1}(a_1) + (d_\alpha)_{G_2}(a_2),
\end{aligned}
$$

$$
\begin{aligned}
(d_\gamma)_{G_1 \,\square\, G_2}(a_1, a_2) &= \sum_{(a_1,a_2)(b_1,b_2)\in E_1 \,\square\, E_2} \Big(\gamma_{Y_1} \,\square\, \gamma_{Y_2}\Big)(a_1, a_2)(b_1, b_2) \\
&= \sum_{a_1=b_1,\ a_2b_2\in E_2} \min\{\gamma_{X_1}(a_1), \gamma_{Y_2}(a_2b_2)\} \\
&\quad + \sum_{a_2=b_2,\ a_1b_1\in E_1} \min\{\gamma_{X_2}(a_2), \gamma_{Y_1}(a_1b_1)\} \\
&= \sum_{a_2b_2\in E_2} \gamma_{Y_2}(a_2b_2) \\
&\quad + \sum_{a_1b_1\in E_1} \gamma_{Y_1}(a_1b_1) \ \ \text{(by using } \gamma_{X_1} \geq \gamma_{Y_2} \text{ and } \gamma_{X_2} \geq \gamma_{Y_1}) \\
&= (d_\gamma)_{G_1}(a_1) + (d_\gamma)_{G_2}(a_2),
\end{aligned}
$$

$$
\begin{aligned}
(d_\beta)_{G_1 \,\square\, G_2}(a_1, a_2) &= \sum_{(a_1,a_2)(b_1,b_2)\in E_1 \,\square\, E_2} \Big(\beta_{Y_1} \,\square\, \beta_{Y_2}\Big)(a_1, a_2)(b_1, b_2) \\
&= \sum_{a_1=b_1,\ a_2b_2\in E_2} \max\{\beta_{X_1}(a_1), \beta_{Y_2}(a_2b_2)\} \\
&\quad + \sum_{a_2=b_2,\ a_1b_1\in E_1} \max\{\beta_{X_2}(a_2), \beta_{Y_1}(a_1b_1)\} \\
&= \sum_{a_2b_2\in E_2} \beta_{Y_2}(a_2b_2) \\
&\quad + \sum_{a_1b_1\in E_1} \beta_{Y_1}(a_1b_1) \ \text{(by using } \beta_{X_1} \leq \beta_{Y_2} \text{ and } \beta_{X_2} \leq \beta_{Y_1}) \\
&= (d_\beta)_{G_1}(a_1) + (d_\beta)_{G_2}(a_2).
\end{aligned}
$$

Hence, $d_{G_1 \,\square\, G_2}(a_1, a_2) = d_{G_1}(a_1) + d_{G_2}(a_2)$.

Definition 10 Let $G_1 = (X_1, Y_1)$ and $G_2 = (X_2, Y_2)$ be two SFGs. Then for any vertex, $(a_1, a_2) \in V_1 \times V_2$,

$$
\begin{aligned}
(td_\alpha)_{G_1 \,\square\, G_2}(a_1, a_2) &= \sum_{(a_1,a_2)(b_1,b_2)\in E_1 \,\square\, E_2} \Big(\alpha_{Y_1} \,\square\, \alpha_{Y_2}\Big)(a_1, a_2)(b_1, b_2) \\
&\quad + \Big(\alpha_{Y_1} \,\square\, \alpha_{Y_2}\Big)(a_1, a_2) \\
&= \sum_{a_1=b_1,\ a_2b_2\in E_2} \min\{\alpha_{X_1}(a_1), \alpha_{Y_2}(a_2b_2)\} \\
&\quad + \sum_{a_2=b_2,\ a_1b_1\in E_1} \min\{\alpha_{X_2}(a_2), \alpha_{Y_1}(a_1b_1)\} \\
&\quad + \min\{\alpha_{X_1}(a_1), \alpha_{X_2}(a_2)\},
\end{aligned}
$$

$$
\begin{aligned}
(td_\gamma)_{G_1 \,\square\, G_2}(a_1, a_2) &= \sum_{(a_1,a_2)(b_1,b_2)\in E_1 \,\square E_2} \Big(\gamma_{Y_1} \,\square\, \gamma_{Y_2}\Big)(a_1, a_2)(b_1, b_2) \\
&\quad + \Big(\gamma_{Y_1} \,\square\, \gamma_{Y_2}\Big)(a_1, a_2) \\
&= \sum_{a_1=b_1,\ a_2b_2\in E_2} \min\{\gamma_{X_1}(a_1), \gamma_{Y_2}(a_2b_2)\} \\
&\quad + \sum_{a_2=b_2,\ a_1b_1\in E_1} \min\{\gamma_{X_2}(a_2), \gamma_{Y_1}(a_1b_1)\} \\
&\quad + \min\{\gamma_{X_1}(a_1), \gamma_{X_2}(a_2)\},
\end{aligned}
$$

$$
\begin{aligned}
(td_\beta)_{G_1 \,\square\, G_2}(a_1,a_2) &= \sum_{(a_1,a_2)(b_1,b_2)\in E_1 \,\square\, E_2} \Big(\beta_{Y_1} \,\square\, \beta_{Y_2}\Big)(a_1,a_2)(b_1,b_2) \\
&\quad + \Big(\beta_{Y_1} \,\square\, \beta_{Y_2}\Big)(a_1,a_2) \\
&= \sum_{a_1=b_1,\ a_2b_2\in E_2} \max\{\beta_{X_1}(a_1), \beta_{Y_2}(a_2b_2)\} \\
&\quad + \sum_{a_2=b_2,\ a_1b_1\in E_1} \max\{\beta_{X_2}(a_2), \beta_{Y_1}(a_1b_1)\} \\
&\quad + \max\{\beta_{X_1}(a_1), \beta_{X_2}(a_2)\}.
\end{aligned}
$$

Theorem 4 *Let $G_1 = (X_1, Y_1)$ and $G_2 = (X_2, Y_2)$ be two SFGs. If*

(i) $\alpha_{X_1} \geq \alpha_{Y_2}$ and $\alpha_{X_2} \geq \alpha_{Y_1}$, then $(td_\alpha)_{G_1 \,\square\, G_2}(a_1,a_2) = (td_\alpha)_{G_1}(a_1) + (td_\alpha)_{G_2}(a_2) - \max\{\alpha_{X_1}(a_1), \alpha_{X_2}(a_2)\}$;
(ii) $\gamma_{X_1} \geq \gamma_{Y_2}$ and $\gamma_{X_2} \geq \gamma_{Y_1}$, then $(td_\gamma)_{G_1 \,\square\, G_2}(a_1,a_2) = (td_\gamma)_{G_1}(a_1) + (td_\gamma)_{G_2}(a_2) - \max\{\gamma_{X_1}(a_1), \gamma_{X_2}(a_2)\}$;
(iii) $\beta_{X_1} \leq \beta_{Y_2}$ and $\beta_{X_2} \leq \beta_{Y_1}$, then $(td_\beta)_{G_1 \,\square\, G_2}(a_1,a_2) = (td_\beta)_{G_1}(a_1) + (td_\beta)_{G_2}(a_2) - \min\{\beta_{X_1}(a_1), \beta_{X_2}(a_2)\}$

for all $(a_1, a_2) \in V_1 \times V_2$.

Proof The proof is obvious using Definition 10 and Theorem 3.

Example 5 Consider two SFGs $G_1 = (X_1, Y_1)$ and $G_2 = (X_2, Y_2)$ as given in Example 4. Their Cartesian product $G_1 \,\square\, G_2$ is shown in Fig. 5.

As, $\alpha_{X_1} \geq \alpha_{Y_2}, \gamma_{X_1} \geq \gamma_{Y_2}, \beta_{X_1} \leq \alpha_{Y_2}$ and $\alpha_{X_2} \geq \alpha_{Y_1}, \gamma_{X_2} \geq \gamma_{Y_1}, \beta_{X_2} \leq \alpha_{Y_1}$. So, by Theorem 3, we have

$$
(d_\alpha)_{G_1 \,\square\, G_2}(b,e) = (d_\alpha)_{G_1}(b) + (d_\alpha)_{G_2}(e) = 0.8,\ (d_\gamma)_{G_1 \,\square\, G_2}(b,e)
$$

$$
= (d_\gamma)_{G_1}(b) + (d_\gamma)_{G_2}(e) = 0.7,
$$

$$
(d_\beta)_{G_1 \,\square\, G_2}(b,e) = (d_\beta)_{G_1}(b) + (d_\beta)_{G_1}(b)
$$

$$
= 1.8. \text{ Hence, } d_{G_1 \,\square\, G_2}(b,e) = (0.8, 0.7, 1.8).
$$

Furthermore, by Theorem 4, we have

$$
(td_\alpha)_{G_1 \,\square\, G_2}(b,e) = (td_\alpha)_{G_1}(b) + (td_\alpha)_{G_2}(e) - \max\{\alpha_{X_1}(b), \alpha_{X_2}(e)\} = 1.1,
$$

$$
(td_\gamma)_{G_1 \,\square\, G_2}(b,e) = (td_\gamma)_{G_1}(b) + (td_\gamma)_{G_2}(e) - \max\{\gamma_{X_1}(b), \gamma_{X_2}(e)\} = 1.1,
$$

$$
(td_\beta)_{G_1 \,\square\, G_2}(b,e) = (td_\beta)_{G_1}(b) + (td_\beta)_{G_2}(e) - \min\{\beta_{X_1}(b), \beta_{X_2}(e)\} = 2.4.
$$

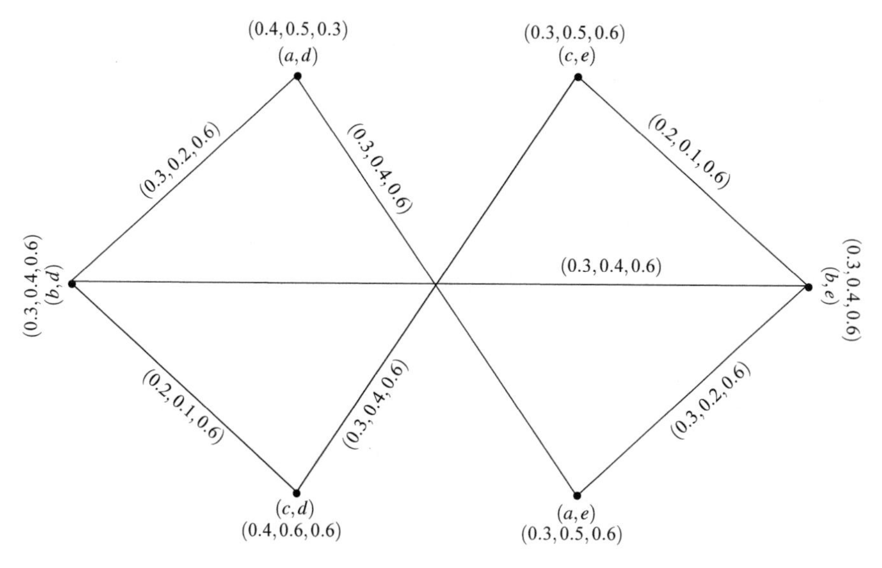

Fig. 5 $G_1 \square G_2$

Hence, $(td)_{G_1 \square G_2}(b, d) = (1.1, 1.1, 2.4)$.
Analogously, we can calculate the degree and total degree of all vertices in $G_1 \square G_2$.

Definition 11 Let $X_1 = (\alpha_{X_1}, \gamma_{X_1}, \beta_{X_1})$ and $X_2 = (\alpha_{X_2}, \gamma_{X_2}, \beta_{X_2})$ be spherical fuzzy subsets defined on V_1 and V_2, and let $Y_1 = (\alpha_{Y_1}, \gamma_{Y_1}, \beta_{Y_1})$ and $Y_2 = (\alpha_{Y_2}, \gamma_{Y_2}, \beta_{Y_2})$ be spherical fuzzy subsets on E_1 and E_2, respectively. Then the semi-strong product of two G_1 and G_2 is a SFG $G_1 \bullet G_2 = (X_1 \bullet X_2, Y_1 \bullet Y_2)$ such that:

(i)
$$\begin{aligned}
(\alpha_{X_1} \bullet \alpha_{X_2})(a_1, a_2) &= \min\{\alpha_{X_1}(a_1), \alpha_{X_2}(a_2)\}\\
(\gamma_{X_1} \bullet \gamma_{X_2})(a_1, a_2) &= \min\{\gamma_{X_1}(a_1), \gamma_{X_2}(a_2)\}\\
(\beta_{X_1} \bullet \beta_{X_2})(a_1, a_2) &= \max\{\beta_{X_1}(a_1), \beta_{X_2}(a_2)\} \quad \text{for all } (a_1, a_2) \in V_1 \times V_2,
\end{aligned}$$

(ii)
$$\begin{aligned}
(\alpha_{Y_1} \bullet \alpha_{Y_2})(a, a_2)(a, b_2) &= \min\{\alpha_{X_1}(a), \alpha_{Y_2}(a_2 b_2)\}\\
(\gamma_{Y_1} \bullet \gamma_{Y_2})(a, a_2)(a, b_2) &= \min\{\gamma_{X_1}(a), \gamma_{Y_2}(a_2 b_2)\}\\
(\beta_{Y_1} \bullet \beta_{Y_2})(a, a_2)(a, b_2) &= \max\{\beta_{X_1}(a), \beta_{Y_2}(a_2 b_2)\}\\
&\quad \text{for all } a \in V_1, \quad \text{for all } a_2 b_2 \in E_2,
\end{aligned}$$

(iii) $\left(\alpha_{Y_1} \bullet \alpha_{Y_2}\right)(a_1, a_2)(b_1, b_2) = \min\{\alpha_{Y_1}(a_1b_1), \alpha_{Y_2}(a_2b_2)\}$

$\left(\gamma_{Y_1} \bullet \gamma_{Y_2}\right)(a_1, a_1)(b_1, b_2) = \min\{\gamma_{Y_1}(a_1b_1), \gamma_{Y_2}(a_2b_2)\}$

$\left(\beta_{Y_1} \bullet \beta_{Y_2}\right)(a_1, a_1)(b_1, b_2) = \max\{\beta_{Y_1}(a_1b_1), \beta_{Y_2}(a_2b_2)\}$ for all $a_1b_1 \in E_1$, for all $a_2b_2 \in E_2$.

Proposition 3 *If G_1 and G_2 are SFGs, then $G_1 \bullet G_2$ is a SFG.*

Proof Let $a \in V_1$ and $a_2b_2 \in E_2$. Then we have

$$\begin{aligned}
\left(\alpha_{Y_1} \bullet \alpha_{Y_2}\right)(a, a_2)(a, b_2) &= \min\left(\alpha_{X_1}(a), \alpha_{Y_2}(a_2b_2)\right) \\
&\leq \min\left(\alpha_{X_1}(a), \min\left(\alpha_{X_2}(a_2), \alpha_{X_2}(b_2)\right)\right) \\
&= \min\left(\min\left(\alpha_{X_1}(a), \alpha_{X_2}(a_2)\right), \min\left(\alpha_{X_1}(a), \alpha_{X_2}(b_2)\right)\right) \\
&= \min\left(\left(\alpha_{X_1} \bullet \alpha_{X_2}\right)(a, a_2), \left(\alpha_{X_1} \bullet \alpha_{X_2}\right)(a, b_2)\right),
\end{aligned}$$

$$\begin{aligned}
\left(\gamma_{Y_1} \bullet \gamma_{Y_2}\right)(a, a_2)(a, b_2) &= \min\left(\gamma_{X_1}(a), \gamma_{Y_2}(a_2b_2)\right) \\
&\leq \min\left(\gamma_{X_1}(a), \min\left(\gamma_{X_2}(a_2), \gamma_{X_2}(b_2)\right)\right) \\
&= \min\left(\min\left(\gamma_{X_1}(a), \gamma_{X_2}(a_2)\right), \min\left(\gamma_{X_1}(a), \gamma_{X_2}(b_2)\right)\right) \\
&= \min\left(\left(\gamma_{X_1} \bullet \gamma_{X_2}\right)(a, a_2), \left(\gamma_{X_1} \bullet \gamma_{X_2}\right)(a, b_2)\right),
\end{aligned}$$

$$\begin{aligned}
\left(\beta_{Y_1} \bullet \beta_{Y_2}\right)(a, a_2)(a, b_2) &= \max\left(\beta_{X_1}(a), \beta_{Y_2}(a_2b_2)\right) \\
&\leq \max\left(\beta_{X_1}(a), \max\left(\beta_{X_2}(a_2), \beta_{X_2}(b_2)\right)\right) \\
&= \max\left(\max\left(\beta_{X_1}(a), \beta_{X_2}(a_2)\right), \max\left(\beta_{X_1}(a), \beta_{X_2}(b_2)\right)\right) \\
&= \max\left(\left(\beta_{X_1} \bullet \beta_{X_2}\right)(a, a_2), \left(\beta_{X_1} \bullet \beta_{X_2}\right)(a, b_2)\right).
\end{aligned}$$

Let $a_1b_1 \in E_1$ and $a_2b_2 \in E_2$. Then we have

$$\begin{aligned}
\left(\alpha_{Y_1} \bullet \alpha_{Y_2}\right)(a_1, a_2)(b_1, b_2) &= \min\left(\alpha_{Y_1}(a_1b_1), \alpha_{Y_2}(a_2b_2)\right) \\
&\leq \min\left(\min\left(\alpha_{X_1}(a_1), \alpha_{X_1}(b_1)\right), \min\left(\alpha_{X_2}(a_2), \alpha_{X_2}(b_2)\right)\right) \\
&= \min\left(\min\left(\alpha_{X_1}(a_1), \alpha_{X_2}(a_2)\right), \min\left(\alpha_{X_1}(b_1), \alpha_{X_2}(b_2)\right)\right) \\
&= \min\left(\left(\alpha_{X_1} \bullet \alpha_{X_2}\right)(a_1, a_2), \left(\alpha_{X_1} \bullet \alpha_{X_2}\right)(b_1, b_2)\right),
\end{aligned}$$

$$\begin{aligned}\Big(\gamma_{Y_1} \bullet \gamma_{Y_2}\Big)(a_1, a_2)(b_1, b_2) &= \min\Big(\gamma_{Y_1}(a_1b_1), \gamma_{Y_2}(a_2b_2)\Big)\\ &\leq \min\Big(\min\big(\gamma_{X_1}(a_1), \gamma_{X_1}(b_1)\big), \min\big(\gamma_{X_2}(a_2), \gamma_{X_2}(b_2)\big)\Big)\\ &= \min\Big(\min\big(\gamma_{X_1}(a_1), \gamma_{X_2}(a_2)\big), \min\big(\gamma_{X_1}(b_1), \gamma_{X_2}(b_2)\big)\Big)\\ &= \min\Big(\big(\gamma_{X_1} \bullet \gamma_{X_2}\big)(a_1, a_2), \big(\gamma_{X_1} \bullet \gamma_{X_2}\big)(b_1, b_2)\Big),\end{aligned}$$

$$\begin{aligned}\Big(\beta_{Y_1} \bullet \beta_{Y_2}\Big)(a_1, a_2)(b_1, b_2) &= \max\Big(\beta_{Y_1}(a_1b_1), \beta_{Y_2}(a_2b_2)\Big)\\ &\leq \max\Big(\max\big(\beta_{X_1}(a_1), \beta_{X_1}(b_1)\big), \max\big(\beta_{X_2}(a_2), \beta_{X_2}(b_2)\big)\Big)\\ &= \max\Big(\max\big(\beta_{X_1}(a_1), \beta_{X_2}(a_2)\big), \max\big(\beta_{X_1}(b_1), \beta_{X_2}(b_2)\big)\Big)\\ &= \max\Big(\big(\beta_{X_1} \bullet \beta_{X_2}\big)(a_1, a_2), \big(\beta_{X_1} \bullet \beta_{X_2}\big)(b_1, b_2)\Big).\end{aligned}$$

This completes the proof.

Definition 12 Let $G_1 = (X_1, Y_1)$ and $G_2 = (X_2, Y_2)$ be two SFGs. Then for any vertex, $(a_1, a_2) \in V_1 \times V_2$,

$$\begin{aligned}(d_\alpha)_{G_1 \bullet G_2}(a_1, a_2) &= \sum_{(a_1,a_2)(b_1,b_2)\in E_1 \bullet E_2} \Big(\alpha_{Y_1} \bullet \alpha_{Y_2}\Big)(a_1, a_2)(b_1, b_2)\\ &= \sum_{a_1=b_1,\ a_2b_2\in E_2} \min\{\alpha_{X_1}(a_1), \alpha_{Y_2}(a_2b_2)\}\\ &\quad + \sum_{a_1b_1\in E_1,\ a_2b_2\in E_2} \min\{\alpha_{Y_1}(a_1b_1), \alpha_{Y_2}(a_2b_2)\},\end{aligned}$$

$$\begin{aligned}(d_\gamma)_{G_1 \bullet G_2}(a_1, a_2) &= \sum_{(a_1,a_2)(b_1,b_2)\in E_1 \bullet E_2} \Big(\gamma_{Y_1} \bullet \gamma_{Y_2}\Big)(a_1, a_2)(b_1, b_2)\\ &= \sum_{a_1=b_1,\ a_2b_2\in E_2} \min\{\gamma_{X_1}(a_1), \gamma_{Y_2}(a_2b_2)\}\\ &\quad + \sum_{a_1b_1\in E_1,\ a_2b_2\in E_2} \min\{\gamma_{Y_1}(a_1b_1), \gamma_{Y_2}(a_2b_2)\},\end{aligned}$$

$$\begin{aligned}(d_\beta)_{G_1 \bullet G_2}(a_1, a_2) &= \sum_{(a_1,a_2)(b_1,b_2)\in E_1 \bullet E_2} \Big(\beta_{Y_1} \bullet \beta_{Y_2}\Big)(a_1, a_2)(b_1, b_2)\\ &= \sum_{a_1=b_1,\ a_2b_2\in E_2} \max\{\beta_{X_1}(a_1), \beta_{Y_2}(a_2b_2)\}\\ &\quad + \sum_{a_1b_1\in E_1,\ a_2b_2\in E_2} \max\{\beta_{Y_1}(a_1b_1), \beta_{Y_2}(a_2b_2)\}.\end{aligned}$$

Theorem 5 *Let $G_1 = (X_1, Y_1)$ and $G_2 = (X_2, Y_2)$ be two SFGs. If $\alpha_{X_1} \geq \alpha_{Y_2}$, $\gamma_{X_1} \geq \gamma_{Y_2}$, $\beta_{X_1} \leq \beta_{Y_2}$ and $\alpha_{Y_1} \leq \alpha_{Y_2}$, $\gamma_{Y_1} \leq \gamma_{Y_2}$, $\beta_{Y_1} \geq \beta_{Y_2}$. Then $d_{G_1 \bullet G_2}(a_1, a_2) = d_{G_1}(a_1) + d_{G_2}(a_2)$ for all $(a_1, a_2) \in V_1 \times V_2$.*

Proof By definition of vertex degree of $G_1 \bullet G_2$, we have

$$\begin{aligned}
(d_\alpha)_{G_1 \bullet G_2}(a_1, a_2) &= \sum_{(a_1,a_2)(b_1,b_2)\in E_1 \bullet E_2} \Big(\alpha_{Y_1} \bullet \alpha_{Y_2}\Big)(a_1, a_2)(b_1, b_2) \\
&= \sum_{a_1=b_1,\ a_2b_2\in E_2} \min\{\alpha_{X_1}(a_1), \alpha_{Y_2}(a_2b_2)\} \\
&\quad + \sum_{a_1b_1\in E_1,\ a_2b_2\in E_2} \min\{\alpha_{Y_1}(ab_1), \alpha_{Y_2}(a_2b_2)\} \\
&= \sum_{a_2b_2\in E_2} \alpha_{Y_2}(a_2b_2) \\
&\quad + \sum_{a_1b_1\in E_1} \alpha_{Y_1}(a_1b_1) \ \text{(by using } \alpha_{X_1} \geq \alpha_{Y_2} \text{ and } \alpha_{Y_1} \leq \alpha_{Y_2}) \\
&= (d_\alpha)_{G_2}(a_2) + (d_\alpha)_{G_1}(a_1).
\end{aligned}$$

Similarly, we can show that $(d_\gamma)_{G_1 \bullet G_2}(a_1, a_2) = (d_\gamma)_{G_1}(a_1) + (d_\gamma)_{G_2}(a_2)$ and $(d_\beta)_{G_1 \bullet G_2}(a_1, a_2) = (d_\beta)_{G_1}(a_1) + (d_\beta)_{G_2}(a_2)$. Hence, $d_{G_1 \bullet G_2}(a_1, a_2) = d_{G_1}(a_1) + d_{G_2}(a_2)$.

Definition 13 Let $G_1 = (X_1, Y_1)$ and $G_2 = (X_2, Y_2)$ be two SFGs. Then for any vertex, $(a_1, a_2) \in V_1 \times V_2$,

$$\begin{aligned}
(td_\alpha)_{G_1 \bullet G_2}(a_1, a_2) &= \sum_{(a_1,a_2)(b_1,b_2)\in E_1 \bullet E_2} \Big(\alpha_{Y_1} \bullet \alpha_{Y_2}\Big)(a_1, a_2)(b_1, b_2) \\
&\quad + \Big(\alpha_{X_1} \bullet \alpha_{X_2}\Big)(a_1, a_2) \\
&= \sum_{a_1=b_1,\ a_2b_2\in E_2} \min\{\alpha_{X_1}(a_1), \alpha_{Y_2}(a_2b_2)\} \\
&\quad + \sum_{a_1b_1\in E_1,\ a_2b_2\in E_2} \min\{\alpha_{Y_1}(a_1b_1), \alpha_{Y_2}(a_2b_2)\} \\
&\quad + \min\{\alpha_{X_1}(a_1), \alpha_{X_2}(a_2)\}
\end{aligned}$$

$$\begin{aligned}(td_\gamma)_{G_1 \bullet G_2}(a_1, a_2) = & \sum_{(a_1,a_2)(b_1,b_2)\in E_1 \bullet E_2} \Big(\gamma_{Y_1} \bullet \gamma_{Y_2}\Big)(a_1, a_2)(b_1, b_2) \\ & + \big(\gamma_{X_1} \bullet \gamma_{X_2}\big)(a_1, a_2) \\ = & \sum_{a_1=b_1,\ a_2b_2\in E_2} \min\{\gamma_{X_1}(a_1), \gamma_{Y_2}(a_2b_2)\} \\ & + \sum_{a_1b_1\in E_1,\ a_2b_2\in E_2} \min\{\gamma_{Y_1}(a_1b_1), \gamma_{Y_2}(a_2b_2)\} \\ & + \min\{\gamma_{X_1}(a_1), \gamma_{X_2}(a_2)\}\end{aligned}$$

$$\begin{aligned}(td_\beta)_{G_1 \bullet G_2}(a_1, a_2) = & \sum_{(a_1,a_2)(b_1,b_2)\in E_1 \bullet E_2} \Big(\beta_{Y_1} \bullet \beta_{Y_2}\Big)(a_1, a_2)(b_1, b_2) \\ & + \big(\beta_{X_1} \bullet \beta_{X_2}\big)(a_1, a_2) \\ = & \sum_{a_1=b_1,\ a_2b_2\in E_2} \max\{\beta_{X_1}(a_1), \beta_{Y_2}(a_2b_2)\} \\ & + \sum_{a_1b_1\in E_1,\ a_2b_2\in E_2} \max\{\beta_{Y_1}(a_1b_1), \beta_{Y_2}(a_2b_2)\} \\ & + \max\{\beta_{X_1}(a_1), \beta_{X_2}(a_2)\}.\end{aligned}$$

Theorem 6 *Let $G_1 = (X_1, Y_1)$ and $G_2 = (X_2, Y_2)$ be two SFGs. If*

(i) $\alpha_{X_1} \geq \alpha_{Y_2}$ and $\alpha_{Y_1} \leq \alpha_{Y_2}$, then $(td_\alpha)_{G_1 \bullet G_2}(a_1, a_2) = (td_\alpha)_{G_1}(a_1) + (td_\alpha)_{G_2}(a_2) - \max\{\alpha_{X_1}(a_1), \alpha_{X_2}(a_2)\}$;

(ii) $\gamma_{X_1} \geq \gamma_{Y_2}$ and $\gamma_{Y_1} \leq \gamma_{Y_2}$, then $(td_\gamma)_{G_1 \bullet G_2}(a_1, a_2) = (td_\gamma)_{G_1}(a_1) + (td_\gamma)_{G_2}(a_2) - \max\{\gamma_{X_1}(a_1), \gamma_{X_2}(a_2)\}$;

(iii) $\beta_{X_1} \leq \beta_{Y_2}$ and $\beta_{Y_1} \geq \beta_{Y_2}$, then $(td_\beta)_{G_1 \bullet G_2}(a_1, a_2) = (td_\beta)_{G_1}(a_1) + (td_\beta)_{G_2}(a_2) - \min\{\beta_{X_1}(a_1), \beta_{X_2}(a_2)\}$

for all $(a_1, a_2) \in V_1 \times V_2$.

Proof (i) If $\alpha_{X_1} \geq \alpha_{Y_2}$ and $\alpha_{Y_1} \leq \alpha_{Y_2}$

$$\begin{aligned}
(td_\alpha)_{G_1 \bullet G_2}(a_1, a_2) &= \sum_{(a_1,a_2)(b_1,b_2)\in E_1 \bullet E_2} \Big(\alpha_{Y_1} \bullet \alpha_{Y_2}\Big)(a_1, a_2)(b_1, b_2) \\
&\quad + \big(\alpha_{X_1} \bullet \alpha_{X_2}\big)(a_1, a_2) \\
&= \sum_{a_1=b_1,\ a_2b_2\in E_2} \min\{\alpha_{X_1}(a_1), \alpha_{Y_2}(a_2b_2)\} \\
&\quad + \sum_{a_1b_1\in E_1,\ a_2b_2\in E_2} \min\{\alpha_{Y_1}(a_1b_1), \alpha_{Y_2}(a_2b_2)\} \\
&\quad + \min\{\alpha_{X_1}(a_1), \alpha_{X_2}(a_2)\} \\
&= \sum_{a_2b_2\in E_2} \alpha_{Y_2}(a_2b_2) + \sum_{a_1b_1\in E_1} \alpha_{Y_1}(a_1b_1) + \alpha_{X_1}(a_1) + \alpha_{X_2}(a_2) \\
&\quad - \max\{\alpha_{X_1}(a_1), \alpha_{X_2}(a_2)\} \\
&= (td_\alpha)_{G_2}(a_2) + (td_\alpha)_{G_1}(a_1) - \max\{\alpha_{X_1}(a_1), \alpha_{X_2}(a_2)\}.
\end{aligned}$$

(ii) If $\gamma_{X_1} \geq \gamma_{Y_2}$ and $\gamma_{Y_1} \leq \gamma_{Y_2}$

$$\begin{aligned}
(td_\gamma)_{G_1 \bullet G_2}(a_1, a_2) &= \sum_{(a_1,a_2)(b_1,b_2)\in E_1 \bullet E_2} \Big(\gamma_{Y_1} \bullet \gamma_{Y_2}\Big)(a_1, a_2)(b_1, b_2) \\
&\quad + \big(\gamma_{X_1} \bullet \gamma_{X_2}\big)(a_1, a_2) \\
&= \sum_{a_1=b_1,\ a_2b_2\in E_2} \min\{\gamma_{X_1}(a_1), \gamma_{Y_2}(a_2b_2)\} \\
&\quad + \sum_{a_1b_1\in E_1,\ a_2b_2\in E_2} \min\{\gamma_{Y_1}(a_1b_1), \gamma_{Y_2}(a_2b_2)\} \\
&\quad + \min\{\gamma_{X_1}(a_1), \gamma_{X_2}(a_2)\} \\
&= \sum_{a_2b_2\in E_2} \gamma_{Y_2}(a_2b_2) \\
&\quad + \sum_{a_1b_1\in E_1} \gamma_{Y_1}(a_1b_1) + \gamma_{X_1}(a_1) + \gamma_{X_2}(a_2) - \max\{\gamma_{X_1}(a_1), \gamma_{X_2}(a_2)\} \\
&= (td_\gamma)_{G_2}(a_2) + (td_\gamma)_{G_1}(a_1) - \max\{\gamma_{X_1}(a_1), \gamma_{X_2}(a_2)\}.
\end{aligned}$$

(iii) If $\beta_{X_1} \le \beta_{Y_2}$ and $\beta_{Y_1} \ge \beta_{Y_2}$

$$\begin{aligned}
(td_\beta)_{G_1 \bullet G_2}(a_1, a_2) &= \sum_{(a_1,a_2)(b_1,b_2)\in E_1 \bullet E_2} \left(\beta_{Y_1} \bullet \beta_{Y_2}\right)(a_1, a_2)(b_1, b_2) \\
&\quad + \left(\beta_{X_1} \bullet \beta_{X_2}\right)(a_1, a_2) \\
&= \sum_{a_1=b_1,\ a_2b_2\in E_2} \max\{\beta_{X_1}(a_1), \beta_{Y_2}(a_2b_2)\} \\
&\quad + \sum_{a_1b_1\in E_1,\ a_2b_2\in E_2} \max\{\beta_{Y_1}(a_1b_1), \beta_{Y_2}(a_2b_2)\} \\
&\quad + \max\{\beta_{X_1}(a_1), \beta_{X_2}(a_2)\} \\
&= \sum_{a_2b_2\in E_2} \beta_{Y_2}(a_2b_2) + \sum_{a_1b_1\in E_1} \beta_{Y_1}(a_1b_1) \\
&\quad + \beta_{X_1}(a_1) + \beta_{X_2}(a_2) - \min\{\beta_{X_1}(a_1), \beta_{X_2}(a_2)\} \\
&= (td_\beta)_{G_2}(a_2) + (td_\beta)_{G_1}(a_1) - \min\{\beta_{X_1}(a_1), \beta_{X_2}(a_2)\}.
\end{aligned}$$

Definition 14 Let $X_1 = (\alpha_{X_1}, \gamma_{X_1}, \beta_{X_1})$ and $X_2 = (\alpha_{X_2}, \gamma_{X_2}, \beta_{X_2})$ be spherical fuzzy subsets defined on V_1 and V_2, and let $Y_1 = (\alpha_{Y_1}, \gamma_{Y_1}, \beta_{Y_1})$ and $Y_2 = (\alpha_{Y_2}, \gamma_{Y_2}, \beta_{Y_2})$ be spherical fuzzy subsets on E_1 and E_2, respectively. Then the strong product of G_1 and G_2 is a SFG $G_1 \boxtimes G_2 = (X_1 \boxtimes X_2, Y_1 \boxtimes Y_2)$ such that:

(i) $$\begin{aligned}
\left(\alpha_{X_1} \boxtimes \alpha_{X_2}\right)(a_1, a_2) &= \min\{\alpha_{X_1}(a_1), \alpha_{X_2}(a_2)\} \\
\left(\gamma_{X_1} \boxtimes \gamma_{X_2}\right)(a_1, a_2) &= \min\{\gamma_{X_1}(a_1), \gamma_{X_2}(a_2)\} \\
\left(\beta_{X_1} \boxtimes \beta_{X_2}\right)(a_1, a_2) &= \max\{\beta_{X_1}(a_1), \beta_{X_2}(a_2)\} \text{ for all } (a_1, a_2) \in V_1 \times V_2,
\end{aligned}$$

(ii) $$\begin{aligned}
\left(\alpha_{Y_1} \boxtimes \alpha_{Y_2}\right)(a, a_2)(a, b_2) &= \min\{\alpha_{X_1}(a), \alpha_{Y_2}(a_2b_2)\} \\
\left(\gamma_{Y_1} \boxtimes \gamma_{Y_2}\right)(a, a_2)(a, b_2) &= \min\{\gamma_{X_1}(a), \gamma_{Y_2}(a_2b_2)\} \\
\left(\beta_{Y_1} \boxtimes \beta_{Y_2}\right)(a, a_2)(a, b_2) &= \max\{\beta_{X_1}(a), \beta_{Y_2}(a_2b_2)\} \\
&\quad \text{for all } a \in V_1, \text{ for all } a_2b_2 \in E_2,
\end{aligned}$$

(iii) $$\begin{aligned}
\left(\alpha_{Y_1} \boxtimes \alpha_{Y_2}\right)(a_1, c)(b_1, c) &= \min\{\alpha_{Y_1}(a_1b_1), \alpha_{X_2}(c)\} \\
\left(\gamma_{Y_1} \boxtimes \gamma_{Y_2}\right)(a_1, c)(b_1, c) &= \min\{\gamma_{Y_1}(a_1b_1), \gamma_{X_2}(c)\} \\
\left(\beta_{Y_1} \boxtimes \beta_{Y_2}\right)(a_1, c)(b_1, c) &= \max\{\beta_{Y_1}(a_1b_1), \beta_{X_2}(c)\} \\
&\quad \text{for all } c \in V_2, \text{ for all } a_1b_1 \in E_1.
\end{aligned}$$

(iv) $\big(\alpha_{Y_1} \boxtimes \alpha_{Y_2}\big)(a_1, a_2)(b_1, b_2) = \min\{\alpha_{Y_1}(a_1b_1), \alpha_{Y_2}(a_2b_2)\}$
$\big(\gamma_{Y_1} \boxtimes \gamma_{Y_2}\big)(a_1, a_1)(b_1, b_2) = \min\{\gamma_{Y_1}(a_1b_1), \gamma_{Y_2}(a_2b_2)\}$
$\big(\beta_{Y_1} \boxtimes \beta_{Y_2}\big)(a_1, a_1)(b_1, b_2) = \max\{\beta_{Y_1}(a_1b_1), \beta_{Y_2}(a_2b_2)\}$ for all $a_1b_1 \in E_1$, for all $a_2b_2 \in E_2$.

Proposition 4 *If G_1 and G_2 are SFGs, then $G_1 \boxtimes G_2$ is a SFG.*

Proof Let $a \in V_1$ and $a_2b_2 \in E_2$. Then we have

$$\begin{aligned}
\big(\alpha_{Y_1} \boxtimes \alpha_{Y_2}\big)(a, a_2)(a, b_2) &= \min\big(\alpha_{X_1}(a), \alpha_{Y_2}(a_2b_2)\big)\\
&\leq \min\Big(\alpha_{X_1}(a), \min\big(\alpha_{X_2}(a_2), \alpha_{X_2}(b_2)\big)\Big)\\
&= \min\Big(\min\big(\alpha_{X_1}(a), \alpha_{X_2}(a_2)\big), \min\big(\alpha_{X_1}(a), \alpha_{X_2}(b_2)\big)\Big)\\
&= \min\Big(\big(\alpha_{X_1} \boxtimes \alpha_{X_2}\big)(a, a_2), \big(\alpha_{X_1} \boxtimes \alpha_{X_2}\big)(a, b_2)\Big),
\end{aligned}$$

$$\begin{aligned}
\big(\gamma_{Y_1} \boxtimes \gamma_{Y_2}\big)(a, a_2)(a, b_2) &= \min\big(\gamma_{X_1}(a), \gamma_{Y_2}(a_2b_2)\big)\\
&\leq \min\Big(\gamma_{X_1}(a), \min\big(\gamma_{X_2}(a_2), \gamma_{X_2}(b_2)\big)\Big)\\
&= \min\Big(\min\big(\gamma_{X_1}(a), \gamma_{X_2}(a_2)\big), \min\big(\gamma_{X_1}(a), \gamma_{X_2}(b_2)\big)\Big)\\
&= \min\Big(\big(\gamma_{X_1} \boxtimes \gamma_{X_2}\big)(a, a_2), \big(\gamma_{X_1} \boxtimes \gamma_{X_2}\big)(a, b_2)\Big),
\end{aligned}$$

$$\begin{aligned}
\big(\beta_{Y_1} \boxtimes \beta_{Y_2}\big)(a, a_2)(a, b_2) &= \max\big(\beta_{X_1}(a), \beta_{Y_2}(a_2b_2)\big)\\
&\leq \max\Big(\beta_{X_1}(a), \max\big(\beta_{X_2}(a_2), \beta_{X_2}(b_2)\big)\Big)\\
&= \max\Big(\max\big(\beta_{X_1}(a), \beta_{X_2}(a_2)\big), \max\big(\beta_{X_1}(a), \beta_{X_2}(b_2)\big)\Big)\\
&= \max\Big(\big(\beta_{X_1} \boxtimes \beta_{X_2}\big)(a, a_2), \big(\beta_{X_1} \boxtimes \beta_{X_2}\big)(a, b_2)\Big).
\end{aligned}$$

Let $c \in V_2$ and $a_1b_1 \in E_1$. Then we have

$$\begin{aligned}
\big(\alpha_{Y_1} \boxtimes \alpha_{Y_2}\big)(a_1, c)(b_1, c) &= \min\big(\alpha_{Y_1}(a_1b_1), \alpha_{X_2}(c)\big)\\
&\leq \min\Big(\min\big(\alpha_{X_1}(a_1), \alpha_{X_1}(b_1)\big), \alpha_{X_2}(c)\Big)\\
&= \min\Big(\min\big(\alpha_{X_1}(a_1), \alpha_{X_2}(c)\big), \min\big(\alpha_{X_1}(b_1), \alpha_{X_2}(c)\big)\Big)\\
&= \min\Big(\big(\alpha_{X_1} \boxtimes \alpha_{X_2}\big)(a_1, c), \big(\alpha_{X_1} \boxtimes \alpha_{X_2}\big)(b_1, c)\Big),
\end{aligned}$$

$$\begin{aligned}(\gamma_{Y_1} \boxtimes \gamma_{Y_2})(a_1, c)(b_1, c) &= \min\big(\gamma_{Y_1}(a_1b_1), \gamma_{X_2}(c)\big)\\ &\leq \min\Big(\min\big(\gamma_{X_1}(a_1), \gamma_{X_1}(b_1)\big), \gamma_{X_2}(c)\Big)\\ &= \min\Big(\min\big(\gamma_{X_1}(a_1), \gamma_{X_2}(c)\big), \min\big(\gamma_{X_1}(b_1), \gamma_{X_2}(c)\big)\Big)\\ &= \min\Big(\big(\gamma_{X_1} \boxtimes \gamma_{X_2}\big)(a_1, c), \big(\gamma_{X_1} \boxtimes \gamma_{X_2}\big)(b_1, c)\Big),\end{aligned}$$

$$\begin{aligned}(\beta_{Y_1} \boxtimes \beta_{Y_2})(a_1, c)(b_1, c) &= \max\big(\beta_{Y_1}(a_1b_1), \beta_{X_2}(c)\big)\\ &\leq \max\Big(\max\big(\beta_{X_1}(a_1), \beta_{X_1}(b_1)\big), \beta_{X_2}(c)\Big)\\ &= \max\Big(\max\big(\beta_{X_1}(a_1), \beta_{X_2}(c)\big), \max\big(\beta_{X_1}(b_1), \beta_{X_2}(c)\big)\Big)\\ &= \max\Big(\big(\beta_{X_1} \boxtimes \beta_{X_2}\big)(a_1, c), \big(\beta_{X_1} \boxtimes \beta_{X_2}\big)(b_1, c)\Big).\end{aligned}$$

Let $a_1b_1 \in E_1$ and $a_2b_2 \in E_2$. Then we have

$$\begin{aligned}\Big(\alpha_{Y_1} \boxtimes \alpha_{Y_2}\Big)(a_1, a_2)(b_1, b_2) &= \min\Big(\alpha_{Y_1}(a_1b_1), \alpha_{Y_2}(a_2b_2)\Big)\\ &\leq \min\Big(\min\big(\alpha_{X_1}(a_1), \alpha_{X_1}(b_1)\big), \min\big(\alpha_{X_2}(a_2), \alpha_{X_2}(b_2)\big)\Big)\\ &= \min\Big(\min\big(\alpha_{X_1}(a_1), \alpha_{X_2}(a_2)\big), \min\big(\alpha_{X_1}(b_1), \alpha_{X_2}(b_2)\big)\Big)\\ &= \min\Big(\big(\alpha_{X_1} \boxtimes \alpha_{X_2}\big)(a_1, a_2), \big(\alpha_{X_1} \boxtimes \alpha_{X_2}\big)(b_1, b_2)\Big),\end{aligned}$$

$$\begin{aligned}\Big(\gamma_{Y_1} \boxtimes \gamma_{Y_2}\Big)(a_1, a_2)(b_1, b_2) &= \min\Big(\gamma_{Y_1}(a_1b_1), \gamma_{Y_2}(a_2b_2)\Big)\\ &\leq \min\Big(\min\big(\gamma_{X_1}(a_1), \gamma_{X_1}(b_1)\big), \min\big(\gamma_{X_2}(a_2), \gamma_{X_2}(b_2)\big)\Big)\\ &= \min\Big(\min\big(\gamma_{X_1}(a_1), \gamma_{X_2}(a_2)\big), \min\big(\gamma_{X_1}(b_1), \gamma_{X_2}(b_2)\big)\Big)\\ &= \min\Big(\big(\gamma_{X_1} \boxtimes \gamma_{X_2}\big)(a_1, a_2), \big(\gamma_{X_1} \boxtimes \gamma_{X_2}\big)(b_1, b_2)\Big),\end{aligned}$$

$$\begin{aligned}\Big(\beta_{Y_1} \boxtimes \beta_{Y_2}\Big)(a_1, a_2)(b_1, b_2) &= \max\Big(\beta_{Y_1}(a_1b_1), \beta_{Y_2}(a_2b_2)\Big)\\ &\leq \max\Big(\max\big(\beta_{X_1}(a_1), \beta_{X_1}(b_1)\big), \max\big(\beta_{X_2}(a_2), \beta_{X_2}(b_2)\big)\Big)\\ &= \max\Big(\max\big(\beta_{X_1}(a_1), \beta_{X_2}(a_2)\big), \max\big(\beta_{X_1}(b_1), \beta_{X_2}(b_2)\big)\Big)\\ &= \max\Big(\big(\beta_{X_1} \boxtimes \beta_{X_2}\big)(a_1, a_2), \big(\beta_{X_1} \boxtimes \beta_{X_2}\big)(b_1, b_2)\Big).\end{aligned}$$

This completes the proof.

Definition 15 Let $G_1 = (X_1, Y_1)$ and $G_2 = (X_2, Y_2)$ be two SFGs. Then for any vertex, $(a_1, a_2) \in V_1 \times V_2$,

$$\begin{aligned}
(d_\alpha)_{G_1 \boxtimes G_2}(a_1, a_2) &= \sum_{(a_1,a_2)(b_1,b_2)\in E_1 \boxtimes E_2} \Big(\alpha_{Y_1} \boxtimes \alpha_{Y_2}\Big)(a_1, a_2)(b_1, b_2) \\
&= \sum_{a_1=b_1,\ a_2b_2\in E_2} \min\{\alpha_{X_1}(a_1), \alpha_{Y_2}(a_2b_2)\} \\
&+ \sum_{a_2=b_2,\ a_1b_1\in E_1} \min\{\alpha_{X_2}(a_2), \alpha_{Y_1}(a_1b_1)\} \\
&+ \sum_{a_1b_1\in E_1,\ a_2b_2\in E_2} \min\{\alpha_{Y_1}(a_1b_1), \alpha_{Y_2}(a_2b_2)\},
\end{aligned}$$

$$\begin{aligned}
(d_\gamma)_{G_1 \boxtimes G_2}(a_1, a_2) &= \sum_{(a_1,a_2)(b_1,b_2)\in E_1 \boxtimes E_2} \Big(\gamma_{Y_1} \boxtimes \gamma_{Y_2}\Big)(a_1, a_2)(b_1, b_2) \\
&= \sum_{a_1=b_1,\ a_2b_2\in E_2} \min\{\gamma_{X_1}(a_1), \gamma_{Y_2}(a_2b_2)\} \\
&+ \sum_{a_2=b_2,\ a_1b_1\in E_1} \min\{\gamma_{X_2}(a_2), \gamma_{Y_1}(a_1b_1)\} \\
&+ \sum_{a_1b_1\in E_1,\ a_2b_2\in E_2} \min\{\gamma_{Y_1}(a_1b_1), \gamma_{Y_2}(a_2b_2)\},
\end{aligned}$$

$$\begin{aligned}
(d_\beta)_{G_1 \boxtimes G_2}(a_1, a_2) &= \sum_{(a_1,a_2)(b_1,b_2)\in E_1 \boxtimes E_2} \Big(\beta_{Y_1} \boxtimes \beta_{Y_2}\Big)(a_1, a_2)(b_1, b_2) \\
&= \sum_{a_1=b_1,\ a_2b_2\in E_2} \max\{\beta_{X_1}(a_1), \beta_{Y_2}(a_2b_2)\} \\
&+ \sum_{a_2=b_2,\ a_1b_1\in E_1} \max\{\beta_{X_2}(a_2), \beta_{Y_1}(a_1b_1)\} \\
&+ \sum_{a_1b_1\in E_1,\ a_2b_2\in E_2} \max\{\beta_{Y_1}(a_1b_1), \beta_{Y_2}(a_2b_2)\}.
\end{aligned}$$

Theorem 7 *Let $G_1 = (X_1, Y_1)$ and $G_2 = (X_2, Y_2)$ be two SFGs. If $\alpha_{X_1} \geq \alpha_{Y_2}$, $\gamma_{X_1} \geq \gamma_{Y_2}$, $\beta_{X_1} \leq \beta_{Y_2}$, $\alpha_{X_2} \geq \alpha_{Y_1}$, $\gamma_{X_2} \geq \gamma_{Y_1}$, $\beta_{X_2} \leq \beta_{Y_1}$ and $\alpha_{Y_1} \leq \alpha_{Y_2}$, $\gamma_{Y_1} \leq \gamma_{Y_2}$, $\beta_{Y_1} \geq \beta_{Y_2}$. Then $d_{G_1 \boxtimes G_2}(a_1, a_2) = x_2 d_{G_1}(a_1) + d_{G_2}(a_2)$, where $x_2 = |V_2|$ for all $(a_1, a_2) \in V_1 \times V_2$.*

Proof By definition of vertex degree of $G_1 \boxtimes G_2$, we have

$$\begin{aligned}
(d_\alpha)_{G_1 \boxtimes G_2}(a_1, a_2) &= \sum_{(a_1,a_2)(b_1,b_2)\in E_1 \boxtimes E_2} \left(\alpha_{Y_1} \boxtimes \alpha_{Y_2}\right)(a_1, a_2)(b_1, b_2) \\
&= \sum_{a_1=b_1,\ a_2b_2\in E_2} \min\{\alpha_{X_1}(a_1), \alpha_{Y_2}(a_2b_2)\} \\
&\quad + \sum_{a_2=b_2,\ a_1b_1\in E_1} \min\{\alpha_{X_2}(a_2), \alpha_{Y_1}(a_1b_1)\} \\
&\quad + \sum_{a_1b_1\in E_1,\ a_2b_2\in E_2} \min\{\alpha_{Y_1}(a_1b_1), \alpha_{Y_2}(a_2b_2)\} \\
&= \sum_{a_2b_2\in E_2} \alpha_{Y_2}(a_2b_2) + \sum_{a_1b_1\in E_1} \alpha_{Y_1}(a_1b_1) \\
&\quad + \sum_{a_1b_1\in E_1} \alpha_{Y_1}(a_1b_1) \\
&\quad \text{(by using } \alpha_{X_1} \geq \alpha_{Y_2},\ \alpha_{X_2} \geq \alpha_{Y_1} \text{ and } \alpha_{Y_1} \leq \alpha_{Y_2}) \\
&= x_2(d_\alpha)_{G_1}(a_1) + (d_\alpha)_{G_2}(a_2).
\end{aligned}$$

Similarly, we can show that $(d_\gamma)_{G_1 \boxtimes G_2}(a_1, a_2) = x_2(d_\gamma)_{G_1}(a_1) + (d_\gamma)_{G_2}(a_2)$ and $(d_\beta)_{G_1 \boxtimes G_2}(a_1, a_2) = x_2(d_\beta)_{G_1}(a_1) + (d_\beta)_{G_2}(a_2)$. Hence, $d_{G_1 \boxtimes G_2}(a_1, a_2) = x_2 d_{G_1}(a_1) + d_{G_2}(a_2)$.

Definition 16 Let $G_1 = (X_1, Y_1)$ and $G_2 = (X_2, Y_2)$ be two SFGs. Then for any vertex, $(a_1, a_2) \in V_1 \times V_2$,

$$\begin{aligned}
(td_\alpha)_{G_1 \boxtimes G_2}(a_1, a_2) &= \sum_{(a_1,a_2)(b_1,b_2)\in E_1 \boxtimes E_2} \left(\alpha_{Y_1} \boxtimes \alpha_{Y_2}\right)(a_1, a_2)(b_1, b_2) \\
&\quad + \left(\alpha_{X_1} \boxtimes \alpha_{X_2}\right)(a_1, a_2) \\
&= \sum_{a_1=b_1,\ a_2b_2\in E_2} \min\{\alpha_{X_1}(a_1), \alpha_{Y_2}(a_2b_2)\} \\
&\quad + \sum_{a_2=b_2,\ a_1b_1\in E_1} \min\{\alpha_{X_2}(a_2), \alpha_{Y_1}(a_1b_1)\} \\
&\quad + \sum_{a_1b_1\in E_1,\ a_2b_2\in E_2} \min\{\alpha_{Y_1}(a_1b_1), \alpha_{Y_2}(a_2b_2)\} \\
&\quad + \min\{\alpha_{X_1}(a_1), \alpha_{X_2}(a_2)\},
\end{aligned}$$

$$\begin{aligned}(td_\gamma)_{G_1 \boxtimes G_2}(a_1,a_2) = &\sum_{(a_1,a_2)(b_1,b_2)\in E_1 \boxtimes E_2} \Big(\gamma_{Y_1} \boxtimes \gamma_{Y_2}\Big)(a_1,a_2)(b_1,b_2)\\ &+ \big(\gamma_{X_1} \boxtimes \gamma_{X_2}\big)(a_1,a_2)\\ = &\sum_{a_1=b_1,\ a_2b_2\in E_2} \min\{\gamma_{X_1}(a_1), \gamma_{Y_2}(a_2b_2)\}\\ &+ \sum_{a_2=b_2,\ a_1b_1\in E_1} \min\{\gamma_{X_2}(a_2), \gamma_{Y_1}(a_1b_1)\}\\ &+ \sum_{a_1b_1\in E_1,\ a_2b_2\in E_2} \min\{\gamma_{Y_1}(a_1b_1), \gamma_{Y_2}(a_2b_2)\}\\ &+ \min\{\gamma_{X_1}(a_1), \gamma_{X_2}(a_2)\},\end{aligned}$$

$$\begin{aligned}(td_\beta)_{G_1 \boxtimes G_2}(a_1,a_2) = &\sum_{(a_1,a_2)(b_1,b_2)\in E_1 \boxtimes E_2} \Big(\beta_{Y_1} \boxtimes \beta_{Y_2}\Big)(a_1,a_2)(b_1,b_2)\\ &+ \big(\beta_{X_1} \boxtimes \beta_{X_2}\big)(a_1,a_2)\\ = &\sum_{a_1=b_1,\ a_2b_2\in E_2} \max\{\beta_{X_1}(a_1), \beta_{Y_2}(a_2b_2)\}\\ &+ \sum_{a_2=b_2,\ a_1b_1\in E_1} \max\{\beta_{X_2}(a_2), \beta_{Y_1}(a_1b_1)\}\\ &+ \sum_{a_1b_1\in E_1,\ a_2b_2\in E_2} \max\{\beta_{Y_1}(a_1b_1), \beta_{Y_2}(a_2b_2)\}\\ &+ \max\{\beta_{X_1}(a_1), \beta_{X_2}(a_2)\}.\end{aligned}$$

Theorem 8 *Let $G_1 = (X_1, Y_1)$ and $G_2 = (X_2, Y_2)$ be two SFGs. If*

(i) $\alpha_{X_1} \geq \alpha_{Y_2}$, $\alpha_{X_2} \geq \alpha_{Y_1}$, and $\alpha_{Y_1} \leq \alpha_{Y_2}$, then $(td_\alpha)_{G_1 \circ G_2}(a_1,a_2)$
$= x_2(td_\alpha)_{G_1}(a_1) - (x_2-1)\alpha_{X_1}(a_1) + (td_\alpha)_{G_2}(a_2) - \max\{\alpha_{X_1}(a_1), \alpha_{X_2}(a_2)\}$;
(ii) $\gamma_{X_1} \geq \gamma_{Y_2}$ and $\gamma_{X_2} \geq \gamma_{Y_1}$, and $\gamma_{Y_1} \leq \gamma_{Y_2}$, then $(td_\gamma)_{G_1 \circ G_2}(a_1,a_2)$
$= x_2(td_\gamma)_{G_1}(a_1) - (x_2-1)\gamma_{X_1}(a_1) + (td_\gamma)_{G_2}(a_2) - \max\{\gamma_{X_1}(a_1), \gamma_{X_2}(a_2)\}$;
(iii) $\beta_{X_1} \leq \beta_{Y_2}$ and $\beta_{X_2} \leq \beta_{Y_1}$, and $\beta_{Y_1} \geq \beta_{Y_2}$, then $(td_\beta)_{G_1 \circ G_2}(a_1,a_2)$
$= x_2(td_\beta)_{G_1}(a_1) - (x_2-1)\beta_{X_1}(a_1)$
$+ (td_\alpha)_{G_2}(a_2) - \min\{\alpha_{X_1}(a_1), \alpha_{X_2}(a_2)\}$,

for all $(a_1,a_2) \in V_1 \times V_2$.

Proof The proof is straightforward.

Definition 17 Let $X_1 = (\alpha_{X_1}, \gamma_{X_1}, \beta_{X_1})$ and $X_2 = (\alpha_{X_2}, \gamma_{X_2}, \beta_{X_2})$ be spherical fuzzy subsets defined on V_1 and V_2, and let $Y_1 = (\alpha_{Y_1}, \gamma_{Y_1}, \beta_{Y_1})$ and $Y_2 = (\alpha_{Y_2}, \gamma_{Y_2}, \beta_{Y_2})$ be spherical fuzzy subsets on E_1 and E_2, respectively. Then, we denote the lexicographic product of two SFGs G_1 and G_2 of the graphs G_1^* and G_2^* by $G_1 \circ G_2 = (X_1 \circ X_2, Y_1 \circ Y_2)$, and define as follows:

$$\begin{aligned}
&\text{(i) } (\alpha_{X_1} \circ \alpha_{X_2})(a_1, a_2) = \min\{\alpha_{X_1}(a_1), \alpha_{X_2}(a_2)\}\\
&(\gamma_{X_1} \circ \gamma_{X_2})(a_1, a_2) = \min\{\gamma_{X_1}(a_1), \gamma_{X_2}(a_2)\}\\
&(\beta_{X_1} \circ \beta_{X_2})(a_1, a_2) = \max\{\beta_{X_1}(a_1), \beta_{X_2}(a_2)\} \text{ for all } a_1, a_2 \in V,
\end{aligned}$$

$$\begin{aligned}
&\text{(ii) } (\alpha_{Y_1} \circ \alpha_{Y_2})(a, a_2)(a, b_2) = \min\{\alpha_{X_1}(a), \alpha_{Y_2}(a_2b_2)\}\\
&(\gamma_{Y_1} \circ \gamma_{Y_2})(a, a_2)(a, b_2) = \min\{\gamma_{X_1}(a), \gamma_{Y_2}(a_2b_2)\}\\
&(\beta_{Y_1} \circ \beta_{Y_2})(a, a_2)(a, b_2) = \max\{\beta_{X_1}(a), \beta_{Y_2}(a_2b_2)\} \text{ for all } a \in V_1,\\
&\text{for all } a_2b_2 \in E_2,
\end{aligned}$$

$$\begin{aligned}
&\text{(iii) } (\alpha_{Y_1} \circ \alpha_{Y_2})(a_1, c)(b_1, c) = \min\{\alpha_{Y_1}(a_1b_1), \alpha_{X_2}(c)\}\\
&(\gamma_{Y_1} \circ \gamma_{Y_2})(a_1, c)(b_1, c) = \min\{\gamma_{Y_1}(a_1b_1), \gamma_{X_2}(c)\}\\
&(\beta_{Y_1} \circ \beta_{Y_2})(a_1, c)(b_1, c) = \max\{\beta_{Y_1}(a_1b_1), \beta_{X_2}(c)\} \text{ for all } c \in V_2,\\
&\text{for all } a_1b_1 \in E_1,
\end{aligned}$$

$$\begin{aligned}
&\text{(iv) } (\alpha_{Y_1} \circ \alpha_{Y_2})(a_1, a_2)(b_1, b_2) = \min\{\alpha_{X_2}(a_2), \alpha_{X_2}(b_2), \alpha_{Y_1}(a_1b_1)\}\\
&(\gamma_{Y_1} \circ \gamma_{Y_2})(a_1, a_2)(b_1, b_2) = \min\{\gamma_{X_2}(a_2), \gamma_{X_2}(b_2), \gamma_{Y_1}(a_1b_1)\}\\
&(\beta_{Y_1} \circ \beta_{Y_2})(a_1, a_2)(b_1, b_2) = \max\{\beta_{X_2}(a_2), \beta_{X_2}(b_2), \beta_{Y_1}(a_1b_1)\}\\
&\text{for all } a_1b_1 \in E_1,\quad a_2 \neq b_2.
\end{aligned}$$

Proposition 5 *If G_1 and G_2 are SFGs, then $G_1 \circ G_2$ is a SFG.*

Proof Let $a \in V_1$ and $a_2b_2 \in E_2$. Then we have

$$\begin{aligned}
(\alpha_{Y_1} \circ \alpha_{Y_2})(a, a_2)(a, b_2) &= \min\left(\alpha_{X_1}(a), \alpha_{Y_2}(a_2b_2)\right)\\
&\leq \min\Big(\alpha_{X_1}(a), \min\left(\alpha_{X_2}(a_2), \alpha_{X_2}(b_2)\right)\Big)\\
&= \min\Big(\min\left(\alpha_{X_1}(a), \alpha_{X_2}(a_2)\right), \min\left(\alpha_{X_1}(a), \alpha_{X_2}(b_2)\right)\Big)\\
&= \min\Big(\left(\alpha_{X_1} \circ \alpha_{X_2}\right)(a, a_2), \left(\alpha_{X_1} \circ \alpha_{X_2}\right)(a, b_2)\Big),
\end{aligned}$$

$$\begin{aligned}
\big(\gamma_{Y_1} \circ \gamma_{Y_2}\big)(a, a_2)(a, b_2) &= \min\big(\gamma_{X_1}(a), \gamma_{Y_2}(a_2 b_2)\big) \\
&\le \min\Big(\gamma_{X_1}(a), \min\big(\gamma_{X_2}(a_2), \gamma_{X_2}(b_2)\big)\Big) \\
&= \min\Big(\min\big(\gamma_{X_1}(a), \gamma_{X_2}(a_2)\big), \min\big(\gamma_{X_1}(a), \gamma_{X_2}(b_2)\big)\Big) \\
&= \min\Big(\big(\gamma_{X_1} \circ \gamma_{X_2}\big)(a, a_2), \big(\gamma_{X_1} \circ \gamma_{X_2}\big)(a, b_2)\Big),
\end{aligned}$$

$$\begin{aligned}
\big(\beta_{Y_1} \circ \beta_{Y_2}\big)(a, a_2)(a, b_2) &= \max\big(\beta_{X_1}(a), \beta_{Y_2}(a_2 b_2)\big) \\
&\le \max\Big(\beta_{X_1}(a), \max\big(\beta_{X_2}(a_2), \beta_{X_2}(b_2)\big)\Big) \\
&= \max\Big(\max\big(\beta_{X_1}(a), \beta_{X_2}(a_2)\big), \max\big(\beta_{X_1}(a), \beta_{X_2}(b_2)\big)\Big) \\
&= \max\Big(\big(\beta_{X_1} \circ \beta_{X_2}\big)(a, a_2), \big(\beta_{X_1} \circ \beta_{X_2}\big)(a, b_2)\Big).
\end{aligned}$$

Let $c \in V_2$ and $a_1 b_1 \in E_1$. Then we have

$$\begin{aligned}
\big(\alpha_{Y_1} \circ \alpha_{Y_2}\big)(a_1, c)(b_1, c) &= \min\big(\alpha_{Y_1}(a_1 b_1), \alpha_{X_2}(c)\big) \\
&\le \min\Big(\min\big(\alpha_{X_1}(a_1), \alpha_{X_1}(b_1)\big), \alpha_{X_2}(c)\Big) \\
&= \min\Big(\min\big(\alpha_{X_1}(a_1), \alpha_{X_2}(c)\big), \min\big(\alpha_{X_1}(b_1), \alpha_{X_2}(c)\big)\Big) \\
&= \min\Big(\big(\alpha_{X_1} \circ \alpha_{X_2}\big)(a_1, c), \big(\alpha_{X_1} \circ \alpha_{X_2}\big)(b_1, c)\Big),
\end{aligned}$$

$$\begin{aligned}
\big(\gamma_{Y_1} \circ \gamma_{Y_2}\big)(a_1, c)(b_1, c) &= \min\big(\gamma_{Y_1}(a_1 b_1), \gamma_{X_2}(c)\big) \\
&\le \min\Big(\min\big(\gamma_{X_1}(a_1), \gamma_{X_1}(b_1)\big), \gamma_{X_2}(c)\Big) \\
&= \min\Big(\min\big(\gamma_{X_1}(a_1), \gamma_{X_2}(c)\big), \min\big(\gamma_{X_1}(b_1), \gamma_{X_2}(c)\big)\Big) \\
&= \min\Big(\big(\gamma_{X_1} \circ \gamma_{X_2}\big)(a_1, c), \big(\gamma_{X_1} \circ \gamma_{X_2}\big)(b_1, c)\Big),
\end{aligned}$$

$$\begin{aligned}
\big(\beta_{Y_1} \circ \beta_{Y_2}\big)(a_1, c)(b_1, c) &= \max\big(\beta_{Y_1}(a_1 b_1), \beta_{X_2}(c)\big) \\
&\le \max\Big(\max\big(\beta_{X_1}(a_1), \beta_{X_1}(b_1)\big), \beta_{X_2}(c)\Big) \\
&= \max\Big(\max\big(\beta_{X_1}(a_1), \beta_{X_2}(c)\big), \max\big(\beta_{X_1}(b_1), \beta_{X_2}(c)\big)\Big) \\
&= \max\Big(\big(\beta_{X_1} \circ \beta_{X_2}\big)(a_1, c), \big(\beta_{X_1} \circ \beta_{X_2}\big)(b_1, c)\Big).
\end{aligned}$$

Let $a_1b_1 \in E_1, \quad a_2 \neq b_2$. Then we have

$$\begin{aligned}\left(\alpha_{Y_1} \circ \alpha_{Y_2}\right)(a_1, a_2)(b_1, b_2) &= \min\left(\alpha_{X_2}(a_2), \alpha_{X_2}(b_2), \alpha_{Y_1}(a_1b_1)\right)\\ &\leq \min\left(\alpha_{X_2}(a_2), \alpha_{X_2}(b_2), \min\left(\alpha_{X_1}(a_1), \alpha_{X_1}(b_1)\right)\right)\\ &= \min\left(\min\left(\alpha_{X_1}(a_1), \alpha_{X_2}(a_2)\right), \min\left(\alpha_{X_1}(b_1), \alpha_{X_2}(b_2)\right)\right)\\ &= \min\left(\left(\alpha_{X_1} \circ \alpha_{X_2}\right)(a_1, a_2), \left(\alpha_{X_1} \circ \alpha_{X_2}\right)(b_1, b_2)\right),\end{aligned}$$

$$\begin{aligned}\left(\gamma_{Y_1} \circ \gamma_{Y_2}\right)(a_1, a_2)(b_1, b_2) &= \min\left(\gamma_{X_2}(a_2), \gamma_{X_2}(b_2), \gamma_{Y_1}(a_1b_1)\right)\\ &\leq \min\left(\gamma_{X_2}(a_2), \gamma_{X_2}(b_2), \min\left(\gamma_{X_1}(a_1), \gamma_{X_1}(b_1)\right)\right)\\ &= \min\left(\min\left(\gamma_{X_1}(a_1), \gamma_{X_2}(a_2)\right), \min\left(\gamma_{X_1}(b_1), \gamma_{X_2}(b_2)\right)\right)\\ &= \min\left(\left(\gamma_{X_1} \circ \gamma_{X_2}\right)(a_1, a_2), \left(\gamma_{X_1} \circ \gamma_{X_2}\right)(b_1, b_2)\right),\end{aligned}$$

$$\begin{aligned}\left(\beta_{Y_1} \circ \beta_{Y_2}\right)(a_1, a_2)(b_1, b_2) &= \max\left(\beta_{X_2}(a_2), \beta_{X_2}(b_2), \beta_{Y_1}(a_1b_1)\right)\\ &\leq \max\left(\beta_{X_2}(a_2), \beta_{X_2}(b_2), \max\left(\beta_{X_1}(a_1), \beta_{X_1}(b_1)\right)\right)\\ &= \max\left(\max\left(\beta_{X_1}(a_1), \beta_{X_2}(a_2)\right), \max\left(\beta_{X_1}(b_1), \beta_{X_2}(b_2)\right)\right)\\ &= \max\left(\left(\beta_{X_1} \circ \beta_{X_2}\right)(a_1, a_2), \left(\beta_{X_1} \circ \beta_{X_2}\right)(b_1, b_2)\right).\end{aligned}$$

This completes the proof.

Definition 18 Let $G_1 = (X_1, Y_1)$ and $G_2 = (X_2, Y_2)$ be two SFGs. Then for any vertex, $(a_1, a_2) \in V_1 \times V_2$,

$$\begin{aligned}(d_\alpha)_{G_1 \circ G_2}(a_1, a_2) &= \sum_{(a_1,a_2)(b_1,b_2)\in E_1 \circ E_2} \left(\alpha_{Y_1} \circ \alpha_{Y_2}\right)(a_1, a_2)(b_1, b_2)\\ &= \sum_{a_1=b_1,\ a_2b_2\in E_2} \min\{\alpha_{X_1}(a_1), \alpha_{Y_2}(a_2b_2)\}\\ &+ \sum_{a_2=b_2,\ a_1b_1\in E_1} \min\{\alpha_{X_2}(a_2), \alpha_{Y_1}(a_1b_1)\}\\ &+ \sum_{a_1b_1\in E_1,\ a_2\neq b_2} \min\{\alpha_{Y_1}(a_1b_1), \alpha_{X_2}(a_2), \alpha_{X_2}(b_2)\},\end{aligned}$$

$$(d_\gamma)_{G_1 \circ G_2}(a_1, a_2) = \sum_{(a_1,a_2)(b_1,b_2)\in E_1 \circ E_2} \Big(\gamma_{Y_1} \circ \gamma_{Y_2}\Big)(a_1, a_2)(b_1, b_2)$$
$$= \sum_{a_1=b_1,\ a_2b_2\in E_2} \min\{\gamma_{X_1}(a_1), \gamma_{Y_2}(a_2b_2)\}$$
$$+ \sum_{a_2=b_2,\ a_1b_1\in E_1} \min\{\gamma_{X_2}(a_2), \gamma_{Y_1}(a_1b_1)\}$$
$$+ \sum_{a_1b_1\in E_1,\ a_2\neq b_2} \min\{\gamma_{Y_1}(a_1b_1), \gamma_{X_2}(a_2), \gamma_{X_2}(b_2)\},$$

$$(d_\beta)_{G_1 \circ G_2}(a_1, a_2) = \sum_{(a_1,a_2)(b_1,b_2)\in E_1 \circ E_2} \Big(\beta_{Y_1} \circ \beta_{Y_2}\Big)(a_1, a_2)(b_1, b_2)$$
$$= \sum_{a_1=b_1,\ a_2b_2\in E_2} \max\{\beta_{X_1}(a_1), \beta_{Y_2}(a_2b_2)\}$$
$$+ \sum_{a_2=b_2,\ a_1b_1\in E_1} \max\{\beta_{X_2}(a_2), \beta_{Y_1}(a_1b_1)\}$$
$$+ \sum_{a_1b_1\in E_1,\ a_2\neq b_2} \max\{\beta_{Y_1}(a_1b_1), \beta_{X_2}(a_2), \beta_{X_2}(b_2)\}.$$

Theorem 9 *Let $G_1 = (X_1, Y_1)$ and $G_2 = (X_2, Y_2)$ be two SFGs. If $\alpha_{X_1} \geq \alpha_{Y_2}$, $\gamma_{X_1} \geq \gamma_{Y_2}$, $\beta_{X_1} \leq \beta_{Y_2}$ and $\alpha_{X_2} \geq \alpha_{Y_1}$, $\gamma_{X_2} \geq \gamma_{Y_1}$, $\beta_{X_2} \leq \beta_{Y_1}$. Then $d_{G_1 \circ G_2}(a_1, a_2) = x_2 d_{G_1}(a_1) + d_{G_2}(a_2)$, where $x_2 = |V_2|$ for all $(a_1, a_2) \in V_1 \times V_2$.*

Proof By definition of vertex degree of $G_1 \circ G_2$, we have

$$(d_\alpha)_{G_1 \circ G_2}(a_1, a_2) = \sum_{(a_1,a_2)(b_1,b_2)\in E_1 \circ E_2} \Big(\alpha_{Y_1} \circ \alpha_{Y_2}\Big)(a_1, a_2)(b_1, b_2)$$
$$= \sum_{a_1=b_1,\ a_2b_2\in E_2} \min\{\alpha_{X_1}(a_1), \alpha_{Y_2}(a_2b_2)\}$$
$$+ \sum_{a_2=b_2,\ a_1b_1\in E_1} \min\{\alpha_{X_2}(a_2), \alpha_{Y_1}(a_1b_1)\}$$
$$+ \sum_{a_1b_1\in E_1,\ a_2\neq b_2} \min\{\alpha_{Y_1}(a_1b_1), \alpha_{X_2}(a_2), \alpha_{X_2}(b_2)\}$$
$$= \sum_{a_2b_2\in E_2} \alpha_{Y_2}(a_2b_2) + \sum_{a_1b_1\in E_1} \alpha_{Y_1}(a_1b_1)$$
$$+ \sum_{a_1b_1\in E_1} \alpha_{Y_1}(a_1b_1) \text{ (by using } \alpha_{X_1} \geq \alpha_{Y_2} \text{ and } \alpha_{X_2} \geq \alpha_{Y_1})$$
$$= x_2(d_\alpha)_{G_1}(a_1) + (d_\alpha)_{G_2}(a_2).$$

Similarly, we can show that $(d_\gamma)_{G_1 \circ G_2}(a_1, a_2) = x_2(d_\gamma)_{G_1}(a_1) + (d_\gamma)_{G_2}(a_2)$ and $(d_\beta)_{G_1 \circ G_2}(a_1, a_2) = x_2(d_\beta)_{G_1}(a_1) + (d_\beta)_{G_2}(a_2)$. Hence, $d_{G_1 \circ G_2}(a_1, a_2) = x_2 d_{G_1}(a_1) + d_{G_2}(a_2)$.

Definition 19 Let $G_1 = (X_1, Y_1)$ and $G_2 = (X_2, Y_2)$ be two SFGs. Then for any vertex, $(a_1, a_2) \in V_1 \times V_2$,

$$\begin{aligned}
(td_\alpha)_{G_1 \circ G_2}(a_1, a_2) = & \sum_{(a_1,a_2)(b_1,b_2)\in E_1 \circ E_2} \Big(\alpha_{Y_1} \circ \alpha_{Y_2}\Big)(a_1, a_2)(b_1, b_2) \\
& + \big(\alpha_{X_1} \circ \alpha_{X_2}\big)(a_1, a_2) \\
= & \sum_{a_1=b_1,\ a_2b_2\in E_2} \min\{\alpha_{X_1}(a_1), \alpha_{Y_2}(a_2b_2)\} \\
& + \sum_{a_2=b_2,\ a_1b_1\in E_1} \min\{\alpha_{X_2}(a_2), \alpha_{Y_1}(a_1b_1)\} \\
& + \sum_{a_1b_1\in E_1,\ a_2\neq b_2} \min\{\alpha_{Y_1}(a_1b_1), \alpha_{X_2}(a_2), \alpha_{X_2}(b_2)\} \\
& + \min\{\alpha_{X_1}(a_1), \alpha_{X_2}(a_2)\},
\end{aligned}$$

$$\begin{aligned}
(td_\gamma)_{G_1 \circ G_2}(a_1, a_2) = & \sum_{(a_1,a_2)(b_1,b_2)\in E_1 \circ E_2} \Big(\gamma_{Y_1} \circ \gamma_{Y_2}\Big)(a_1, a_2)(b_1, b_2) \\
& + \big(\gamma_{X_1} \circ \gamma_{X_2}\big)(a_1, a_2) \\
= & \sum_{a_1=b_1,\ a_2b_2\in E_2} \min\{\gamma_{X_1}(a_1), \gamma_{Y_2}(a_2b_2)\} \\
& + \sum_{a_2=b_2,\ a_1b_1\in E_1} \min\{\gamma_{X_2}(a_2), \gamma_{Y_1}(a_1b_1)\} \\
& + \sum_{a_1b_1\in E_1,\ a_2\neq b_2} \min\{\gamma_{Y_1}(a_1b_1), \gamma_{X_2}(a_2), \gamma_{X_2}(b_2)\} \\
& + \min\{\gamma_{X_1}(a_1), \gamma_{X_2}(a_2)\},
\end{aligned}$$

$$\begin{aligned}
(td_\beta)_{G_1 \circ G_2}(a_1, a_2) = & \sum_{(a_1,a_2)(b_1,b_2)\in E_1 \circ E_2} \Big(\beta_{Y_1} \circ \beta_{Y_2}\Big)(a_1, a_2)(b_1, b_2) \\
& + \big(\beta_{X_1} \circ \beta_{X_2}\big)(a_1, a_2) \\
= & \sum_{a_1=b_1,\ a_2b_2\in E_2} \max\{\beta_{X_1}(a_1), \beta_{Y_2}(a_2b_2)\} \\
& + \sum_{a_2=b_2,\ a_1b_1\in E_1} \max\{\beta_{X_2}(a_2), \beta_{Y_1}(a_1b_1)\} \\
& + \sum_{a_1b_1\in E_1,\ a_2\neq b_2} \max\{\beta_{Y_1}(a_1b_1), \beta_{X_2}(a_2), \beta_{X_2}(b_2)\} \\
& + \max\{\beta_{X_1}(a_1), \beta_{X_2}(a_2)\}.
\end{aligned}$$

Theorem 10 *Let $G_1 = (X_1, Y_1)$ and $G_2 = (X_2, Y_2)$ be two SFGs. If*

(i) $\alpha_{X_1} \geq \alpha_{Y_2}$ and $\alpha_{X_2} \geq \alpha_{Y_1}$, then $(td_\alpha)_{G_1 \circ G_2}(a_1, a_2) = x_2(td_\alpha)_{G_1}(a_1) - (x_2 - 1)\alpha_{X_1}(a_1) + (td_\alpha)_{G_2}(a_2) - \max\{\alpha_{X_1}(a_1), \alpha_{X_2}(a_2)\}$;
(ii) $\gamma_{X_1} \geq \gamma_{Y_2}$ and $\gamma_{X_2} \geq \gamma_{Y_1}$, then $(td_\gamma)_{G_1 \circ G_2}(a_1, a_2) = x_2(td_\gamma)_{G_1}(a_1) - (x_2 - 1)\gamma_{X_1}(a_1) + (td_\gamma)_{G_2}(a_2) - \max\{\gamma_{X_1}(a_1), \gamma_{X_2}(a_2)\}$;
(iii) $\beta_{X_1} \leq \beta_{Y_2}$ and $\beta_{X_2} \leq \beta_{Y_1}$, then $(td_\beta)_{G_1 \circ G_2}(a_1, a_2) = x_2(td_\beta)_{G_1}(a_1) - (x_2 - 1)\beta_{X_1}(a_1) + (td_\alpha)_{G_2}(a_2) - \min\{\alpha_{X_1}(a_1), \alpha_{X_2}(a_2)\}$,

for all $(a_1, a_2) \in V_1 \times V_2$.

Proof The proof is straightforward.

3 Applications to Decision Making

Decision-making can be regarded as a problem-solving activity terminated by a solution deemed to be optimal, or at least satisfactory. The group decision-making problems regarding the 'selection of a manager' and 'selection of a player' are presented to illustrate the applicability of proposed notion of SFGs depend on spherical fuzzy preference relations. An algorithm for this selection within the context of spherical fuzzy preference relation is also explored. We present steps of our proposed method for solving the decision making problems.

Step 1: Form a committee of decision-makers, then identify the evaluation criteria.
Step 2: Construct the SFPRs matrices according to decision makers.
Step 3: Define a SFWA operator to fuse the individual SFPRs.
Step 4: Construct the collective SFPR matrix.
Step 5: Draw a directed model relating to a collective SFPR.
Step 7: Draw a partial directed model relating to a collective SFPR by using the condition $\alpha_{jk} \geq 0.5$.
Step 8: Calculate the out-degrees of each alternative a_j in a partial directed network.
Step 9: According to the truthness degrees of out-$d(a_i)$, the ranking order of all choices can be determined.
Step 10: Then, we get the best alternative.

Definition 20 A spherical fuzzy preference relation R on a set of choices $V = \{v_1, v_2, ..., , v_n\}$ is described by a matrix $R = (r_{jk})_{n\times n} \subset V \times V$, where $r_{jk} = (\alpha(v_jv_k), \gamma(v_jv_k), \beta(v_jv_k))$ for all $j, k = 1, 2, ..., n$. Let $r_{jk} = (\alpha_{jk}, \gamma_{jk}, \beta_{jk})$ be a spherical fuzzy value, possessed by the truthness degree α_{jk} to which v_j is preferred to v_k, the falseness degree β_{jk} to which v_j is not preferred to v_k and γ_{jk} indicates the indeterminacy-membership degree, with $\alpha_{jk}, \gamma_{jk}, \beta_{jk} \in [0, 1]$, $0 \leq \alpha_{jk}^2 + \gamma_{jk}^2 + \beta_{jk}^2 \leq 1$, $\alpha_{jk} = \beta_{kj}$, $\beta_{jk} = \alpha_{kj}$, $\gamma_{jk} = \gamma_{kj}$ and $\alpha_{jj} = \gamma_{jj} = \beta_{jj} = 0.5$ for all $j, k = 1, 2, ..., n$.

3.1 Selection of a Manager

Syngenta is a Swiss-based worldwide organization that produces agrochemicals and seeds. As a biotechnology organization, it conducts genomic research. Syngenta require a zonal manager in sales department. Four sales officers are working in the zone and company want to select one of them. These sales officers are Amir a_1, Rashid a_2, Ahmed a_3 and Anwar a_4. To select the attractive zonal manager, three experts e_t $(t = 1, 2, 3)$ are requested to coordinate in the decision analysis, who comes from different departments of Syngenta. The experts compare given choices and give personal judgments using the following spherical fuzzy preference relations (SFPRs) $R_t = (r_{jk}^{(t)})_{4\times 4}$ $(t = 1, 2, 3)$:
The spherical fuzzy digraphs (SFDs) G_t corresponding to SFPRs R_t $(t = 1, 2, 3)$ given in the following matrices, and shown in Fig. 6.

$$R_1 = \begin{bmatrix} (0.5, 0.5, 0.5) & (0.8, 0.2, 0.4) & (0.3, 0.4, 0.7) & (0.6, 0.4, 0.2) \\ (0.4, 0.2, 0.8) & (0.5, 0.5, 0.5) & (0.4, 0.3, 0.8) & (0.7, 0.1, 0.6) \\ (0.7, 0.4, 0.3) & (0.8, 0.3, 0.4) & (0.5, 0.5, 0.5) & (0.2, 0.6, 0.7) \\ (0.2, 0.4, 0.6) & (0.6, 0.1, 0.7) & (0.7, 0.6, 0.2) & (0.5, 0.5, 0.5) \end{bmatrix},$$

$$R_2 = \begin{bmatrix} (0.5, 0.5, 0.5) & (0.7, 0.4, 0.3) & (0.5, 0.2, 0.8) & (0.4, 0.7, 0.2) \\ (0.3, 0.4, 0.7) & (0.5, 0.5, 0.5) & (0.8, 0.1, 0.5) & (0.3, 0.4, 0.7) \\ (0.8, 0.2, 0.5) & (0.5, 0.1, 0.8) & (0.5, 0.5, 0.5) & (0.6, 0.3, 0.4) \\ (0.2, 0.7, 0.4) & (0.7, 0.4, 0.3) & (0.4, 0.3, 0.6) & (0.5, 0.5, 0.5) \end{bmatrix},$$

$$R_3 = \begin{bmatrix} (0.5, 0.5, 0.5) & (0.4, 0.2, 0.7) & (0.3, 0.5, 0.6) & (0.8, 0.2, 0.3) \\ (0.7, 0.2, 0.4) & (0.5, 0.5, 0.5) & (0.2, 0.4, 0.6) & (0.6, 0.3, 0.4) \\ (0.6, 0.5, 0.3) & (0.6, 0.4, 0.2) & (0.5, 0.5, 0.5) & (0.9, 0.1, 0.2) \\ (0.3, 0.2, 0.8) & (0.4, 0.3, 0.6) & (0.2, 0.1, 0.9) & (0.5, 0.5, 0.5) \end{bmatrix}.$$

Use the accumulation operator to combine all the individual SFPRs $R_t = (r_{jk}^{(t)})_{4\times 4}$ $(t = 1, 2, 3)$ into the collective $R = (r_{jk})_{4\times 4}$. We define spherical fuzzy weighted averaging operator on the basis of spherical fuzzy weighted averaging operator. Here, we use the spherical fuzzy weighted averaging (SFWA) operator to combine the individual SFPR. Thus, we have

$$SFWA(r_{jk}^{(1)}, r_{jk}^{(2)}, ..., r_{jk}^{(t)})$$

$$= \left(\sqrt{1 - \left(\prod_{s=1}^{t} (1 - (\alpha_{jk}^{2})^{(s)}) \right)^{\frac{1}{t}}}, \left(\prod_{s=1}^{t} (\gamma_{jk}^{(s)}) \right)^{\frac{1}{t}}, \left(\prod_{s=1}^{t} (\beta_{jk}^{(s)} + \gamma_{jk}^{(s)}) \right)^{\frac{1}{t}} - \left(\prod_{s=1}^{t} (\gamma_{jk}^{(s)}) \right)^{\frac{1}{t}} \right),$$

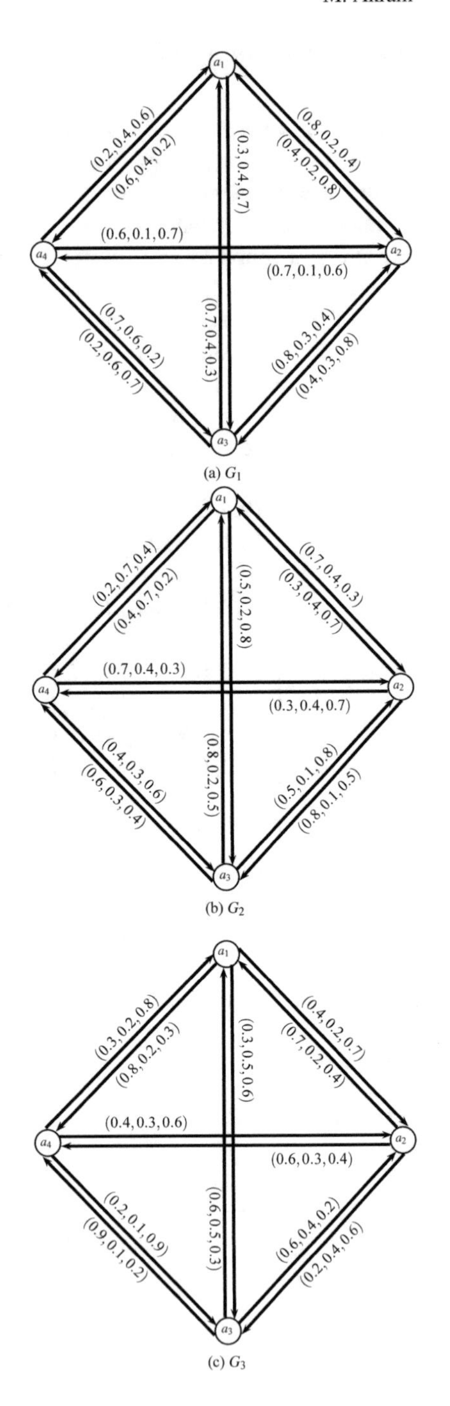

Fig. 6 Spherical fuzzy digraphs

we get the collective SFPR;

$$R = \begin{bmatrix} (0.5000, 0.5000, 0.5000) & (0.6785, 0.2555, 0.4699) & (0.3814, 0.3457, 0.7192) & (0.6468, 0.3863, 0.2629) \\ (0.5169, 0.2555, 0.6164) & (0.5000, 0.5000, 0.5000) & (0.5789, 0.2323, 0.6396) & (0.5745, 0.2323, 0.5832) \\ (0.7118, 0.3457, 0.3884) & (0.6631, 0.2323, 0.4931) & (0.5000, 0.5000, 0.5000) & (0.7126, 0.2656, 0.3859) \\ (0.2375, 0.3863, 0.6455) & (0.5895, 0.2323, 0.5653) & (0.5041, 0.2656, 0.6317) & (0.5000, 0.5000, 0.5000) \end{bmatrix}.$$

In the directed model relating to a collective SFPR above, as shown in Fig. 7, we select those spherical fuzzy numbers whose truthness degrees $\alpha_{jk} \geq 0.5$ $(j, k = 1, 2, 3, 4)$, and resulting partial model is shown in Fig. 8.

Compute the out-degrees out-$d(a_j)$ $(j = 1, 2, 3, 4)$ in a partial directed model as follows:

$$out - d(a_1) = (1.3253, 0.6418, 0.7328), \quad out - d(a_2) = (1.6703, 0.7201, 1.8392),$$
$$out - d(a_3) = (2.0875, 0.8436, 1.2674), \quad out - d(a_4) = (1.0936, 0.4979, 1.1970).$$

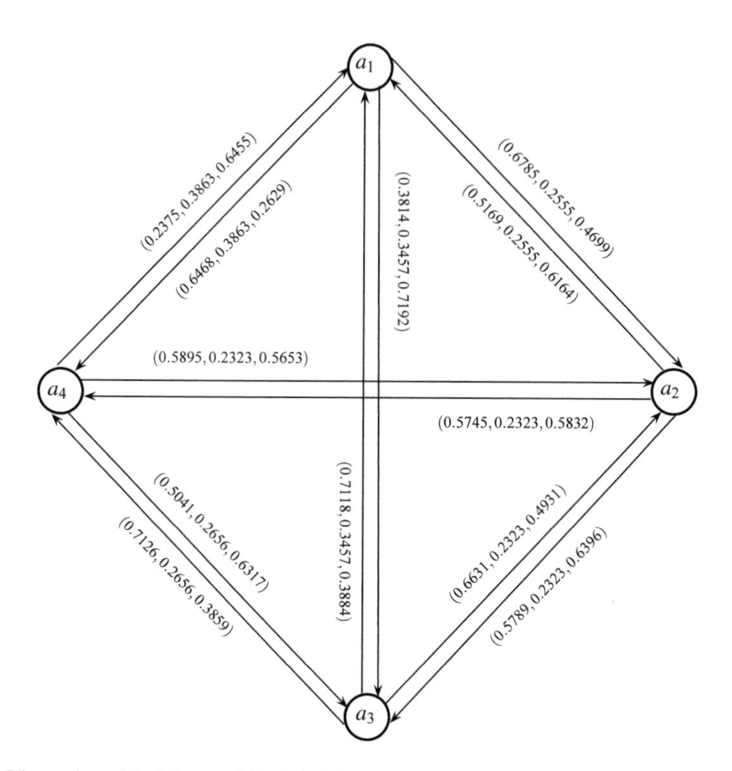

Fig. 7 Directed model of the combined SFPR

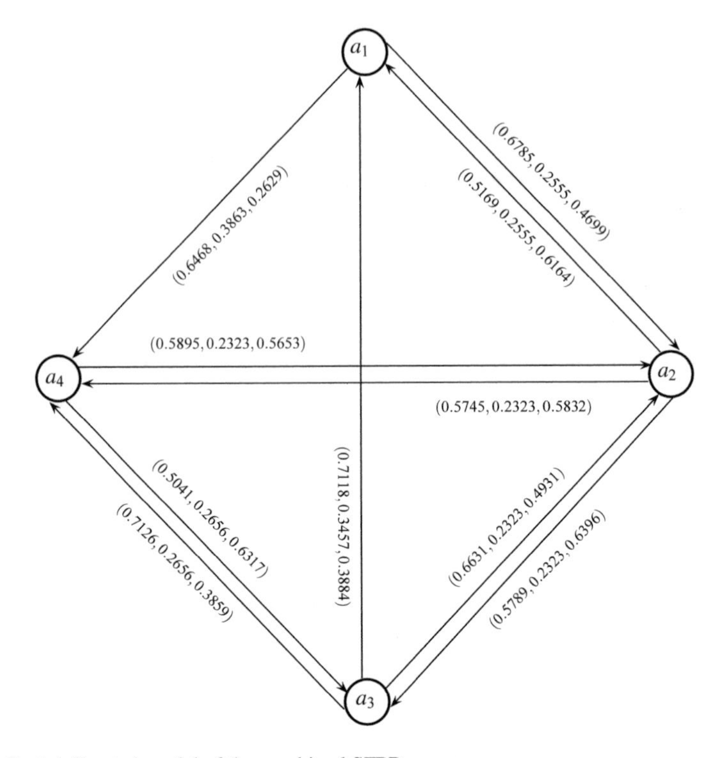

Fig. 8 Partial directed model of the combined SFPR

As indicated by truthness degrees of out-$d(a_j)$ $(j = 1, 2, 3, 4)$, we get the ranking of the variables $a_j (j = 1, 2, 3, 4)$ as:

$$a_3 \succ a_2 \succ a_1 \succ a_4.$$

Thus, the best choice is Ahmed a_3.
Now, using spherical fuzzy weighted geometric (SFWG) operator (Mahmood et al. 2018),

$$SFWG(r_{jk}^{(1)}, r_{jk}^{(2)}, ..., r_{jk}^{(t)}) = \Bigg(\Big(\prod_{s=1}^{t} (\alpha_{kj}^{(s)} + \gamma_{jk}^{(s)}) \Big)^{\frac{1}{t}} - \Big(\prod_{s=1}^{t} (\gamma_{jk}^{(s)}) \Big)^{\frac{1}{t}}, \Big(\prod_{s=1}^{t} (\gamma_{jk}^{(s)}) \Big)^{\frac{1}{t}},$$
$$\sqrt{1 - \Big(\prod_{s=1}^{t} (1 - (\beta_{jk}^2)^{(s)}) \Big)^{\frac{1}{t}}} \Bigg),$$

we get the collective SFPR;

$$R = \begin{bmatrix} (0.5000, 0.5000, 0.5000) & (0.6164, 0.2555, 0.5169) & (0.3884, 0.3457, 0.7118) & (0.6455, 0.3863, 0.2375) \\ (0.4699, 0.2555, 0.6785) & (0.5000, 0.5000, 0.5000) & (0.4931, 0.2323, 0.6658) & (0.5653, 0.2323, 0.5895) \\ (0.7192, 0.3457, 0.3814) & (0.6396, 0.2323, 0.5789) & (0.5000, 0.5000, 0.5000) & (0.6317, 0.2656, 0.5041) \\ (0.2629, 0.3863, 0.6468) & (0.5832, 0.2323, 0.5745) & (0.3859, 0.2656, 0.7126) & (0.5000, 0.5000, 0.5000) \end{bmatrix}.$$

In the directed model relating to a collective SFPR above, as shown in Fig. 9, we select those spherical fuzzy numbers whose truthness degrees $\alpha_{jk} \geq 0.5$ $(j, k = 1, 2, 3, 4)$, and resulting partial model is shown in Fig. 10.

Compute the out-degrees out-$d(a_j)$ $(j = 1, 2, 3, 4)$ in a partial directed model as follows:

$$out - d(a_1) = (1.2619, 0.6418, 0.7544), \quad out - d(a_2) = (0.5653, 0.2323, 0.5895),$$
$$out - d(a_3) = (1.9905, 0.8436, 1.4644), \quad out - d(a_4) = (0.5832, 0.2323, 0.5745).$$

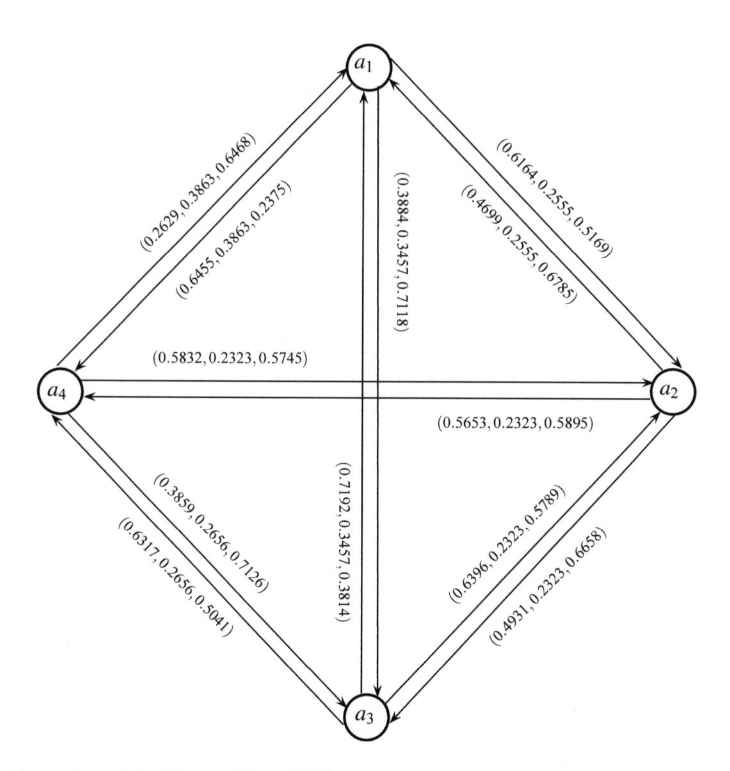

Fig. 9 Directed model of the combined SFPR

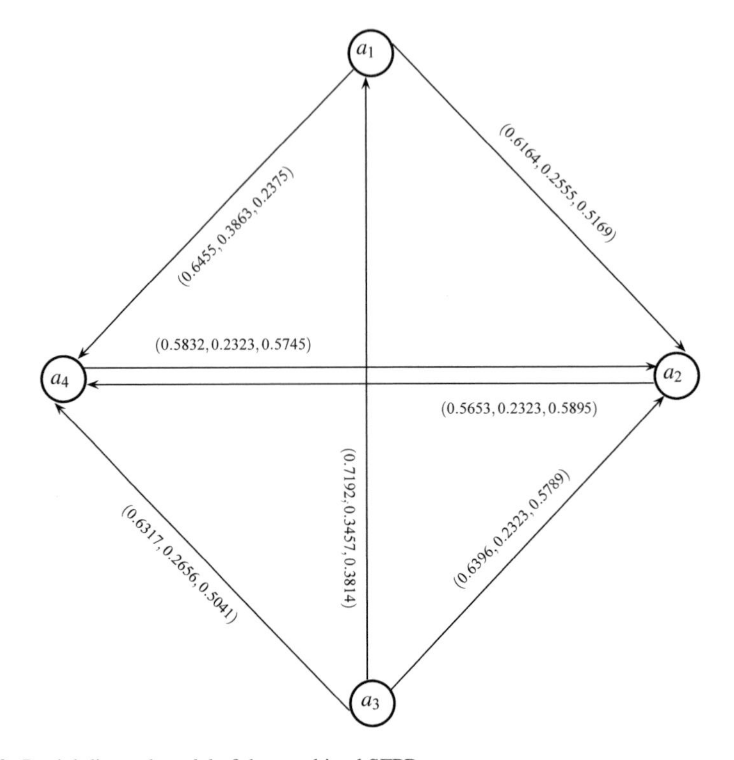

Fig. 10 Partial directed model of the combined SFPR

As indicated by truthness degrees of out-$d(a_j)$ $(j = 1, 2, 3, 4)$, we get the ranking of the variables $a_j (j = 1, 2, 3, 4)$ as:

$$a_3 \succ a_1 \succ a_4 \succ a_2.$$

Thus, the best choice is Ahmed a_3.

3.2 Selection of a Player

The Pakistan Billiard and Snooker Federation (PBSF) is the national overseeing body to create and advance the games of Billiards and snooker in Pakistan. The Federation was established in 1958. PBSF wants to select a best player to play in World Snooker Championship. World Snooker Championship was a professional snooker tournament, held in England. There are five players which are short listed after preliminary screening. They are Abrar a_1, Yousaf a_2, Sajjad a_3, Hamza a_4, Asif a_5. To select the best player, four experts e_t $(t = 1, 2, 3, 4)$ are requested to cooperate in the decision analysis, who comes from various departments of PBSF. The experts compare given choices and give personal judgments using the following spherical fuzzy preference relations (SFPRs) $R_t = (r_{jk}^{(t)})_{5\times 5}$ $(t = 1, 2, 3, 4)$:

The SFDGs G_t corresponding to SFPRs R_t $(t = 1, 2, 3, 4)$ given in following matrices, are shown in Fig. 11.

$$R_1 = \begin{bmatrix} (0.5, 0.5, 0.5) & (0.6, 0.4, 0.3) & (0.4, 0.5, 0.7) & (0.4, 0.2, 0.8) & (0.5, 0.3, 0.6) \\ (0.3, 0.4, 0.6) & (0.5, 0.5, 0.5) & (0.8, 0.4, 0.3) & (0.3, 0.7, 0.6) & (0.4, 0.5, 0.7) \\ (0.7, 0.5, 0.4) & (0.3, 0.4, 0.8) & (0.5, 0.5, 0.5) & (0.7, 0.3, 0.4) & (0.3, 0.6, 0.5) \\ (0.8, 0.2, 0.4) & (0.6, 0.7, 0.3) & (0.4, 0.3, 0.7) & (0.5, 0.5, 0.5) & (0.5, 0.4, 0.7) \\ (0.6, 0.3, 0.5) & (0.7, 0.5, 0.4) & (0.5, 0.6, 0.3) & (0.7, 0.4, 0.5) & (0.5, 0.5, 0.5) \end{bmatrix},$$

$$R_2 = \begin{bmatrix} (0.5, 0.5, 0.5) & (0.3, 0.7, 0.5) & (0.6, 0.3, 0.7) & (0.4, 0.5, 0.6) & (0.7, 0.4, 0.3) \\ (0.5, 0.7, 0.3) & (0.5, 0.5, 0.5) & (0.3, 0.6, 0.5) & (0.5, 0.2, 0.7) & (0.4, 0.8, 0.3) \\ (0.7, 0.3, 0.6) & (0.5, 0.6, 0.3) & (0.5, 0.5, 0.5) & (0.8, 0.2, 0.4) & (0.3, 0.9, 0.1) \\ (0.6, 0.5, 0.4) & (0.7, 0.2, 0.5) & (0.4, 0.2, 0.8) & (0.5, 0.5, 0.5) & (0.2, 0.9, 0.3) \\ (0.3, 0.4, 0.7) & (0.3, 0.8, 0.4) & (0.1, 0.9, 0.3) & (0.3, 0.9, 0.2) & (0.5, 0.5, 0.5) \end{bmatrix},$$

$$R_3 = \begin{bmatrix} (0.5, 0.5, 0.5) & (0.8, 0.2, 0.5) & (0.4, 0.6, 0.3) & (0.2, 0.8, 0.4) & (0.7, 0.3, 0.6) \\ (0.5, 0.2, 0.8) & (0.5, 0.5, 0.5) & (0.6, 0.4, 0.5) & (0.5, 0.3, 0.8) & (0.4, 0.6, 0.2) \\ (0.3, 0.6, 0.4) & (0.5, 0.4, 0.6) & (0.5, 0.5, 0.5) & (0.7, 0.4, 0.3) & (0.5, 0.4, 0.7) \\ (0.4, 0.8, 0.2) & (0.8, 0.3, 0.5) & (0.3, 0.4, 0.7) & (0.5, 0.5, 0.5) & (0.6, 0.3, 0.4) \\ (0.6, 0.3, 0.7) & (0.2, 0.6, 0.4) & (0.7, 0.4, 0.5) & (0.4, 0.3, 0.6) & (0.5, 0.5, 0.5) \end{bmatrix},$$

$$R_4 = \begin{bmatrix} (0.5, 0.5, 0.5) & (0.3, 0.9, 0.1) & (0.7, 0.4, 0.3) & (0.4, 0.2, 0.8) & (0.6, 0.3, 0.4) \\ (0.1, 0.9, 0.3) & (0.5, 0.5, 0.5) & (0.6, 0.3, 0.2) & (0.2, 0.4, 0.7) & (0.4, 0.5, 0.1) \\ (0.3, 0.4, 0.7) & (0.2, 0.3, 0.6) & (0.5, 0.5, 0.5) & (0.3, 0.7, 0.4) & (0.7, 0.2, 0.4) \\ (0.8, 0.2, 0.4) & (0.7, 0.4, 0.2) & (0.4, 0.7, 0.3) & (0.5, 0.5, 0.5) & (0.1, 0.9, 0.2) \\ (0.4, 0.3, 0.6) & (0.1, 0.5, 0.4) & (0.4, 0.2, 0.7) & (0.2, 0.9, 0.1) & (0.5, 0.5, 0.5) \end{bmatrix}.$$

Use the accumulation operator to combine all the individual SFPRs $R_t = (r_{jk}^{(t)})_{5\times 5}$ $(t = 1, 2, 3, 4)$ into the collective $R = (r_{jk})_{5\times 5}$. Here, we use the SFWA operator to combine the individual SFPR. Thus, we have

$$SFWA(r_{jk}^{(1)}, r_{jk}^{(2)}, \ldots, r_{jk}^{(t)})$$
$$= \left(\sqrt{1 - \left(\prod_{s=1}^{t}(1 - (\alpha_{jk}^{2})^{(s)})\right)^{\frac{1}{t}}}, \left(\prod_{s=1}^{t}(\gamma_{jk}^{(s)})\right)^{\frac{1}{t}}, \left(\prod_{s=1}^{t}(\beta_{jk}^{(s)} + \gamma_{jk}^{(s)})\right)^{\frac{1}{t}} - \left(\prod_{s=1}^{t}(\gamma_{jk}^{(s)})\right)^{\frac{1}{t}}\right),$$

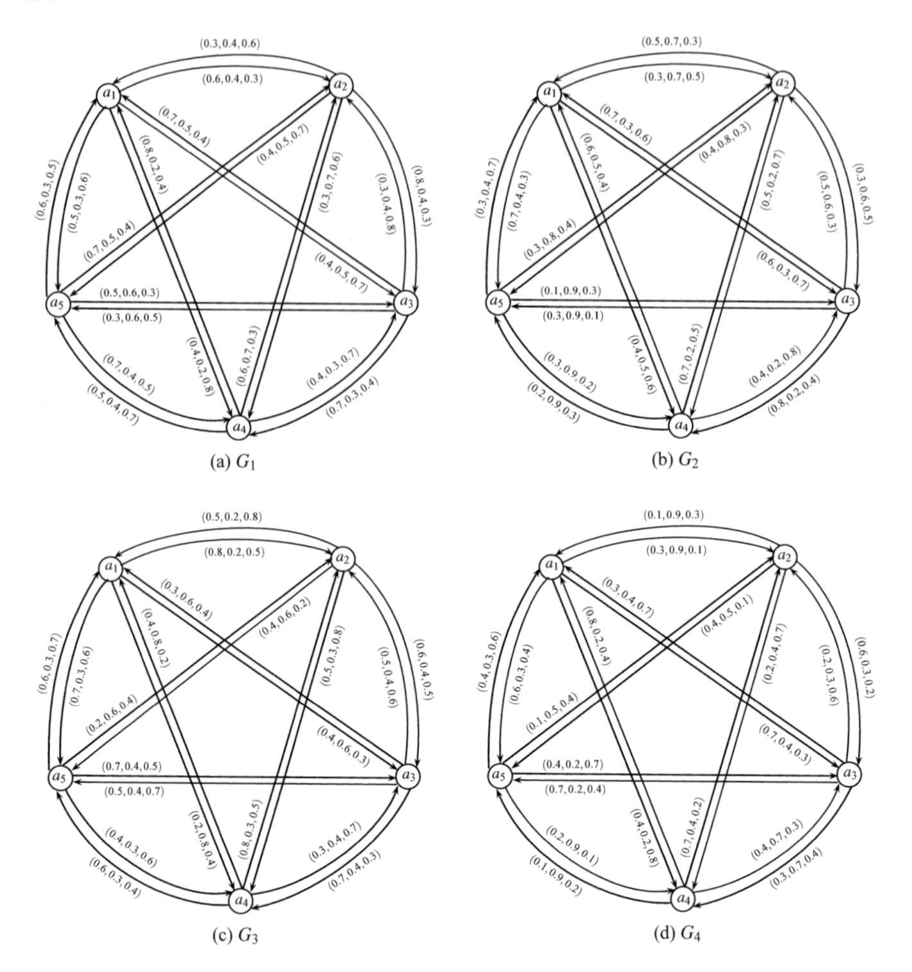

Fig. 11 Spherical fuzzy digraphs

we get the collective SFPR:

$$R = \begin{bmatrix} (0.5000,0.5000,0.5000) & (0.5823,0.4738,0.4019) & (0.5543,0.4356,0.4969) & (0.3626,0.3557,0.7162) & (0.6368,0.3224,0.4713) \\ (0.3953,0.4738,0.6110) & (0.5000,0.5000,0.5000) & (0.6283,0.4120,0.3552) & (0.4034,0.3600,0.7308) & (0.4000,0.5886,0.3036) \\ (0.5645,0.4356,0.5360) & (0.4034,0.4120,0.5809) & (0.5000,0.5000,0.5000) & (0.6780,0.3600,0.3941) & (0.4998,0.4559,0.4672) \\ (0.6973,0.3557,0.3988) & (0.7108,0.3600,0.4041) & (0.3781,0.3600,0.6641) & (0.5000,0.5000,0.5000) & (0.4221,0.5584,0.4457) \\ (0.5020,0.3224,0.6210) & (0.4302,0.5886,0.4013) & (0.4990,0.4559,0.5112) & (0.4667,0.5584,0.4132) & (0.5000,0.5000,0.5000) \end{bmatrix}.$$

In the directed model relating to a collective SFPR above, as shown in Fig. 12, we select those spherical fuzzy numbers whose truthness degrees $\alpha_{jk} \geq 0.5$ $(j, k = 1, 2, 3, 4, 5)$, and resulting partial model is shown in Fig. 13.

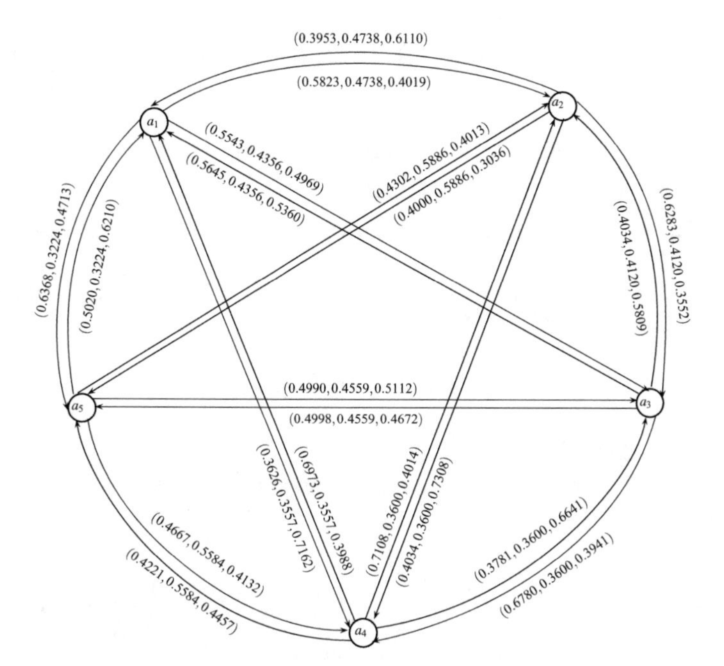

Fig. 12 Directed model of the combined SFPR

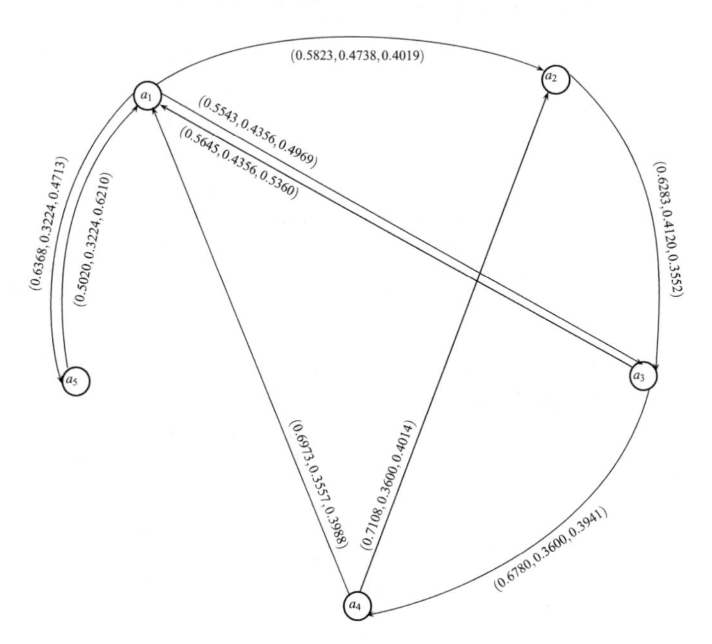

Fig. 13 Partial directed model of the combined SFPR

Compute the out-degrees out-$d(a_j)$ $(j = 1, 2, 3, 4, 5)$ in a partial directed model as follows:

$$\begin{aligned}
&\text{out}-d(a_1) = (1.7734, 1.2318, 1.3701),\ \text{out}-d(a_2) = (0.6283, 0.4120, 0.3552),\\
&\text{out}-d(a_3) = (1.2425, 0.7956, 0.9301),\\
&\text{out}-d(a_4) = (1.4081, 0.7157, 0.8002),\ \text{out}-d(a_5) = (0.5020, 0.3224, 0.6210).
\end{aligned}$$

As indicated by truthness degrees of out-$d(a_j)$ $(j = 1, 2, 3, 4, 5)$, we get the ranking of the variables $a_j (j = 1, 2, 3, 4, 5)$ as:

$$a_1 \succ a_4 \succ a_3 \succ a_2 \succ a_5.$$

Thus, the best choice is Abrar a_1.

Now, using spherical fuzzy weighted geometric (SFWG) operator (Mahmood et al. 2018),

$$SFWG(r_{jk}^{(1)}, r_{jk}^{(2)}, ..., r_{jk}^{(t)}) = \Big(\Big(\prod_{s=1}^{t}(\alpha_{kj}^{(s)} + \gamma_{jk}^{(s)})\Big)^{\frac{1}{t}} - \Big(\prod_{s=1}^{t}(\gamma_{jk}^{(s)})\Big)^{\frac{1}{t}},$$
$$\Big(\prod_{s=1}^{t}(\gamma_{jk}^{(s)})\Big)^{\frac{1}{t}}, \sqrt{1 - \Big(\prod_{s=1}^{t}(1 - (\beta_{jk}^{2})^{(s)})\Big)^{\frac{1}{t}}}\Big),$$

we get the collective SFPR:

$$R = \begin{bmatrix}
(0.5000,0.5000,0.5000) & (0.6110,0.4738,0.3953) & (0.5360,0.4356,0.5645) & (0.3988,0.3557,0.6973) & (0.6210,0.3224,0.5020)\\
(0.4019,0.4738,0.5823) & (0.5000,0.5000,0.5000) & (0.5809,0.4120,0.4034) & (0.4014,0.3600,0.7108) & (0.4013,0.5886,0.4302)\\
(0.4969,0.4356,0.5543) & (0.3552,0.4120,0.6283) & (0.5000,0.5000,0.5000) & (0.6641,0.3600,0.3781) & (0.5112,0.4559,0.6990)\\
(0.7162,0.3557,0.3626) & (0.7308,0.3600,0.4034) & (0.3941,0.3600,0.6780) & (0.5000,0.5000,0.5000) & (0.4132,0.5584,0.4667)\\
(0.4713,0.3224,0.6368) & (0.3036,0.5886,0.4000) & (0.4672,0.4559,0.4998) & (0.4457,0.5584,0.4221) & (0.5000,0.5000,0.5000)
\end{bmatrix}.$$

In the directed model relating to a collective SFPR above, as shown in Fig. 14, we select those spherical fuzzy numbers whose truthness degrees $\alpha_{jk} \geq 0.5$ $(j, k = 1, 2, 3, 4, 5)$, and resulting partial model is shown in Fig. 15.

Compute the out-degrees out-$d(a_j)$ $(j = 1, 2, 3, 4, 5)$ in a partial directed model as follows:

$$\begin{aligned}
&\text{out} - d(a_1) = (1.7680, 1.2318, 1.4618),\ \text{out} - d(a_2) = (0.5809, 04120, 0.4034),\\
&\text{out} - d(a_3) = (1.5713, 0.8159, 0.8771),\\
&\text{out} - d(a_4) = (1.4470, 0.7157, 0.7660),\ \text{out} - d(a_5) = (0.0000, 0.0000, 0.0000).
\end{aligned}$$

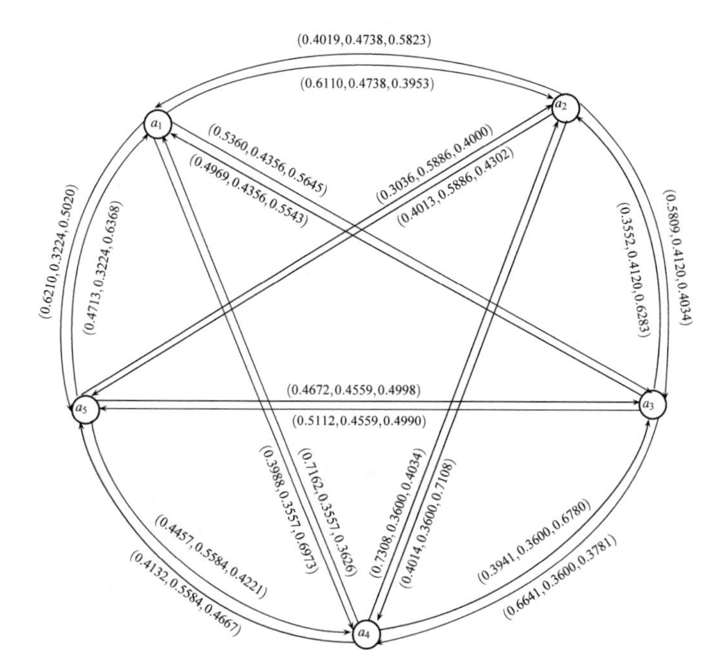

Fig. 14 Directed model of the combined SFPR

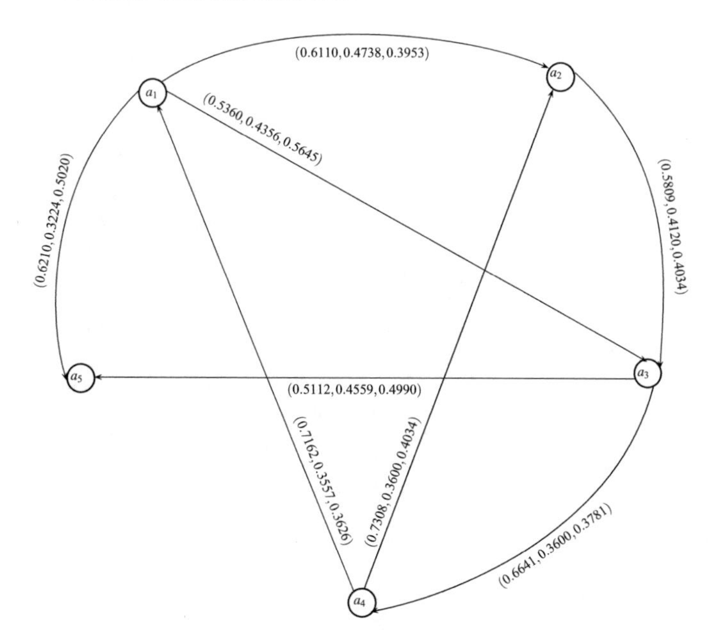

Fig. 15 Partial directed model of the combined SFPR

As indicated by truthness degrees of out-$d(a_j)$ $(j = 1, 2, 3, 4, 5)$, we get the ranking of the variables $a_j (j = 1, 2, 3, 4, 5)$ as:

$$a_1 \succ a_4 \succ a_3 \succ a_2 \succ a_5.$$

Thus, the best choice is Abrar a_1. A similar process can be applied for SFWG operator to choose the best alternative.

4 Conclusions

Spherical fuzzy model deals with uncertainty problems more efficiently with the constraint $0 \leq \alpha^2 + \gamma^2 + \beta^2 \leq 1$, providing a vast space to appoint degrees of onefs own choice as compared to picture fuzzy model. This paper has considered concept of SFGs and some operations on SFGs. The study has provided a novel approach to decision-making in order to highlight the applicability of proposed concept in real world. In future, we will extend this concept to (1) Interval-valued spherical fuzzy graphs; (2) Complex spherical fuzzy graphs; and (3) Hesitant spherical fuzzy graphs.

Conflict of interest The authors declare that they have no conflict of interest regarding the publication of the research article.

References

Akram M, Ali G (2018) Hybrid models for decision-making based on rough Pythagorean fuzzy bipolar soft information. Granular Comput. https://doi.org/10.1007/s41066-018-0132-3
Akram M, Davvaz B (2012) Strong intuitionistic fuzzy graphs. Filomat 26:177–196
Akram M, Habib A (2019) q-Rung picture fuzzy graphs: a creative view on regularity with applications. J Appl Math Comput 61:235–280
Akram M, Naz S (2018) Energy of Pythagorean fuzzy graphs with applications. Mathematics 6:136. https://doi.org/10.3390/math6080136
Akram M, Ashraf A, Sarwar M (2014) Novel applications of intuitionistic fuzzy digraphs in decision support systems. Scientific World J. Article ID 904606, 11 pages
Akram M, Habib A, Ilyas F, Dar JM (2018a) Specific types of Pythagorean fyzzy graphs and application to decision-making. Math Comput Appl 23:42. https://doi.org/10.3390/mca23030042
Akram M, Dar JM, Farooq A (2018b) Planar graphs under Pythagorean fuzzy environment. Mathematics 6:278. https://doi.org/10.3390/math6120278
Akram M, Dar JM, Naz S (2019a) Certain graphs under Pythagorean fuzzy environment. Complex Intell Syst 5:127–144
Akram M, Habib A, Koam AN (2019b) A novel description on edge-regular q-rung picture fuzzy graphs with application. Symmetry 11:489. https://doi.org/10.3390/sym11040489
Akram M, Habib A, Davvaz B (2019c) Direct sum of n Pythagorean fuzzy graphs with application to group decision-making. J Multiple-Valued Logic Soft Comput 33:75–115
Ashraf S, Abdullah S, Mahmood T, Ghani F (2019) Spherical fuzzy sets and their applications in multi-attribute decision making problems. J Intell Fuzzy Syst 36:2829–2844
Atanassov KT (1986) Intuitionistic fuzzy sets. Fuzzy Sets Syst 20:87–96

Cuong BC (2013a) Picture fuzzy sets: first results, Part 1. In: Seminar neuro-fuzzy systems with applications, preprint 03/2013, Institute of Mathematics, Vietnam Academy of Science and Technology, Hanoi-Vietnam
Cuong BC (2013b) Picture fuzzy sets: first results, Part 2. In: Seminar neuro-fuzzy systems with applications, preprint 04/2013, Institute of Mathematics, Vietnam Academy of Science and Technology, Hanoi-Vietnam
Cuong BC (2014) Picture fuzzy sets. J Comput Sci Cybern 30:409–420
Cuong BC, Kreinovich V (2013) Picture fuzzy sets: a new concept for computational intelligence problems. In: Proceedings of the 3rd world congress on information and communication technologies (WICT 2013), Hanoi, Vietnam, IEEE CS, pp 1–6. https://doi.org/10.1109/WICT.2013.7113099.
Cuong BC, Cuong C, Hai PV (2015) Some fuzzy logic operators for picture fuzzy sets. In: Seventh international conference on knowledge and systems engineering (KSE). IEEE. https://doi.org/10.1109/KSE.2015.20
Garg H (2017) Some picture fuzzy aggregation operators and their applications to multicriteria decision-making. Arab J Sci Eng 42:5275–5290
Gündo FK, Kahraman C (2019) Spherical fuzzy sets and spherical fuzzy TOPSIS method. J Intell Fuzzy Syst 36:337–352
Habib A, Akram M, Farooq A (2019) q-Rung orthopair fuzzy competition graphs with application in soil ecosystem. Mathematics 7: 91. https://doi.org/10.3390/math7010091.
Kaufmann A (1973) Introduction a la Theorie des Sour-ensembles Flous. Masson et Cie, Vol. 1
Mahmood T, Ullah K, Khan Q, Jan N (2018) An approach toward decision-making and medical diagnosis problems using the concept of spherical fuzzy sets. J Neutral Comput Appl 31:7041–7053
Naz S, Ashraf S, Akram M (2018) A novel approach to decision-making with Pythagorean fuzzy information. Mathematics 6: 95. https://doi.org/10.3390/math6060095
Parvathi R, Karunambigai MG (2006) Intuitionistic fuzzy graphs. Computational intelligence, theory and applications. Springer, Berlin, Heidelberg, pp 139–150
Rosenfeld A(1975) Fuzzy graphs. In: Zadeh LA, Fu KS, Shimura M (eds) Fuzzy sets and their applications. Academic Press, New York, pp 77–95
Sarwar M, Akram M (2016) An algorithm for computing certain metrics in intuitionistic fuzzy graphs. J Intell Fuzzy Syst 30:2405–2416
Singh P (2015) Correlation coefficients for picture fuzzy sets. J Intell Fuzzy Syst 28:591–604
Wang CY, Zhou XQ, Tu XN, Tao SD (2017) Some geometric aggregation operators based on picture fuzzy sets and their application in multiple attribute decision making. Ital J Pure Appl Math 37:477–492
Yager RR (2013) Pythagorean fuzzy subsets. In: 2013 joint IFSA world congress and NAFIPS annual meeting (IFSA/NAFIPS). https://doi.org/10.1109/IFSA-NAFIPS.2013.6608375
Zadeh LA (1965) Fuzzy sets. Inf Control 8:338–353
Zadeh LA (1971) Similarity relations and fuzzy orderings. Inf Sci 3:177–200
Zhang H, Zhang R, Huang H, Wang J (2018) Some picture fuzzy Dombi Heronian mean operators with their application to multi-attribute decision-making. Symmetry 10:593. https://doi.org/10.3390/sym10110593

Spherical Fuzzy Multi-criteria Decision Making

Optimal Site Selection of Electric Vehicle Charging Station by Using Spherical Fuzzy TOPSIS Method

Fatma Kutlu Gündoğdu and Cengiz Kahraman

Abstract TOPSIS method is a distance based multi-criteria decision making tool which is one of the most used methods. It has been extended by using each of the recently proposed fuzzy set types such as type-2 fuzzy TOPSIS, hesitant fuzzy TOPSIS, intuitionistic fuzzy TOPSIS, and Pythagorean fuzzy TOPSIS. As a new extension of Picture fuzzy sets, the emerging spherical fuzzy sets (SFS) and spherical fuzzy TOPSIS method have been proposed by Kutlu Gündoğdu and Kahraman (2019a). In spherical fuzzy sets, the sum of membership, non-membership and hesitancy degrees must satisfy the condition $0 \leq \mu^2 + v^2 + \pi^2 \leq 1$ in which these parameters are assigned independently. In this chapter, spherical fuzzy TOPSIS method is used in solving a multiple criteria selection problem for optimal site selection of electric vehicle charging station to verify the developed approach and to demonstrate its practicality and effectiveness. A comparative analysis with single-valued spherical fuzzy CODAS method is also performed.

1 Introduction

After the demonstration of ordinary fuzzy sets by Zadeh (1965), they have been very popular in almost all branches of science. Various researchers (Zadeh 1965; Smarandache 1998; Grattan-Guinness 1976; Sambuc 1975; Zadeh 1975; Jahn 1975; Atanassov 1986; Torra 2010; Yager 1986, 2013, 2016; Garibaldi and Ozen 2007; Cuong and Kreinovich 2013) have developed several extensions of ordinary fuzzy sets. The spherical fuzzy sets introduced in the literature are based on the fact that

F. Kutlu Gündoğdu (✉)
Industrial Engineering Department, National Defence University, Turkish Air Force Academy, 34149 Istanbul, Turkey
e-mail: fatmakutlugundogdu@gmail.com

C. Kahraman
Industrial Engineering Department, Istanbul Technical University, 34367 Besiktas, Istanbul, Turkey
e-mail: kahramanc@itu.edu.tr

© Springer Nature Switzerland AG 2021

C. Kahraman and F. Kutlu Gündoğdu (eds.), *Decision Making with Spherical Fuzzy Sets*, Studies in Fuzziness and Soft Computing 392,
https://doi.org/10.1007/978-3-030-45461-6_8

the hesitancy of a decision maker can be defined independently from membership and non-membership degrees. The novel concept of Spherical Fuzzy Sets provides a larger preference domain for decision makers to assign membership degrees since the squared sum of the spherical parameters is allowed to be at most 1.0. DMs can determine their hesitancy degree independently under spherical fuzzy environment. Spherical fuzzy Sets (SFS) are a generalization of Pythagorean Fuzzy Sets, and picture fuzzy sets (Kahraman and Kutlu Gündoğdu 2018; Kutlu Gündoğdu and Kahraman 2019a, b, c; Kahraman et al. 2019; Kutlu Gündoğdu 2019). There are several articles on spherical fuzzy sets (Ashraf et al. 2019; Ulah et al. 2019; Mahmood et al. 2019).

The crisp TOPSIS method proposed by Hwang and Yoon (1981) is a simple and useful method to solve multi-criteria decision making (MCDM) problems aiming at choosing the best alternative with the shortest distance from the positive ideal solution (PIS) and the farthest distance from the negative ideal solution (NIS). Some researchers have extended the TOPSIS method for solving MCDM problems under a variety of different fuzzy environments over the last decades such as ordinary fuzzy TOPSIS (Chen 2000), interval-valued fuzzy TOPSIS (Chen and Tsao 2008; Ye 2010) IF-TOPSIS (Boran et al. 2009, 2012; Vahdani et al. 2013; Büyüközkan and Güleryüz 2016) hesitant fuzzy TOPSIS (Xu and Zhang 2013), hesitant fuzzy linguistic TOPSIS (Beg and Rashid 2013), Neutrosophic TOPSIS (Broumi et al. 2015; Biswas et al. 2016; Elhassouny and Smarandache 2016), Pythagorean fuzzy TOPSIS (Cevik Onar et al. 2018), Interval-valued PF-TOPSIS (Sajjad et al. 2018), Spherical fuzzy TOPSIS (Kutlu Gündoğdu and Kahraman 2019a, b) and interval-valued spherical fuzzy TOPSIS (Kutlu Gündoğdu and Kahraman 2019b).

In this chapter, spherical fuzzy TOPSIS method is used in solving a multiple criteria selection problem for optimal site selection of electric vehicle charging station.

Rest of the chapter is organized as follows: In Sect. 2, preliminaries of spherical fuzzy sets are presented. In Sect. 3, spherical fuzzy TOPSIS method is given. In Sect. 4, a descriptive application is conducted. In Sect. 5, a comparative analysis is performed and results are discussed. The paper ends with conclusion and suggestions for further researches.

2 Spherical Fuzzy Sets: Preliminaries

In the following, preliminaries of SFS are presented (Kutlu Gündoğdu and Kahraman 2019a):

Definition 2.1 Single valued Spherical Fuzzy Sets (SFS) $\tilde{A}_S$ of the universe of discourse U is given by

$$\tilde{A}_S = \left\{ \langle u, (\mu_{\tilde{A}_S}(u), \nu_{\tilde{A}_S}(u), \pi_{\tilde{A}_S}(u)) \middle| u \in U \right\} \quad (1)$$

where

$$\mu_{\tilde{A}_S}(u) : U \to [0, 1], \nu_{\tilde{A}_S}(u) : U \to [0, 1], \pi_{\tilde{A}_S}(u) : U \to [0, 1]$$

and

$$0 \leq \mu^2_{\tilde{A}_S}(u) + \nu^2_{\tilde{A}_S}(u) + \pi^2_{\tilde{A}_S}(u) \leq 1 \quad \forall u \in U \tag{2}$$

For each u, the numbers $\mu_{\tilde{A}_S}(u)$, $\nu_{\tilde{A}_S}(u)$ and $\pi_{\tilde{A}_S}(u)$ are the degree of membership, non-membership and hesitancy of u to $\tilde{A}_S$, respectively. $\chi_{\tilde{A}_S}(u) = \sqrt{1 - \mu^2_{\tilde{A}_S}(u) - \nu^2_{\tilde{A}_S}(u) - \pi^2_{\tilde{A}_S}(u)}$ is called as a refusal degree.

Definition 2.2 Basic operators of Single-valued SFS;

i. $$\tilde{A}_S \oplus \tilde{B}_S = \left\{ \begin{array}{l} \left(\mu^2_{\tilde{A}_S} + \mu^2_{\tilde{B}_S} - \mu^2_{\tilde{A}_S}\mu^2_{\tilde{B}_S}\right)^{1/2}, v_{\tilde{A}_S}v_{\tilde{B}_S}, \\ \left(\left(1 - \mu^2_{\tilde{B}_S}\right)\pi^2_{\tilde{A}_S} + \left(1 - \mu^2_{\tilde{A}_S}\right)\pi^2_{\tilde{B}_S} - \pi^2_{\tilde{A}_S}\pi^2_{\tilde{B}_S}\right)^{1/2} \end{array} \right\} \tag{3}$$

ii. $$\tilde{A}_S \otimes \tilde{B}_S = \left\{ \begin{array}{l} \mu_{\tilde{A}_S}\mu_{\tilde{B}_S}, \left(v^2_{\tilde{A}_S} + v^2_{\tilde{B}_S} - v^2_{\tilde{A}_S}v^2_{\tilde{B}_S}\right)^{1/2}, \\ \left(\left(1 - v^2_{\tilde{B}_S}\right)\pi^2_{\tilde{A}_S} + \left(1 - v^2_{\tilde{A}_S}\right)\pi^2_{\tilde{B}_S} - \pi^2_{\tilde{A}_S}\pi^2_{\tilde{B}_S}\right)^{1/2} \end{array} \right\} \tag{4}$$

iii. $$\lambda \cdot \tilde{A}_S = \left\{ \begin{array}{l} \left(1 - \left(1 - \mu^2_{\tilde{A}_S}\right)^{\lambda}\right)^{1/2}, v^{\lambda}_{\tilde{A}_S}, \\ \left(\left(1 - \mu^2_{\tilde{A}_S}\right)^{\lambda} - \left(1 - \mu^2_{\tilde{A}_S} - \pi^2_{\tilde{A}_S}\right)^{\lambda}\right)^{1/2} \end{array} \right\} \quad for\, \lambda \geq 0 \tag{5}$$

iv. $$\tilde{A}^{\lambda}_S = \left\{ \begin{array}{l} \mu^{\lambda}_{\tilde{A}_S}, \left(1 - \left(1 - v^2_{\tilde{A}_S}\right)^{\lambda}\right)^{1/2}, \\ \left(\left(1 - v^2_{\tilde{A}_S}\right)^{\lambda} - \left(1 - v^2_{\tilde{A}_S} - \pi^2_{\tilde{A}_S}\right)^{\lambda}\right)^{1/2} \end{array} \right\} \quad for\, \lambda \geq 0 \tag{6}$$

Definition 2.3 For these SFS $\tilde{A}_S = (\mu_{\tilde{A}_S}, v_{\tilde{A}_S}, \pi_{\tilde{A}_S})$ and $\tilde{B}_S = (\mu_{\tilde{B}_S}, v_{\tilde{B}_S}, \pi_{\tilde{B}_S})$, the followings are valid under the condition $\lambda, \lambda_1, \lambda_2 \geq 0$.

i. $$\tilde{A}_S \oplus \tilde{B}_S = \tilde{B}_S \oplus \tilde{A}_S \tag{7}$$

ii. $$\tilde{A}_S \otimes \tilde{B}_S = \tilde{B}_S \otimes \tilde{A}_S \tag{8}$$

iii. $$\lambda(\tilde{A}_S \oplus \tilde{B}_S) = \lambda\tilde{A}_S \oplus \lambda\tilde{B}_S \tag{9}$$

iv. $$\lambda_1\tilde{A}_S \oplus \lambda_2\tilde{A}_S = (\lambda_1 + \lambda_2)\tilde{A}_S \tag{10}$$

v. $(\tilde{A}_S \otimes \tilde{B}_S)^\lambda = \tilde{A}_S^\lambda \otimes \tilde{B}_S^\lambda$ (11)

vi. $\tilde{A}_S^{\lambda_1} \otimes \tilde{A}_S^{\lambda_2} = \tilde{A}_S^{\lambda_1+\lambda_2}$ (12)

Definition 2.4 Single-valued Spherical Weighted Arithmetic Mean (SWAM) with respect to, $w = (w_1, w_2 \ldots, w_n)$; $w_i \in [0, 1]$; $\sum_{i=1}^n w_i = 1$, SWAM is defined as;

$$SWAM_w(\tilde{A}_{S1}, \ldots, \tilde{A}_{Sn}) = w_1\tilde{A}_{S1} + w_2\tilde{A}_{S2} + \cdots + w_n\tilde{A}_{Sn}$$
$$= \left\{ \begin{array}{l} \left[1 - \prod_{i=1}^{n}(1 - \mu_{\tilde{A}_{Si}}^2)^{w_i}\right]^{1/2}, \\ \prod_{i=1}^{n} v_{\tilde{A}_{Si}}^{w_i}, \left[\prod_{i=1}^{n}(1 - \mu_{\tilde{A}_{Si}}^2)^{w_i} - \prod_{i=1}^{n}(1 - \mu_{\tilde{A}_{Si}}^2 - \pi_{\tilde{A}_{Si}}^2)^{w_i}\right]^{1/2} \end{array} \right\} \tag{13}$$

Definition 2.5 Single-valued Spherical Weighted Geometric Mean (SWGM) with respect to,$w = (w_1, w_2 \ldots, w_n)$; $w_i \in [0, 1]$; $\sum_{i=1}^n w_i = 1$, SWGM is defined as;

$$SWAM_w(\tilde{A}_1, \ldots, \tilde{A}_n) = \tilde{A}_{S1}^{w_1} + \tilde{A}_{S2}^{w_2} + \cdots + \tilde{A}_{Sn}^{w_n}$$
$$= \left\{ \begin{array}{l} \prod_{i=1}^{n} \mu_{\tilde{A}_{Si}}^{w_i} \left[1 - \prod_{i=1}^{n}(1 - v_{\tilde{A}_{Si}}^2)^{w_i}\right]^{1/2}, \\ \left[\prod_{i=1}^{n}(1 - v_{\tilde{A}_{Si}}^2)^{w_i} - \prod_{i=1}^{n}(1 - v_{\tilde{A}_{Si}}^2 - \pi_{\tilde{A}_{Si}}^2)^{w_i}\right]^{\frac{1}{2}} \end{array} \right\} \tag{14}$$

Definition 2.6 Score functions and Accuracy functions of sorting SFS are defined by;

$$Score\left(\tilde{A}_S\right) = \left(2\mu_{\tilde{A}_S} - \frac{\pi_{\tilde{A}_S}}{2}\right)^2 - \left(v_{\tilde{A}_S} - \frac{\pi_{\tilde{A}_S}}{2}\right)^2 \tag{15}$$

$$Accuracy\left(\tilde{A}_S\right) = \mu_{\tilde{A}_S}^2 + v_{\tilde{A}_S}^2 + \pi_{\tilde{A}_S}^2 \tag{16}$$

Note that: $\tilde{A}_S < \tilde{B}_S$ if and only if

i. $Score(\tilde{A}_S) < Score(\tilde{B}_S)$

or

ii. $Score(\tilde{A}_S) = Score(\tilde{B}_S)$ and $Accuracy(\tilde{A}_S) < Accuracy(\tilde{B}_S)$

2.1 Spherical Fuzzy TOPSIS Method

The proposed single-valued spherical fuzzy TOPSIS (SF-TOPSIS) method is composed of several steps as given in this section. Before giving these steps, we present the flow chart of the SF-TOPSIS in Fig. 1.

A MCDM problem can be presented as a decision matrix whose elements show the evaluation values of all alternatives with respect to each criterion under spherical fuzzy environment. Let $X = \{x_1, x_2, \ldots x_m\}$ $(m \geq 2)$ be a discrete set of m feasible alternatives and $C = \{C_1, C_2, \ldots C_n\}$ be a finite set of criteria, and $w = \{w_1, w_2, \ldots w_n\}$ be the weights vector of all criteria which satisfies $0 \leq w_j \leq 1$ and $\sum_{j=1}^{n} w_j = 1$. The following methodology is proposed by Kutlu Gündoğdu in 2019.

Step 1: Let Decision makers fill in the decision matrices using the linguistic terms given in Table 1. DMs can also evaluate the criteria in order to define their weights by using Table 1.

Step 2: Aggregate the judgments of each decision maker (DM) using Spherical Weighted Arithmetic Mean (SWAM) by Eq. (17). Based on their experiences, each DM may have a different importance weight, w_j, $j = 1, 2, \ldots, n$

$$SWAM_w(\tilde{A}_{S1}, \ldots, \tilde{A}_{Sn}) = w_1\tilde{A}_{S1} + w_2\tilde{A}_{S2} + \cdots + w_n\tilde{A}_{Sn}$$
$$= \left\{ \begin{array}{c} \left[1 - \prod_{i=1}^{n}(1 - \mu_{\tilde{A}_{Si}}^2)^{w_i}\right]^{1/2}, \\ \prod_{i=1}^{n} v_{\tilde{A}_{Si}}^{w_i}, \left[\prod_{i=1}^{n}(1 - \mu_{\tilde{A}_{Si}}^2)^{w_i} - \prod_{i=1}^{n}(1 - \mu_{\tilde{A}_{Si}}^2 - \pi_{\tilde{A}_{Si}}^2)^{w_i}\right]^{1/2} \end{array} \right\} \tag{17}$$

Step 3: Aggregate the criteria weights.

Step 4: Construct the aggregated spherical fuzzy decision matrix based on the individual decision matrices.

Denote the evaluation values of alternative $X_i (i = 1, 2 \ldots m)$ with respect to criterion $C_j (j = 1, 2 \ldots n)$ by $C_j(X_i) = (\mu_{ij}, v_{ij}, \pi_{ij}).D = \left(C_j(X_i)\right)_{mxn}$ is a spherical fuzzy decision matrix. For a MCDM problem with SFSs, the aggregated decision matrix $D = \left(C_j(X_i)\right)_{mxn}$ should be constructed as in Eq. (18).

$$D = \left(C_j(X_i)\right)_{mxn} = \begin{pmatrix} (\mu_{11}, v_{11}, \pi_{11}) & (\mu_{12}, v_{12}, \pi_{12}) & \ldots & (\mu_{1n}, v_{1n}, \pi_{1n}) \\ (\mu_{21}, v_{21}, \pi_{21}) & (\mu_{22}, v_{22}, \pi_{22}) & \ldots & (\mu_{2n}, v_{2n}, \pi_{2n}) \\ \vdots & \vdots & & \vdots \\ (\mu_{m1}, v_{m1}, \pi_{m1}) & (\mu_{m2}, v_{m2}, \pi_{m2}) & \ldots & (\mu_{mn}, v_{mn}, \pi_{mn}) \end{pmatrix} \tag{18}$$

Decision makers evaluate the alternatives with respect to the criteria as if they were all benefit criteria such that they assign a lower linguistic term if it is a cost criterion.

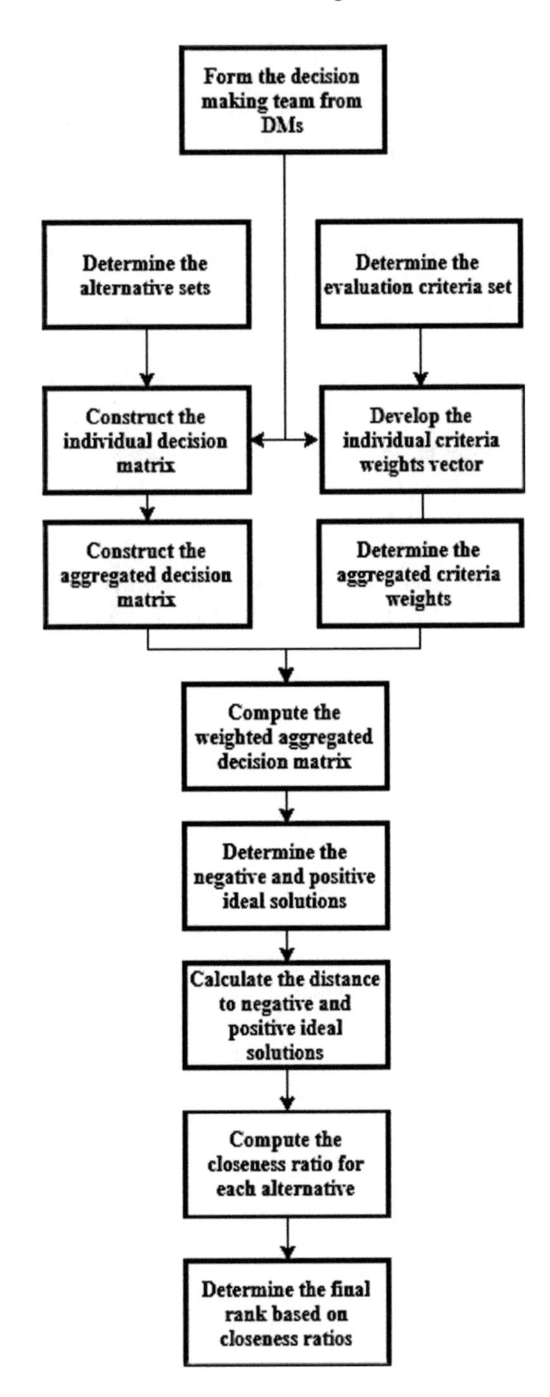

Fig. 1 Flow chart of the proposed SF-TOPSIS method

Table 1 Linguistic terms and their corresponding spherical fuzzy numbers (Kahraman et al. 2019)

Linguistic terms	(μ, v, π)
Absolutely more importance (AMI)	(0.9, 0.1, 0.1)
Very high importance (VHI)	(0.8, 0.2, 0.2)
High importance (HI)	(0.7, 0.3, 0.3)
Slightly more importance (SMI)	(0.6, 0.4, 0.4)
Equally importance (EI)	(0.5, 0.5, 0.5)
Slightly low importance (SLI)	(0.4, 0.6, 0.4)
Low importance (LI)	(0.3, 0.7, 0.3)
Very low importance (VLI)	(0.2, 0.8, 0.2)
Absolutely low importance (ALI)	(0.1, 0.9, 0.1)

Step 5: Construct the aggregated weighted spherical fuzzy decision matrix as given in Eq. (19). After the aggregated weights of the criteria and the aggregated rating of the alternatives are determined, the aggregated weighted spherical fuzzy decision matrix is constructed by utilizing Eq. (4) and then the aggregated weighted spherical fuzzy decision matrix can be defined as follows:

$$D = \left(C_j(X_{iw})\right)_{mxn} = \begin{pmatrix} (\mu_{11w}, v_{11w}, \pi_{11w}) & (\mu_{12w}, v_{12w}, \pi_{12w}) & \dots & (\mu_{1nw}, v_{1nw}, \pi_{1nw}) \\ (\mu_{21w}, v_{21w}, \pi_{21w}) & (\mu_{22w}, v_{22w}, \pi_{22w}) & \dots & (\mu_{2nw}, v_{2nw}, \pi_{2nw}) \\ \vdots & \vdots & & \vdots \\ (\mu_{m1w}, v_{m1w}, \pi_{m1w}) & (\mu_{m2w}, v_{m2w}, \pi_{m2w}) & \dots & (\mu_{mnw}, v_{mnw}, \pi_{mnw}) \end{pmatrix} \tag{19}$$

Step 6: Defuzzify the aggregated weighted decision matrix by using the proposed score function.

$$Score\left(C_j(X_{iw})\right) = \left(2\mu_{ijw} - \frac{\pi_{ijw}}{2}\right)^2 - \left(v_{ijw} - \frac{\pi_{ijw}}{2}\right)^2 \tag{20}$$

Step 7: Determine the Spherical Fuzzy Negative Ideal Solution (SF-NIS) and Spherical Fuzzy Positive Ideal Solution (SF-PIS), based on the score values obtained in Step 6.
For the SF-NIS:

$$X^- = \left\{ C_j, \min_i < Score(C_j(X_{iw})) > \middle| j = 1, 2 \dots n \right\} \tag{21}$$

or

$$X^- = \left\{\left\langle C_1, (\mu_1^-, v_1^-, \pi_1^-)\right\rangle, \left\langle C_2, (\mu_2^-, v_2^-, \pi_2^-)\right\rangle \dots \left\langle C_n, (\mu_n^-, v_n^-, \pi_n^-)\right\rangle\right\} \tag{22}$$

For the SF-PIS:

$$X^* = \left\{ C_j, \max_i < Score(C_j(X_{iw})) > \middle| j = 1, 2 \ldots n \right\} \tag{23}$$

or

$$X^* = \left\{ \langle C_1, (\mu_1^*, v_1^*, \pi_1^*) \rangle, \langle C_2, (\mu_2^*, v_2^*, \pi_2^*) \rangle \ldots \langle C_n, (\mu_n^*, v_n^*, \pi_n^*) \rangle \right\} \tag{24}$$

Step 8: Calculate the distances between alternative X_i and the SF-NIS as well as SF-NIS, respectively. Normalized Euclidean distance (Szmidt and Kacprzyk 2000) or normalized Zhang and Xu's distance equations will be used for this step (Ejegwa 2019).
Normalized Euclidean distance to SF-NIS:

$$D_E(\tilde{X}_{ij}, \tilde{X}_j^-) = \sqrt{\frac{1}{2n} \sum_{i=1}^{n} \left(\mu_{\tilde{X}_{ij}} - \mu_{\tilde{X}_j^-} \right)^2 + \left(v_{\tilde{X}_{ij}} - v_{\tilde{X}_j^-} \right)^2 + \left(\pi_{\tilde{X}_{ij}} - \pi_{\tilde{X}_j^-} \right)^2} \quad \forall i \tag{25}$$

Normalized Euclidean distance to SF-PIS:

$$D_E(\tilde{X}_{ij}, \tilde{X}_j^*) = \sqrt{\frac{1}{2n} \sum_{i=1}^{n} \left(\mu_{\tilde{X}_{ij}} - \mu_{\tilde{X}_j^*} \right)^2 + \left(v_{\tilde{X}_{ij}} - v_{\tilde{X}_j^*} \right)^2 + \left(\pi_{\tilde{X}_{ij}} - \pi_{\tilde{X}_j^*} \right)^2} \quad \forall i \tag{26}$$

or
Normalized Zhang and Xu's distance to SF-NIS:

$$D_{ZX}\left(\tilde{X}_{ij}, \tilde{X}_j^-\right) = \frac{1}{2n} \sum_{j=1}^{n} \left(\left| \mu_{\tilde{X}_{ij}}^2 - \mu_{\tilde{X}_j^-}^2 \right| + \left| v_{\tilde{X}_{ij}}^2 - v_{\tilde{X}_j^-}^2 \right| + \left| \pi_{\tilde{X}_{ij}}^2 - \pi_{\tilde{X}_j^-}^2 \right| \right) \quad \forall i \tag{27}$$

Normalized Zhang and Xu's distance to SF-PIS:

$$D_{ZX}\left(\tilde{X}_{ij}, \tilde{X}_j^*\right) = \frac{1}{2n} \sum_{j=1}^{n} \left(\left| \mu_{\tilde{X}_{ij}}^2 - \mu_{\tilde{X}_j^*}^2 \right| + \left| v_{\tilde{X}_{ij}}^2 - v_{\tilde{X}_j^*}^2 \right| + \left| \pi_{\tilde{X}_{ij}}^2 - \pi_{\tilde{X}_j^*}^2 \right| \right) \quad \forall i \tag{28}$$

Step 9: Calculate the closeness ratio of each alternative based on Eq. (29). Determine the optimal ranking order of the alternatives and identify the optimal alternative. We put the alternatives into orders with respect to the decreasing values of closeness ratio.

$$Closeness\ Ratio_i = \xi_i = \frac{D\left(X_{ij}, X_j^-\right)}{D\left(X_{ij}, X_j^-\right) + D\left(X_{ij}, X_j^+\right)} \tag{29}$$

3 Application: Site Selection of Electric Vehicle Charge Station

A manager would like to select the most proper site in İzmir in Turkey for electric vehicle charge station. A linguistic evaluation process will be applied since experts give their judgments to the decision matrix with linguistic terms. The alternatives will be evaluated under some criteria by three decision makers. There are three decision makers (DM1, DM2, and DM3) are included who are three experienced engineers in facility design and electric vehicles charge stations management. The weights of these decision makers who have different experience levels are 0.5, 0.3 and 0.2, respectively. In this application we have 7 criteria and 5 alternatives. The alternative sites are as follows: Alternative 1 (X_1)—Goztepe, Alternative 2(X_2)—Karsıyaka, Alternative 3(X_3)—Konak, Alternative 4(X_4)—Guzelyalı, and Alternative 5(X_5)—Bornova.

The criteria are as follows (Guo and Zhao 2015):

Criterion 1 (C_1)—Construction Cost (Cost Criterion)
Criterion 2 (C_2)—Operation and maintenance cost (Cost Criterion)
Criterion 3 (C_3)—Traffic Condition (Benefit Criterion)
Criterion 4 (C_4)—Waste disposal (Cost Criterion)
Criterion 5 (C_5)—Destruction degree on green area (Cost Criterion)
Criterion 6 (C_6)—Service capability (Benefit Criterion)
Criterion 7 (C_7)—Harmonization of electric vehicle charging station with the development planning of urban road network (Benefit Criterion)

Firstly, the judgments are gathered from decision makers as linguistic terms given in Table 1 with respect to the goal. All judgments are given in Table 2.

The linguistic importance weights of the criteria assigned by DMs and the spherical fuzzy weight of each criterion obtained by using SWAM operator is presented in Table 3.

The aggregated weighted spherical fuzzy decision matrix is constructed by utilizing Eq. (4) as given in Table 4.

Based on Table 4 and Eq. (20), the score function values are calculated as in Table 5. The values in bold represent PIS (negative ideal solution) and the values in *italic* represent NIS.

By utilizing Eqs. (21–24), the corresponding Spherical Fuzzy Negative Ideal Solution (SF-NIS) and Spherical Fuzzy Positive Ideal Solution (SF-PIS) are given in Table 6.

In the next step based on Eqs. (25–28), we can calculate the distances between alternative X_i and the SF-NIS as well as SF-PIS. They are given in Table 7.

From Table 7, based on Eq. (29) closeness ratios are obtained for Euclidean distance and presented in Table 8.

From Table 7, based on Eq. (29) closeness ratios are obtained for Zhang and Xu's distance and presented in Table 9.

Table 2 Judgments of decision makers

Decision makers	Alternatives	C_1	C_2	C_3	C_4	C_5	C_6	C_7
DM1	X_1	SMI	HI	HI	VHI	EI	HI	SMI
	X_2	SMII	HI	HI	SMI	SMI	LI	HI
	X_3	VHI	AMI	AMI	VHI	VHI	AMI	VHI
	X_4	VHI	VHI	HI	VHI	AMI	HI	HI
	X_5	HI	HI	LI	HI	HI	EI	AMI
DM2	X_1	HI	VHI	ALI	SMI	SMI	LI	ALI
	X_2	HI	LI	SMI	HI	HI	EI	SMI
	X_3	AMI	HI	VHI	HI	VHI	VHI	VHI
	X_4	AMI	SLI	LI	SLI	EI	VHI	HI
	X_5	VHI	SLI	LI	SLI	HI	VHI	VHI
DM3	X_1	HI	VHI	VHI	VHI	AMI	VHI	VHI
	X_2	SMI	SMI	HI	SMI	SMI	LI	SMI
	X_3	HI	VHI	AMI	VHI	VHI	HI	AMI
	X_4	HI	SMI	LI	SMI	VHI	SLI	HI
	X_5	HI	ALI	HI	ALI	LI	SLI	SMI

Table 3 Importance weights of the main criteria

Criteria	DM1	DM2	DM3	Aggregated criteria weights
C_1	SMI	HI	HI	(0.65, 0.35, 0.35)
C_2	HI	LI	SMI	(0.60, 0.41, 0.33)
C_3	AMI	VHI	AHI	(0.88, 0.12, 0.13)
C_4	VHI	SLI	SMI	(0.69, 0.32, 0.29)
C_5	HI	HI	LI	(0.65, 0.36, 0.30)
C_6	EI	VHI	SLI	(0.62, 0.39, 0.39)
C_7	AMI	VHI	SMI	(0.84, 0.16, 0.18)

The closeness ratios indicate that the best alternative is third and overall ranking is $X_3 > X_4 > X_1 > X_5 > X_2$. The best alternative is Konak and the worst place is Karsiyaka.

4 Comparative Analyses

We compare the proposed SF-TOPSIS with SF-CODAS (spherical fuzzy combinative distance-based assessment) in this section (Kutlu Gündoğdu and Kahraman 2019c). Table 1 presents the SF linguistic scale, which we use for the comparison purpose. In this comparison, the same judgments as given in Table 2 was used and

Table 4 Aggregated weighted spherical fuzzy decision matrix

Alternatives	C_1	C_2	C_3	C_4	C_5	C_6	C_7
X_1	(0.43, 0.47, 0.20)	(0.45, 0.47, 0.15)	(0.57, 0.40, 0.08)	(0.52, 0.40, 0.13)	(0.44, 0.48, 0.19)	(0.41, 0.51, 0.19)	(0.50, 0.47, 0.12)
X_2	(0.42, 0.49, 0.21)	(0.37, 0.55, 0.17)	(0.59, 0.35, 0.12)	(0.44, 0.47, 0.19)	(0.41, 0.49, 0.19)	(0.23, 0.70, 0.20)	(0.55, 0.38, 0.14)
X_3	(0.54, 0.38, 0.14)	(0.51, 0.43, 0.13)	(0.77, 0.17, 0.03)	(0.54, 0.38, 0.12)	(0.52, 0.40, 0.12)	(0.53, 0.42, 0.17)	(0.69, 0.24, 0.06)
X_4	(0.54, 0.38, 0.14)	(0.42, 0.50, 0.16)	(0.50, 0.47, 0.10)	(0.48, 0.44, 0.15)	(0.54, 0.40, 0.13)	(0.44, 0.48, 0.19)	(0.59, 0.34, 0.11)
X_5	(0.48, 0.43, 0.17)	(0.34, 0.59, 0.16)	(0.38, 0.60, 0.10)	(0.39, 0.54, 0.15)	(0.43, 0.49, 0.15)	(0.38, 0.54, 0.23)	(0.71, 0.23, 0.06)

Table 5 Score function values

Alternatives	C_1	C_2	C_3	C_4	C_5	C_6	C_7
X_1	0.433	0.551	1.073	0.853	0.476	0.348	*0.706*
X_2	*0.379*	0.196	1.178	0.473	*0.381*	*−0.230*	0.964
X_3	**0.920**	**0.773**	**2.297**	**0.918**	0.855	**0.827**	1.804
X_4	0.920	0.394	0.705	0.649	**0.908**	0.449	1.175
X_5	0.655	*0.113*	*0.208*	*0.291*	0.438	0.251	**1.870**

Table 6 Spherical fuzzy negative ideal solution (SF-NIS)

	C_1	C_2	C_3	C_4	C_5	C_6	C_7
SF-NIS	(0.42, 0.49, 0.21)	(0.34, 0.59, 0.16)	(0.38, 0.60, 0.10)	(0.39, 0.54, 0.15)	(0.41, 0.49, 0.19)	(0.23, 0.70, 0.20)	(0.50, 0.47, 0.12)
SF-PIS	(0.54, 0.38, 0.14)	(0.51, 0.43, 0.13)	(0.77, 0.17, 0.03)	(0.54, 0.38, 0.12)	(0.54, 0.40, 0.13)	(0.53, 0.42, 0.17)	(0.71, 0.23, 0.06)

Table 7 Euclidean and spherical distances to negative ideal solution

Alternatives	Euclidean distance		Zhang and Xu's distance	
	$D_E(\tilde{X}_{ij}, \tilde{X}_j^-)$	$D_E(\tilde{X}_{ij}, \tilde{X}_j^*)$	$D_{ZX}(\tilde{X}_{ij}, \tilde{X}_j^-)$	$D_{ZX}(\tilde{X}_{ij}, \tilde{X}_j^*)$
X_1	0.122	0.139	0.091	0.110
X_2	0.096	0.171	0.054	0.150
X_3	0.232	0.006	0.198	0.003
X_4	0.129	0.129	0.115	0.085
X_5	0.110	0.189	0.065	0.138

Table 8 Closeness ratios based on Euclidean distance

Alternatives	Closeness ratio	Rank
X_1	0.4676	3
X_2	0.3597	5
X_3	0.9743	1
X_4	0.5000	2
X_5	0.3675	4

Table 9 Closeness ratios based on Zhang and Xu's distance

Alternatives	Closeness Ratio	Rank
X_1	0.4534	3
X_2	0.2645	5
X_3	0.9850	1
X_4	0.5747	2
X_5	0.3190	4

aggregated using SWAM operator given in Eq. (13). Aggregated weighted decision matrix is given in Table 4. Score values and spherical fuzzy negative ideal solution are the same with SF-TOPSIS method.

We calculated the distances between alternative X_i and the SF-NIS as given in Table 10. Normalized Euclidean distance (Szmidt and Kacprzyk 2000) and spherical distance (Kutlu Gundogdu and Kahraman 2019a) formulas will be used for this step.

Normalized Euclidean distance to SF-NIS:

$$D_E(\tilde{X}_{ij}, \tilde{X}_j^-) = \sqrt{\frac{1}{2n}\sum_{i=1}^{n}\left(\mu_{\tilde{X}_{ij}} - \mu_{\tilde{X}_j^-}\right)^2 + \left(v_{\tilde{X}_{ij}} - v_{\tilde{X}_j^-}\right)^2 + \left(\pi_{\tilde{X}_{ij}} - \pi_{\tilde{X}_j^-}\right)^2} \quad \forall i \tag{30}$$

Normalized Spherical distance to SF-NIS:

$$D_S(\tilde{X}_{ij}, \tilde{X}_j^-) = \frac{2}{n\pi}\sum_{i=1}^{n}\arccos\left(\mu_{\tilde{X}_{ij}} \cdot \mu_{\tilde{X}_j^-} + v_{\tilde{X}_{ij}} \cdot v_{\tilde{X}_j^-} + \pi_{\tilde{X}_{ij}} \cdot \pi_{\tilde{X}_j^-}\right) \quad \forall i \tag{31}$$

Table 10 Euclidean and spherical distances to negative ideal solution

Alternatives	Euclidean distance	Spherical distance
X_1	0.122	3.424
X_2	0.096	3.370
X_3	0.232	3.503
X_4	0.129	3.432
X_5	0.110	3.372

Table 11 Relative assessment matrix (RA) based on Euclidean and spherical distances

Alternatives	X_1	X_2	X_3	X_4	X_5
X_1	0.00	0.08	−0.19	−0.01	0.01
X_2	−0.08	0.00	−0.27	−0.09	−0.01
X_3	0.19	0.27	0.00	0.17	0.25
X_4	0.01	0.09	−0.17	0.00	0.02
X_5	−0.01	0.01	−0.25	−0.02	0.00

Relative assessment matrix (RA) was computed based on Eq. (32) is given in Table 11.

$$RA = [s_{ik}]_{nxn};$$

$$s_{ik} = \left(D_{E_i} - D_{E_k}\right) + \left(\Phi(D_{E_i} - D_{E_k}) \cdot (D_{S_i} - D_{S_k})\right) \quad (32)$$

where $k \in \{1, 2, \ldots, n\}$ and Φ is a threshold function that is defined as follows:

$$\Phi(x) = \begin{cases} 1 \; if \; |x| \geq \alpha \\ 0 \; if \; |x| < \alpha \end{cases} \quad (33)$$

Decision maker can set the threshold parameter of this function. In this study, we use $\alpha = 0.02$.

We determined the optimal ranking order of the alternatives and identify the optimal alternative based on the appraisal scores obtained in Eq. (34) from Table 11. The largest value of AS_i indicates the best alternative as seen from Table 12.

$$AS_i = \sum_{k=1}^{n} s_{ik} \quad (34)$$

The appraisal scores indicate that the best alternative is X_3 and overall ranking is the same with SF-TOPSIS method. When we used Zhang and Xu's distance formula instead of Euclidean distance formula, fourth and fifth alternatives were replaced each other in the final ranking.

Table 12 Appraisal scores and ranking based on Euclidean and spherical distances

Alternatives	Appraisal score	Ranking
X_1	−0.103	3
X_2	−0.457	5
X_3	0.884	1
X_4	−0.053	2
X_5	−0.270	4

5 Conclusions

Spherical fuzzy sets present an excellent tool to handle the vagueness and uncertainty existing in this problem. The literature review shows that SFSs present a new way to extend MCDM methods under fuzzy environment. Gradually, the number of MCDM papers with SFS is increasing.

TOPSIS method is proposed by Hwang and Yoon (1981) is an effective and useful technique to overcome multi-criteria decision making (MCDM) problems targeting at choosing the best alternative with the shortest distance from the positive ideal solution (PIS) and the farthest distance from the negative ideal solution (NIS). The TOPSIS method is extended to its different fuzzy extensions in the literature. We have extended TOPSIS method to Spherical fuzzy TOPSIS in order to select the best location for electric vehicle charge station in Izmir under spherical fuzzy environment. Spherical fuzzy TOPSIS method gives significant results and can be used as an alternative MCDM method under fuzziness. In order to see the robustness of the method, comparative and sensitivity analyses are shown. For further research, we suggest other extensions of ordinary fuzzy sets to be used for the same case study such as Pythagorean fuzzy sets, and picture fuzzy sets.

References

Ashraf S, Abdullah S, Aslam M, Qiyas M, Kutbi MA (2019) Spherical fuzzy sets and its representation of spherical fuzzy t-norms and t-conorms. J Intell Fuzzy Syst 36(6):6089–6102

Atanassov KT (1986) Intuitionistic fuzzy sets. Fuzzy Sets Syst 20(1):87–96

Beg I, Rashid T (2013) TOPSIS for hesitant fuzzy linguistic term sets. Int J Intell Syst 28(12):1162–1171

Biswas P, Pramanik S, Giri BC (2016) TOPSIS method for multi-attribute group decision-making under single-valued neutrosophic environment. Neural Comput Appl 27(3):727–737

Boran FE, Genç S, Kurt M, Akay D (2009) A multi-criteria intuitionistic fuzzy group decision making for supplier selection with TOPSIS method. Expert Syst Appl 36(8):11363–11368

Boran FE, Boran K, Menlik T (2012) The evaluation of renewable energy technologies for electricity generation in Turkey using intuitionistic fuzzy TOPSIS. Energy Sour Part B 7(1):81–90

Broumi S, Ye J, Smarandache F (2015) An extended TOPSIS method for multiple attribute decision making based on interval neutrosophic uncertain linguistic variables. Infinite Study

Büyüközkan G, Güleryüz S (2016) Multi criteria group decision making approach for smart phone selection using intuitionistic fuzzy TOPSIS. Int J Comput Intell Syst, 709–725

Cevik Onar S, Oztaysi B, Kahraman C (2018) Multicriteria evaluation of cloud service providers using pythagorean fuzzy TOPSIS. J Multiple-Valued Logic Soft Comput 30:263–283

Chen CT (2000) Extensions of the TOPSIS for group decision-making under fuzzy environment. Fuzzy Sets Syst 114(1):1–9

Chen TY, Tsao CY (2008) The interval-valued fuzzy TOPSIS method and experimental analysis. Fuzzy Sets Syst 159(11):1410–1428

Cuong BC, Kreinovich V (2013) Picture fuzzy sets-a new concept for computational intelligence problems. In: 2013 third world congress on information and communication technologies (WICT 2013). IEEE, pp 1–6

Ejegwa PA (2019) Modified Zhang and Xu's distance measure for Pythagorean fuzzy sets and its application to pattern recognition problems. Neural Comput Appl, 1–10

Elhassouny A, Smarandache F (2016) Neutrosophic-simplified-TOPSIS multi-criteria decision-making using combined simplified-TOPSIS method and neutrosophics. In: 2016 IEEE international conference fuzzy systems (FUZZ-IEEE), pp 2468–2474
Garibaldi JM, Ozen T (2007) Uncertain fuzzy reasoning: a case study in modelling expert decision making. IEEE Trans Fuzzy Syst 15(1):16–30
Grattan-Guinness I (1976) Fuzzy membership mapped onto interval and many-valued quantities. Z fur math Logik und Grundladen der Math 22(1):149–160
Guo S, Zhao H (2015) Optimal site selection of electric vehicle charging station by using fuzzy TOPSIS based on sustainability perspective. Appl Energy 158:390–402
Hwang CL, Yoon K (1981) Multiple attributes decision making methods and applications. Springer, Berlin Heidelberg
Jahn KU (1975) Intervall-wertige Mengen. Math Nachr 68(1):115–132
Kahraman C, Kutlu Gündoğdu F (2018) From 1D to 3D membership: spherical fuzzy sets. In: BOS/SOR2018 conference, Warsaw, Poland
Kahraman C, Gundogdu FK, Onar SC, Oztaysi B (2019) Hospital location selection using spherical fuzzy TOPSIS. In: 2019 Conference of the international fuzzy systems association and the European society for fuzzy logic and technology (EUSFLAT 2019). Atlantis Press
Kutlu Gündoğdu F (2019) Principals of spherical fuzzy sets. In: International conference on intelligent and fuzzy systems. Springer, Cham, pp 15–23
Kutlu Gündoğdu F, Kahraman C (2019a) Spherical fuzzy sets and spherical fuzzy TOPSIS method. J Intell Fuzzy Syst 36(1):337–352
Kutlu Gündoğdu F, Kahraman C (2019b) A novel fuzzy TOPSIS method using emerging interval-valued spherical fuzzy sets. Eng Appl Artif Intell 85:307–323
Kutlu Gündoğdu FK, Kahraman C (2019c) Extension of CODAS with spherical fuzzy sets. J Multi-Valued Logic Soft Comput 33:481–505
Mahmood T, Ullah K, Khan Q, Jan N (2019) An approach toward decision-making and medical diagnosis problems using the concept of spherical fuzzy sets. Neural Comput Appl 31(11):7041–7053
Sajjad A, Khan M, Abdullah S, Yousaf Ali M (2018) Extension of TOPSIS method base on Choquet integral under interval-valued Pythagorean fuzzy environment. J Intell Fuzzy Syst 34(1):267–282
Sambuc R (1975) Function Φ-Flous, Application a l'aide au Diagnostic en Pathologie Thyroidienne. University of Marseille
Smarandache F (1998) Neutrosophy: neutrosophic probability, set, and logic: analytic synthesis & synthetic analysis
Szmidt E, Kacprzyk J (2000) Distances between intuitionistic fuzzy sets. Fuzzy Sets Syst 114(3):505–518
Torra V (2010) Hesitant fuzzy sets. Int J Intell Syst 25(6):529–539
Ullah K, Hassan N, Mahmood T, Jan N, Hassan M (2019) Evaluation of investment policy based on multi-attribute decision-making using interval valued T-spherical fuzzy aggregation operators. Symmetry 11(3):357
Vahdani B, Mousavi SM, Tavakkoli-Moghaddam R, Hashemi H (2013) A new design of the elimination and choice translating reality method for multi-criteria group decision-making in an intuitionistic fuzzy environment. Appl Math Model 37(4):1781–1799
Xu Z, Zhang X (2013) Hesitant fuzzy multi-attribute decision making based on TOPSIS with incomplete weight information. Knowl-Based Syst 52:53–64
Yager R (1986) On the theory of bags. Int J Gener Syst 13(1):23–37
Yager RR (2013) Pythagorean fuzzy subsets. In: Joint IFSA world congress and NAFIPS annual meeting, Edmonton, Canada, pp 57–61
Yager RR (2016) Generalized orthopair fuzzy sets. IEEE Trans Fuzzy Syst 25(5):1222–1230

Ye F (2010) An extended TOPSIS method with interval-valued intuitionistic fuzzy numbers for virtual enterprise partner selection. Expert Syst Appl 37(10):7050–7055
Zadeh LA (1965) Fuzzy Sets. Inf Control 8:338–353
Zadeh LA (1975) The concept of a linguistic variable and its application to approximate reasoning. Inf Sci 8:199–249

Spherical Fuzzy VIKOR with SWAM and SWGM Operators for MCDM

Iman Mohamad Sharaf

Abstract In this study, VIseKriterijumska Optimizacija I Kompromisno Resenje (VIKOR), multi-criteria optimization and compromise solution, is extended using spherical fuzzy sets (SFSs). SFSs were recently introduced as a novel type of fuzzy sets that allows the decision-makers to express their degree of hesitation explicitly. Therefore, SFSs provide a greater preference domain for decision-makers. Spherical fuzzy VIKOR (SFVIKOR) is implemented in four different ways based on two aggregation operators and two proposed ideal solutions. The used aggregation operators are the spherical weighted averaging mean (SWAM) and the spherical weighted geometric mean (SWGM). The proposed ideal solutions are the absolute ideal solution and the relative ideal solution. Both ideal solutions don't rely on the score function and take the degree of hesitation into account. An illustrative example in supplier selection is solved using SFVIKOR with the four implementations and the results are compared with the results of the spherical fuzzy technique of order preference by similarity to an ideal solution (SFTOPSIS). Then, the stability of the solution is examined under these different implementations.

1 Introduction

Multi-criteria decision making (MCDM) proved to be a powerful tool in handling models with imperfectly defined facts and imprecise knowledge since the introduction of fuzzy set theory. Finding fuzzy sets that can efficiently address subjective human judgement and fuzzy objective evaluations in decision making led to the introduction of many types of fuzzy sets, each is better than its predecessor in modelling vagueness and uncertainty.

The starting point was type-1 fuzzy sets (T1FSs) which have a crisp membership function value. Yet, in an uncertain environment, it is hard to estimate the exact membership function of a fuzzy set. As a result, type-2 fuzzy sets (T2FSs) were introduced

---

I. M. Sharaf (✉)
Department of Basic Sciences, Higher Technological Institute, Tenth of Ramadan City, Egypt
e-mail: iman_sharaf@hotmail.com

© Springer Nature Switzerland AG 2021

C. Kahraman and F. Kutlu Gündoğdu (eds.), *Decision Making with Spherical Fuzzy Sets*, Studies in Fuzziness and Soft Computing 392,
https://doi.org/10.1007/978-3-030-45461-6_9

that have a three-dimensional membership function that captures the linguistic assessment and lessens the effect of fuzziness. However, the computational complexity of T2FSs led to the introduction of interval type-2 fuzzy sets (IT2FSs) and interval-valued fuzzy sets (IVFSs) that possess reasonable computational complexity (Türk et al. 2014). Nonetheless, there was still a need for fuzzy sets that can describe human judgements in a more effective manner. This led to the introduction of intuitionistic fuzzy sets (IFSs) that provide a wider preference scope for decision-makers having a membership degree a non-membership degree and a hesitancy degree, where the sum of the three degrees is equal to one. In some cases, the decision-makers might fail to fulfill this condition. Therefore, type-2 intuitionistic fuzzy sets (T2IFS) [also known as Pythagorean fuzzy sets (PFSs)] were introduced in which the sum of the squares of the three degrees is equal to one. In either IFSs or PFSs, the hesitancy degree depends on both the membership degree and the non-membership degree. A natural extension of IFSs and PFSs are spherical fuzzy sets (SFSs) in which the hesitancy degree is independent of the membership degree and the non-membership degree. This new type of FSs is based on the spherical fuzzy distances previously defined in the literature. This provides a greater preference domain for decision-makers to express their hesitance about a criterion (Gündoğdu and Kahraman 2019a).

Multi-criteria optimization and compromise solution (VIKOR, acronyms from the Serbian name), is a known MCDM method. The method sorts a set of alternatives for the decision makers' preferences and proposes a compromise solution for a problem with conflicting criteria (Ju and Wang 2013). The technique is based on an aggregating function that represents the closeness of the alternatives to the best solution (Mehbodniya et al. 2013). VIKOR is characterized by the trade-off between the maximum group utility of the majority and the minimum individual regret of the opponent. Its main advantage is the simple calculations (Ju and Wang 2013). VIKOR is suitable for cautious decision-makers who aim at maximizing the profits and minimizing the risks (Sayadi et al. 2009). VIKOR was initially proposed for MCDM problems in which the weights of the criteria and the performance rating of the alternatives are precisely known, and then it was extended to include the fuzzy domain resulting from using subjective or fuzzy objective evaluations (Mehbodniya et al. 2013).

In this study, VIKOR is extended using spherical fuzzy sets (SFSs). SFSs were recently introduced as a new type of fuzzy sets that allows the decision-makers to express their hesitance explicitly. Therefore, provide the decision-maker with a robust framework to convey uncertainty and imprecision of knowledge. SFVIKOR is implemented in four different ways. Two aggregation operators are used, the spherical weighted averaging mean (SWAM) and the spherical weighted geometric mean (SWGM). Also, two different ideal solutions are proposed that takes the degree of hesitation into account and are not based on the score function. An illustrative example in supplier selection is solved using the proposed SFVIKOR with the four implementations and the results are compared with the results of spherical fuzzy TOPSIS (SFTOPSIS). Then, the stability of the solution under these different implementations is tested.

From the previous view, the originality of this study comes from proposing a new version of the SFVIKOR using different implementations, thus gaining two main advantages. The decision-makers can reflect their hesitation explicitly, hence providing a richer structure to express the fuzziness and uncertainty while attaining the best and most robust implementation of the method.

The chapter is organized as follows. A literature review is given in Sect. 2. SFSs with their basic operators and the conventional VIKOR are presented in Sect. 3. SFVIKOR is introduced in Sect. 4. An illustrative example is solved using the SFVIKOR with its four different implementations and the comparative analysis is given in Sect. 5. Finally, the conclusion is given in Sect. 6.

2 A Literature Review

VIKOR is one of the most prominent and widely used MCDM techniques. Opricovic (1998) introduced VIKOR to solve discrete multi-criteria problems with incommensurable and conflicting criteria. The VIKOR method determines a compromise solution based on these criteria and established by mutual concessions. It is based on an aggregating function that represents "closeness to the ideal" originating from the compromise programming method (Zhang and Wei 2013). A compromise solution is a feasible solution close to the ideal one and provides a maximum group utility for the majority and a minimum of an individual regret for the opponent. The classical VIKOR method has been extended using various fuzzy sets, e.g. fuzzy VIKOR (FVIKOR), intuitionistic fuzzy VIKOR (IFVIKOR), hesitant fuzzy VIKOR (HFVIKOR), and Pythagorean fuzzy VIKOR (PFVIKOR). The various fuzzy extensions of VIKOR in the past decade with its diverse applications can be summarized chronologically as follows.

Sayadi et al. (2009) extended the VIKOR method with interval numbers. Sanayei et al. (2010) presented an FVIKOR using trapezoidal or triangular T1FSs to deal with supplier selection problems in the supply chain system. Vahdani et al. (2010) proposed an FVIKOR method based on triangular IVFSs to select a maintenance strategy. Devi (2011) proposed an IFVIKOR using triangular IFSs for robot selection problem for material handling task. Lu and Tang (2011) utilized an IFVIKOR based on intuitionistic language multi-criteria for evaluating auto parts suppliers. Opricovic (2011) utilized an FVIKOR using triangular T1FSs to plan water resources. Shemshadi et al. (2011) developed an FVIKOR method using trapezoidal T1FSs with a mechanism to extract and deploy objective weights based on Shannon entropy concept.

Roostaee et al. (2012) used IFVIKOR for supplier selection under incomplete and uncertain information environment. Ju and Wang (2013) extended VIKOR using the 2-tuple linguistic approach for emergency alternative selection problem. Liao and Xu (2013) proposed a VIKOR based method for MCDM using hesitant fuzzy sets to evaluate and rank the service quality among domestic airlines. Park et al. (2013) developed IFVIKOR for dynamic intuitionistic fuzzy MCDM to evaluate university faculty for tenure and promotion. Peng et al. (2013) proposed an optimization procedure that utilizes IFVIKOR in solving a multi-response optimization problem. Wan et al. (2013) applied IFVIKOR with triangular IFSs in which the weights of the criteria and decision-makers are completely unknown for personnel selection. Zhang and Wei (2013) developed an HFVIKOR to develop large projects within a five-year plan. Zhao et al. (2013) extended VIKOR under the interval-valued intuitionistic fuzzy (IVIF) environment. Furthermore, they proposed a fuzzy cross-entropy approach to state the discrimination measure between optional and optimal interval-valued IFSs. Hashemi et al. (2014) proposed a new IFVIKOR based technique. They introduced two new basic IVIF operations and a collective index to sort alternatives applied to reservoir flood control operation.

Ghorabaee et al. (2015) proposed a new version of FVIKOR with IT2FSs to handle the project selection problems. You et al. (2015) proposed an extended VIKOR method for group multi-criteria supplier selection with interval 2-tuple linguistic (ITL) information. The feasibility and practicability of the proposed ITLVIKOR were demonstrated through three realistic supplier selection examples. Awasthi and Kannan (2016) addressed the problem of evaluating green supplier development programs and proposed a fuzzy nominal group technique (NGT) and FVIKOR using triangular T1FS. Ghorabaee (2016) developed an FVIKOR using trapezoidal IT2FSs to select robots for a particular application and manufacturing environment. Gupta et al. (2016) presented a novel IFVIKOR to solve plant location selection problem with IF information captured through trapezoidal intuitionistic fuzzy numbers (TrIFNs). Mousavi et al. (2016) introduced an intuitionistic ranking index for IFVIKOR for the portfolio selection in the stock exchange problem and for the material handling selection problem. Wu et al. (2016) proposed an extended VIKOR under linguistic information using the cloud model to handle imprecise numerical quantities applied to supplier selection in the nuclear power industry. For extensive surveys on VIKOR and its fuzzy extensions for diverse applications until the year 2016, the reader is referred to Yazdani and Graeml (2014) and Gul et al. (2016).

Chatterjee and Kar (2017) proposed an MCDM framework integrating the analytical hierarchy process (AHP) and VIKOR in prioritizing risk responses to manage green supply chain risks from the perspective of a plastic manufacturing company. Luo and Wang (2017) proposed a new distance measure within IFVIKOR context for enterprise resource planning system. Sari (2017) integrated AHP, VIKOR, and Monte Carlo simulation to evaluate green supply chain management practices. Soner et al. (2017) proposed a hybrid approach that integrates the AHP method into VIKOR technique under IT2F environment. The proposed approach was demonstrated with a hatch cover design selection problem since it is important in the structure of bulk carrier ships to prevent water ingress and protect cargo from the external damages. Wang and Cai (2017) extended the classical VIKOR to a generalized distance-based VIKOR to handle heterogeneous information containing a crisp number, interval number, intuitionistic fuzzy number, and hesitant fuzzy linguistic value to solve emergency supplier selection appropriately and flexibly. Zhao et al. (2017) develop a new IF hybrid VIKOR for solving supplier selection problems.

Chen (2018) proposed a remoteness index-based PFVIKOR methodology that considers uncertain information represented by PF values and conducted two types of problems; service quality and internet stock performance evaluation, and internet stock and R&D project investment problems. Cui et al. (2018) developed a PFVIKOR approach to solving the electric vehicle charging stations (EVCSs) site selection problems. Emeç and Akkaya (2018) developed a stochastic MCDM approach to solving the warehouse location problem in the stochastic environment which contains an uncertain condition. In this approach, the weights of the used criteria were calculated by using the stochastic AHP. The alternatives' ranking was evaluated by FVIKOR. Gul et al. (2018) proposed a new risk assessment method incorporating a FAHP with FVIKOR. FAHP is used in weighting the risk parameters derived from the Fine–Kinney method. The priority orders of hazards were determined by an FVIKOR method. Gul (2018) proposed a new approach for risk assessment in the field of occupational health and safety (OHS) by integrating the PFAHP and the FVIKOR into a risk assessment process. Liu and Liang (2018) extended the VIKOR method to multi-granularity unbalanced linguistic term sets with incomplete attribute weight in green supplier selection. Liu et al. (2018) presented an integrated ANP (analytical network process) and FVIKOR method using IT2FSs to solve supplier selection problem in sustainable supply chain management. Zhou et al. (2018) developed an extended FVIKOR based model for the construction of a robotic automation system in the healthcare industry.

Gul et al. (2019) provided a PFVIKOR approach to improve the overall safety levels of underground mining by considering and advising on the potential hazards of risk management. Khan et al. (2019) proposed a broad new extension of classical VIKOR method for decision-making problems with Pythagorean hesitant fuzzy information. Phochanikorn et al. (2019) also proposed a method based on ANP and VIKOR to help business analysis and supply chain managers formulate both short-term and long-term flexible decision strategies for successfully managing and implementing reverse logistics adoption in the supply chain scenarios. Sharaf (2019) modified FVIKOR using IT2FSs to reduce the computations, and allow the participation of the decision-makers through the optimism level.

Regarding SFSs, Gündoğdu and Kahraman (2019b) extended the VIKOR method to the spherical fuzzy environment to handle warehouse location selection problem. Then, Gündoğdu et al. (2020) employed SFVIKOR to tackle a waste disposal site selection problem.

3 Fundamentals

3.1 Basic Operators of SFSs

An SFS with a membership degree $\mu_{\tilde{A}_s}$, a non-membership degree $\upsilon_{\tilde{A}_s}$, and a hesitancy degree $\pi_{\tilde{A}_s}$, is given by

$$\tilde{A}_s = \{u, (\mu_{\tilde{A}_s}(u), \upsilon_{\tilde{A}_s}(u), \pi_{\tilde{A}_s}(u)) | u \in U\}, \tag{1}$$

where

$$\mu_{\tilde{A}_s} : U \to [0, 1], \upsilon_{\tilde{A}_s} : U \to [0, 1], \pi_{\tilde{A}_s} : U \to [0, 1]$$
$$\text{and } 0 \le \mu^2_{\tilde{A}_s}(u) + \upsilon^2_{\tilde{A}_s}(u) + \pi^2_{\tilde{A}_s}(u) \le 1, \quad \forall u \in U. \tag{2}$$

The basic operations on SFSs, the SWAM and the SWGM operators are given as follows (Gündoğdu and Kahraman 2019a).

Addition

$$\tilde{A}_s \oplus \tilde{B}_s = \left\{ \left(\mu^2_{\tilde{A}_s} + \mu^2_{\tilde{B}_s} - \mu^2_{\tilde{A}_s}\mu^2_{\tilde{B}_s} \right)^{1/2}, \upsilon_{\tilde{A}_s}\upsilon_{\tilde{B}_s}, \left(\left(1 - \mu^2_{\tilde{B}_s}\right)\pi^2_{\tilde{A}_s} + \left(1 - \mu^2_{\tilde{A}_s}\right)\pi^2_{\tilde{B}_s} - \pi^2_{\tilde{A}_s}\pi^2_{\tilde{B}_s} \right)^{1/2} \right\}. \tag{3}$$

Multiplication

$$\tilde{A}_s \otimes \tilde{B}_s = \left\{ \mu_{\tilde{A}_s}\mu_{\tilde{B}_s}, \left(\upsilon^2_{\tilde{A}_s} + \upsilon^2_{\tilde{B}_s} - \upsilon^2_{\tilde{A}_s}\upsilon^2_{\tilde{B}_s} \right)^{1/2}, \left(\left(1 - \upsilon^2_{\tilde{B}_s}\right)\pi^2_{\tilde{A}_s} + \left(1 - \upsilon^2_{\tilde{A}_s}\right)\pi^2_{\tilde{B}_s} - \pi^2_{\tilde{A}_s}\pi^2_{\tilde{B}_s} \right)^{1/2} \right\}. \tag{4}$$

Multiplication by a scalar $\lambda > 0$

$$\lambda \cdot \tilde{A}_s = \left\{ \left(1 - \left(1 - \mu^2_{\tilde{A}_s}\right)^{\lambda} \right)^{1/2}, \upsilon^{\lambda}_{\tilde{A}_s}, \left(\left(1 - \mu^2_{\tilde{A}_s}\right)^{\lambda} - \left(1 - \mu^2_{\tilde{A}_s} - \pi^2_{\tilde{A}_s}\right)^{\lambda} \right)^{1/2} \right\}. \tag{5}$$

Liu (2010) generalized IFSs and defined the operation at $\lambda = 0$, and the result is (0, 1). Ashraf et al. (2019) also defined the operation for SFSs when the scalar is zero. Similarly, the operation of multiplication by a scalar can be defined at $\lambda = 0$, to give (0, 1, 0). Then the result of multiplying an SFS by zero can be interpreted as a total disagreement or strict objection.

$$\text{SWAM}_w\left(\tilde{A}_{S_1}, \tilde{A}_{S_2}, \ldots, \tilde{A}_{S_n}\right) = w_1\tilde{A}_{S_1} + w_2\tilde{A}_{S_2} + \cdots + w_n\tilde{A}_{S_n}$$
$$= \left\{ \left[1 - \prod_{i=1}^{n}\left(1 - \mu^2_{\tilde{A}_{S_i}}\right)^{w_i} \right]^{1/2}, \prod_{i=1}^{n}\upsilon^{w_i}_{\tilde{A}_{S_i}}, \left[\prod_{i=1}^{n}\left(1 - \mu^2_{\tilde{A}_{S_i}}\right)^{w_i} - \prod_{i=1}^{n}\left(1 - \mu^2_{\tilde{A}_{S_i}} - \pi^2_{\tilde{A}_{S_i}}\right)^{w_i} \right]^{1/2} \right\}, \tag{6}$$

where $w_i \in [0, 1]$; $\sum_{i=1}^{n} w_i = 1$.

$$\text{SWGM}_w\left(\tilde{A}_{S_1}, \tilde{A}_{S_2}, \ldots, \tilde{A}_{S_n}\right) = \tilde{A}^{w_1}_{S_1} + \tilde{A}^{w_2}_{S_2} + \cdots + \tilde{A}^{w_n}_{S_n}$$
$$= \left\{ \prod_{i=1}^{n}\mu^{w_i}_{\tilde{A}_{S_i}}, \left[1 - \prod_{i=1}^{n}\left(1 - \upsilon^2_{\tilde{A}_{S_i}}\right)^{w_i} \right]^{1/2}, \left[\prod_{i=1}^{n}\left(1 - \upsilon^2_{\tilde{A}_{S_i}}\right)^{w_i} - \prod_{i=1}^{n}\left(1 - \upsilon^2_{\tilde{A}_{S_i}} - \pi^2_{\tilde{A}_{S_i}}\right)^{w_i} \right]^{1/2} \right\}, \tag{7}$$

where $w_i \in [0, 1]$; $\sum_{i=1}^{n} w_i = 1$.

The Score and Accuracy functions for sorting SFSs are given by

$$\text{Score}\left(\tilde{A}_s\right) = \left(\mu_{\tilde{A}_s} - \pi_{\tilde{A}_s}\right)^2 - \left(\upsilon_{\tilde{A}_s} - \pi_{\tilde{A}_s}\right)^2, \tag{8}$$

and

$$\text{Accuracy}\left(\tilde{A}_s\right) = \mu_{\tilde{A}_s}^2 + \upsilon_{\tilde{A}_s}^2 + \pi_{\tilde{A}_s}^2, \tag{9}$$

$$\tilde{A}_s < \tilde{B}_s \quad \text{iff} \quad \text{Score}\left(\tilde{A}_s\right) < \text{Score}\left(\tilde{B}_s\right),$$

$$or \ \text{Score}\left(\tilde{A}_s\right) = \text{Score}\left(\tilde{B}_s\right) \text{ and Accuracy}\left(\tilde{A}_s\right) < \text{Accuracy}\left(\tilde{B}_s\right).$$

The normalized Euclidean distance is given by

$$d\left(\tilde{A}_s, \tilde{B}_s\right) = \sqrt{\frac{1}{2n}\sum_{i=1}^{n}\left(\left(\mu_{\tilde{A}_s} - \mu_{\tilde{B}_s}\right)^2 + \left(\upsilon_{\tilde{A}_s} - \upsilon_{\tilde{B}_s}\right)^2 + \left(\pi_{\tilde{A}_s} - \pi_{\tilde{B}_s}\right)^2\right)} \tag{10}$$

When using the score function for sorting SFSs, sometimes it fails regarding the distances. Consider the following SFSs:

$\tilde{F}_1 = (0.73, 0.3, 0.31)$, $\tilde{F}_2 = (0.71, 0.54, 0.45)$, and $\tilde{F}_3 = (0.81, 0.46, 0.26)$. Their scores are $Sc\left(\tilde{F}_1\right) = 0.1763$, $Sc\left(\tilde{F}_2\right) = 0.0595$, and $Sc\left(\tilde{F}_3\right) = 0.2625$. According to the score values the positive ideal solution (PIS) is $\tilde{F}_3$ and the negative ideal solution (NIS) is $\tilde{F}_2$.

The distance between the PIS and the NIS should be the largest. Yet, $d\left(\tilde{F}_1, \tilde{F}_2\right) = 0.1970$, while $d\left(\tilde{F}_2, \tilde{F}_3\right) = 0.1620$. Then the score values are not suitable for choosing ideal solutions (ISs). Accordingly, two ISs will be proposed in the context of SFVIKOR.

It will be assumed that the cost criteria are positively evaluated in the same manner as benefit criteria. For example, instead of assessing the price as low, the price offer is evaluated as good. Else, the conjugate is directly used. The conjugate as given by (Ashraf et al. 2019)

$$\tilde{A}_s^c = \left(\upsilon_{\tilde{A}_s}, \mu_{\tilde{A}_s}, \pi_{\tilde{A}_s}\right). \tag{11}$$

3.2 The VIKOR Method

VIKOR is one of the known MCDM methods which optimize multi-criteria based complicated systems (Mehbodniya et al. 2013). Opricovic (1998) introduced this method to solve decision-making problems with conflicting and incommensurable criteria by proposing a compromise solution (Shemshadi et al. 2011).

For an MCDM problem, let $X = \{X_1, X_2, \ldots, X_n\}$ be the set of n alternatives, $C = \{C_1, C_2, \ldots, C_m\}$ be the set of m criteria, and $D = \{D_1, D_2, \ldots, D_k\}$ be the set of k decision-makers of weights $\{\omega_1, \omega_2, \ldots, \omega_k\}$. The compromise ranking technique VIKOR is summarized as follows (Opricovic and Tzeng 2004).

Step 1 Form the decision matrices, the average decision matrix, the weighting matrices, and the average weighting matrix.

The decision matrix and the weighting matrix of the pth decision maker are given by

$$\mathbf{D}_p = \begin{array}{c} \\ X_1 \\ X_2 \\ \vdots \\ X_n \end{array} \begin{array}{c} \begin{array}{cccc} C_1 & C_2 & & C_m \end{array} \\ \begin{bmatrix} X^p_{11} & X^p_{12} & \cdots & X^p_{1m} \\ X^p_{21} & X^p_{22} & & X^p_{2m} \\ & \vdots & \ddots & \vdots \\ X^p_{n1} & X^p_{n2} & \cdots & X^p_{nm} \end{bmatrix} \end{array}, \quad \text{and } \mathbf{W}_p = \begin{bmatrix} w^p_1 & w^p_2 & \ldots & w^p_m \end{bmatrix}, \tag{12}$$

where X^p_{ij} is the rating of the ith alternative for the jth criterion as evaluated by the pth decision-maker. The weights of the criteria satisfy $w_j > 0$ and $\sum_{j=1}^{m} w_j = 1$. The average decision matrix and the average weighting matrix are computed as follows

$$\mathbf{D} = (\omega_1 \mathbf{D}_1 + \omega_2 \mathbf{D}_2 + \cdots + \omega_k \mathbf{D}_k)/k,$$

and

$$\mathbf{W} = (\omega_1 \mathbf{W}_1 + \omega_2 \mathbf{W}_2 + \cdots + \omega_k \mathbf{W}_k)/k. \tag{13}$$

The weights of the decision-makers satisfy $\omega_j > 0$ and $\sum_{j=1}^{m} \omega_j = 1$.

Step 2 Set the PIS $\left(X_j^+\right)$ and the NIS $\left(X_j^-\right)$.

$$X_j^+ = \max_i X_{ij}, \text{ and } X_j^- = \min_i X_{ij}, \text{ for benefit criteria,}$$
$$\text{while } X_j^+ = \min_i X_{ij}, \text{ and } X_j^- = \max_i X_{ij}, \text{ for cost criteria.} \tag{14}$$

Step 3 Compute the separation measures.

This step measures the closeness of an alternative to the positive ideal solution using the L_P metric to formulate the group utility and the individual regret of the opponent (Liao and Xu 2013).

$$L_{P,i} = \left\{ \sum_{j=1}^{m} \left[w_j \left(X_j^+ - X_{ij} \right) / \left(X_j^+ - X_j^- \right)^P \right] \right\}^{1/P}, 1 \leq P \leq \infty, \quad i = 1, 2, \ldots, n. \tag{15}$$

The ranking measure is based on the variation of the parameter P from 1 to ∞, hence moving from minimizing the sum of individual regrets to minimizing the maximum regret. For $p = 1$, the deviations are weighted equally. For $P = \infty$, the deviations can be interpreted as the maximum individual regret.

$$S_i = L_{1,i} = \sum_{j=1}^{m} w_j \left(X_j^+ - X_{ij} \right) / \left(X_j^+ - X_j^- \right), \tag{16}$$

$$R_i = L_{\infty,i} = \max_j \left[w_j \left(X_j^+ - X_{ij} \right) / \left(X_j^+ - X_j^- \right) \right], \tag{17}$$

where w_j is the weight of the jth criteria, S_i is the utility measure, and R_i is the regret measure.

Step 4 Determine S^+, S^-, R^+, and R^-.

$S^+ = \min_i S_i$ is the maximum group utility ('majority rule index"), and

$S^- = \max_i S_i$ is the minimum group utility.

$R^+ = \min_i R_i$ is the minimum individual regret of the opponent, and

$R^- = \max_i R_i$ is the maximum individual regret of the opponent.

Step 5 Calculate the index Q_i

$$Q_i = \gamma \left(S_i - S^+ \right) / \left(S^- - S^+ \right) + (1 - \gamma) \left(R_i - R^+ \right) / \left(R^- - R^+ \right), \tag{18}$$

where γ is the weight of "the maximum group utility" strategy, while $(1 - \gamma)$ is the weight of "the individual regret".

Step 6 Propose a compromise solution.

Form three ranking lists of the alternatives by the values of S, R and Q.

The alternative with minimum Q is the best choice, $X^{(1)}$, if the following two conditions are satisfied.

i. **Acceptable advantage**

$$Q\left(X^{(2)} \right) - Q\left(X^{(1)} \right) \geq 1/(n - 1), \tag{19}$$

where $X^{(2)}$ is the 2nd alternative in the Q rank.
If this condition is not satisfied, then
$X^{(1)}, X^{(2)}, \ldots, X^{(r)}$ are the compromised solutions having

$$Q\left(X^{(r)}\right) - Q\left(X^{(1)}\right) < 1/(n-1). \quad (20)$$

ii. **Acceptable stability in decision making**
The alternative $X^{(1)}$ is also ranked 1st with minimum S and/or R
If this condition is not satisfied, then $X^{(1)}$ and $X^{(2)}$ only are the compromised solutions.

4 The Spherical Fuzzy VIKOR

In an MCDM problem, k decision-makers of different weights $\{\omega_1, \omega_2, \ldots, \omega_k\}$ evaluate a set of n alternatives $X = \{X_1, X_2, \ldots, X_n\}$ according to a set of m criterion $C = \{C_1, C_2, \ldots, C_m\}$. The decision matrix of the pth decision-maker $\mathbf{D}_p = \left[X_{ij}^{p}\right]$ expresses his evaluation of the alternative X_i for the criterion C_j using SFSs. The weighting vector $\mathbf{W}_p = \left[w_j^{p}\right]$ expresses the pth decision-maker's ratings of the importance of the jth criterion using also SFSs. The steps of the technique are as follows.

Step 1 Form the decision matrices, hence the average decision matrix.
For the decision matrices $\mathbf{D}_1, \mathbf{D}_2, \ldots, \mathbf{D}_k$ the average decision matrix is computed by

$$\tilde{\mathbf{D}} = \omega_1 * \mathbf{D}_1 + \omega_2 * \mathbf{D}_2 + \cdots + \omega_k * \mathbf{D}_k. \quad (21)$$

Aggregation can be formed using the SWAM operator or the SWGM operator.
Step 2 Form the weighting matrices, hence the average weighting matrix.
For the weighting matrices $\mathbf{W}_1, \mathbf{W}_2, \ldots, \mathbf{W}_k$, the average weighting matrix is computed by

$$\tilde{\mathbf{W}} = \omega_1 * \mathbf{W}_1 + \omega_2 * \mathbf{W}_2 + \cdots + \omega_k * \mathbf{W}_k. \quad (22)$$

Aggregation can be formed using the SWAM operator or the SWGM operator.
Step 3 Set the PIS and the NIS
Two ideal solutions are proposed. The first is the absolute ideal solution which is independent of the decision variablesThe average decision matrix obtained by using the SWAM operator is

$$X^{+} = (1, 0, 0), \text{ and } X^{-} = (0, 0, 1). \quad (23)$$

Since the hesitancy degree is independent of the membership degree and the non-membership degree, being hesitant and unable to decide is considered the worst case that might happen. In this case, $d\left(X^{+}, X^{-}\right) = 1$.
The second is the relative ideal solution that depends on the ratings of the alternatives for the assessment criteria.
Let $\mu_P = \max\limits_i \mu_i$, $\nu_P = \min\limits_i \upsilon_i$, and $\pi_P = \min\limits_i \pi_i$,

$$X_j^{+} = (\mu_P, \upsilon_P, \pi_P). \tag{24}$$

Let $\mu_N = \min_i \mu_i$, $\upsilon_N = \max_i \upsilon_i$, and $\pi_N = \max_i \pi_i$,

$$X_j^{-} = \begin{cases} (\mu_N, \upsilon_N, \pi_N) & if\ \mu_N^2 + \upsilon_N^2 + \pi_N^2 \le 1, \\ \left(\mu_N, \upsilon_N, \sqrt{1 - \left(\mu_N^2 + \upsilon_N^2\right)}\right) & \text{otherwise.} \end{cases} \tag{25}$$

Step 4 Compute the weighted distances $\tilde{R}_{ij}$ and $\tilde{S}_i$.
The closeness of X_{ij} (the rating of the ith alternative for the jth criterion) to X_j^{+} (the PIS of the jth criterion) is calculated

$$\tilde{R}_{ij} = w_j * \left(\frac{d\left(X_j^{+}, X_{ij}\right)}{d\left(X_j^{+}, X_j^{-}\right)} \right). \tag{26}$$

Then, the total weighted distance is calculated for each alternative

$$\tilde{S}_i = \sum_{j=1}^{m} \tilde{R}_{ij}. \tag{27}$$

Step 5 Use the score function to form the score matrices, hence determine the separation measures.

$$\mathbf{R} = \left[\text{Score}\left(\tilde{R}_{ij}\right)\right], \tag{28}$$

and

$$R_i = \max_j \left[\text{Score}\left(\tilde{R}_{ij}\right)\right]. \tag{29}$$

$$\mathbf{S} = [S_i] = \left[\text{Score}\left(\tilde{S}_i\right)\right]. \tag{30}$$

Step 6 Sort R_i and S_i, then determine S^{+}, S^{-}, R^{+}, and R^{-}.

$$S^{+} = \min_i S_i,\ S^{-} = \max_i S_i,\ R^{+} = \min_i R_i \text{ and } R^{-} = \max_i R_i. \tag{31}$$

Step 7 Compute and rank Q_i.
Use (18) to calculate Q_i, and then rank in ascending order.
Step 8 Propose a compromise solution.

The compromise solution is chosen using (19) and (20) as previously mentioned in the VIKOR method.

5 An Illustrative Example

The example in this section is due to Gündoğdu and Kahraman (2019a). It is resolved using the proposed SFVIKOR to illustrate its application and compare the results with the SFTOPSIS. This example is solved using four different implementations. The first implementation SFVIKOR1 uses the SWAM operator and the absolute ideal solution (23). The second implementation SFVIKOR2 uses the SWGM operator and the absolute ideal solution (23). The third implementation SFVIKOR3 uses the SWAM operator and the relative ideal solution (24) and (25). The fourth implementation SFVIKOR4 uses the SWGM operator and the relative ideal solution (24) and (25). The example is solved for $\gamma = 0.5$ (voting by consensus). Then, the stability of the solution for the different implementations is tested

5.1 Example

A company purchases air conditions from four suppliers $\{X_1, X_2, X_3, X_4\}$. Evaluation is based on four criteria: price (C_1), quality (C_2), delivery (C_3), and performance (C_4). Three decision-makers $\{D_1, D_2, D_3\}$ are involved with different levels of experience. Their weights are $\{0.3, 0.2, 0.5\}$ respectively. The linguistic terms and their corresponding spherical fuzzy numbers and the judgements of the decision makers can be found in Gündoğdu and Kahraman (2019a). The decision matrices are given directly.

SFVIKOR1: Using the SWAM operator and the absolute ideal solution (23).

Step 1 Form the decision matrices, hence the average decision matrix.

$$\widetilde{\mathbf{D}}_1 = \begin{array}{c} \\ X_1 \\ X_2 \\ X_3 \\ X_4 \end{array} \begin{array}{c} \begin{array}{cccc} C_1 & C_2 & C_3 & C_4 \end{array} \\ \begin{bmatrix} \text{AMI} & \text{HI} & \text{EI} & \text{SMI} \\ \text{SLI} & \text{VHI} & \text{VHI} & \text{EI} \\ \text{LI} & \text{VLI} & \text{SMI} & \text{HI} \\ \text{HI} & \text{VH} & \text{VHI} & \text{SMI} \end{bmatrix} \end{array}, \quad \widetilde{\mathbf{D}}_2 = \begin{array}{c} \\ X_1 \\ X_2 \\ X_3 \\ X_4 \end{array} \begin{array}{c} \begin{array}{cccc} C_1 & C_2 & C_3 & C_4 \end{array} \\ \begin{bmatrix} \text{VHI} & \text{HI} & \text{VHI} & \text{SMI} \\ \text{SLI} & \text{AMI} & \text{VHI} & \text{EI} \\ \text{SLI} & \text{SMI} & \text{HI} & \text{SMI} \\ \text{VLI} & \text{SLI} & \text{VHI} & \text{LI} \end{bmatrix} \end{array}$$

$$\text{and } \widetilde{\mathbf{D}}_3 = \begin{array}{c} \\ X_1 \\ X_2 \\ X_3 \\ X_4 \end{array} \begin{array}{c} \begin{array}{cccc} C_1 & C_2 & C_3 & C_4 \end{array} \\ \begin{bmatrix} \text{HI} & \text{HI} & \text{AMI} & \text{AMI} \\ \text{AMI} & \text{EI} & \text{HI} & \text{SLI} \\ \text{HI} & \text{SMI} & \text{VHI} & \text{VLI} \\ \text{LI} & \text{EI} & \text{LI} & \text{HI} \end{bmatrix} \end{array}.$$

The average decision matrix obtained by using the SWAM operator is

$$\widetilde{\mathbf{D}}_A = \begin{array}{c} \\ X_1 \\ X_2 \\ X_3 \\ X_4 \end{array} \begin{array}{c} \begin{array}{cccc} C_1 & C_2 & C_3 & C_4 \end{array} \\ \begin{bmatrix} (0.80, 0.20, 0.21) & (0.70, 0.30, 0.30) & (0.82, 0.19, 0.23) & (0.81, 0.20, 0.23) \\ (0.77, 0.24, 0.22) & (0.74, 0.28, 0.32) & (0.74, 0.27, 0.27) & (0.45, 0.55, 0.46) \\ (0.57, 0.44, 0.32) & (0.53, 0.49, 0.37) & (0.74, 0.27, 0.28) & (0.52, 0.52, 0.31) \\ (0.48, 0.56, 0.29) & (0.62, 0.39, 0.39) & (0.65, 0.37, 0.25) & (0.62, 0.39, 0.34) \end{bmatrix} \end{array}$$

Step 2 Form the weighting matrices, hence the average weighting matrix.

$$\mathbf{W}_1 = \begin{array}{cccc} C_1 & C_2 & C_3 & C_4 \\ [\text{LI} & \text{EI} & \text{AMI} & \text{SMI}] \end{array}, \quad \mathbf{W}_2 = \begin{array}{cccc} C_1 & C_2 & C_3 & C_4 \\ [\text{SLI} & \text{VHI} & \text{VHI} & \text{SMI}] \end{array},$$

$$\text{and } \mathbf{W}_3 = \begin{array}{cccc} C_1 & C_2 & C_3 & C_4 \\ [\text{HI} & \text{AMI} & \text{HI} & \text{AMI}] \end{array}.$$

Using the SWAM operator, the average weighting matrix is

$$\widetilde{\mathbf{W}}_A = \begin{array}{cccc} C_1 & C_2 & C_3 & C_4 \\ [(0.57, 0.44, 0.32) & (0.82, 0.19, 0.23) & (0.80, 0.20, 0.21) & (0.81, 0.20, 0.23)] \end{array}$$

Step 3 Set the PIS and the NIS
$X^+ = (1, 0, 0)$, and $X^- = (0, 0, 1)$. In this case, $d(X^+, X^-) = 1$.
Step 4 Compute the weighted distances $\tilde{R}_{ij}$ and $\tilde{S}_i$.
The distances between each alternative and the PIS of each criterion are given in Table 1.

$$\tilde{R}_{11} = w_1 d(X_{11}, X^+) = (0.3053, 0.8150, 0.1908),$$
$$\tilde{R}_{12} = w_2 d(X_{12}, X^+) = (0.5800, 0.5433, 0.2039),$$

Table 1 Distance between each alternative and the PIS of each criterion for SFVIKOR1

	C_1	C_2	C_3	C_4
X_1	0.2491	0.3674	0.2464	0.2540
X_2	0.2819	0.3524	0.3266	0.6390
X_3	0.4904	0.5468	0.3308	0.5463
X_4	0.5780	0.4736	0.4012	0.4539

$$\begin{aligned}\tilde{R}_{13} &= w_3 d\left(X_{13}, X^+\right) = (0.4718, 0.6726, 0.1570),\\ \tilde{R}_{14} &= w_4 d\left(X_{14}, X^+\right) = (0.4873, 0.6644, 0.1780),\\ \text{and } \tilde{S}_1 &= (0.8020, 0.1979, 0.2426).\end{aligned}$$

$$\begin{aligned}\tilde{R}_{21} &= w_1 d\left(X_{21}, X^+\right) = (0.3238, 0.7934, 0.2014),\\ \tilde{R}_{22} &= w_2 d\left(X_{22}, X^+\right) = (0.5702, 0.5570, 0.2015),\\ \tilde{R}_{23} &= w_3 d\left(X_{23}, X^+\right) = (0.5326, 0.5912, 0.1730),\\ \tilde{R}_{24} &= w_4 d\left(X_{24}, X^+\right) = (0.7032, 0.3576, 0.2262),\\ \text{and } \tilde{S}_2 &= (0.8839, 0.0934, 0.2229).\end{aligned}$$

$$\begin{aligned}\tilde{R}_{31} &= w_1 d\left(X_{31}, X^+\right) = (0.4186, 0.6686, 0.2528),\\ \tilde{R}_{32} &= w_2 d\left(X_{32}, X^+\right) = (0.6758, 0.4033, 0.2233),\\ \tilde{R}_{33} &= w_3 d\left(X_{33}, X^+\right) = (0.5355, 0.5872, 0.1737),\\ \tilde{R}_{34} &= w_4 d\left(X_{34}, X^+\right) = (0.6647, 0.4151, 0.2206),\\ \text{and } \tilde{S}_3 &= (0.9064, 0.0657, 0.2185).\end{aligned}$$

$$\begin{aligned}\tilde{R}_{41} &= w_1 d\left(X_{31}, X^+\right) = (0.4507, 0.6222, 0.2688),\\ \tilde{R}_{42} &= w_2 d\left(X_{32}, X^+\right) = (0.6407, 0.4554, 0.2172),\\ \tilde{R}_{43} &= w_3 d\left(X_{33}, X^+\right) = (0.5799, 0.5243, 0.18410),\\ \tilde{R}_{44} &= w_4 d\left(X_{34}, X^+\right) = (0.6197, 0.4817, 0.2121),\\ \text{and } \tilde{S}_4 &= (0.8989, 0.0715, 0.2252).\end{aligned}$$

Step 5 Use the score function to form the score matrices, hence determine the separation measures.
Using Eq. (28) to get the **R** score matrix, we get

$$\mathbf{R} = \begin{array}{c} \\ X_1 \\ X_2 \\ X_3 \\ X_4 \end{array} \begin{array}{c} \begin{array}{cccc} C_1 & C_2 & C_3 & C_4 \end{array} \\ \begin{bmatrix} -0.3765 & 0.0263 & -0.1667 & -0.1409 \\ -0.3355 & 0.0096 & -0.0456 & +0.2103 \\ -0.1454 & 0.1724 & -0.0401 & +0.1594 \\ -0.0918 & 0.1226 & +0.0409 & +0.0935 \end{bmatrix} \end{array}.$$

Using Eq. (29) to find the regret measure R_i, $R_i = \max_j \left[\text{Score}\left(\tilde{R}_{ij} \right) \right]$, we get $R_1 = 0.0263$, $R_2 = 0.2103$, $R_3 = 0.1724$, and $R_4 = 0.1226$.
Using (30) to find the utility measure S_i, $\mathbf{S} = \left[\text{Score}\left(\tilde{S}_i \right) \right]$, we get

$$\mathbf{S} = \begin{array}{cccc} X_1 & X_2 & X_3 & X_4 \\ [0.3109 & 0.4204 & 0.4499 & 0.4302] \end{array}.$$

Step 6 Sort S_i and R_i, then determine S^+, S^-, R^+, and R^-.
Using (31), we have

$$S^+ = 0.3109,\ S^- = 0.4499,\ R^+ = 0.0263,\ \text{and}\ R^- = 0.2103.$$

Step 7 Compute and rank Q_i

$$Q_i = \gamma\left(S_i - S^+\right)/\left(S^- - S^+\right) + (1-\gamma)\left(R_i - R^+\right)/\left(R^- - R^+\right).$$

$$Q_1 = 0,\ Q_2 = 0.8939,\ Q_3 = 0.897 \text{ and } Q_4 = 0.6908.$$

The ranking is Q_1, Q_4, Q_2 and Q_3.
Step 8 Propose a compromise solution.
X_1 is the best alternative since it has the minimum S and R and satisfies the acceptable advantage condition $Q_4 - Q_1 = 0.6908 > 0.33$.
The best alternative X_1 is the same as that obtained by SFTOPSIS with the SWAM operator. The overall ranking is also the same X_1, X_4, X_2 and X_3.

SFVIKOR2: Using the SWGM operator and the absolute ideal solution (23).
The results can be summarized as follows.

$$\widetilde{\mathbf{D}}_G = \begin{array}{c} \\ X_1 \\ X_2 \\ X_3 \\ X_4 \end{array} \begin{array}{c} \begin{array}{cccc} C_1 & C_2 & C_3 & C_4 \end{array} \\ \begin{bmatrix} (0.78, 0.24, 0.24) & (0.70, 0.30, 0.30) & (0.74, 0.31, 0.34) & (0.73, 0.30, 0.31) \\ (0.66, 0.45, 0.33) & (0.65, 0.38, 0.41) & (0.87, 0.27, 0.28) & (0.71, 0.55, 0.45) \\ (0.49, 0.54, 0.33) & (0.43, 0.59, 0.33) & (0.80, 0.30, 0.30) & (0.81, 0.66, 0.26) \\ (0.36, 0.66, 0.28) & (0.55, 0.47, 0.43) & (0.62, 0.55, 0.28) & (0.61, 0.46, 0.34) \end{bmatrix} \end{array},$$

$$\widetilde{\mathbf{W}}_G = \begin{array}{cccc} C_1 & C_2 & C_3 & C_4 \\ [(0.49, 0.54, 0.33) & (0.74, 0.31, 0.34) & (0.78, 0.24, 0.24) & (0.73, 0.30, 0.31)] \end{array}.$$

Table 2 Distance between each alternative and the PIS of each criterion for SFVIKOR2

	C_1	C_2	C_3	C_4
X_1	0.2860	0.3674	0.3737	0.3599
X_2	0.4855	0.4664	0.2900	0.5427
X_3	0.5747	0.6253	0.3317	0.5193
X_4	0.6796	0.5515	0.5125	0.4895

The distances between each alternative and the PIS of each criterion are given in Table 2.

$$\mathbf{R} = \begin{matrix} \\ X_1 \\ X_2 \\ X_3 \\ X_4 \end{matrix} \begin{matrix} \mathrm{C}_1 & \mathrm{C}_2 & \mathrm{C}_3 & \mathrm{C}_4 \\ \left[\begin{matrix} -0.4020 & -0.0883 & -0.0313 & -0.1026 \\ -0.2296 & -0.0124 & -0.1348 & 0.0334 \\ -0.1797 & 0.0673 & -0.0796 & 0.0202 \\ -0.1214 & 0.0353 & 0.0899 & 0.0017 \end{matrix}\right] \end{matrix}.$$

$R_1 = -0.0313$, $R_2 = 0.0334$, $R_3 = 0.0673$, and $R_4 = 0.0899$, and

$$\mathbf{S} = \begin{matrix} \mathrm{X}_1 & \mathrm{X}_2 & \mathrm{X}_3 & \mathrm{X}_4 \\ [0.2265 & 0.2457 & 0.2756 & 0.3061] \end{matrix}.$$

$S^+ = 0.2265$, $S^- = 0.3061$, $R^+ = -0.0313$, and $R^- = 0.0899$.

$$Q_1 = 0,\ Q_2 = 0.3875,\ Q_3 = 0.7238 \text{ and } Q_4 = 1.$$

The ranking is Q_1, Q_2, Q_3 and Q_4. Then, X_1 is the best alternative having the minimum S and R and satisfy the acceptable advantage condition $Q_2 - Q_1 = 0.3875 > 0.33$.

The overall ranking of the SFTOPSIS with the SWGM operator is $\mathrm{X}_1, \mathrm{X}_2, \mathrm{X}_4$ and X_3. Then, the ranking is quite similar having the third and fourth alternatives exchanging ranks.

SFVIKOR3: Using the SWAM operator and the relative ideal solution (24) **and** (25).

The results can be summarized as follows.

$$\widetilde{\mathbf{D}}_{\mathrm{A}} = \begin{matrix} \\ \mathrm{X}_1 \\ \mathrm{X}_2 \\ \mathrm{X}_3 \\ \mathrm{X}_4 \end{matrix} \begin{matrix} \mathrm{C}_1 & \mathrm{C}_2 & \mathrm{C}_3 & \mathrm{C}_4 \\ \left[\begin{matrix} (0.80,0.20,0.21) & (0.70, 0.30, 0.30) & (0.82, 0.19, 0.23) & (0.81, 0.20, 0.23) \\ (0.77, 0.24, 0.22) & (0.74, 0.28, 0.32) & (0.74, 0.27, 0.27) & (0.45, 0.55, 0.46) \\ (0.57, 0.44, 0.32) & (0.53, 0.49, 0.37) & (0.74, 0.27, 0.28) & (0.52, 0.52, 0.31) \\ (0.48, 0.56, 0.29) & (0.62, 0.39, 0.39) & (0.65, 0.37, 0.25) & (0.62, 0.39, 0.34) \end{matrix}\right] \end{matrix}.$$

The ISs are

Table 3 Distance between each alternative and the PIS of each criterion for SFVIKOR3

	C_1	C_2	C_3	C_4
X_1	0	0.0316	0	0
X_2	0.0361	0.0141	0.0849	0.3905
X_3	0.2476	0.2158	0.0972	0.3106
X_4	0.3453	0.1315	0.1756	0.2007
$d\left(X_j^+, X_j^-\right)$	0.3494	0.2194	0.1863	0.3905

$$\begin{array}{c} \\ \text{PIS} \\ \text{NIS} \end{array} \begin{array}{cccc} C_1 & C_2 & C_3 & C_4 \\ \left[\begin{array}{c} (0.80, 0.20, 0.21) \\ (0.48, 0.56, 0.32) \end{array}\right. & \begin{array}{c} (0.74, 0.28, 0.30) \\ (0.53, 0.49, 0.39) \end{array} & \begin{array}{c} (0.82, 0.19, 0.23) \\ (0.65, 0.37, 0.28) \end{array} & \left.\begin{array}{c} (0.81, 0.20, 0.23) \\ (0.45, 0.55, 0.46) \end{array}\right] \end{array}$$

The distances between each alternative and the PIS of each criterion are given in Table 3.

$$\mathbf{R} = \begin{array}{c} \\ X_1 \\ X_2 \\ X_3 \\ X_4 \end{array} \begin{array}{c} \begin{array}{cccc} C_1 & C_2 & C_3 & C_4 \end{array} \\ \begin{bmatrix} -1.0000 & -0.3540 & -1.0000 & -1.0000 \\ -0.6211 & -0.6083 & +0.0921 & +0.3355 \\ -0.0313 & +0.3420 & +0.1432 & +0.2741 \\ +0.0460 & +0.1681 & +0.3309 & +0.1384 \end{bmatrix} \end{array},$$

$R_1 = -0.3540$, $R_2 = 0.3355$, $R_3 = 0.3420$, and $R_4 = 0.3309$, and

$$\mathbf{S} = \begin{array}{cccc} X_1 & X_2 & X_3 & X_4 \\ [-0.3540 & 0.4594 & 0.6371 & 0.5855] \end{array}.$$

$S^+ = -0.3540$, $S^- = 0.6371$, $R^+ = -0.3540$, and $R^- = 0.3420$.

$$Q_1 = 0,\ Q_2 = 0.9057,\ Q_3 = 1 \text{ and } Q_4 = 0.9660.$$

The ranking is Q_1, Q_2, Q_4 and Q_3. Then, X_1 is the best alternative having the minimum S and R and satisfy the acceptable advantage condition $Q_2 - Q_1 = 0.9057 > 0.33$.

The best alternative X_1 is the same as that obtained by SFTOPSIS with the SWAM operator X_1, X_4, X_2 and X_3. Regarding the overall ranking, the best and worst solutions are the same, the second and third alternative exchange ranks.

SFVIKOR4: Using the SWGM operator and the relative ideal solution (24) **and** (25).

The results can be summarized as follows.

$$\widetilde{\mathbf{D}}_{\mathrm{G}} = \begin{array}{c} \\ \mathrm{X}_1 \\ \mathrm{X}_2 \\ \mathrm{X}_3 \\ \mathrm{X}_4 \end{array} \begin{array}{cccc} \mathrm{C}_1 & \mathrm{C}_2 & \mathrm{C}_3 & \mathrm{C}_4 \end{array} \begin{bmatrix} (0.78,0.24,0.24) & (0.70, 0.30, 0.30) & (0.74, 0.31, 0.34) & (0.73, 0.30, 0.31) \\ (0.66, 0.45, 0.33) & (0.65, 0.38, 0.41) & (0.87, 0.27, 0.28) & (0.71, 0.55, 0.45) \\ (0.49, 0.54, 0.33) & (0.43, 0.59, 0.33) & (0.80, 0.30, 0.30) & (0.81, 0.66, 0.26) \\ (0.36, 0.66, 0.28) & (0.55, 0.47, 0.43) & (0.62, 0.55, 0.28) & (0.61, 0.46, 0.34) \end{bmatrix}$$

The ISs are

$$\begin{array}{c} \\ \mathrm{PIS} \\ \mathrm{NIS} \end{array} \begin{array}{cccc} \mathrm{C}_1 & \mathrm{C}_2 & \mathrm{C}_3 & \mathrm{C}_4 \end{array} \begin{bmatrix} (0.78, 0.24, 0.24) & (0.70, 0.30, 0.30) & (0.87, 0.27, 0.28) & (0.81, 0.30, 0.26) \\ (0.36, 0.66, 0.33) & (0.43, 0.59, 0.43) & (0.62, 0.55, 0.34) & (0.61, 0.66, 0.43) \end{bmatrix}$$

The distances between each alternative and the IS of each criterion are given in Table 4.

$$\mathbf{R} = \begin{array}{c} X_1 \\ X_2 \\ X_3 \\ X_4 \end{array} \overset{\begin{array}{cccc} \mathrm{C}_1 & \mathrm{C}_2 & \mathrm{C}_3 & \mathrm{C}_4 \end{array}}{\begin{bmatrix} -1.0000 & -1.0000 & -0.0132 & -0.2951 \\ -0.2304 & -0.1070 & -1.0000 & +0.1153 \\ -0.1081 & +0.1512 & -0.2712 & +0.1349 \\ +0.0205 & +0.0678 & +0.2883 & +0.0633 \end{bmatrix}},$$

$R_1 = -0.0132$, $R_2 = 0.1153$, $R_3 = 0.1512$, and $R_4 = 0.2883$,
And

$$\mathbf{S} = \begin{array}{cccc} \mathrm{X}_1 & \mathrm{X}_2 & \mathrm{X}_3 & \mathrm{X}_4 \\ [0.1153 & 0.1881 & 0.3479 & 0.4582] \end{array}.$$

$S^+ = 0.1153, S^- = 0.4582, R^+ = -0.0132$, and $R^- = 0.2883$.

$$Q_1 = 0,\ Q_2 = 0.3193,\ Q_3 = 0.6118, \text{ and } Q_4 = 1.$$

The ranking is Q_1, Q_2, Q_3 and Q_4.

Table 4 Distance between each alternative and the PIS of each criterion for SFVIKOR4

	C_1	C_2	C_3	C_4
X_1	0	0	0.1051	0.0667
X_2	0.2057	0.1025	0	0.2330
X_3	0.3018	0.2810	0.0557	0.2546
X_4	0.4210	0.1848	0.2654	0.1897
$d\left(X_j^+, X_j^-\right)$	0.4248	0.2949	0.2688	0.3150

Table 5 Summary of the SFVIKOR results

	R	S	Q	Compromise solution
SFVIKOR1	R_1, R_4, R_3, R_2	S_1, S_2, S_4, S_3	Q_1, Q_4, Q_2, Q_3	X_1
SFVIKOR2	R_1, R_2, R_3, R_4	S_1, S_2, S_3, S_4	Q_1, Q_2, Q_3, Q_4	X_1
SFVIKOR3	R_1, R_4, R_2, R_3	S_1, S_2, S_4, S_3	Q_1, Q_2, Q_4, Q_3	X_1
SFVIKOR4	R_1, R_2, R_3, R_4	S_1, S_2, S_3, S_4	Q_1, Q_2, Q_3, Q_4	X_1 and X_2

The acceptable advantage condition is not satisfied $Q_2 - Q_1 = 0.3193 < 0.33$. Then, X_1 and X_2 are the compromised solutions, since they are the only alternatives that satisfy $Q(X^{(r)}) - Q(X^{(1)}) < 1/(n-1)$.

The compromise solution list is half the list obtained by the SFTOPSIS with the SWGM operator X_1, X_2, X_4 and X_3.

The results of the different implementations are given in Table 5.

5.2 *The Stability of Solution*

In this section, the performance of the SFVIKOR using the four implementations is examined to test the stability of the proposed compromised solutions when changing some data in the problem. Wang and Triantaphyllou (2006) established three test criteria to evaluate the performance of MCDM methods.

Test criteria 1: When a non-optimal alternative is replaced by another worse alternative, keeping the decision criteria unchanged, the best solution remains the same.

Applying test criteria 1, and changing a non-optimal alternative by another worse one.

For SFVIKOR1 and SFVIKOR2, the performance of the alternative X_3 for the criteria is changed by switching the membership degree and the non-membership degree whenever the membership degree has a larger value than the non-membership degree. In SFVIKOR1, the best alternative X_1 is unchanged. Meanwhile, in SFVIKOR2, the compromise solution set becomes $\{X_1, X_2\}$.

For SFVIKOR3 and SFVIKOR4, the performance of the alternative X_3 for the criteria is replaced by the NIS. In SFVIKOR3, the best alternative X_1 is unchanged. In SFVIKOR4, the compromise solution set becomes $\{X_1, X_2\}$.

Test criteria 2: When the problem is decomposed to smaller sub-problems, the rankings of the sub-problems should follow the property of transitivity.

Test criteria 3: When combining the rankings of the smaller sub-problems, the overall ranking should be identical to the ranking of the original problem.

Applying test criteria 2 and 3, three smaller sub-problems are solved. The results are summarized in Tables 6, 7, 8 and 9.

SFVIKOR results using the SWAM/SWGM operator and the absolute ideal solution (23) are not affected when omitting one of the alternatives. It maintained

Table 6 Results of subproblems using SFVIKOR1

Subproblem	R	S	Q
(X_1, X_2, X_3)	$R_1 = 0.03$, $R_3 = 0.17$, $R_2 = 0.21$.	$S_1 = 0.31$, $S_2 = 0.42$, $S_3 = 0.45$.	$Q_1 = 0.00$, $Q_2 = 0.89$, $Q_3 = 0.90$.
(X_2, X_3, X_4)	$R_4 = 0.12$, $R_3 = 0.17$, $R_2 = 0.21$.	$S_2 = 0.42$, $S_4 = 0.43$, $S_3 = 0.45$.	$Q_4 = 0.17$, $Q_2 = 0.50$, $Q_3 = 0.78$
(X_1, X_2, X_4)	$R_1 = 0.03$, $R_4 = 0.12$, $R_2 = 0.21$.	$S_1 = 0.31$, $S_2 = 0.42$, $S_4 = 0.43$.	$Q_1 = 0.00$, $Q_4 = 0.76$, $Q_2 = 0.96$.
(X_1, X_3, X_4)	$R_1 = 0.03$, $R_4 = 0.12$, $R_3 = 0.17$.	$S_1 = 0.31$, $S_4 = 0.43$, $S_3 = 0.45$.	$Q_1 = 0.00$ $Q_4 = 0.76$, $Q_3 = 1.00$.
Combined ranking	R_1, R_4, R_3 and R_2	S_1, S_2, $S_4$and S_3	Q_1, Q_4, $Q_2$and Q_3

Table 7 Results of subproblems using SFVIKOR2

Subproblem	R	S	Q
(X_1, X_2, X_3)	$R_1 = -0.03$, $R_2 = 0.03$, $R_3 = 0.07$.	$S_1 = 0.23$, $S_2 = 0.25$, $S_3 = 0.28$.	$Q_1 = 0.00$, $Q_2 = 0.52$, $Q_3 = 1.00$.
(X_2, X_3, X_4)	$R_2 = 0.03$, $R_3 = 0.07$, $R_4 = 0.09$.	$S_2 = 0.25$, $S_3 = 0.28$, $S_4 = 0.31$.	$Q_2 = 0.00$, $Q_3 = 0.55$, $Q_4 = 1.00$.
(X_1, X_2, X_4)	$R_1 = -0.03$, $R_2 = 0.03$, $R_4 = 0.09$.	$S_1 = 0.23$, $S_2 = 0.25$, $S_4 = 0.31$.	$Q_1 = 0.00$, $Q_2 = 0.38$, $Q_4 = 1.00$.
(X_1, X_3, X_4)	$R_1 = -0.03$, $R_3 = 0.07$, $R_4 = 0.09$.	$S_1 = 0.23$, $S_3 = 0.28$, $S_4 = 0.31$.	$Q_1 = 0.00$, $Q_3 = 0.42$, $Q_4 = 1.00$.
Combined ranking	R_1,R_2, R_3 and R_4	S_1, S_2, S_3 and S_4	Q_1, Q_2,Q_3 and Q_4

Table 8 Results of subproblems using SFVIKOR3

Subproblem	R	S	Q
(X_1, X_2, X_3)	$R_1 = -0.35$, $R_2 = 0.34$, $R_3 = 0.35$.	$S_1 = -0.35$, $S_2 = 0.56$, $S_3 = 0.72$.	$Q_1 = 0.00$, $Q_2 = 0.92$, $Q_3 = 1.00$
(X_2, X_3, X_4)	$R_2 = 0.34$, $R_4 = 0.34$, $R_3 = 0.35$.	$S_2 = 0.37$, $S_4 = 0.52$, $S_3 = 0.55$.	$Q_2 = 0.00$, $Q_4 = 0.41$, $Q_3 = 1.00$
(X_1, X_3, X_4)	$R_1 = -1.00$, $R_3 = 0.34$, $R_4 = 0.35$.	$S_1 = -1.00$, $S_4 = 0.62$, $S_3 = 0.66$.	$Q_1 = 0.00$, $Q_4 = 0.42$, $Q_3 = 1.00$
Combined ranking	Inconsistent	S_1, S_2, $S_4$and S_3	Q_1, Q_2, $Q_4$and Q_3

Table 9 Results of subproblems using SFVIKOR4

Subproblem	R	S	Q
(X_1, X_2, X_3)	$R_2 = 0.15$, $R_3 = 0.17$, $R_1 = 0.29$.	$S_2 = 0.21$, $S_1 = 0.32$, $S_3 = 0.47$.	$Q_2 = 0.00$, $Q_3 = 0.56$, $Q_1 = 0.71$.
(X_2, X_3, X_4)	$R_2 = 0.10$, $R_3 = 0.15$, $R_4 = 0.29$.	$S_2 = 0.17$, $S_3 = 0.31$, $S_4 = 0.45$.	$Q_2 = 0.00$, $Q_3 = 0.37$, $Q_4 = 1.00$.
(X_1, X_3, X_4)	$R_1 = 0.16$, $R_3 = 0.17$, $R_4 = 0.29$.	$S_1 = 0.23$, $S_3 = 0.34$, $S_4 = 0.44$.	$Q_1 = 0.00$, $Q_3 = 0.42$, $Q_4 = 1.00$.
Combined ranking	Inconsistent	S_2, S_1, S_3 and S_4	Inconsistent

the three ranking lists R, S and Q unchanged. Meanwhile, the results using the SWAM/SWGM operator and the relative ideal solution (24) and (25) are affected by omitting an alternative and especially with the SWGM operator. The ranking of the subgroups changed from the ranking of the original problem.

The compromised solution obtained by the absolute ideal solution (23) is more stable when used in SFVIKOR. Since the relative ideal solution (24) and (25) rely on three independent membership functions, then omitting one of the alternatives can completely change the ISs which may change the ranking.

Regarding the acceptable advantage condition, the results showed that using the SWAM operator in SFVIKOR gives a better acceptable advantage than the SWGM operator, and the compromise solution was unique.

Although the different implementations of the SFVIKOR gave the same best compromise solution, yet the implementation of the SFVIKOR using the SWAM and the absolute ideal solution is the most robust.

6 Conclusion

In this chapter, the VIKOR method was extended using SFSs with four different implementations. The SWAM and the SWGM operators were used with two different ideal solutions. The proposed ideal solutions are the absolute ideal solution and the relative ideal solution. Both ideal solutions don't rely on the score function and take the degree of hesitation into consideration. The relative ideal solution depends on the ratings of the alternatives for different decision criteria, while the absolute ideal solution is independent of these ratings. An illustrative example in supplier selection was solved utilizing the four implementations. The four different implementations of the SFVIKOR gave the same best solution which is also the best solution using the SFTOPSIS. The overall ranking of the SFVIKOR1 and the SFVIKOR3 (using the SWAM operator) is different. However, the best and worst alternatives are the same in the two implementations. On the other hand, the overall ranking of the SFVIKOR2 and the SFVIKOR4 (using the SWGM operator) was the same.

The stability of the solution under the different implementations of SFVIKOR was tested. The results show that the performance of the SFVIKOR technique using the absolute ideal solution is more stable than its performance using the relative ideal solution. Meanwhile, the performance of the SFVIKOR with the SWAM operator gives a better acceptable advantage than its performance with the SWGM operator. In general, the SFVIKOR implemented by the SWAM operator and the absolute ideal solution is the most robust implementation. Thus, in MCDM problems in which the evaluation of different sets of alternatives may require the inclusion or exclusion of an alternative, this implementation is recommended.

The contributions of this chapter are as follows. VIKOR is extended to the spherical fuzzy environment which makes the approach more powerful in representing vagueness and uncertainty. Two ideal solutions are proposed apart from the score function. The hesitancy degree plays a role in defining these ideal solutions being independent of the membership and non-membership degrees. Different implementations are examined to obtain the best and most robust implementation of the method.

A comparative study among the performance of IFVIKOR, PFVIKOR, and SFVIKOR with different implementations will be considered in future research.

References

Ashraf S, Abdullah S, Mahmood T, Ghani F, Mahmood T (2019) Spherical fuzzy sets and their applications in multi-attribute decision making problems. J Intell Fuzzy Syst 36(3):2829–2844

Awasthi A, Kannan G (2016) Green supplier development program selection using NGT and VIKOR under fuzzy environment. Comput Ind Eng 91:100–108

Chatterjee K, Kar S (2017) Unified granular-number-based AHP-VIKOR multi-criteria decision framework. Granul Comput 2:199–221

Chen T-Y (2018) Remoteness index-based Pythagorean fuzzy VIKOR methods with a generalized distance measure for multiple criteria decision analysis. Inf Fusion 41:129–150

Cui F-B, You X-Y, Shi H, Liu H-C (2018) Optimal siting of electric vehicle charging stations using Pythagorean fuzzy VIKOR approach. Math Prob Eng. https://doi.org/10.1155/2018/9262067

Devi K (2011) Extension of VIKOR method in intuitionistic fuzzy environment for robot selection. Expert Syst Appl 38:14163–14168

Emeç S, Akkaya G (2018) Stochastic AHP and fuzzy VIKOR approach for warehouse location selection problem. J Enterp Inf Manag 31(6):950–962

Ghorabaee MK (2016) Developing an MCDM method for robot selection with interval type-2 fuzzy sets. Robot Comput Integr Manuf 37:221–232

Ghorabaee MK, Amiri M, Sadaghiani JS, Zavadskas EK (2015) Multi-criteria project selection using an extended VIKOR method with interval type-2 fuzzy sets. Int J Inf Technol Decis Mak 14:993–1016

Gul M (2018) Application of Pythagorean fuzzy AHP and VIKOR methods in occupational health and safety risk assessment: the case of a gun and rifle barrel external surface oxidation and colouring unit. Int J Occup Saf Ergon. https://doi.org/10.1080/10803548.2018.1492251

Gul M, Ak MF, Guneri AF (2019) Pythagorean fuzzy VIKOR-based approach for safety risk assessment in mine industry. J Saf Res 69:135–153

Gul M, Celik E, Aydin N, Gumus AT, Guneri AF (2016) A state of the art literature review of VIKOR and its fuzzy extensions on applications. Appl Soft Comput 46:60–89
Gul M, Guven B, Guneri AF (2018) A new fine-Kinney-based risk assessment framework using FAHP-FVIKOR incorporation. J Loss Prev Process Ind 53:3–16
Gündoğdu FK, Kahraman C (2019a) Spherical fuzzy sets and spherical fuzzy TOPSIS method. J Intell Fuzzy Syst 36(1):337–352
Gündoğdu FK, Kahraman C (2019b) A novel VIKOR method using spherical fuzzy sets and its application to warehouse site selection. J Intell Fuzzy Syst 37:1197–1211
Gündoğdu FK, Kahraman C, Karaşan Ali (2020) Spherical fuzzy VIKOR method and its application to waste management. In: Kahraman C, Cebi S, Cevik Onar S, Oztaysi B, Tolga A, Sari I (eds) Intelligent and fuzzy techniques in big data analytics and decision making. INFUS 2019. Advances in intelligent systems and computing, vol 1029. Springer, Cham, pp 997–1005
Gupta P, Mehlawat MK, Grover N (2016) Intuitionistic fuzzy multi-attribute group decision-making with an application to plant location selection based on a new extended VIKOR method. Inf Sci 370–371:184–203
Hashemi H, Bazargan J, Mousavi SM, Vahdani B (2014) An extended compromise ratio model with an application to reservoir flood control operation under an interval-valued intuitionistic fuzzy environment. Appl Math Model 38:3495–3511
Ju Y, Wang A (2013) Extension of VIKOR method for multi-criteria group decision making problem with linguistic information. Appl Math Model 37:3112–3125
Khan MSA, Abdullah S, Ali A, Amin F (2019) An extension of VIKOR method for multi-attribute decision-making under Pythagorean hesitant fuzzy setting. Granular Comput 4(3):421–434
Liao H, Xu Z (2013) A VIKOR-based method for hesitant fuzzy multi-criteria decision making. Fuzzy Optim Decis Mak 12:373–392
Liu H-C (2010) Generalized addition and scalar multiplication operations on Liu's generalized intuitionistic fuzzy numbers. In: Proceedings of the ninth international conference on machine learning and cybernetics, Qingdao, 11–14 July 2010, pp 2736–2741
Liu J, Liang Y (2018) Multi-granularity unbalanced linguistic group decision-making with incomplete weight information based on VIKOR method. Granular Comput 3(3):219–228
Liu K, Liu Y, Qin J (2018) An integrated ANP-VIKOR methodology for supplier selection with interval type-2 fuzzy sets. Granul Comput 3(3):193–208
Lu S, Tang J (2011) Research on evaluation of auto parts suppliers by VIKOR method based on intuitionistic language multi-criteria. Key Eng Mater 467–469:31–35
Luo X, Wang X (2017) Extended VIKOR method for intuitionistic fuzzy multi-attribute decision-making based on a new distance measure. Math Prob Eng. https://doi.org/10.1155/2017/4072486
Mehbodniya A, Kaleem F, Yen KK, Adachi F (2013) A fuzzy extension of VIKOR for target network selection in heterogeneous wireless environments. Phys Commun 7:145–155
Mousavi SM, Vahdani B, Behzadi SS (2016) Designing a model of intuitionistic fuzzy VIKOR in multi-attribute group decision—making problems. Iran J Fuzzy Syst 13(1):45–65
Opricovic S (1998) Multicriteria optimization of civil engineering systems. PhD thesis, Faculty of Civil Engineering, Belgrade
Opricovic S (2011) Fuzzy VIKOR with an application to water resources planning. Expert Syst Appl 38:12983–12990
Opricovic S, Tzeng G-H (2004) Compromise solution by MCDM methods: a comparative analysis of VIKOR and TOPSIS. Eur J Oper Res 156:445–455
Park J-H, Cho H-J, Kwun Y-C (2013) Extension of the VIKOR method to dynamic intuitionistic fuzzy multiple attribute decision making. Comput Math Appl 65:731–744
Peng J-P, Yeh W-C, Lai T-C, Hsu C-P (2013) Similarity-based method for multi-response optimization problems with intuitionistic fuzzy sets. Proc IMechE Part B: J Eng Manuf 227(6):908–916
Phochanikorn P, Tan C, Chen W (2019) Barriers analysis for reverse logistics in Thailand's palm oil industry using fuzzy multi-criteria decision-making method for prioritizing the solutions. Granular Comput. https://doi.org/10.1007/s41066-019-00155-9

Roostaee R, Izadikhah M, Lotfi FH, Rostamy-Malkhalifeh M (2012) A multi-criteria intuitionistic fuzzy group decision making method for supplier selection with VIKOR method. Int J Fuzzy Syst Appl 2(1):1–17

Sanayei A, Farid SM, Yazdankhah A (2010) Group decision-making process for supplier selection with VIKOR under fuzzy environment. Expert Syst Appl 37(1):24–30

Sari K (2017) A novel multi-criteria decision framework for evaluating green supply chain management practices. Comput Ind Eng 105:338–347

Sayadi MK, Heydari M, Shahanaghi K (2009) Extension of VIKOR method for decision making problem with interval numbers. Appl Math Model 33:2257–2262

Sharaf IM (2019) Supplier selection using a flexible interval-valued fuzzy VIKOR. Comput, Granul. https://doi.org/10.1007/s41066-019-00169-3

Shemshadi A, Shirazi H, Toreihi M, Tarokh MJ (2011) A fuzzy VIKOR method for supplier selection based on entropy measure for objective weighting. Expert Syst Appl 38:12160–12167

Soner O, Celik E, Akyuz E (2017) Application of AHP and VIKOR methods under interval type 2 fuzzy environment in maritime transportation. Ocean Eng 129:107–116

Türk S, John R, Özkan E (2014) Interval Type-2 fuzzy sets in supplier selection. 14th UK workshop on computational intelligence (UKCI), IEEE, Bradford, UK, 8–10 Sept 2014

Vahdani B, Hadipour H, Sadaghiani JS, Amiri M (2010) Extension of VIKOR method based on interval-valued fuzzy sets. Int Adv Manuf Technol 47:1231–1239

Wan S-P, Wang Q-Y, Dong J-Y (2013) The extended VIKOR method for multi-attribute group decision making with triangular intuitionistic fuzzy numbers. Knowl-Based Syst 52:65–77

Wang X, Cai J (2017) A group decision-making model based on distance-based VIKOR with incomplete heterogeneous information and its application to emergency supplier selection. Kybernetes 46:501–529

Wang X, Triantaphyllou E (2006) Ranking irregularities when evaluating alternatives by using some multi-criteria decision analysis methods. In: Badiru A (ed) Handbook of industrial and systems engineering. CRC Press, Taylor & Francis Group, Boca Raton, Chapter 27, pp 27-1 to 27-12

Wu Y, Chen K, Zeng B, Xu H, Yang Y (2016) Supplier selection in nuclear power industry with extended VIKOR method under linguistic information. Appl Soft Comput 48:444–457

Yazdani M, Graeml FR (2014) VIKOR and its applications: a state-of-the-art survey. Int J Strat Decis Sci 5(2):56–83

You X-Y, You J-X, Liu H-C, Zhen L (2015) Group multi-criteria supplier selection using an extended VIKOR method with interval 2-tuple linguistic information. Expert Syst Appl 42:1906–1916

Zhang N, Wei G (2013) Extension of VIKOR method for decision making problem based on hesitant fuzzy set. Appl Math Model 37:4938–4947

Zhao X, Tang S, Yang S, Huang K (2013) Extended VIKOR method based on cross-entropy for interval-valued intuitionistic fuzzy multiple criteria group decision making. J Intell Fuzzy Syst 25(4):1053–1066

Zhao J, You X-Y, Liu H-C, Wu S-M (2017) An extended VIKOR method using intuitionistic fuzzy sets and combination weights for supplier selection. Symmetry 9(9):169. https://doi.org/10.3390/sym9090169

Zhou F, Wang X, Goh M (2018) Fuzzy extended VIKOR based mobile robot selection model for hospital pharmacy. Int J Adv Rob Syst 15(4):1–11

Simple Additive Weighting and Weighted Product Methods Using Spherical Fuzzy Sets

Fatma Kutlu Gündoğdu and Mehmet Yörükoğlu

Abstract Scoring methods are the most frequently used multi attribute decision-making methods because of their easiness and effectiveness. In this chapter, Simple Additive Weighting (SAW) and Weighted Product Methods (WPM) are extended to their fuzzy versions by using recently developed fuzzy extension which is entitled spherical fuzzy sets. These sets not only overcome the vagueness but also clarify hesitancy of decision makers' judgments. This also presents larger preference domain to decision makers. In this chapter, we apply single-valued and interval-valued spherical SAW and WPM methods for the selection of insurance options.

1 Introduction

In Aristotelian philosophy, logic involved with two different values. George Boole formed an algebra system in the 19th century, and adjusted the theory that could cover mathematically with this two-valued logic, matching true and false with 1 and 0, respectively. Zadeh had figured out that traditional two-valued logic could not enough for data representing include personal or imprecise opinions, and so he proposed fuzzy logic to let people identify uncertain thoughts (Zadeh 1965). Fuzzy set is a mathematical model that is generated by using multi-valued logic. After the publication of Zadeh's inspiring study, fuzzy set studies and its extensions are significantly increased. Fuzzy set studies were highly successful and revolutionary in the in the field of control theory in 1980s. After the 1900s, it began to be used in data analysis studies and then extensively developed in computational intelligence, machine learning, and decision making applications (Kahraman et al. 2016).

F. Kutlu Gündoğdu (✉) · M. Yörükoğlu
Industrial Engineering Department, National Defence University, Turkish Air Force Academy, 34149, Istanbul, Turkey
e-mail: fatmakutlugundogdu@gmail.com

M. Yörükoğlu
e-mail: myorukoglu@hho.edu.tr

© Springer Nature Switzerland AG 2021

C. Kahraman and F. Kutlu Gündoğdu (eds.), *Decision Making with Spherical Fuzzy Sets*, Studies in Fuzziness and Soft Computing 392,
https://doi.org/10.1007/978-3-030-45461-6_10

Spherical fuzzy sets (SFS) are a recent extension of previous fuzzy sets such as picture fuzzy sets and Pythagorean fuzzy sets (Ashraf et al. 2019). Spherical fuzzy set is a new notion that has been proposed to handle some limitations encountered in previous fuzzy sets (Kutlu Gündoğdu 2019). In the recent SFS notion, decision makers can define hesitancy degree according to a certain criteria for an alternative just as membership and non-membership dimensions, so that a wider range of preferences is possible thanks to SFS. Simply, SFS are characterized by the degree of membership, the degree of non-membership, and the degree of hesitancy, the square sum of these degrees is equal to or less than one (Kutlu Gündoğdu and Kahraman 2019a). Jin et al. (2019) handled the vague and defective information in decision making by using linguistic Spherical Fuzzy Aggregation Operators. Kutlu Gündoğdu and Kahraman (2019c) introduced the spherical fuzzy sets and used the spherical fuzzy CODAS method for the selection of hospital location. They also presented novel interval-valued spherical fuzzy sets and employed it to develop the extension of TOPSIS under spherical fuzzy environment and used in solving a multiple criteria selection problem for 3D printers (Kutlu Gündoğdu and Kahraman 2019b). Kutlu Gündoğdu and Kahraman (2020) proposed spherical fuzzy quality function deployment (SF-QFD) approach for designing linear delta robot technology. Liu et al. (2020) proposed the linguistic spherical fuzzy numbers (Lt-SFNs) to suggest the public's knowledge of language valuation, then they proposed the linguistic spherical fuzzy weighted averaging (Lt-SFSWA) operator for integrating the language assessment information.

In the literature, many multi-criteria decision-making approaches have been detected in many application areas (Kahraman et al. 2016). To handle impreciseness, ordinary fuzzy sets was applied to simple additive weighting model by Chen and Hwang (1992) and Chen and Klein (1997) and also was employed to weighted product model (Triantaphyllou and Lin 1996). Later, because of the ambiguous and uncertain environment of real life applications, SAW and WPM methods have been extended to other extensions of fuzzy sets such as Type 2 fuzzy SAW (Chen 2014), intuitionistic fuzzy SAW (Kaur and Kumar 2013), hesitant fuzzy SAW (Zhang 2017) and neutrosophic SAW and WPM models (Boltürk et al. 2019). Zavadskas et al. (2014) proposed weighted aggregated sum product assessment with interval-valued intuitionistic fuzzy numbers by integrating SAW and WPM models. To the best our knowledge, there is no any study on SAW and WPM models under spherical fuzzy environment.

In this chapter, single valued spherical SAW, WPM methods and also interval-valued spherical SAW and WPM methods are developed and they applied to the selection of insurance options. The criteria and alternatives are determined by an expert group and weights of criteria and the scores in the decision matrices are aggregated through their opinions.

For the sake of clarity, this chapter is organized as follows. In Sect. 2, preliminaries for spherical fuzzy sets are introduced. In Sect. 3, the proposed single-valued spherical SAW and WPM methods and interval-valued neutrosophic SAW and WPM methods are presented. In Sect. 4, an application for the selection of insurance options is given by using the proposed methods and compared them to check the robustness of the decisions. The paper ends with the conclusions and suggestions for further research.

2 Spherical Fuzzy Sets: Preliminaries

The novel concept of SFS (Spherical Fuzzy Sets) provides a larger preference domain for decision makers to assign membership degrees since the squared sum of the spherical parameters is allowed to be at most 1.0 (Kutlu Gündoğdu and Kahraman 2019a; 2019b; 2019c; 2019d; Kahraman et al. 2019). In the following, definition of SFS is presented:

Definition 2.1 Single valued Spherical Fuzzy Sets (SFS) $\tilde{A}_S$ of the universe of discourse U is given by

$$\tilde{A}_S = \left\{\langle u,\ (\mu_{\tilde{A}_S}(u),\ v_{\tilde{A}_S}(u),\ \pi_{\tilde{A}_S}(u))\middle| u \in U\right\} \tag{1}$$

where

$$\mu_{\tilde{A}_S}(u) : U \rightarrow [0, 1],\ v_{\tilde{A}_S}(u) : U \rightarrow [0, 1],\ \pi_{\tilde{A}_S}(u) : U \rightarrow [0, 1]$$

and

$$0 \leq \mu^2_{\tilde{A}_S}(u) + v^2_{\tilde{A}_S}(u) + \pi^2_{\tilde{A}_S}(u) \leq 1 \quad \forall u \in U \tag{2}$$

For each u, the numbers $\mu_{\tilde{A}_S}(u)$, $v_{\tilde{A}_S}(u)$ and $\pi_{\tilde{A}_S}(u)$ are the degree of membership, non-membership and hesitancy of u to $\tilde{A}_S$, respectively. $\chi_{\tilde{A}_S}(u) = \sqrt{1 - \mu^2_{\tilde{A}_S}(u) - v^2_{\tilde{A}_S}(u) - \pi^2_{\tilde{A}_S}(u)}$ is called as a refusal degree (Kutlu Gündoğdu and Kahraman 2019a; Ashraf and Abdullah 2019).

Definition 2.2 Basic operators of Single-valued SFS;

i. $$\tilde{A}_S \oplus \tilde{B}_S = \left\{ \begin{array}{c} \left(\mu^2_{\tilde{A}_S} + \mu^2_{\tilde{B}_S} - \mu^2_{\tilde{A}_S}\mu^2_{\tilde{B}_S}\right)^{1/2},\ v_{\tilde{A}_S}v_{\tilde{B}_S}, \\ \left(\left(1 - \mu^2_{\tilde{B}_S}\right)\pi^2_{\tilde{A}_S} + \left(1 - \mu^2_{\tilde{A}_S}\right)\pi^2_{\tilde{B}_S} - \pi^2_{\tilde{A}_S}\pi^2_{\tilde{B}_S}\right)^{1/2} \end{array} \right\} \tag{3}$$

ii. $$\tilde{A}_S \otimes \tilde{B}_S = \left\{ \begin{array}{c} \mu_{\tilde{A}_S}\mu_{\tilde{B}_S},\ \left(v^2_{\tilde{A}_S} + v^2_{\tilde{B}_S} - v^2_{\tilde{A}_S}v^2_{\tilde{B}_S}\right)^{1/2}, \\ \left(\left(1 - v^2_{\tilde{B}_S}\right)\pi^2_{\tilde{A}_S} + \left(1 - v^2_{\tilde{A}_S}\right)\pi^2_{\tilde{B}_S} - \pi^2_{\tilde{A}_S}\pi^2_{\tilde{B}_S}\right)^{1/2} \end{array} \right\} \tag{4}$$

$$\text{iii. } \lambda \cdot \tilde{A}_S = \left\{ \begin{array}{c} \left(1 - \left(1 - \mu_{\tilde{A}_S}^2\right)^{\lambda}\right)^{1/2}, \ v_{\tilde{A}_S}^{\lambda}, \\ \left(\left(1 - \mu_{\tilde{A}_S}^2\right)^{\lambda} - \left(1 - \mu_{\tilde{A}_S}^2 - \pi_{\tilde{A}_S}^2\right)^{\lambda}\right)^{1/2} \end{array} \right\} \text{ for } \lambda > 0 \tag{5}$$

$$\text{iv. } \tilde{A}_S^{\lambda} = \left\{ \begin{array}{c} \mu_{\tilde{A}_S}^{\lambda}, \left(1 - \left(1 - v_{\tilde{A}_S}^2\right)^{\lambda}\right)^{1/2}, \\ \left(\left(1 - v_{\tilde{A}_S}^2\right)^{\lambda} - \left(1 - v_{\tilde{A}_S}^2 - \pi_{\tilde{A}_S}^2\right)^{\lambda}\right)^{1/2} \end{array} \right\} \ for\ \lambda > 0 \tag{6}$$

Definition 2.3 Single-valued Spherical Weighted Arithmetic Mean (SWAM) with respect to, $w = (w_1, w_2 \ldots, w_n)$; $w_i \in [0, 1]$; $\sum_{i=1}^{n} w_i = 1$, SWAM is defined as;

$$\begin{aligned} SWAM_w(\tilde{A}_{S1}, \ldots, \tilde{A}_{Sn}) &= w_1\tilde{A}_{S1} + w_2\tilde{A}_{S2} + \cdots + w_n\tilde{A}_{Sn} \\ &= \left\{ \begin{array}{c} \left[1 - \prod_{i=1}^{n} (1 - \mu_{\tilde{A}_{Si}}^2)^{w_i}\right]^{1/2}, \\ \prod_{i=1}^{n} v_{\tilde{A}_{Si}}^{w_i}, \left[\prod_{i=1}^{n} (1 - \mu_{\tilde{A}_{Si}}^2)^{w_i} - \prod_{i=1}^{n} (1 - \mu_{\tilde{A}_{Si}}^2 - \pi_{\tilde{A}_{Si}}^2)^{w_i}\right]^{1/2} \end{array} \right\} \end{aligned} \tag{7}$$

Definition 2.4 Single-valued Spherical Weighted Geometric Mean (SWGM) with respect to, $w = (w_1, w_2 \ldots, w_n)$; $w_i \in [0, 1]$; $\sum_{i=1}^{n} w_i = 1$, SWGM is defined as;

$$\begin{aligned} SWGM_w(\tilde{A}_1, \ldots, \tilde{A}_n) &= \tilde{A}_{S1}^{w_1} + \tilde{A}_{S2}^{w_2} + \cdots + \tilde{A}_{Sn}^{w_n} \\ &= \left\{ \begin{array}{c} \prod_{i=1}^{n} \mu_{\tilde{A}_{Si}}^{w_i}, \left[1 - \prod_{i=1}^{n} (1 - v_{\tilde{A}_{Si}}^2)^{w_i}\right]^{1/2}, \\ \left[\prod_{i=1}^{n} (1 - v_{\tilde{A}_{Si}}^2)^{w_i} - \prod_{i=1}^{n} (1 - v_{\tilde{A}_{Si}}^2 - \pi_{\tilde{A}_{Si}}^2)^{w_i}\right]^{1/2} \end{array} \right\} \end{aligned} \tag{8}$$

Definition 2.5 Score functions and Accuracy functions of sorting SFS are defined by;

$$Score\left(\tilde{A}_S\right) = (\mu_{\tilde{A}_S} - \pi_{\tilde{A}_S}/2)^2 - \left(v_{\tilde{A}_S} - \pi_{\tilde{A}_S}/2\right)^2 \tag{9}$$

$$Accuracy\left(\tilde{A}_S\right) = \mu_{\tilde{A}_S}^2 + v_{\tilde{A}_S}^2 + \pi_{\tilde{A}_S}^2 \tag{10}$$

Note that: $\tilde{A}_S < \tilde{B}_S$ if and only if

i. $Score(\tilde{A}_S) < Score(\tilde{B}_S)$ or
ii. $Score(\tilde{A}_S) = Score(\tilde{B}_S)$ and $Accuracy(\tilde{A}_S) < Accuracy(\tilde{B}_S)$

In this section, we give the definition of Interval-valued spherical fuzzy sets (IVSFS) and summarize arithmetic operations, aggregation and defuzzification operations.

Definition 2.6 An Interval-Valued Spherical Fuzzy Set $\tilde{A}_S$ of the universe of discourse U is defined as in Eq. (11).

$$\tilde{A}_S = \left\{\left\langle u, \left(\left[\mu^L_{\tilde{A}_S}(u), \mu^U_{\tilde{A}_S}(u)\right], \left[v^L_{\tilde{A}_S}(u), v^U_{\tilde{A}_S}(u)\right], \left[\pi^L_{\tilde{A}_S}(u), \pi^U_{\tilde{A}_S}(u)\right]\right)\right| u \in U\right\} \tag{11}$$

where $0 \leq \mu^L_{\tilde{A}_S}(u) \leq \mu^U_{\tilde{A}_S}(u) \leq 1$, $0 \leq v^L_{\tilde{A}_S}(u) \leq v^U_{\tilde{A}_S}(u) \leq 1$ and $0 \leq \left(\mu^U_{\tilde{A}_S}(u)\right)^2 + \left(v^U_{\tilde{A}_S}(u)\right)^2 + \left(\pi^U_{\tilde{A}_S}(u)\right)^2 \leq 1$. For each $u \in U$, $\mu^U_{\tilde{A}_S}(u)$, $v^U_{\tilde{A}_S}(u)$ *and* $\pi^U_{\tilde{A}_S}(u)$ are the upper degrees of membership, non-membership and hesitancy of u to $\tilde{A}_S$, respectively. For an IVSFS $\tilde{A}_S$, the pair $\left\langle\left[\mu^L_{\tilde{A}_S}(u), \mu^U_{\tilde{A}_S}(u)\right], \left[v^L_{\tilde{A}_S}(u), v^U_{\tilde{A}_S}(u)\right], \left[\pi^L_{\tilde{A}_S}(u), \pi^U_{\tilde{A}_S}(u)\right]\right\rangle$ is called an interval-valued spherical fuzzy number. For convenience, the pair $\left\langle\left[\mu^L_{\tilde{A}_S}(u), \mu^U_{\tilde{A}_S}(u)\right], \left[v^L_{\tilde{A}_S}(u), v^U_{\tilde{A}_S}(u)\right], \left[\pi^L_{\tilde{A}_S}(u), \pi^U_{\tilde{A}_S}(u)\right]\right\rangle$ is denoted by $\tilde{\alpha} = \langle [a, b], [c, d], [e, f]\rangle$ where $[a, b] \subset [0, 1]$, $[c, d] \subset [0, 1]$, $[e, f] \subset [0, 1]$ and $b^2 + d^2 + f^2 \leq 1$.

Definition 2.7 Let $\tilde{\alpha} = \langle [a, b], [c, d], [e, f]\rangle$, $\tilde{\alpha}_1 = \langle [a_1, b_1], [c_1, d_1], [e_1, f_1]\rangle$, and $\tilde{\alpha}_2 = \langle [a_2, b_2], [c_2, d_2], [e_2, f_2]\rangle$ be IVSFS then;

$$\tilde{\alpha}_1 \cup \tilde{\alpha}_2 = \left\{\begin{array}{r} [\max\{a_1, a_2\}, \max\{b_1, b_2\}], [\min\{c_1, c_2\}, \min\{d_1, d_2\}], \\ [\min\{e_1, e_2\}, \min\{f_1, f_2\}] \end{array}\right\} \tag{12}$$

$$\tilde{\alpha}_1 \cap \tilde{\alpha}_2 = \left\{\begin{array}{r} [\min\{a_1, a_2\}, \min\{b_1, b_2\}], [\max\{c_1, c_2\}, \max\{d_1, d_2\}], \\ [\min\{e_1, e_2\}, \min\{f_1, f_2\}] \end{array}\right\} \tag{13}$$

$$\tilde{\alpha}_1 \oplus \tilde{\alpha}_2 = \left\{\begin{array}{c} \left[\left((a_1)^2 + (a_2)^2 - (a_1)^2(a_2)^2\right)^{1/2}, \left((b_1)^2 + (b_2)^2 - (b_1)^2(b_2)^2\right)^{1/2}\right], [c_1c_2, d_1d_2], \\ \left[\begin{array}{c} \left(\left(1 - (a_2)^2\right)(e_1)^2 + \left(1 - (a_1)^2\right)(e_2)^2 - (e_1)^2(e_2)^2\right)^{1/2}, \\ \left(\left(1 - (b_2)^2\right)(f_1)^2 + \left(1 - (b_1)^2\right)(f_2)^2 - (f_1)^2(f_2)^2\right)^{1/2} \end{array}\right] \end{array}\right\} \tag{14}$$

$$\tilde{\alpha}_1 \otimes \tilde{\alpha}_2 = \left\{\begin{array}{c} [a_1a_2, b_1b_2], \left[\left((c_1)^2 + (c_2)^2 - (c_1)^2(c_2)^2\right)^{1/2}, \left((d_1)^2 + (d_2)^2 - (d_1)^2(d_2)^2\right)^{1/2}\right], \\ \left[\begin{array}{c} \left(\left(1 - (c_2)^2\right)(e_1)^2 + \left(1 - (c_1)^2\right)(e_2)^2 - (e_1)^2(e_2)^2\right)^{1/2}, \\ \left(\left(1 - (d_2)^2\right)(f_1)^2 + \left(1 - (d_1)^2\right)(f_2)^2 - (f_1)^2(f_2)^2\right)^{1/2} \end{array}\right] \end{array}\right\} \tag{15}$$

$$\lambda \cdot \tilde{\alpha} = \left\{ \begin{array}{c} \left[\left(1-\left(1-a^2\right)^{\lambda}\right)^{1/2}, \left(1-\left(1-b^2\right)^{\lambda}\right)^{1/2} \right], \left[c^{\lambda}, d^{\lambda}\right], \\ \left[\left(\left(1-a^2\right)^{\lambda} - \left(1-a^2-e^2\right)^{\lambda}\right)^{1/2}, \left(\left(1-b^2\right)^{\lambda} - \left(1-b^2-f^2\right)^{\lambda}\right)^{1/2} \right] \end{array} \right\} \; for\ \lambda > 0 \tag{16}$$

$$\tilde{\alpha}^{\lambda} = \left\{ \begin{array}{c} \left[a^{\lambda}, b^{\lambda}\right], \left[\left(1-\left(1-c^2\right)^{\lambda}\right)^{1/2}, \left(1-\left(1-d^2\right)^{\lambda}\right)^{1/2} \right], \\ \left[\left(\left(1-c^2\right)^{\lambda} - \left(1-c^2-e^2\right)^{\lambda}\right)^{1/2}, \left(\left(1-d^2\right)^{\lambda} - \left(1-d^2-f^2\right)^{\lambda}\right)^{1/2} \right] \end{array} \right\} \; for\ \lambda > 0 \tag{17}$$

Definition 2.8 Let $\lambda,\ \lambda_1,\ \lambda_2 \geq 0$, then

i. $\tilde{\alpha}_1 \oplus \tilde{\alpha}_2 = \tilde{\alpha}_2 \oplus \tilde{\alpha}_1$ (18)

ii. $\tilde{\alpha}_1 \otimes \tilde{\alpha}_2 = \tilde{\alpha}_2 \otimes \tilde{\alpha}_1$ (19)

iii. $\lambda(\tilde{\alpha}_1 \oplus \tilde{\alpha}_2) = \lambda \cdot \tilde{\alpha}_1 \oplus \lambda \cdot \tilde{\alpha}_2$ (20)

iv. $(\tilde{\alpha}_1 \otimes \tilde{\alpha}_2)^{\lambda} = \tilde{\alpha}_1^{\lambda} \otimes \tilde{\alpha}_2^{\lambda}$ (21)

v. $\lambda_1 \cdot \tilde{\alpha} \oplus \lambda_2 \cdot \tilde{\alpha} = (\lambda_1 + \lambda_2) \cdot \tilde{\alpha}$ (22)

vi. $\tilde{\alpha}^{\lambda_1} \otimes \tilde{\alpha}^{\lambda_2} = \tilde{\alpha}^{\lambda_1 + \lambda_2}$ (23)

Definition 2.9 Let $\tilde{\alpha}_j = \langle [a_j, b_j], [c_j, d_j], [e_j, f_j] \rangle$ be a collection of Interval-valued Spherical Weighted Arithmetic Mean (IVSWAM) with respect to, $w_j = (w_1, w_2 \ldots, w_n)$; $\quad w_j \in [0, 1]\ and\ \sum_{j=1}^{n} w_j = 1$, IVSWAM is defined as;

$$IVSWAM_w(\tilde{\alpha}_1, \tilde{\alpha}_2 \ldots, \tilde{\alpha}_n) = w_1 \cdot \tilde{\alpha}_1 \oplus w_2 \cdot \tilde{\alpha}_2 \oplus \cdots \oplus w_n \cdot \tilde{\alpha}_n$$

$$= \left\{ \begin{array}{c} \left[\left(1 - \prod_{j=1}^{n} (1-a_j^2)^{w_j}\right)^{1/2}, \left(1 - \prod_{j=1}^{n} (1-b_j^2)^{w_j}\right)^{1/2} \right], \left[\prod_{j=1}^{n} c_j^{w_j}, \prod_{j=1}^{n} d_j^{w_j} \right], \\ \left[\left(\prod_{j=1}^{n} (1-a_j^2)^{w_j} - \prod_{j=1}^{n} (1-a_j^2-e_j^2)^{w_j} \right)^{1/2}, \left(\prod_{j=1}^{n} (1-b_j^2)^{w_j} - \prod_{j=1}^{n} (1-b_j^2-f_j^2)^{w_j} \right)^{1/2} \right] \end{array} \right\} \tag{24}$$

Definition 2.10 Let $\tilde{\alpha}_j = \langle [a_j, b_j], [c_j, d_j], [e_j, f_j] \rangle$ be a collection of Interval-valued Spherical Geometric Mean (IVSWGM) with respect to, $w_j = (w_1, w_2 \ldots, w_n)$; $w_j \in [0, 1]\ and\ \sum_{j=1}^{n} w_j = 1$, IVSWGM is defined as;

$$IVSWGM_w(\tilde{\alpha}_1, \tilde{\alpha}_2 \ldots, \tilde{\alpha}_n) = \tilde{\alpha}_1^{w_1} \otimes \tilde{\alpha}_2^{w_2} \otimes \cdots \otimes \tilde{\alpha}_n^{w_n}$$
$$= \left\{ \begin{array}{c} \left[\prod_{j=1}^{n} a_j^{w_j}, \prod_{j=1}^{n} b_j^{w_j}\right], \left[\left(1 - \prod_{j=1}^{n} (1 - c_j^2)^{w_j}\right)^{1/2}, \left(1 - \prod_{j=1}^{n} (1 - d_j^2)^{w_j}\right)^{1/2}\right], \\ \left[\left(\prod_{j=1}^{n} (1 - c_j^2)^{w_j} - \prod_{j=1}^{n} (1 - c_j^2 - e_j^2)^{w_j}\right)^{1/2}, \left(\prod_{j=1}^{n} (1 - d_j^2)^{w_j} - \prod_{j=1}^{n} (1 - d_j^2 - f_j^2)^{w_j}\right)^{1/2}\right] \end{array} \right\} \tag{25}$$

Definition 2.11 The score function of IVSFS number α is defined as;

$$Score(\tilde{\alpha}) = S(\tilde{\alpha}) = \frac{a^2 + b^2 - c^2 - d^2 - (e/2)^2 - (f/2)^2}{2} \tag{26}$$

The accuracy function of IVSFS number α is defined as;

$$Accuracy(\tilde{\alpha}) = H(\tilde{\alpha}) = \frac{a^2 + b^2 + c^2 + d^2 + e^2 + f^2}{2} \tag{27}$$

Note that: $\tilde{\alpha}_1 < \tilde{\alpha}_2$ if and only if

i. $S(\tilde{\alpha}_1) < S(\tilde{\alpha}_2)$ or
ii. $S(\tilde{\alpha}_1) = S(\tilde{\alpha}_2)$ and $H(\tilde{\alpha}_1) < H(\tilde{\alpha}_2)$

3 Simple Additive Weighting and Simple Weighted Product Methods Using Single-Valued Spherical Fuzzy Sets

Scoring methods are the most often used multi attribute decision-making methods because of their easiness and effectiveness. In this section, Simple Additive Weighting (SAW) and Weighted Product Methods (WPM) are extended to their fuzzy versions by using single-valued and interval-valued spherical fuzzy sets.

3.1 Simple Additive Weighting Using Single-Valued Spherical Fuzzy Sets

A multi-criteria decision-making problem can be expressed as a decision matrix whose elements show the evaluation values of all alternatives with respect to each criterion under Spherical fuzzy environment. Let $X = \{x_1, x_2, \ldots x_m\}$ $(m \geq 2)$ be a discrete set of m feasible alternatives and $C = \{C_1, C_2, \ldots C_n\}$ be a finite set of criteria and $w = \{w_1, w_2, \ldots w_n\}$ be the weight vector of all criteria which satisfies $0 \leq w_j \leq 1$ and $\sum_{j=1}^{n} w_j = 1$.

Step 1 Let Decision Makers (DMs) fill in the decision and criteria evaluation matrices based on the linguistic terms given in Table 1.

Step 2 Aggregate the judgments of each decision maker (DM) using Spherical Weighted Arithmetic Mean (SWAM) as given in Eq. (28). Construct aggregated spherical fuzzy decision matrix based on the opinions of decision makers. Signify the evaluation values of Alternative $x_i(i = 1, 2 \ldots m)$ with respect to criterion $C_j(j = 1, 2 \ldots n)$ by $C_j(\tilde{x}_i) = (\mu_{ij}, v_{ij}, \pi_{ij})$ and $\tilde{x}_{ij} = \left(C_j(\tilde{x}_i)\right)_{mxn}$ is a spherical fuzzy decision matrix. For a MCDM problem with SFS, decision matrix $\tilde{x}_{ij} = \left(C_j(\tilde{x}_i)\right)_{mxn}$ should be written as in Eq. (29).

$$SWAM_w(\tilde{A}_{S1}, \ldots, \tilde{A}_{Sn}) = w_1\tilde{A}_{S1} + w_2\tilde{A}_{S2} + \cdots + w_n\tilde{A}_{Sn}$$
$$= \left\{ \begin{array}{c} \left[1 - \prod_{i=1}^{n} (1 - \mu_{\tilde{A}_{Si}}^2)^{w_i}\right]^{1/2}, \\ \prod_{i=1}^{n} v_{\tilde{A}_{Si}}^{w_i}, \left[\prod_{i=1}^{n} (1 - \mu_{\tilde{A}_{Si}}^2)^{w_i} - \prod_{i=1}^{n} (1 - \mu_{\tilde{A}_{Si}}^2 - \pi_{\tilde{A}_{Si}}^2)^{w_i}\right]^{1/2} \end{array} \right\} \tag{28}$$

$$\tilde{x}_{ij} = \left(C_j(\tilde{x}_i)\right)_{mxn}$$
$$= \begin{pmatrix} (\mu_{11}, v_{11}, \pi_{11}) & (\mu_{12}, v_{12}, \pi_{12}) & \cdots & (\mu_{1n}, v_{1n}, \pi_{1n}) \\ (\mu_{21}, v_{21}, \pi_{21}) & (\mu_{22}, v_{22}, \pi_{22}) & \cdots & (\mu_{2n}, v_{2n}, \pi_{2n}) \\ \cdot & \cdot & & \cdot \\ \cdot & \cdot & & \cdot \\ (\mu_{m1}, v_{m1}, \pi_{m1}) & (\mu_{m2}, v_{m2}, \pi_{m2}) & \cdots & (\mu_{mn}, v_{mn}, \pi_{mn}) \end{pmatrix} \tag{29}$$

Step 3 Decision makers also assess the decision criteria as given in Step 1. Decision makers evaluate the alternatives with respect to the criteria as if they were benefit criteria such that they assign a lower linguistic term if it is a cost criterion.

Table 1 Linguistic terms (Kutlu Gündoğdu and Kahraman 2019a)

Linguistic terms	(μ, v, π)
Absolutely more importance (AMI)	(0.9, 0.1, 0.1)
Very high importance (VHI)	(0.8, 0.2, 0.2)
High importance (HI)	(0.7, 0.3, 0.3)
Slightly more importance (SMI)	(0.6, 0.4, 0.4)
Equally importance (EI)	(0.5, 0.5, 0.5)
Slightly low importance (SLI)	(0.4, 0.6, 0.4)
Low importance (LI)	(0.3, 0.7, 0.3)
Very low importance (VLI)	(0.2, 0.8, 0.2)
Absolutely low importance (ALI)	(0.1, 0.9, 0.1)

Aggregate the judgments of each decision maker (DM) with respect to each criterion using Spherical Weighted Arithmetic Mean (SWAM) operator as given in Eq. (28).

Step 4 Calculate the score function value of each criterion and then normalize these values.

Step 4.1 Defuzzify the aggregated criteria weights based on the modified score function given in Eq. (30).

$$S(\tilde{w}_j^s) = (3\mu_j - \pi_j/2)^2 - (v_j - \pi_j/2)^2 \tag{30}$$

Step 4.2 Normalize the aggregated criteria weights by using Eq. (31).

$$\bar{w}_j^s = \frac{S\left(\tilde{w}_j^s\right)}{\sum_{J=1}^{n} S\left(\tilde{w}_j^s\right)} \tag{31}$$

Step 5 Compute the fuzzy results of Single-Valued Simple Additive Weighting (SVSAW) as given in Eq. (32) by utilizing Eqs. (3) and (5).

$$SVSAW_i = \sum_{j=1}^{n} \tilde{x}_{ijw} = \sum_{j=1}^{n} \tilde{x}_{ij} \cdot \bar{w}_j^s \quad \forall i \tag{32}$$

Step 6 Calculate the score of each alternative by using Eq. (33) and put the alternatives into order with respect to the decreasing values of score values.

$$S(SVSAW_i) = (3\mu_i - \pi_i/2)^2 - (v_i - \pi_i/2)^2 \tag{33}$$

3.2 Simple Weighted Product Methods Using Single-Valued Spherical Fuzzy Sets

The steps 1–4 applied in Sect. 4.1 are also used in this method in the same way.

Step 1 Identical to SVSAW method.

Step 2 Identical to SVSAW method.

Step 3 Identical to SVSAW method.

Step 4 Identical to SVSAW method.

Step 5 Calculate the results of Single-Valued Weighted Product Model (SVWPM) as presented in Eq. (34) by utilizing Eqs. (4) and (6).

$$SVWPM_i = \prod_{j=1}^{n} \tilde{x}_{ij}^{\bar{w}_j^s} \quad \forall i \tag{34}$$

Step 6 Calculate the score of each alternative by using Eq. (35) which is adopted from Eq. (9) and put the alternatives into order with respect to the decreasing values of score values.

$$S(SVWPM_i) = (3\mu_i - \pi_i/2)^2 - (v_i - \pi_i/2)^2 \tag{35}$$

3.3 Simple Additive Weighting Using Interval-Valued Spherical Fuzzy Sets

In this section, simple additive weighting and simple weighted product models are extended to interval-valued spherical fuzzy versions.

Step 1 Use the scale in Table 2 for data input. Intermediate values can be used if DMs do not prefer using the given linguistic terms. This provides a large spherical volume for the assignment of membership, non-membership, and hesitancy degrees.

Step 1.1 Collect the assessments of DMs for the criteria weighting.

Step 1.2 Let DMs fill in the decision matrices using the linguistic terms.

Table 2 Linguistic terms and their corresponding interval-valued spherical fuzzy numbers

Linguistic terms	$\left(\left[\mu^L_{\tilde{A}_S}(u), \mu^U_{\tilde{A}_S}(u)\right], \left[v^L_{\tilde{A}_S}(u), v^U_{\tilde{A}_S}(u)\right], \left[\pi^L_{\tilde{A}_S}(u), \pi^U_{\tilde{A}_S}(u)\right]\right)$
Absolutely more importance (AMI)	([0.85,0.95], [0.10,0.15], [0.05,0.15])
Very high importance (VHI)	([0.75,0.85], [0.15,0.20], [0.15,0.20])
High importance (HI)	([0.65,0.75], [0.20,0.25], [0.20,0.25])
Slightly more importance (SMI)	([0.55,0.65], [0.25,0.30], [0.25,0.30])
Equally importance (EI)	([0.50,0.55], [0.45,0.55], [0.30,0.40])
Slightly low importance (SLI)	([0.25,0.30], [0.55,0.65], [0.25,0.30])
Low importance (LI)	([0.20,0.25], [0.65,0.75], [0.20,0.25])
Very low importance (VLI)	([0.15,0.20], [0.75,0.85], [0.15,0.20])
Absolutely low importance (ALI)	([0.10,0.15], [0.85,0.95], [0.05,0.15])

A decision matrix is defined as follows:

Let $C_j(X_i) = \left(\left[\mu^L_{ij}(u), \mu^U_{ij}(u)\right], \left[v^L_{ij}(u), v^U_{ij}(u)\right], \left[\pi^L_{ij}(u), \pi^U_{ij}(u)\right]\right)$ denote the assessments of each alternative $X_i(i = 1, 2 \ldots m)$ with respect to each criterion $C_j(j = 1, 2 \ldots n)$. $D = \left(C_j(X_i)\right)_{mxn}$ is a spherical fuzzy decision matrix. Then, the decision matrix $D = \left(C_j(X_i)\right)_{mxn}$ is constructed as in Eq. (36).

$$\tilde{x}^{IV}_{ij} = D = \left(C_j(X_i)\right)_{mxn} = \begin{pmatrix} \left(\left[\mu^L_{11}(u),\mu^U_{11}(u)\right],\left[v^L_{11}(u),v^U_{11}(u)\right],\left[\pi^L_{11}(u),\pi^U_{11}(u)\right]\right) & \cdots & \left(\left[\mu^L_{1n}(u),\mu^U_{1n}(u)\right],\left[v^L_{1n}(u),v^U_{1n}(u)\right],\left[\pi^L_{1n}(u),\pi^U_{1n}(u)\right]\right) \\ \left(\left[\mu^L_{22}(u),\mu^U_{22}(u)\right],\left[v^L_{22}(u),v^U_{22}(u)\right],\left[\pi^L_{22}(u),\pi^U_{22}(u)\right]\right) & \cdots & \left(\left[\mu^L_{2n}(u),\mu^U_{2n}(u)\right],\left[v^L_{2n}(u),v^U_{2n}(u)\right],\left[\pi^L_{2n}(u),\pi^U_{2n}(u)\right]\right) \\ \vdots & & \vdots \\ \left(\left[\mu^L_{m1}(u),\mu^U_{m1}(u)\right],\left[v^L_{m1}(u),v^U_{m1}(u)\right],\left[\pi^L_{m1}(u),\pi^U_{m1}(u)\right]\right) & \cdots & \left[\mu^L_{mn}(u),\mu^U_{mn}(u)\right],\left[v^L_{mn}(u),v^U_{mn}(u)\right],\left[\pi^L_{mn}(u),\pi^U_{mn}(u)\right] \end{pmatrix} \tag{36}$$

Step 2 Aggregate the decision matrices using Interval-valued Spherical Weighted Arithmetic Mean (IVSWAM) operator as given in Eq. (24).

Step 2.1 Aggregate the assessments on criteria weights.

Step 2.2 Aggregate the interval-valued spherical fuzzy decision matrices.

Step 3 Defuzzify the criteria weights by using Eq. (37) which is adopted from Eq. (26), and normalize them based on Eq. (31).

$$s\left(\tilde{w}^{IV}_j\right) = \frac{\left(\mu^L_j(u)\right)^2 + \left(\mu^U_j(u)\right)^2 - \left(v^L_j(u)\right)^2 - \left(v^U_j(u)\right)^2 - \left(\pi^L_j(u)/2\right)^2 - \left(\pi^U_j(u)/2\right)^2}{2} + 1 \tag{37}$$

Step 4 Compute the fuzzy results of Interval-valued Simple Additive Weighting (IVSAW) as given in Eq. (38) by utilizing Eqs. (14) and (16).

$$IVSAW_i = \sum_{j=1}^{n} \tilde{x}_{ijw} = \sum_{j=1}^{n} \tilde{x}_{ij}^{IV} \cdot \bar{w}_j^{IV} \quad \forall i \tag{38}$$

Step 5 Calculate the score of each alternative by using Eq. (39) and put the alternatives into order with respect to the decreasing values of score values.

$$S(IVSAW_i) = \frac{\left(\mu_i^L(u)\right)^2 + \left(\mu_i^U(u)\right)^2 - \left(v_i^L(u)\right)^2 - \left(v_i^U(u)\right)^2 - \left(\pi_i^L(u)/2\right)^2 - \left(\pi_i^U(u)/2\right)^2}{2} + 1 \tag{39}$$

3.4 *Simple Weighted Product Methods Using Interval-Valued Spherical Fuzzy Sets*

The steps 1–4 applied in Sect. 4.3 are also used in this method in the same way.

Step 1 Identical to IVSAW method.

Step 2 Identical to IVSAW method.

Step 3 Identical to IVSAW method.

Step 4 Identical to IVSAW method.

Step 5 Calculate the results of Interval-valued Weighted Product Model (IVWPM) as presented in Eq. (40) by utilizing Eqs. (15) and (17).

$$IVWPM_i = \prod_{j=1}^{n} \tilde{x}_{ij}^{\bar{w}_j^{IV}} \quad \forall i \tag{40}$$

Step 6 Calculate the score of each alternative by using Eq. (41) which is adopted from Eq. (26) and put the alternatives into order with respect to the decreasing values of score values.

$$S(IVWPM_i) = \frac{\left(\mu_i^L(u)\right)^2 + \left(\mu_i^U(u)\right)^2 - \left(v_i^L(u)\right)^2 - \left(v_i^U(u)\right)^2 - \left(\pi_i^L(u)/2\right)^2 - \left(\pi_i^U(u)/2\right)^2}{2} + 1 \tag{41}$$

4 An Illustrative Example: Choosing Insurance Options

Selection among several insurance options is a very popular problem for most of the people. Suppose that the following insurance options are presented to three people (DM1, DM2, and DM3) whose ages are 25, 50 and 75. The weights of these decision makers who have different important levels are 0.3, 0.5 and 0.2, respectively. These people request a help from you for the selection among the following options: Dental

Insurance (X_1), Hospital and Doctor Insurance (X_2), Vision Insurance (X_3), Travel Insurance (X_4), Accident Insurance (X_5). The selection criteria can be age (C_1), health status of person (C_2), work status of person (C_3), and economic conditions (C_4). First of all, the assessments for the criteria are collected from decision makers with respect to the goal, based on the linguistic terms given in Table 1. All assessments are given in Table 3.

These judgments are aggregated using single-valued SWAM operator by considering the importance levels of decision makers. Aggregated decision matrix is obtained as in Table 4.

The linguistic importance weights of the criteria assigned by DMs are shown in Table 5.

Table 3 Assessments of DM1, DM2, and DM3

DM1	C_1	C_2	C_3	C_4
X_1	HI	SMI	SLI	EI
X_2	AMI	SMI	VHI	SMI
X_3	SLI	LI	LI	EI
X_4	SMI	HI	LI	HI
X_5	HI	SLI	HI	EI
DM2	C_1	C_2	C_3	C_4
X_1	EI	HI	LI	HI
X_2	AMI	AMI	HI	VHI
X_3	HI	LI	LI	SMI
X_4	HI	EI	EI	EI
X_5	HI	SLI	HI	EI
DM3	C_1	C_2	C_3	C_4
X_1	HI	LI	VLI	SLI
X_2	SMI	HI	VHI	HI
X_3	HI	LI	VLI	HI
X_4	SMI	HI	SLI	HI
X_5	HI	EI	HI	EI

Table 4 Aggregated decision matrix

Alternatives	C_1	C_2	C_3	C_4
X_1	(0.62, 0.39, 0.40)	(0.62, 0.39, 0.34)	(0.32, 0.69, 0.32)	(0.61, 0.40, 0.38)
X_2	(0.87, 0.13, 0.15)	(0.82, 0.19, 0.22)	(0.76, 0.24, 0.25)	(0.74, 0.27, 0.28)
X_3	(0.64, 0.37, 0.33)	(0.30, 0.70, 0.30)	(0.28, 0.72, 0.28)	(0.60, 0.40, 0.41)
X_4	(0.65, 0.35, 0.35)	(0.62, 0.39, 0.40)	(0.43, 0.57, 0.44)	(0.62, 0.39, 0.40)
X_5	(0.70, 0.30, 0.30)	(0.42, 0.58, 0.43)	(0.70, 0.30, 0.30)	(0.50, 0.50, 0.50)

Table 5 Importance weights of the criteria

Criteria	DM1	DM2	DM3
C_1	HI	EI	HI
C_2	SMI	AMI	HI
C_3	LI	LI	VLI
C_4	EI	EI	EI

Table 6 Aggregation of criteria weights

Criteria	Weight of each criterion
C_1	(0.62, 0.39, 0.40)
C_2	(0.82, 0.19, 0.22)
C_3	(0.28, 0.72, 0.28)
C_4	(0.50, 0.50, 0.50)

The weight of each criterion obtained by using single-valued SWAM operator is presented in Table 6.

After the weights of the criteria have been determined, the defuzzified and normalized criteria weights are calculated by using Eqs. (30) and (31) as given in Table 7.

By utilizing Eqs. (32), and (33) and $S(SVSAW_i)$ is calculated as in Table 8. The final ranking is $X_2 > X_4 > X_1 > X_5 > X_3$.

Hospital and doctor insurance is the best alternative for decision makers with respect to single-valued spherical fuzzy simple additive weighting model. Based on Eqs. (34) and (35), $SVWPM_i$ is calculated as in Table 9. The final ranking is the same with $SVSAW$ method.

Table 7 Defuzzified and normalized criteria weights

Criteria	Weight of each criterion
C_1	0.274
C_2	0.556
C_3	0.017
C_4	0.152

Table 8 Single-valued simple additive weighting model and ranking

Alternatives	$SVSAW_i$	$S(SVSAW_i)$	Ranking
X_1	(0.62, 0.39, 0.36)	2.725	3
X_2	(0.82, 0.18, 0.21)	5.595	1
X_3	(0.48, 0.54, 0.34)	1.498	5
X_4	(0.63, 0.38, 0.39)	2.804	2
X_5	(0.54, 0.47, 0.40)	1.972	4

Table 9 Single-valued simple weighted product model and ranking

Alternatives	$SVWPM_i$	$S(SVWPM_i)$	Ranking
X_1	(0.61, 0.40, 0.36)	2.686	3
X_2	(0.82, 0.19, 0.21)	5.485	1
X_3	(0.41, 0.60, 0.32)	0.942	5
X_4	(0.41, 0.60, 0.32)	2.777	2
X_5	(0.50, 0.51, 0.42)	1.595	4

The score values indicate that the best alternative is X_2 and overall ranking is $X_2 > X_4 > X_1 > X_3 > X_5$.

In order to test the reliability and robustness, we compared the proposed single-valued spherical fuzzy SAW and WPM models with interval-valued spherical fuzzy SAW and WPM models. Table 2 presents the interval-valued spherical fuzzy linguistic scale, which we use for the comparison purpose. In this comparison, the same judgments as given in Table 3 were used and aggregated using IVSWAM (Interval-valued Spherical Fuzzy Weighted Arithmetic Mean) operator given in Eq. (24). The aggregated interval-valued spherical fuzzy decision matrix is constructed by utilizing Eq. (36) as given in Table 10.

After the weights of the criteria have been determined, the defuzzified and normalized criteria weights are calculated by using Eqs. (31) and (37) as given in Table 11.

By utilizing Eqs. (38) and (39) and $S(IVSAW_i)$ is calculated as in Table 12. The final ranking is $X_2 > X_4 > X_1 > X_5 > X_3$. Hospital and doctor insurance is the

Table 10 Aggregated interval-valued spherical fuzzy decision matrix

Alternatives	C_1	C_2	C_3	C_4
X_1	([0.58, 0.67], [0.30, 0.37], [0.25,0.32])	([0.57, 0.67], [0.27, 0.33], [0.22, 0.27])	([0.21, 0.26], [0.64, 0.74], [0.21, 0.26])	([0.56, 0.65], [0.31, 0.38], [0.24, 0.31])
X_2	([0.82, 0.93], [0.12, 0.17], [0.09, 0.17])	([0.76, 0.88], [0.15, 0.20], [0.14, 0.21])	([0.71, 0.81],[0.17, 0.22], [0.17, 0.22])	([0.68, 0.79], [0.19, 0.24], [0.19, 0.24])
X_3	([0.58, 0.67], [0.27, 0.33], [0.21, 0.27])	([0.20, 0.25],[0.65, 0.75], [0.20, 0.25])	([0.19, 0.24], [0.67, 0.77], [0.19, 0.24])	([0.56, 0.65],[0.29, 0.35], [0.26, 0.32])
X_4	([0.60, 0.71], [0.22, 0.27], [0.22, 0.27])	([0.58, 0.67], [0.30, 0.37], [0.25, 0.32])	([0.39, 0.44], [0.52, 0.62], [0.27, 0.36])	([0.58, 0.67], [0.30, 0.37], [0.25, 0.32])
X_5	([0.65, 0.75], [0.20, 0.25], [0.20, 0.25])	([0.32, 0.37], [0.53, 0.63], [0.26, 0.33])	([0.65, 0.75], [0.20, 0.25], [0.20, 0.25])	([0.50, 0.55], [0.45, 0.55], [0.30, 0.40])

Table 11 Defuzzified and normalized criteria weights

Criteria	C_1	Weight of each criterion
C_1	([0.58, 0.67], [0.30, 0.37], [0.25,0.32])	1.239
C_2	([0.76, 0.88], [0.15, 0.20], [0.14, 0.21])	1.628
C_3	([0.19, 0.24], [0.67, 0.77], [0.19, 0.24])	0.504
C_4	([0.50, 0.55], [0.45, 0.55], [0.30, 0.40])	0.961

Table 12 Interval-valued simple additive weighting model and ranking

Alternatives	$IVSAW_i$	$S(IVSAW_i)$	Ranking
X_1	([0.56, 0.66], [0.29, 0.36], [0.24,0.30])	1.232	3
X_2	([0.78, 0.89], [0.13, 0.18], [0.14, 0.20])	1.660	1
X_3	([0.46, 0.55], [0.39, 0.47], [0.23, 0.29])	1.031	5
X_4	([0.59, 0.68], [0.27, 0.33], [0.25, 0.32])	1.277	2
X_5	([0.55, 0.63], [0.32, 0.39], [0.25, 0.32]	1.179	4

Table 13 Interval-valued simple weighted product model and ranking

Alternatives	$IVWPM_i$	$S(IVWPM_i)$	Ranking
X_1	([0.48, 0.57], [0.37, 0.45], [0.24,0.30])	1.067	3
X_2	([0.73, 0.85], [0.16, 0.21], [0.15, 0.22])	1.579	1
X_3	([0.31, 0.38], [0.54, 0.64], [0.22, 0.28])	0.742	5
X_4	([0.54, 0.62], [0.33, 0.41], [0.26, 0.33])	1.158	2
X_5	([0.44, 0.51], [0.43, 0.52], [0.26, 0.34]	0.952	4

best alternative for decision makers with respect to interval-valued spherical fuzzy simple additive weighting model.

Based on Eqs. (40) and (41), $IVWPM_i$ is calculated as in Table 13. The final ranking is the same with $IVSAW$ method.

The score values indicate that the best alternative is X_2 and overall ranking is $X_2 > X_4 > X_1 > X_3 > X_5$ in all approaches. Hospital and doctor insurance is the best alternative and vision insurance is the worst alternative for decision makers with respects all methods under spherical fuzzy environment.

5 Conclusions

Scoring methods are comparatively simple and mostly used multi-criteria decision-making methods in the literature. In the real life problems, these methods have tangible and intangible criteria which mean hard to assign a single numerical value. Fuzzy sets can be useful for these kind situations. Spherical fuzzy sets, one of the recent extensions of ordinary fuzzy sets, let experts express their judgments for a considered problem with a larger preference domain about the membership, non-membership, and hesitancy, individualistically. In this chapter, simple additive weighting and weighted product methods have been extended to single-valued and interval-valued spherical fuzzy versions. Both spherical fuzzy SAW and spherical fuzzy WPM methods have been applied to an insurance selection problem. The obtained results are consistent with some slight differences.

For further research, we suggest other types of spherical fuzzy sets such as triangular SFS, and trapezoidal SFS to be employed instead of single-valued and interval-valued spherical fuzzy sets in the developed spherical SAW and WPM methods.

References

Ashraf S, Abdullah S, Mahmood T, Ghani F, Mahmood T (2019) Spherical fuzzy sets and their applications in multi-attribute decision making problems. J Intell Fuzzy Syst 36(3):2829–2844

Boltürk E, Karaşan A, Kahraman C (2019) Simple additive weighting and weighted product methods using neutrosophic sets. In Fuzzy multi-criteria decision-making using neutrosophic sets. Springer, Cham, pp 647–676

Chen CB, Klein CM (1997) An efficient approach to solving fuzzy MADM problems. Fuzzy Sets Syst 88(1):51–67

Chen SJ, Hwang CL (1992) Fuzzy multiple attribute decision making methods. In Fuzzy Multiple Attribute Decision Making. Springer, Berlin, Heidelberg

Chen T (2014) An interactive signed distance approach for multiple criteria group decision-making based on simple additive weighting method with incomplete preference information defined by interval type-2 fuzzy sets. Int J Inf Technol Decis Mak 13(5):979–1012

Jin H, Ashraf S, Abdullah S, Qiyas M, Bano M, Zeng S (2019) Linguistic spherical fuzzy aggregation operators and their applications in multi-attribute decision making problems. Mathematics 7(5):413

Kahraman C, Gundogdu FK, Onar SC, Oztaysi B (2019, August) Hospital location selection using spherical fuzzy TOPSIS. In: 2019 conference of the international fuzzy systems association and the European society for fuzzy logic and technology (EUSFLAT 2019). Atlantis Press

Kahraman C, Öztayşi B, Çevik Onar S (2016) A comprehensive literature review of 50 years of fuzzy set theory. Int J Comput Intell Syst 9(sup1):3–24

Kaur P, Kumar S (2013) An intuitionistic fuzzy simple additive weighting (IFSAW) method for selection of vendor. IOSR J Bus Manag 15(2):78–81

Kutlu Gündoğdu F (2019) Principals of spherical fuzzy sets. In: International conference on intelligent and fuzzy systems. Springer, Cham, pp 15–23

Kutlu Gündoğdu F, Kahraman C (2019a) Spherical fuzzy sets and spherical fuzzy TOPSIS method. J Intell Fuzzy Syst 36(1):337–352

Kutlu Gündoğdu F, Kahraman C (2019b) A novel fuzzy TOPSIS method using emerging interval-valued spherical fuzzy sets. Eng Appl Artif Intell 85:307–323

Kutlu Gündoğdu F, Kahraman C (2019c) Extension of CODAS with spherical fuzzy sets. J Mult-Valued Log Soft Comput 33:481–505

Kutlu Gündoğdu F, Kahraman C (2019d) A novel VIKOR method using spherical fuzzy sets and its application to warehouse site selection. J Intell Fuzzy Syst 37(1):1197–1211

Kutlu Gündoğdu F, Kahraman C (2020) A novel spherical fuzzy QFD method and its application to the linear delta robot technology development. Eng Appl Artif Intell 87:103348

Liu P, Zhu B, Wang P, Shen M (2020) An approach based on linguistic spherical fuzzy sets for public evaluation of shared bicycles in China. Eng Appl Artif Intell 87:103295
Triantaphyllou E, Lin CT (1996) Development and evaluation of five fuzzy multiattribute decision-making method. Int J Approximate Reasoning 14(4):281–310
Zadeh L (1965) Fuzzy sets. Inf Control 8(1965):338–353
Zavadskas EK, Antucheviciene J, Hajiagha SHR, Hashemi SS (2014) Extension of weighted aggregated sum product assessment with interval-valued intuitionistic fuzzy numbers (WASPAS-IVIF). Appl Soft Comput 24:1013–1021
Zhang Z (2017) Hesitant fuzzy multi-criteria group decision making with unknown weight information. Int J Fuzzy Syst 19(3):615–636

Prioritizing Manufacturing Challenges of a Contract Manufacturing Company for Personal Auto by Using Spherical WASPAS Method

Eda Boltürk and Fatma Kutlu Gündoğdu

Abstract Since the development of manufacturing systems provides many opportunities such as standardizing the tasks, increasing productivity, it also brings some challenges for the companies to deal with. These challenges can be sorted as manual handling and safety, skilled labor shortage, internet of things, maintaining the right inventory levels, robotics and automation change, labor shortage, new technologies, cyber security, global competition, attracting qualified leads, etc. Since these challenges contain both qualitative and quantitative evaluations for the measurement together with the solution alternatives, it can be considered as a multi criteria decision making (MCDM) problem. For the qualitative evaluations, process includes uncertainties and vagueness as a result of human evaluations. To overcome uncertainty and vagueness in the decision-making process, the spherical fuzzy sets (SFSs) are utilized for the applicability of the available data for the WASPAS method. In this chapter, a ranking method, WASPAS, based on single-valued spherical fuzzy sets is applied for the prioritization of the manufacturing challenges of a contract manufacturing company by considering evaluations of a group of experts.

1 Introduction

With the new technology methods, the challenges and problems in manufacturing is growing day by day. Every manufacturing system have some advantages and disadvantages but not every type of them are appropriate for every company's systems. There are some manufacturing challenges in the world such as manual handling

E. Boltürk (✉)
Industrial Engineering Department, Istanbul Technical University, Besiktas, Istanbul 34367, Turkey
e-mail: bolturk@itu.edu.tr

F. Kutlu Gündoğdu
Industrial Engineering Department, National Defence University, Turkish Air Force Academy, Istanbul 34149, Turkey
e-mail: fatmakutlugundogdu@gmail.com

© Springer Nature Switzerland AG 2021

C. Kahraman and F. Kutlu Gündoğdu (eds.), *Decision Making with Spherical Fuzzy Sets*, Studies in Fuzziness and Soft Computing 392,
https://doi.org/10.1007/978-3-030-45461-6_11

and safety, skilled labor shortage, internet of things, maintaining the right inventory levels, robotics and automation change, labor shortage, new technologies, cyber security, global competition, attracting qualified leads, etc. In the literature review, some solution ways for different difficulties are discussed while discussing system requirements.

Agile manufacturing is one of the opportunities in order to minimize the manufacturing challenges in automotive industry. Because of the flexibility, speed and responsively of agile manufacturing, it is appropriate for personal auto demands. Agile manufacturing systems will allow quick cost-efficient responses to irregular product demand, and strength rapid product launches for unintentional products tailored to meet changing customer desires (Elkins et al. 2004). Elkins et al. (2004) presented two decision models for providing initial insights and industry perspective into the business case for investment in agile manufacturing systems and these models are applied to study hypothetical decision. Purohit et al. (2006) took rapid prototyping/tooling/manufacturing as primary enablers of agile manufacturing. They showed the role of prototype parts in automobile development programs and the necessity to accept rapid prototyping/tooling/manufacturing in agile manufacturing, as well as the tools available for implementing rapid prototyping/tooling/manufacturing, and current limitations of rapid techniques for realizing agile manufacturing. Fritzsche (2018) supplied empirical data about the constraints under that scheduling and sequencing takes place for agile manufacturing in the automotive industry. Qamar et al. (2019) handled the difference between agile and lean manufacturing with quality and flexibility. They also aimed to identify the relationship between quality and flexibility.

In the Scopus database, the papers which related to agile manufacturing systems requirements are presented. In Fig. 1, the percentage of published papers per year is

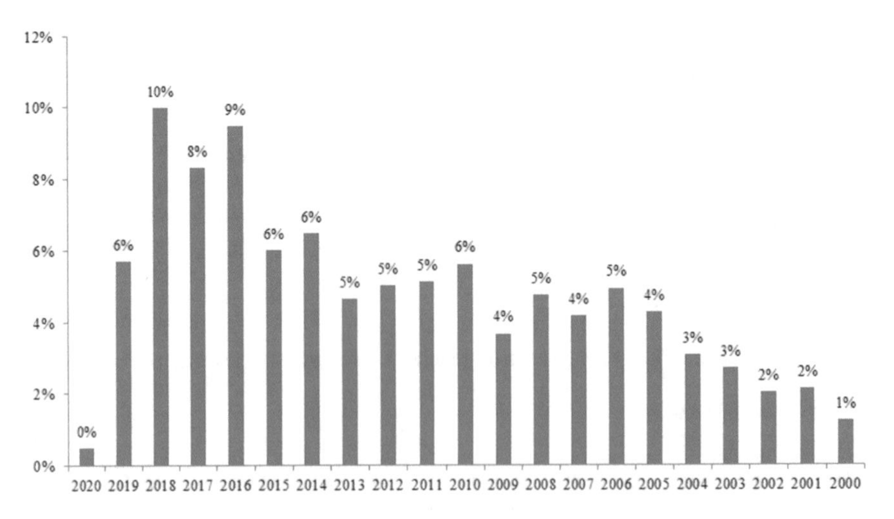

Fig. 1 The percentage of documents per year

given. The highest percentage is belonged to 2018. The distribution of authors who has published in agile manufacturing systems requirements are given in Fig. 2. In Fig. 3, the paper distribution by their territory is given about agile manufacturing system challenge. United States, China and Germany are in the first three ranking. In Fig. 4, the percentage of document type is given about agile manufacturing and related system requirements are illustrated. The most published type of agile manufacturing system requirements is conference papers.

There are some criteria, parameters and questions for companies in order to determine the appropriate technology for business processes and parameters to be assessed. In order to find the best suitable method in manufacturing, decision making methods can be used. Multi criteria decision making (MCDM) methods are significant tools in order to solve complex and big problems which have different

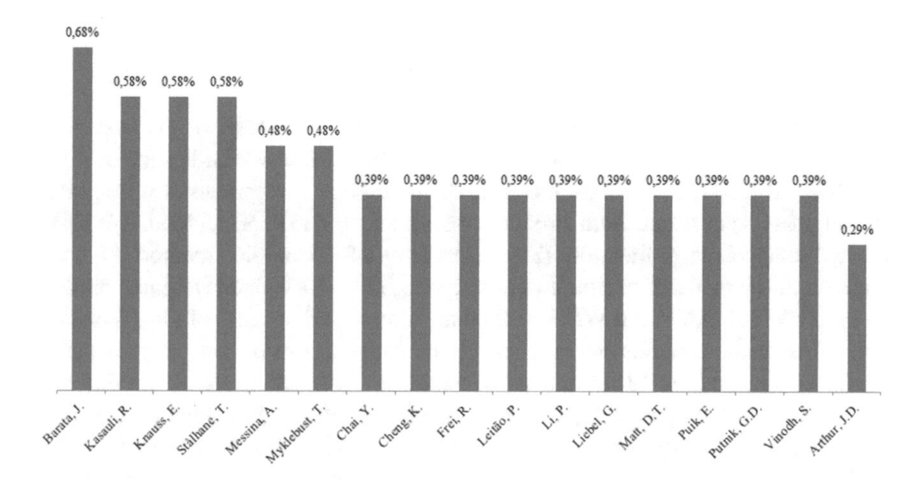

Fig. 2 The percentage of number of authors per year

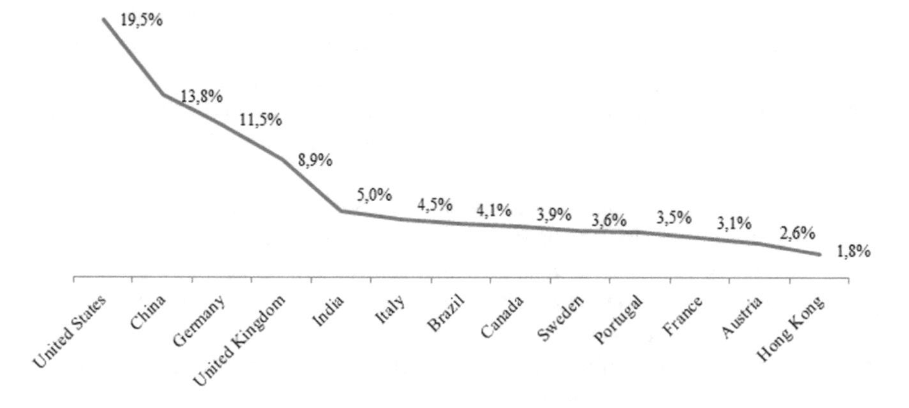

Fig. 3 The percentage of documents by territory

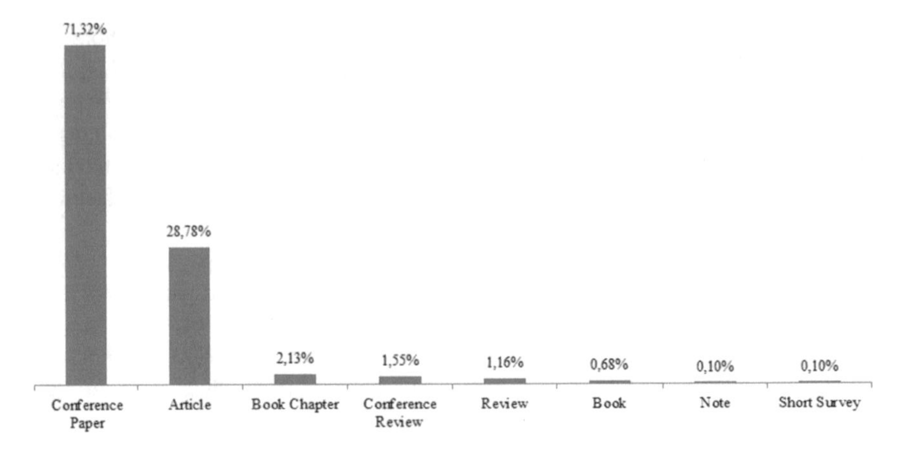

Fig. 4 The percentage of documents type

alternatives with respect to diversified criteria in order to select of the best alternative. In the literature, there are various different MCDM methods such as AHP, TOPSIS, ELECTRE, CODAS, etc. One of the current methods used in literature is Weighted Aggregated Sum Product Assessment (WASPAS) method. WASPAS method is appeared in literature (Zavadskas et al. 2012) and this method is a combination of the methods of Simple additive weighting (SAW) and weighted product method (WPM). SAW and WPM method are simple and because of this, WASPAS method is mathematically uncomplicated and capability to provide more accurate results as compared to WSM and WPM methods. This method used with different fuzzy sets like Pythagorean fuzzy sets, type 2 fuzzy sets, neutrosophic sets, etc.

Ordinary fuzzy sets are introduced by Zadeh (1965). The fuzzy sets has been diversified its different extensions like hesitant fuzzy sets (Torra 2010), intuitionistic fuzzy sets (Atanassov 1986), neutrosophic sets (Smarandache 1998), Pythagorean fuzzy sets (Yager 2014), picture fuzzy sets (Cường 2014), spherical fuzzy sets and etc. Every type of fuzzy sets has explained the membership and non-membership functions in different ways with or without hesitancy. Spherical fuzzy sets are developed by Kutlu Gündoğdu and Kahraman (2019d). These fuzzy sets are used in multi criteria decision making methods like TOPSIS and AHP. In spherical fuzzy sets, the sum of membership, non-membership and hesitancy degrees must be between 0 and 1 and three parameters are assigned independently.

In this chapter, we use spherical fuzzy sets with WASPAS method in order to prioritizing manufacturing challenges of a contract manufacturing company for personal auto. The method is based on the arithmetic operations of Spherical fuzzy sets such as addition, multiplication, and division. In Sect. 2, the preliminaries of Spherical fuzzy sets are given. In Sect. 3, the Spherical fuzzy WASPAS method is given step by step. In Sect. 4, the application is presented. In Sect. 5, the conclusion is given, and the future research suggestions are presented.

2 Spherical Fuzzy Sets: Preliminaries

The novel concept of SFS (Spherical Fuzzy Sets) provides a larger preference domain for decision makers to assign membership degrees since the squared sum of the spherical parameters is allowed to be at most 1.0. DMs can define their hesitancy information independently under spherical fuzzy environment. SFS are a generalization of Pythagorean Fuzzy Sets (PFS), picture fuzzy sets and neutrosophic sets (Kutlu Gündoğdu and Kahraman 2019a, b, c, d, e; Kahraman et al. 2019, Kutlu Gündogdu & Kahraman 2020). In the following, definition of SFS is presented:

Definition 2.1 Single valued Spherical Fuzzy Sets (SFS) $\tilde{A}_S$ of the universe of discourse U is given by

$$\tilde{A}_S = \left\{\left\langle u, \left(\mu_{\tilde{A}_S}(u), v_{\tilde{A}_S}(u), \pi_{\tilde{A}_S}(u)\right)\right| u \in U\right\} \quad (1)$$

where

$$\mu_{\tilde{A}_S}(u) : U \to [0, 1], \quad v_{\tilde{A}_S}(u) : U \to [0, 1], \quad \pi_{\tilde{A}_S}(u) : U \to [0, 1]$$

and

$$0 \le \mu^2_{\tilde{A}_S}(u) + v^2_{\tilde{A}_S}(u) + \pi^2_{\tilde{A}_S}(u) \le 1 \qquad \forall u \in U \quad (2)$$

For each u, the numbers $\mu_{\tilde{A}_S}(u)$, $v_{\tilde{A}_S}(u)$ and $\pi_{\tilde{A}_S}(u)$ are the degree of membership, non-membership and hesitancy of u to $\tilde{A}_S$, respectively. $\chi_{\tilde{A}_S}(u) = \sqrt{1 - \mu^2_{\tilde{A}_S}(u) - v^2_{\tilde{A}_S}(u) - \pi^2_{\tilde{A}_S}(u)}$ is called as a refusal degree.

Definition 2.2 Basic operators of Single-valued SFS;

$$\text{i. } \tilde{A}_S \oplus \tilde{B}_S = \left\{ \begin{array}{c} \left(\mu^2_{\tilde{A}_S} + \mu^2_{\tilde{B}_S} - \mu^2_{\tilde{A}_S}\mu^2_{\tilde{B}_S}\right)^{1/2}, \; v_{\tilde{A}_S} v_{\tilde{B}_S}, \\ \left(\left(1 - \mu^2_{\tilde{B}_S}\right)\pi^2_{\tilde{A}_S} + \left(1 - \mu^2_{\tilde{A}_S}\right)\pi^2_{\tilde{B}_S} - \pi^2_{\tilde{A}_S}\pi^2_{\tilde{B}_S}\right)^{1/2} \end{array} \right\} \quad (3)$$

$$\text{ii. } \tilde{A}_S \otimes \tilde{B}_S = \left\{ \begin{array}{c} \mu_{\tilde{A}_S}\mu_{\tilde{B}_S}, \; \left(v^2_{\tilde{A}_S} + v^2_{\tilde{B}_S} - v^2_{\tilde{A}_S} v^2_{\tilde{B}_S}\right)^{1/2}, \\ \left(\left(1 - v^2_{\tilde{B}_S}\right)\pi^2_{\tilde{A}_S} + \left(1 - v^2_{\tilde{A}_S}\right)\pi^2_{\tilde{B}_S} - \pi^2_{\tilde{A}_S}\pi^2_{\tilde{B}_S}\right)^{1/2} \end{array} \right\} \quad (4)$$

$$\text{iii. } \lambda \cdot \tilde{A}_S = \left\{ \begin{array}{c} \left(1 - \left(1 - \mu^2_{\tilde{A}_S}\right)^{\lambda}\right)^{1/2}, \; v^{\lambda}_{\tilde{A}_S}, \\ \left(\left(1 - \mu^2_{\tilde{A}_S}\right)^{\lambda} - \left(1 - \mu^2_{\tilde{A}_S} - \pi^2_{\tilde{A}_S}\right)^{\lambda}\right)^{1/2} \end{array} \right\} \quad for\ \lambda > 0 \quad (5)$$

$$\text{iv. } \tilde{A}_S^{\lambda} = \left\{ \begin{array}{l} \mu_{\tilde{A}_S}^{\lambda}, \left(1 - \left(1 - v_{\tilde{A}_S}^2\right)^{\lambda}\right)^{1/2}, \\ \left(\left(1 - v_{\tilde{A}_S}^2\right)^{\lambda} - \left(1 - v_{\tilde{A}_S}^2 - \pi_{\tilde{A}_S}^2\right)^{\lambda}\right)^{1/2} \end{array} \right\} \quad for\, \lambda > 0 \tag{6}$$

Definition 2.3 Single-valued Spherical Weighted Arithmetic Mean (SWAM) with respect to, $w = (w_1, w_2, \ldots, w_n)$; $w_i \in [0, 1]$; $\sum_{i=1}^{n} w_i = 1$, SWAM is defined as;

$$SWAM_w(\tilde{A}_{S1}, \ldots, \tilde{A}_{Sn}) = w_1\tilde{A}_{S1} + w_2\tilde{A}_{S2} + \ldots + w_n\tilde{A}_{Sn}$$
$$= \left\{ \begin{array}{l} \left[1 - \prod_{i=1}^{n} (1 - \mu_{\tilde{A}_{Si}}^2)^{w_i}\right]^{1/2}, \\ \prod_{i=1}^{n} v_{\tilde{A}_{Si}}^{w_i}, \left[\prod_{i=1}^{n} (1 - \mu_{\tilde{A}_{Si}}^2)^{w_i} - \prod_{i=1}^{n} (1 - \mu_{\tilde{A}_{Si}}^2 - \pi_{\tilde{A}_{Si}}^2)^{w_i}\right]^{1/2} \end{array} \right\} \tag{7}$$

Definition 2.4 Single-valued Spherical Weighted Geometric Mean (SWGM) with respect to, $w = (w_1, w_2, \ldots, w_n)$; $w_i \in [0, 1]$; $\sum_{i=1}^{n} w_i = 1$, SWGM is defined as;

$$SWGM_w(\tilde{A}_1, \ldots, \tilde{A}_n) = \tilde{A}_{S1}^{w_1} + \tilde{A}_{S2}^{w_2} + \ldots + \tilde{A}_{Sn}^{w_n}$$
$$= \left\{ \begin{array}{l} \prod_{i=1}^{n} \mu_{\tilde{A}_{Si}}^{w_i}, \left[1 - \prod_{i=1}^{n} (1 - v_{\tilde{A}_{Si}}^2)^{w_i}\right]^{1/2}, \\ \left[\prod_{i=1}^{n} (1 - v_{\tilde{A}_{Si}}^2)^{w_i} - \prod_{i=1}^{n} (1 - v_{\tilde{A}_{Si}}^2 - \pi_{\tilde{A}_{Si}}^2)^{w_i}\right]^{1/2} \end{array} \right\} \tag{8}$$

Definition 2.5 Score functions and Accuracy functions of sorting SFS are defined by;

$$Score\left(\tilde{A}_S\right) = \left(2\mu_{\tilde{A}_S} - \pi_{\tilde{A}_S}/2\right)^2 - \left(v_{\tilde{A}_S} - \pi_{\tilde{A}_S}/2\right)^2 \tag{9}$$

$$Accuracy\left(\tilde{A}_S\right) = \mu_{\tilde{A}_S}^2 + v_{\tilde{A}_S}^2 + \pi_{\tilde{A}_S}^2 \tag{10}$$

Note that: $\tilde{A}_S < \tilde{B}_S$ if and only if

i. $Score(\tilde{A}_S) < Score(\tilde{B}_S)$ or
ii. $Score(\tilde{A}_S) = Score(\tilde{B}_S)$ and $Accuracy(\tilde{A}_S) < Accuracy(\tilde{B}_S)$.

3 Spherical Fuzzy WASPAS Method

A MCDM problem can be expressed as a decision matrix whose elements indicate the evaluation values of all alternatives with respect to each criterion under Spherical fuzzy environment. Let $X = \{x_1, x_2, \ldots, x_m\}$ $(m \geq 2)$ be a discrete set of m feasible alternatives and $C = \{C_1, C_2, \ldots, C_n\}$ be a finite set of criteria and $w = \{w_1, w_2, \ldots, w_n\}$ be the weight vector of all criteria which satisfies $0 \leq w_j \leq 1$ and $\sum_{j=1}^{n} w_j = 1$.

The proposed spherical fuzzy WASPAS method is consist of several steps as given in this section. Before giving these steps, the flow chart of the SF-WASPAS method is presented in Fig. 5 in order to make it easily comprehensible.

Step 1: Let Decision Makers (DMs) fill in the decision and criteria evaluation matrices based on the linguistic terms given in Table 1.

Step 2: Aggregate the judgments of each decision maker (DM) using Spherical Weighted Arithmetic Mean (SWAM) or Spherical Weighted Geometric Mean (SWGM) as given in Eqs. (11) and (12). Aggregate the criteria weights and construct aggregated spherical fuzzy decision matrix based on the opinions of decision makers. Signify the evaluation values of Alternative $x_i (i = 1, 2, \ldots, m)$ with respect to criterion $C_j (j = 1, 2, \ldots, n)$ by $C_j(\tilde{x}_i) = (\mu_{ij}, v_{ij}, \pi_{ij})$ and $\tilde{x}_{ij} = \left(C_j(\tilde{x}_i)\right)_{mxn}$ is a spherical fuzzy decision matrix. For a MCDM problem with SFS, decision matrix $\tilde{x}_{ij} = \left(C_j(\tilde{x}_i)\right)_{mxn}$ should be constructed as in Eq. (13).

$$SWAM_w(\tilde{A}_{S1}, \ldots, \tilde{A}_{Sn}) = w_1\tilde{A}_{S1} + w_2\tilde{A}_{S2} + \ldots + w_n\tilde{A}_{Sn}$$
$$= \left\{ \begin{array}{l} \left[1 - \prod_{i=1}^{n} (1 - \mu^2_{\tilde{A}_{Si}})^{w_i}\right]^{1/2}, \\ \prod_{i=1}^{n} v^{w_i}_{\tilde{A}_{Si}}, \left[\prod_{i=1}^{n} (1 - \mu^2_{\tilde{A}_{Si}})^{w_i} - \prod_{i=1}^{n} (1 - \mu^2_{\tilde{A}_{Si}} - \pi^2_{\tilde{A}_{Si}})^{w_i}\right]^{1/2} \end{array} \right\} \tag{11}$$

or

$$SWGM_w(\tilde{A}_1, \ldots, \tilde{A}_n) = \tilde{A}^{w_1}_{S1} + \tilde{A}^{w_2}_{S2} + \ldots + \tilde{A}^{w_n}_{Sn}$$
$$= \left\{ \begin{array}{l} \prod_{i=1}^{n} \mu^{w_i}_{\tilde{A}_{Si}}, \left[1 - \prod_{i=1}^{n} (1 - v^2_{\tilde{A}_{Si}})^{w_i}\right]^{1/2}, \\ \left[\prod_{i=1}^{n} (1 - v^2_{\tilde{A}_{Si}})^{w_i} - \prod_{i=1}^{n} (1 - v^2_{\tilde{A}_{Si}} - \pi^2_{\tilde{A}_{Si}})^{w_i}\right]^{1/2} \end{array} \right\} \tag{12}$$

$$\tilde{x}_{ij} = \left(C_j(\tilde{x}_i)\right)_{mxn} = \begin{array}{llll} (\mu_{11}, v_{11}, \pi_{11}) & (\mu_{12}, v_{12}, \pi_{12}) & \ldots & (\mu_{1n}, v_{1n}, \pi_{1n}) \\ (\mu_{21}, v_{21}, \pi_{21}) & (\mu_{22}, v_{22}, \pi_{22}) & \ldots & (\mu_{2n}, v_{2n}, \pi_{2n}) \\ . & . & & . \\ . & . & & . \\ (\mu_{m1}, v_{m1}, \pi_{m1}) & (\mu_{m2}, v_{m2}, \pi_{m2}) & \ldots & (\mu_{mn}, v_{mn}, \pi_{mn}) \end{array} \tag{13}$$

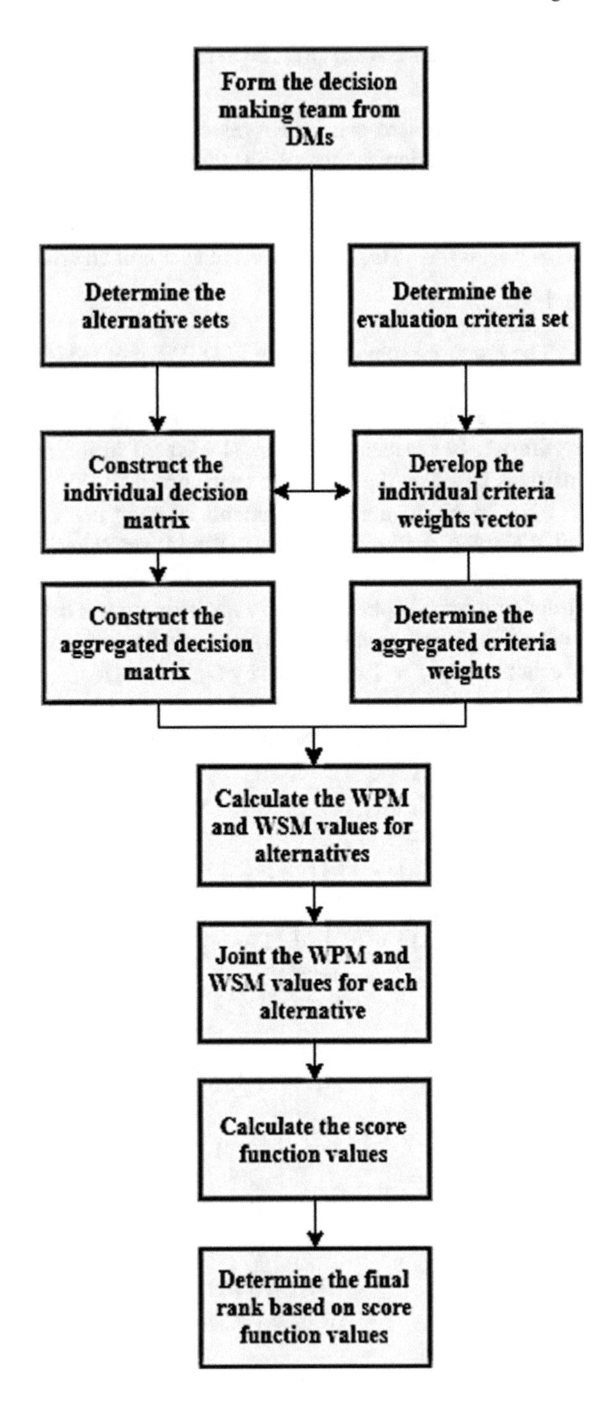

Fig. 5 SF-WASPAS proposed methodology

Table 1 Linguistic terms

Linguistic Terms	(μ, v, π)
Absolutely more importance (AMI)	(0.9, 0.1, 0.1)
Very high importance (VHI)	(0.8, 0.2, 0.2)
High importance (HI)	(0.7, 0.3, 0.3)
Slightly more importance (SMI)	(0.6, 0.4, 0.4)
Equally importance (EI)	(0.5, 0.5, 0.5)
Slightly low importance (SLI)	(0.4, 0.6, 0.4)
Low importance (LI)	(0.3, 0.7, 0.3)
Very low importance (VLI)	(0.2, 0.8, 0.2)
Absolutely low importance (ALI)	(0.1, 0.9, 0.1)

Table 2 Assessment of criteria by DMs

Criteria	DM1	DM2	…	DMk
C_1	$(\mu_{11}, v_{11}, \pi_{11})$	$(\mu_{12}, v_{12}, \pi_{12})$	…	$(\mu_{1k}, v_{1k}, \pi_{1k})$
C_2	$(\mu_{21}, v_{21}, \pi_{21})$	$(\mu_{22}, v_{22}, \pi_{22})$	…	$(\mu_{2k}, v_{2k}, \pi_{2k})$
⋮	⋮	⋮	⋱	⋮
C_j	$(\mu_{j1}, v_{j2}, \pi_{j1})$	$(\mu_{j2}, v_{j2}, \pi_{j2})$	…	$(\mu_{jk}, v_{jk}, \pi_{jk})$

Step 3: Decision makers also assess the decision criteria as given in Table 2.

Decision makers evaluate the alternatives with respect to the criteria as if they were benefit criteria such that they assign a lower linguistic term if it is a cost criterion.

Step 4: Calculate the score function value of each criterion in Table 3 and then normalize these values.

Step 4.1: Defuzzify the aggregated criteria weights based on the modified score function given in Eq. (14).

$$w_j^s = \left(2\mu_j - \pi_j/2\right)^2 - \left(v_j - \pi_j/2\right)^2 \tag{14}$$

Step 4.2: Normalize the aggregated criteria weights by using Eq. (15).

$$\bar{w}_j^s = \frac{w_j^s}{\sum_{J=1}^{n} w_j^s} \tag{15}$$

Step 5: Compute the results of Weighted Sum Model (WSM) as given in Eq. (16).

$$\tilde{Q}_i^{(1)} = \sum_{j=1}^{n} \tilde{x}_{ijw} = \sum_{j=1}^{n} \tilde{x}_{ij} \cdot \bar{w}_j^s \tag{16}$$

Table 3 Assessments of DM1, DM2, and DM3

DM1	C_1	C_2	C_3	C_4
X_1	SMI	VHI	HI	EI
X_2	VHI	AMI	HI	HI
X_3	EI	HI	VHI	HI
X_4	HI	AMI	HI	VHI
X_5	HI	LI	LI	SMI
DM2	C_1	C_2	C_3	C_4
X_1	HI	SMI	VHI	EI
X_2	AMI	SMI	VHI	SMI
X_3	SLI	LI	HI	EI
X_4	SMI	HI	LI	HI
X_5	HI	SLI	HI	EI
DM3	C_1	C_2	C_3	C_4
X_1	VHI	HI	HI	HI
X_2	SMI	HI	SMI	HI
X_3	HI	LI	VLI	HI
X_4	SMI	HI	VHI	HI
X_5	HI	EI	EI	EI

Equation (16) can be separated into two parts for ease of operations. First, the multiplication operator, then the addition operator is used.

Step 5.1: Determine the multiplication part of Eq. (16) based on Eq. (17).

$$\tilde{x}_{ijw} = \tilde{x}_{ij} \cdot \bar{w}_j^s = \left\langle \left(\left(1 - \left(1 - \mu_{\tilde{x}_{ij}}^2 \right)^{\bar{w}_j^s} \right) \right)^{1/2}, v_{\tilde{x}_{ij}}^{\bar{w}_j^s}, \left(\left(1 - \mu_{\tilde{x}_{ij}}^2 \right)^{\bar{w}_j^s} - \left(1 - \mu_{\tilde{x}_{ij}}^2 - \pi_{\tilde{x}_{ij}}^2 \right)^{\bar{w}_j^s} \right)^{1/2} \right\rangle \tag{17}$$

Step 5.2: Compute each addition term in Eq. (16) by using Eq. (18).

$$\tilde{x}_{i1w} \oplus \tilde{x}_{i2w} = \left\langle \begin{matrix} \left(\mu_{\tilde{x}_{i1w}}^2 + \mu_{\tilde{x}_{i2w}}^2 - \mu_{\tilde{x}_{i1w}}^2 \mu_{\tilde{x}_{i2w}}^2 \right)^{1/2}, v_{\tilde{x}_{i1w}} v_{\tilde{x}_{i2w}}, \\ \left(\left(1 - \mu_{\tilde{x}_{i2w}}^2 \right) \pi_{\tilde{x}_{i1w}}^2 + \left(1 - \mu_{\tilde{x}_{i1w}}^2 \right) \pi_{\tilde{x}_{i2w}}^2 - \pi_{\tilde{x}_{i1w}}^2 \pi_{\tilde{x}_{i2w}}^2 \right)^{1/2} \end{matrix} \right\rangle \tag{18}$$

Step 6: Calculate the results of Weighted Product Model (WPM) as presented in Eq. (19);

$$\tilde{Q}_i^{(2)} = \prod_{j=1}^{n} \tilde{x}_{ij}^{\bar{w}_j^s} \tag{19}$$

Equation (19) can be also separated into two parts for ease of operations. First, the exponential operator and then the multiplication operator is used.

Step 6.1: Determine the exponential part of Eq. (19) by using Eq. (20).

$$\tilde{x}_{ij}^{\bar{w}_j^s} = \left\langle \left(\mu_{\tilde{x}_{ij}}^{\bar{w}_j^s}, \left(1 - \left(1 - v_{\tilde{x}_{ij}}^2\right)^{\bar{w}_j^s}\right)^{1/2}, \left(\left(1 - v_{\tilde{x}_{ij}}^2\right)^{\bar{w}_j^s} - \left(1 - v_{\tilde{x}_{ij}}^2 - \pi_{\tilde{x}_{ij}}^2\right)^{\bar{w}_j^s}\right)^{1/2} \right) \right\rangle \tag{20}$$

Step 6.2: Calculate each multiplication term in Eq. (19) by using Eq. (21).

$$\tilde{x}_{i1}^{\bar{w}_1^s} \otimes \tilde{x}_{i2}^{\bar{w}_2^s} = \left\langle \begin{array}{l} \mu_{\tilde{x}_{i1}^{\bar{w}_1^s}} \mu_{\tilde{x}_{i2}^{\bar{w}_2^s}}, \left(v^2_{\tilde{x}_{i1}^{\bar{w}_1^s}} + v^2_{\tilde{x}_{i2}^{\bar{w}_2^s}} - v^2_{\tilde{x}_{i1}^{\bar{w}_1^s}} v^2_{\tilde{x}_{i2}^{\bar{w}_2^s}} \right)^{1/2}, \\ \left(\left(1 - v^2_{\tilde{x}_{i2}^{\bar{w}_2^s}}\right) \pi^2_{\tilde{x}_{i1}^{\bar{w}_1^s}} + \left(1 - v^2_{\tilde{x}_{i1}^{\bar{w}_1^s}}\right) \pi^2_{\tilde{x}_{i2}^{\bar{w}_2^s}} - \pi^2_{\tilde{x}_{i1}^{\bar{w}_1^s}} \pi^2_{\tilde{x}_{i2}^{\bar{w}_2^s}} \right)^{1/2} \end{array} \right\rangle \tag{21}$$

Step 7: Define the threshold number λ and calculate Eqs. (22) and (23):

$$\lambda \cdot \tilde{Q}_i^{(1)} = \left\langle \begin{array}{l} \left(1 - \left(1 - \mu^2_{\tilde{Q}_i^{(1)}}\right)^{\lambda}\right)^{1/2}, v^{\lambda}_{\tilde{Q}_i^{(1)}}, \\ \left(\left(1 - \mu^2_{\tilde{Q}_i^{(1)}}\right)^{\lambda} - \left(1 - \mu^2_{\tilde{Q}_i^{(1)}} - \pi^2_{\tilde{Q}_i^{(1)}}\right)^{\lambda}\right)^{1/2} \end{array} \right\rangle \tag{22}$$

$$(1 - \lambda) \cdot \tilde{Q}_i^{(2)} = \left\langle \begin{array}{l} \left(1 - \left(1 - \mu^2_{\tilde{Q}_i^{(2)}}\right)^{1-\lambda}\right)^{1/2}, v^{1-\lambda}_{\tilde{Q}_i^{(2)}}, \\ \left(\left(1 - \mu^2_{\tilde{Q}_i^{(2)}}\right)^{1-\lambda} - \left(1 - \mu^2_{\tilde{Q}_i^{(2)}} - \pi^2_{\tilde{Q}_i^{(2)}}\right)^{1-\lambda}\right)^{1/2} \end{array} \right\rangle \tag{23}$$

Step 8: Sum Eqs. (22) and (23) as given by Eq. (24).

$$\tilde{Q}_i = \lambda \cdot \tilde{Q}_i^{(1)} + (1 - \lambda) \cdot \tilde{Q}_i^{(2)} \tag{24}$$

Step 9: Defuzzify $\tilde{Q}_i$ by using the score function as given in Eq. (9). We put the alternatives into order with respect to the decreasing values of score values. If the score values of two alternatives are equal, their accuracy function values might be taken into account as given in Eq. (10).

4 Application

A company in Turkey wants to prioritize its most faced challenges during their manufacturing line. To make a more productive system, an agreement with the experts is performed which aim to start a preparatory work about sorting of the challenges in the manufacturing plant. Through this work, 4 criteria are determined ad possible 6 alternatives are suggested as the solution. 3 decision makers consist of production department manager, firm's technology officer, and an academician who has expertise on MCDM methods under uncertain environments are selected and agreed to perform the study to determine which alternative robot is the most appropriate one for the solution. Our proposed methodology is applied to an industrial robot selection problem. For this goal, mostly used five robots which are 6-axis robots X_1, Scara robots X_2, Dual-arm robots X_3, Redundant robots X_4, Cartesian robots X_5. These alternatives are evaluated under 4 criteria which are efficiency (C_1), suitability (C_2), automation (C_3), and ergonomics (C_4) The criteria are determined after a comprehensive literature review. The weights of three decision makers (DM1, DM2, DM3) having different experience levels are 0.4, 0.4 and 0.2, respectively.

First of all, the assessments for the criteria are collected from decision makers with respect to the goal, based on the linguistic terms given in Table 1. All assessments are given in Table 3.

These judgments are aggregated using SWAM operator by considering the importance levels of decision makers. Aggregated decision matrix is achieved as in Table 4.

The linguistic importance weights of the criteria assigned by DMs are shown in Table 5.

The weight of each criterion obtained by using SWGM operator is presented in Table 6.

Table 4 Aggregated decision matrix

Alternatives	C_1	C_2	C_3	C_4
X_1	(0.68, 0.33, 0.34)	(0.69, 0.32, 0.32)	(0.77, 0.27, 0.27)	(0.53, 0.47, 0.48)
X_2	(0.79, 0.23, 0.24)	(0.73, 0.30, 0.31)	(0.72, 0.29, 0.30)	(0.66, 0.34, 0.35)
X_3	(0.49, 0.52, 0.43)	(0.42, 0.60, 0.31)	(0.57, 0.48, 0.25)	(0.61, 0.40, 0.41)
X_4	(0.64, 0.36, 0.37)	(0.77, 0.24, 0.25)	(0.51, 0.52, 0.30)	(0.74, 0.27, 0.27)
X_5	(0.70, 0.30, 0.30)	(0.37, 0.63, 0.38)	(0.47, 0.55, 0.36)	(0.54, 0.46, 0.47)

Table 5 Importance weights of the criteria

Criteria	DM1	DM2	DM3
C_1	SMI	HI	VHI
C_2	AMI	SMI	HI
C_3	VHI	HI	VLI
C_4	SMI	EI	EI

Table 6 Aggregation of criteria weights

Criteria	Weight of each criterion
C_1	(0.68, 0.33, 0.34)
C_2	(0.73, 0.30, 0.31)
C_3	(0.57, 0.48, 0.25)
C_4	(0.54, 0.46, 0.47)

Table 7 Defuzzified and normalized criteria weights

Criteria	Weight of each criterion
C_1	0.297
C_2	0.359
C_3	0.200
C_4	0.141

After the weights of the criteria have been determined, the defuzzified and normalized criteria weights are calculated by using Eqs. (14) and (15) as given in Table 7.

By utilizing Eqs. (16), (17), and (18) and, $\tilde{Q}_i^{(1)}$ is calculated as in Table 8. Based on first column of Table 8 and Eq. (22), $\lambda \cdot \tilde{Q}_i^{(1)}$ is obtained as given in the second column of Table 8.

Based on Eqs. (19), (20) and (21), $\tilde{Q}_i^{(2)}$ is calculated as in Table 9. Based on the first column of Table 9 and Eq. (23), $(1 - \lambda) \cdot \tilde{Q}_i^{(2)}$ is obtained as given in the second column of Table 9.The threshold value λ is accepted as 0.5 in this chapter.

In the next step, based on Tables 8 and 9, we can calculate the final spherical fuzzy value of SF-WASPAS by utilizing Eq. (24). They are given in Table 10.

Table 8 Weighted sum product ($\tilde{Q}_i^{(1)}$)

Alternatives	$\tilde{Q}_i^{(1)}$	$\lambda \cdot \tilde{Q}_i^{(1)}$
X_1	(0.70, 0.31, 0.32)	(0.53, 0.55, 0.27)
X_2	(0.78, 0.23, 0.25)	(0.61, 0.48, 0.23)
X_3	(0.59, 0.43, 0.35)	(0.44, 0.66, 0.28)
X_4	(0.73, 0.28, 0.27)	(0.56, 0.53, 0.24)
X_5	(0.56, 0.45, 0.37)	(0.42, 0.67, 0.30)

Table 9 Weighted product model ($\tilde{Q}_i^{(2)}$)

Alternatives	$\tilde{Q}_i^{(2)}$	$(1 - \lambda) \cdot \tilde{Q}_i^{(2)}$
X_1	(0.69, 0.32, 0.33)	(0.53, 0.56, 0.28)
X_2	(0.77, 0.24, 0.26)	(0.60, 0.49, 0.23)
X_3	(0.57, 0.45, 0.35)	(0.43, 0.67, 0.28)
X_4	(0.71, 0.30, 0.29)	(0.55, 0.55, 0.25)
X_5	(0.52, 0.49, 0.38)	(0.38, 0.70, 0.30)

Table 10 $\tilde{Q}_i$ spherical fuzzy values

Alternatives	$\tilde{Q}_i$
X_1	(0.69, 0.31, 0.33)
X_2	(0.78, 0.23, 0.25)
X_3	(0.58, 0.44, 0.35)
X_4	(0.72, 0.29, 0.28)
X_5	(0.54, 0.47, 0.38)

Table 11 Score values and final ranking

Alternatives	Score	Ranking
X_1	1.479	3
X_2	2.014	1
X_3	0.902	4
X_4	1.673	2
X_5	0.728	5

From Table 10, the score value of each alternative is obtained based on Eq. (9) and given in Table 11.

The score values indicate that the best alternative is X_1 and overall ranking is $X_2 > X_4 > X_1 > X_3 > X_5$. Now, in order to test the reliability and robustness, we compare the proposed single-valued SF-WASPAS with single-valued SF-TOPSIS. Table 1 presents the single-valued SF linguistic scale, which also we use for the comparison purpose.

In this comparison, the same judgments as given in Table 3 were used and aggregated using SWAM (Spherical fuzzy Weighted Arithmetic Mean) operator given in Eq. (7). After the weights of the criteria and the rating of the alternatives have been determined, the aggregated weighted spherical fuzzy decision matrix is constructed by utilizing Eq. (4) as given in Table 12.

Spherical positive and negative ideal solutions (SF-PIS and SF-NIS) are given in Table 13. They are calculated by using Eq. (9). According to the best and worst scores, the corresponding Spherical Fuzzy Positive Ideal Solution (SF-PIS) and Spherical Fuzzy Negative Ideal Solution (SF-NIS) are given in Table 13.

Table 12 Aggregated weighted decision matrix

Alternatives	C_1	C_2	C_3	C_4
X_1	(0.40, 0.54, 0.39)	(0.53, 0.42, 0.32)	(0.44, 0.51, 0.32)	(0.26, 0.69, 0.48)
X_2	(0.48, 0.47, 0.35)	(0.59, 0.36, 0.29)	(0.43, 0.52, 0.33)	(0.31, 0.63, 0.45)
X_3	(0.32, 0.63, 0.44)	(0.39, 0.58, 0.33)	(0.42, 0.54, 0.29)	(0.31, 0.64, 0.46)
X_4	(0.38, 0.56, 0.40)	(0.60, 0.35, 0.28)	(0.41, 0.55, 0.31)	(0.35, 0.59, 0.44)
X_5	(0.41, 0.53, 0.38)	(0.27, 0.71, 0.38)	(0.30, 0.67, 0.36)	(0.25, 0.69, 0.48)

Table 13 Positive and negative ideal solutions

	C_1	C_2	C_3	C_4
X^*	(0.48, 0.47, 0.35)	(0.60, 0.35, 0.28)	(0.44, 0.51, 0.32)	(0.35, 0.59, 0.44)
X^-	(0.32, 0.63, 0.44)	(0.27, 0.71, 0.38)	(0.30, 0.67, 0.36)	(0.25, 0.69, 0.48)

Table 14 Distances to positive and negative ideal solutions

Alternatives	$d\left(X_{ij}, X_j^+\right)$	$d\left(X_{ij}, X_j^-\right)$
X_1	0.0313	0.1390
X_2	0.0111	0.1887
X_3	0.0344	0.0882
X_4	0.0072	0.1710
X_5	0.0637	0.0293

Based on Eqs. (25) and (26), we have calculated the Zhang and Xu's (2014) distances between alternative X_i and SF-PIS as well as X_i and SF-NIS. They are given in Table 14.

$$D_{ZX}\left(X_{ij}, X_j^*\right) = \frac{1}{2n}\sum_{j=1}^{n}\left(\left|\mu_{X_{ij}}^2 - \mu_{X_j^*}^2\right| + \left|v_{X_{ij}}^2 - v_{X_j^*}^2\right| + \left|\pi_{X_{ij}}^2 - \pi_{X_j^*}^2\right|\right) \forall i \quad (25)$$

$$D_{ZX}\left(X_{ij}, X_j^-\right) = \frac{1}{2n}\sum_{j=1}^{n}\left(\left|\mu_{X_{ij}}^2 - \mu_{X_j^-}^2\right| + \left|v_{X_{ij}}^2 - v_{X_j^-}^2\right| + \left|\pi_{X_{ij}}^2 - \pi_{X_j^-}^2\right|\right) \forall i \quad (26)$$

Based on Eq. (27), closeness ratios are calculated and presented in Table 15.

$$Closeness\ Ratio_i = \frac{d\left(X_{ij}, X_j^-\right)}{d\left(X_{ij}, X_j^-\right) + d\left(X_{ij}, X_j^+\right)} \quad (27)$$

The closeness ratios based on SF-TOPSIS method indicate that the best alternative is X_4 and the overall ranking ($X_4 > X_2 > X_1 > X_3 > X_5$) is slightly different than

Table 15 Closeness ratio of each alternative i

Alternatives	Closeness ratio	Rank
X_1	0.8162	3
X_2	0.9446	2
X_3	0.7196	4
X_4	0.9597	1
X_5	0.3149	5

SF-WASPAS method. When we changed the distance formula from Zhang and Xu's distance to Euclidean distance, the final ranking will be same with SF-WASPAS.

5 Conclusions

In this chapter, an extended WASPAS method is proposed for the prioritization of the solutions of manufacturing challenges for a contract manufacturing company. Since the challenges contain both qualitative and quantitative data, spherical fuzzy sets are applied for the application. Also, a comparative analysis is applied to check the applicability of the proposed extension. Based on the results, it is observed that, the results of the compared methods are slightly different. For further researches, the proposed technique can be applied to different production problems by considering labor force perspective. Through that, the results can be compared with the other types of fuzzy extensions such as intuitionistic fuzzy sets, q-rank fuzzy sets, neutrosophic sets.

References

Atanassov KT (1986) Intuitionistic fuzzy sets. Fuzzy Sets Syst 20(1):87–96

Cường BC (2014) Picture fuzzy sets. J Comput Sci Cybern 30(4):409

Elkins DA, Huang N, Alden JM (2004) Agile manufacturing systems in the automotive industry. Int J Prod Econ 91(3):201–214

Fritzsche A (2018) Implications of agile manufacturing in the automotive industry for order management in the factories-evidence from the practitioner's perspective. Procedia CIRP 72:369–374

Kahraman C, Gundogdu FK, Onar SC, Oztaysi B (2019) Hospital Location Selection Using Spherical Fuzzy TOPSIS. In: 2019 conference of the international fuzzy systems association and the European society for fuzzy logic and technology (EUSFLAT 2019). Atlantis Press

Kutlu Gündoğdu F, Kahraman C (2019a) Extension of WASPAS with spherical Fuzzy Sets. Informatica 30(2):269–292

Kutlu Gündoğdu F, Kahraman C (2019b) A novel spherical fuzzy analytic hierarchy process and its renewable energy application. Soft Computing in-press

Kutlu Gündoğdu F, Kahraman C (2019c) A novel VIKOR method using spherical fuzzy sets and its application to warehouse site selection. J Intell Fuzzy Syst 37(1):1197–1211

Kutlu Gündoğdu F, Kahraman C (2019d) Spherical fuzzy sets and spherical fuzzy TOPSIS method. J Intell Fuzzy Syst 36(1):337–352

Kutlu Gündoğdu F, Kahraman C (2019e) A novel fuzzy TOPSIS method using emerging interval-valued spherical fuzzy sets. Eng Appl Artif Intell 85:307–323

Kutlu Gündoğdu F, Kahraman C (2020) A novel spherical fuzzy QFD method and its application to the linear delta robot technology development. Eng Appl Artif Intell 87:103348

Smarandache F (1998) Neutrosophy neutrosophic probability, set, and logic. Amer Res Press Rehoboth:12–20

Qamar A, Hall MA, Chicksand D, Collinson S (2019) Quality and flexibility performance trade-offs between lean and agile manufacturing firms in the automotive industry. Prod Plan Control. https://doi.org/10.1080/09537287.2019.1681534

Purohit A, Pant R, Deb A (2006) Role of rapid technologies as enablers for agile manufacturing in the automotive industry. Int J Agile Manuf 9(2):91–97
Torra V (2010) Hesitant fuzzy sets. Int J Intell Syst 25(6):529–539
Yager RR (2014) Pythagorean membership grades in multicriteria decision making. IEEE Trans Fuzzy Syst 22(4):958–965
Zadeh LA (1965) Fuzzy Sets. Inf Control 338–353
Zavadskas E, Turskis Z, Antucheviciene J, Zakarevicius A (2012) Optimization of weighted aggregated sum product assessment. Elektron Elektrotechnika 122(6):3–6

Assessment of Livability Indices of Suburban Places of Istanbul by Using Spherical Fuzzy CODAS Method

Ali Karaşan, Eda Boltürk, and Fatma Kutlu Gündoğdu

Abstract Spherical Fuzzy Sets (SFSs) are characterized by three degree are a membership degree, a non-membership degree and a hesitancy degree satisfying the condition that the square sum of its membership degree and non-membership degree is equal to or less than one, which is a generalization of Pythagorean Fuzzy Sets (PFSs) (Kutlu Gündoğdu and Kahraman 2019a). Livability indices help to understand how a place is livable and COmbinative Distance-based ASsessment (CODAS) method computes the Euclidean as the primary measure and Taxicab distances as the secondary measure to assess alternatives based on predetermined criteria. In this paper, CODAS method is extended to its Spherical Fuzzy CODAS version for handling the impreciseness and vagueness in human thoughts. The applicability of the methodology is illustrated through the assessment of livability index of suburban districts. Sensitivity analysis demonstrates the robustness of the decision-making methodology.

1 Introduction

Quality of life for the societies becomes a vital priority since the population, density in cities, is increasing exponentially. Istanbul, the most crowded city of Turkey, is in north-western Turkey within the Marmara Region and is divided by the Bosporus

A. Karaşan
Graduate School of Science and Engineering, Yildiz Technical University, 34220 Davutpasa, Istanbul, Turkey

F. Kutlu Gündoğdu
Industrial Engineering Department, National Defence University, Turkish Air Force Academy, 34149 Istanbul, Turkey
e-mail: fatmakutlugundogdu@gmail.com

E. Boltürk (✉)
Industrial Engineering Department, Istanbul Technical University, 34367 Macka Besiktas, Istanbul, Turkey
e-mail: bolturk@itu.edu.tr

© Springer Nature Switzerland AG 2021

C. Kahraman and F. Kutlu Gündoğdu (eds.), *Decision Making with Spherical Fuzzy Sets*, Studies in Fuzziness and Soft Computing 392,
https://doi.org/10.1007/978-3-030-45461-6_12

which connects the Sea of Marmara to the Black Sea. Since the city is the largest investment area both for manufacturing and service plants, instability in number of populations is increasing due to new investment opportunities, bankruptcy. Based on that, there are 385,000 people who immigrated to Istanbul and 210,000 people who emigrated from there in 2018 based on the data of Turkish Statistical Institute. This fluctuation negatively affects the social stability and decreases the level of being a livable city for Istanbul. Livability index is a measurement for the quality of life for the cities, regions or even nations based on multiple parameters which are including both qualitative and quantitative data (Scerri 2014). The index can be calculated by the methodologies consist of simulation approaches (Giap et al. 2014), statistical models (Higgs et al. 2019), integrated methodologies (Karasan et al. 2019). The aim of the usage of different methodologies is based on data type which can be based on cardinal, statistical or/and linguistic information and aim to improve the ability to reflect the data by using the appropriate methodology. Since the problem evaluates a list of alternative areas with respect to multi criteria based on group decision, it can be handled by using Multi Criteria Decision Making (MCDM) techniques.

MCDM methods enable a system for decision-makers to subjectively evaluate the performance of alternatives with respect to the predetermined criteria (Zavadskas et al. 2014). In the literature, there are many MCDM methods such as Analytic Hierarchy Process (AHP) (Saaty 1980), Technique for Order of Preference by Similarity to Ideal Solution (TOPSIS) (Hwang 1981), ELimination Et Choix Traduisant la REalité (ELECTRE) (Roy 1991), Analytic Network Process (ANP) (Saaty 1996), Evaluation Based on Distance from Average Solution (EDAS) (Keshavarz Ghorabaee et al. 2015) and Combinative Distance-Based Assessment (CODAS) (Keshavarz Ghorabaee et al. 2016). The constructed methods aim to handle the environmental, human, and social aspects of the problem based on the available data. Since the data may consist both quantitative and qualitative variables, reflecting them in the problem is a vital challenge. Fuzzy sets theory was introduced by Zadeh to capture the uncertainties in human thoughts through the degree of memberships of the elements in a set (Zadeh 1965). In order to increase the capability of handling vagueness and impreciseness in the problems, ordinary fuzzy sets have been extended to many types. Type-n fuzzy sets were developed by Zadeh to reduce the uncertainty of the membership functions in the ordinary fuzzy sets (Zadeh 1975). Interval-valued fuzzy sets were introduced independently by Zadeh (1975), Grattan-Guiness (1975), Sambuc (1975), Jahn (1975). Intuitionistic fuzzy sets were introduced by Atanassov to show how the hesitancy degree of a decision maker can be handled (Atanassov 1986). Smarandache developed neutrosophic sets for demonstrating the differences between relativity and absoluteness in the decision makers' preferences (Smarandache 1995). Hesitant fuzzy sets (HFSs), introduced by Torra, are the extensions of ordinary fuzzy sets where a set of values are possible for the membership of a single element (Torra 2010). In 2013, Yager introduced Pythagorean fuzzy sets (PFSs) which are an extension of IFS to offer a larger domain than IFS for the decision makers to express their judgments (Yager 2013). In 2019, Kutlu Gündoğdu and Kahraman (2019a) introduced Spherical fuzzy sets.

In this study, spherical fuzzy CODAS method is introduced and then applied for the problem of assessment of livability indices of suburban places in Istanbul. Through that, the evaluation criteria are collected based on literature review and alternatives are listed based on the highlighted suburban places of Istanbul. The aim of this study is to present applicability of spherical fuzzy CODAS method by applying descriptive illustrative examples under different scenarios. Also, a sensitivity analysis based on criteria weights is conducted to check the alterations of decisions of the method.

Rest of the chapter is organized as follows: In Sect. 2, a literature review on the studies based on methods which are used for determining the livability index is given. In Sect. 3, preliminaries of spherical fuzzy sets and the extended method is presented. In Sect. 4, a descriptive application is conducted, and results are discussed. The paper ends with conclusion and suggestions for further researches.

2 Literature Review: Livability Index

The Spherical fuzzy sets (SFS) have been recently introduced by Kutlu Gündoğdu and Kahraman (2019a) and they proposed a spherical fuzzy TOPSIS method. Kutlu Gündoğdu et al. (2019b) proposed Spherical fuzzy interval valued Spherical fuzzy TOPSIS. Kutlu Gündoğdu and Kahraman (2019c) proposed a novel approach based on spherical fuzzy VIKOR and they gave an application to warehouse selection. Kutlu Gündoğdu and Kahraman (2020a) developed spherical fuzzy AHP for renewable energy application.

In order to determine the appropriate criteria and to analyze the applied methodologies, a systematic literature review is carried out. Norouzian-Maleki et al. (2015) aimed to develop a practical method for assessing the liveability of a residential neighbourhood, tested in two contrasting countries, Iran and Estonia with respect to the criteria based on sustainable urban environment. Tan et al. (2016) studied to apply a comprehensive Liveability Cities index to rank the liveability of 100 cities in the Greater China Region based on 96 indicators across five environments, namely, economic vibrancy and competitiveness; environmental friendliness and sustainability; domestic security and stability; socio-cultural conditions; and political governance. Antognelli and Vizzari (2017) studied on the influences of Ecosystem Services and Urban Services to the landscape liveability in a comparable manner with context to landscape planning and policy-making purposes. Yin and Yin (2009) carried out a study for assessing the cities in Shandong Province in China involving the assessment indicator system and the assessment method by combining with the statistical data of 18 cities. Wang et al. (2011) studied a comparison analysis of liveability index with respect to analyses of four metropolitan cities especially based on the criteria social development and environmental quality.

Zanella et al. (2015) introduced an assessment tool to evaluate livability in European cities covering two components of livability: human wellbeing and environmental impact. Ruth and Franklin (2014) investigated the stock of the current discourse on livability, identifies two central elements that have yet to shape the assessments of livability and policies to promote it, and introduced strategies for research and practice to transform the livability concept, and with it the places in which the lives and livelihoods of people unfold. Kashef (2016) studied on comparative analytical assessment of diverse approaches and lays out a nuanced understanding of urban livability that draws on the richness and diversity embedded in design, planning, and current ranking tools.

3 Spherical Fuzzy Sets

The novel concept of Spherical Fuzzy Sets provides a larger preference domain for decision makers to assign membership degrees since the squared sum of the spherical parameters is allowed to be at most 1.0. DMs can determine their hesitancy degree independently under spherical fuzzy environment. Spherical fuzzy sets (SFS) are a generalization of Pythagorean fuzzy sets, picture fuzzy sets and neutrosophic sets (Kutlu Gündoğdu and Kahraman 2019a, b, c; Kutlu Gündoğdu and Kahraman 2020a, b).

3.1 Preliminaries: Spherical Fuzzy Sets

In the following, definition of SFS is presented:

Definition 3.1 Single valued Spherical Fuzzy Sets (SFS) $\tilde{A}_S$ of the universe of discourse U is given by

$$\tilde{A}_S = \left\{\langle u, (\mu_{\tilde{A}_S}(u), \nu_{\tilde{A}_S}(u), \pi_{\tilde{A}_S}(u)) \middle| u \in U\right\} \tag{1}$$

where

$$\mu_{\tilde{A}_S}(u) : U \to [0, 1],\ \ \nu_{\tilde{A}_S}(u) : U \to [0, 1],\ \ \pi_{\tilde{A}_S}(u) : U \to [0, 1]$$

and

$$0 \le \mu^2_{\tilde{A}_S}(u) + \nu^2_{\tilde{A}_S}(u) + \pi^2_{\tilde{A}_S}(u) \le 1 \quad \forall u \in U \tag{2}$$

For each u, the numbers $\mu_{\tilde{A}_S}(u)$, $\nu_{\tilde{A}_S}(u)$ and $\pi_{\tilde{A}_S}(u)$ are the degree of membership, non-membership and hesitancy of u to $\tilde{A}_S$, respectively.

Definition 3.2 Basic operators of single-valued SFS;

$$1.\ \tilde{A}_S \oplus \tilde{B}_S = \left\{ \begin{array}{l} \left(\mu^2_{\tilde{A}_S} + \mu^2_{\tilde{B}_S} - \mu^2_{\tilde{A}_S}\mu^2_{\tilde{B}_S}\right)^{1/2}, \ \nu_{\tilde{A}_S}\nu_{\tilde{B}_S}, \\ \left(\left(1 - \mu^2_{\tilde{B}_S}\right)\pi^2_{\tilde{A}_S} + \left(1 - \mu^2_{\tilde{A}_S}\right)\pi^2_{\tilde{B}_S} - \pi^2_{\tilde{A}_S}\pi^2_{\tilde{B}_S}\right)^{1/2} \end{array} \right\} \tag{3}$$

$$2.\ \tilde{A}_S \otimes \tilde{B}_S = \left\{ \begin{array}{l} \mu_{\tilde{A}_S}\mu_{\tilde{B}_S},\ \left(v^2_{\tilde{A}_S} + v^2_{\tilde{B}_S} - v^2_{\tilde{A}_S}v^2_{\tilde{B}_S}\right)^{1/2}, \\ \left(\left(1 - v^2_{\tilde{B}_S}\right)\pi^2_{\tilde{A}_S} + \left(1 - v^2_{\tilde{A}_S}\right)\pi^2_{\tilde{B}_S} - \pi^2_{\tilde{A}_S}\pi^2_{\tilde{B}_S}\right)^{1/2} \end{array} \right\} \quad (4)$$

$$3.\ \lambda \cdot \tilde{A}_S = \left\{ \begin{array}{l} \left(1 - \left(1 - \mu^2_{\tilde{A}_S}\right)^{\lambda}\right)^{1/2},\ v^{\lambda}_{\tilde{A}_S}, \\ \left(\left(1 - \mu^2_{\tilde{A}_S}\right)^{\lambda} - \left(1 - \mu^2_{\tilde{A}_S} - \pi^2_{\tilde{A}_S}\right)^{\lambda}\right)^{1/2} \end{array} \right\} \quad for\, \lambda \geq 0 \quad (5)$$

$$4.\ \tilde{A}_S^{\lambda} = \left\{ \begin{array}{l} \mu^{\lambda}_{\tilde{A}_S},\ \left(1 - \left(1 - v^2_{\tilde{A}_S}\right)^{\lambda}\right)^{1/2}, \\ \left(\left(1 - v^2_{\tilde{A}_S}\right)^{\lambda} - \left(1 - v^2_{\tilde{A}_S} - \pi^2_{\tilde{A}_S}\right)^{\lambda}\right)^{1/2} \end{array} \right\} \quad for\, \lambda > 0 \quad (6)$$

Definition 3.3 For these SFS $\tilde{A}_S = (\mu_{\tilde{A}_S}, v_{\tilde{A}_S}, \pi_{\tilde{A}_S})$ and $\tilde{B}_S = (\mu_{\tilde{B}_S}, v_{\tilde{B}_S}, \pi_{\tilde{B}_S})$, the followings are valid under the condition $\lambda, \lambda_1, \lambda_2 \geq 0$.

$$1.\ \tilde{A}_S \oplus \tilde{B}_S = \tilde{B}_S \oplus \tilde{A}_S \quad (7)$$

$$2.\ \tilde{A}_S \otimes \tilde{B}_S = \tilde{B}_S \otimes \tilde{A}_S \quad (8)$$

$$3.\ \lambda(\tilde{A}_S \oplus \tilde{B}_S) = \lambda\tilde{A}_S \oplus \lambda\tilde{B}_S \quad (9)$$

$$4.\ \lambda_1\tilde{A}_S \oplus \lambda_2\tilde{A}_S = (\lambda_1 + \lambda_2)\tilde{A}_S \quad (10)$$

$$5.\ (\tilde{A}_S \otimes \tilde{B}_S)^{\lambda} = \tilde{A}_S^{\lambda} \otimes \tilde{B}_S^{\lambda} \quad (11)$$

$$6.\ \tilde{A}_S^{\lambda_1} \otimes \tilde{A}_S^{\lambda_2} = \tilde{A}_S^{\lambda_1+\lambda_2} \quad (12)$$

Definition 3.4 Single-valued Spherical Weighted Arithmetic Mean (SWAM) with respect to, $w = (w_1, w_2 \ldots, w_n)$; $w_i \in [0, 1]$; $\sum_{i=1}^{n} w_i = 1$, SWAM is defined as;

$$SWAM_w(\tilde{A}_{S1}, \ldots, \tilde{A}_{Sn}) = w_1\tilde{A}_{S1} + w_2\tilde{A}_{S2} + \cdots + w_n\tilde{A}_{Sn}$$

$$= \left\{ \begin{array}{l} \left[1 - \prod_{i=1}^{n}(1 - \mu^2_{\tilde{A}_{Si}})^{w_i}\right]^{1/2},\ \prod_{i=1}^{n} v^{w_i}_{\tilde{A}_{Si}}, \\ \left[\prod_{i=1}^{n}(1 - \mu^2_{\tilde{A}_{Si}})^{w_i} - \prod_{i=1}^{n}(1 - \mu^2_{\tilde{A}_{Si}} - \pi^2_{\tilde{A}_{Si}})^{w_i}\right]^{1/2} \end{array} \right\} \quad (13)$$

Definition 3.5 Single-valued Spherical Weighted Geometric Mean (SWGM) with respect to, $w = (w_1, w_2 \ldots, w_n)$; $w_i \in [0, 1]$; $\sum_{i=1}^{n} w_i = 1$, SWGM is defined as;

$$SWGM_w(\tilde{A}_1, \ldots, \tilde{A}_n) = \tilde{A}_{S1}^{w_1} + \tilde{A}_{S2}^{w_2} + \cdots + \tilde{A}_{Sn}^{w_n} = \left\{ \begin{array}{l} \prod_{i=1}^{n} \mu_{\tilde{A}_{Si}}^{w_i}, \left[1 - \prod_{i=1}^{n} (1 - v_{\tilde{A}_{Si}}^2)^{w_i}\right]^{1/2}, \\ \left[\prod_{i=1}^{n} (1 - v_{\tilde{A}_{Si}}^2)^{w_i} - \prod_{i=1}^{n} (1 - v_{\tilde{A}_{Si}}^2 - \pi_{\tilde{A}_{Si}}^2)^{w_i}\right]^{1/2} \end{array} \right\} \tag{14}$$

Definition 3.6 Score functions and accuracy functions of sorting SFS are defined by;

$$Score\left(\tilde{A}_S\right) = \left(\mu_{\tilde{A}_S} - \pi_{\tilde{A}_S}/2\right)^2 - \left(v_{\tilde{A}_S} - \pi_{\tilde{A}_S}/2\right)^2 \tag{15}$$

$$Accuracy\left(\tilde{A}_S\right) = \mu_{\tilde{A}_S}^2 + v_{\tilde{A}_S}^2 + \pi_{\tilde{A}_S}^2 \tag{16}$$

Note that: $\tilde{A}_S < \tilde{B}_S$ if and only if

1. $Score(\tilde{A}_S) < Score(\tilde{B}_S)$ or
2. $Score(\tilde{A}_S) = Score(\tilde{B}_S)$ and $Accuracy(\tilde{A}_S) < Accuracy(\tilde{B}_S)$.

3.2 *Extension of CODAS with Spherical Fuzzy Sets*

The proposed single-valued spherical fuzzy CODAS (SF-CODAS) method is composed of several steps as given in this section. Before giving these steps, we present the flow chart of the SF-CODAS in Fig. 1 in order to make it easily understandable.

A MCDM problem can be expressed as a decision matrix whose elements indicate the evaluation values of all alternatives with respect to each criterion under spherical fuzzy environment. Let $X = \{x_1, x_2, \ldots x_m\}$ $(m \geq 2)$ be a discrete set of m feasible alternatives and $C = \{C_1, C_2, \ldots C_n\}$ be a finite set of criteria, and $w = \{w_1, w_2, \ldots w_n\}$ be the weights vector of all criteria which satisfies $0 \leq w_j \leq 1$ and $\sum_{j=1}^{n} w_j = 1$. The following methodology is proposed by Kutlu Gündoğdu (2019a).

Step 1: Let Decision makers fill in the decision matrices using the linguistic terms given in Table 1. DMs can also evaluate the criteria in order to define their weights by using Table 1.

Step 2: Aggregate the judgments of each decision maker (DM) using Spherical Weighted Arithmetic Mean (SWAM) by Eq. (17). Based on their experiences, each DM may have a different importance weight, w_j, $j = 1, 2, \ldots, n$

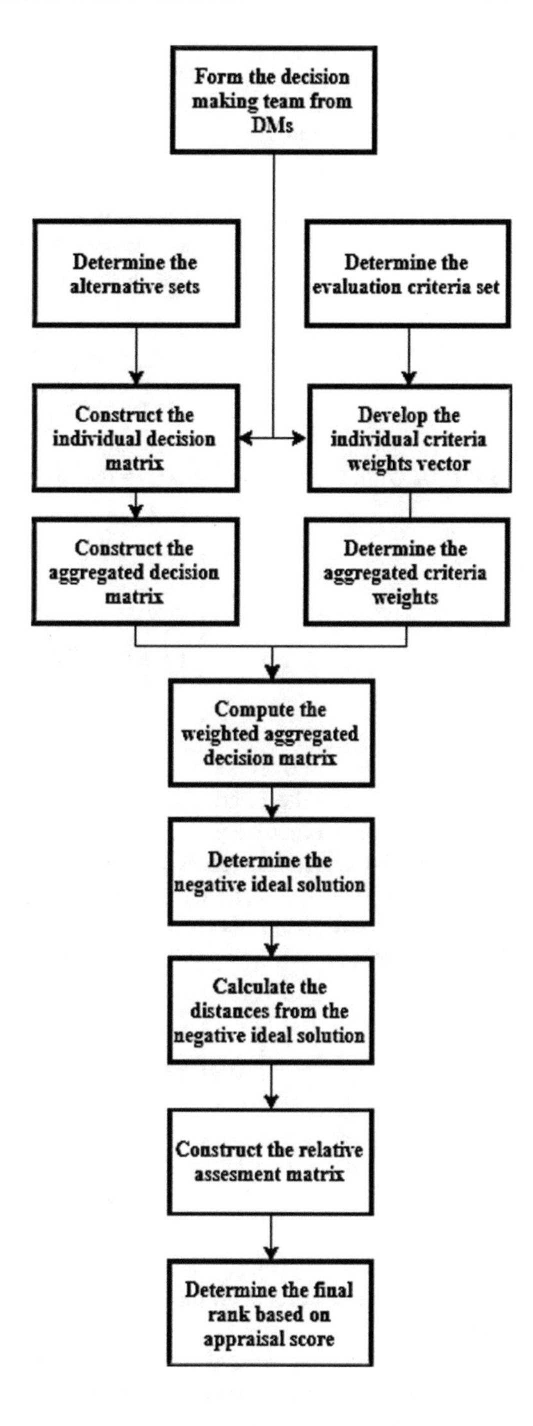

Fig. 1 Flow chart of the proposed SF-CODAS method

Table 1 Linguistic terms and their corresponding spherical fuzzy numbers (Kutlu Gündoğdu and Kahraman 2019c)

Linguistic terms	(μ, v, π)
Absolutely more importance (AMI)	(0.9, 0.1, 0.1)
Very high importance (VHI)	(0.8, 0.2, 0.2)
High importance (HI)	(0.7, 0.3, 0.3)
Slightly more importance (SMI)	(0.6, 0.4, 0.4)
Equally importance (EI)	(0.5, 0.5, 0.5)
Slightly low importance (SLI)	(0.4, 0.6, 0.4)
Low importance (LI)	(0.3, 0.7, 0.3)
Very low importance (VLI)	(0.2, 0.8, 0.2)
Absolutely low importance (ALI)	(0.1, 0.9, 0.1)

$$SWAM_w(\tilde{A}_{S1}, \ldots, \tilde{A}_{Sn}) = w_1\tilde{A}_{S1} + w_2\tilde{A}_{S2} + \cdots + w_n\tilde{A}_{Sn} = \left\{ \begin{array}{l} \left[1 - \prod_{i=1}^{n}(1 - \mu_{\tilde{A}_{Si}}^2)^{w_i}\right]^{1/2}, \prod_{i=1}^{n} v_{\tilde{A}_{Si}}^{w_i}, \\ \left[\prod_{i=1}^{n}(1 - \mu_{\tilde{A}_{Si}}^2)^{w_i} - \prod_{i=1}^{n}(1 - \mu_{\tilde{A}_{Si}}^2 - \pi_{\tilde{A}_{Si}}^2)^{w_i}\right]^{1/2} \end{array} \right\} \tag{17}$$

Step 3: Aggregate the criteria weights.
Step 4: Construct the aggregated spherical fuzzy decision matrix based on the individual decision matrices.

Denote the evaluation values of alternative $X_i (i = 1, 2 \ldots m)$ with respect to criterion $C_j (j = 1, 2 \ldots n)$ by $C_j(X_i) = (\mu_{ij}, v_{ij}, \pi_{ij}) \cdot D = \left(C_j(X_i)\right)_{m \times n}$ is a spherical fuzzy decision matrix. For a MCDM problem with SFSs, the aggregated decision matrix $D = \left(C_j(X_i)\right)_{m \times n}$ should be constructed as in Eq. (18).

$$D = \left(C_j(X_i)\right)_{m \times n} = \begin{pmatrix} (\mu_{11}, v_{11}, \pi_{11}) & (\mu_{12}, v_{12}, \pi_{12}) & \ldots & (\mu_{1n}, v_{1n}, \pi_{1n}) \\ (\mu_{21}, v_{21}, \pi_{21}) & (\mu_{22}, v_{22}, \pi_{22}) & \ldots & (\mu_{2n}, v_{2n}, \pi_{2n}) \\ . & . & & . \\ . & . & & . \\ (\mu_{m1}, v_{m1}, \pi_{m1}) & (\mu_{m2}, v_{m2}, \pi_{m2}) & \ldots & (\mu_{mn}, v_{mn}, \pi_{mn}) \end{pmatrix} \tag{18}$$

Decision makers evaluate the alternatives with respect to the criteria as if they were all benefit criteria such that they assign a lower linguistic term if it is a cost criterion.

Step 5: Construct the aggregated weighted spherical fuzzy decision matrix as given in Eq. (19). After the aggregated weights of the criteria and the aggregated rating of the alternatives are determined, the aggregated weighted spherical fuzzy decision matrix is constructed by utilizing Eq. (4) and then the aggregated weighted spherical fuzzy decision matrix can be defined as follows:

$$D = \left(C_j(X_{iw})\right)_{m\times n} = \begin{pmatrix} (\mu_{11w}, v_{11w}, \pi_{11w}) & (\mu_{12w}, v_{12w}, \pi_{12w}) & \ldots & (\mu_{1nw}, v_{1nw}, \pi_{1nw}) \\ (\mu_{21w}, v_{21w}, \pi_{21w}) & (\mu_{22w}, v_{22w}, \pi_{22w}) & \ldots & (\mu_{2nw}, v_{2nw}, \pi_{2nw}) \\ . & . & & . \\ . & . & & . \\ (\mu_{m1w}, v_{m1w}, \pi_{m1w}) & (\mu_{m2w}, v_{m2w}, \pi_{m2w}) & \ldots & (\mu_{mnw}, v_{mnw}, \pi_{mnw}) \end{pmatrix} \tag{19}$$

Step 6: Defuzzify the aggregated weighted decision matrix by using the proposed score function.

$$Score\left(C_j(X_{iw})\right) = \left(\mu_{ijw} - \frac{\pi_{ijw}}{2}\right)^2 - \left(v_{ijw} - \frac{\pi_{ijw}}{2}\right)^2 \tag{20}$$

Step 7: Determine the Spherical Fuzzy Negative Ideal Solution (SF-NIS), based on the score values obtained in Step 6.

For the SF-NIS:

$$X^- = \left\{ C_j, \min_i < Score(C_j(X_{iw})) > \middle| \; j = 1, 2 \ldots n \right\} \tag{21}$$

or

$$X^- = \left\{ \langle C_1, (\mu_1^-, v_1^-, \pi_1^-) \rangle, \langle C_2, (\mu_2^-, v_2^-, \pi_2^-) \rangle \ldots \langle C_n, (\mu_n^-, v_n^-, \pi_n^-) \rangle \right\} \tag{22}$$

Step 8: Calculate the distances between alternative X_i and the SF-NIS. Normalized Euclidean distance (Szmidt and Kacprzyk 2000) and spherical distance (Kutlu Gundogdu and Kahraman 2019a) equations will be used for this step.

Normalized Euclidean distance to SF-NIS:

$$D_E(\tilde{X}_{ij}, \tilde{X}_j^-) = \sqrt{\frac{1}{2n} \sum_{i=1}^{n} \left(\mu_{\tilde{X}_{ij}} - \mu_{\tilde{X}_j^-}\right)^2 + \left(v_{\tilde{X}_{ij}} - v_{\tilde{X}_j^-}\right)^2 + \left(\pi_{\tilde{X}_{ij}} - \pi_{\tilde{X}_j^-}\right)^2} \quad \forall i \tag{23}$$

Normalized spherical distance to SF-NIS:

$$D_S(\tilde{X}_{ij}, \tilde{X}_j^-) = \frac{2}{n\pi} \sum_{i=1}^{n} \arccos\left(\mu_{\tilde{X}_{ij}} \cdot \mu_{\tilde{X}_j^-} + v_{\tilde{X}_{ij}} \cdot v_{\tilde{X}_j^-} + \pi_{\tilde{X}_{ij}} \cdot \pi_{\tilde{X}_j^-}\right) \; \forall i \tag{24}$$

Step 9: Determine relative assessment matrix (RA) as follows:

$$RA = [s_{ik}]_{n\times n};$$

$$s_{ik} = \left(D_{E_i} - D_{E_k}\right) + \left(\Phi(D_{E_i} - D_{E_k}) \cdot (D_{S_i} - D_{S_k})\right) \quad (25)$$

where $k \in \{1, 2, \ldots, n\}$ and Φ is a threshold function that is defined as follows:

$$\Phi(x) = \begin{Bmatrix} 1\ if\ |x| \geq \alpha \\ 0\ if\ |x| < \alpha \end{Bmatrix} \quad (26)$$

Decision maker can set the threshold parameter of this function. In this study, we use $\alpha = 0.02$.

Step 10: Calculate the appraisal score of each alternative as follows:

$$AS_i = \sum_{k=1}^{n} s_{ik} \quad (27)$$

Step 11: Determine the optimal ranking order of the alternatives and identify the optimal alternative based on the appraisal scores obtained in Step 10. The largest value of AS_i indicates the best alternative.

4 Applications to Livability Index

A person wants to select the most suitable suburban place in Istanbul in Turkey by using livability indices. A linguistic evaluation process will be applied since experts give their assessments to the decision matrix with linguistic terms. The alternatives will be evaluated under some criteria by three decision makers. There are three decision makers are DM1, DM2 and DM3 and their weights are 0.5, 0.25 and 0.25, respectively. In this application we have 7 criteria and 5 alternatives.

The alternatives are:

Alternative 1 (X_1)—Beşiktaş
Alternative 2 (X_2)—Kadıköy
Alternative 3 (X_3)—Sarıyer
Alternative 4 (X_4)—Beykoz
Alternative 5 (X_5)—Arnavutköy.

The criteria are:

Criteria 1 (C_1)—Air Condition (Cost Criterion)
Criteria 2 (C_2)—Available Health Opportunity (Benefit Criterion)
Criteria 3 (C_3)—Traffic Condition (Benefit Criterion)
Criteria 4 (C_4)—Closeness to City Centre (Benefit Criterion)
Criteria 5 (C_5)—Green Area (Benefit Criterion)
Criteria 6 (C_6)—Earthquake Resistance (Benefit Criterion)
Criteria 7 (C_7)—Number of Apartments (Benefit Criterion).

4.1 Application of Spherical Fuzzy CODAS

The weights of these decision makers who have different experience levels are 0.5, 0.25 and 0.25, respectively. All judgments are given in Table 2. These judgments are aggregated using SWAM operator by considering the significance levels of decision makers.

The linguistic importance weights of the criteria assigned by DMs and the spherical fuzzy weight of each criterion obtained by using SWAM operator is presented in Table 3.

The aggregated weighted spherical fuzzy decision matrix is constructed by utilizing Eq. (4) as given in Table 4.

Table 2 Judgments of decision makers

Decision makers	Alternatives	C_1	C_2	C_3	C_4	C_5	C_6	C_7
DM1	X_1	SMI	ALI	LI	SMI	EI	LI	SMI
	X_2	SMII	HI	HI	ALI	SMI	LI	HI
	X_3	VHI	AMI	AMI	VHI	VHI	AMI	HI
	X_4	VHI	VHI	HI	VHI	AMI	HI	HI
	X_5	HI	HI	LI	HI	HI	EI	HI
DM2	X_1	HI	LI	ALI	SMI	SMI	LI	ALI
	X_2	HI	EI	SMI	HI	HI	EI	SMI
	X_3	AMI	AMI	VHI	VHI	VHI	VHI	VHI
	X_4	AMI	AMI	VHI	VHI	VHI	VHI	VHI
	X_5	VHI	AMI	LI	SLI	HI	VHI	VHI
DM3	X_1	HI	LI	ALI	SMI	SMI	LI	ALI
	X_2	HI	EI	SMI	HI	HI	EI	SMI
	X_3	AMI	AMI	VHI	VHI	VHI	VHI	VHI
	X_4	AMI	AMI	VHI	VHI	VHI	VHI	VHI
	X_5	VHI	AMI	LI	SLI	HI	VHI	VHI

Table 3 Importance weights of the main criteria

Criteria	DM1	DM2	DM3	Aggregated criteria weights
C_1	SMI	HI	HI	(0.65, 0.35, 0.35)
C_2	HI	EI	EI	(0.62, 0.39, 0.40)
C_3	AMI	VHI	VHI	(0.86, 0.14, 0.15)
C_4	VHI	VHI	VHI	(0.80, 0.20, 0.20)
C_5	HI	HI	HI	(0.70, 0.30, 0.30)
C_6	EI	VHI	VHI	(0.30, 0.70, 0.30)
C_7	HI	VHI	VHI	(0.54, 0.52, 0.26)

Table 4 Aggregated weighted spherical fuzzy decision matrix

Alternatives	C_1	C_2	C_3	C_4	C_5	C_6	C_7
X_1	(0.43, 0.47, 0.20)	(0.14, 0.83, 0.10)	(0.19, 0.80, 0.06)	(0.48, 0.44, 0.18)	(0.39, 0.52, 0.24)	(0.21, 0.74, 0.13)	(0.34, 0.63, 0.14)
X_2	(0.43, 0.47, 0.20)	(0.38, 0.53, 0.25)	(0.56, 0.37, 0.14)	(0.43, 0.55, 0.09)	(0.46, 0.45, 0.18)	(0.29, 0.64, 0.22)	(0.49, 0.42, 0.16)
X_3	(0.56, 0.37, 0.14)	(0.56, 0.40, 0.17)	(0.74, 0.20, 0.04)	(0.64, 0.28, 0.08)	(0.56, 0.36, 0.12)	(0.60, 0.34, 0.13)	(0.57, 0.34, 0.11)
X_4	(0.56, 0.37, 0.14)	(0.53, 0.41, 0.17)	(0.65, 0.28, 0.08)	(0.64, 0.28, 0.08)	(0.60, 0.33, 0.11)	(0.52, 0.39, 0.16)	(0.57, 0.34, 0.11)
X_5	(0.49, 0.42, 0.16)	(0.51, 0.42, 0.18)	(0.26, 0.71, 0.10)	(0.47, 0.46, 0.14)	(0.49, 0.41, 0.16)	(0.48, 0.44, 0.20)	(0.57, 0.34, 0.11)

Based on Table 4 and Eq. (20), the score function values are calculated as in Table 5. The values in bold represent NIS (negative ideal solution).

Based on Tables 4 and 5, by utilizing Eqs. (21) and (22), the corresponding Spherical Fuzzy Negative Ideal Solution (SF-NIS) is given in Table 6.

In the next step based on Eqs. (23) and (24), we can calculate the distances between alternative X_i and the SF-NIS. They are given in Table 7.

From Table 7, we determine the maximum distance to the SF-NIS. Based on Eqs. (25) and (26), relative assessment matrix (RA) is obtained and presented in Table 8.

Table 5 Score function values

Alternatives	C_1	C_2	C_3	C_4	C_5	C_6	C_7
X_1	**−0.033**	**−0.600**	**−0.564**	0.030	**−0.090**	**−0.429**	**−0.242**
X_2	−0.033	−0.096	0.153	**−0.103**	0.009	−0.252	0.059
X_3	0.153	0.124	0.483	0.304	0.163	0.203	0.184
X_4	0.153	0.094	0.314	0.304	0.225	0.099	0.184
X_5	0.059	0.070	−0.390	0.007	0.056	0.032	0.184

Table 6 Spherical Fuzzy Negative Ideal Solution (SF-NIS)

	C_1	C_2	C_3	C_4	C_5	C_6	C_7
SF-NIS	(0.43, 0.47, 0.20)	(0.14, 0.83, 0.10)	(0.19, 0.80, 0.06)	(0.43, 0.55, 0.09)	(0.39, 0.52, 0.24)	(0.21, 0.74, 0.13)	(0.34, 0.63, 0.14)

Table 7 Euclidean and Spherical distances to negative ideal solution

Alternatives	Euclidean Distance	Spherical Distance
X_1	0.039	3.051
X_2	0.208	3.370
X_3	0.346	3.612
X_4	0.318	3.573
X_5	0.221	3.397

Table 8 Relative assessment matrix (RA)

Alternatives	X_1	X_2	X_3	X_4	X_5
X_1	0.00	−0.49	−0.87	−0.80	−0.53
X_2	0.49	0.00	−0.38	−0.31	−0.01
X_3	0.87	0.38	0.00	0.07	0.34
X_4	0.80	0.31	−0.07	0.00	0.27
X_5	0.53	0.01	−0.34	−0.27	0.00

Table 9 Appraisal scores and ranking

Alternatives	Appraisal Score	Ranking
X_1	−2.684	5
X_2	−0.218	4
X_3	1.655	1
X_4	1.321	2
X_5	−0.074	3

From Table 8, the appraisal scores are determined based on Eq. (27) and given in Table 9.

The appraisal scores indicate that the best alternative is third and overall ranking is $X_3 > X_4 > X_5 > X_2 > X_1$. The best alternative is Sarıyer and the worst place is Beşiktaş.

4.2 Sensitivity Analyses

For the sensitivity analysis, we have changed the spherical distance formula with Zhang and Xu's formula as given in Eq. (28). Distances calculate the distances between alternative X_i and the SF-NIS. They are given in Table 10.

$$d\left(\tilde{X}_{ij}, \tilde{X}_j^-\right) = \frac{1}{2n}\sum_{j=1}^{n}\left(\left|\mu_{\tilde{X}_{ij}}^2 - \mu_{\tilde{X}_j^-}^2\right| + \left|v_{\tilde{X}_{ij}}^2 - v_{\tilde{X}_j^-}^2\right| + \left|\pi_{\tilde{X}_{ij}}^2 - \pi_{\tilde{X}_j^-}^2\right|\right) \quad \forall i \tag{28}$$

Table 10 Euclidean and Zhang and Xu's distances to negative ideal solution

Alternatives	Euclidean Distance	Spherical Distance
X_1	0.039	0.012
X_2	0.208	0.150
X_3	0.346	0.301
X_4	0.318	0.285
X_5	0.221	0.178

Table 11 Relative assessment matrix (RA) based on Euclidean and Zhang and Xu's distances

Alternatives	X_1	X_2	X_3	X_4	X_5
X_1	0.00	−0.31	−0.60	−0.55	−0.35
X_2	0.31	0.00	−0.14	−0.11	−0.01
X_3	0.60	0.14	0.00	0.05	0.25
X_4	0.55	0.11	−0.05	0.00	0.20
X_5	0.35	0.01	−0.25	−0.20	0.00

From Table 10, we determine the maximum distance to the SF-NIS. Based on Eqs. (25) and (26), relative assessment matrix (RA) is obtained and given in Table 11.

From Table 11, the appraisal scores are determined based on Eq. (27) and given in Table 12.

The appraisal scores indicate that the best alternative is X_3 and overall ranking is $X_3 > X_4 > X_2 > X_5 > X_1$. Third and fourth alternatives are replaced each other when compared with the previous result.

We applied sensitivity analysis by changing the threshold value (α) observed similar results of the given decisions. Sensitivity analysis showed that very robust decisions have been obtained from SF-CODAS as given in Fig. 2 which is based on Euclidean and Zhang and Xu's distances. Although the appraisal scores changed, the ranking of alternatives remained the same when the threshold value is greater than 0.125. Alternative 3 is the best alternative in all cases whereas Alternative 1 is always the worst. However, the general ranking is $X_3 > X_4 > X_2 > X_5 > X_1$ when the

Table 12 Appraisal scores and ranking based on Euclidean and Zhang and Xu's distances

Alternatives	Appraisal Score	Ranking
X_1	−1.800	5
X_2	0.044	3
X_3	1.028	1
X_4	0.819	2
X_5	−0.092	4

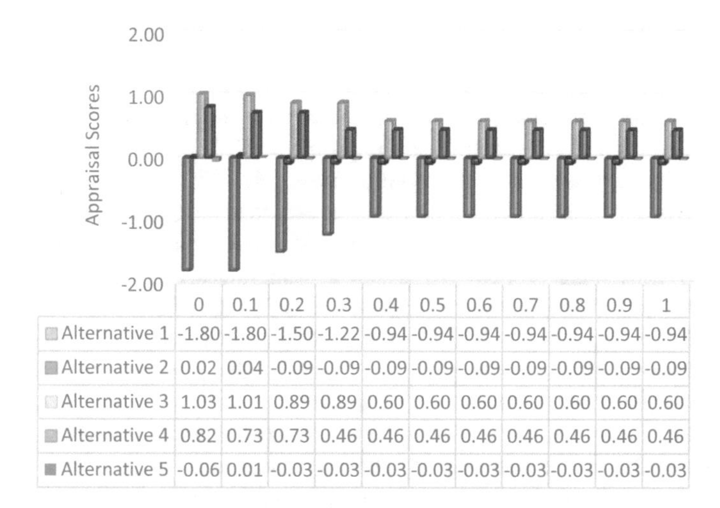

	0	0.1	0.2	0.3	0.4	0.5	0.6	0.7	0.8	0.9	1
Alternative 1	-1.80	-1.80	-1.50	-1.22	-0.94	-0.94	-0.94	-0.94	-0.94	-0.94	-0.94
Alternative 2	0.02	0.04	-0.09	-0.09	-0.09	-0.09	-0.09	-0.09	-0.09	-0.09	-0.09
Alternative 3	1.03	1.01	0.89	0.89	0.60	0.60	0.60	0.60	0.60	0.60	0.60
Alternative 4	0.82	0.73	0.73	0.46	0.46	0.46	0.46	0.46	0.46	0.46	0.46
Alternative 5	-0.06	0.01	-0.03	-0.03	-0.03	-0.03	-0.03	-0.03	-0.03	-0.03	-0.03

Fig. 2 Sensitivity analysis by changing the threshold value (α) based on Euclidean and Zhang and Xu's distances

threshold value is less than 0.125. After this critical value, the general ranking will be $X_3 > X_4 > X_5 > X_2 > X_1$ (Sarıyer > Beykoz > Arnavutköy > Kadıköy > Beşiktaş).

Sensitivity analysis showed that very robust decisions have been obtained from SF-CODAS as given in Fig. 3 which is based on Euclidean and Spherical distances. Alternative 3 is the best alternative in all cases whereas Alternative 1 is always the worst. The general ranking is $X_3 > X_4 > X_5 > X_2 > X_1$ (Sarıyer > Beykoz > Arnavutköy > Kadıköy > Beşiktaş).

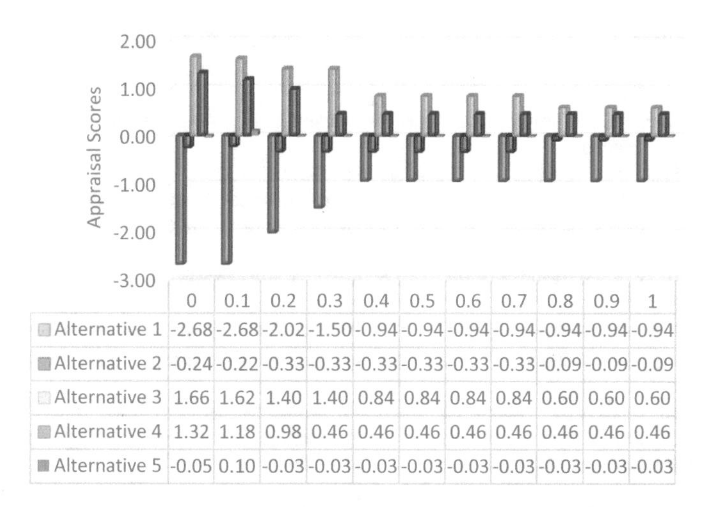

	0	0.1	0.2	0.3	0.4	0.5	0.6	0.7	0.8	0.9	1
Alternative 1	-2.68	-2.68	-2.02	-1.50	-0.94	-0.94	-0.94	-0.94	-0.94	-0.94	-0.94
Alternative 2	-0.24	-0.22	-0.33	-0.33	-0.33	-0.33	-0.33	-0.33	-0.09	-0.09	-0.09
Alternative 3	1.66	1.62	1.40	1.40	0.84	0.84	0.84	0.84	0.60	0.60	0.60
Alternative 4	1.32	1.18	0.98	0.46	0.46	0.46	0.46	0.46	0.46	0.46	0.46
Alternative 5	-0.05	0.10	-0.03	-0.03	-0.03	-0.03	-0.03	-0.03	-0.03	-0.03	-0.03

Fig. 3 Sensitivity analysis by changing the threshold value (α) based on Euclidean and spherical distance

5 Conclusions

It is a tough decision that which suburban place type should be selected since several uncertainties and criteria exist in the problem. SFSs present an excellent tool to capture the vagueness and uncertainty existing in this problem. The literature review shows that SFSs provide a new way to extend MCDM methods under fuzzy environment. Day by day, the number of MCDM papers with SFSs is increasing.

CODAS method is a proposed by Keshavarz Ghorabaee et al. (2016) method for MCDM problems. The CODAS method is extended to its different extensions in the literature. We have extended CODAS method to Spherical fuzzy CODAS in order to select the best suitable suburban place in Istanbul under fuzziness. Spherical fuzzy CODAS method gives significant results and can be used as an alternative MCDM method under fuzziness. We applied Spherical Fuzzy CODAS method and observed successful results for the selection of suburban places in this paper. In order to see the robustness of the method, a sensitivity analysis is shown. For further research, we suggest other extensions of ordinary fuzzy sets to be used such as Pythagorean fuzzy sets, picture fuzzy sets, or sets.

References

Antognelli S, Vizzari M (2017) Landscape liveability spatial assessment integrating ecosystem and urban services with their perceived importance by stakeholders. Ecol Indic 72:703–725

Atanassov K (1986) Intuitionistic fuzzy sets. Fuzzy Sets Syst 20(1):87–96

Giap TK, Thye WW, Aw G (2014) A new approach to measuring the liveability of cities: the Global Liveable Cities Index. World Rev Sci Technol Sustain Dev 11(2):176–196

Grattan-Guiness I (1975) Fuzzy membership mapped onto interval and many-valued quantities. Z Math Logik Grundladen Math 22(1):149–160

Higgs C, Badland H, Simons K, Knibbs LD, Giles-Corti B (2019) The Urban Liveability Index: developing a policy-relevant urban liveability composite measure and evaluating associations with transport mode choice. Int J Health Geograph 18(1)

Hwang CLYK (1981) Multiple criteria decision making. In: Lecture notes in economics and mathematical systems

Jahn K (1975) Intervall-wertige Mengen. Math Nach 68:115–132

Karasan A, Bolturk E, Kahraman C (2019) An integrated methodology using neutrosophic CODAS & fuzzy inference system: assessment of livability index of urban districts. J Intell Fuzzy Syst 36(6):5443–5455

Kashef M (2016) Urban livability across disciplinary and professional boundaries. Front Arch Res 5(2):239–253

Keshavarz Ghorabaee M, Zavadskas EK, Olfat L, Turskis Z (2015) Multi-criteria inventory classification using a new method of Evaluation Based on Distance from Average Solution (EDAS). Informatica 26(3):435–451

Keshavarz Ghorabaee M, Zavadskas EK, Turskis Z, Antucheviciene J (2016) A new combinative distance-based assessment (CODAS) method for multi-criteria decision-making. Econ Comput Econ Cybern Stud Res 50(3):25–44

Kutlu Gündoğdu F, Kahraman C (2019a) Spherical fuzzy sets and spherical fuzzy TOPSIS method. J Intell Fuzzy Syst 36(1):337–352

Kutlu Gündoğdu F, Kahraman C (2019b) A novel fuzzy TOPSIS method using emerging interval-valued spherical fuzzy sets. Eng Appl Artif Intell 85:307–323

Kutlu Gündoğdu F, Kahraman C (2019c) A novel VIKOR method using spherical fuzzy sets and its application to warehouse site selection. J Intell Fuzzy Syst 37(1):1197–1211
Kutlu Gündoğdu F, Kahraman C (2020a) A novel spherical fuzzy analytic hierarchy process and its renewable energy application. Soft Comput 24, 4607–4621
Kutlu Gündoğdu F, Kahraman C (2020b) Spherical fuzzy sets and decision making applications. In: Kahraman C, Cebi S, Cevik Onar S, Oztaysi B, Tolga A, Sari I (eds) Intelligent and fuzzy techniques in big data analytics and decision making. INFUS 2019. Advances in intelligent systems and computing, vol 1029. Springer, Cham
Norouzian-Maleki S, Bell S, Hosseini SB, Faizi M (2015) Developing and testing a framework for the assessment of neighbourhood liveability in two contrasting countries: Iran and Estonia. Ecol Indic 48:263–271
Roy B (1991) The outranking approach and the foundations of ELECTRE methods. Theor Decis 31(1):49–73
Ruth M, Franklin RS (2014) Livability for all? Conceptual limits and practical implications. Appl Geogr 49:18–23
Saaty TL (1980) The analytic hierarchy process: planning, priority setting, resource allocation. McGraw-Hill International Book Company, New York
Saaty TL (1996) Decision making with dependence and feedback: the analytic network process. RWS Publications, Pittsburgh
Sambuc R (1975) Fonctions ϕ-floues. Application l'aide au diagnostic en pathologie thyroidienne. University of Marseille, France
Scerri A (2014) Livability index. In: Michalos AC (ed) Encyclopedia of quality of life and well-being research. Springer, Dordrecht
Smarandache F (1995) Neutrosophic logic and set, mss
Szmidt E, Kacprzyk J (2000) Distances between intuitionistic fuzzy sets. Fuzzy Sets Syst 114(3):505–518
Tan KG, Nie T, Baek S (2016) Empirical assessment on the liveability of cities in the Greater China Region. Compet Rev 26(1):2–24
Torra V (2010) Hesitant fuzzy sets. Int J Intell Syst 25:529–539
Wang J, Su M, Chen B, Chen S, Liang C (2011) A comparative study of Beijing and three global cities: a perspective on urban livability. Front Earth Sci 5(3):323–329
Yager RR (2013) Pythagorean fuzzy subsets. In: 2013 joint IFSA world congress and NAFIPS annual meeting (IFSA/NAFIPS): 57–61
Yin L, Yin Y (2009) Research on assessment of city livability based on principle component analysis—taking Shandong Province for example. In: 2009 international conference on management and service science. IEEE:1–4
Zadeh L (1965) Fuzzy sets. Inf Control 8(3):338–353
Zadeh L (1975) The concept of a linguistic variable and its application to approximate reasoning-1. Inf Sci 8:199–249
Zanella A, Camanho AS, Dias TG (2015) The assessment of cities' livability integrating human wellbeing and environmental impact. Ann Oper Res 226(1):695–726
Zavadskas EK, Turskis Z, Kildienė S (2014) State of art surveys of overviews on MCDM/MADM methods. Technol Econ Dev Econ 20:165–179

Interval-Valued Spherical Fuzzy MULTIMOORA Method and Its Application to Industry 4.0

Serhat Aydın and Fatma Kutlu Gündoğdu

Abstract Industry 4.0 represents the fourth industrial revolution in production; manufacturing and industry. Industry 4.0 connects new technologies providing flexibility in manufacturing where the conditions change rapidly. Companies need to use Industry 4.0 technologies to take their place in a competitive environment. Evaluating companies' performance based on Industry 4.0 is a complex multi criteria problem including both quantitative and qualitative factors. In this chapter, we propose a novel fuzzy MULTIMOORA method based on interval-valued spherical fuzzy sets to evaluate companies that are using Industry 4.0 technologies. Five different alternatives are evaluated according to seven conflicting criteria by three decision makers. The results are compared with single-valued spherical fuzzy MULTIMOORA method. The results indicate that Industry 4.0 performance evaluating problem can be tackled by using the proposed methodology effectively.

1 Introduction

Decision making is the selection process of the best alternative among available alternatives. To realize this process decision makers need collecting information from various sources, model the process, analyze the results and make a final decision about alternatives.

Multicriteria decision making (MCDM) is defined as prioritizing, ranking, or selecting a set of alternatives according to independent and conflicting criteria (Fenton and Wang 2006). MCDM includes both multi attribute decision making (MADM) and multi objective decision making (MODM) problems. MADM refers to making

S. Aydın (✉) · F. Kutlu Gündoğdu
Industrial Engineering Department, Turkish Air Force Academy, National Defence University, 34149 Istanbul, Turkey
e-mail: saydin3@hho.edu.tr

F. Kutlu Gündoğdu
e-mail: fatmakutlugundogdu@gmail.com

© Springer Nature Switzerland AG 2021

C. Kahraman and F. Kutlu Gündoğdu (eds.), *Decision Making with Spherical Fuzzy Sets*, Studies in Fuzziness and Soft Computing 392,
https://doi.org/10.1007/978-3-030-45461-6_13

preference decision among available alternatives according to the multiple and conflicting attributes. MODM methods optimize many conflicting criteria which can be both beneficial and non-beneficial and it is known as the continuous type of the MCDM. The aim of the MODM methods is achieving multiple objectives which are conflicting with each other (Lu et al. 2007).

MADM can be also used in many field organizations to solve the problems of project selection, human resource evaluation, facility planning, production material selection, etc. There are many methods to overcome MADM problems; Analytic Hierarchy Process (AHP) (Saaty 1980), Analytic Network Process (Saaty 2001), Technique for Order of Preference by Similarity to Ideal Solution (TOPSIS) (Hwang and Yoon 1981), Ornasition, Rangement Et Synthese De donnees Relationnelles (ORESTE) (Roubens 1982), VIKOR (Opricovic 1998), Elimination Et choix Traduisant la Realite (ELECTRE) (Roy 1991), Complex Proportional Assessment (COPRAS) (Zavadskas et al. 1994), Evaluation based on distance from average solution (EDAS) (Ghorabaee et al. 2015).

MOORA method is one of the effective MCDM methods developed by Brauers and Zavadskas (2006). MOORA method analyzes the complete throughout for each alternative as a variance between the sum of cost criteria and benefit criteria. The method involves two different methods as the ratio system method and the reference point method. Brauers and Zavadskas (2011) added the full multiplicative form into steps of MOORA and they named the new method under the name of MULTIMOORA. MULTIMOORA method has many advantages as follows; it doesn't require complex mathematical calculations and using full multiplicative form doesn't require normalization procedure.

In many real-life problems, it is not easy to get crisp data from decision makers. Therefore, decision makers usually need to take into account the vagueness of the information. The fuzzy set theory developed, by Zadeh (1965) for solving problems in which description of activities and observations are imprecise, vague, and uncertain. MADM problems with uncertain information can be modeled by using fuzzy set theory in decision making process.

In recent years, fuzzy set theory has been extended to new types. Zadeh (1975) introduced Type-2 fuzzy sets in 1975. Atanassov (1986) developed intuitionistic fuzzy sets which include the membership value as well as the non-membership value, and satisfy that the sum of membership and non-membership functions is less than or equal to 1. Yager (1986) proposed fuzzy multi-sets theory. The theory represents multiple occurrences of a subject item with degrees of relevance. Smarandache (1998) proposed neutrosophic sets (NS) by adding an independent indeterminacy-membership function to intuitionistic fuzzy sets. Garibaldi and Ozen (2007) proposed Nonstationary fuzzy sets that are evolved to handle uncertainties in fuzzy systems with reducing the computational burden of Type-2 sets. Torra (2010) developed hesitant fuzzy sets which handle the situations where a set of values are possible for the membership of a single element. Yager (2013) proposed Pythagorean fuzzy sets which satisfy the squared sum of membership and non-membership degrees is less than or equal to 1. Finally, Kutlu Gündoğdu and Kahraman (2019a) proposed Spherical fuzzy sets that satisfy the squared sum of membership, non-membership and hesitancy degree is less than or equal to 1. Spherical fuzzy sets are based on the spherical fuzzy distances and hesitancy of a decision maker can be defined independently from membership and non-membership degrees, unlike the Pythagorean fuzzy sets. Spherical fuzzy sets satisfy the condition as follows: $0 \leq \mu_{\tilde{A}}^2(u) + v_{\tilde{A}}^2(u) + \pi_{\tilde{A}}^2(u) \leq 1$. At this point, decision makers can identify their hesitancy judgment about alternatives according to the determined criteria more easily in decision making process.

It is hard to reach crisp data to model real life problems; therefore fuzzy sets and MCDM methods have been combined in many research fields. Decision makers also require fuzzy set theory in MULTIMOORA method due to uncertain and insufficient information about alternatives. In this study, an interval-valued Spherical fuzzy sets and MULTIMOORA are combined and an application is given for evaluation Industry 4.0 companies. The originality of the paper is, it is the first study of combined with interval-valued Spherical fuzzy set theory and MULTIMOORA method in the literature.

The rest of the chapter is organized as follows: the literature review on spherical fuzzy sets and fuzzy extension of MULTIMOORA has given in Sect. 2. Preliminaries of spherical fuzzy sets are given in Sect. 3. The Methodology is given in Sect. 4. A case study and sensitivity analysis are given in Sect. 5. Finally, conclusions are presented in Sect. 6.

2 Literature Review

In the literature review section we focus on researches about spherical fuzzy sets and fuzzy extensions of MULTIMOORA method. Spherical fuzzy sets are recently developed one of the extensions of ordinary fuzzy sets; therefore, there are a little number of applications in the literature.

The spherical fuzzy sets (SFS) have been recently introduced by Kutlu Gündoğdu and Kahraman (2019a) and they proposed a spherical fuzzy TOPSIS method. Ashraf et al. (2019) defined spherical fuzzy sets and spherical t-norms and t-conorms. Kutlu Gündogdu et al. (2020) proposed Spherical fuzzy VIKOR method and gave an application to waste management. Kutlu Gündoğdu and Kahraman (2019b) proposed a novel approach based on spherical fuzzy VIKOR and they gave an application to warehouse selection. Kutlu Gündoğdu and Kahraman (2019c) developed spherical fuzzy AHP for renewable energy application.

After a glance at spherical fuzzy sets, we focus on the fuzzy extension of MULTIMORA method. Brauers et al. proposed an ordinary fuzzy MULTIMOORA method to rank the EU Member States according to their performance (2011). Baležentis and Zeng (2013) developed a fuzzy MULTIMOORA method using interval-valued trapezoidal fuzzy number. Datta et al. (2013) developed a grey MULTIMOORA method in robot selection. Li (2014) proposed a hesitant MULTIMOORA method for software selection. Baležentis and Zeng (2013) improved an extension

MULTIMOORA method based on interval-valued fuzzy numbers and they gave a numerical example of human resource management. Zavadskas et al. (2015) proposed a MULTIMOORA method by using interval-valued intuitionistic fuzzy sets. They presented a case study on civil engineering problem and used the method for ranking alternatives. Hafezalkotob and Hafezalkotob (2016) presented the ordinary fuzzy entropy-weighted MULTIMOORA method for a practical materials selection problem related to automotive industry. Stanujkic et al. (2017) proposed a MULTIMOORA method using a neutrosophic set they assessed residential house projects. Gou et al. (2017) evaluated the implementation status of haze controlling measures by using hesitant fuzzy MULTIMOORA. Tian et al. (2017) presented an implemented MULTIMOORA method based on simplified neutrosophic sets. Zhao et al. (2017) proposed a novel approach based on interval-valued intuitionistic fuzzy sets and MULTIMOORA method to handle the uncertainty and vagueness from FMEA team members' subjective assessments. Zavadskas et al. (2017) developed single valued neutrosophic extension of MULTIMOORA method to select the material of single-family residential houses. Chen et al. (2018) proposed a novel method based on the MULTIMOORA with triangular fuzzy numbers. Aydın (2018) presented the augmented reality (AR) eyeglass selection problem based on neutrosophic MULTIMOORA method. Dorfeshan et al. (2018) proposed interval-valued type-2 fuzzy MULTIMMORA method, and they showed the applicability of the proposed method on construction project evaluation. Liang et al. proposed an extended MULTIMOORA based on linguistic neutrosophic numbers for selecting the optimal mining method (2019). Zhang et al. (2019) proposed intuitionistic fuzzy MULTIMOORA method and they gave the empirical example considers the case of energy storage technology selection.

Based on the literature review there is no research on interval-valued spherical fuzzy MULTIMOORA method. Therefore, we developed a new MULTIMOORA method based on interval-valued spherical fuzzy sets and give an illustrative example in order to demonstrate the applicability of developed methodology.

3 Spherical Fuzzy Sets: Preliminaries

The novel concept of Spherical Fuzzy Sets provides a larger preference domain for decision makers to assign membership degrees since the squared sum of the spherical parameters is allowed to be at most 1.0. DMs can define their hesitancy information independently under spherical fuzzy environment. Spherical fuzzy Sets (SFS) are a generalization of Pythagorean Fuzzy Sets, picture fuzzy sets and neutrosophic sets. In the following, definition of SFS is presented:

Definition 3.1 Single valued Spherical Fuzzy Sets (SFS) $\tilde{A}_S$ of the universe of discourse U is given by

$$\tilde{A}_S = \left\{ \langle u, (\mu_{\tilde{A}_S}(u), \nu_{\tilde{A}_S}(u), \pi_{\tilde{A}_S}(u)) \middle| u \in U \right\} \quad (1)$$

where

$$\mu_{\tilde{A}_S}(u) : U \rightarrow [0, 1],\ v_{\tilde{A}_S}(u) : U \rightarrow [0, 1],\ \pi_{\tilde{A}_S}(u) : U \rightarrow [0, 1]$$

and

$$0 \le \mu^2_{\tilde{A}_S}(u) + v^2_{\tilde{A}_S}(u) + \pi^2_{\tilde{A}_S}(u) \le 1 \quad \forall u \in U \tag{2}$$

For each u, the numbers $\mu_{\tilde{A}_S}(u)$, $v_{\tilde{A}_S}(u)$ and $\pi_{\tilde{A}_S}(u)$ are the degree of membership, non-membership and hesitancy of u to $\tilde{A}_S$, respectively.

Definition 3.2 Basic operators of Single-valued SFS;

1. $$\tilde{A}_S \oplus \tilde{B}_S = \left\{ \begin{array}{l} \left(\mu^2_{\tilde{A}_S} + \mu^2_{\tilde{B}_S} - \mu^2_{\tilde{A}_S}\mu^2_{\tilde{B}_S}\right)^{1/2},\ v_{\tilde{A}_S}v_{\tilde{B}_S}, \\ \left(\left(1 - \mu^2_{\tilde{B}_S}\right)\pi^2_{\tilde{A}_S} + \left(1 - \mu^2_{\tilde{A}_S}\right)\pi^2_{\tilde{B}_S} - \pi^2_{\tilde{A}_S}\pi^2_{\tilde{B}_S}\right)^{1/2} \end{array} \right\} \tag{3}$$

2. $$\tilde{A}_S \otimes \tilde{B}_S = \left\{ \begin{array}{l} \mu_{\tilde{A}_S}\mu_{\tilde{B}_S},\ \left(v^2_{\tilde{A}_S} + v^2_{\tilde{B}_S} - v^2_{\tilde{A}_S}v^2_{\tilde{B}_S}\right)^{1/2}, \\ \left(\left(1 - v^2_{\tilde{B}_S}\right)\pi^2_{\tilde{A}_S} + \left(1 - v^2_{\tilde{A}_S}\right)\pi^2_{\tilde{B}_S} - \pi^2_{\tilde{A}_S}\pi^2_{\tilde{B}_S}\right)^{1/2} \end{array} \right\} \tag{4}$$

3. $$\lambda \cdot \tilde{A}_S = \left\{ \begin{array}{l} \left(1 - \left(1 - \mu^2_{\tilde{A}_S}\right)^{\lambda}\right)^{1/2},\ v^{\lambda}_{\tilde{A}_S}, \\ \left(\left(1 - \mu^2_{\tilde{A}_S}\right)^{\lambda} - \left(1 - \mu^2_{\tilde{A}_S} - \pi^2_{\tilde{A}_S}\right)^{\lambda}\right)^{1/2} \end{array} \right\} \quad for\ \lambda > 0 \tag{5}$$

4. $$\tilde{A}^{\lambda}_S = \left\{ \begin{array}{l} \mu^{\lambda}_{\tilde{A}_S},\ \left(1 - \left(1 - v^2_{\tilde{A}_S}\right)^{\lambda}\right)^{1/2}, \\ \left(\left(1 - v^2_{\tilde{A}_S}\right)^{\lambda} - \left(1 - v^2_{\tilde{A}_S} - \pi^2_{\tilde{A}_S}\right)^{\lambda}\right)^{1/2} \end{array} \right\} \quad for\ \lambda > 0 \tag{6}$$

Definition 3.3 Single-valued Spherical Weighted Arithmetic Mean (SWAM) with respect to, $w = (w_1, w_2 \ldots, w_n)$; $w_i \in [0, 1]$; $\sum_{i=1}^{n} w_i = 1$, SWAM is defined as;

$$SWAM_w(\tilde{A}_{S1}, \ldots, \tilde{A}_{Sn}) = w_1\tilde{A}_{S1} + w_2\tilde{A}_{S2} + \cdots + w_n\tilde{A}_{Sn}$$
$$= \left\{ \begin{array}{l} \left[1 - \prod_{i=1}^{n} (1 - \mu_{\tilde{A}_{Si}}^2)^{w_i}\right]^{1/2}, \prod_{i=1}^{n} v_{\tilde{A}_{Si}}^{w_i}, \\ \left[\prod_{i=1}^{n} (1 - \mu_{\tilde{A}_{Si}}^2)^{w_i} - \prod_{i=1}^{n} (1 - \mu_{\tilde{A}_{Si}}^2 - \pi_{\tilde{A}_{Si}}^2)^{w_i}\right]^{1/2} \end{array} \right\} \quad (7)$$

Definition 3.4 Single-valued Spherical Weighted Geometric Mean (SWGM) with respect to, $w = (w_1, w_2 \ldots, w_n)$; $w_i \in [0, 1]$; $\sum_{i=1}^{n} w_i = 1$, SWGM is defined as;

$$SWGM_w(\tilde{A}_1, \ldots, \tilde{A}_n) = \tilde{A}_{S1}^{w_1} + \tilde{A}_{S2}^{w_2} + \cdots + \tilde{A}_{Sn}^{w_n}$$
$$= \left\{ \begin{array}{l} \prod_{i=1}^{n} \mu_{\tilde{A}_{Si}}^{w_i}, \left[1 - \prod_{i=1}^{n} (1 - v_{\tilde{A}_{Si}}^2)^{w_i}\right]^{1/2}, \\ \left[\prod_{i=1}^{n} (1 - v_{\tilde{A}_{Si}}^2)^{w_i} - \prod_{i=1}^{n} (1 - v_{\tilde{A}_{Si}}^2 - \pi_{\tilde{A}_{Si}}^2)^{w_i}\right]^{1/2} \end{array} \right\} \quad (8)$$

Definition 3.5 Score functions and Accuracy functions of sorting SFS are defined by;

$$Score\left(\tilde{A}_S\right) = (\mu_{\tilde{A}_S} - \pi_{\tilde{A}_S}/2)^2 - (v_{\tilde{A}_S} - \pi_{\tilde{A}_S}/2)^2 \quad (9)$$

$$Accuracy\left(\tilde{A}_S\right) = \mu_{\tilde{A}_S}^2 + v_{\tilde{A}_S}^2 + \pi_{\tilde{A}_S}^2 \quad (10)$$

Note that: $\tilde{A}_S < \tilde{B}_S$ if and only if

1. $Score(\tilde{A}_S) < Score(\tilde{B}_S)$ or
2. $Score(\tilde{A}_S) = Score(\tilde{B}_S)$ and $Accuracy(\tilde{A}_S) < Accuracy(\tilde{B}_S)$.

In this section, we give the definition of Interval-valued spherical fuzzy sets (IVSFS) and summarize arithmetic operations, aggregation and defuzzification operations.

Definition 3.6 An Interval-Valued Spherical Fuzzy Set $\tilde{A}_S$ of the universe of discourse U is defined as in Eq. (11).

$$\tilde{A}_S = \left\{\left\langle u, \left(\left[\mu_{\tilde{A}_S}^L(u), \mu_{\tilde{A}_S}^U(u)\right], \left[v_{\tilde{A}_S}^L(u), v_{\tilde{A}_S}^U(u)\right], \left[\pi_{\tilde{A}_S}^L(u), \pi_{\tilde{A}_S}^U(u)\right]\right) \middle| u \in U\right.\right\} \quad (11)$$

where $0 \leq \mu^L_{\tilde{A}_S}(u) \leq \mu^U_{\tilde{A}_S}(u) \leq 1$, $0 \leq v^L_{\tilde{A}_S}(u) \leq v^U_{\tilde{A}_S}(u) \leq 1$ and $0 \leq \left(\mu^U_{\tilde{A}_S}(u)\right)^2 + \left(v^U_{\tilde{A}_S}(u)\right)^2 + \left(\pi^U_{\tilde{A}_S}(u)\right)^2 \leq 1$. For each $u \in U$, $\mu^U_{\tilde{A}_S}(u)$, $v^U_{\tilde{A}_S}(u)\, and\, \pi^U_{\tilde{A}_S}(u)$ are the upper degrees of membership, non-membership and hesitancy of u to $\tilde{A}_S$, respectively. For an IVSFS $\tilde{A}_S$, the pair $\left\langle\left[\mu^L_{\tilde{A}_S}(u), \mu^U_{\tilde{A}_S}(u)\right], \left[v^L_{\tilde{A}_S}(u), v^U_{\tilde{A}_S}(u)\right], \left[\pi^L_{\tilde{A}_S}(u), \pi^U_{\tilde{A}_S}(u)\right]\right\rangle$ is called an interval-valued spherical fuzzy number. For convenience, the pair $\left\langle\left[\mu^L_{\tilde{A}_S}(u), \mu^U_{\tilde{A}_S}(u)\right], \left[v^L_{\tilde{A}_S}(u), v^U_{\tilde{A}_S}(u)\right], \left[\pi^L_{\tilde{A}_S}(u), \pi^U_{\tilde{A}_S}(u)\right]\right\rangle$ is denoted by $\tilde{\alpha} = \langle [a, b], [c, d], [e, f]\rangle$ where $[a, b] \subset [0, 1]$, $[c, d] \subset [0, 1]$, $[e, f] \subset [0, 1]$ and $b^2 + d^2 + f^2 \leq 1$.

Definition 3.7 Let $\tilde{\alpha} = \langle [a, b], [c, d], [e, f]\rangle, \tilde{\alpha}_1 = \langle [a_1, b_1], [c_1, d_1], [e_1, f_1]\rangle$, and $\tilde{\alpha}_2 = \langle [a_2, b_2], [c_2, d_2], [e_2, f_2]\rangle$ be IVSFS then;

$$\tilde{\alpha}_1 \cup \tilde{\alpha}_2 = \left\{\begin{array}{l}[\max\{a_1, a_2\}, \max\{b_1, b_2\}],\ [\min\{c_1, c_2\}, \min\{d_1, d_2\}], \\ [\min\{e_1, e_2\}, \min\{f_1, f_2\}]\end{array}\right\} \tag{12}$$

$$\tilde{\alpha}_1 \cap \tilde{\alpha}_2 = \left\{\begin{array}{l}[\min\{a_1, a_2\}, \min\{b_1, b_2\}], [\max\{c_1, c_2\}, \max\{d_1, d_2\}], \\ [\min\{e_1, e_2\}, \min\{f_1, f_2\}]\end{array}\right\} \tag{13}$$

$$\tilde{\alpha}_1 \oplus \tilde{\alpha}_2 = \left\{\begin{array}{l}\left[\left((a_1)^2 + (a_2)^2 - (a_1)^2(a_2)^2\right)^{1/2}, \left((b_1)^2 + (b_2)^2 - (b_1)^2(b_2)^2\right)^{1/2}\right], \\ [c_1c_2, d_1d_2], \left[\begin{array}{l}\left(\left(1 - (a_2)^2\right)(e_1)^2 + \left(1 - (a_1)^2\right)(e_2)^2 - (e_1)^2(e_2)^2\right)^{1/2}, \\ \left(\left(1 - (b_2)^2\right)(f_1)^2 + \left(1 - (b_1)^2\right)(f_2)^2 - (f_1)^2(f_2)^2\right)^{1/2}\end{array}\right]\end{array}\right\} \tag{14}$$

$$\tilde{\alpha}_1 \otimes \tilde{\alpha}_2 = \left\{\begin{array}{l}[a_1a_2, b_1b_2], \left[\left((c_1)^2 + (c_2)^2 - (c_1)^2(c_2)^2\right)^{1/2}, \left((d_1)^2 + (d_2)^2 - (d_1)^2(d_2)^2\right)^{1/2}\right], \\ \left[\begin{array}{l}\left(\left(1 - (c_2)^2\right)(e_1)^2 + \left(1 - (c_1)^2\right)(e_2)^2 - (e_1)^2(e_2)^2\right)^{1/2}, \\ \left(\left(1 - (d_2)^2\right)(f_1)^2 + \left(1 - (d_1)^2\right)(f_2)^2 - (f_1)^2(f_2)^2\right)^{1/2}\end{array}\right]\end{array}\right\} \tag{15}$$

$$\lambda \cdot \tilde{\alpha} = \left\{\begin{array}{l}\left[\left(1 - \left(1 - a^2\right)^\lambda\right)^{1/2}, \left(1 - \left(1 - b^2\right)^\lambda\right)^{1/2}\right], \left[c^\lambda, d^\lambda\right], \\ \left[\left(\left(1 - a^2\right)^\lambda - \left(1 - a^2 - e^2\right)^\lambda\right)^{1/2}, \left(\left(1 - b^2\right)^\lambda - \left(1 - b^2 - f^2\right)^\lambda\right)^{1/2}\right]\end{array}\right\} \quad for\ \lambda > 0 \tag{16}$$

$$\tilde{\alpha}^\lambda = \left\{\begin{array}{l}\left[a^\lambda, b^\lambda\right], \left[\left(1 - \left(1 - c^2\right)^\lambda\right)^{1/2}, \left(1 - \left(1 - d^2\right)^\lambda\right)^{1/2}\right], \\ \left[\left(\left(1 - c^2\right)^\lambda - \left(1 - c^2 - e^2\right)^\lambda\right)^{1/2}, \left(\left(1 - d^2\right)^\lambda - \left(1 - d^2 - f^2\right)^\lambda\right)^{1/2}\right]\end{array}\right\} \quad for\ \lambda > 0 \tag{17}$$

Definition 3.8 Let $\lambda, \lambda_1, \lambda_2 \geq 0$, then

1. $\tilde{\alpha}_1 \oplus \tilde{\alpha}_2 = \tilde{\alpha}_2 \oplus \tilde{\alpha}_1$ (18)

2. $\tilde{\alpha}_1 \otimes \tilde{\alpha}_2 = \tilde{\alpha}_2 \otimes \tilde{\alpha}_1$ (19)

3. $\lambda(\tilde{\alpha}_1 \oplus \tilde{\alpha}_2) = \lambda \cdot \tilde{\alpha}_1 \oplus \lambda \cdot \tilde{\alpha}_2$ (20)

4. $(\tilde{\alpha}_1 \otimes \tilde{\alpha}_2)^{\lambda} = \tilde{\alpha}_1^{\lambda} \otimes \tilde{\alpha}_2^{\lambda}$ (21)

5. $\lambda_1 \cdot \tilde{\alpha} \oplus \lambda_2 \cdot \tilde{\alpha} = (\lambda_1 + \lambda_2) \cdot \tilde{\alpha}$ (22)

6. $\tilde{\alpha}^{\lambda_1} \otimes \tilde{\alpha}^{\lambda_2} = \tilde{\alpha}^{\lambda_1 + \lambda_2}$ (23)

Definition 3.9 Let $\tilde{\alpha}_j = \langle [a_j, b_j], [c_j, d_j], [e_j, f_j] \rangle$ be a collection of Interval-valued Spherical Weighted Arithmetic Mean (IVSWAM) with respect to, $w_j = (w_1, w_2 \ldots, w_n)$; $w_j \in [0, 1]$ *and* $\sum_{j=1}^{n} w_j = 1$, IVSWAM is defined as;

$$IVSWAM_w(\tilde{\alpha}_1, \tilde{\alpha}_2 \ldots, \tilde{\alpha}_n) = w_1 \cdot \tilde{\alpha}_1 \oplus w_2 \cdot \tilde{\alpha}_2 \oplus \cdots \oplus w_n \cdot \tilde{\alpha}_n$$
$$= \left\{ \begin{array}{l} \left[\left(1 - \prod_{j=1}^{n} (1 - a_j^2)^{w_j}\right)^{1/2}, \left(1 - \prod_{j=1}^{n} (1 - b_j^2)^{w_j}\right)^{1/2} \right], \left[\prod_{j=1}^{n} c_j^{w_j}, \prod_{j=1}^{n} d_j^{w_j} \right], \\ \left[\left(\prod_{j=1}^{n} (1 - a_j^2)^{w_j} - \prod_{j=1}^{n} (1 - a_j^2 - e_j^2)^{w_j} \right)^{1/2}, \left(\prod_{j=1}^{n} (1 - b_j^2)^{w_j} - \prod_{j=1}^{n} (1 - b_j^2 - f_j^2)^{w_j} \right)^{1/2} \right] \end{array} \right\} \tag{24}$$

Definition 3.10 Let $\tilde{\alpha}_j = \langle [a_j, b_j], [c_j, d_j], [e_j, f_j] \rangle$ be a collection of Interval-valued Spherical Geometric Mean (IVSWGM) with respect to,$w_j = (w_1, w_2 \ldots, w_n)$; $w_j \in [0, 1]$ *and* $\sum_{j=1}^{n} w_j = 1$, IVSWGM is defined as;

$$IVSWGM_w(\tilde{\alpha}_1, \tilde{\alpha}_2 \ldots, \tilde{\alpha}_n) = \tilde{\alpha}_1^{w_1} \otimes \tilde{\alpha}_2^{w_2} \otimes \cdots \otimes \tilde{\alpha}_n^{w_n}$$
$$= \left\{ \begin{array}{l} \left[\prod_{j=1}^{n} a_j^{w_j}, \prod_{j=1}^{n} b_j^{w_j} \right], \left[\left(1 - \prod_{j=1}^{n} (1 - c_j^2)^{w_j}\right)^{1/2}, \left(1 - \prod_{j=1}^{n} (1 - d_j^2)^{w_j}\right)^{1/2} \right], \\ \left[\left(\prod_{j=1}^{n} (1 - c_j^2)^{w_j} - \prod_{j=1}^{n} (1 - c_j^2 - e_j^2)^{w_j} \right)^{1/2}, \left(\prod_{j=1}^{n} (1 - d_j^2)^{w_j} - \prod_{j=1}^{n} (1 - d_j^2 - f_j^2)^{w_j} \right)^{1/2} \right] \end{array} \right\} \tag{25}$$

Definition 3.11 The score function of IVSFS number α is defined as;

$$Score(\tilde{\alpha}) = S(\tilde{\alpha}) = \frac{a^2 + b^2 - c^2 - d^2 - \left(e/2\right)^2 - \left(f/2\right)^2}{2} \tag{26}$$

The accuracy function of IVSFS number α is defined as;

$$Accuracy(\tilde{\alpha}) = H(\tilde{\alpha}) = \frac{a^2 + b^2 + c^2 + d^2 + e^2 + f^2}{2} \tag{27}$$

Note that: $\tilde{\alpha}_1 < \tilde{\alpha}_2$ if and only if

1. $S(\tilde{\alpha}_1) < S(\tilde{\alpha}_2)$ or
2. $S(\tilde{\alpha}_1) = S(\tilde{\alpha}_2)$ and $H(\tilde{\alpha}_1) < H(\tilde{\alpha}_2)$.

4 Extension of MULTIMOORA with Spherical Fuzzy Sets

The proposed single-valued spherical fuzzy (SF-MULTIMOORA) and interval-valued spherical fuzzy MULTIMOORA (IVSF-MULTIMOORA) methods are composed of several steps as given in this section. Before giving these steps, we present the flow chart of the SF-MULTIMOORA and IVSF-MULTIMOORA methods in Fig. 1 in order to make it easily understandable.

4.1 Extension of MULTIMOORA with Single-Valued Spherical Fuzzy Sets

The general MULTIMOORA method includes three approaches: the Ratio System (RS) approach, the Reference Point (RP) approach and the Full Multiplicative Form (FMF) approach. In our methodology, first four steps are the same as all three approaches.

A MCDM problem can be expressed as a decision matrix whose elements indicate the evaluation values of all alternatives with respect to each criterion under spherical fuzzy environment. Let $X = \{x_1, x_2, \ldots x_m\}$ $(m \geq 2)$ be a discrete set of m feasible alternatives and $C = \{C_1, C_2, \ldots C_n\}$ be a finite set of criteria, and $w = \{w_1, w_2, \ldots w_n\}$ be the weights vector of all criteria which satisfies $0 \leq w_j \leq 1$ and $\sum_{j=1}^{n} w_j = 1$. The following methodology is proposed by Kutlu Gündoğdu in 2019.

Step 1: Let DMs fill in the criteria evaluation matrix and their individual decision matrix using the linguistic terms given in Table 1.

Step 2: Aggregate the judgments of each decision maker (DM) using Single-valued Spherical Weighted Arithmetic Mean (SWAM).

$$SWAM_w(\tilde{A}_{S1}, \ldots, \tilde{A}_{Sn}) = w_1\tilde{A}_{S1} + w_2\tilde{A}_{S2} + \cdots + w_n\tilde{A}_{Sn}$$
$$= \left\{ \begin{array}{l} \left[1 - \prod_{i=1}^{n}(1 - \mu_{\tilde{A}_{Si}}^2)^{w_i}\right]^{1/2}, \prod_{i=1}^{n} v_{\tilde{A}_{Si}}^{w_i}, \\ \left[\prod_{i=1}^{n}(1 - \mu_{\tilde{A}_{Si}}^2)^{w_i} - \prod_{i=1}^{n}(1 - \mu_{\tilde{A}_{Si}}^2 - \pi_{\tilde{A}_{Si}}^2)^{w_i}\right]^{1/2} \end{array} \right\} \quad (28)$$

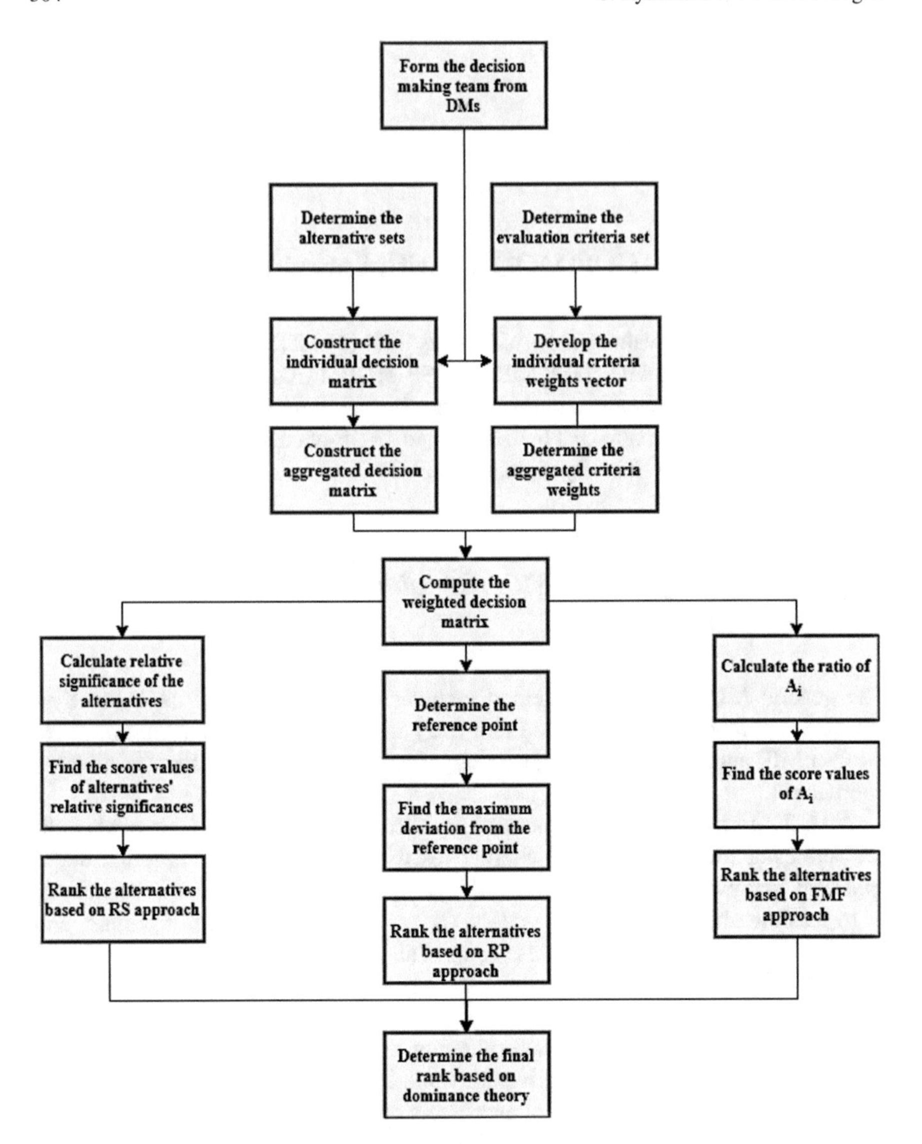

Fig. 1 Flow chart of the proposed SF and IVSF-MULTIMOORA methods

Step 3: Aggregate the criteria weights.
Step 4: Construct the aggregated spherical fuzzy decision matrix based on the individual decision matrices.

Denote the evaluation values of alternative $X_i(i = 1, 2 \ldots m)$ with respect to criterion $C_j(j = 1, 2 \ldots n)$ by $C_j(X_i) = (\mu_{ij}, \upsilon_{ij}, \pi_{ij}) \cdot D = \left(C_j(X_i)\right)_{m \times n}$ is a spherical fuzzy decision matrix. For a MCDM problem with SFSs, the aggregated decision matrix $D = \left(C_j(X_i)\right)_{m \times n}$ should be constructed as in Eq. (29).

Table 1 Linguistic terms and their corresponding spherical fuzzy numbers (Kutlu Gündoğdu and Kahraman 2019a)

Linguistic terms	(μ, v, π)
Absolutely more importance (AMI)	(0.9, 0.1, 0.1)
Very high importance (VHI)	(0.8, 0.2, 0.2)
High importance (HI)	(0.7, 0.3, 0.3)
Slightly more importance (SMI)	(0.6, 0.4, 0.4)
Equally importance (EI)	(0.5, 0.5, 0.5)
Slightly low importance (SLI)	(0.4, 0.6, 0.4)
Low importance (LI)	(0.3, 0.7, 0.3)
Very low importance (VLI)	(0.2, 0.8, 0.2)
Absolutely low importance (ALI)	(0.1, 0.9, 0.1)

$$D = \left(C_j(X_i)\right)_{m \times n} = \begin{pmatrix} (\mu_{11}, v_{11}, \pi_{11}) & (\mu_{12}, v_{12}, \pi_{12}) & \ldots & (\mu_{1n}, v_{1n}, \pi_{1n}) \\ (\mu_{21}, v_{21}, \pi_{21}) & (\mu_{22}, v_{22}, \pi_{22}) & \ldots & (\mu_{2n}, v_{2n}, \pi_{2n}) \\ . & . & & . \\ . & . & & . \\ (\mu_{m1}, v_{m1}, \pi_{m1}) & (\mu_{m2}, v_{m2}, \pi_{m2}) & \ldots & (\mu_{mn}, v_{mn}, \pi_{mn}) \end{pmatrix} \tag{29}$$

Decision makers evaluate the alternatives with respect to the criteria as if they were all benefit criteria such that they assign a lower linguistic term if it is a cost criterion. For example, for an alternative which has a higher first cost between two investment alternatives will be assigned a lower linguistic term than *equally importance* (EI).
Step 5: Constitute the aggregated weighted spherical fuzzy decision matrix. After the aggregated weights of the criteria and the aggregated rating of the alternatives are determined, the aggregated weighted spherical fuzzy decision matrix is constructed by utilizing Eq. (4) and then the aggregated weighted spherical fuzzy decision matrix can be defined as follows:

$$D = \left(C_j(X_{iw})\right)_{m \times n} = \begin{pmatrix} (\mu_{11w}, v_{11w}, \pi_{11w}) & (\mu_{12w}, v_{12w}, \pi_{12w}) & \ldots & (\mu_{1nw}, v_{1nw}, \pi_{1nw}) \\ (\mu_{21w}, v_{21w}, \pi_{21w}) & (\mu_{22w}, v_{22w}, \pi_{22w}) & \ldots & (\mu_{2nw}, v_{2nw}, \pi_{2nw}) \\ . & . & & . \\ . & . & & . \\ (\mu_{m1w}, v_{m1w}, \pi_{m1w}) & (\mu_{m2w}, v_{m2w}, \pi_{m2w}) & \ldots & (\mu_{mnw}, v_{mnw}, \pi_{mnw}) \end{pmatrix} \tag{30}$$

1st Approach: The Ratio System (RS) Approach
Step 1.6: Calculate Y_i^+ based on single-valued SWAM operator, as follows:

$$\tilde{Y}_i^+ = \left\{ \begin{aligned} & \left[1 - \prod_{j=1}^{n} (1 - \mu_{A_{Sj}}^2)^{1/n}\right]^{1/2}, \prod_{j=1}^{n} v_{A_{Sj}}^{1/n}, \\ & \left[\prod_{j=1}^{n} (1 - \mu_{A_{Sj}}^2)^{1/n} - \prod_{j=1}^{n} (1 - \mu_{A_{Sj}}^2 - \pi_{A_{Sj}}^2)^{1/n}\right]^{1/2} \end{aligned} \right\} \tag{31}$$

Defuzzify the $\tilde{Y}_i^+$ by using Eq. (32)

$$y_i^+ = Score\left(\tilde{Y}_i^+\right) = \left(2\mu_{\tilde{Y}_i^+} - \pi_{\tilde{Y}_i^+}/2\right)^2 - \left(v_{\tilde{Y}_i^+} - \pi_{\tilde{Y}_i^+}/2\right)^2 \tag{32}$$

Step 1.7: Rank the alternatives based on y_i^+ in descending order and the alternative with the highest values of y_i^+ is considered the best ranked.
2nd Approach: The Reference Point (RP) Approach
Step 2.6: Determine the reference point based on the score function values of

$$X_j^* = \left\{ C_j, \max_i < Score(C_j(X_{iw})) > \middle| j = 1, 2 \ldots n \right\} \tag{33}$$

or

$$X_j^* = \left\{\langle C_1, (\mu_1^*, v_1^*, \pi_1^*)\rangle, \ldots \langle C_n, (\mu_n^*, v_n^*, \pi_n^*)\rangle\right\} \tag{34}$$

Step 2.7: Calculate the distance from each alternative to all the coordinates of the reference point by using Eq. (35).

$$d\left(\tilde{X}_{ij}, \tilde{X}_j^*\right) = \frac{1}{2}\left(\left|\mu_{\tilde{X}_{ij}}^2 - \mu_{\tilde{X}_j^*}^2\right| + \left|v_{\tilde{X}_{ij}}^2 - v_{\tilde{X}_j^*}^2\right| + \left|\pi_{\tilde{X}_{ij}}^2 - \pi_{\tilde{X}_j^*}^2\right|\right) \tag{35}$$

Step 2.8: Rank the alternatives. Final rank is given based on the deviation from the reference point according to Eq. (36). In this approach, the compared alternatives are ranked based on $\max_j d\left(\tilde{X}_{ij}, \tilde{X}_j^*\right)$ in ascending order and the alternative with the lowest value is considered the best ranked.

$$\min_i \left\{ \max_j d\left(\tilde{X}_{ij}, \tilde{X}_j^*\right) \right\} \tag{36}$$

3rd Approach: The Fuzzy Full Multiplicative Form (FMF) Approach
Step 3.6: Calculate the ratio of $\tilde{A}_i$ for maximized decision criteria by using Eq. (37).

$$\tilde{A}_i = \left\{ \begin{array}{c} \prod_{j=1}^{m} \mu_{\tilde{A}_{ij}}, \left[1 - \prod_{j=1}^{m} (1 - v_{\tilde{A}_{ij}}^2) \right]^{1/2}, \\ \left[\prod_{j=1}^{m} (1 - v_{\tilde{A}_{ij}}^2) - \prod_{j=1}^{m} (1 - v_{\tilde{A}_{ij}}^2 - \pi_{\tilde{A}_{ij}}^2) \right]^{1/2} \end{array} \right\} \tag{37}$$

Step 3.7: Defuzzify the $\tilde{A}_i$ by using Eq. (38).

$$a_i = Score\left(\tilde{A}_i\right) = \left(2\mu_{\tilde{A}_i} - \pi_{\tilde{A}_i}/2\right)^2 - \left(v_{\tilde{A}_i} - \pi_{\tilde{A}_i}/2\right)^2 \tag{38}$$

Step 3.8: Rank the alternatives based on a_i values in descending order and the alternative with the highest value of a_i is accepted the best ranked.

The considered alternatives are ranked based on all three approaches and the final ranking order and the final decision is made according to the theory of dominance. In other words, the alternative with the highest number of appearances in the first position on all ranking lists is the best-ranked alternative.

4.2 *Extension of MULTIMOORA with Interval-Valued Spherical Fuzzy Sets*

In this chapter, steps of the novel interval-valued spherical MULTIMOORA method are given as follows;

Step 1: Let DMs fill in the criteria evaluation matrix and their individual decision matrix using the interval-valued linguistic terms given in Table 2.

Step 2: Aggregate the judgments of each decision maker (DM) using Interval-valued Spherical Weighted Arithmetic Mean (IVSWAM).

$$IVSWAM_w(\tilde{\alpha}_1, \tilde{\alpha}_2 \ldots, \tilde{\alpha}_n) = w_1 \cdot \tilde{\alpha}_1 \oplus w_2 \cdot \tilde{\alpha}_2 \oplus \cdots \oplus w_n \cdot \tilde{\alpha}_n$$
$$= \left\{ \begin{array}{c} \left[\left(1 - \prod_{j=1}^{n} (1 - a_j^2)^{w_j} \right)^{1/2}, \left(1 - \prod_{j=1}^{n} (1 - b_j^2)^{w_j} \right)^{1/2} \right], \left[\prod_{j=1}^{n} c_j^{w_j}, \prod_{j=1}^{n} d_j^{w_j} \right], \\ \left[\left(\prod_{j=1}^{n} (1 - a_j^2)^{w_j} - \prod_{j=1}^{n} (1 - a_j^2 - e_j^2)^{w_j} \right)^{1/2}, \left(\prod_{j=1}^{n} (1 - b_j^2)^{w_j} - \prod_{j=1}^{n} (1 - b_j^2 - f_j^2)^{w_j} \right)^{1/2} \right] \end{array} \right\} \tag{39}$$

Step 3: Aggregate the criteria weights based on IVSWAM operator.

Step 4: Construct the aggregated interval-valued spherical fuzzy decision matrix.

Let $C_j(X_i) = \left(\left[\mu_{ij}^L(u), \mu_{ij}^U(u)\right], \left[v_{ij}^L(u), v_{ij}^U(u)\right], \left[\pi_{ij}^L(u), \pi_{ij}^U(u)\right]\right)$ denote the assessments of each alternative $X_i (i = 1, 2 \ldots m)$ with respect to each criterion $C_j (j = 1, 2 \ldots n) \cdot D = \left(C_j(X_i)\right)_{m \times n}$ is a spherical fuzzy decision matrix. Then,

Table 2 Interval-valued spherical fuzzy linguistic terms

Linguistic terms	$\left(\left[\mu^L_{\tilde{A}_S}(u), \mu^U_{\tilde{A}_S}(u)\right], \left[v^L_{\tilde{A}_S}(u), v^U_{\tilde{A}_S}(u)\right], \left[\pi^L_{\tilde{A}_S}(u), \pi^U_{\tilde{A}_S}(u)\right]\right)$ *or* $([a, b], [c, d], [e, f])$
Absolutely more importance (AMI)	([0.85, 0.95], [0.10, 0.15], [0.05, 0.15])
Very high importance (VHI)	([0.75, 0.85], [0.15, 0.20], [0.15, 0.20])
High importance (HI)	([0.65, 0.75], [0.20, 0.25], [0.20, 0.25])
Slightly more importance (SMI)	([0.55, 0.65], [0.25, 0.30], [0.25, 0.30])
Equally importance (EI)	([0.50, 0.55], [0.45, 0.55], [0.30, 0.40])
Slightly low importance (SLI)	([0.25, 0.30], [0.55, 0.65], [0.25, 0.30])
Low importance (LI)	([0.20, 0.25], [0.65, 0.75], [0.20, 0.25])
Very low importance (VLI)	([0.15, 0.20], [0.75, 0.85], [0.15, 0.20])
Absolutely low importance (ALI)	([0.10, 0.15], [0.85, 0.95], [0.05, 0.15])

the decision matrix $D = \left(C_j(X_i)\right)_{m \times n}$ is constructed as in (40).

$$D = \left(C_j(X_i)\right)_{m\times n} = \begin{pmatrix} \left(\left[\mu^L_{11}(u), \mu^U_{11}(u)\right], \left[v^L_{11}(u), v^U_{11}(u)\right], \left[\pi^L_{11}(u), \pi^U_{11}(u)\right]\right) & \cdots & \left(\left[\mu^L_{1n}(u), \mu^U_{1n}(u)\right], \left[v^L_{1n}(u), v^U_{1n}(u)\right], \left[\pi^L_{1n}(u), \pi^U_{1n}(u)\right]\right) \\ \left(\left[\mu^L_{22}(u), \mu^U_{22}(u)\right], \left[v^L_{22}(u), v^U_{22}(u)\right], \left[\pi^L_{22}(u), \pi^U_{22}(u)\right]\right) & \cdots & \left(\left[\mu^L_{2n}(u), \mu^U_{2n}(u)\right], \left[v^L_{2n}(u), v^U_{2n}(u)\right], \left[\pi^L_{2n}(u), \pi^U_{2n}(u)\right]\right) \\ \vdots & & \vdots \\ \left(\left[\mu^L_{m1}(u), \mu^U_{m1}(u)\right], \left[v^L_{m1}(u), v^U_{m1}(u)\right], \left[\pi^L_{m1}(u), \pi^U_{m1}(u)\right]\right) & \cdots & \left[\mu^L_{mn}(u), \mu^U_{mn}(u)\right], \left[v^L_{mn}(u), v^U_{mn}(u)\right], \left[\pi^L_{mn}(u), \pi^U_{mn}(u)\right] \end{pmatrix} \tag{40}$$

Step 5: Constitute the aggregated weighted interval-valued spherical fuzzy decision matrix. After the aggregated weights of the criteria and the aggregated rating of the alternatives are determined, the aggregated weighted interval-valued spherical fuzzy decision matrix is constructed by utilizing Eq. (15) and then the aggregated weighted interval-valued spherical fuzzy decision matrix can be defined as follows:

$$D = \left(C_j(X_{iw})\right)_{m\times n} = \begin{pmatrix} \left(\left[\mu^L_{11w}(u), \mu^U_{11w}(u)\right], \left[v^L_{11w}(u), v^U_{11w}(u)\right], \left[\pi^L_{11w}(u), \pi^U_{11w}(u)\right]\right) & \cdots & \left(\left[\mu^L_{1nw}(u), \mu^U_{1nw}(u)\right], \left[v^L_{1nw}(u), v^U_{1nw}(u)\right], \left[\pi^L_{1nw}(u), \pi^U_{1nw}(u)\right]\right) \\ \left(\left[\mu^L_{22w}(u), \mu^U_{22w}(u)\right], \left[v^L_{22w}(u), v^U_{22w}(u)\right], \left[\pi^L_{22w}(u), \pi^U_{22w}(u)\right]\right) & \cdots & \left(\left[\mu^L_{2nw}(u), \mu^U_{2nw}(u)\right], \left[v^L_{2nw}(u), v^U_{2nw}(u)\right], \left[\pi^L_{2nw}(u), \pi^U_{2nw}(u)\right]\right) \\ \vdots & & \vdots \\ \left(\left[\mu^L_{m1w}(u), \mu^U_{m1w}(u)\right], \left[v^L_{m1w}(u), v^U_{m1w}(u)\right], \left[\pi^L_{m1w}(u), \pi^U_{m1w}(u)\right]\right) & \cdots & \left[\mu^L_{mnw}(u), \mu^U_{mnw}(u)\right], \left[v^L_{mnw}(u), v^U_{mnw}(u)\right], \left[\pi^L_{mnw}(u), \pi^U_{mnw}(u)\right] \end{pmatrix} \tag{41}$$

1st Approach: The Ratio System (RS) Approach for IVSF-MULTIMOORA

Step 1.6: Calculate Y_i^+ by using the IVSWAM operator, as follows:

$$\tilde{Y}_i^+ = \left\{ \begin{array}{l} \left[\left(1 - \prod_{j=1}^{n} (1 - a_j^2)^{1/n}\right)^{1/2}, \left(1 - \prod_{j=1}^{n} (1 - b_j^2)^{1/n}\right)^{1/2} \right], \left[\prod_{j=1}^{n} c_j^{1/n}, \prod_{j=1}^{n} d_j^{1/n} \right], \\ \left[\left(\prod_{j=1}^{n} (1 - a_j^2)^{1/n} - \prod_{j=1}^{n} (1 - a_j^2 - e_j^2)^{1/n} \right)^{1/2}, \left(\prod_{j=1}^{n} (1 - b_j^2)^{1/n} - \prod_{j=1}^{n} (1 - b_j^2 - f_j^2)^{1/n} \right)^{1/2} \right] \end{array} \right\} \tag{42}$$

Defuzzify the $\tilde{Y}_i^+$ based on score function is given in Eq. (43)

$$y_i^+ = Score\left(\tilde{Y}_i^+\right) = \frac{a^2 + b^2 - c^2 - d^2 - \left(e/2\right)^2 - \left(f/2\right)^2}{2} \tag{43}$$

Step 1.7: Rank the alternatives based on y_i^+ in descending order and the alternative with the highest values of y_i^+ is considered the best ranked.

2nd Approach: The Reference Point (RP) Approach for IVSF-MULTIMOORA

Step 2.6: Determine the interval-valued reference point according to the score function values;

$$X^* = \left\{ C_j, \max_i < S(C_j(X_{iw})) > \middle| j = 1, 2 \ldots n \right\} \tag{44}$$

or

$$\begin{aligned} X^* = \{ & \langle C_1, ([\mu_1^{L*}, \mu_1^{U*}], [v_1^{L*}, v_1^{U*}], [\pi_1^{L*}, \pi_1^{U*}]) \rangle, \ldots \\ & \langle C_n, ([\mu_n^{L*}, \mu_n^{U*}], [v_n^{L*}, v_n^{U*}], [\pi_n^{L*}, \pi_n^{U*}]) \rangle \} \end{aligned} \tag{45}$$

Step 2.7: Calculate the distance from each alternative to all the coordinates of the reference point by using Eq. (46).

$$\begin{aligned} d\left(X_{ij}, X_j^*\right) = \frac{1}{4} \Big(& \left| \left(\mu_{ij}^L\right)^2 - \left(\mu_j^*\right)^2 \right| + \left| \left(\mu_{ij}^U\right)^2 - \left(\mu_j^*\right)^2 \right| + \left| \left(v_{ij}^L\right)^2 - \left(v_j^*\right)^2 \right| + \left| \left(v_{ij}^U\right)^2 - \left(v_j^*\right)^2 \right| \\ & + \left| \left(\pi_{ij}^L\right)^2 - \left(\pi_j^*\right)^2 \right| + \left| \left(\pi_{ij}^U\right)^2 - \left(\pi_j^*\right)^2 \right| \Big) \ldots \forall i \end{aligned} \tag{46}$$

Step 2.8: Rank the alternatives. Final rank is given based on the deviation from the reference point according to Eq. (47). In this approach, the compared alternatives are ranked based on $\max_j d\left(\tilde{X}_{ij}, \tilde{X}_j^*\right)$ in ascending order and the alternative with the lowest value is considered the best ranked.

$$\min_i \left\{ \max_j d\left(\tilde{X}_{ij}, \tilde{X}_j^*\right) \right\} \tag{47}$$

3rd Approach: The Fuzzy Full Multiplicative Form (FMF) Approach for IVSF-MULTIMOORA Method

Step 3.6: Calculate the value of $\tilde{A}_i$ for maximized decision criteria by using IVSWGM operator

$$\tilde{A}_i = \left\{ \begin{array}{l} \left[\prod_{j=1}^{n} a_j^{1/n}, \prod_{j=1}^{n} b_j^{1/n} \right], \left[\left(1 - \prod_{j=1}^{n} (1 - c_j^2)^{1/n} \right)^{1/2}, \left(1 - \prod_{j=1}^{n} (1 - d_j^2)^{1/n} \right)^{1/2} \right], \\ \left[\left(\prod_{j=1}^{n} (1 - c_j^2)^{1/n} - \prod_{j=1}^{n} (1 - c_j^2 - e_j^2)^{1/n} \right)^{1/2}, \left(\prod_{j=1}^{n} (1 - d_j^2)^{1/n} - \prod_{j=1}^{n} (1 - d_j^2 - f_j^2)^{1/n} \right)^{1/2} \right] \end{array} \right\} \tag{48}$$

Step 3.7: Defuzzify the $\tilde{A}_i$ by using Eq. (49).

$$a_i = Score\left(\tilde{A}_i\right) = \frac{a^2 + b^2 - c^2 - d^2 - \left(e/2\right)^2 - \left(f/2\right)^2}{2} \tag{49}$$

Step 3.8: Rank the alternatives based on a_i values in descending order. The last decision ranking order is made based on the theory of dominance.

5 Application to Industry 4.0

Industry 4.0 refers to a new stage in the industrial revolution that focuses deeply on interconnectivity, automation, artificial intelligence, cloud computing, and real-time data as shown in Fig. 2. Companies need to be investing in technology and solutions that help you develop and optimize your own operation. To stay competitive, they have to have the systems and processes in place to allow you to provide the better level of service to their customers that they could be getting from a company. Companies also that invest in modern, innovative Industry 4.0 technologies are better positioned to attract and retain new workers and customers. However, these investments should be proper to company's targets. It is important that the company gives priority to investments because of the cost management.

In our descriptive example, we evaluate five companies/alternatives in terms of adaptation to Industry 4.0 applications. We assumed that criteria are cloud computing, virtual reality, 3D printing, automation, data analytics, artificial intelligence, and internet of things as criteria. There are three decision makers who expert in Industry 4.0. The weights of these decision makers who have different experience levels are 0.5, 0.25 and 0.25, respectively.

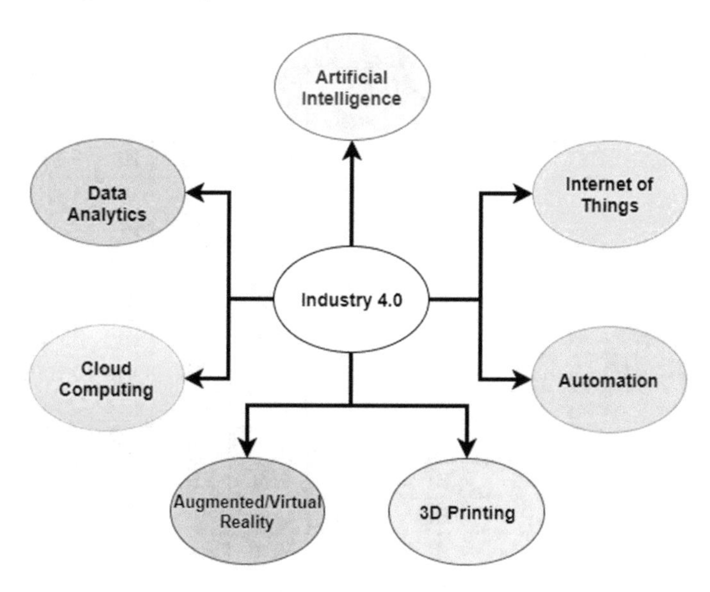

Fig. 2 Industry 4.0 implementations

5.1 Single-Valued Spherical Fuzzy MULTIMOORA Application

In this section, single-valued spherical fuzzy MULTIMOORA method is applied to Industry 4.0. All assessments are given in Table 3. These judgments are aggregated using single-valued SWAM operator by considering the significance levels of decision makers.

The linguistic importance weights of the criteria assigned by DMs and the spherical fuzzy weight of each criterion obtained by using SWAM operator is presented in Table 4.

The aggregated weighted spherical fuzzy decision matrix is constructed by utilizing Eq. (4) as given in Table 5.

In the ratio system (RS) approach, $\tilde{Y}_i^+$ values can be calculated by using Eqs. (31) and (32) as given in Table 6. Besides, $Score\left(\tilde{Y}_i^+\right)$ values and ranking of the alternatives are given in Table 6.

Reference points in the weighted decision matrix can be calculated based on Eqs. (33) and (34) and the corresponding spherical fuzzy sets are given in Table 7.

Modified Zhang and Xu's distances from each alternative to all the coordinates of the reference point by using Eq. (35) are presented as given in Table 8 (Ejegwa 2019). The rank based on 2nd approach is also given in Table 8.

For the sensitivity analysis, the distance formula can be changed with the Euclidean distance. While distance values may be changed, the rank will be remained same as given in Table 9.

Table 3 Judgments of decision makers

Decision makers	Alternatives	C1	C2	C3	C4	C5	C6	C7
DM1	X_1	VHI	VHI	HI	VHI	AMI	HI	HI
	X_2	HI	HI	LI	SMI	HI	EI	HI
	X_3	VHI	AMI	AMI	VHI	VHI	AMI	HI
	X_4	SMI	ALI	LI	SMI	EI	LI	SMI
	X_5	SMI	HI	HI	ALI	SMI	LI	HI
DM2	X_1	VHI	AMI	VHI	AMI	HI	VHI	VHI
	X_2	HI	EI	SMI	HI	HI	EI	SMI
	X_3	AMI	AMI	VHI	VHI	VHI	VHI	VHI
	X_4	ALI	SLI	SMI	EI	SLI	LI	SMI
	X_5	HI	LI	ALI	SMI	SMI	LI	ALI
DM3	X_1	HI	VHI	AMI	VHI	HI	HI	VHI
	X_2	HI	SMI	LI	HI	EI	SLI	HI
	X_3	HI	VHI	VHI	AMI	AMI	VHI	VHI
	X_4	SMI	SMI	HI	ALI	LI	LI	SMI
	X_5	HI	ALI	HI	HI	HI	SLI	SMI

Table 4 Importance weights of the criteria

Criteria	DM1	DM2	DM3	Aggregated criteria weights
C1	VHI	VHI	HI	(0.80, 0.20, 0.20)
C2	HI	EI	SMI	(0.62, 0.39, 0.40)
C3	AMI	VHI	VHI	(0.86, 0.14, 0.15)
C4	SMI	EI	ALI	(0.55, 0.45, 0.45)
C5	SMI	SMI	HI	(0.60, 0.40, 0.40)
C6	LI	LI	SLI	(0.30, 0.70, 0.30)
C7	HI	ALI	SMI	(0.54, 0.52, 0.26)

Based on the spherical fuzzy Full Multiplicative Form (FMF) approach, we can calculate the ratio of $\tilde{A}_i$ and $Score\left(\tilde{A}_i\right)$ values by using Eqs. (37) and (38) as given in Table 10, respectively.

As it can be seen from Tables 6, 8, and 10, first and third approaches, integrated in the single-valued SF-MULTIMOORA, have resulted in the same ranking orders. However, in the second approach, third and fourth alternatives replaced each other. Based on the dominance theory, Alternative 3 is the best alternative and alternative 1 is in the second rank.

Table 5 Aggregated weighted spherical fuzzy decision matrix

Alternatives	C1	C2	C3	C4	C5	C6	C7
X_1	(0.64, 0.28, 0.08)	(0.53, 0.41, 0.17)	(0.65, 0.28, 0.08)	(0.48, 0.46, 0.21)	(0.50, 0.43, 0.18)	(0.23, 0.72, 0.11)	(0.41, 0.56, 0.11)
X_2	(0.56, 0.36, 0.12)	(0.38, 0.53, 0.25)	(0.42, 0.54, 0.15)	(0.36, 0.54, 0.25)	(0.42, 0.49, 0.21)	(0.15, 0.79, 0.17)	(0.35, 0.60, 0.14)
X_3	(0.69, 0.24, 0.06)	(0.56, 0.40, 0.17)	(0.74, 0.20, 0.04)	(0.44, 0.48, 0.22)	(0.48, 0.44, 0.18)	(0.26, 0.71, 0.10)	(0.41, 0.56, 0.11)
X_4	(0.36, 0.62, 0.13)	(0.18, 0.78, 0.14)	(0.42, 0.54, 0.15)	(0.31, 0.60, 0.28)	(0.27, 0.64, 0.26)	(0.09, 0.86, 0.08)	(0.32, 0.62, 0.16)
X_5	(0.52, 0.39, 0.15)	(0.35, 0.57, 0.19)	(0.46, 0.53, 0.08)	(0.25, 0.70, 0.20)	(0.36, 0.54, 0.24)	(0.09, 0.86, 0.08)	(0.29, 0.68, 0.10)

Table 6 The ranking of the alternatives based on the RS approach

Alternatives	$\tilde{Y}_i^+$ *values*	$Score\left(\tilde{Y}_i^+\right)$	Ranking
X_1	(0.52, 0.43, 0.14)	0.80	2
X_2	(0.40, 0.54, 0.19)	0.30	3
X_3	(0.55, 0.40, 0.13)	0.96	1
X_4	(0.30, 0.66, 0.19)	−0.06	5
X_5	(0.36, 0.60, 0.16)	0.15	4

Table 7 Reference points

Reference point	C1	C2	C3	C4	C5	C6	C7
X_j^*	(0.69, 0.24, 0.06)	(0.56, 0.40, 0.17)	(0.74, 0.20, 0.04)	(0.48, 0.46, 0.21)	(0.50, 0.43, 0.18)	(0.26, 0.71, 0.10)	(0.41, 0.56, 0.11)

5.2 Interval-Valued Spherical Fuzzy MULTIMOORA Application

In this chapter, single-valued spherical fuzzy MULTIMOORA and interval-valued spherical fuzzy MULTIMOORA methods are compared. Table 2 presents interval-valued spherical fuzzy linguistic scale, which we use for the comparison purpose.

In this comparison, the same decision matrix and same criteria evaluations are used as given in Tables 3 and 4. These judgments are aggregated using IVSWAM

Table 8 The ranking of the alternatives based on Zhang and Xu's distance

X_i	$d\left(\tilde{X}_{i1}, \tilde{X}_j^*\right)$	$d\left(\tilde{X}_{2j}, \tilde{X}_j^*\right)$	$d\left(\tilde{X}_{3j}, \tilde{X}_j^*\right)$	$d\left(\tilde{X}_{4j}, \tilde{X}_j^*\right)$	$d\left(\tilde{X}_{5j}, \tilde{X}_j^*\right)$	$d\left(\tilde{X}_{6j}, \tilde{X}_j^*\right)$	$d\left(\tilde{X}_{7j}, \tilde{X}_j^*\right)$	$\max_j d\left(\tilde{X}_{ij}, \tilde{X}_j^*\right)$	Rank
X_1	0.04	0.02	0.08	0.00	0.00	0.02	0.00	0.08	2
X_2	0.12	0.16	0.32	0.10	0.07	0.09	0.05	0.32	4
X_3	0.00	0.00	0.00	0.02	0.01	0.00	0.00	0.02	1
X_4	0.34	0.37	0.32	0.16	0.22	0.15	0.07	0.37	5
X_5	0.16	0.18	0.29	0.22	0.13	0.15	0.12	0.29	3

Table 9 The ranking of the alternatives based on Euclidean distance

X_i	$d\left(\tilde{X}_{i1}, \tilde{X}_j^*\right)$	$d\left(\tilde{X}_{2j}, \tilde{X}_j^*\right)$	$d\left(\tilde{X}_{3j}, \tilde{X}_j^*\right)$	$d\left(\tilde{X}_{4j}, \tilde{X}_j^*\right)$	$d\left(\tilde{X}_{5j}, \tilde{X}_j^*\right)$	$d\left(\tilde{X}_{6j}, \tilde{X}_j^*\right)$	$d\left(\tilde{X}_{7j}, \tilde{X}_j^*\right)$	$\max_j d\left(\tilde{X}_{ij}, \tilde{X}_j^*\right)$	Rank
X_1	0.04	0.02	0.09	0.00	0.00	0.03	0.00	0.09	2
X_2	0.13	0.16	0.34	0.10	0.07	0.11	0.05	0.34	4
X_3	0.00	0.00	0.00	0.03	0.01	0.00	0.00	0.03	1
X_4	0.36	0.38	0.34	0.16	0.22	0.16	0.08	0.38	5
X_5	0.17	0.19	0.31	0.23	0.13	0.16	0.12	0.31	3

Table 10 The ranking of the alternatives based on FMF approach

Alternatives	$\tilde{A}_i$ *values*	$Score\left(\tilde{A}_i\right)$	Ranking
X_1	(0.47, 0.49, 0.14)	0.57	2
X_2	(0.36, 0.58, 0.19)	0.15	3
X_3	(0.48, 0.48, 0.14)	0.64	1
X_4	(0.25, 0.69, 0.18)	−0.19	5
X_5	(0.30, 0.65, 0.16)	−0.07	4

operator by considering the importance levels of decision makers. The aggregated weighted interval-valued spherical fuzzy decision matrix is constructed by utilizing Eq. (15) as given in Table 11.

Table 11 Aggregated weighted interval-valued spherical fuzzy decision matrix

Alternatives	C1	C2	C3	C4	C5	C6	C7
X_1	([0.53, 0.69], [0.23, 0.30], [0.22, 0.29])	([0.46, 0.61], [0.29, 0.36], [0.26, 0.34])	([0.60, 0.78], [0.20, 0.27], [0.18, 0.27])	([0.37, 0.49], [0.41, 0.49], [0.27, 0.35])	([0.45, 0.60], [0.27, 0.34], [0.26, 0.33])	([0.15, 0.21], [0.64, 0.74], [0.25, 0.30])	([0.39, 0.53], [0.35, 0.42], [0.26, 0.32])
X_2	([0.47, 0.62], [0.25, 0.32], [0.25, 0.31])	([0.35, 0.47], [0.36, 0.44], [0.32, 0.39])	([0.27, 0.38], [0.52, 0.61], [0.23, 0.30])	([0.29, 0.39], [0.44, 0.52], [0.31, 0.38])	([0.36, 0.48], [0.34, 0.41], [0.31, 0.38])	([0.10, 0.13], [0.72, 0.82], [0.29, 0.33])	([0.35, 0.48], [0.36, 0.44], [0.28, 0.34])
X_3	([0.56, 0.72], [0.22, 0.29], [0.21, 0.28])	([0.49, 0.64], [0.28, 0.35], [0.25, 0.33])	([0.65, 0.84], [0.17, 0.24], [0.14, 0.24])	([0.37, 0.49], [0.41, 0.49], [0.27, 0.35])	([0.45, 0.60], [0.27, 0.34], [0.26, 0.33])	([0.17, 0.24], [0.63, 0.73], [0.23, 0.28])	([0.39, 0.53], [0.35, 0.42], [0.26, 0.32])
X_4	([0.36, 0.48], [0.37, 0.44], [0.27, 0.33])	([0.19, 0.27], [0.60, 0.69], [0.27, 0.32])	([0.38, 0.51], [0.40, 0.48], [0.24, 0.31])	([0.22, 0.31], [0.53, 0.62], [0.31, 0.38])	([0.23, 0.30], [0.56, 0.66], [0.33, 0.40])	([0.04, 0.07], [0.80, 0.89], [0.22, 0.24])	([0.30, 0.42], [0.39, 0.46], [0.30, 0.37])
X_5	([0.44, 0.59], [0.27, 0.34], [0.27, 0.33])	([0.30, 0.41], [0.45, 0.54], [0.28, 0.35])	([0.47, 0.62], [0.31, 0.38], [0.21, 0.29])	([0.22, 0.30], [0.56, 0.65], [0.27, 0.34])	([0.33, 0.46], [0.33, 0.40], [0.32, 0.38])	([0.05, 0.07], [0.79, 0.88], [0.23, 0.25])	([0.31, 0.43], [0.42, 0.50], [0.27, 0.34])

Table 12 The ranking of the alternatives based on the RS approach

Alternatives	$\tilde{Y}_i^+$ *values*	$Score\left(\tilde{Y}_i^+\right)$	Ranking
X_1	([0.45, 0.60], [0.32, 0.40], [0.24, 0.32])	0.11	2
X_2	([0.33, 0.45], [0.41, 0.49], [0.29, 0.35])	−0.10	3
X_3	([0.47, 0.63], [0.31, 0.39], [0.23, 0.31])	0.15	1
X_4	([0.27, 0.37], [0.50, 0.59], [0.28, 0.34])	−0.24	5
X_5	([0.33, 0.46], [0.42, 0.50], [0.27, 0.33])	−0.10	4

Table 13 Reference points

Reference point	C1	C2	C3	C4	C5	C6	C7
X_j^*	([0.56, 0.72], [0.22, 0.29], [0.21, 0.28])	([0.49, 0.64], [0.28, 0.35], [0.25, 0.33])	([0.65, 0.84], [0.17, 0.24], [0.14, 0.24])	([0.37, 0.49], [0.41, 0.49], [0.27, 0.35])	([0.45, 0.60], [0.27, 0.34], [0.26, 0.33])	([0.17, 0.24], [0.63, 0.73], [0.23, 0.28])	([0.39, 0.53], [0.35, 0.42], [0.26, 0.32])

In the ratio system (RS) approach, $\tilde{Y}_i^+$ values can be calculated by using Eqs. (42) and (43) as given in Table 12. Besides, $Score\left(\tilde{Y}_i^+\right)$ values and ranking of the alternatives are given in Table 12.

Reference points in the aggregated weighted decision matrix can be calculated based on Eqs. (44) and (45) and the corresponding interval-valued spherical fuzzy sets are given in Table 13.

Distances from each alternative to all the coordinates of the reference point by using Eq. (46) are presented as given in Table 14 (Ejegwa 2019). The rank based on 2nd approach is also given in Table 14.

For the sensitivity analysis, the same distance calculation can be performed based on the Euclidean distance (Szmidt and Kacprzyk 2000) is given in Table 15.

Based on Eqs. (48) and (49), $\tilde{A}_i$ and $Score\left(\tilde{A}_i\right)$ values can be calculated as given in Table 16, respectively.

As it can be seen from Tables 13, 14, and 16, Alternative 3 is the best alternative and Alternative 1 is in the second rank.

6 Conclusions

The MULTIMOORA method has been used for solving different decision-making and engineering problems in the literature. Several fuzzy extensions have been introduced for the MULTIMOORA method such as intuitionistic fuzzy MULTIMOORA,

Table 14 The ranking of the alternatives based on Zhang and Xu's distance

X_i	$d\left(\tilde{X}_{i1}, \tilde{X}_j^*\right)$	$d\left(\tilde{X}_{2j}, \tilde{X}_j^*\right)$	$d\left(\tilde{X}_{3j}, \tilde{X}_j^*\right)$	$d\left(\tilde{X}_{4j}, \tilde{X}_j^*\right)$	$d\left(\tilde{X}_{5j}, \tilde{X}_j^*\right)$	$d\left(\tilde{X}_{6j}, \tilde{X}_j^*\right)$	$d\left(\tilde{X}_{7j}, \tilde{X}_j^*\right)$	$\max_j d\left(\tilde{X}_{ij}, \tilde{X}_j^*\right)$	Rank
X_1	0.02	0.02	0.05	0.00	0.00	0.02	0.00	0.05	2
X_2	0.08	0.13	0.38	0.06	0.09	0.10	0.03	0.38	4
X_3	0.00	0.00	0.00	0.00	0.00	0.00	0.00	0.00	1
X_4	0.19	0.30	0.27	0.14	0.27	0.15	0.07	0.30	5
X_5	0.10	0.18	0.18	0.14	0.10	0.14	0.07	0.18	3

Table 15 The ranking of the alternatives based on Euclidean distance

X_i	$d\left(\tilde{X}_{i1}, \tilde{X}_j^*\right)$	$d\left(\tilde{X}_{2j}, \tilde{X}_j^*\right)$	$d\left(\tilde{X}_{3j}, \tilde{X}_j^*\right)$	$d\left(\tilde{X}_{4j}, \tilde{X}_j^*\right)$	$d\left(\tilde{X}_{5j}, \tilde{X}_j^*\right)$	$d\left(\tilde{X}_{6j}, \tilde{X}_j^*\right)$	$d\left(\tilde{X}_{7j}, \tilde{X}_j^*\right)$	$\max_j d\left(\tilde{X}_{ij}, \tilde{X}_j^*\right)$	Rank
X_1	0.02	0.03	0.05	0.00	0.00	0.03	0.00	0.05	2
X_2	0.08	0.13	0.40	0.07	0.10	0.10	0.04	0.40	4
X_3	0.00	0.00	0.00	0.00	0.00	0.00	0.00	0.00	1
X_4	0.20	0.33	0.27	0.15	0.29	0.16	0.08	0.33	5
X_5	0.11	0.20	0.18	0.16	0.11	0.15	0.09	0.20	3

Table 16 The ranking of the alternatives based on FMF approach for IVSF-MULTIMOORA

Alternatives	$\tilde{A}_i$ *values*	$Score\left(\tilde{A}_i\right)$	Ranking
X_1	([0.39, 0.52], [0.38, 0.47], [0.25, 0.32])	−0.01	2
X_2	([0.29, 0.39], [0.47, 0.56], [0.29, 0.36])	−0.21	3
X_3	([0.41, 0.55], [0.37, 0.46], [0.24, 0.31])	0.02	1
X_4	([0.21, 0.29], [0.56, 0.66], [0.28, 0.34])	−0.36	5
X_5	([0.25, 0.35], [0.51, 0.60], [0.27, 0.33])	−0.26	4

hesitant fuzzy MULTIMOORA, and neutrosophic MULTIMOORA. Compared with these fuzzy extensions, spherical fuzzy sets provide a larger domain for solving complex multi-criteria decision-making problems. The parameters of a SFS, membership, non-membership, and hesitancy degrees are independently assigned to satisfy that their squared sum is at most equal to 1. The proposed single-valued and interval-valued spherical fuzzy MULTIMOORA methods include new aggregation operations and score functions for spherical fuzzy sets.

A group decision-making Industry 4.0 strategy selection problem has been solved by single-valued spherical fuzzy MULTIMOORA and interval-valued spherical fuzzy MULTIMOORA and compared each other. Both approaches have given same ranking for the same problem. However, interval-valued spherical fuzzy MULTIMOORA method has given larger preference domain to decision makers for assigning membership functions than single-valued spherical fuzzy MULTIMOORA and other MULTIMOORA extensions of fuzzy sets.

References

Ashraf A, Abdullah S, Aslam M, Qiyas M, Kutbi MA (2019) Spherical fuzzy sets and its representation of spherical fuzzy t-norms and t-conorms. J Intell Fuzzy Syst 36(6):6089–6102

Atanassov KT (1986) Intuitionistic fuzzy sets. Fuzzy Sets Syst 20:87–96

Aydın S (2018) Augmented reality goggles selection by using neutrosophic MULTIMOORA method. J Enterp Inf Manag 31(4):565–576

Baležentis T, Zeng S (2013) Group multi-criteria decision making based upon interval-valued fuzzy numbers: an extension of the MULTIMOORA method. Expert Syst Appl 40(2):543–550

Brauers WKM, Zavadskas EK (2006) The MOORA method and its application to privatization in a transition economy. Control Cybernet 35(2):445–469

Brauers WKM, Zavadskas EK (2011) MULTIMOORA optimization used to decide on a bank loan to buy property. Technol Econ Dev Econ 17(1):174–188

Brauers WKM, Baležentis A, Baležentis T (2011) Multimoora for the EU member states updated with fuzzy number theory. Technol Econ Dev Econ 17(2):259–290

Chen X, Zhao L, Liang H (2018) A novel multi-attribute group decision-making method based on the MULTIMOORA with linguistic evaluations. Soft Comput 22:5347–5361

Datta S, Sahu N, Mahapatra S (2013) Robot selection based on grey-MULTIMOORA approach. Grey Syst Theory Appl 3(2):201–232

Dorfeshan Y, Mousavi MS, Mohagheghi V, Vahdanib B (2018) Selecting project-critical path by a new interval type-2 fuzzy decision methodology based on MULTIMOORA, MOOSRA and TOPSIS methods. Comput Ind Eng 120:160–178
Ejegwa PA (2019) Modified Zhang and Xu's distance measure for Pythagorean fuzzy sets and its application to pattern recognition problems. Neural Comput Appl 1–10
Fenton N, Wang W (2006) Risk and confidence analysis for fuzzy multicriteria decision making. Knowl Based Syst 19:430–437
Garibaldi JM, Ozen T (2007) Uncertain fuzzy reasoning: a case study in modelling expert decision making. IEEE Trans Fuzzy Syst 15:16–30
Ghorabaee MK, Zavadskas EK, Olfat L, Turskis Z (2015) Multi-criteria inventory classification using a new method of evaluation based on distance from average solution (EDAS). Informatica 26(3):435–451
Gou X, Liao H, Xu Z, Herrera F (2017) Double hierarchy hesitant fuzzy linguistic term set and MULTIMOORA method: a case of study to evaluate the implementation status of haze controlling measures. Inf Fusion 38:22–34
Hafezalkotob A, Hafezalkotob A (2016) Fuzzy entropy-weighted MULTIMOORA method for materials selection. J Intell Fuzzy Syst 31(3):1211–1226
Hwang CL, Yoon K (1981) Multiple attribute decision making: methods and applications. Springer, New York
Kutlu Gündoğdu F, Kahraman C (2019a) Spherical fuzzy sets and spherical fuzzy TOPSIS method. J Intell Fuzzy Syst 36(1):337–352
Kutlu Gündoğdu F, Kahraman C (2019b) A novel VIKOR method using spherical fuzzy sets and its application to warehouse site selection. J Intell Fuzzy Syst 37(1):1197–1211
Kutlu Gündoğdu F, Kahraman C (2019c) A novel spherical fuzzy analytic hierarchy process and its renewable energy application. Soft Comput (in press)
Kutlu Gündoğdu F, Kahraman C, Karaşan A (2020) Spherical fuzzy VIKOR method and its application to waste management. In: Kahraman C, Cebi S, Cevik Onar S, Oztaysi B, Tolga A, Sari I (eds) Intelligent and fuzzy techniques in big data analytics and decision making. INFUS 2019. Advances in intelligent systems and computing, vol 1029. Springer, Cham
Li Z-H (2014) An extension of the MULTIMOORA method for multiple criteria group decision making based upon hesitant fuzzy sets. J Appl Math 2014, 16 pages. Article ID 527836
Liang W, Zhao G, Hong C (2019) Selecting the optimal mining method with extended multi-objective optimization by ratio analysis plus the full multiplicative form (MULTIMOORA) approach. Neural Comput Appl 31(10):5871–5886
Lu J, Zhang G, Ruan D, Wu F (2007) Multi-objective group decision making—methods, software and applications with fuzzy set techniques. Imperial College Press, London
Opricovic S (1998) Multicriteria optimization of civil engineering systems. Dissertation, Faculty of Civil Engineering, Belgrade
Roubens M (1982) Preference relations on actions and criteria in multi criteria decision making. Eur J Oper Res 10(1):51–55
Roy B (1991) The outranking approach and the foundations of ELECTRE methods. Theory Decis 31:49–73
Saaty TL (1980) The analytic hierarchy process. McGraw-Hill, New York
Saaty TL (2001) Decision making with dependence and feedback: the analytic network process, 2nd edn. RWS Publications, Pittsburgh
Smarandache FA (1998) Unifying field in logics. Neutrosophy: neutrosophic probability, set and logic. American Research Press, Rehoboth
Stanujkic D, Zavadskas EK, Smarandache F, Brauers WKM, Karabasevic D (2017) A neutrosophic extension of the MULTIMOORA method. Informatica 28(1):181–192
Szmidt E, Kacprzyk J (2000) Distances between intuitionistic fuzzy sets. Fuzzy Sets Syst 114(3):505–518

Tian ZP, Wang J, Wang JQ, Zhang HY (2017) An improved MULTIMOORA approach for multi-criteria decision-making based on interdependent inputs of simplified neutrosophic linguistic information. Neural Comput Appl 28(1):585–597

Torra V (2010) Hesitant fuzzy sets. Int J Intell Syst 25:529–539

Yager RR (1986) On the theory of bags. Int J Gen Syst 13:23–37

Yager RR (2013) Pythagorean membership grades in multicriteria decision making. In: Technical report MII-3301. Machine Intelligence Institute, Iona College, New Rochelle

Zadeh LA (1965) Fuzzy sets. Inf Control 8(3):338–353

Zadeh LA (1975) The concept of a linguistic variable and its application to approximate reasoning-1. Inf Sci 8:199–249

Zavadskas EK, Kaklauskas A, Sarka V (1994) The new method of multicriteria complex proportional assessment of projects. Technol Econ Dev Econ 1(3):131–139

Zavadskas EK, Antucheviciene J, Hajiagha SHR, Hashemi SS (2015) The interval-valued intuitionistic fuzzy MULTIMOORA method for group decision making in engineering. Math Probl Eng 2015:1–13

Zavadskas EK, Bausys R, Juodagalviene B, Sapranaviciene GI (2017) Model for residential house element and material selection by neutrosophic MULTIMOORA method. Eng Appl Artif Intell 64:315–324

Zhang C, Chen C, Streimikiene D, Balezentis T (2019) Intuitionistic fuzzy MULTIMOORA approach for multi-criteria assessment of the energy storage technologies. Appl Soft Comput 79:410–423

Zhao H, You JX, Liu HC (2017) Failure mode and effect analysis using MULTIMOORA method with continuous weighted entropy under interval-valued intuitionistic fuzzy environment. Soft Comput 21(18):5355–5367

Global Supplier Selection with Spherical Fuzzy Analytic Hierarchy Process

Iman Mohamad Sharaf

Abstract Spherical fuzzy sets (SFSs) have emerged lately as a new type of fuzzy sets in which the hesitancy degree is independent of the membership degree and the non-membership degree, thus providing a broader preference domain. This study extends the analytic hierarchy process (AHP) using (SFSs). In the proposed method, SFSs are used to construct the pairwise comparison matrices. Therefore, a spherical fuzzy preference scale is presented. Then, the consistency of the spherical fuzzy pairwise comparison matrices is checked to ensure obtaining a reasonable solution. To achieve this, the spherical fuzzy pairwise comparison matrices are converted to crisp matrices, and then Saaty's eigenvalue method is applied to check the consistency. A prioritization function is proposed to determine the importance weights, hence construct the priority vector of the criteria and the priority vector of the alternatives for each criterion. Finally, the priorities of the alternatives obtained are combined in a weighted sum to form the global priority of each alternative. The alternative with the highest global priority is the best. The proposed method is used to solve an example in global supplier selection using SFSs and intuitionistic fuzzy sets (IFSs) to compare the results with the results of an intuitionistic fuzzy AHP.

1 Introduction

The analytic hierarchy process (AHP) is one of the widely used methods in Operations Research and Management Science to determine priorities in multi-criteria decision-making (MCDM) (Xu and Liao 2014). Saaty (1977, 1980) introduced AHP as a structured approach in decision making. As defined by Saaty (2008) "The AHP is a theory of measurement through pairwise comparisons and relies on the judgements of experts to derive priority scales". It is an efficient tool that helps decision-makers to solve complex problems. The procedure of an AHP can be summarized in three phases: decomposition, comparative judgements, and synthesis of priorities

I. M. Sharaf (✉)
Department of Basic Sciences, Higher Technological Institute, Tenth of Ramadan City, Egypt
e-mail: iman_sharaf@hotmail.com

© Springer Nature Switzerland AG 2021

C. Kahraman and F. Kutlu Gündoğdu (eds.), *Decision Making with Spherical Fuzzy Sets*, Studies in Fuzziness and Soft Computing 392,
https://doi.org/10.1007/978-3-030-45461-6_14

(Xu and Liao 2014). Any complex problem can be decomposed into a multi-level hierarchic structure of objectives, criteria, sub-criteria and alternatives. Comparative judgements identify the relation between two stimuli both present to the observer (Blumenthal 1977). Comparisons use a scale of numbers that indicates how many times one element is dominant over another with respect to the criterion they are compared (Saaty 2008). To rank the alternatives, an overall ratio scale of priorities is synthesized after deriving a ratio scale of relative magnitudes (Xu and Liao 2014).

In most real-life problems, the decision-makers are unable to provide an exact numerical value to their evaluations. This is due to their limited knowledge, reluctance, the imprecision of data, or the subjective nature of the qualitative evaluation criteria. In this case, using fuzzy sets is essential to capture the uncertainties associated with the human cognitive process (Kahraman et al. 2014).

Zadeh (1965) introduced the first known fuzzy sets; type 1 fuzzy sets (T1FSs). T1FSs have a membership function that assigns to each element in the set a grade of membership in the interval [0, 1]. Nevertheless, T1FSs were not satisfactory in expressing linguistic expressions. Having a crisp membership function, they didn't provide a flexible framework to incorporate intra-uncertainty and inter-uncertainty related to linguistic terms.

Later, Zadeh (1975) proposed type-2 fuzzy sets (T2FSs) in which the membership degree of each element in the set is defined by another fuzzy set over the interval [0, 1]. The three-dimensional membership function of T2FSs has an advantage and a disadvantage. The advantage is minimizing the effect of uncertainty. The disadvantage is the heavy computations. T2FSs are difficult in understanding and applications (Türk et al. 2014).

Some researchers proposed simpler variants of T2FSs. Liang and Mendel (2000) presented interval type-2 fuzzy sets (IT2FSs) as a special case of T2FSs in which the secondary membership degree is equal to one. Sambuc (1975) proposed interval-valued fuzzy sets (IVFSs) in which membership values are intervals. Thus, IT2FSs and IVFSs are defined by two T1FSs: an upper membership function and a lower membership function. In spite of the apparent similarity between an IVFS and an IT2FS, they are quite different. The information capacity of the two sets is not identical (Niewiadomski 2007). An IT2FS permits the representation of concepts that cannot be represented by an IVFS (Niewiadomski 2007; Sola et al. 2015). Actually an IT2FS can be viewed as a generalization of an IVFS (Sola et al. 2015).

However, since type-1 and type-2 fuzzy sets are single-valued functions expressing the degree of membership, they cannot be used to express the support and the objection in many practical situations (Xu and Liao 2014). In many cases, decision-makers do not have a precise or sufficient knowledge of the problem domain, or they are unable to explicitly discriminate one alternative from the others due to the complicated socio-economic environment. Herein, preference evaluations manifest three characteristics: affirmation, negation and hesitation. Atanassov (1986) introduced intuitionistic fuzzy sets (IFSs) to express the support, the opposition, and the indeterminacy of the decision-maker concurrently. Intuitionistic fuzzy sets (IFSs) can express these situations and model human perception more effectively whenever both imprecision and uncertainty are salient in the problem domain. IFSs have three dimensions: the membership degree, the non-membership degree, and the hesitancy degree. The sum of these dimensions must be equal to one.

Smarandache (1998) proposed neutrosophic fuzzy sets (NFSs) as an extension to the concept of IFSs. NFSs provide a new perspective to uncertainty, impreciseness, inconsistency and vagueness. Smarandache (1998) defined an NFS by three parameters: truth membership, indeterminacy membership, and falsity membership. Thus, he introduced the degree of indeterminacy as a new and independent component of fuzzy sets. Since indeterminacy parameter gives more details in the membership functions, using NFSs in decision making can produce better results. Nonetheless, an NFS is more complex to apply in real scientific and engineering fields (Boltürk and Kahraman 2018).

Torra and Narukawa (2009) and Torra (2010) proposed a novel type of fuzzy sets named hesitant fuzzy sets (HFSs). HFSs don't have an exact membership function, and the possible values of the membership degree are random. HFSs are close to the human cognition and more natural in representing the fuzziness and vagueness than all the other forms of fuzzy sets (Liao and Xu 2013).

Zadeh (2011) proposed the concept of Z-number to provide a basis for computation with numbers which are not reliable. An uncertain variable "Z", Z-number, is an ordered pair of fuzzy numbers (A, B). The first component "A" plays the role of a fuzzy restriction, while the second component "B" is the reliability of the first component. A Z-number can describe the knowledge of human, it can describe both restraint and reliability (Azadeh et al. 2013).

Yager (2014) proposed Pythagorean fuzzy sets (PFSs), also known as type-2 intuitionistic fuzzy sets (T2IFSs). In some situations when decision-makers evaluate the preference of the alternatives, the degrees of membership and non-membership are greater than one. The main difference between the PFSs and the IFSs is the condition on the sum of the three dimensions. The condition is modified to make the sum of the squares of the membership degree and non-membership degree equal to or less than one. This makes PFSs better than IFSs in modelling the lack of certainty and doubtfulness of decision-makers when making a decision when confronted with imprecise, inconsistent, or incomplete information.

In both IFSs and PFSs the hesitancy degree depends on the degrees of membership and non-membership. Since the hesitation degree represents the ignorance or indeterminacy resulting from the lack of information, it is more appropriate to be expressed unrelated to the membership degree and non-membership degree. Recently, Gündoğdu and Kahraman (2019a) introduced the three-dimensional spherical fuzzy sets (SFSs) as a new extension of IFSs, NFSs, and PFSs in which the hesitancy of a decision-maker can be expressed explicitly. Accordingly, the decision-makers can express their hesitance regardless of their degree of agreement or disagreement. Thus, SFSs can reveal experts' judgements and preferences efficiently.

In this study, the AHP method is extended using SFSs. SFSs are used to construct the pairwise comparison matrices of the criteria and the alternatives with respect to the criteria. Therefore, a spherical fuzzy preference scale is presented in accordance with Saaty's 1-9 scale and in line with that proposed by Gündoğdu and Kahraman (2019a). Then, the consistency of the spherical fuzzy pairwise comparison matrices is checked to ensure obtaining a reasonable solution. This is accomplished by converting the fuzzy pairwise comparison matrices into crisp matrices, and then Saaty's eigenvalue method is applied to check the consistency. A prioritization function is proposed to determine the importance weights, hence construct the priority vector of the criteria, and the priority vector of the alternatives with respect to each criterion. Finally, the priorities of the alternatives obtained are combined in a weighted sum to form the global priority of each alternative. The alternative with the highest global priority is the best.

The originality of the study comes from the implementation of the AHP using SFSs that provide the decision-makers with a reliable structure that reflects their cognition. A spherical fuzzy preference scale is proposed. A novel technique is presented to defuzzify the spherical pairwise comparison matrices, and a new function is proposed to prioritize the criteria and alternatives.

The chapter is organized as follows. A literature review is given in Sect. 2. The SFSs with their basic operators and the conventional AHP are presented in Sect. 3. The proposed SFAHP is introduced in Sect. 4. An example is solved in Sect. 5 to demonstrate the algorithm and the results are compared with the results of an IFAHP. Finally, the conclusion is given in Sect. 6.

2 A Literature Review

Saaty (1977, 1980) introduced the AHP method to handle complex problems in MCDM. The AHP method determines the relative priorities for a set of alternatives for the assessment criteria using the decision-makers' pairwise judgments. The AHP constructs the decision-making criteria as a hierarchy and then calculates the weights of the criteria and the available alternatives. The consistencies of the alternatives' comparisons for the used criteria are checked. In the classical AHP method, the decision-makers' evaluations are crisp numbers. However, when the decision-makers cannot express the assessments by crisp numbers, fuzzy logic can be used. This provides a mathematical strength to capture the uncertainties accompanying human cognition (Gündoğdu and Kahraman 2019b). The conventional AHP method has been extended using the various types of fuzzy sets. These extensions are classified according to the type of the used fuzzy set as follows.

Van Laarhoven and Pedrycz (1983) proposed the first fuzzy AHP (FAHP) using type-1 triangular fuzzy sets. They used the logarithmic least square method to derive fuzzy weights and fuzzy performance scores. Buckley (1985) proposed a FAHP using trapezoidal fuzzy numbers and used the geometric mean method to get the fuzzy weights and the fuzzy performance scores. Boender et al. (1989) presented a modification of Van Laarhoven and Pedrycz's method using a logarithmic regression function and geometric ratio scales to develop an efficient procedure. Chang (1996) proposed a new approach for handling fuzzy AHP with the use of triangular fuzzy numbers for pairwise comparison scale and the use of the extent analysis method. Wang et al. (2008) showed by examples that the priority vectors determined by the extent analysis method did not represent the relative importance of the decision criteria or the alternatives and that the misapplication of the extent analysis method to FAHP problems may lead to wrong decisions. Wang (2018) combined FAHP with representative utility functions under fuzzy environment to resolve the drawbacks and difficulties in Chang's method. Gim and Kim (2014) evaluated hydrogen storage systems for automobiles using the FAHP. Kahraman et al. (2014) developed a FAHP using IT2FSs with a new ranking method. Rezaei et al. (2014) designed a supplier selection methodology in the airline retail industry. In the first phase, a conjunctive screening method is used to reduce the initial set of potential suppliers. In the second phase, a FAHP is used to evaluate the suppliers against the main criteria and sub-criteria. Tan et al. (2015) proposed a FAHP to evaluate the options for the harvesting and drying process. Öztaysi et al. (2018) proposed a prioritization method for possible business analytics projects using Type-2 fuzzy AHP. Since business analytics projects require considerable financial investment and have potential risks and benefits, the problem can be formulated as an MCDM problem. FAHP was also integrated with other decision-making methods. For example, Chatterjee and Kar (2017) proposed an MCDM framework using the AHP and VIseKriterijumska Optimizacija I Kompromisno Resenje (VIKOR), multi-criteria optimization and compromise solution, in a granular domain. Also, Sari (2017) proposed a hybrid method using the AHP, VIKOR and Monte Carlo simulation.

Sadiq and Tasfamariam (2009) applied intuitionistic FAHP (IFAHP) to select the best drilling fluid (mud) for drilling operations under multiple environmental criteria. Wu et al. (2013) proposed an AHP approach using interval-valued intuitionistic fuzzy sets (IVIFs) and applied the approach in E-commerce. Abdullah and Najib (2014) proposed a new IFAHP characterized by a new preference scale. Xu and Liao (2014) proposed a new technique to check the consistency of an intuitionistic preference relation in IFAHP and introduced an automatic procedure to repair an inconsistent one. Buyukozkan et al. (2016) provided a new framework for evaluating the web service performance of healthcare institutions to improve their reputation and recognition. The proposed framework is based on an integrated MCDM methodology that exploits the IFAHP and IFVIKOR. Deepika and Kanaan (2016) proposed an IFAHP and developed an efficient method to check the consistency and automatic repairing procedure. Tooranloo and Iranpour (2007) proposed an AHP method using IVIFSs to select the appropriate supplier in group decision-making environments.

After Smarandache (1998) proposed neutrosophic fuzzy sets (NFSs), several types of NFSs have been proposed. Wang et al. (2005a, b) developed interval-valued neutrosophic fuzzy sets (IVNFSs) and single-valued NFSs (SVNFSs), respectively. Bhowmik and Pal (2010) proposed intuitionistic neutrosophic fuzzy sets (INFSs). Peng et al. (2014) presented multi-valued neutrosophic fuzzy sets (MVNFS) which allow the truth-membership, indeterminacy membership and falsity-membership degree to have a set of crisp values between zero and one. Ye (2015) proposed trapezoidal neutrosophic fuzzy sets (TrNFSs) based on the combination of trapezoidal fuzzy numbers and SVNFSs.

Abdel-Basset et al. (2018) studied the integration of the AHP into Delphi framework in the neutrosophic environment and presented a new technique to check the consistency and calculate the consensus degree of expert's opinions. They also developed a real-life example based on expert opinions about the evaluation process of many international search engines. Boltürk and Kahraman (2018) introduced two novel AHP methods with IVNFSs: IVNF-AHP alone and IVNF-AHP based on cosine similarity (IVNFAHP-CS) measure that provides an objective scoring procedure for pairwise comparison matrices under neutrosophic uncertainty. They gave an application in energy alternative selection to illustrate the developed methods. Kahraman et al. (2018) used an NFAHP for comparing the performances of law offices that can be comparatively measured by MCDM methods in which linguistic assessments can be used rather than exact numerical evaluations.

The AHP was extended in the hesitant fuzzy environment since HFSs permit the membership degree to have a set of possible values. HFSs are ideal to reflect hesitancy and to handle conditions where decision-makers are hesitant to provide their preferences over objects, rather than a margin of error. Zhu et al. (2016) proposed an HFAHP to assess the strategic positions of islands and reefs regarding national security, military deployments, resource and economy. Öztaysi et al. (2015) developed an HFAHP method involving multi-experts' linguistic evaluations aggregated by ordered weighted averaging (OWA) operator, and the method is applied to the supplier selection problem. Boltürk et al. (2016) proposed a new HFAHP to solve the humanitarian logistics warehouse location selection problem which is an important issue for humanitarian relief organizations to improve their relief aid capability and rescue plan. Senvar (2018) proposed an evaluation framework for customer-oriented performance rankings of departments (sales, delivery, quality, maintenance) of an organization using HFAHP.

Z-number fuzzy sets didn't receive much attention like other fuzzy sets. Azadeh et al. (2013) proposed a frame of AHP using Z-number. They applied the Z-AHP model to evaluate universities.

Pythagorean FAHP (PFAHP) has been extensively used in the context of occupational health and safety (OHS). Ilbahar et al. (2018) developed a novel integrated approach, using Pythagorean fuzzy proportional risk assessment, PFAHP, and a fuzzy inference system for risk assessment. Gul and Ak (2018) subjectively assessed two parameters, likelihood and severity by occupational health and safety experts and determined their importance levels using the PFAHP. Gul (2018) proposed a new approach for risk assessment that integrates the PFAHP and the FVIKOR. PFAHP is used in weighing the risk parameters, and then FVIKOR is applied to prioritize the hazards. Karaşan et al. (2018) introduced a novel approach, Safety and Critical Effect Analysis (SCEA), and its extension with PFSs to provide a more comprehensive and accurate risk assessment.

Recently, Gündoğdu and Kahraman (2019b) extended the AHP in the spherical fuzzy environment and applied the method in the field of renewable energy. Gündoğdu and Kahraman (2020) also applied the SFAHP method to the industrial robot selection problem.

3 Fundamentals

3.1 Basic Operators of SFSs

An SFS is defined by

$$\tilde{A}_s = \left\{ \left\langle u, \left(\mu_{\tilde{A}_s}(u), v_{\tilde{A}_s}(u), \pi_{\tilde{A}_s}(u) \right) | u \in U \right\rangle \right\} \tag{1}$$

where $\mu_{\tilde{A}_s}: U \rightarrow [0, 1]$ is the membership degree, $v_{\tilde{A}_s}: U \rightarrow [0, 1]$ is the non-membership degree, $\pi_{\tilde{A}_s}: U \rightarrow [0, 1]$ is the hesitancy degree, and

$$0 \leq \mu^2_{\tilde{A}_s}(u) + v^2_{\tilde{A}_s}(u) + \pi^2_{\tilde{A}_s}(u) \leq 1, \quad \forall u \in U. \tag{2}$$

The basic operations on SFSs are given as follows (Gündoğdu and Kahraman 2019a).

The addition operation

$$\tilde{A}_s \oplus \tilde{B}_s = \left\{ \left(\mu^2_{\tilde{A}_s} + \mu^2_{\tilde{B}_s} - \mu^2_{\tilde{A}_s}\mu^2_{\tilde{B}_s} \right)^{1/2}, v_{\tilde{A}_s}v_{\tilde{B}_s}, \left(\left(1 - \mu^2_{\tilde{B}_s}\right)\pi^2_{\tilde{A}_s} + \left(1 - \mu^2_{\tilde{A}_s}\right)\pi^2_{\tilde{B}_s} - \pi^2_{\tilde{A}_s}\pi^2_{\tilde{B}_s} \right)^{1/2} \right\}. \tag{3}$$

The multiplication operation

$$\tilde{A}_s \otimes \tilde{B}_s = \left\{ \mu_{\tilde{A}_s}\mu_{\tilde{B}_s}, \left(v^2_{\tilde{A}_s} + v^2_{\tilde{B}_s} - v^2_{\tilde{A}_s}v^2_{\tilde{B}_s} \right)^{1/2}, \left(\left(1 - v^2_{\tilde{B}_s}\right)\pi^2_{\tilde{A}_s} + \left(1 - v^2_{\tilde{A}_s}\right)\pi^2_{\tilde{B}_s} - \pi^2_{\tilde{A}_s}\pi^2_{\tilde{B}_s} \right)^{1/2} \right\}. \tag{4}$$

Multiplication by a scalar $\lambda > 0$,

$$\lambda \cdot \tilde{A}_s = \left\{ \left(1 - \left(1 - \mu^2_{\tilde{A}_s}\right)^{\lambda} \right)^{1/2}, v^{\lambda}_{\tilde{A}'_s}, \left(\left(1 - \mu^2_{\tilde{A}_s}\right)^{\lambda} - \left(1 - \mu^2_{\tilde{A}_s} - \pi^2_{\tilde{A}_s}\right)^{\lambda} \right)^{1/2} \right\}. \tag{5}$$

Raising $\widetilde{A}_s$ to the Power of $\lambda > 0$,

$$\tilde{A}^{\lambda}_s = \left\{ \mu^{\lambda}_{\tilde{A}_s}, \left(1 - \left(1 - v^2_{\tilde{A}_s}\right)^{\lambda} \right)^{1/2}, \left(\left(1 - v^2_{\tilde{A}_s}\right)^{\lambda} - \left(1 - v^2_{\tilde{A}_s} - \pi^2_{\tilde{A}_s}\right)^{\lambda} \right)^{1/2} \right\}. \tag{6}$$

3.2 The AHP

The decision modelling is the basic step in an AHP, i.e. constructing a hierarchy to analyze the decision (Mu and Pereyra-Rojas 2017). Structuring the problem as a hierarchy leads to a better understanding of the required decision, the used criteria and the available alternatives. The AHP is summarized as follows (Saaty 2008; Mu and Pereyra-Rojas 2017).

1. Define the problem, and then determine the type of required information.
2. Decomposition: construct the decision hierarchy starting from "the goal" at the top level, followed by "the criteria" and "the sub-criteria" at the intermediate level, and finally "the alternatives" at the lowest level.
3. Comparative judgements: construct a set of pairwise comparison matrices. Each element in an upper level is used to compare the elements in the next lower level with respect to it, e.g. derive the relative priority of criteria with respect to the overall objective, and the priorities of the alternatives with respect to each criterion separately (local priorities). This is done according to a proposed numerical scale.
4. Consistency Check: The judgements are reviewed to ensure a rational level of consistency regarding proportionality and transitivity.
5. Prioritization: derive the priority vector of each preference relation (local priority).
6. Synthesis of priorities: the priorities of the alternatives obtained are combined in a weighted sum to form the global (overall) priority of each alternative. The alternative with the highest global priority is the best.

4 The SFAHP Method

4.1 The Spherical Fuzzy Preference Relation

A spherical fuzzy preference relation (SFPR) on the set of alternatives $X = \{x_1, x_2, \ldots, x_n\}$ is given by a matrix

$$R = \left[r_{ij}\right]_{n \times n}, \tag{7}$$

where

$$r_{ij} = \left\langle \left(x_i, x_j\right), \mu\left(x_i, x_j\right), \upsilon\left(x_i, x_j\right), \pi\left(x_i, x_j\right) \right\rangle, \quad \forall i, j = 1, 2, \ldots, n,$$

or simply

$$r_{ij} = \left(\mu_{ij}, \upsilon_{ij}, \pi_{ij}\right). \tag{8}$$

The degree to which the alternative x_i is preferred to the alternative x_j is represented by $\mu(x_i, x_j)$, while $\upsilon(x_i, x_j)$ indicates the degree to which the alternative x_i is not preferred to the alternative x_j, and $\pi(x_i, x_j)$ denotes the inability to express the preference, with the conditions

$$\mu_{ij},\ \upsilon_{ij},\ \text{ and } \pi_{ij} \in [0, 1];\quad \mu_{ij}^2(u) + \upsilon_{ij}^2(u) + \pi_{ij}^2(u) \leq 1;$$

$$\mu_{ij} = \upsilon_{ji}, \upsilon_{ij} = \mu_{ji},\quad \text{and } \pi_{ij} = \pi_{ji},$$

$$\text{and } \mu_{ii} = \upsilon_{ii} = \pi_{ii} = 0.5,\quad \forall i, j = 1, 2, \ldots, n. \tag{9}$$

The following properties of SFPR are similar to the properties of intuitionistic fuzzy preference relation IFPR given by Liao and Xu (2014) with slight adjustments to take the hesitancy degree into consideration:

(a) Weak transitivity: if $r_{ik} \geq (0.5, 0.5, 0.5)$ and $r_{kj} \geq (0.5, 0.5, 0.5)$, then $r_{ij} \geq (0.5, 0.5, 0.5), \forall i, j, k = 1, 2, \ldots, n$. (i.e. if the alternative x_i is preferred to x_k, and x_k is preferred to x_j, then x_i should be preferred to x_j).
(b) Max–min transitivity: if $r_{ij} \geq \min\{r_{ik}, r_{kj}\}$, $\forall i, j, k = 1, 2, \ldots, n$. (i.e. the preference value obtained by a direct comparison between two alternatives x_i and x_j should be equal to or greater than the minimum preference value obtained from comparing both alternatives with an intermediate one x_i, x_k and x_k, x_j).
(c) Max–max transitivity: if $r_{ij} \geq \max\{r_{ik}, r_{kj}\}$, $\forall i, j, k = 1, 2, \ldots, n$. (i.e. The preference value obtained by a direct comparison between two alternatives x_i and x_j should be equal to or greater than the maximum preference value obtained by comparing two alternatives with an intermediate one x_i, x_k and x_k, x_j).
(d) Restricted max–min transitivity: If $r_{ik} \geq (0.5, 0.5, 0.5)$ and $r_{kj} \geq (0.5, 0.5, 0.5)$, then $r_{ij} \geq \min\{r_{ik}, r_{kj}\}$, $\forall i, j, k = 1, 2, \ldots, n$. (i.e. if the alternative x_i is preferred to x_k with a preference value r_{ik}, and x_k is preferred to x_j with a preference value r_{kj}, then x_i should be preferred to x_j with at least a preference value r_{ij} which is equal to the minimum of these values.
(e) Restricted max–max transitivity: If $r_{ik} \geq (0.5, 0.5, 0.5)$ and $r_{kj} \geq (0.5, 0.5, 0.5)$, then $r_{ij} \geq \max\{r_{ik}, r_{kj}\}$, $\forall i, j, k = 1, 2, \ldots, n$. (i.e. if the alternative x_i is preferred to x_k with a preference value r_{ik}, and x_k is preferred to x_j with a preference value r_{kj}, then x_i should be preferred to x_j with at least a preference value r_{ij} which is equal to the maximum of these values.
(f) Multiplicative transitivity: $r_{ij}/r_{ji} = (r_{ik}/r_{ki})(r_{kj}/r_{jk})$, $\forall i, j, k = 1, 2, \ldots, n$.
    (i.e. the ratio of the preference intensity (r_{ij}/r_{ji}) for the alternative x_i to that of x_j should be equal to the multiplication of the ratios of preferences using an intermediate alternative x_k).

From the above-mentioned transitivity properties, only the multiplicative transitivity implies reciprocity. Therefore multiplicative transitivity is used to verify the consistency of an SFPR.

An intuitionistic fuzzy preference relation (IFPR) is called multiplicative consistent if the following multiplicative transitivity is satisfied (Jin et al. 2016)

$$\mu_{ij}\mu_{jk}\mu_{ki} = \mu_{ik}\mu_{kj}\mu_{ji}, \quad \forall\ i, j, k = 1, 2, \ldots, n. \tag{10}$$

From (9), since $\mu_{ij} = \upsilon_{ji}, \mu_{jk} = \upsilon_{kj},$ and $\mu_{ki} = \upsilon_{ik},$

$$\forall i, j, k = 1, 2, \ldots, n,$$

substituting in (10) we get

$$\mu_{ij}\upsilon_{kj}\upsilon_{ik} = \mu_{ik}\mu_{kj}\upsilon_{ij},$$

$$\frac{\mu_{ij}}{\upsilon_{ij}} = \frac{\mu_{ik}}{\upsilon_{ik}}\frac{\mu_{kj}}{\upsilon_{kj}}, \quad \forall i, j, k = 1, 2, \ldots, n. \tag{11}$$

Or simply,

$$\rho_{ij} = \rho_{ik}\rho_{kj}, \quad \forall i, j, k = 1, 2, \ldots, n \tag{12}$$

where

$$\rho_{ij} = \frac{\mu_{ij}}{\upsilon_{ij}}, \quad \rho_{ik} = \frac{\mu_{ik}}{\upsilon_{ik}}, \text{ and } \rho_{kj} = \frac{\mu_{kj}}{\upsilon_{kj}}.$$

The relation (12) is valid for SFSs regardless of the hesitancy degree since the hesitancy degree is independent of the membership degree and the non-membership degree. In addition, the hesitancy degree is multiplicative consistent since the multiplicative transitivity $\pi_{ij}\pi_{jk}\pi_{ki} = \pi_{ik}\pi_{kj}\pi_{ji}$ is always satisfied since $\pi_{ij} = \pi_{ji}, \pi_{jk} = \pi_{kj},$ and $\pi_{ki} = \pi_{ik}$.

From the previous demonstration, we are only concerned with the membership degree and the non-membership degree in checking the consistency. A spherical fuzzy comparison matrix is converted to a crisp matrix by replacing "r_{ij}" with "ρ_{ij}" which makes the form of the matrix similar to the positive reciprocal pairwise comparison matrix proposed by Saaty $\left(a_{ij} = 1/a_{ji}, a_{ii} = 1\right)$. The crisp pairwise comparison matrix is given by:

$$R = \left[\rho_{ij}\right]. \tag{13}$$

Then Saaty's eigenvalue method is applied to check the consistency.

4.2 Consistency Check

In the context of FAHP, Kwong and Bai (2003) converted the FPR to its corresponding crisp multiplicative preference relation and then used Saaty's method to check the consistency. Still, many researchers neglected the consistency check process (Xu and Liao 2014). In the context of IFAHP, Xu and Liao (2014) proposed a method which not only checks the consistency but also repairs an inconsistent IFPR and changes it into a consistent one.

A spherical fuzzy pairwise comparison matrix is converted to a crisp pairwise comparison matrix $R = \left[\rho_{ij}\right]$, as previously explained, and then Saaty's eigenvalue method is applied to check the consistency.

The conventional AHP method of Saaty provides a consistency index to measure the consistency within a pairwise comparison matrix. The method can be summarized as follows (Mu and Pereyra-Rojas 2017).

The consistency index (CI), and the consistency ratio (CR) for a pairwise comparison matrix can be computed by

$$\mathrm{CI} = (\lambda_{max} - n)/(n - 1), \quad \mathrm{CR} = (\mathrm{CI}/\mathrm{RI}(n)), \tag{14}$$

where λ_{max} is the largest eigenvalue of the pairwise comparison matrix $\left(a_{ij} = 1/a_{ji}, a_{ii} = 1\right)$ of size n and $\mathrm{RI}(n)$ is a random index that depends on n. For the consistency indices for different values of n, the reader is referred to Mu and Pereyra-Rojas (2017). If the CR is less than 0.1, although a ratio of less than 0.2 is considered tolerable (Wedley 1993), the consistency of the pairwise judgment is acceptable. Otherwise, the judgments expressed by the experts are considered to be inconsistent, and the decision-maker has to repeat the pairwise comparison matrix. This method proved to be powerful in checking consistency (Xu and Liao 2014).

4.3 Prioritization Function

A scale of priorities is an n-dimensional vector $\mathrm{W} = [w_1, w_2, \ldots, w_n]$ obtained from an SFPR. A weight w_i represents the relative dominance of a criterion among other criteria, or an alternative among other alternatives with respect to a certain criterion. In this section, finding a function to derive the weights of the criteria and alternatives is investigated.

Gündoğdu and Kahraman (2019a) defined the score function and the accuracy function of an SFS as follows:

$$\mathrm{Score}\left(\widetilde{A}_s\right) = \left(\mu_{\widetilde{A}_s} - \pi_{\widetilde{A}_s}\right)^2 - \left(\upsilon_{\widetilde{A}_s} - \pi_{\widetilde{A}_s}\right)^2, \tag{15}$$

$$\mathrm{Accuracy}\left(\widetilde{A}_s\right) = \mu_{\widetilde{A}_s}^2 + \upsilon_{\widetilde{A}_s}^2 + \pi_{\widetilde{A}_s}^2 \tag{16}$$

Each function solely might be inappropriate for deriving the relative weights. The score function may give a zero or a negative value. For example, the SFS (0.5 0.5 0.2) has a zero score, while the SFS (0.2 0.8 0.2) has a negative score "0.36". In addition, the score function may produce the same value for two different SFSs. For example, consider the SFSs $\widetilde{A}_1 = (0.5, 0.4, 0.1)$ and $\widetilde{A}_2 = (0.6, 0.5, 0.2)$, both sets have the same score function "0.07". The accuracy function can also give the same value for two different SFSs. Consider the SFSs $\widetilde{A}_1 = (0.5, 0.4, 0.1)$ and $\widetilde{A}_2 = (0.4, 0.1, 0.5)$, the accuracy function produces the same value "0.42" for both sets. Then, an alternative function is required to derive the weights of the assessment criteria and alternatives.

Szmidt and Kacprzyk (2009) proposed a function based on the amount of information associated with an alternative, and the reliability of this information to rank IFSs. The amount of information is measured by the distance from the positive ideal alternative, while the reliability of information is measured by the hesitation margin. The function is given by $\mathrm{F}(\widetilde{A}_I) = 0.5(1+\pi)(1-\mu)$. The smaller the value of $\mathrm{F}(\widetilde{A}_I)$, the greater the IFS. Based on this idea, the weighting function is proposed as follows.

The SFS representing the ideal alternative IA $= (1, 0, 0)$, i.e. 100% preferred, 0% not preferred, and 0% hesitancy. The smaller the distance of an alternative from the IA, the better this alternative than other considered alternatives. Therefore, the weight of an alternative increases as the distance between this alternative and the IA decreases. For any distance formula, e.g. the Hamming distance, the Euclidean distance, the main building blocks are the difference between the membership degrees, the non-membership degrees, and the hesitancy degrees of an alternative and their peers in the IA. Then, for an alternative (μ, υ, π) the basic terms for its distance from the ideal alternative IA $= (1, 0, 0)$ are $|\mu - 1|$, (υ), and (π). Then, the distance decreases as the membership degree increases, and both the non-membership degree and the hesitancy degree decreases. The higher the agreement (affirmation), the lower the disagreement (negation) and the indeterminacy or abstention (hesitation), the SFS increases in weight. Consequently, the prioritization function is given by

$$\mathcal{F}(\widetilde{A}_s) = \mu(1-\upsilon)(1-\pi). \tag{17}$$

4.4 The Proposed SFAHP

The steps of the proposed SFAHP can be summarized as follows.

Step 1: Problem definition and hierarchy construction.
This step is similar to the conventional AHP, FAHP, and IFAHP.
Step 2: Formation of the spherical fuzzy pairwise comparison matrices.

Table 1 The spherical fuzzy preference scale

Linguistic term	(μ, υ, π)
Extremely preferred (EP)	(0.9, 0.1, 0.1)
Very strongly preferred (VSP)	(0.8, 0.2, 0.2)
Strongly preferred (SP)	(0.7, 0.3, 0.3)
Moderately preferred (MP)	(0.6, 0.4, 0.4)
Equally preferred (EP)	(0.5, 0.5, 0.5)
Moderately not preferred (MNP)	(0.4, 0.6, 0.4)
Strongly not preferred (SNP)	(0.3, 0.7, 0.3)
Very strongly not preferred (VSNP)	(0.2, 0.8, 0.2)
Extremely not preferred (ENP)	(0.1, 0.9, 0.1)

The proposed spherical fuzzy preference scale, given in Table 1, is used to set up the SFPRs via the pairwise comparison between the elements in a level with respect to the elements in the level immediately above it, e.g. the pairwise comparison between each criterion and sub-criterion, and the pairwise comparison between the alternatives under each criterion or sub-criterion. Thus, the spherical fuzzy pairwise comparison matrices are formed.

Step 3: Check the consistency.

The spherical fuzzy pairwise comparison matrices are converted to crisp positive reciprocal pairwise comparison matrices, using $R = \left[\rho_{ij}\right]$, where $\rho_{ij} = \frac{\mu_{ij}}{\upsilon_{ij}}$ as previously explained, and then Saaty's eigenvalue method is applied to check the consistency.

Step 4: Derive the local priorities.

To weigh the criteria and the alternatives, for each comparison matrix do:

(i) compute the spherical fuzzy geometric mean for the ith row using the aggregation operations of SFSs:

$$r^i = \left(\prod_{j=1}^{n} r_{ij}\right)^{1/n}.$$

(ii) find the weight of each aggregated row $\mathcal{F}\left(r^i\right)$ using (17), the proposed prioritization function, $\mathcal{F}\left(\widetilde{A}_s\right) = \mu(1 - \upsilon)(1 - \pi)$. Then, compute the total weight: $\sum_{i=1}^{n} \mathcal{F}\left(r^i\right)$.

(iii) determine the priority vector $\mathrm{W} = [w_1, w_2, \ldots, w_n]$, where

$$w_i = \mathcal{F}\left(r^i\right) / \sum_{i=1}^{n} \mathcal{F}\left(r^i\right).$$

Step 5: Derive global priorities and rank.

The performance score of each alternative is given by $W_i = \sum w_j w_{ij}$, where W_i is the utility of the ith alternative, w_j is the weight of the jth criterion, and w_{ij} is the performance score of the ith alternative for the jth criterion. The alternative with the highest utility score is the best.

5 Examples

The following example is adapted from Xu and Liao (2014). The example is solved twice using the proposed SFAHP. First the example is solved using the spherical fuzzy preference scale given in Table 1. Then, the example is resolved as given by Xu and Liao (2014) using the intuitionistic fuzzy preference scale. The hierarchy construction and the comparison matrices are given in details in Chan and Kumar (2007) and Xu and Liao (2014). Therefore, we skip step 1 and 2 and start the technique from step 3.

The establishment of a long-term business relationship with competitive global suppliers worldwide became a necessary trend in markets. The selection of diverse international suppliers on broad comparison basis is critical and has a significant impact on the performance of an organization. Global supplier selection is a more complex MCDM problem than a domestic one due to additional factors and needs a thorough analysis. The analysis requires disintegrating this complex problem into simpler sub-problems and thus constructing the hierarchy. Chan and Kumar (2007) developed a four-level hierarchy for this problem; with the top-level "the goal" or "the objective" is to select the best supplier. The criteria and sub-criteria considered are:

$$C_1: \textbf{the overall cost of the product}\begin{cases} S_1: \textbf{product price} \\ S_2: \textbf{freight cost} \\ S_3: \textbf{traffic} \end{cases},$$

$$C_2: \textbf{quality of the product}\begin{cases} S_4: \textbf{rejection rate of the product} \\ S_5: \textbf{increased lead-time} \\ S_6: \textbf{quality assessment} \\ S_7: \textbf{the remedy for quality problems} \end{cases}$$

$$C_3: \textbf{service performance}\begin{cases} S_8: \textbf{delivery schedule} \\ S_9: \textbf{technological support} \\ S_{10}: \textbf{response to changes} \\ S_{11}: \textbf{ease of communication} \end{cases},$$

$$C_4: \textbf{supplier's profile}\begin{cases} S_{12}: \textbf{financial status} \\ S_{13}: \textbf{customer base} \\ S_{14}: \textbf{performance history} \\ S_{15}: \textbf{production facility and capacity} \end{cases},$$

$$C_5: \textbf{risk factor}\begin{cases} S_{16}: \textbf{geographical location} \\ S_{17}: \textbf{political stability} \\ S_{18}: \textbf{economy} \\ S_{19}: \textbf{terrorism} \end{cases}.$$

Suppose that there are three suppliers. The pairwise comparison matrices are constructed conveying the preference of a criterion, sub-criterion, or an alternative with respect to the others by a questionnaire or by the experience of the experts. For simplification, the criteria are compared as a whole. The pairwise comparison matrices are given as follows.

The preference relation of the criteria with respect to the overall objective is given by:

$$\tilde{R}_o = \begin{matrix} & C_1 & C_2 & C_3 & C_4 & C_5 \\ C_1 & \textbf{EP} & \textbf{MP} & \textbf{MP} & \textbf{SP} & \textbf{SP} \\ C_2 & \textbf{MNP} & \textbf{EP} & \textbf{MP} & \textbf{SP} & \textbf{SP} \\ C_3 & \textbf{MNP} & \textbf{MNP} & \textbf{EP} & \textbf{MP} & \textbf{MP} \\ C_4 & \textbf{SNP} & \textbf{SNP} & \textbf{MNP} & \textbf{EP} & \textbf{MP} \\ C_5 & \textbf{SNP} & \textbf{SNP} & \textbf{MNP} & \textbf{MNP} & \textbf{EP} \end{matrix}.$$

The preference relations of the alternatives with respect to the criteria are given by:

$$\tilde{R}_{C_1} = \begin{matrix} C_1 & A_1 & A_2 & A_3 \\ A_1 & \text{EP} & \text{MP} & \text{MNP} \\ A_2 & \text{MNP} & \text{EP} & \text{SNP} \\ A_3 & \text{MP} & \text{SP} & \text{EP} \end{matrix}, \tilde{R}_{C_2} = \begin{matrix} C_2 & A_1 & A_2 & A_3 \\ A_1 & \text{EP} & \text{MP} & \text{MNP} \\ A_2 & \text{MNP} & \text{EP} & \text{SNP} \\ A_3 & \text{MP} & \text{SP} & \text{EP} \end{matrix},$$

$$\tilde{R}_{C_3} = \begin{array}{c} C_3 \\ A_1 \\ A_2 \\ A_3 \end{array} \begin{array}{c} \begin{array}{ccc} A_1 & A_2 & A_3 \end{array} \\ \begin{bmatrix} EP & MP & SP \\ MNP & EP & MP \\ SNP & MNP & EP \end{bmatrix} \end{array}, \tilde{R}_{C_4} = \begin{array}{c} C_4 \\ A_1 \\ A_2 \\ A_3 \end{array} \begin{array}{c} \begin{array}{ccc} A_1 & A_2 & A_3 \end{array} \\ \begin{bmatrix} EP & VSP & SP \\ VSNP & EP & MP \\ SNP & MNP & EP \end{bmatrix} \end{array},$$

and

$$\tilde{R}_{C_5} = \begin{array}{c} C_5 \\ A_1 \\ A_2 \\ A_3 \end{array} \begin{array}{c} \begin{array}{ccc} A_1 & A_2 & A_3 \end{array} \\ \begin{bmatrix} EP & SP & SP \\ SNP & EP & MP \\ SNP & MNP & EP \end{bmatrix} \end{array}.$$

Step 3: Check the consistency.

Form the crisp pairwise comparison matrix $R = [\rho_{ij}]$, where $\rho_{ij} = \frac{\mu_{ij}}{\upsilon_{ij}}$, then apply the eigenvalue method (14)

$CI = (\lambda_{max} - n)/(n - 1)$ and $CR = CI/RI$.

The random indices are RI(5) = 1.12, and RI(3) = 0.58 (Mu and Pereyra-Rojas 2017).

$$R_o = \begin{bmatrix} 1 & 3/2 & 3/2 & 7/3 & 7/3 \\ 2/3 & 1 & 3/2 & 7/3 & 7/3 \\ 2/3 & 2/3 & 1 & 3/2 & 3/2 \\ 3/7 & 3/7 & 2/3 & 1 & 3/2 \\ 3/7 & 3/7 & 2/3 & 2/3 & 1 \end{bmatrix}, \lambda_{max} = 5.0398, \text{ and } CR = 0.009.$$

$$R_{C_1} = R_{C_2} = \begin{bmatrix} 1 & 3/2 & 2/3 \\ 2/3 & 1 & 3/7 \\ 3/2 & 7/3 & 1 \end{bmatrix}, \lambda_{max} = 3.0001 \text{ and } CR = 8.6 * 10^{-5}.$$

$$R_{C_3} = \begin{bmatrix} 1 & 3/2 & 7/3 \\ 2/3 & 1 & 3/2 \\ 3/7 & 2/3 & 1 \end{bmatrix}, \lambda_{max} = 3.0001 \text{ and } CR = 8.6 * 10^{-5}.$$

$$R_{C_4} = \begin{bmatrix} 1 & 4 & 7/3 \\ 1/4 & 1 & 3/2 \\ 3/7 & 2/3 & 1 \end{bmatrix}, \lambda_{max} = 3.0999 \text{ and } CR = 0.09.$$

$$R_{C_5} = \begin{bmatrix} 1 & 7/3 & 7/3 \\ 3/7 & 1 & 3/2 \\ 3/7 & 2/3 & 1 \end{bmatrix}, \lambda_{max} = 3.0183 \text{ and } CR = 0.02.$$

The consistency ratio is less than 0.1, for all pairwise comparison matrices, which is acceptable to continue the SFAHP analysis.

Step 4: Derive the local priorities.

For the comparison matrix $\widetilde{R}_o$

$$r_o^1 = \left(\prod_{j=1}^{5} r_{1j}\right)^{1/5} = (0.6153, 0.3901, 0.3985),\ \mathcal{F}\left(r_o^1\right) = 0.2257.$$
$$r_o^2 = \left(\prod_{j=1}^{5} r_{2j}\right)^{1/5} = (0.5674, 0.4438, 0.4003),\ \mathcal{F}\left(r_o^2\right) = 0.1893.$$
$$r_o^3 = \left(\prod_{j=1}^{5} r_{3j}\right)^{1/5} = (0.4919, 0.5133, 0.4245),\ \mathcal{F}\left(r_o^3\right) = 0.1378.$$
$$r_o^4 = \left(\prod_{j=1}^{5} r_{4j}\right)^{1/5} = (0.4043, 0.6025, 0.3808),\ \mathcal{F}\left(r_o^4\right) = 0.0995.$$
$$r_o^5 = \left(\prod_{j=1}^{5} r_{5j}\right)^{1/5} = (0.3728, 0.6181, 0.3713),\ \mathcal{F}\left(r_o^5\right) = 0.0895.$$
$$\mathcal{F}_o = \sum_{i=1}^{5} \mathcal{F}\left(r_o^i\right) = 0.7418.$$

Then, the priority weight vector of the criteria is

$$\mathrm{W_o} = \left(0.3043\ 0.2552\ 0.1858\ 0.1341\ 0.1207\right).$$

The priority of the alternatives with respect to C_1 using $\widetilde{R}_{C_1}$:

$$r_{C_1}^1 = \left(\prod_{j=1}^{3} r_{1j}\right)^{1/3} = (0.4932, 0.5111, 0.4385),\ \mathcal{F}\left(r_{C_1}^1\right) = 0.1354.$$
$$r_{C_1}^2 = \left(\prod_{j=1}^{3} r_{2j}\right)^{1/3} = (0.4309, 0.5703, 0.4338),\ \mathcal{F}\left(r_{C_1}^2\right) = 0.1048.$$
$$r_{C_1}^3 = \left(\prod_{j=1}^{3} r_{3j}\right)^{1/3} = (0.5944, 0.4114, 0.4213),\ \mathcal{F}\left(r_{C_1}^3\right) = 0.2025.$$

Then, $\mathcal{F}_{C_1} = \sum_{i=1}^{3} \mathcal{F}\left(r_{C_1}^i\right) = 0.4427$ and $\mathrm{W}_{C_1} = \left(0.3059\ 0.2367\ 0.4574\right)$.
The priority of the alternatives with respect to C_3 using $\widetilde{R}_{C_3}$:

$$r_{C_3}^1 = (0.5944, 0.4114, 0.4213),\ \mathcal{F}\left(r_{C_3}^1\right) = 0.2025.$$
$$r_{C_3}^2 = (0.5646, 0.4372, 0.4417),\ \mathcal{F}\left(r_{C_3}^2\right) = 0.1774.$$
$$r_{C_3}^3 = (0.3915, 0.6119, 0.4002),\ \mathcal{F}\left(r_{C_3}^3\right) = 0.0911.$$

Then, $\mathcal{F}_{C_3} = \sum_{i=1}^{3} \mathcal{F}\left(r_{C_2}^i\right) = 0.4710$ and $\mathrm{W}_{C_3} = \left(0.4299\ 0.3766\ 0.1934\right)$.
The priority of the alternatives with respect to C_4 using $\widetilde{R}_{C_4}$:

Table 2 Weights of the alternatives with respect to the criteria using SFSs

	C_1	C_2	C_3	C_4	C_5	W
	0.3043	0.2552	0.1858	0.1341	0.1207	
A_1	0.3059	0.3059	0.4299	0.64	0.5258	0.3298
A_2	0.2367	0.2367	0.3766	0.16	0.2114	0.2629
A_3	0.4574	0.4574	0.1934	0.2	0.2060	0.3438

$$r_{C_4}^1 = (0.6542, 0.3626, 0.3822),\ \mathcal{F}\left(r_{C_4}^1\right) = 0.2576.$$
$$r_{C_4}^2 = (0.3915, 0.6246, 0.3639),\ \mathcal{F}\left(r_{C_4}^2\right) = 0.0935.$$
$$r_{C_4}^3 = (0.3915, 0.6119, 0.4002),\ \mathcal{F}\left(r_{C_4}^3\right) = 0.0911.$$

Then, $\mathcal{F}_{C_4} = \sum_{i=1}^3 \mathcal{F}\left(r_{C_2}^i\right) = 0.6384$ and $\mathrm{W}_{C_4} = \left(0.5825\ 0.2114\ 0.2060\right)$.
The priority of the alternatives with respect to the fifth criterion using $\widetilde{R}_{C_5}$:

$$r_{C_5}^1 = (0.6257, 0.3832, 0.3973),\ \mathcal{F}\left(r_{C_5}^1\right) = 0.2326.$$
$$r_{C_5}^2 = (0.4481, 0.5613, 0.4027),\ \mathcal{F}\left(r_{C_5}^2\right) = 0.1174.$$
$$r_{C_5}^3(0.3915, 0.6119, 0.4002),\ \mathcal{F}\left(r_{C_5}^3\right) = 0.0911.$$

Then, $\mathcal{F}_{C_5} = \sum_{i=1}^3 \mathcal{F}\left(r_{C_2}^i\right) = 0.4411$ and $\mathrm{W}_{C_5} = \left(0.5273\ 0.2661\ 0.2065\right)$.

Step 5: Derive global priorities and rank.

$$W_i = \sum w_j w_{ij}$$

$\mathrm{W}_1 = 0.3928,\ \mathrm{W}_2 = 0.2629,$ and $\mathrm{W}_3 = 0.3438$..
The results are summarized in Table 2.
The ranking of the suppliers $A_1 \succ A_3 \succ A_2$.
The proposed method is used to solve the original problem in which IFSs are used for comparative judgements.
The preference relation of the criteria with respect to the overall objective.

$$\tilde{R}_o = \begin{matrix} \\ C_1 \\ C_2 \\ C_3 \\ C_4 \\ C_5 \end{matrix}\begin{bmatrix} C_1 & C_2 & C_3 & C_4 & C_5 \\ (0.5, 0.5, 0) & (0.6, 0.2, 0.2) & (0.6, 0.2, 0.2) & (0.65, 0.25, 0.1) & (0.65, 0.25, 0.1) \\ (0.2, 0.6, 0.2) & (0.5, 0.5, 0) & (0.6, 0.2, 0.2) & (0.65, 0.25, 0.1) & (0.65, 0.25, 0.1) \\ (0.2, 0.6, 0.2) & (0.2, 0.6, 0.2) & (0.5, 0.5, 0) & (0.6, 0.2, 0.2) & (0.6, 0.2, 0.2) \\ (0.25, 0.65, 0.1) & (0.25, 0.65, 0.1) & (0.2, 0.6, 0.2) & (0.5, 0.5, 0) & (0.6, 0.2, 0.2) \\ (0.25, 0.65, 0.1) & (0.25, 0.65, 0.1) & (0.2, 0.6, 0.2) & (0.5, 0.5, 0) & (0.2, 0.6, 0.2) \end{bmatrix}$$

The preference relation of the alternatives with respect to C_1

$$\widetilde{R}_{C_1} = \begin{array}{c} \\ A_1 \\ A_2 \\ A_3 \end{array} \begin{array}{c} \begin{array}{ccc} A_1 & A_2 & A_3 \end{array} \\ \begin{bmatrix} (0.5, 0.5, 0) & (0.6, 0.2, 0.2) & (0.4, 0.5, 0.1) \\ (0.2, 0.6, 0.2) & (0.5, 0.5, 0) & (0.3, 0.4, 0.3) \\ (0.5, 0.4, 0.1) & (0.4, 0.3, 0.3) & (0.5, 0.5, 0) \end{bmatrix} \end{array}.$$

The preference relation of the alternatives with respect to C_2

$$\widetilde{R}_{C_2} = \begin{array}{c} \\ A_1 \\ A_2 \\ A_3 \end{array} \begin{array}{c} \begin{array}{ccc} A_1 & A_2 & A_3 \end{array} \\ \begin{bmatrix} (0.5, 0.5, 0) & (0.6, 0.2, 0.2) & (0.45, 0.4, 0.15) \\ (0.2, 0.6, 0.2) & (0.5, 0.5, 0) & (0.35, 0.2, 0.45) \\ (0.4, 0.45, 0.15) & (0.2, 0.35, 0.45) & (0.5, 0.5, 0) \end{bmatrix} \end{array}.$$

The preference relation of the alternatives with respect to C_3

$$\widetilde{R}_{C_3} = \begin{array}{c} \\ A_1 \\ A_2 \\ A_3 \end{array} \begin{array}{c} \begin{array}{ccc} A_1 & A_2 & A_3 \end{array} \\ \begin{bmatrix} (0.5, 0.5, 0) & (0.55, 0.25, 0.2) & (0.65, 0.1, 0.25) \\ (0.25, 0.55, 0.2) & (0.5, 0.5, 0) & (0.6, 0.3, 0.1) \\ (0.1, 0.65, 0.25) & (0.3, 0.6, 0.1) & (0.5, 0.5, 0) \end{bmatrix} \end{array}.$$

The preference relation of the alternatives with respect to C_4

$$\widetilde{R}_{C_4} = \begin{array}{c} \\ A_1 \\ A_2 \\ A_3 \end{array} \begin{array}{c} \begin{array}{ccc} A_1 & A_2 & A_3 \end{array} \\ \begin{bmatrix} (0.5, 0.5, 0) & (0.8, 0.1, 0.1) & (0.7, 0.2, 0.1) \\ (0.1, 0.8, 0.1) & (0.5, 0.5, 0) & (0.55, 0.4, 0.05) \\ (0.2, 0.7, 0.1) & (0.4, 0.55, 0.05) & (0.5, 0.5, 0) \end{bmatrix} \end{array}.$$

The preference relation of the alternatives with respect to C_5

$$\widetilde{R}_{C_5} = \begin{array}{c} \\ A_1 \\ A_2 \\ A_3 \end{array} \begin{array}{c} \begin{array}{ccc} A_1 & A_2 & A_3 \end{array} \\ \begin{bmatrix} (0.5, 0.5, 0) & (0.7, 0.2, 0.1) & (0.65, 0.2, 0.15) \\ (0.2, 0.7, 0.1) & (0.5, 0.5, 0) & (0.55, 0.25, 0.2) \\ (0.2, 0.65, 0.15) & (0.25, 0.55, 0.2) & (0.5, 0.5, 0) \end{bmatrix} \end{array}.$$

Step 3: Check the consistency.

Form the crisp pairwise comparison matrix $R = [\rho_{ij}]$, where $\rho_{ij} = \frac{\mu_{ij}}{\upsilon_{ij}}$, then apply the maximum eigenvalue method (14).

$$R_o = \begin{bmatrix} 1 & 6/2 & 6/2 & 65/25 & 65/25 \\ 2/6 & 1 & 6/2 & 65/25 & 65/25 \\ 2/6 & 2/6 & 1 & 6/2 & 6/2 \\ 25/65 & 25/65 & 2/6 & 1 & 6/2 \\ 25/65 & 25/65 & 2/6 & 2/6 & 1 \end{bmatrix}$$

$\lambda_{max} = 5.5359$, CI $= 0.134$ and CR $= 0.134/1.12 = 0.1196$. Since CR > 0.1, then the relation is inconsistent and needs modification, which is also the case in Xu and Liao (2014) IFAHP.

Taking the modified R_o as given by Xu and Liao (2014)

$$R_o = \begin{array}{c} \\ C_1 \\ C_2 \\ C_3 \\ C_4 \\ C_5 \end{array} \left[\begin{array}{ccccc} C_1 & C_2 & C_3 & C_4 & C_5 \\ (0.5, 0.5, 0) & (0.6, 0.2, 0.2) & (0.67, 0.08, 0.25) & (0.7, 0.09, 0.21) & (0.78, 0.04, 0.18) \\ (0.2, 0.6, 0.2) & (0.5, 0.5, 0) & (0.6, 0.2, 0.2) & (0.67, 0.08, 0.25) & (0.7, 0.09, 0.21) \\ (0.08, 0.67, 0.25) & (0.2, 0.6, 0.2) & (0.5, 0.5, 0) & (0.6, 0.2, 0.2) & (0.67, 0.08, 0.25) \\ (0.09, 0.7, 0.21) & (0.08, 0.67, 0.25) & (0.2, 0.6, 0.2) & (0.5, 0.5, 0) & (0.6, 0.2, 0.2) \\ (0.04, 0.78, 0.18) & (0.09, 0.7, 0.21) & (0.08, 0.67, 0.25) & (0.2, 0.6, 0.2) & (0.5, 0.5, 0) \end{array} \right]$$

$$\text{Then, } R_o = \begin{bmatrix} 1 & 6/2 & 67/8 & 70/9 & 78/4 \\ 2/6 & 1 & 6/2 & 67/8 & 70/9 \\ 8/67 & 2/6 & 1 & 6/2 & 67/8 \\ 9/70 & 8/67 & 2/6 & 1 & 6/2 \\ 4/78 & 9/70 & 8/67 & 2/6 & 1 \end{bmatrix}$$

$\lambda_{max} = 5.2930$, CI $= 0.0733$ and CR $= 0.0654$. Since CR < 0.1, then the relation is consistent and the priority vector can be calculated.

Similarly, the rest of the preference matrices are checked for consistency

$$R_{C_1} = \begin{bmatrix} 1 & 6/2 & 4/5 \\ 2/6 & 1 & 3/4 \\ 5/4 & 4/3 & 1 \end{bmatrix}, \lambda_{max} = 3.12 \text{ and CR} = 0.1.$$

$$R_{C_2} = \begin{bmatrix} 1 & 6/2 & 45/40 \\ 2/6 & 1 & 35/20 \\ 40/45 & 20/35 & 1 \end{bmatrix}, \lambda_{max} = 3.2695 \text{ and CR} = 0.23.$$

$$R_{C_3} = \begin{bmatrix} 1 & 55/25 & 65/10 \\ 25/55 & 1 & 6/3 \\ 10/65 & 3/6 & 1 \end{bmatrix}, \lambda_{max} = 3.0169 \text{ and CR} = 0.014.$$

$$R_{C_4} = \begin{bmatrix} 1 & 8 & 7/2 \\ 1/8 & 1 & 55/40 \\ 2/7 & 40/55 & 1 \end{bmatrix}, \lambda_{max} = 3.1475 \text{ and CR} = 0.12.$$

$$R_{C_4} = \begin{bmatrix} 1 & 7/2 & 65/20 \\ 2/7 & 1 & 55/25 \\ 20/65 & 25/55 & 1 \end{bmatrix}, \lambda_{max} = 3.0832 \text{ and CR} = 0.07.$$

From the previous consistency ratios, the preference relations of the alternatives with respect to the second and fourth criteria need to be revised. This is different from the results obtained by Xu and Liao method in which all the preference relations are consistent. Yet, since the main aim is to compare the results, the solution will be continued with the same comparison matrices.

Step 4: Derive the local priorities.

Starting with the comparison matrix $\widetilde{R}_o$

$$r_o^1 = \left(\prod_{j=1}^{5} r_{1j}\right)^{1/5} = (0.6427, 0.2581, 0.1846),\ \mathcal{F}\left(r_o^1\right) = 0.3888.$$

$$r_o^2 = \left(\prod_{j=1}^{5} r_{2j}\right)^{1/5} = (0.4895, 0.3821, 0.1907),\ \mathcal{F}\left(r_o^2\right) = 0.2448.$$

$$r_o^3 = \left(\prod_{j=1}^{5} r_{3j}\right)^{1/5} = (0.3170, 0.4906, 0.2076),\ \mathcal{F}\left(r_o^3\right) = 0.1280.$$

$$r_o^4 = \left(\prod_{j=1}^{5} r_{4j}\right)^{1/5} = (0.2091, 0.5793, 0.2023),\ \mathcal{F}\left(r_o^4\right) = 0.0702.$$

$$r_o^5 = \left(\prod_{j=1}^{5} r_{5j}\right)^{1/5} = (0.1585, 0.6670, 0.1971),\ \mathcal{F}\left(r_o^5\right) = 0.0424.$$

$$\mathcal{F}_o = \sum_{i=1}^{5} \mathcal{F}\left(r_o^i\right) = 0.8742.$$

The priority weight vector of the criteria is $\mathrm{W}_o = \begin{pmatrix} 0.44 & 0.28 & 0.15 & 0.08 & 0.05 \end{pmatrix}$.
The priority of the alternatives with respect to the first criterion using $\widetilde{R}_{C_1}$:

$$r_{C_1}^1 = \left(\prod_{j=1}^{3} r_{1j}\right)^{1/3} = (0.4932, 0.4310, 0.1227),\ \mathcal{F}\left(r_{C_1}^1\right) = 0.2462.$$

$$r_{C_1}^2 = \left(\prod_{j=1}^{3} r_{2j}\right)^{1/3} = (0.3107, 0.5111, 0.2062),\ \mathcal{F}\left(r_{C_1}^2\right) = 0.1206.$$

$$r_{C_1}^3 = \left(\prod_{j=1}^{3} r_{3j}\right)^{1/3} = (0.4642, 0.4114, 0.1776),\ \mathcal{F}\left(r_{C_1}^3\right) = 0.2247.$$

Then, $\mathcal{F}_{C_1} = \sum_{i=1}^{3} \mathcal{F}\left(r_{C_1}^i\right) = 0.5915$ and $\mathrm{W}_{C_1} = \begin{pmatrix} 0.42 & 0.20 & 0.38 \end{pmatrix}$.
The priority of the alternatives with respect to the second criterion using $\tilde{R}_{C_2}$:

$$r_{C_2}^1 = (0.5130, 0.3929, 0.1394),\ \mathcal{F}\left(r_{C_2}^1\right) = 0.2680.$$

$$r^2_{C_2} = (0.3271, 0.4771, 0.2717),\ \mathcal{F}\left(r^2_{C_2}\right) = 0.1246.$$
$$r^3_{C_2} = (0.3420, 0.4397, 0.2730),\ \mathcal{F}\left(r^3_{C_2}\right) = 0.1393.$$

Then, $\mathcal{F}_{C_2} = \sum_{i=1}^{3} \mathcal{F}\left(r^i_{C_2}\right) = 0.5319$ and $\mathrm{W}_{C_2} = \left(0.5\ 0.23\ 0.26\right)$.

The priority of the alternatives with respect to the third criterion using $\tilde{R}_{C_3}$:

$$r^1_{C_3} = (0.5634, 0.3373, 0.1777),\ \mathcal{F}\left(r^1_{C_3}\right) = 0.3070.$$
$$r^2_{C_3} = (0.4217, 0.4682, 0.1343),\ \mathcal{F}\left(r^2_{C_3}\right) = 0.1941.$$
$$r^3_{C_3} = (0.2466, 0.5899, 0.1665),\ \mathcal{F}\left(r^3_{C_3}\right) = 0.0843.$$

Then, $\mathcal{F}_{C_3} = \sum_{i=1}^{3} \mathcal{F}\left(r^i_{C_2}\right) = 0.5854$ and $\mathrm{W}_{C_3} = \left(0.52\ 0.33\ 0.14\right)$.

The priority of the alternatives with respect to the fourth criterion using $\tilde{R}_{C_4}$:

$$r^1_{C_4} = (0.6542, 0.3267, 0.0782), \mathcal{F}\left(r^1_{C_4}\right) = 0.4060.$$
$$r^2_{C_4} = (0.3018, 0.6246, 0.0794), \mathcal{F}\left(r^2_{C_4}\right) = 0.1043.$$
$$r^3_{C_4} = (0.3420, 0.5969, 0.0707),\ \mathcal{F}\left(r^3_{C_4}\right) = 0.1281.$$

Then, $\mathcal{F}_{C_4} = \sum_{i=1}^{3} \mathcal{F}\left(r^i_{C_2}\right) = 0.6384$ and $\mathrm{W}_{C_4} = \left(0.64\ 0.16\ 0.2\right)$.

The priority of the alternatives with respect to the fifth criterion using $\tilde{R}_{C_5}$:

$$r^1_{C_5} = (0.6105, 0.3403, 0.1001),\ \mathcal{F}\left(r^1_{C_5}\right) = 0.3624.$$
$$r^2_{C_5} = (0.3803, 0.5094, 0.1424),\ \mathcal{F}\left(r^2_{C_5}\right) = 0.16.$$
$$r^3_{C_5} = (0.2924, 0.5736, 0.1475),\ \mathcal{F}\left(r^3_{C_5}\right) = 0.1063.$$

Then, $\mathcal{F}_{C_5} = \sum_{i=1}^{3} \mathcal{F}\left(r^i_{C_2}\right) = 0.6287$ and $\mathrm{W}_{C_5} = \left(0.58\ 0.25\ 0.17\right)$.

Step 5: Derive global priorities and rank.

$$W_i = \sum w_i w_{ij}$$

$\mathrm{W}_1 = 0.4830$, $\mathrm{W}_2 = 0.2552$, and $\mathrm{W}_3 = 0.2855$. The ranking of the suppliers $A_1 \succ A_3 \succ A_2$. The results are summarized in Table 3. The ranking is the same as the ranking of Xu and Liao (2014).

Table 3 Weights of the alternatives with respect to the criteria using IFSs

	C_1	C_2	C_3	C_4	C_5	W
	0.44	0.28	0.15	0.08	0.05	
A_1	0.42	0.5	0.52	0.64	0.58	0.4830
A_2	0.2	0.23	0.33	0.16	0.25	0.2552
A_3	0.38	0.26	0.14	0.2	0.17	0.2855

6 Conclusion

The contribution of this study is the extension of the AHP method using SFSs. Using three-dimensional membership functions, with an independent degree of hesitation, increases the decision-makers preference domain and provides them with a robust structure that reflects their hesitancy. SFSs are used to form the pairwise comparison matrices of the criteria with respect to the objective and the alternatives with respect to the criteria. Therefore, a spherical fuzzy preference scale is proposed in accordance with Saaty's 1-9 scale and in line with that proposed by Gündoğdu and Kahraman (2019a). To check the consistency of the pairwise comparison matrices, they are converted to crisp matrices having the form of Saaty's positive reciprocal pairwise comparison matrix, and then the eigenvalue method is applied. To overcome the deficiencies of the score function, a prioritization function is presented. The prioritization function is used to determine the importance weights, hence the priority vector of the criteria and the priority vectors of the alternatives with respect to the criteria. Finally, the priorities of the alternatives obtained are combined in a weighted sum to form the global priority of each alternative. The alternative with the highest global priority is the best. The proposed technique was used to solve an example in global supplier selection using SFSs and IFSs. The ranking obtained using both sets are identical. Moreover, the ranking coincides with the ranking of the method proposed by Xu and Liao (2014) using IFSs.

For future research, due to the deficiencies of the score function (15) in evaluating the weights of the criteria and the performance of the alternatives with respect to the criteria, it is required to propose other score functions that convey the essence of an SFS. Also, a comparison between the performance of the AHP in the spherical fuzzy environment and its performance in the Pythagorean fuzzy environment is suggested.

References

Abdel-Basset M, Mohamed M, Sangaiah AK (2018) Neutrosophic AHP-Delphi Group decision making model based on trapezoidal neutrosophic numbers. J Ambient Intell Humaniz Comput 9(5):1427–1443

Abdullah L, Najib L (2014) A new preference scale of intuitionistic fuzzy analytic hierarchy process in multi-criteria decision making problems. J Intell Fuzzy Syst 26:1039–1049

Atanassov KT (1986) Intuitionistic fuzzy sets. Fuzzy Sets Syst 20(1):87–96

Azadeh A, Saberi M, Atashbar NZ, Chang E, Pazhoheshfar P (2013) Z-AHP: a Z-number extension of fuzzy analytical hierarchy process. In: 7th IEEE international conference on digital ecosystems and technologies (DEST), Menlo Park, CA, USA, 24–26 July 2013
Bhowmik M, Pal M (2010) Intuitionistic neutrosophic set relations and some of its properties. J Inf Comput Sci 5(3):183–192
Blumenthal AL (1977) The process of cognition. Prentice Hall, Englewood Cliffs
Boender CGE, De Graan JG, Lootsma FA (1989) Multicriteria decision analysis with fuzzy pairwise comparison. Fuzzy Sets Syst 29:133–143
Boltürk E, Kahraman C (2018) A novel interval-valued neutrosophic AHP with cosine similarity measure. Soft Comput 22:4941–4958
Boltürk E, Onar SÇ, Öztayşi B, Kahraman C, Goztepe K (2016) Multi-attribute warehouse location selection in humanitarian logistics using hesitant fuzzy AHP. Int J Anal Hierarchy Process 8(2):271–298
Buckley JJ (1985) Fuzzy hierarchical analysis. Fuzzy Sets Syst 17:233–247
Buyukozkan G, Feyzioglu O, Gocer F (2016) Evaluation of hospital web services using intuitionistic fuzzy AHP and intuitionistic fuzzy VIKOR. IEEE Int Conf Ind Eng Eng Manag (IEEM), Bali, Indonesia, 4–7 Dec 2016
Chan FTS, Kumar N (2007) Global supplier development considering risk factors using fuzzy extended AHP-based approach. Omega 35:417–431
Chang DY (1996) Applications of the extent analysis method on fuzzy AHP. Eur J Oper Res 95:649–655
Chatterjee K, Kar S (2017) Unified granular-number-based AHP-VIKOR multi-criteria decision framework. Granul Comput 2:199–221
Deepika M, Kannan ASK (2016) Global supplier selection using intuitionistic fuzzy analytic hierarchy process. In: International conference on electrical, electronics and optimization techniques (ICEEOT), Chennai, India, 3–5 Mar 2016
Gim B, Kim JW (2014) Multi-criteria evaluation of hydrogen storage systems for automobiles in Korea using the fuzzy analytic hierarchy process. Int J Hydrog Energy 39(15):7852–7858
Gul M (2018) Application of pythagorean fuzzy AHP and VIKOR methods in occupational health and safety risk assessment: the case of a gun and rifle barrel external surface oxidation and colouring unit. Int J Occup Saf Ergon. https://doi.org/10.1080/10803548.2018.1492251
Gul M, Ak MF (2018) A comparative outline for quantifying risk ratings in occupational health and safety risk assessment. J Clean Prod 196:653–664
Gündoğdu FK, Kahraman C (2019a) Spherical fuzzy sets and spherical fuzzy TOPSIS method. J Intell Fuzzy Syst 36(1):337–352
Gündoğdu FK, Kahraman C (2019b) A novel spherical fuzzy analytic hierarchy process and its renewable energy application. Soft Comput. https://doi.org/10.1007/s00500-019-04222-w
Gündoğdu FK, Kahraman C (2020) Spherical fuzzy analytic hierarchy process (AHP) and its application to industrial robot selection. In: Kahraman C, Cebi S, Cevik Onar S, Oztaysi B, Tolga A, Sari I (eds) Intelligent and fuzzy techniques in big data analytics and decision making. In: INFUS 2019. Advances in Intelligent Systems and Computing, vol 1029, pp 988–996. Springer, Cham (2020)
Ilbahar E, Karaşan A, Cebi S, Kahraman C (2018) A novel approach to risk assessment for occupational health and safety using Pythagorean fuzzy AHP & fuzzy inference system. Saf Sci 103:124–136
Jin F, Ni Z, Chen H, Li Y (2016) Approaches to group decision making with intuitionistic fuzzy preference relations based on multiplicative consistency. Knowl-Based Syst 97:48–59
Kahraman C, Öztayşi B, Onar SÇ, Boltürk E (2018) Neutrosophic AHP and prioritization of legal service outsourcing firms/law offices. In: World scientific proceedings series on computer engineering and information science. Data science and knowledge engineering for sensing decision support, pp 1176–1183
Kahraman C, Öztayşi B, Sari IU, Turanoğlu E (2014) Fuzzy analytic hierarchy process with interval type-2 fuzzy sets. Knowl-Based Syst 59:48–57

Karaşan A, Ilbahar E, Cebi S, Kahraman C (2018) A new risk assessment approach: safety and critical effect analysis (SCEA) and its extension with Pythagorean fuzzy sets. Saf Sci 108:173–187
Kwong CK, Bai H (2003) Determining the importance weights for the customer requirements in QFD using a fuzzy AHP with an extent analysis approach. IIE Trans 35:619–626
Liang Q, Mendel J (2000) Interval-type 2 fuzzy logic systems: theory and design. IEEE Trans Fuzzy Syst 8(5):535–550
Liao H, Xu Z (2013) A VIKOR-based method for hesitant fuzzy multi-criteria decision making. Fuzzy Optim Decis Making 12:373–392
Liao H, Xu Z (2014) Priorities of intuitionistic fuzzy preference relation based on multiplicative consistency. IEEE Trans Fuzzy Syst 22(6):1669–1681
Mu E, Pereyra-Rojas M (2017) Understanding the analytic hierarchy process (Chap. 2). In: Practical decision making. Springer briefs in operations research, pp 7–22
Niewiadomski A (2007) Interval-valued and interval type-2 fuzzy sets: a subjective comparison. In: IEEE international fuzzy systems conference, London, UK, 23–26 July 2007
Öztaysi B, Onar SÇ, Boltürk E, Kahraman C (2015) Hesitant fuzzy analytic hierarchy process. In: IEEE international conference on fuzzy systems (FUZZ-IEEE). IEEE, Istanbul, Turkey, 2–5 Aug 2015
Öztaysi B, Onar SÇ, Kahraman C (2018) Prioritization of business analytics projects using interval type-2 fuzzy AHP. Adv Intell Syst Comput 643:106–117
Peng J-J, Wang J-Q, Wu X-H, Wang J, Chen X-H (2014) Multi-valued neutrosophic sets and power aggregation operators with their applications in multi-criteria group decision-making problems. Int J Comput Intell Syst 8:345–363
Rezaei J, Fahim PBM, Tavasszy L (2014) Supplier selection in the airline retail industry using a funnel methodology: conjunctive screening method and fuzzy AHP. Expert Syst Appl 41(18):8165–8179
Saaty TL (1977) A scaling method for priorities in a hierarchical structure. J Math Psychol 15:234–281
Saaty TL (1980) The analytic hierarchy process: planning, priority setting, resource allocation. McGraw-Hill, New York
Saaty TL (2008) Decision making with the analytic hierarchy process. Int J Serv Sci 1(1):83–97
Sadiq R, Tasfamariam S (2009) Environmental decision making under uncertainty using intuitionistic fuzzy analytic hierarchy process (IF-AHP). Stoch Environ Res Risk Assess 23:75–91
Sambuc R (1975) Function U-Flous, Application a l'aide au diagnostic en pathologie thyroidienne. Thèse de Doctorate en Medicine, Séction Medecine University of Marseille, Marseille, France
Sari K (2017) A novel multi-criteria decision framework for evaluating green supply chain management practices. Comput Ind Eng 105:338–347
Senvar OA (2018) Systematic customer oriented approach based on hesitant fuzzy AHP for performance assessments of service departments. Adv Intell Syst Comput 643:289–300
Smarandache F (1998) Neutrosophy neutrosophic probability: set, and logic. American Research Press, Rehoboth
Sola HB, Fernandez J, Hagras H, Herrera F, Pagola M, Barrenechea E (2015) Interval type 2 fuzzy sets: toward a wider view on their relationship. IEEE Trans Fuzzy Syst 23(5):1876–1882
Szmidt E, Kacprzyk J (2009) Amount of information and its reliability in the ranking of Atanassov's intuitionistic fuzzy alternatives. In: Rakus-Andersson E, Yager RR, Ichalkaranje N, Jain LC (eds) Recent advances in decision making. Studies in computational intelligence, vol 222. Springer, Berlin, pp 7–19
Tan J, Low KY, Sulaiman NMN, Tan RR, Promentilla MAB (2015) Fuzzy analytical hierarchy process (AHP) for multi-criteria selection in drying and harvesting process of microalgae system. Chem Eng Trans 45:829–834
Tooranloo HS, Iranpour A (2007) Supplier selection and evaluation using interval-valued intuitionistic fuzzy AHP method. Int J Procure Manag 10(5):539–554
Torra V (2010) Hesitant fuzzy sets. Int J Intell Syst 25:529–539

Torra V, Narukawa Y (2009) On hesitant fuzzy sets and decision. IEEE international conference on fuzzy systems, Jeju Island, Korea, 20–24 Aug 2009
Türk S, John R, Özcan E (2014) Interval type-2 fuzzy sets in supplier selection. In: 14th UK workshop on computational intelligence (UKCI), Bradford, UK, 8–10 Sept 2014
Van Laarhoven PJM, Pedrycz W (1983) A fuzzy extension of Saaty's priority theory. Fuzzy Sets Syst 11:229–241
Wang H, Smarandache F, Sunderraman R, Zhang Y-Q (2005a) Interval neutrosophic sets and logic: theory and applications in computing. Hexis, Phoenix Az
Wang H, Smarandache F, Zhang Y, Sunderraman R (2005) Single valued neutrosophic sets. In: Proceedings of 10th 476 international conference on fuzzy theory and technology, Salt Lake City, 477 Utah
Wang Y-J (2018) Fuzzy multi-criteria decision making on combining fuzzy analytic hierarchy process with representative utility functions under fuzzy environment. Soft Comput 22:1641–1650
Wang YM, Luo Y, Hua ZS (2008) On the extent analysis method for fuzzy AHP and its applications. Eur J Oper Res 186:735–747
Wedley WC (1993) Consistency prediction for incomplete AHP matrices. Mathl Comput Model 17(4/5):151–161
Wu J, Huang H-B, Cao Q-W (2013) Research on AHP with interval valued intuitionistic fuzzy sets and its application in multicriteria decision making problems. Appl Math Model 37(24):9898–9906
Xu Z, Liao H (2014) Intuitionistic fuzzy analytic hierarchy process. IEEE Trans Fuzzy Sets Syst 22(4):749–761
Yager RR (2014) Pythagorean membership grades in multi-criteria decision-making. IEEE Trans Fuzzy Syst 22(4):958–965
Ye J (2015) Trapezoidal neutrosophic set and its application to multiple attribute decision-making. Neural Comput Appl 26(5):1157–1166
Zadeh LH (1965) Fuzzy sets. Inf. Control 8(3):338–353
Zadeh LH (1975) The concept of a linguistic variable and its applications to approximate reasoning. Inf Sci 8:199–249
Zadeh LA (2011) A note on Z-numbers. Inf Sci 181:2923–2932
Zhu B, Xu Z, Zhang R, Hong M (2016) Hesitant analytic hierarchy process. Eur J Oper Res 250(2):602–614

Hospital Performance Assessment Using Interval-Valued Spherical Fuzzy Analytic Hierarchy Process

Fatma Kutlu Gündoğdu and Cengiz Kahraman

Abstract Health-care service quality is the core of the medical institution's management. However, the measurement of service quality is difficult by classical measurement approaches. The fuzzy set theory can capture this difficulty through its linguistic approaches. The extensions of ordinary fuzzy sets such as intuitionistic fuzzy sets (IFS), Pythagorean fuzzy sets (PFS), and neutrosophic sets (NS), whose membership functions are based on three dimensions, aim at collecting experts' judgments more informatively and explicitly. In the literature, generalized three-dimensional spherical fuzzy sets have been introduced by Kutlu Gündoğdu and Kahraman (2019a) and then they proposed spherical fuzzy analytic hierarchy process (SF-AHP) method. In this chapter, this method is extended to interval-valued spherical fuzzy AHP (IVSF-AHP) method and the proposed method is used to compare the service performances of several hospitals. For this purpose, the method has been designed to analyze the service quality in the health-care industry based on SERVQUAL dimensions. Additionally, we present a comparative analysis with neutrosophic AHP to show robustness and validity of the proposed method. Hospital managers can use the results of this evaluation as a basis of strategies that would ensure the quality services to patients.

Keywords Spherical fuzzy sets · Interval-valued spherical fuzzy sets · Hospital performance evaluation · Interval-valued spherical fuzzy AHP · Neutrosophic AHP

F. Kutlu Gündoğdu
Industrial Engineering Department, National Defence University, Turkish Air Force Academy, Istanbul 34149, Turkey
e-mail: fatmakutlugundogdu@gmail.com

C. Kahraman (✉)
Industrial Engineering Department, Istanbul Technical University, Beşiktaş, Istanbul 34367, Turkey
e-mail: kahramanc@itu.edu.tr

© Springer Nature Switzerland AG 2021

C. Kahraman and F. Kutlu Gündoğdu (eds.), *Decision Making with Spherical Fuzzy Sets*, Studies in Fuzziness and Soft Computing 392,
https://doi.org/10.1007/978-3-030-45461-6_15

1 Introduction

Assessment of hospital performance based on several criteria by hospital managers helps to determine its position among competitive hospitals. Through hospital performance assessment based on realistic and effective tools, hospitals can identify where they are weak and strong, and in which aspects they are advantages and disadvantages to improve their medical standards. Conducting hospital performance evaluation regularly and applying the assessment results into hospital's strategic management can contribute to their quality improvement (Pinnarelli et al. 2012). Thus, a remarkable hospital performance evaluation system can be created.

Analytic Hierarchy Process (AHP) proposed by Saaty (1980) is one of the most used multi-criteria decision making methods to evaluate, prioritize, and rank decision choices. AHP creates a hierarchy based on the objective, decision criteria, sub-criteria and alternatives and sorts the numerous alternatives according to their relative priorities. Individual experts' experiences are utilized to estimate the relative magnitudes of factors through pairwise comparisons. Each of the respondents has to compare the relative importance between the two items under special designed questionnaire.

The ordinary fuzzy set theory introduced by Zadeh (1965) is a way of dealing with uncertainty by representing every parameter in the set together with a membership degree. In the literature, ordinary fuzzy set theory has been used to obtain efficient solutions under vagueness for the research areas such as engineering, mathematics, computer sciences, medical sciences, business, economics, and social sciences. The ordinary fuzzy sets have been extended to many new extensions: Type-2 fuzzy sets (Zadeh 1975), interval-valued fuzzy sets (Sambuc 1975; Jahn 1975, Grattan-Guinness 1976), fuzzy multisets (Yager 1986), intuitionistic fuzzy sets (Atanassov 1986), neutronsophic fuzzy sets (Smarandache 1998), non-stationary fuzzy sets (Garibaldi and Ozen 2007), hesitant fuzzy sets (Torra 2010), Pythagorean fuzzy sets (Yager and Abbasov 2013; Yager 2013), orthopair fuzzy sets (Yager 2016), picture fuzzy sets (Cuong 2014) and spherical fuzzy sets (Kutlu Gundogdu and Kahraman 2019a, c, d).

Kutlu Gündoğdu and Kahraman (2019a, d, e) have recently introduced the spherical fuzzy sets (SFS). SFS are described by a membership degree, a non-membership degree, and a hesitancy degree satisfying the condition that their squared sum is equal to or less than one:

$$0 \leq \mu_{\tilde{A}_S}^2(u) + \nu_{\tilde{A}_S}^2(u) + \pi_{\tilde{A}_S}^2(u) \leq 1 \quad \forall u \in U \tag{1}$$

SFS give a larger domain for experts to assign membership function parameters. This is the superiority of SFS with respect to Pythagorean fuzzy, intuitionistic fuzzy and neutrosophic sets.

In this chapter, a decision making model for performance evaluation of hospitals is developed based on interval-valued spherical fuzzy sets since the expert evaluations of hospital alternatives involve vagueness, impreciseness and hesitancy. This decision model integrates Analytic Hierarchy Process (AHP) with interval-valued spherical fuzzy sets. We apply the developed interval-valued spherical fuzzy AHP method to hospital performance evaluation for organ transplantation. The advantages of the spherical fuzzy sets are that it enables decision makers to independently reflect their hesitancies in the decision process and gives larger domain to decision makers for membership assignment (Kutlu Gündoğdu and Kahraman 2019b). Interval-valued fuzzy sets are used for defining the parameters of a fuzzy set with intervals instead of single values (Kutlu Gündoğdu and Kahraman 2019d). The originality of the chapter comes from the presentation of a novel interval-valued spherical fuzzy AHP method and its application to hospital performance evaluation as the first time. A case study containing four main criteria, two or more sub-criteria under each main criterion and four alternatives is presented.

The rest of this chapter is organized as follows. Section 2 summarizes a literature review. Section 3 includes the introductory definitions and the preliminaries on interval-valued SFS. Section 4 includes interval-valued spherical fuzzy AHP method (IVSF-AHP). Section 5 applies IVSF-AHP method to a high-qualified hospital selection problem and includes a comparative analysis of IVSF-AHP and neutrosophic AHP in Sect. 6. Finally, the study is concluded in the last section.

2 Literature Review

Analytic Hierarchy Process (AHP) introduced by Saaty (1980) is a structured approach used for complex decision making problems. This method lets us construct the decision making criteria as a hierarchy, calculate the weights of the criteria and alternatives and measure the consistency of pairwise comparisons. In the traditional AHP method, decision makers' judgments are represented by exact numbers such as 1, 3, 5, 7, and 9. However, in cases where decision makers cannot express their assessments by exact numbers, fuzzy sets can be employed, providing a mathematical tool to deal with the uncertainties existing in human perceptions (Kahraman and Kaya 2010). Based on this fact, traditional AHP method has been extended to several fuzzy versions because of vague and imprecise data. The first fuzzy AHP method based on triangular fuzzy membership functions was proposed by Van Laarhoven and Pedrycz (1983). Buckley (1985) developed a trapezoidal fuzzy AHP to derive fuzzy weights and performance scores. Chang (1996) proposed an extent analysis method for obtaining synthetic extent values of the pairwise comparisons by utilizing triangular fuzzy numbers. Zeng et al. (2007) developed a new fuzzy AHP method by using arithmetic averaging method to get performance scores and extended the method with different scales including triangular, trapezoidal, and crisp numbers.

Some studies on other fuzzy extensions of AHP are as follows: Fuzzy AHP with type-2 fuzzy sets (Kahraman et al. 2014), fuzzy AHP with intuitionistic fuzzy sets (Büyüközkan and Güleryüz 2016), fuzzy AHP with hesitant fuzzy sets (Oztaysi et al. 2015; Boltürk et al. 2016), and fuzzy AHP with interval-valued intuitionistic fuzzy sets (Oztaysi et al. 2017). A neutrosophic AHP method was developed by Radwan et al. (2016) and applied to the learning management system selection. A neutrosophic

AHP-Delphi group decision-making model based on trapezoidal neutrosophic numbers in order to handle experts' non-deterministic evaluation values was presented by Abdel-Basset et al. (2017). Pythagorean fuzzy AHP method developed by Ilbahar et al. (2018) presented a novel approach to risk assessment for occupational health and safety based on Pythagorean fuzzy AHP and a fuzzy inference system. Boltürk and Kahraman (2018) proposed an AHP method with interval-valued neutrosophic sets (IVN-AHP) and an IVN-AHP method with cosine similarity (IVNAHP-CS). As the number of publications implies, fuzzy AHP method has been used as a research area in numerous problems such as supplier selection, airline retail industry, facility site selection, public transportation leisure travel industry, hydrogen storage systems and performance evaluations.

Fuzzy MCDM methods have often been employed in the evaluation of health-care systems. Dursun et al. (2011) proposed a fuzzy multi-criteria decision making technique for the evaluation of health-care waste (HCW) treatment alternatives. Kuo et al. (2012) used the fuzzy TOPSIS method to rank the severity of the failure modes in health-care systems. Thokala and Duenas (2012) analyzed the possible applications of multi-criteria decision analysis approaches in health technology assessment. Afkham et al. (2012) evaluated health-care service quality by using fuzzy MCDM in Iranian health-care centers. Büyüközkan and Çifçi (2012) presented a hybrid AHP–TOPSIS method for health-care e-service quality strategy analysis. Zeng et al. (2013) modified the VIKOR method for increasing accuracy and validity in health-care management. Diaby et al. (2013) documented MCDM applications in health-care and aimed at identifying publication patterns. Akdag et al. (2014) applied Multiple Criteria Decision Making (MCDM) to evaluate the service quality of Turkish hospitals. They combined the fuzzy set theory and some MCDM methods such as TOPSIS and AHP. Chang (2014) proposed a framework based on the concept of fuzzy set theory and the VIKOR method to provide a rational, scientific and systematic process for evaluating the hospital service quality. Chui et al. (2015) identified heart failure using an electrocardiogram identifier based on fuzzy MCDM. Kulak et al. (2015) developed the fuzzy axiomatic design with risk factors (RFAD) approach and used it in multi attribute comparisons of medical imaging systems in a university hospital.

Ahmadi et al. (2015) employed fuzzy multi-criteria decision making using a hospital information system to study Malaysian public hospitals. Lupo (2016) constructed a fuzzy framework using AHP with triangular fuzzy numbers to assess healthcare service quality for Sicilian public hospitals. Islam et al. (2016) determined the weights of five healthcare quality dimensions in the SERVQUAL approach using fuzzy AHP. Ren et al. (2017) developed a weighted averaging operator based TOPSIS for assessing China's hierarchical medical system. Ravangard et al. (2017) weighted the criteria to rank military hospitals using fuzzy AHP. Yu et al. (2018) determined related criteria concerning drugs in treating asthmatic children and ranked medications using AHP. Wu et al. (2018) presented a hesitant fuzzy linguistic projection model to help hospital decision support systems. Torkzad and Beheshtinia (2019) used four hybrid multi criteria decision-making (MCDM) methods for evaluating hospital service quality.

The literature review above indicates that there are a lot of studies on the measurement of health-care system performances using various fuzzy multi-criteria decision approaches. In the following, a novel comprehensive spherical fuzzy MCDM approach is used to evaluate the performances of several hospitals.

3 Interval-Valued Spherical Fuzzy Sets: Preliminaries

The novel concept of SFS (Spherical Fuzzy Sets) provides a larger preference domain for decision makers to assign membership degrees since the squared sum of the spherical parameters is allowed to be at most 1.0. DMs can define their hesitancy information independently under spherical fuzzy environment. For instance; a decision maker may assign his/her preference for an alternative with respect to a criterion with (0.7, 0.3, 0.4). It is clearly seen that the sum of the parameters is larger than 1 whereas the squared sum is 0.74. SFS are a generalization of Pythagorean Fuzzy Sets (PFS), picture fuzzy sets and neutrosophic sets (Kutlu Gündoğdu and Kahraman 2019a, b, e, f; Kutlu Gündogdu 2019; Kahraman et al. 2019a, b). In the following, definition of SFS is presented:

Definition 3.1 Single valued Spherical Fuzzy Sets (SFS) $\overline{A}_S$ of the universe of discourse U is given by

$$\tilde{A}_S = \left\{ \langle u, (\mu_{\tilde{A}_S}(u), v_{\tilde{A}_S}(u), \pi_{\tilde{A}_S}(u)) \middle| u \in U \right\} \tag{2}$$

where

$$\mu_{\tilde{A}_S}: U \to [0, 1], \quad v_{\tilde{A}_S}(u): U \to [0, 1], \quad \pi_{\tilde{A}_S}: U \to [0, 1]$$

and

$$0 \le \mu^2_{\tilde{A}_S}(u) + v^2_{\tilde{A}_S}(u) + \pi^2_{\tilde{A}_S}(u) \le 1 \quad \forall u \in U \tag{3}$$

For each u, the numbers $\mu_{\tilde{A}_S}(u)$, $v_{\tilde{A}_S}(u)$ and $\pi_{\tilde{A}_S}(u)$ are the degrees of membership, non-membership and hesitancy of u to $\overline{A}_S$, respectively.

In this section, we give the definition of Interval-valued spherical fuzzy sets (IVSFS) and summarize arithmetic operations, aggregation and defuzzification operations.

Definition 3.2 An Interval-Valued Spherical Fuzzy Set $\tilde{A}_S$ of the universe of discourse U is defined as in Eq. (4).

$$\tilde{A}_S = \left\{ \left\langle u, \left(\left[\mu^L_{\tilde{A}_S}(u), \mu^U_{\tilde{A}_S}(u) \right], \left[v^L_{\tilde{A}_S}(u), v^U_{\tilde{A}_S}(u) \right], \left[\pi^L_{\tilde{A}_S}(u), \pi^U_{\tilde{A}_S}(u) \right] \right) \right| u \in U \right\} \tag{4}$$

where $0 \leq \mu^L_{\tilde{A}_S}(u) \leq \mu^U_{\tilde{A}_S}(u) \leq 1, 0 \leq v^L_{\tilde{A}_S}(u) \leq v^U_{\tilde{A}_S}(u) \leq 1$ and $0 \leq \left(\mu^U_{\tilde{A}_S}(u)\right)^2 + \left(v^U_{\tilde{A}_S}(u)\right)^2 + \left(\pi^U_{\tilde{A}_S}(u)\right)^2 \leq 1$. For each $u \in U$, $\mu^U_{\tilde{A}_S}(u)$, $v^U_{\tilde{A}_S}(u)$ and $\pi^U_{\tilde{A}_S}(u)$ are the upper degrees of membership, non-membership and hesitancy of u to $\tilde{A}_S$, respectively.

For an IVSFS $\overline{A}_S$, the pair $\left\langle\left[\mu^L_{\tilde{A}_S}(u), \mu^U_{\tilde{A}_S}(u)\right], \left[v^L_{\tilde{A}_S}(u), v^U_{\tilde{A}_S}(u)\right], \left[\pi^L_{\tilde{A}_S}(u), \pi^U_{\tilde{A}_S}(u)\right]\right\rangle$ is called an interval-valued spherical fuzzy number. For convenience, the pair $\left\langle\left[\mu^L_{\tilde{A}_S}(u), \mu^U_{\tilde{A}_S}(u)\right], \left[v^L_{\tilde{A}_S}(u), v^U_{\tilde{A}_S}(u)\right], \left[\pi^L_{\tilde{A}_S}(u), \pi^U_{\tilde{A}_S}(u)\right]\right\rangle$ is denoted by $\tilde{\alpha} = \langle[a, b], [c, d], [e, f]\rangle$ where $[a, b] \subset [0, 1]$, $[c, d] \subset [0, 1]$, $[e, f] \subset [0, 1]$ and $b^2 + d^2 + f^2 \leq 1$.

Definition 3.3 Let $\tilde{\alpha} = \langle[a, b], [c, d], [e, f]\rangle, \tilde{\alpha}_1 = \langle[a_1, b_1], [c_1, d_1], [e_1, f_1]\rangle$, and $\tilde{\alpha}_2 = \langle[a_2, b_2], [c_2, d_2], [e_2, f_2]\rangle$ be IVSFS then

$$\tilde{\alpha}_1 \cup \tilde{\alpha}_2 = \{[\max\{a_1, a_2\}, \max\{b_1, b_2\}], [\min\{c_1, c_2\}, \min\{d_1, d_2\}], [\min\{e_1, e_2\}, \min\{f_1, f_2\}]\} \tag{5}$$

$$\tilde{\alpha}_1 \cap \tilde{\alpha}_2 = \{[\min\{a_1, a_2\}, \min\{b_1, b_2\}], [\max\{c_1, c_2\}, \max\{d_1, d_2\}], [\min\{e_1, e_2\}, \min\{f_1, f_2\}]\} \tag{6}$$

$$\tilde{\alpha}_1 \oplus \tilde{\alpha}_2 = \left\{\begin{array}{l} \left[\left((a_1)^2 + (a_2)^2 - (a_1)^2(a_2)^2\right)^{1/2}, \left((b_1)^2 + (b_2)^2 - (b_1)^2(b_2)^2\right)^{1/2}\right], [c_1c_2, d_1d_2], \\ \left[\left(\left(1 - (a_2)^2\right)(e_1)^2 + \left(1 - (a_1)^2\right)(e_2)^2 - (e_1)^2(e_2)^2\right)^{1/2}, \right. \\ \left. \left(\left(1 - (b_2)^2\right)(f_1)^2 + \left(1 - (b_1)^2\right)(f_2)^2 - (f_1)^2(f_2)^2\right)^{1/2}\right] \end{array}\right\} \tag{7}$$

$$\tilde{\alpha}_1 \otimes \tilde{\alpha}_2 = \left\{\begin{array}{l} [a_1a_2, b_1b_2], \left[\left((c_1)^2 + (c_2)^2 - (c_1)^2(c_2)^2\right)^{1/2}, \left((d_1)^2 + (d_2)^2 - (d_1)^2(d_2)^2\right)^{1/2}\right], \\ \left[\left(\left(1 - (c_2)^2\right)(e_1)^2 + \left(1 - (c_1)^2\right)(e_2)^2 - (e_1)^2(e_2)^2\right)^{1/2}, \right. \\ \left. \left(\left(1 - (d_2)^2\right)(f_1)^2 + \left(1 - (d_1)^2\right)(f_2)^2 - (f_1)^2(f_2)^2\right)^{1/2}\right] \end{array}\right\} \tag{8}$$

Multiplication by a scalar; $\lambda \geq 0$

$$\lambda \cdot \tilde{\alpha} = \left\{\begin{array}{l} \left[\left(1 - \left(1 - a^2\right)^{\lambda}\right)^{1/2}, \left(1 - \left(1 - b^2\right)^{\lambda}\right)^{1/2}\right], \left[c^{\lambda}, d^{\lambda}\right], \\ \left[\left(\left(1 - a^2\right)^{\lambda} - \left(1 - a^2 - e^2\right)^{\lambda}\right)^{1/2}, \left(\left(1 - b^2\right)^{\lambda} - \left(1 - b^2 - f^2\right)^{\lambda}\right)^{1/2}\right] \end{array}\right\} \tag{9}$$

λth **Power of** $\tilde{\alpha}$; $\lambda \geq 0$

$$\tilde{\alpha}^{\lambda} = \left\{\begin{array}{l} \left[a^{\lambda}, b^{\lambda}\right], \left[\left(1 - \left(1 - c^2\right)^{\lambda}\right)^{1/2}, \left(1 - \left(1 - d^2\right)^{\lambda}\right)^{1/2}\right], \\ \left[\left(\left(1 - c^2\right)^{\lambda} - \left(1 - c^2 - e^2\right)^{\lambda}\right)^{1/2}, \left(\left(1 - d^2\right)^{\lambda} - \left(1 - d^2 - f^2\right)^{\lambda}\right)^{1/2}\right] \end{array}\right\} \tag{10}$$

Definition 3.4 Let $\lambda, \lambda_1, \lambda_2 \geq 0$, then

$$\tilde{\alpha}_1 \oplus \tilde{\alpha}_2 = \tilde{\alpha}_2 \oplus \tilde{\alpha}_1 \tag{11}$$

$$\tilde{\alpha}_1 \otimes \tilde{\alpha}_2 = \tilde{\alpha}_2 \otimes \tilde{\alpha}_1 \tag{12}$$

$$\lambda(\tilde{\alpha}_1 \oplus \tilde{\alpha}_2) = \lambda \cdot \tilde{\alpha}_1 \oplus \lambda \cdot \tilde{\alpha}_2 \tag{13}$$

$$(\tilde{\alpha}_1 \otimes \tilde{\alpha}_2)^{\lambda} = \tilde{\alpha}_1^{\lambda} \otimes \tilde{\alpha}_2^{\lambda} \tag{14}$$

$$\lambda_1 \cdot \tilde{\alpha} \oplus \lambda_2 \cdot \tilde{\alpha} = (\lambda_1 + \lambda_2) \cdot \tilde{\alpha} \tag{15}$$

$$\tilde{\alpha}^{\lambda_1} \otimes \tilde{\alpha}^{\lambda_2} = \tilde{\alpha}^{\lambda_1 + \lambda_2} \tag{16}$$

Definition 3.5 Let $\tilde{\alpha}_j = \langle [a_j, b_j], [c_j, d_j], [e_j, f_j] \rangle$ be a collection of IVSFS. Interval-valued Spherical Weighted Arithmetic Mean (IVSWAM) with respect to, $w_j = (w_1, w_2 \ldots, w_n)$; $w_j \in [0, 1]$ and $\sum_{j=1}^{n} w_j = 1$, IVSWAM is defined as;

$$IVSWAM_w(\tilde{\alpha}_1, \tilde{\alpha}_2 \ldots, \tilde{\alpha}_n) = w_1 \cdot \tilde{\alpha}_1 \oplus w_2 \cdot \tilde{\alpha}_2 \oplus \cdots \oplus w_n \cdot \tilde{\alpha}_n$$
$$= \left\{ \begin{gathered} \left[\left(1 - \prod_{j=1}^{n} (1 - a_j^2)^{w_j}\right)^{1/2}, \left(1 - \prod_{j=1}^{n} (1 - b_j^2)^{w_j}\right)^{1/2} \right], \left[\prod_{j=1}^{n} c_j^{w_j}, \prod_{j=1}^{n} d_j^{w_j} \right], \\ \left[\left(\prod_{j=1}^{n} (1 - a_j^2)^{w_j} - \prod_{j=1}^{n} (1 - a_j^2 - e_j^2)^{w_j} \right)^{1/2}, \left(\prod_{j=1}^{n} (1 - b_j^2)^{w_j} - \prod_{j=1}^{n} (1 - b_j^2 - f_j^2)^{w_j} \right)^{1/2} \right] \end{gathered} \right\} \tag{17}$$

Definition 3.6 Let $\tilde{\alpha}_j = \langle [a_j, b_j], [c_j, d_j], [e_j, f_j] \rangle$ be a collection of IVSFS. Interval-valued Spherical Geometric Mean (IVSWGM) with respect to, $w_j = (w_1, w_2 \ldots, w_n)$; $w_j \in [0, 1]$ and $\sum_{j=1}^{n} w_j = 1$, IVSWGM is defined as;

$$IVSWGM_w(\tilde{\alpha}_1, \tilde{\alpha}_2 \ldots, \tilde{\alpha}_n) = \tilde{\alpha}_1^{w_1} \otimes \tilde{\alpha}_2^{w_2} \otimes \cdots \otimes \tilde{\alpha}_n^{w_n}$$
$$= \left\{ \begin{gathered} \left[\prod_{j=1}^{n} a_j^{w_j}, \prod_{j=1}^{n} b_j^{w_j} \right], \left[\left(1 - \prod_{j=1}^{n} (1 - c_j^2)^{w_j}\right)^{1/2}, \left(1 - \prod_{j=1}^{n} (1 - d_j^2)^{w_j}\right)^{1/2} \right], \\ \left[\left(\prod_{j=1}^{n} (1 - c_j^2)^{w_j} - \prod_{j=1}^{n} (1 - c_j^2 - e_j^2)^{w_j} \right)^{1/2}, \left(\prod_{j=1}^{n} (1 - d_j^2)^{w_j} - \prod_{j=1}^{n} (1 - d_j^2 - f_j^2)^{w_j} \right)^{1/2} \right] \end{gathered} \right\} \tag{18}$$

Definition 3.7 The score function of IVSF number $\tilde{\alpha}$ is defined as

$$Score(\tilde{\alpha}) = S(\tilde{\alpha}) = \frac{a^2 + b^2 - c^2 - d^2 - \left(e/2\right)^2 - \left(f/2\right)^2}{2} \tag{19}$$

Definition 3.8 The accuracy function of IVSF number $\tilde{\alpha}$ is defined as;

$$Accuracy(\tilde{\alpha}) = H(\tilde{\alpha}) = \frac{a^2 + b^2 + c^2 + d^2 + e^2 + f^2}{2} \tag{20}$$

Note that: $\tilde{\alpha}_1 < \tilde{\alpha}_2$ if and only if
$S(\tilde{\alpha}_1) < S(\tilde{\alpha}_2)$ or $S(\tilde{\alpha}_1) = S(\tilde{\alpha}_2)$ and $H(\tilde{\alpha}_1) < H(\tilde{\alpha}_2)$.

4 Extension of AHP with Spherical Fuzzy Sets

The proposed spherical fuzzy AHP method is composed of several steps as given in this section.

Step 1. Construct the hierarchical structure

In this step, a hierarchical structure consisting of at least three levels is developed (Fig. 1). Level 1 represents an objective (selecting the best alternative) based on score index. The score index is estimated based on a finite set of criteria $C = \{C_1, C_2, \ldots C_n\}$, which are shown at Level 2. There are many sub-criteria which are at Level 3 defined for any criterion C in this hierarchical structure. Therefore, at Level 4, a discrete set of m feasible alternative $X = \{x_1, x_2, \ldots x_m\}$ $(m \geq 2)$ is defined.

Step 2. Form pairwise comparisons using interval-valued spherical fuzzy judgment matrices based on the linguistic measure of importance given in Table 1.

Step 3. Calculate the consistency of each pairwise comparison matrix. To do that, switch the linguistic terms in the pairwise comparison matrix to their corresponding score indices given in Table 1. Then, apply the classical consistency check. The threshold of the CR must be less than 10%. If the crisp pairwise comparison matrix is consistent, we continue with its corresponding spherical fuzzy pairwise comparison matrix. For instance,

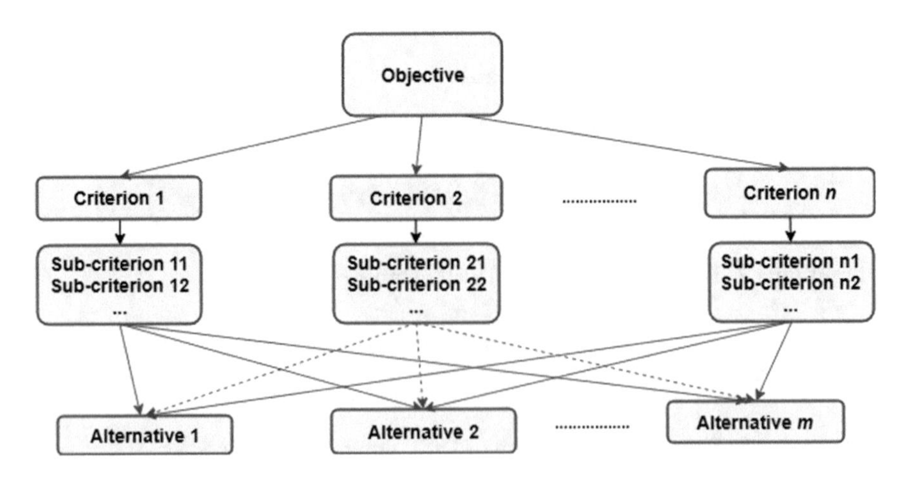

Fig. 1 A hierarchical structure

Table 1 Linguistic terms used for pairwise comparisons (Kutlu Gündoğdu and Kahraman 2019d)

Linguistic terms	$\tilde{A}_S = \left(\left[\mu^L_{\tilde{A}_S}(u), \mu^U_{\tilde{A}_S}(u)\right], \left[v^L_{\tilde{A}_S}(u), v^U_{\tilde{A}_S}(u)\right], \left[\pi^L_{\tilde{A}_S}(u), \pi^U_{\tilde{A}_S}(u)\right]\right)$ *or* $\tilde{\alpha} = \langle [a, b], [c, d], [e, f] \rangle$	Score index
Absolutely more importance (AMI)	([0.85, 0.95], [0.10, 0.15], [0.05, 0.15])	9
Very high importance (VHI)	([0.75, 0.85], [0.15, 0.20], [0.15, 0.20])	7
High importance (HI)	([0.65, 0.75], [0.20, 0.25], [0.20, 0.25])	5
Slightly more importance (SMI)	([0.55, 0.65], [0.25, 0.30], [0.25, 0.30])	3
Equally importance (EI)	([0.50, 0.55], [0.45, 0.55], [0.30, 0.40])	1
Slightly low importance (SLI)	([0.25, 0.30], [0.55, 0.65], [0.25, 0.30])	1/3
Low importance (LI)	([0.20, 0.25], [0.65, 0.75], [0.20, 0.25])	1/5
Very low importance (VLI)	([0.15, 0.20], [0.75, 0.85], [0.15, 0.20])	1/7
Absolutely low importance (ALI)	([0.10, 0.15], [0.85, 0.95], [0.05, 0.15])	1/9

the pairwise comparison matrix $J = \begin{matrix} C_1 \\ C_2 \\ C_3 \end{matrix}\begin{vmatrix} EI & SLI & HI \\ SMI & EI & VHI \\ LI & VLI & EI \end{vmatrix}$ is converted to $J = \begin{matrix} C_1 \\ C_2 \\ C_3 \end{matrix}\begin{vmatrix} 1 & 1/3 & 5 \\ 3 & 1 & 7 \\ 1/5 & 1/7 & 1 \end{vmatrix}$ and the consistency ratio is calculated by using the classical way and found to be 0.048, which indicates that the pairwise comparison matrix is consistent.

Step 4. Calculate the interval-valued spherical fuzzy local weights of criteria and alternatives.

Determine the weight of each alternative using IVSWAM operator given in Eq. (20) with respect to each criterion. The weighted arithmetic mean is used to compute the interval-valued spherical fuzzy weights.

$$IVSWAM_w(\tilde{\alpha}_1, \tilde{\alpha}_2 \ldots, \tilde{\alpha}_k) = w_1 \cdot \tilde{\alpha}_1 \oplus w_2 \cdot \tilde{\alpha}_2 \oplus \cdots \oplus w_k \cdot \tilde{\alpha}_k$$
$$= \left\{ \begin{matrix} \left[\left(1 - \prod_{j=1}^{k} (1 - a_j^2)^{w_j}\right)^{1/2}, \left(1 - \prod_{j=1}^{k} (1 - b_j^2)^{w_j}\right)^{1/2} \right], \left[\prod_{j=1}^{k} c_j^{w_j}, \prod_{j=1}^{k} d_j^{w_j} \right], \\ \left[\left(\prod_{j=1}^{k} (1 - a_j^2)^{w_j} - \prod_{j=1}^{k} (1 - a_j^2 - e_j^2)^{w_j} \right)^{1/2}, \left(\prod_{j=1}^{k} (1 - b_j^2)^{w_j} - \prod_{j=1}^{k} (1 - b_j^2 - f_j^2)^{w_j} \right)^{1/2} \right] \end{matrix} \right\} \quad (21)$$

where $w = 1/n$.

Step 5. Constitute the hierarchical form to obtain global weights

The interval-valued spherical fuzzy weights at each level are aggregated to estimate final ranking orders for the alternatives. This computation is performed from bottom level (alternatives) to top level (objective) as shown in Fig. 1.

We present different IVSF-AHP approaches. In the following, Eqs. (22)–(25) are used in our partially IVSF-AHP approach while Eqs. (25)–(26) are used in completely IVSF-AHP approach.

Equation (22) defuzzifies the criteria weights by using a modified score function. We add 1.0 to the previous definition of score function since a positive score value may be more practical for spherical calculations.

$$Score\left(\tilde{w}_j^s\right) = S\left(\tilde{w}_j^s\right) = \frac{a^2 + b^2 - c^2 - d^2 - \left(e/2\right)^2 - \left(f/2\right)^2}{2} + 1 \quad (22)$$

Equation (23) normalizes the criteria weights:

$$\bar{w}_j^s = \frac{S(\tilde{w}_j^s)}{\sum_{J=1}^{n} S(\tilde{w}_j^s)} \quad (23)$$

Equation (24) is used for weighting the decision matrix.

$$\tilde{\alpha}_{S_{ij}} = \bar{w}_j^s \cdot \tilde{\alpha}_{S_i} = \left\{ \begin{matrix} \left[\left(1 - \left(1 - a_{S_i}^2\right)^{\bar{w}_j^s}\right)^{1/2}, \left(1 - \left(1 - b_{S_i}^2\right)^{\bar{w}_j^s}\right)^{1/2} \right], \left[c_{S_i}^{\bar{w}_j^s}, d_{S_i}^{\bar{w}_j^s} \right], \\ \left[\left(\left(1 - a_{S_i}^2\right)^{\bar{w}_j^s} - \left(1 - a_{S_i}^2 - e_{S_i}^2\right)^{\bar{w}_j^s} \right)^{1/2}, \left(\left(1 - b_{S_i}^2\right)^{\bar{w}_j^s} - \left(1 - b_{S_i}^2 - f_{S_i}^2\right)^{\bar{w}_j^s} \right)^{1/2} \right] \end{matrix} \right\} \quad (24)$$

The final spherical fuzzy AHP score for each alternative is obtained by carrying out the spherical fuzzy addition operator over each global preference weights as given in Eq. (25).

$$\tilde{F} = \sum_{j=1}^{n} \tilde{\alpha}_{S_{ij}} = \tilde{\alpha}_{S_{i1}} \oplus \tilde{\alpha}_{S_{i2}} \cdots \oplus \tilde{\alpha}_{S_{in}} \forall i$$

$$\text{i.e. } \tilde{\alpha}_{S_{11}} \oplus \tilde{\alpha}_{S_{12}} = \left\{ \begin{array}{l} \left[\begin{array}{l} \left(\left(a_{\tilde{\alpha}_{S_{11}}}\right)^2 + \left(a_{\tilde{\alpha}_{S_{12}}}\right)^2 - \left(a_{\tilde{\alpha}_{S_{11}}}\right)^2 \left(a_{\tilde{\alpha}_{S_{12}}}\right)^2 \right)^{1/2}, \\ \left(\left(b_{\tilde{\alpha}_{S_{11}}}\right)^2 + \left(b_{\tilde{\alpha}_{S_{12}}}\right)^2 - \left(b_{\tilde{\alpha}_{S_{11}}}\right)^2 \left(b_{\tilde{\alpha}_{S_{12}}}\right)^2 \right)^{1/2} \end{array} \right], \left[c_{\tilde{\alpha}_{S_{11}}} c_2, d_{\tilde{\alpha}_{S_{11}}} d_2 \right], \\ \left[\begin{array}{l} \left(\left(1 - \left(a_{\tilde{\alpha}_{S_{12}}}\right)^2\right) \left(e_{\tilde{\alpha}_{S_{11}}}\right)^2 + \left(1 - \left(a_{\tilde{\alpha}_{S_{11}}}\right)^2\right) \left(e_{\tilde{\alpha}_{S_{12}}}\right)^2 - \left(e_{\tilde{\alpha}_{S_{11}}}\right)^2 \left(e_{\tilde{\alpha}_{S_{12}}}\right)^2 \right)^{1/2}, \\ \left(\left(1 - \left(b_{\tilde{\alpha}_{S_{12}}}\right)^2\right) \left(f_{\tilde{\alpha}_{S_{11}}}\right)^2 + \left(1 - \left(b_{\tilde{\alpha}_{S_{11}}}\right)^2\right) \left(f_{\tilde{\alpha}_{S_{12}}}\right)^2 - \left(f_{\tilde{\alpha}_{S_{11}}}\right)^2 \left(f_{\tilde{\alpha}_{S_{12}}}\right)^2 \right)^{1/2} \end{array} \right] \end{array} \right\} \quad (25)$$

The second way to follow is to maintain without defuzzification operation. In this case, spherical fuzzy global preference weights are computed based on Eq. (26).

$$\prod_{j=1}^{n} \tilde{\alpha}_{S_{ij}} = \tilde{\alpha}_{S_{i1}} \otimes \tilde{\alpha}_{S_{i2}} \cdots \otimes \tilde{\alpha}_{S_{in}} \forall i$$

$$\text{i.e. } \tilde{\alpha}_{S_{11}} \otimes \tilde{\alpha}_{S_{12}} = \left\{ \begin{array}{l} \left[a_{\tilde{\alpha}_{S_{11}}} a_2, b_{\tilde{\alpha}_{S_{11}}} b_2 \right], \left[\begin{array}{l} \left(\left(c_{\tilde{\alpha}_{S_{11}}}\right)^2 + \left(c_{\tilde{\alpha}_{S_{12}}}\right)^2 - \left(c_{\tilde{\alpha}_{S_{11}}}\right)^2 \left(c_{\tilde{\alpha}_{S_{12}}}\right)^2 \right)^{1/2}, \\ \left(\left(d_{\tilde{\alpha}_{S_{11}}}\right)^2 + \left(d_{\tilde{\alpha}_{S_{12}}}\right)^2 - \left(d_{\tilde{\alpha}_{S_{11}}}\right)^2 \left(d_{\tilde{\alpha}_{S_{12}}}\right)^2 \right)^{1/2} \end{array} \right], \\ \left[\begin{array}{l} \left(\left(1 - \left(c_{\tilde{\alpha}_{S_{12}}}\right)^2\right) \left(e_{\tilde{\alpha}_{S_{11}}}\right)^2 + \left(1 - \left(c_{\tilde{\alpha}_{S_{11}}}\right)^2\right) \left(e_{\tilde{\alpha}_{S_{12}}}\right)^2 - \left(e_{\tilde{\alpha}_{S_{11}}}\right)^2 \left(e_{\tilde{\alpha}_{S_{12}}}\right)^2 \right)^{1/2}, \\ \left(\left(1 - \left(d_{\tilde{\alpha}_{S_{12}}}\right)^2\right) \left(f_{\tilde{\alpha}_{S_{11}}}\right)^2 + \left(1 - \left(d_{\tilde{\alpha}_{S_{11}}}\right)^2\right) \left(f_{\tilde{\alpha}_{S_{12}}}\right)^2 - \left(f_{\tilde{\alpha}_{S_{11}}}\right)^2 \left(f_{\tilde{\alpha}_{S_{12}}}\right)^2 \right)^{1/2} \end{array} \right] \end{array} \right\} \quad (26)$$

Finally, the final score ($\widetilde{F}$) is calculated by using Eq. (21) operator.

Step 6. Defuzzify the final score of each alternative by using the score function given in Eq. (22).

Step 7. Rank the alternatives with respect to the defuzzified final scores. The largest value provides the best alternative.

5 A Case Study: Selecting a High-Qualified Hospital

In Turkey, the hospitals are of high quality but the financial cost is relatively low and this creates a large selection area. Indeed, the Turkish health-care services have a very competitive environment. Therefore, it is necessary to evaluate the health-care services provided by some private hospitals from the patients' viewpoint to make continuous progress. To demonstrate our proposed novel spherical fuzzy AHP methodology, we utilize a case study given by Kutlu Gündoğdu et al. (2018). The present study is implemented for selecting high quality hospital for organ transplantation in Istanbul. For this goal, four popular hospitals in Istanbul are assessed based on service quality dimensions as given in Fig. 2. According to the literature, a variety of criteria and sub-criteria are examined. In this study, four alternative hospitals, four main criteria and twelve sub-criteria are evaluated by the consensus of three decision makers.

The consistency ratio of an interval-valued spherical fuzzy pairwise comparison matrix is calculated by using the corresponding score indices in Table 1 and

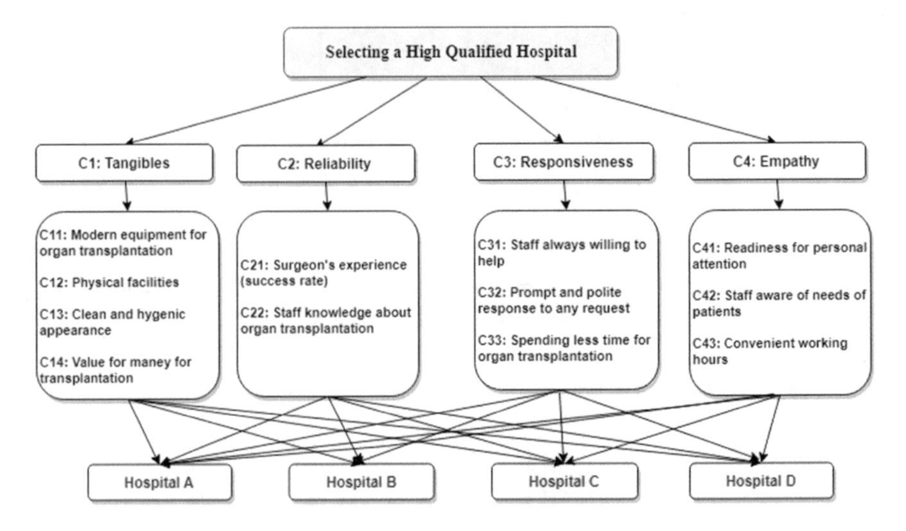

Fig. 2 Hierarchical structure for the problem (Kutlu Gündoğdu et al. 2018)

then applying the classical consistency measurement. Pairwise comparisons and the obtained interval-valued spherical weights ($\tilde{w}^s$) and crisp weights ($\bar{w}^s$) are presented in Tables 2, 3, 4, 5, 6, 7, 8, 9, 10, 11, 12, 13, 14, 15, 16, 17 and 18 together with their consistency ratios (CR).

Based on Eq. (26) in Step 5 of the completely fuzzy approach, Table 19 includes the final spherical fuzzy global priority weights of the competitive hospitals with respect to the considered criteria. Table 20 includes the spherical fuzzy priority weights of

Table 2 Pairwise comparison of main criteria

Criteria	C1	C2	C3	C4	$\tilde{w}^s$	$\bar{w}^s$
C1	EI	LI	ALI	SLI	([0.31, 0.36], [0.61, 0.71], [0.23, 0.31])	0.151
C2	HI	EI	SLI	HI	([0.55, 0.64], [0.32, 0.39], [0.23, 0.30])	0.284
C3	AMI	SMI	EI	VHI	([0.71, 0.82], [0.20, 0.27], [0.18, 0.25])	0.358
C4	SMI	LI	VLI	EI	([0.40, 0.47], [0.48, 0.57], [0.24, 0.31])	0.208

CR = 0.067

Table 3 Pairwise comparison of hospital performance with respect to tangibles

C1	C11	C12	C13	C14	$\tilde{w}^s$	$\bar{w}^s$
C11	EI	SMI	VHI	SMI	([0.61, 0.70], [0.25, 0.32], [0.24, 0.29])	0.320
C12	SLI	EI	SMI	SLI	([0.42, 0.49], [0.43, 0.51], [0.27,0.34])	0.229
C13	VLI	SLI	EI	VLI	([0.31, 0.35], [0.61, 0.71], [0.23,0.31])	0.154
C14	SLI	SMI	VHI	EI	([0.57, 0.66], [0.31, 0.38], [0.24, 0.30])	0.297

CR = 0.060

Table 4 Pairwise comparison of hospital performance with respect to reliability

C2	C21	C22	$\tilde{w}^s$	$\bar{w}^s$
C21	EI	ALI	([0.37, 0.42], [0.64, 0.73], [0.23, 0.33])	0.29
C22	AMI	EI	([0.74, 0.86], [0.26, 0.33], [0.17, 0.25])	0.71

Table 5 Pairwise comparison of hospital performance with respect to responsiveness

C3	C31	C32	C33	$\tilde{w}^s$	$\bar{w}^s$
C31	EI	AMI	VHI	([0.74, 0.86], [0.19, 0.25], [0.17, 0.23])	0.490
C32	ALI	EI	SLI	([0.34, 0.38], [0.59, 0.70], [0.24, 0.32])	0.209
C33	VLI	SMI	EI	([0.45, 0.52], [0.44, 0.52], [0.25, 0.33])	0.300

CR = 0.092

Table 6 Pairwise comparison of hospital performance with respect to empathy

C4	C41	C42	C43	$\tilde{w}^s$	$\bar{w}^s$
C41	EI	VHI	HI	([0.65, 0.75], [0.24, 0.30], [0.22, 0.28])	0.451
C42	VLI	EI	SLI	([0.34, 0.39], [0.57, 0.67], [0.25, 0.33])	0.229
C43	LI	SMI	EI	([0.45, 0.53], [0.42, 0.50], [0.26, 0.33])	0.320

CR = 0.057

Table 7 Pairwise comparison of alternatives with respect to sub-criterion C11

C11	A	B	C	D	$\tilde{w}^s$	$\bar{w}^s$
A	EI	SMI	VHI	SMI	([0.61, 0.70], [0.25, 0.32], [0.24, 0.29])	0.320
B	SLI	EI	SMI	SLI	([0.42, 0.49], [0.43, 0.51], [0.27, 0.33])	0.229
C	VLI	SLI	EI	VLI	([0.31, 0.35], [0.61, 0.71], [0.23, 0.30])	0.154
D	SLI	SMI	VHI	EI	([0.57, 0.66], [0.31, 0.38], [0.24, 0.30])	0.297

CR = 0.060

the hospitals based on Eq. (24) in Step 5 of the partially fuzzy approach. Table 21 presents the results of total spherical fuzzy weights of the hospitals and their score

Table 8 Pairwise comparison of alternatives with respect to sub-criterion C12

C12	A	B	C	D	$\tilde{w}^s$	$\bar{w}^s$
A	EI	SLI	ALI	ALI	([0.30, 0.34], [0.65, 0.75], [0.21, 0.29])	0.134
B	SMI	EI	SLI	LI	([0.41, 0.48], [0.45, 0.53], [0.26, 0.33])	0.213
C	AMI	SMI	EI	SMI	([0.66, 0.78], [0.23, 0.29], [0.21, 0.27])	0.331
D	AMI	HI	SLI	EI	([0.65, 0.77], [0.27, 0.34], [0.20, 0.27])	0.322

CR = 0.098

Table 9 Pairwise comparison of alternatives with respect to sub-criterion C13

C13	A	B	C	D	$\tilde{w}^s$	$\bar{w}^s$
A	EI	SMI	VLI	SLI	([0.41, 0.48], [0.46, 0.55], [0.25, 0.32])	0.214
B	SLI	EI	ALI	LI	([0.31, 0.36], [0.61, 0.71], [0.23, 0.31])	0.152
C	VHI	AMI	EI	HI	([0.72, 0.84], [0.19, 0.25], [0.17, 0.24])	0.366
D	SMI	HI	LI	EI	([0.52, 0.60], [0.35, 0.42], [0.24, 0.31])	0.268

CR = 0.064

Table 10 Pairwise comparison of alternatives with respect to sub-criterion C14

C14	A	B	C	D	$\tilde{w}^s$	$\bar{w}^s$
A	EI	SMI	VLI	SLI	([0.41, 0.48], [0.46, 0.55], [0.25, 0.32])	0.214
B	SLI	EI	ALI	LI	([0.31, 0.36], [0.61, 0.71], [0.23, 0.31])	0.152
C	VHI	AMI	EI	HI	([0.72, 0.84], [0.19, 0.25], [0.17, 0.24])	0.366
D	SMI	HI	LI	EI	([0.52, 0.60], [0.35, 0.42], [0.24, 0.31])	0.268

CR = 0.064

Table 11 Pairwise comparison of alternatives with respect to sub-criterion C21

C21	A	B	C	D	$\tilde{w}^s$	$\bar{w}^s$
A	EI	LI	SLI	ALI	([0.31, 0.36], [0.61, 0.71], [0.23, 0.31])	0.152
B	HI	EI	SMI	LI	([0.52, 0.60], [0.35, 0.42], [0.24, 0.31])	0.268
C	SMI	SLI	EI	VLI	([0.41, 0.48], [0.46, 0.55], [0.25, 0.32])	0.214
D	AMI	HI	VHI	EI	([0.72, 0.84], [0.19, 0.25], [0.17, 0.24])	0.366

CR = 0.064

Table 12 Pairwise comparison of alternatives with respect to sub-criterion C22

C22	A	B	C	D	$\tilde{w}^s$	$\bar{w}^s$
A	EI	HI	AMI	SMI	([0.68, 0.80], [0.22, 0.28], [0.20, 0.26])	0.345
B	LI	EI	SMI	LI	([0.41, 0.48], [0.47, 0.55], [0.24, 0.32])	0.212
C	ALI	SLI	EI	VLI	([0.30, 0.35], [0.63, 0.73], [0.22, 0.30])	0.143
D	SLI	HI	VHI	EI	([0.59, 0.69], [0.29, 0.37], [0.22, 0.29])	0.300

CR = 0.066

Table 13 Pairwise comparison of alternatives with respect to sub-criterion C31

C31	A	B	C	D	$\tilde{w}^s$	$\bar{w}^s$
A	EI	HI	AMI	SMI	([0.68, 0.80], [0.22, 0.28], [0.20, 0.26])	0.345
B	LI	EI	SMI	LI	([0.41, 0.48], [0.47, 0.55], [0.24, 0.32])	0.212
C	ALI	SLI	EI	VLI	([0.30, 0.35], [0.63, 0.73], [0.22, 0.30])	0.143
D	SLI	HI	VHI	EI	([0.59, 0.69], [0.29, 0.37], [0.22, 0.29])	0.300

CR = 0.066

Table 14 Pairwise comparison of alternatives with respect to sub-criterion C32

C32	A	B	C	D	$\tilde{w}^s$	$\bar{w}^s$
A	EI	SMI	AMI	VHI	([0.71, 0.82], [0.20, 0.27], [0.19, 0.25])	0.359
B	SLI	EI	SMI	HI	([0.52, 0.61], [0.33, 0.40], [0.25, 0.32])	0.272
C	ALI	SLI	EI	SLI	([0.32, 0.36], [0.58, 0.69], [0.24, 0.32])	0.160
D	VLI	LI	SMI	EI	([0.40, 0.47], [0.48, 0.57], [0.24, 0.31])	0.209

CR = 0.096

Table 15 Pairwise comparison of alternatives with respect to sub-criterion C33

C33	A	B	C	D	$\tilde{w}^s$	$\bar{w}^s$
A	EI	SMI	VHI	SMI	([0.61, 0.70], [0.25, 0.32], [0.24, 0.29])	0.320
B	SLI	EI	SMI	SLI	([0.42, 0.49], [0.43, 0.51], [0.27, 0.34])	0.229
C	VLI	SLI	EI	VLI	([0.31, 0.35], [0.61, 0.71], [0.23, 0.31])	0.154
D	SLI	SMI	VHI	EI	([0.57, 0.66], [0.31, 0.38], [0.24, 0.30])	0.297

CR = 0.060

Table 16 Pairwise comparison of alternatives with respect to sub-criterion C41

C41	A	B	C	D	$\tilde{w}^s$	$\bar{w}^s$
A	EI	SLI	ALI	ALI	([0.30, 0.34], [0.65, 0.75], [0.21, 0.29])	0.134
B	SMI	EI	SLI	LI	([0.41, 0.48], [0.45, 0.53], [0.26, 0.33])	0.213
C	AMI	SMI	EI	SMI	([0.66, 0.78], [0.23, 0.29], [0.21, 0.27])	0.331
D	AMI	HI	SLI	EI	([0.65, 0.77], [0.27, 0.34], [0.20, 0.27])	0.322

CR = 0.098

Table 17 Pairwise comparison of alternatives with respect to sub-criterion C42

C42	A	B	C	D	$\tilde{w}^s$	$\bar{w}^s$
A	EI	LI	ALI	SLI	([0.31, 0.36], [0.61, 0.71], [0.23, 0.31])	0.151
B	HI	EI	SLI	HI	([0.55, 0.64], [0.32, 0.39], [0.24, 0.30])	0.284
C	AMI	SMI	EI	VHI	([0.71, 0.82], [0.20, 0.27], [0.19, 0.25])	0.358
D	SMI	LI	VLI	EI	([0.40, 0.47], [0.48, 0.57], [0.24, 0.31])	0.208

CR = 0.064

Table 18 Pairwise comparison of alternatives with respect to sub-criterion C43

C43	A	B	C	D	$\tilde{w}^s$	$\bar{w}^s$
A	EI	HI	SMI	VHI	([0.63, 0.73], [0.24, 0.30], [0.22, 0.28])	0.329
B	LI	EI	SLI	SMI	([0.41, 0.48], [0.45, 0.53], [0.26, 0.33])	0.222
C	SLI	SMI	EI	VHI	([0.57, 0.66], [0.31, 0.38], [0.24, 0.30])	0.296
D	VLI	SLI	VLI	EI	([0.31, 0.35], [0.61, 0.71], [0.23, 0.31])	0.153

CR = 0.052

Table 19 Spherical fuzzy weight matrix based on completely fuzzy approach

Alternatives	C11	C12	C13	C14	C21	C22	C31	C32	C33	C41	C42	C43
A	([0.11, 0.18], [0.67, 0.77], [0.32, 0.37])	([0.04, 0.06], [0.84, 0.92], [0.26, 0.27])	([0.04, 0.06], [0.83, 0.91], [0.26, 0.27])	([0.07, 0.11], [0.74, 0.84], [0.31, 0.34])	([0.06, 0.10], [0.82, 0.90], [0.27, 0.30])	([0.28, 0.44], [0.45, 0.55], [0.32, 0.40])	([0.36, 0.56], [0.35, 0.45], [0.30, 0.39])	([0.17, 0.26], [0.64, 0.75], [0.30, 0.37])	([0.19, 0.30], [0.53, 0.62], [0.34, 0.41])	([0.08, 0.12], [0.76, 0.86], [0.28, 0.31])	([0.04, 0.07], [0.82, 0.90], [0.27, 0.29])	([0.12, 0.18], [0.64, 0.73], [0.34, 0.39])
B	([0.08, 0.12], [0.72, 0.82], [0.32, 0.35])	([0.05, 0.08], [0.77, 0.86], [0.31, 0.33])	([0.03, 0.04], [0.87, 0.94], [0.24, 0.24])	([0.05, 0.08], [0.80, 0.89], [0.28, 0.29])	([0.11, 0.16], [0.73, 0.82], [0.31, 0.36])	([0.17, 0.26], [0.59, 0.69], [0.33, 0.39])	([0.21, 0.34], [0.53, 0.63], [0.31, 0.39])	([0.12, 0.19], [0.67, 0.78], [0.32, 0.37])	([0.13, 0.21], [0.61, 0.71], [0.34, 0.40])	([0.11, 0.17], [0.65, 0.75], [0.33, 0.38])	([0.08, 0.12], [0.73, 0.83], [0.32, 0.35])	([0.08, 0.12], [0.70, 0.80], [0.33, 0.37])
C	([0.06, 0.09], [0.79, 0.88], [0.28, 0.30])	([0.09, 0.14], [0.72, 0.82], [0.31, 0.35])	([0.07, 0.11], [0.79, 0.88], [0.27, 0.30])	([0.13, 0.20], [0.67, 0.78], [0.30, 0.36])	([0.08, 0.13], [0.77, 0.85], [0.30, 0.34])	([0.12, 0.19], [0.70, 0.81], [0.29, 0.34])	([0.16, 0.25], [0.67, 0.77], [0.28, 0.34])	([0.08, 0.12], [0.77, 0.86], [0.28, 0.32])	([0.10, 0.15], [0.72, 0.82], [0.30, 0.34])	([0.17, 0.28], [0.56, 0.66], [0.33, 0.40])	([0.10, 0.15], [0.71, 0.81], [0.31, 0.35])	([0.10, 0.16], [0.65, 0.75], [0.34, 0.39])
D	([0.11, 0.17], [0.68, 0.79], [0.32, 0.36])	([0.08, 0.13], [0.72, 0.82], [0.31, 0.35])	([0.05, 0.08], [0.81, 0.89], [0.28, 0.29])	([0.09, 0.14], [0.71, 0.81], [0.32, 0.35])	([0.15, 0.22], [0.70, 0.79], [0.30, 0.36])	([0.24, 0.38], [0.48, 0.59], [0.33, 0.40])	([0.31, 0.48], [0.39, 0.50], [0.31, 0.39])	([0.10, 0.15], [0.72, 0.82], [0.30, 0.34])	([0.18, 0.28], [0.55, 0.65], [0.34, 0.41])	([0.17, 0.27], [0.57, 0.68], [0.33, 0.39])	([0.06, 0.09], [0.78, 0.87], [0.30, 0.32])	([0.06, 0.09], [0.78, 0.87], [0.30, 0.32])

Table 20 Spherical fuzzy weight matrix partially fuzzy approach

Alternatives	C11	C12	C13	C14	C21	C22	C31	C32	C33	C41	C42	C43
A	([0.15, 0.18], [0.94, 0.95], [0.07, 0.09])	([0.06, 0.07], [0.99, 0.99], [0.04, 0.06])	([0.07, 0.08], [0.98, 0.99], [0.04, 0.06])	([0.09, 0.11], [0.97, 0.97], [0.06, 0.08])	([0.09, 0.11], [0.96, 0.97], [0.07, 0.10])	([0.34, 0.43], [0.74, 0.78], [0.12, 0.18])	([0.32, 0.40], [0.77, 0.80], [0.11, 0.17])	([0.22, 0.28], [0.89, 0.91], [0.07, 0.12])	([0.22, 0.27], [0.86, 0.88], [0.10, 0.14])	([0.09, 0.11], [0.96, 0.97], [0.07, 0.10])	([0.07, 0.08], [0.98, 0.98], [0.05, 0.07])	([0.18, 0.22], [0.91, 0.92], [0.07, 0.11])
B	([0.10, 0.11], [0.96, 0.97], [0.07, 0.09])	([0.08, 0.10], [0.97, 0.98], [0.05, 0.07])	([0.05, 0.06], [0.99, 0.99], [0.04, 0.05])	([0.07, 0.08], [0.98, 0.98], [0.05, 0.07])	([0.16, 0.19], [0.92, 0.93], [0.08, 0.11])	([0.19, 0.22], [0.86, 0.89], [0.12, 0.16])	([0.18, 0.21], [0.87, 0.90], [0.11, 0.15])	([0.15, 0.18], [0.92, 0.93], [0.08, 0.11])	([0.14, 0.17], [0.91, 0.93], [0.10, 0.13])	([0.13, 0.16], [0.93, 0.94], [0.09, 0.12])	([0.13, 0.16], [0.95, 0.96], [0.06, 0.09])	([0.11, 0.13], [0.95, 0.96], [0.07, 0.10])
C	([0.07, 0.08], [0.98, 0.98], [0.05, 0.07])	([0.14, 0.18], [0.95, 0.96], [0.05, 0.08])	([0.13, 0.17], [0.96, 0.97], [0.04, 0.07])	([0.18, 0.23], [0.93, 0.94], [0.05, 0.09])	([0.12, 0.15], [0.94, 0.95], [0.08, 0.11])	([0.14, 0.16], [0.91, 0.94], [0.10, 0.14])	([0.13, 0.15], [0.92, 0.95], [0.10, 0.14])	([0.09, 0.10], [0.96, 0.97], [0.07, 0.10])	([0.10, 0.12], [0.95, 0.96], [0.08, 0.11])	([0.23, 0.29], [0.87, 0.89], [0.08, 0.14])	([0.18, 0.23], [0.93, 0.94], [0.06, 0.10])	([0.16, 0.19], [0.93, 0.94], [0.07, 0.10])
D	([0.14, 0.17], [0.94, 0.95], [0.06, 0.09])	([0.14, 0.18], [0.96, 0.96], [0.05, 0.08])	([0.08, 0.10], [0.98, 0.98], [0.04, 0.06])	([0.12, 0.14], [0.95, 0.96], [0.06, 0.08])	([0.24, 0.31], [0.87, 0.89], [0.07, 0.12])	([0.29, 0.35], [0.78, 0.82], [0.12, 0.17])	([0.27, 0.33], [0.81, 0.84], [0.11, 0.16])	([0.11, 0.14], [0.95, 0.96], [0.07, 0.10])	([0.20, 0.24], [0.88, 0.90], [0.09, 0.13])	([0.22, 0.28], [0.88, 0.90], [0.08, 0.13])	([0.09, 0.11], [0.97, 0.97], [0.06, 0.08])	([0.08, 0.09], [0.97, 0.98], [0.06, 0.09])

Table 21 Score values and ranking based on completely fuzzy approach

Alternatives	Score values	Ranking
A	0.502	1
B	0.418	3
C	0.394	4
D	0.499	2

Table 22 Score values and ranking based on partially fuzzy approach

Alternatives	Score values	Ranking
A	1.265	1
B	0.953	4
C	1.000	3
D	1.234	2

values for the completely fuzzy approach. Based on Eq. (25) in Step 5, Table 22 gives the results of total SF weights of the hospitals for the partially fuzzy approach.

As seen in Tables 21 and 22, the first and second alternatives are same in both approaches. However, alternative 3 and alternative 4 replaced each other. We suggest the completely fuzzy approach to be preferred since it does not let any information be lost.

6 A Comparative Analysis

Buckley's fuzzy AHP was extended using interval neutrosophic sets by Kahraman et al. (2018). Neutrosophic AHP has been presented in their paper and performed to the performance comparison of law firms successfully. In this study, our proposed methodology compared with neutrosophic AHP for the hospital performance assessment. Table 23 presents the neutrosophic linguistic scale, which we use for the comparison purpose (Kahraman et al. 2018; Kutlu Gündoğdu and Kahraman 2019b).

In this section, the same pairwise comparisons are used as given in Tables 2, 3, 4, 5, 6, 7, 8, 9, 10, 11, 12, 13, 14, 15, 16, 17 and 18.

We checked the consistency of pairwise comparison matrices by using Eq. (27).

$$dv = 0.6 + 0.4T - 0.2I - 0.4F \tag{27}$$

where dv is the deneutrosophicated value. For the next step, geometric mean for main criteria and sub-criteria will be calculated based on Eqs. (28), (29), and (30).

Table 23 Neutrosophic AHP linguistic scale

Linguistic terms	(T, I, F)
Absolutely more importance (AMI)	(1, 0.07, 0.015)
Very high importance (VHI)	(0.9, 0.2, 0.1)
High importance (HI)	(0.8, 0.3, 0.2)
Slightly more importance (SMI)	(0.7, 0.4, 0.3)
Equally importance (EI)	(1, 1, 1)
Slightly low importance (SLI)	(0.02, 0.226, 0.623)
Low importance (LI)	(0.016, 0.145, 0.679)
Very low importance (VLI)	(0.013, 0.100, 0.711)
Absolutely low importance (ALI)	(0.009, 0.005, 0.765)

$$\begin{aligned} T_1 &= [1 \times T_{12} \times \cdots \times T_{1n}]^{1/n} \\ &\vdots \\ T_n &= [T_{n1} \times T_{n2} \times \cdots \times 1]^{1/n} \end{aligned} \tag{28}$$

$$\begin{aligned} I_{1m} &= [1 \times I_{12m} \times \cdots \times I_{1nm}]^{1/n} \\ &\vdots \\ I_{nm} &= [I_{n1m} \times I_{n2m} \times \cdots \times 1]^{1/n} \end{aligned} \tag{29}$$

$$\begin{aligned} F_{1m} &= [1 \times F_{12u} \times \cdots \times F_{1nu}]^{1/n} \\ &\vdots \\ F_{im} &= [F_{n1m} \times F_{n2m} \times \cdots \times 1]^{1/n} \end{aligned} \tag{30}$$

The results of Eqs. (28), (29), and (30) for the criteria are given in Table 24. Because of the space restrictions, the rest of the geometric means of the sub-criteria and alternatives is not given.

Calculated neutrosophic weights of main criteria and sub-criteria by dividing T, I, F values with the sum of the geometric means in the row for lower, medium and upper parameters are given in Eq. (31). Assume that the sums of the geometric mean values in the row are a_{1s} for lower parameters, a_{2s} for medium parameters, and a_{3s} for upper parameters.

Table 24 Geometric means of the main criteria

Criteria	Geometric mean		
	T	I	F
C1	0.041	0.113	0.754
C2	0.336	0.378	0.397
C3	0.891	0.274	0.146
C4	0.110	0.276	0.617
Total	1.378	1.041	1.914

Table 25 Relative fuzzy priorities of the main criteria

Criteria	$\tilde{r}_{ij}$			Deneutrosophicated value	Normalized value
	T	I	F		
C1	0.021	0.109	0.547	0.368	0.190
C2	0.176	0.363	0.288	0.483	0.250
C3	0.466	0.263	0.106	0.691	0.358
C4	0.057	0.265	0.448	0.391	0.202

$$\tilde{r}_{ij} = \left\{ \begin{array}{c} \left(\frac{a_{1l}}{a_{3s}}, \frac{b_{1m}}{a_{2s}}, \frac{c_{1u}}{a_{1s}} \right) \\ \left(\frac{a_{2l}}{a_{3s}}, \frac{b_{2m}}{a_{2s}}, \frac{c_{2u}}{a_{1s}} \right) \\ \vdots \\ \left(\frac{a_{il}}{a_{3s}}, \frac{b_{im}}{a_{2s}}, \frac{c_{iu}}{a_{1s}} \right) \end{array} \right\} \tag{31}$$

The result of Eqs. (28)–(31) for the criteria is given in Table 25.

By using Eq. (27), we defuzzified and normalized the neutrosophic weights of criteria with the performance scores of the alternatives and calculated the overall performance score to aggregate them based on Eq. (31) as follows:

$$\bar{U}_i = \sum_{j=1}^{n} \bar{w}_j \bar{r}_{ij}, \quad \forall i \tag{32}$$

Table 26 indicates the local and global weights of each sub-criterion.

Table 27 shows that neutrosophic performance scores of each alternative based on Eqs. (28), (29), (30) and (31).

The result of Eq. (32) for each alternative is given in Table 28.

When compared with IVSF-AHP, the ranking is the same as the partially fuzzy AHP approach. However, the ranking is slightly different from the completely fuzzy AHP approach. The best alternative and the second best alternative is the same for all approaches but third and fourth alternatives are replaced each other. The main reason of this difference is the defuzzification step that may cause a little bit information loss.

7 Conclusion

New extensions of ordinary fuzzy sets such as intuitionistic fuzzy sets, hesitant fuzzy sets, Pythagorean fuzzy sets, picture fuzzy sets, and spherical fuzzy sets have been often used in obtaining the new fuzzy versions of almost every approach or method in the literature. In this study, we employed the interval-valued spherical fuzzy sets

Table 26 Local and global weights of each sub-criterion

	C11	C12	C13	C14	C21	C22	C31	C32	C33	C41	C42	C43
Local weights	0.341	0.205	0.187	0.267	0.257	0.743	0.520	0.229	0.251	0.488	0.227	0.285
Global weights	0.065	0.039	0.036	0.051	0.064	0.186	0.186	0.082	0.090	0.099	0.046	0.058

Table 27 Neutrosophic performance scores of each alternative

Alternatives	C11	C12	C13	C14	C21	C22	C31	C32	C33	C41	C42	C43
A	0.328	0.186	0.203	0.203	0.192	0.348	0.348	0.36	0.328	0.186	0.322	0.334
B	0.213	0.193	0.192	0.192	0.241	0.200	0.200	0.242	0.213	0.193	0.187	0.211
C	0.189	0.336	0.365	0.365	0.203	0.188	0.188	0.193	0.189	0.336	0.200	0.267
D	0.271	0.285	0.241	0.241	0.365	0.263	0.263	0.205	0.271	0.285	0.291	0.188

Table 28 Overall performance scores of each alternative and ranking

Overall performance score ($\bar{U}_i$)	Ranking
0.299	1
0.206	4
0.230	3
0.264	2

in the extension of AHP method to incorporate a larger domain for the assignment of membership, non-membership and hesitancy degrees. These parameters can be independently assigned in IVSFS. An interval-valued spherical fuzzy AHP analysis has been used to compare hospital performances in terms of service quality. IVSF-AHP has been performed to measure the SERVQUAL of four hospitals in Istanbul by computing the priority of each of health-care SERVQUAL dimensions. Then, these priorities have been used for determining the best hospital from the patient's perspective. Consistencies of spherical fuzzy pairwise comparison matrices have been checked and continued to operations when their consistencies are satisfied. Hospital performance evaluation problem has been successfully solved by IVSF-AHP and compared with neutrosophic AHP. The ranking results in these approaches are almost same. Hospital managers can use the proposed approach under vagueness and impreciseness in order to ensure their service quality.

For further research, we suggest interval-valued SF-AHP to be compared with the single valued single valued SF-AHP. Besides, other fuzzy extensions of AHP such as hesitant fuzzy AHP and Pythagorean fuzzy AHP can be employed in order to compare with IVSF-AHP.

References

Abdel-Basset M, Mohamed M, Zhou Y, Hezam I (2017) Multi-criteria group decision making based on neutrosophic analytic hierarchy process. J Intell Fuzzy Syst 33(6):4055–4066

Afkham L, Abdi F, Komijan A (2012) Evaluation of service quality by using fuzzy MCDM: a case study in Iranian health-care centers. Manag Sci Lett 2(1):291–300

Ahmadi H, Nilashi M, Ibrahim O (2015) Organizational decision to adopt hospital information system: an empirical investigation in the case of Malaysian public hospitals. Int J Med Inform 84(3):166–188

Akdag H, Kalaycı T, Karagöz S, Zülfikar H, Giz D (2014) The evaluation of hospital service quality by fuzzy MCDM. Appl Soft Comput 23:239–248

Atanassov KT (1986) Intuitionistic fuzzy sets. Fuzzy Sets Syst 20(1):87–96

Bolturk E, Kahraman C (2018) A novel interval-valued neutrosophic AHP with cosine similarity measure. Soft Comput 1:1–8

Boltürk E, Çevik Onar S, Öztayşi B, Kahraman C, Goztepe K (2016) Multi-attribute warehouse location selection in humanitarian logistics using hesitant fuzzy AHP. Int J Anal Hierarchy Process 8:271–298

Buckley JJ (1985) Fuzzy hierarchical analysis. Fuzzy Sets Syst 17(3):233–247

Büyüközkan G, Çifçi G (2012) A combined fuzzy AHP and fuzzy TOPSIS based strategic analysis of electronic service quality in healthcare industry. Expert Syst Appl 39(3):2341–2354

Büyüközkan G, Güleryüz S (2016) A new integrated intuitionistic fuzzy group decision making approach for product development partner selection. Comput Ind Eng 102:383–395
Chang DY (1996) Applications of the extent analysis method on fuzzy AHP. Eur J Oper Res 95(3):649–655
Chang TH (2014) Fuzzy VIKOR method: a case study of the hospital service evaluation in Taiwan. Inf Sci 271:196–212
Chui KT, Tsang KF, Wu CK, Hung FH, Chi HR, Chung HSH, Ko KT et al (2015) Cardiovascular diseases identification using electrocardiogram health identifier based on multiple criteria decision making. Exp Syst Appl 42(13):5684–5695
Cuong BC (2014) Picture fuzzy sets. J Comput Sci Cybern 30(4):409
Diaby V, Campbell K, Goeree R (2013) Multi-criteria decision analysis (MCDA) in health care: a bibliometric analysis. Oper Res Health Care 2(1–2):20–24
Dursun M, Karsak EE, Karadayi MA (2011) Assessment of health-care waste treatment alternatives using fuzzy multi-criteria decision making approaches. Resour Conserv Recycl 57:98–107
Garibaldi JM, Ozen T (2007) Uncertain fuzzy reasoning: a case study in modelling expert decision making. IEEE Trans Fuzzy Syst 15(1):16–30
Grattan-Guinness I (1976) Fuzzy membership mapped onto intervals and many-valued quantities. Math Logic Q 22(1):149–160
Ilbahar E, Karaşan A, Cebi S, Kahraman C (2018) A novel approach to risk assessment for occupational health and safety using Pythagorean fuzzy AHP & fuzzy inference system. Saf Sci 103:124–136
Islam R, Ahmed S, Tarique KM (2016) Prioritisation of service quality dimensions for healthcare sector. Int J Med Eng Inf 8(2):108–123
Jahn KU (1975) Intervall-wertige Mengen. Mathematische Nachrichten 68(1):115–132
Kahraman C, Kaya İ (2010) A fuzzy multicriteria methodology for selection among energy alternatives. Expert Syst Appl 37(9):6270–6281
Kahraman C, Öztayşi B, Sarı İU, Turanoğlu E (2014) Fuzzy analytic hierarchy process with interval type-2 fuzzy sets. Knowl-Based Syst 59:48–57
Kahraman C, Öztayşi B, Onar SÇ, Boltürk E (2018) Neutrosophic AHP and prioritization of legal service outsourcing firms/law offices. World Scientific Proceedings Series on Computer Engineering and Information Science, Data Science and Knowledge Engineering for Sensing Decision Support, pp 1176–1183. https://doi.org/10.1142/9789813273238_0148
Kahraman C, Gundogdu FK, Onar SC, Oztaysi B (2019). Hospital location selection using spherical fuzzy TOPSIS. In: 2019 conference of the international fuzzy systems association and the European Society for fuzzy logic and technology (EUSFLAT 2019). Atlantis Press, Aug 2019
Kulak O, Goren HG, Supciller AA (2015) A new multi criteria decision making approach for medical imaging systems considering risk factors. Appl Soft Comput 35:931–941
Kuo RJ, Wu YH, Hsu TS (2012) Integration of fuzzy set theory and TOPSIS into HFMEA to improve outpatient service for elderly patients in Taiwan. J Chin Med Assoc 75(7):341–348
Kutlu Gündoğdu F (2019) Principals of spherical fuzzy sets. In: International conference on intelligent and fuzzy systems. Springer, Cham, pp 15–23
Kutlu Gündoğdu F, Kahraman C (2019a) Spherical fuzzy sets and spherical fuzzy TOPSIS method. J Intell Fuzzy Syst 36(1):337–352
Kutlu Gundogdu F, Kahraman C (2019b) A novel spherical fuzzy analytic hierarchy process and its renewable energy application. Soft Comput 1–15
Kutlu Gundogdu F, Kahraman C (2019a) Extension of WASPAS with spherical fuzzy sets. Informatica 30(2):269–292
Kutlu Gündoğdu F, Kahraman C (2019b) A novel fuzzy TOPSIS method using emerging interval-valued spherical fuzzy sets. Eng Appl Artif Intell 85:307–323
Kutlu Gündoğdu F, Kahraman C (2019e) A novel VIKOR method using spherical fuzzy sets and its application to warehouse site selection. J Intell Fuzzy Syst, pp 1–15 (Preprint)

Kutlu Gündoğdu F, Kahraman C (2019f). Spherical fuzzy analytic hierarchy process (AHP) and its application to industrial robot selection. In: International conference on intelligent and fuzzy systems. Springer, Cham, pp 988–996
Kutlu Gündoğdu F, Kahraman C, Civan HN (2018) A novel hesitant fuzzy EDAS method and its application to hospital selection. J Intell Fuzzy Syst, pp 1–13 (Preprint)
Lupo T (2016) A fuzzy framework to evaluate service quality in the healthcare industry: an empirical case of public hospital service evaluation in Sicily. Appl Soft Comput 40:468–478
Oztaysi B, Onar SÇ, Boltürk E, Kahraman C (2015). Hesitant fuzzy analytic hierarchy process. In: 2015 IEEE international conference on fuzzy systems (FUZZ-IEEE). IEEE, Aug 2015, pp 1–7
Oztaysi B, Onar SC, Kahraman C (2017) Prioritization of business analytics projects using interval type-2 fuzzy AHP. In: Advances in fuzzy logic and technology 2017, pp 106–117. Springer, Cham
Pinnarelli L, Nuti S, Sorge C, Davoli M, Fusco D, Agabiti N, Perucci CA, et al (2012) What drives hospital performance? The impact of comparative outcome evaluation of patients admitted for hip fracture in two Italian regions. BMJ Qual Saf 21(2):127–134
Radwan NM, Senousy MB, Alaa El Din MR (2016) Neutrosophic AHP multi criteria decision making method applied on the selection of learning management system. Infinite Study
Ravangard R, Bahadori M, Raadabadi M, Teymourzadeh E, Alimomohammadzadeh K, Mehrabian F (2017) A model for the development of hospital beds using fuzzy analytical hierarchy process (fuzzy AHP). Iranian J Publ Health 46(11):1555
Ren P, Xu Z, Liao H, Zeng XJ (2017) A thermodynamic method of intuitionistic fuzzy MCDM to assist the hierarchical medical system in China. Inf Sci 420:490–504
Saaty TL (1980) The analytic hierarchy process: planning, priority setting, resources allocation. McGraw-Hill Inc. Retrieved from. http://www.mendeley.com/research/the-analytic-hierarchy-process
Sambuc R (1975) Function ϕ-flous, Application a l'aide au Diagnostic en Pathologie Thyroidienne. University of Marseille
Smarandache F (1998) Neutrosophy: neutrosophic probability, set, and logic: analytic synthesis and synthetic analysis
Thokala P, Duenas A (2012) Multiple criteria decision analysis for health technology assessment. Value Health 15(8):1172–1181
Torkzad A, Beheshtinia MA (2019) Evaluating and prioritizing hospital service quality. Int J Health Care Qual Assur 32(2):332–346
Torra V (2010) Hesitant fuzzy sets. Int J Intell Syst 25(6):529–539
Van Laarhoven PJ, Pedrycz W (1983) A fuzzy extension of Saaty's priority theory. Fuzzy Sets Syst 11(1–3):229–241
Wu H, Xu Z, Ren P, Liao H (2018) Hesitant fuzzy linguistic projection model to multi-criteria decision making for hospital decision support systems. Comput Ind Eng 115:449–458
Yager RR (1986) On the theory of bags. Int J General Syst 13(1):23–37
Yager RR (2013) Pythagorean fuzzy subsets. In: 2013 joint IFSA world congress and NAFIPS annual meeting (IFSA/NAFIPS). IEEE, June, pp 57–61
Yager RR (2016) Generalized orthopair fuzzy sets. IEEE Trans Fuzzy Syst 25(5):1222–1230
Yager RR, Abbasov AM (2013) Pythagorean membership grades, complex numbers, and decision-making. Int J Intell Syst 28(5):436–452
Yu Y, Jia L, Meng Y, Hu L, Liu Y, Nie X, Wang X et al (2018) Method development for clinical comprehensive evaluation of pediatric drugs based on multi-criteria decision analysis: application to inhaled corticosteroids for children with asthma. Pediatric Drugs 20(2):195–204
Zadeh LA (1965) Fuzzy sets. Inf Control 8(3):338–353
Zadeh LA (1975) The concept of a linguistic variable and its applications to approximate reasoning. Part I: Inf Sci 8:199–249
Zeng J, An M, Smith NJ (2007) Application of a fuzzy based decision making methodology to construction project risk assessment. Int J Project Manag 25(6):589–600
Zeng QL, Li DD, Yang YB (2013) VIKOR method with enhanced accuracy for multiple criteria decision making in healthcare management. J Med Syst 37(2):9908

Evaluating Geothermal Energy Systems Using Spherical Fuzzy PROMETHEE

Iman Mohamad Sharaf

Abstract Spherical fuzzy sets (SFSs) have been recently introduced to provide a larger preference domain for decision-makers. The preference ranking organization method of enrichment evaluation (PROMETHEE) is one of the known methods for multi-criteria decision making (MCDM). This study extends the conventional PROMETHEE into the spherical fuzzy environment. In the proposed method, both the weights of the criteria and the preference relations are SFSs. The relative degree of closeness is used to determine the preference indices. Since it is more appropriate to express the preference using linguistic terms, a linguistic preference function is used. Two examples are solved to illustrate the applicability and efficiency of the proposed method in solving MCDM problems. In the first example, the weights of the criteria and the ratings of the alternatives are SFSs. In the second example, the evaluation of the alternative energy exploitation projects, the weights of the criteria and the performance of the scenarios are crisp values. The results of the second example are compared with the results of the conventional PROMETHEE, the fuzzy PROMETHEE, the intuitionistic fuzzy PROMETHEE, and the Pythagorean fuzzy PROMETHEE.

1 Introduction

Multi-criteria decision making (MCDM) is the process of evaluation and assessment of several alternatives that behave differently under several diverse, even conflicting, criteria to select the optimal alternative. The theory of MCDM has been extensively used in various practical applications, e.g. selection of investment (Garg 2016a, b), enterprise resource planning (Liang et al. 2017; Wei 2017), and occupational health and safety risk assessment (Gul 2018; Gul and Ak 2018).

Several approaches have been proposed to handle MCDM problems. These methods can be classified into two main classes: utility theory-based methods (UBMs)

I. M. Sharaf (✉)
Department of Basic Sciences, Higher Technological Institute, Tenth of Ramadan City, Egypt
e-mail: iman_sharaf@hotmail.com

© Springer Nature Switzerland AG 2021

C. Kahraman and F. Kutlu Gündoğdu (eds.), *Decision Making with Spherical Fuzzy Sets*, Studies in Fuzziness and Soft Computing 392,
https://doi.org/10.1007/978-3-030-45461-6_16

and outranking methods (OMs) (Liao and Xu 2014). UBMs depend on merging the evaluation values of each alternative over various criteria into an overall index, hence ranking the alternatives according to the overall values. They emphasize on the decision-makers' ability to formulate a convenient utility function (Zhang et al. 2019). On the other hand, OMs depend on pairwise comparisons and dominance relations. They identify whether an alternative is indifferent, preferable, or incomparable to another alternative over the criteria (Zhang et al. 2019). They maintain the original verbal meaning of the criteria since they use the original ordinal scales without conversion to abstract ones within a certain range. Besides, they can model incomplete knowledge of data by taking the indifference and preference thresholds into consideration (Liao and Xu 2014).

OMs can be classified into two families: elimination et choix traduisant la realité (ELECTRE), i.e. elimination and choice expressing reality, and Preference ranking organization method of enrichment evaluation (PROMETHEE).

ELECTRE was initially developed by Roy (1968) to overcome the difficulties faced when using UBMs in handling practical problems (Bouyssou 2001). Later, he expanded the method to the well-known family of ELECTRE (Roy and Bertier 1973; Roy 1978, 1986, 1991). ELECTRE is based on the concordance-discordance principle. An alternative A_1 is at least as good as an alternative A_2 if the majority of criteria are in favour of this assertion (concordance condition), and if the opposition of the other minority criteria is not too strong (non-discordance condition) (Bouyssou 2001). These two conditions must be satisfied simultaneously for validating the outranking (Zhang et al. 2019).

PROMETHEE is one of the illustrious methods for MCDM. It is an effective, comprehensive and applicable technique for a wide range of MCDM problems (Zhang et al. 2019). It was introduced by Brans (1982) and later developed by Brans et al. (1984); Brans and Vincke (1985); and Brans et al. (1986); Brans and Mareschal (1992, 1995). PROMETHEE ranking is based on the positive and negative flows or net outranking flows according to selected preferences. A positive outranking flow indicates the extent an alternative is superior to other alternatives, a negative outranking flow indicates the extent an alternative is inferior to other alternatives, while the net outranking flow indicates the balance between the positive and the negative outranking flows. The method is simple in conception and application compared to other MCDM methods (Goumas and Lygerou 2000). Moreover, it easy to implement and all the required information is clear and easy to define (Liao and Xu 2014). It has been successfully applied in various fields, e.g. Environment Management, Hydrology and Water Management, Business and Financial Management, Chemistry, Logistics and Transportation, Manufacturing and Assembly, and Energy Management (Behzadian et al. 2010).

Several variants of PROMETHEE were introduced to handle different situations. PROMETHEE I (partial ranking), PROMETHEE II (complete ranking), PROMETHEE III (ranking based on intervals), PROMETHEE IV (continuous case), PROMETHEE V (for problems with segmentation constraints), and PROMETHEE VI (for the representation of the human brain). PROMETHEE/GDSS (for group decision support system) was developed to provide decision aid to a group of decision-makers. An iteration of the PROMETHEE/ GDSS procedure consists of 11 steps grouped in three phases. Phase I: generation of alternatives and criteria, phase II: individual evaluation by each decision-maker, and phase III: global evaluation by the group (Brans and Mareschal 2005). Further, PROMETHEE/GAIA (Geometrical Analysis for Interactive Aid) was proposed to provide a graphical representation supporting the PROMETHEE methodology to help in more complicated decision-making situations (Liao and Xu 2014).

PROMETHEE I and II are closely related. PROMETHEE I aggregates the preference indices over different criteria to get the positive and negative outranking flows, hence attaining a partial order. Although PROMETHEE I results in partial ranking which might aid the decision-maker in choosing the best alternative, sometimes the results can be incomparable. PROMETHEE II eliminates incomparability using the net outranking flow by calculating the difference between the positive and negative outranking flows to obtain a complete ranking. The complete ranking is comprehensive and efficient (Zhang et al. 2019).

Leyva-López and Fernández-González (2003) extended ELECTRE III to assist a group of decision-makers with different value systems and compared their method with PROMETHEE II for group decision-making. Dulmin and Mininno (2003) investigated using PROMETHEE/GAIA in supplier selection decision given the financial importance and the multi-objective nature of such decision. Elevli and Demirci (2004) successfully used PROMETHEE in solving mining engineering problems. Wang and Yang (2007) proposed using the analytic hierarchy process (AHP) and PROMETHEE to make information systems outsourcing decisions. The AHP is used to analyze the structure of the outsourcing problem and to determine the weights of the criteria, while PROMETHEE is used for final ranking. Araz et al. (2007) developed an outsourcer evaluation and management system for a textile company using fuzzy goal programming (FGP). First, the existing outsourcers are evaluated by PROMETHEE, and then the FGP model selects the most strategic outsourcers with the company. Shih et al. (2016) generalized PROMETHEE III by introducing risk preferences of decision-makers. The risk preferences are expressed by an S-shaped value function with gain and loss parts. The method was used in environmental evaluation of waste treatment plants for waste electrical and electronic equipment in Taiwan.

The classical PROMETHEE methods are modelled to address crisp data which is not common in MCDM problems which relies mainly on human cognition and data. These factors are subject to imprecision, fuzziness, uncertainty, and vagueness. Accordingly, several versions of fuzzy PROMETHEE (F-PROMETHEE) were introduced to encompass fuzzy data. Radojevic and Petrovic (1997) introduced a fuzzy system for preference function modelling to give the decision-maker a wider application of the PROMETHEE method. Le Téno and Mareschal (1998) proposed a new version using interval numbers to measure products environmental quality to help in the selection and the design of environmentally friendly products. Goumas and Lygerou (2000) introduced another version using triangular fuzzy numbers for the evaluation and ranking of alternative energy exploitation schemes of a low-temperature geothermal field. Both versions considered fuzziness in the values of the alternatives only and neglected fuzziness in the decision maker's preferences and the weights of the criteria. Therefore, Geldermann et al. (2000) extended the fuzziness of data to preferences, scores and weights to enhance the flexibility of the method to the users for environmental assessment in iron and steel making industry. Fernández-Castro and Jiménez (2005) used PROMETHEE III scoring and fuzzy integer linear programming to select the best subset of alternatives subject to a set of flexible constraints where some parameters may be fuzzy numbers. Bilsel et al. (2006) utilized an F-PROMETHEE to develop a quality evaluation model for measuring the performance of hospital Websites. Zhang et al. (2009) developed an F-PROMETHEE using trapezoidal fuzzy sets based on comparative risk methodology. Halouani et al. (2009) integrated a 2-tuple linguistic representation model dealing with non-homogeneous and imprecise information data made up by valued intervals, numerical and linguistic values into the aggregation operators of PROMETHEE to apply in project selection problems. Tuzkaya et al. (2010) proposed an integrated approach utilized from fuzzy sets, analytic network process (ANP) and F-PROMETHEE for material handling equipment selection problem. Li and Li (2010) proposed an F-PROMETHEE based on generalized fuzzy numbers. The proposed method considers the difference between corresponding points in two interval numbers based on the α-cut of generalized fuzzy numbers. Rao and Patel (2010) integrated PROMETHEE with AHP and the fuzzy logic for decision-making in various real-life situations of the manufacturing environment considering both crisp and fuzzy criteria. Chen et al. (2011) presented an F-PROMETHEE technique to select an appropriate information system (IS)/ information technology (IT) partner using a realistic case study to enhance the organization competitiveness, reduce costs, increase the focus on internal resources and core activities, and sustain competitive advantage. Shirinfar and Haleh (2011) developed an outsourcer evaluation and management system for a manufacturing company by integrating FGP and F-PROMETHEE. Gupta et al. (2012) proposed an F-PROMETHEE/GDSS approach to rank various logistics service providers (LSPs) for a cement company in the Northern part of India. Senvar et al. (2014) utilized a method for multi-criteria supplier selection problem based on F-PROMETHEE. Adali et al. (2016) proposed an alternative version of F-PROMETHEE in which the preference functions are handled in terms of fuzzy distances between the alternatives for each criterion. Gul et al. (2018) presented an F-PROMETHEE method based on trapezoidal fuzzy interval numbers to select the materials for an automotive instrument panel. The method is illustrated, validated, and compared against three different fuzzy MCDM methods (F-VIKOR, F-TOPSIS, and F-ELECTRE) in terms of its ranking performance.

The variants of F-PROMETHEE may experience some limitations since fuzzy sets themselves do not convey the support and the opposition of the decision-maker simultaneously. Whenever imprecision and uncertainty accompany the problem domain, the judgement of cognitive performance manifest the characteristics of affirmation, negation and hesitation (Liao and Xu 2014). Since Atanassov (1986) introduced intuitionistic fuzzy sets (IFSs) with its power to express these characteristics in humans' perception, PROMETHEE has been extended using IFSs. Liao and Xu (2014) developed an intuitionistic fuzzy PROMETHEE (IF-PROMETHEE), taking not only intuitionistic fuzzy preferences but also intuitionistic fuzzy weights into account. Rani and Jain (2017) proposed an IF-PROMETHEE based on entropy measure. Krishankumar et al. (2017) developed a new IF-PROMETHEE for solving the supplier selection problem with linguistic preferences.

On the other hand, IFSs have a disadvantage. The sum of the membership degree and the non-membership degree of IFSs must be equal to or less than one.

This may be intractable in many practical applications since human cognition might not satisfy this constraint. Yager (2014) proposed Pythagorean fuzzy sets (PFSs) as a generalization of IFSs with the sum of the squares of the membership degree and non-membership degree is equal to or less than one. This makes PFSs more robust than IFSs in modelling the haziness of human perception in MCDM problems. Zhang et al. (2019) extended PROMETHEE into the Pythagorean fuzzy environment (PF-PROMETHEE). The proposed method takes both the weights of different criteria and the preference relations as PFSs, therefore providing a broader range of choices for the decision-makers to express their preferences.

In both IFSs and PFSs the third dimension, i.e. hesitancy degree, depends on the membership degree and non-membership degree. Gündoğdu and Kahraman (2019) introduced spherical fuzzy sets (SFSs) in which the hesitancy degree of the decision-maker is independent of the membership degree and the non-membership degree. Since the hesitancy degree reflects the associated uncertainty, the smaller its value the more certain we are about the information we get. Therefore, the independence of the hesitancy degree is closer to reality. Thus, SFSs can characterize uncertain information sufficiently and more accurately and can describe the experts' judgements and preferences explicitly and in a more informative way.

This study extends the conventional PROMETHEE into the spherical fuzzy environment. In the proposed method, both the weights of the criteria and the preference relations are expressed by SFSs. The relative degree of closeness, which is based on the distances from the positive ideal solution and the negative ideal solution, is used to determine the preference index between the alternatives for each criterion. Since it is more convenient for decision-makers to express their preference using linguistic terms to indicate the intensity of "preferred", "not preferred" and "indeterminate", a linguistic preference function is used. Two practical examples are solved to illustrate the applicability and efficiency of the proposed method in solving MCDM problems. In the first example, the weights of the criteria and the ratings of the alternatives are SFSs. In the second example, the ranking of the alternative energy exploitation projects, the weights of the criteria and the performance of the scenarios are crisp values, and then the results are compared with the results of the conventional PROMETHEE, the F-PROMETHEE, the IF-PROMETHEE, and the PF-PROMETHEE.

The originality of this chapter stems from implementing the PROMETHEE method in a spherical fuzzy environment, thus allowing the decision-makers to express their indeterminacy explicitly and independent of the degrees of affirmation and negation. Also, a new preference function is presented differently from the known types. The proposed preference function is based on linguistic terms which are closer to human cognition. Therefore, the proposed SF-PROMETHEE can handle the decision-makers' uncertainty efficiently.

The chapter is organized as follows. The SFSs with their basic operations and the conventional PROMETHEE are given in Sect. 2. The SF-PROMETHEE is presented in Sect. 3. Two practical examples are solved in Sect. 4 and the results are compared with previously used techniques. Finally, the conclusion and discussion are given in Sect. 5.

2 Preliminaries

2.1 Basic Operators of SFSs

An SFS is denoted by

$$\tilde{A}_s = \left\{\left\langle u, \left(\mu_{\tilde{A}_s}(u), \upsilon_{\tilde{A}_s}(u), \pi_{\tilde{A}_s}(u)\right) | u \in U\right\rangle\right\}, \tag{1}$$

where $\mu_{\tilde{A}_s} : U \rightarrow [0, 1]$ is the membership degree, $\upsilon_{\tilde{A}_s} : U \rightarrow [0, 1]$ is the non-membership degree, $\pi_{\tilde{A}_s} : U \rightarrow [0, 1]$ is the hesitancy degree, and

$$0 \leq \mu^2_{\tilde{A}_s}(u) + \upsilon^2_{\tilde{A}_s}(u) + \pi^2_{\tilde{A}_s}(u) \leq 1, \quad \forall u \in U. \tag{2}$$

The basic operations on SFSs are given as follows (Gündoğdu and Kahraman, 2019).

$$\tilde{A}_s \oplus \tilde{B}_s = \left\{ \begin{array}{l} \left(\mu^2_{\tilde{A}_s} + \mu^2_{\tilde{B}_s} - \mu^2_{\tilde{A}_s}\mu^2_{\tilde{B}_s}\right)^{(1/2)}, \upsilon_{\tilde{A}_s}\upsilon_{\tilde{B}_s}, \\ \left((1-\mu^2_{\tilde{B}_s})\pi^2_{\tilde{A}_s} + (1-\mu^2_{\tilde{A}_s})\pi^2_{\tilde{B}_s} - \pi^2_{\tilde{A}_s}\pi^2_{\tilde{B}_s}\right)^{1/2} \end{array} \right\}. \tag{3}$$

$$\tilde{A}_s \otimes \tilde{B}_s = \left\{ \begin{array}{l} \mu_{\tilde{A}_s}\mu_{\tilde{B}_s}, \left(\upsilon^2_{\tilde{A}_s} + \upsilon^2_{\tilde{B}_s} - \upsilon^2_{\tilde{A}_s}\upsilon^2_{\tilde{B}_s}\right)^{1/2}, \\ \left(\left(1-\upsilon^2_{\tilde{B}_s}\right)\pi^2_{\tilde{A}_s} + \left(1-\upsilon^2_{\tilde{A}_s}\right)\pi^2_{\tilde{B}_s} - \pi^2_{\tilde{A}_s}\pi^2_{\tilde{B}_s}\right)^{1/2} \end{array} \right\}. \tag{4}$$

$$\lambda \cdot \tilde{A}_s = \left\{ \begin{array}{l} \left(1 - \left(1 - \mu^2_{\tilde{A}_s}\right)^{\lambda}\right)^{1/2}, \\ \upsilon^{\lambda}_{\tilde{A}_s}, \left(\left(1 - \mu^2_{\tilde{A}_s}\right)^{\lambda} - \left(1 - \mu^2_{\tilde{A}_s} - \pi^2_{\tilde{A}_s}\right)^{\lambda}\right)^{1/2} \end{array} \right\}, \quad \lambda > 0. \tag{5}$$

$$\tilde{A}^{\lambda}_s = \left\{\mu^{\lambda}_{\tilde{A}_s}, \left(1 - \left(1 - \upsilon^2_{\tilde{A}_s}\right)^{\lambda}\right)^{1/2}, \left(\left(1 - \upsilon^2_{\tilde{A}_s}\right)^{\lambda} - \left(1 - \upsilon^2_{\tilde{A}_s} - \pi^2_{\tilde{A}_s}\right)^{\lambda}\right)^{1/2}\right\}. \tag{6}$$

The Score and Accuracy functions for ranking SFSs are calculated by

$$\text{Score}\left(\tilde{A}_s\right) = \left(\mu_{\tilde{A}_s} - \pi_{\tilde{A}_s}\right)^2 - \left(\upsilon_{\tilde{A}_s} - \pi_{\tilde{A}_s}\right)^2, \tag{7}$$

$$\text{Accuracy}(\tilde{A}_s) = \mu^2_{(\tilde{A}_s)} + \upsilon^2_{(\tilde{A}_s)} + \pi^2_{(\tilde{A}_s)}, \tag{8}$$

$$\begin{aligned} \tilde{A}_s < \tilde{B}_s \quad \text{iff } & \text{Score}\left(\tilde{A}_s\right) < \text{Score}\left(\tilde{B}_s\right), \\ & \text{or Score}\left(\tilde{A}_s\right) = \text{Score}\left(\tilde{B}_s\right) \\ & \text{and Accuracy}\left(\tilde{A}_s\right) < \text{Accuracy}\left(\tilde{B}_s\right). \end{aligned}$$

The normalized Euclidean distance is given by

$$D\left(\tilde{A}_s, \tilde{B}_s\right) = \sqrt{\frac{1}{2n}\sum_{i=1}^{n}\left(\left(\mu_{\tilde{A}_s} - \mu_{\tilde{B}_s}\right)^2 + \left(\upsilon_{\tilde{A}_s} - \upsilon_{\tilde{B}_s}\right)^2 + \left(\pi_{\tilde{A}_s} - \pi_{\tilde{B}_s}\right)^2\right)}. \tag{9}$$

2.2 *The Conventional PROMETHEE*

For a MCDM problem, let $\mathrm{X} = \{x_1, x_2, \ldots, x_n\}$ be the set of alternatives and $\mathrm{C} = \{c_1, c_2, \ldots, c_\mathrm{m}\}$ be the set of criteria. The conventional PROMETHEE can be summarized as follows (Liao and Xu 2014; Zhang et al. 2019).

Step 1. Evaluate the used criteria $c_j (j = 1, \ldots m)$ and assign a weight to each criterion w_j such that $\sum_{j=1}^{m} w_j = 1$.

Step 2. Evaluate the alternatives $\{x_1, x_2, \ldots, x_n\}$ for each criterion c_j and determine the deviations based on pairwise comparisons.

$$d_j(x_i, x_k) = c_j(x_i) - c_j(x_k). \tag{10}$$

Step 3. Compute the preference index of the alternative x_i over the alternative x_k for each criterion c_j.

$$P_j(x_i, x_k) = f_j\left[d_j(x_i, x_k)\right], \quad \forall x_i, x_k \in \mathrm{X}, \tag{11}$$

where f_j is a preference function (generalized criteria) that maps the difference between the evaluation of the alternatives x_i and x_k for a criterion c_j onto a preference index $P_j(x_i, x_k) \in [0, 1]$.

The preference function has six types (Brans et al. 1986): usual criterion, quasi-criterion (U-shape), criterion with linear preference (V-shape), level criterion, criterion with linear preference and indifference area (V-shape with indifference), and Gaussian criterion.

Step 4. Compute the global preference index.

$$\Pi(x_i, x_k) = \sum_{j=1}^{m} w_j P_j(x_i, x_k). \tag{12}$$

Step 5. Compute the positive and negative outranking flows of the alternative x_i.

$$\varphi^+(x_i) = \frac{1}{n-1} \sum_{x \in X} \Pi(x_i, x), \varphi^-(x_i) = \frac{1}{n-1} \sum_{x \in X} \Pi(x, x_i). \tag{13}$$

Step 6. Rank the alternatives.
Partial ranking can be obtained as follows:

(i) If $\varphi^+(x_i) \geq \varphi^+(x_k)$ and $\varphi^-(x_i) \leq \varphi^-(x_k)$, then x_i outranks x_k.
(ii) If $\varphi^+(x_i) = \varphi^+(x_k)$ and $\varphi^-(x_i) = \varphi^-(x_k)$, then x_i and x_k are indifferent.
(iii) If $\varphi^+(x_i) > \varphi^+(x_k)$ and $\varphi^-(x_i) > \varphi^-(x_k)$, or $\varphi^+(x_i) < \varphi^+(x_k)$ and $\varphi^-(x_i) < \varphi^-(x_k)$, then x_i and x_k are incomparable.

When necessary compute the net outranking flow to get a complete ranking:

$$\varphi(x_i) = \varphi^+(x_i) - \varphi^-(x_i) \tag{14}$$

3 The SF-PROMETHEE

3.1 The Spherical Fuzzy Preference Function

The preference function is the cornerstone in PROMETHEE. It is the key step that maps the deviation of the alternatives to a preference index. In the conventional PROMETHEE, the preference function yields crisp preference indices that can only express the intensity of "preferred". However, a decision-maker may be unable to accurately decide how much an alternative is preferred over the other. The decision-maker judgement manifests affirmation, negation and hesitation characteristics. Therefore, it is more proper to represent the preference indices using fuzzy sets to express the intensity of "preferred", "not preferred" and "indeterminate".

Liao and Xu (2014) proposed an IF-PROMETHEE in which they utilized one of the conventional PROMETHEE preference function, the V-shape with indifference criterion to construct the preference degree μ_{ik} of the alternative x_i over the alternative x_k for the criterion c_j. Using the condition $\mu_{ik} = \upsilon_{ki}$ and $\mu_{ki} = \upsilon_{ik}$, the preference index is given by $r_{ik}^j = \left(\mu_{ik}^j, \upsilon_{ik}^j\right)$, where μ_{ik} is the degree to which x_i is preferred to x_k, and υ_{ik} is the degree to which x_i is not preferred to x_k. Zhang et al. (2019)

proposed a PF preference function similar to the IF preference function with a slight adjustment to develop a PF-PROMETHEE.

The IF preference function is characterized by two parameters: the strict preference threshold p, and the indifference threshold q. The strict preference threshold indicates strict preference whenever the deviation exceeds this critical point. While the indifference threshold indicates the region in which the deviation is not significant enough to guarantee any preference. Zhang et al. (2019) added the "blind" confidence parameter r to a PF preference function which indicates the case when the deviation is zero.

Within the context of SFSs subtraction operation is undefined, thus the deviation between any two alternatives for each criterion cannot be calculated similarly to the conventional PROMETHEE $d_j(x_i, x_k) = c_j(x_i) - c_j(x_k)$. Therefore, the relative degree of closeness (RC) will be used instead

$$d_j(x_i, x_k) = RC\left[\tilde{c}_j(x_i)\right] - RC\left[\tilde{c}_j(x_k)\right], \tag{15}$$

where $\tilde{c}_j(x)$ is the spherical fuzzy evaluation of the alternative x for the criterion c_j, $R\left[\tilde{c}_j(x)\right]$ is the known relative degree of closeness used by the technique of order preference by similarity to an ideal solution (TOPSIS). It is computed by:

$$RC\left[\tilde{c}_j(x_i)\right] = \frac{D^-(x_i)}{D^+(x_i) + D^-(x_i)}, \tag{16}$$

where $D^+(x)$ is the distance between an alternative and the positive ideal solution (x^+)

$$D^+(x_i) = \sqrt{\frac{1}{2n}\sum_{i=1}^{n}\left(\left(\mu_{x_i} - \mu_{x^+}\right)^2 + \left(\upsilon_{x_i} - \upsilon_{x^+}\right)^2 + \left(\pi_{x_i} - \pi_{x^+}\right)^2\right)}, \tag{17}$$

and D$^-(x)$ is the distance between an alternative and the negative ideal solution(x^-)

$$D^-(x_i) = \sqrt{\frac{1}{2n}\sum_{i=1}^{n}\left(\left(\mu_{x_i} - \mu_{x^-}\right)^2 + \left(\upsilon_{x_i} - \upsilon_{x^-}\right)^2 + \left(\pi_{x_i} - \pi_{x^-}\right)^2\right)}. \tag{18}$$

A linguistic preference function is used for SFSs, and it is defined as follows:

$$f_j\left[d_j(x_i, x_k)\right] = \begin{cases} \text{equal preference} & d < 0.1, \\ \text{weak preference} & 0.1 \le d < 0.2, \\ \text{moderate preference} & 0.2 \le d < 0.3, \\ \text{strong preference} & 0.3 \le d < 0.4, \\ \text{very strong preference} & 0.4 \le d < 0.5, \\ \text{strict preference} & d \ge 0.5. \end{cases} \tag{19}$$

Table 1 Linguistic preference and their corresponding SFSs

Linguistic preference	(μ, υ, π)
Equal preference (EP)	(0.5, 0.5, 0.5)
weak preference (WP)	(0.6, 0.4, 0.4)
moderate preference (MP)	(0.7, 0.3, 0.3)
Strong preference(SP)	(0.8, 0.2, 0.2)
Very strong preference (VSP)	(0.9, 0.1, 0.1)
Strict preference (STP)	(1, 0, 0)

The preference threshold $p = 0.5$, and the indifference threshold $q = 0.1$. For negative d, the complement of the SFS is used. The SFSs corresponding the linguistic term is given in Table 1.

The SFSs corresponding to the linguistic preference function satisfy the main properties as IFSs and PFSs with some adjustments to encompass the hesitancy degree:

(i) $\mu(d) : [0, 1] \rightarrow [0, 1], \upsilon(d) : [0, 1] \rightarrow [0, 1], \pi(d) : [0, 1] \rightarrow [0, 1], \mu^2 + \upsilon^2 + \pi^2 \leq 1$;
(ii) $\mu(d_{ik}) = \upsilon(d_{ki}), \mu(d_{ki}) = \upsilon(d_{ik})$, and $\pi(d_{ki}) = \pi(d_{ik})$;
(iii) If $0 \leq |d_{ik}| \leq q$, then $\mu(d) = \upsilon(d) = \pi(d) = 0.5$;
(iv) If $d_{ik} \geq p$, then $\mu(d) = 1, \upsilon(d) = 0,\ and\ \pi(d) = 0$;
(v) μ is monotonically increasing, while υ and π are monotonically decreasing.

3.2 The Proposed SF-PROMETHEE

The steps of the SF-PROMETHEE are illustrated as follows.

Step 1. Define the positive and negative ideal solutions for each criterion.

Let $\mu_P = \max_i \mu_i$, $\upsilon_P = \min_i \upsilon_i$, and $\pi_P = \min_i \pi_i$,

$$x_j^+ = (\mu_P, \upsilon_P, \pi_P). \tag{20}$$

Let $\mu_N = \min_i \mu_i$, $\upsilon_N = \max_i \upsilon_i$, and $\pi_N = \max_i \pi_i$.

$$x_j^- = \begin{cases} (\mu_N, \upsilon_N, \pi_N) & if\, \mu_N^2 + \upsilon_N^2 + \pi_N^2 \leq 1, \\ \left(\mu_N, \upsilon_N, \sqrt{1 - \left(\mu_N^2 + \upsilon_N^2\right)}\right) & \text{otherwise.} \end{cases} \tag{21}$$

Step 2. Calculate the relative degree of closeness of each alternative for each criterion.

Use (16), (17), and (18) to find $RC\left[\tilde{c}_j(x_i)\right]$.

Step 3. Calculate the deviation between every two alternatives for each criterion $d_j(x_i, x_k)$ using (15).

Step 4. Construct the preference matrix of the alternatives for each criterion using the linguistic preference function (19), and then transform to the equivalent SFSs.

$$\mathbf{R}_L^j = \left[f_j\left[d_j(x_i, x_k)\right]\right], \tag{22}$$

and

$$\tilde{\mathbf{R}}_s^j = \left[\tilde{r}_{ik}^j\right] = \begin{bmatrix} - & \left(\mu_{12}^j, \upsilon_{12}^j, \pi_{12}^j\right) & \cdots & \left(\mu_{1n}^j, \upsilon_{1n}^j, \pi_{1n}^j\right) \\ \left(\mu_{21}^j, \upsilon_{21}^j, \pi_{21}^j\right) & - & \cdots & \left(\mu_{2n}^j, \upsilon_{2n}^j, \pi_{2n}^j\right) \\ \vdots & \vdots & \ddots & \vdots \\ \left(\mu_{n1}^j, \upsilon_{n1}^j, \pi_{n1}^j\right) & \left(\mu_{n2}^j, \upsilon_{n2}^j, \pi_{n2}^j\right) & \cdots & - \end{bmatrix}. \tag{23}$$

Step 5. Construct the overall spherical fuzzy preference matrix using the global preference indices.

$$\tilde{\mathbf{R}}_s = \left[\tilde{r}_{ik}\right], \quad \text{where } \tilde{r}_{ik} = \underset{j=1}{\overset{m}{\oplus}} \left(\tilde{w}_j \otimes \tilde{r}_{ik}^j\right). \tag{24}$$

Step 6. Derive the positive and negative outranking flows.

$$\varphi^+(x_i) = \frac{1}{n-1} \underset{k=1, k \neq i}{\overset{n}{\oplus}} \tilde{r}_{ik}, \varphi^-(x_i) = \frac{1}{n-1} \underset{k=1, k \neq i}{\overset{n}{\oplus}} \tilde{r}_{ki}. \tag{25}$$

Step 7. Use the score function values for ranking.

Find $\text{Score}\left(\varphi^+(x_i)\right)$ and $Score(\varphi^-(x_i))$, then apply the comparison rules of the conventional PROMETHEE.

(i) If $\text{Score}\left(\varphi^+(x_i)\right) \geq \text{Score}\left(\varphi^+(x_k)\right)$ and $\text{Score}\left(\varphi^-(x_i)\right) \leq \text{Score}\left(\varphi^-(x_k)\right)$, then x_i outranks x_k.

(ii) If $\text{Score}\left(\varphi^+(x_i)\right) = \text{Score}\left(\varphi^+(x_k)\right)$ and $\text{Score}\left(\varphi^-(x_i)\right) = \text{Score}\left(\varphi^-(x_k)\right)$, then x_i and x_k are indifferent.

(iii) If $\text{Score}\left(\varphi^+(x_i)\right) > \text{Score}\left(\varphi^+(x_k)\right)$ and $\text{Score}\left(\varphi^-(x_i)\right) > \text{Score}\left(\varphi^-(x_k)\right)$, or $\text{Score}\left(\varphi^+(x_i)\right) < \text{Score}\left(\varphi^+(x_k)\right)$ and $\text{Score}\left(\varphi^-(x_i)\right) < \text{Score}\left(\varphi^-(x_k)\right)$, then x_i and x_k are incomparable.

Since the subtraction operation is undefined for SFSs, the score function values are also used to calculate the net outranking flow for complete ranking if incomparability occurs,

$$\varphi(x_i) = \text{Score}\left(\varphi^+(x_i)\right) - \text{Score}\left(\varphi^-(x_i)\right).$$

4 Practical Examples

In this section, two practical examples are solved. The first example is adapted from Gündoğdu and Kahraman (2019) for supplier selection in which the weights of the criteria and the ratings of the alternatives are SFSs. The second example is due to Goumas and Lygerou (2000) for ranking of the alternative energy exploitation projects, in which the weights of the criteria and the performance of the scenarios are crisp values.

Example 1 Three decision-makers $\{D_1, D_2, D_3\}$ assess four suppliers $\{x_1, x_2, x_3, x_4\}$ according to four criteria: price (c_1), quality (c_2), delivery (c_3), and performance (c_4). The decision matrix and the weights of the criteria are given in Table 2 (Gündoğdu and Kahraman, 2019). The solution steps using the SF-PROMETHEE are demonstrated as follows.

Step 1. Define the positive and negative ideal solutions using (20) and (21) for each criterion.

The ideal solutions are given in Table 3.

Step 2. Calculate the relative degree of closeness of each alternative for each criterion. The results are given in Table 4.

Step 3. Calculate the deviation between every two alternatives for each criterion. The deviations are given in Table 5.

Step 4. Construct the preference matrix of the alternatives for each criterion using the linguistic preference function (19), then transform to the corresponding SFSs.

Table 2 The decision matrix and the weights of the criteria

	c_1	c_2	c_3	c_4
Alternatives				
x_1	(0.80, 0.20, 0.21)	(0.70, 0.30, 0.30)	(0.82, 0.19, 0.23)	(0.81, 0.20, 0.23)
x_2	(0.77, 0.24, 0.22)	(0.74, 0.28, 0.32)	(0.74, 0.27, 0.28)	(0.45, 0.55, 0.46)
x_3	(0.57, 0.44, 0.32)	(0.53, 0.49, 0.37)	(0.74, 0.27, 0.28)	(0.52, 0.52, 0.31)
x_4	(0.48, 0.20, 0.29)	(0.62, 0.39, 0.39)	(0.65, 0.37, 0.25)	(0.62, 0.39, 0.34)
Weights	(0.57, 0.44, 0.32)	(0.82, 0.19, 0.23)	(0.80, 0.20, 0.21)	(0.81, 0.20, 0.23)

Table 3 The ideal solutions for each criterion

Criteria	PIS (x^+)	NIS (x^-)
c_1	(0.80, 0.20, 0.21)	(0.48, 0.56, 0.32)
c_2	(0.74, 0.28, 0.30)	(0.53, 0.49, 0.39)
c_3	(0.82, 0.19, 0.23)	(0.65, 0.37, 0.28)
c_4	(0.81, 0.20, 0.23)	(0.45, 0.55, 0.46)

Table 4 RC of the alternatives for c_i

	$D^+(x_i)$	$D^-(x_i)$	$RC[\tilde{c}_i(x_i)]$		$D^+(x_i)$	$D^-(x_i)$	$RC[\tilde{c}_i(x_i)]$
c_1				c_3			
x_1	0	0.3494	1	x_1	0	0.1786	1
x_2	0.0361	0.3134	0.896	x_2	0.0849	0.0954	0.5219
x_3	0.2476	0.1061	0.3	x_3	0.0875	0.0951	0.5208
x_4	0.3456	0.0212	0.0578	x_4	0.1756	0.0212	0.1077
c_2				c_4			
x_1	0.0316	0.1912	0.8582	x_1	0	0.3905	1
x_2	0.0141	0.2158	0.9387	x_2	0.3905	0	0
x_3	0.2158	0.0141	0.0613	x_3	0.3460	0.1190	0.2559
x_4	0.1315	0.0951	0.4197	x_4	0.2053	0.1856	0.4748

Table 5 The deviations between the alternatives for c_i

	x_1	x_2	x_3	x_4
$d_1(x_i, x_k)$				
x_1	–	0.1033	0.7	0.9422
x_2	−0.1033	–	0.5967	0.8389
x_3	−0.7	−0.5967	–	0.2422
x_4	−0.9422	−0.8389	−0.2422	–
$d_2(x_i, x_k)$				
x_1	–	−0.0805	0.7969	0.4385
x_2	0.0805	–	0.8774	0.5190
x_3	−0.7969	−0.8774	–	−0.3584
x_4	−0.4385	−0.5190	0.3584	–
$d_3(x_i, x_k)$				
x_1	–	0.4709	0.4792	0.8923
x_2	−0.4709	–	$8.3 * 10^{-3}$	0.4114
x_3	−0.4792	$-8.3 * 10^{-3}$	–	0.4131
x_4	−0.8923	−0.4114	−0.4131	–
$d_4(x_i, x_k)$				
x_1	–	1	0.7441	0.5252
x_2	−1	–	−0.2359	−0.4748
x_3	−0.7441	0.2359	–	−0.2189
x_4	−0.5252	0.4748	0.2189	–

$$\mathbf{R}_L^1 = \begin{bmatrix} - & \text{WP} & \text{STP} & \text{STP} \\ \sim\text{WP} & - & \text{STP} & \text{STP} \\ \sim\text{STP} & \sim\text{STP} & - & \text{MP} \\ \sim\text{STP} & \sim\text{STP} & \sim\text{MP} & - \end{bmatrix}$$

$$\tilde{\mathbf{R}}_S^1 = \begin{bmatrix} - & (0.6, 0.4, 0.4) & (1, 0, 0) & (1, 0, 0) \\ (0.4, 0.6, 0.4) & - & (1, 0, 0) & (1, 0, 0) \\ (0, 1, 0) & (0, 1, 0) & - & (0.7, 0.3, 0.3) \\ (0, 1, 0) & (0, 1, 0) & (0.3, 0.7, 0.3) & - \end{bmatrix}$$

$$\mathbf{R}_L^2 = \begin{bmatrix} - & \text{EP} & \text{STP} & \text{VSP} \\ \text{EP} & - & \text{STP} & \text{STP} \\ \sim\text{STP} & \sim\text{STP} & - & \sim\text{SP} \\ \sim\text{VSP} & \sim\text{STP} & \text{SP} & - \end{bmatrix},$$

$$\tilde{\mathbf{R}}_S^2 = \begin{bmatrix} - & (0.5, 0.5, 0.5) & (1, 0, 0) & (0.9, 0.1, 0.1) \\ (0.5, 0.5, 0.5) & - & (1, 0, 0) & (1, 0, 0) \\ (0, 1, 0) & (0, 1, 0) & - & (0.2, 0.8, 0.2) \\ (0.1, 0.9, 0.1) & (0, 1, 0) & (0.8, 0.2, 0.2) & - \end{bmatrix}.$$

$$\mathbf{R}_L^3 = \begin{bmatrix} - & \text{SP} & \text{SP} & \text{STP} \\ \sim\text{SP} & - & \text{EP} & \text{SP} \\ \sim\text{SP} & \text{EP} & - & \text{SP} \\ \sim\text{STP} & \sim\text{SP} & \sim\text{SP} & - \end{bmatrix},$$

$$\tilde{\mathbf{R}}_S^3 = \begin{bmatrix} - & (0.8, 0.2, 0.2) & (0.8, 0.2, 0.2) & (1, 0, 0) \\ (0.2, 0.8, 0.2) & - & (0.5, 0.5, 0.5) & (0.8, 0.2, 0.2) \\ (0.2, 0.8, 0.2) & (0.5, 0.5, 0.5) & - & (0.8, 0.2, 0.2) \\ (0, 1, 0) & (0.2, 0.8, 0.2) & (0.2, 0.8, 0.2) & - \end{bmatrix}$$

$$\mathbf{R}_L^4 = \begin{bmatrix} - & \text{STP} & \text{STP} & \text{STP} \\ \sim\text{STP} & - & \sim\text{MP} & \sim\text{VSP} \\ \sim\text{STP} & \text{MP} & - & \sim\text{MP} \\ \sim\text{STP} & \text{VSP} & \text{MP} & - \end{bmatrix},$$

$$\tilde{\mathbf{R}}_S^4 = \begin{bmatrix} - & (1, 0, 0) & (1, 0, 0) & (1, 0, 0) \\ (0, 1, 0) & - & (0.3, 0.7, 0.3) & (0.1, 0.9, 0.1) \\ (0, 1, 0) & (0.7, 0.3, 0.3) & - & (0.3, 0.7, 0.3) \\ (0, 1, 0) & (0.9, 0.1, 0.1) & (0.7, 0.3, 0.3) & - \end{bmatrix}.$$

Step 5. Construct the overall spherical fuzzy preference matrix using (24).

Table 6 Positive and negative outranking flows

Alternative	$\varphi^{+}(x_i)$	$\varphi^{-}(x_i)$
x_1	(0.9703, 0.0464, 0.1100)	(0.3065, 0.6006, 0.4414)
x_2	(0.8604, 0.0585, 0.3530)	(0.8133, 0.0843, 0.3856)
x_3	(0.6785, 0.1744, 0.4156)	(0.9118, 0.0000, 0.2896)
x_4	(0.5859, 0.3072, 0.5622)	(0.9424, 0.0000, 0.2172)

$$\tilde{\mathbf{R}}_s = \begin{bmatrix} - & (0.92, 0.02, 0.30) & (0.98, 0.0047, 0.15) & (0.98, 0.0038, 0.13) \\ (0.48, 0.30, 0.61) & - & (0.91, 0.03, 0.31) & (0.93, 0.02, 0.23) \\ (0.16, 0.81, 0.23) & (0.66, 0.19, 0.50) & - & (0.85, 0.04, 0.34) \\ (0.08, 0.90, 0.14) & (0.74, 0.18, 0.29) & (0.61, 0.18, 0.48) & - \end{bmatrix}$$

Step 6. Derive the positive and negative outranking flows using (25).
The results are shown in Table 6.

Step 7. Use the score function values for ranking.

$$\text{Score}\big(\varphi^{+}(x_1)\big) = 0.7361, \text{Score}\ \big(\varphi^{-}(x_1)\big) = -0.0071.$$

$$\text{Score}\big(\varphi^{+}(x_2)\big) = 0.1707, \text{Score}\ \big(\varphi^{-}(x_2)\big) = 0.0921.$$

$$\text{Score}\big(\varphi^{+}(x_3)\big) = 0.0109, \text{Score}\ \big(\varphi^{-}(x_3)\big) = 0.3033.$$

$$\text{Score}\big(\varphi^{+}(x_4)\big) = 0.0470, \text{Score}\ \big(\varphi^{-}(x_4)\big) = 0.4789.$$

From the score function values it can be concluded that $\varphi^{+}(x_1) > \varphi^{+}(x_2) > \varphi^{+}(x_4) > \varphi^{+}(x_3)$ and $\varphi^{-}(x_1) < \varphi^{-}(x_2) < \varphi^{-}(x_3) < \varphi^{-}(x_4)$, which implies that x_1 is the best alternative followed by x_2, while x_3 and x_4 are incomparable.

$$\varphi(x_3) = \text{Score}\big(\varphi^{+}(x_3)\big) - \text{Score}\big(\varphi^{-}(x_3)\big) = -0.2924.$$

$$\varphi(x_4) = \text{Score}\big(\varphi^{+}(x_4)\big) - \text{Score}\big(\varphi^{-}(x_4)\big) = -0.4319.$$

The complete ranking is $x_1 > x_2 > x_3 > x_4$.

The best alternative x_1 is the same as the best alternative obtained by Gündoğdu and Kahraman (2019) using SF-TOPSIS and IF-TOPSIS. Yet, the overall ranking differs. The ranking using SF-TOPSIS with spherical weighted arithmetic mean (SWAM) operator is $x_1 > x_4 > x_2 > x_3$. Meanwhile, the ranking using SF-TOPSIS with spherical weighted geometric mean (SWGM) operator and IF-TOPSIS is $x_1 > x_2 > x_4 > x_3$.

Example 2 To exploit the low enthalpy geothermal energy in a rural community in Greece, Nea Kessani, four alternative exploitation scenarios have been proposed

Table 7 Performance of the different scenarios and the PIS and NIS

	s_1	s_2	s_3	s_4	PIS	NIS
c_1	373	3706	3809	3860	3860	373
c_2	44	99	100	104	104	44
c_3	2240	5100	4930	4750	5100	2240
c_4	3	6	8	9	3	9

$\{s_1, s_2, s_3, s_4\}$. Four evaluation criteria have been selected, the net present value of the investment (NPV) (c_1), the creation of new jobs (c_2), the energy consumed (c_3), and the risk index (c_4). According to the importance of these criteria, relative weights have been assigned, $w_1 = 0.3$, $w_2 = 0.3$, $w_3 = 0.1$, and$w_4 = 0.3$.

For the four different exploitation scenarios s_1, s_2, s_3, s_4 the performance for the four criteria is given in Table 7 with the PIS and NIS. The NPV for the four scenarios is given in million Greek Drachmas. The risk index is represented on a scale from 1 to 20, where 1 indicates no risk and 20 indicates very risky.

Goumas (1995) proposed this example and used the conventional PROMETHEE to rank the alternative scenarios. Later, it was resolved by Goumas and Lygerou (2000) using F-PROMETHEE, Liao and Xu (2014) using IF-PROMETHEE, and by Zhang et al. (2019) using PF-PROMETHEE. The results of the SF-PROMETHEE are compared with the results of these techniques.

The SF-PROMETHEE is applied as follows.

Step 1. Define the positive and negative ideal solutions for each criterion.

Refer to Table 7.

Step 2. Calculate the relative degree of closeness of each scenario for each criterion. Since the data is crisp, the difference between the value of the scenario for each criterion and the ideal solutions $\left(\Delta^+(s_i), \Delta^-(s_i)\right)$ are used directly to calculate the relative degree of closeness.

$$\Delta_j^+(x_i) = c_j\left(x^+\right) - c_j(x_i), \text{ and } \Delta_j^-(x_i) = c_j(x_i) - c_j\left(x^-\right), \text{ and}$$

$$RC\left[c_j(x_i)\right] = \frac{\Delta^-(x_i)}{\Delta^+(x_i) + \Delta^-(x_i)}.$$

The results are given in Table 8.

Step 3. Calculate the deviation between every two scenarios for each criterion. The results are given in Table 9.

Step 4. Construct the preference matrix of the alternatives for each criterion using the linguistic preference function (19), then transform to the corresponding SFSs.

Table 8 RC of the scenarios for c_i

	$\Delta^{+}(s_i)$	$\Delta^{-}(s_i)$	$RC[c_i(s_i)]$		$\Delta^{+}(s_i)$	$\Delta^{-}(s_i)$	$RC[c_i(s_i)]$
c_1				c_3			
s_1	3487	0	0	s_1	2860	0	0
s_2	154	3333	0.96	s_2	0	2860	1
s_3	51	3436	0.99	s_3	170	2690	0.94
s_4	0	3487	1	s_4	350	2510	0.88
c_2				c_4			
s_1	60	0	0	s_1	0	6	1
s_2	5	55	0.92	s_2	3	3	0.5
s_3	4	56	0.93	s_3	5	1	0.17
s_4	0	60	1	s_4	6	0	0

Table 9 The deviations between the scenarios

	s_1	s_2	s_3	s_4
$d_1(s_i, s_k)$				
s_1	–	−0.96	−0.99	−1
s_2	0.96	–	−0.03	−0.04
s_3	0.99	0.03	–	−0.01
s_4	1	0.04	0.01	–
$d_2(s_i, s_k)$				
s_1	–	−0.92	−0.93	−1
s_2	0.92	–	−0.01	−0.08
s_3	0.93	0.01	–	−0.07
s_4	1	0.08	0.07	–
$d_3(s_i, s_k)$				
s_1	–	−1	−0.94	−0.88
s_2	1	–	0.06	0.12
s_3	0.94	−0.06	–	0.06
s_4	0.88	−0.12	−0.06	–
$d_4(s_i, s_k)$				
s_1	–	0.5	0.83	1
s_2	−0.5	–	0.33	0.5
s_3	−0.83	−0.33	–	0.17
s_4	−1	−0.5	−0.17	–

$$\mathbf{R}_L^1 = \begin{bmatrix} - & \sim \text{STP} & \sim \text{STP} & \sim \text{STP} \\ \text{STP} & - & \text{EP} & \text{EP} \\ \text{STP} & \text{EP} & - & \text{EP} \\ \text{STP} & \text{EP} & \text{EP} & - \end{bmatrix},$$

$$\tilde{\mathbf{R}}_S^1 = \begin{bmatrix} - & (0, 1, 0) & (0, 1, 0) & (0, 1, 0) \\ (1, 0, 0) & - & (0.5, 0.5, 0.5) & (0.5, 0.5, 0.5) \\ (1, 0, 0) & (0.5, 0.5, 0.5) & - & (0.5, 0.5, 0.5) \\ (1, 0, 0) & (0.5, 0.5, 0.5) & (0.5, 0.5, 0.5) & - \end{bmatrix}.$$

$$\mathbf{R}_L^2 = \begin{bmatrix} - & \sim \text{STP} & \sim \text{STP} & \sim \text{STP} \\ \text{STP} & - & \text{EP} & \text{EP} \\ \text{STP} & \text{EP} & - & \text{EP} \\ \text{STP} & \text{EP} & \text{EP} & - \end{bmatrix},$$

$$\tilde{\mathbf{R}}_S^2 = \begin{bmatrix} - & (0, 1, 0) & (0, 1, 0) & (0, 1, 0) \\ (1, 0, 0) & - & (0.5, 0.5, 0.5) & (0.5, 0.5, 0.5) \\ (1, 0, 0) & (0.5, 0.5, 0.5) & - & (0.5, 0.5, 0.5) \\ (1, 0, 0) & (0.5, 0.5, 0.5) & (0.5, 0.5, 0.5) & - \end{bmatrix}.$$

$$\mathbf{R}_L^3 = \begin{bmatrix} - & \sim \text{STP} & \sim \text{STP} & \sim \text{STP} \\ \text{STP} & - & \text{EP} & \text{WP} \\ \text{STP} & \text{EP} & - & \text{EP} \\ \text{STP} & \sim \text{WP} & \text{EP} & - \end{bmatrix},$$

$$\tilde{\mathbf{R}}_S^3 = \begin{bmatrix} - & (0, 1, 0) & (0, 1, 0) & (0, 1, 0) \\ (1, 0, 0) & - & (0.5, 0.5, 0.5) & (0.6, 0.4, 0.4) \\ (1, 0, 0) & (0.5, 0.5, 0.5) & - & (0.5, 0.5, 0.5) \\ (1, 0, 0) & (0.4, 0.6, 0.4) & (0.5, 0.5, 0.5) & - \end{bmatrix}.$$

$$\mathbf{R}_L^4 = \begin{bmatrix} - & \text{STP} & \text{STP} & \text{STP} \\ \sim \text{STP} & - & \text{SP} & \text{STP} \\ \sim \text{STP} & \sim \text{SP} & - & \text{WP} \\ \sim \text{STP} & \sim \text{STP} & \sim \text{WP} & - \end{bmatrix},$$

$$\tilde{\mathbf{R}}_S^4 = \begin{bmatrix} - & (1, 0, 0) & (1, 0, 0) & (1, 0, 0) \\ (0, 1, 0) & - & (0.8, 0.2, 0.2) & (1, 0, 0) \\ (0, 1, 0) & (0.2, 0.8, 0.2) & - & (0.6, 0.4, 0.4) \\ (0, 1, 0) & (0, 1, 0) & (0.4, 0.6, 0.4) & - \end{bmatrix}.$$

Step 5. Construct the overall spherical fuzzy preference matrix using (24). Since the relative weights of the criteria are crisp values, they are transformed to an equivalent SFS as follows: $(0.3, 0, 0)$, $(0.3, 0, 0)$, $(0.1, 0, 0)$, and$(0.3, 0, 0)$, which indicates the percentage of agreement, with no disagreement and no hesitancy.

Table 10 Positive and negative outranking flows

Alternative	$\varphi^{+}(s_i)$	$\varphi^{-}(s_i)$
s_1	(0.3, 0, 0)	(0.4245, 0, 0)
s_2	(0.3729, 0, 0.6095)	(0.2496, 0, 0.6350)
s_3	(0.3233, 0, 0.6578)	(0.2903, 0, 0.6651)
s_4	(0.3125, 0, 0.6409)	(0.3175, 0, 0.6403)

$$\tilde{\mathbf{R}}_s = \begin{bmatrix} - & (0.3, 0, 0) & (0.3, 0, 0) & (0.3, 0, 0) \\ (0.4245, 0, 0) & - & (0.3190, 0.0250, 0.7357) & (0.3655, 0, 0.6808) \\ (0.4245, 0, 0) & (0.2243, 0.1, 0.7559) & - & (0.2788, 0.05, 0.7764) \\ (0.4245, 0, 0) & (0.2145, 0.15, 0.7143) & (0.2462, 0.0750, 0.728) & - \end{bmatrix}$$

Step 6. Derive the positive and negative outranking flows using (25).
The results are shown in Table 10.
Step 7. Use the score function values for ranking.

$$\text{Score}\left(\varphi^{+}(s_1)\right) = 0.0900, \text{Score}\left(\varphi^{-}(s_1)\right) = 0.1802.$$

$$\text{Score}\left(\varphi^{+}(s_2)\right) = -0.3379, \text{Score}\left(\varphi^{-}(s_2)\right) = -0.2547.$$

$$\text{Score}\left(\varphi^{+}(s_3)\right) = -0.3208, \text{Score}\left(\varphi^{-}(s_3)\right) = -0.3019.$$

$$\text{Score}\left(\varphi^{+}(s_4)\right) = -0.3029, \text{Score}\left(\varphi^{-}(s_4)\right) = -0.3058.$$

Since $\varphi^{+}(s_3) > \varphi^{+}(s_2)$ and $\varphi^{-}(s_3) < \varphi^{-}(s_2)$, then $x_3 > x_2$.

The best scenario cannot be determined using partial ranking. Therefore, a complete ranking is required.

$$\varphi(s_1) = \text{Score}\big(\varphi^{+}(s_1)\big) - \text{Score}\big(\varphi^{-}(s_1)\big) = -0.0902.$$

$$\varphi(s_2) = \text{Score}\big(\varphi^{+}(s_2)\big) - \text{Score}\big(\varphi^{-}(s_2)\big) = -0.0832.$$

$$\varphi(s_3) = \text{Score}\big(\varphi^{+}(s_3)\big) - \text{Score}\big(\varphi^{-}(s_3)\big) = -0.0189.$$

$$\varphi(s_4) = \text{Score}\big(\varphi^{+}(s_4)\big) - \text{Score}\big(\varphi^{-}(s_4)\big) = 0.0029.$$

The complete ranking is $s_4 > s_3 > s_2 > s_1$. Then, the best scenario is s_4.

This ranking is consistent with the results of the conventional PROMETHEE, the IF-PROMETHEE, and the PF-PROMETHEE.

Note that the IF-PROMETHEE and the PF-PROMETHEE rankings are the same despite the difference in the hesitancy degrees of the used criteria weights in the two methods. The used criteria weights were: $\{(0.3, 0.5, 0.2), (0.3, 0.1, 0.6), (0.1, 0.3, 0.6), (0.3, 0.3, 0.4)\}$ for the IF-PROMETHEE and $\{(0.3, 0.5, 0.81), (0.3, 0.1, 0.95), (0.1, 0.3, 0.95), (0.3, 0.3, 0.91)\}$ for the PF-PROMETHEE. This can be attributed to dealing with only two degrees in the ranking process due to the dependency of the hesitancy degree on the membership degree and the non-membership degree. The IF-PROMETHEE used the membership degree and the hesitancy degree, while the PF-PROMETHEE used the membership degree and the non-membership degree.

When solving the problem with the proposed SF-PROMETHEE using the PF weights, the ranking changes to $s_2 > s_3 > s_1 > s_4$. This ranking is quite similar to the ranking of the F-PROMETHEE $s_2 > s_3 > s_4 > s_1$. According to Liao and Xu (2014), the difference in ranking is due to the type of fuzzy data used. The F-ROMETHEE presumes fuzzy performance values and crisp preference values, while the IF-PROMETHEE presumes IF performance values and IF preference values. Regarding the SF-PROMETHEE, the result is affected by the hesitancy degree of the weights, being treated as an independent degree which is taken into consideration in the solution procedure and the ranking.

5 Conclusion and Discussion

In this study, the conventional PROMETHEE was extended to the spherical fuzzy environment. The contribution of this study is the extension of the PROMRTHEE method within the context of SFSs, gaining the advantage of using three-dimensional membership functions to increase the decision-makers preference domain. Furthermore, using a linguistic preference function is more convenient for the decision-makers as it is closer to human perception. Therefore, the proposed SF-PROMETHEE provides the decision-makers with a flexible framework to express the fuzziness and uncertainty. SFSs are used to express both the weights of the criteria and the preference relations. The relative degree of closeness, which is based on the distances from the positive ideal solution and the negative ideal solution, is used to determine the preference index between the alternatives for each criterion. A linguistic preference function is used to construct the preference matrix. The linguistic terms are then transformed to their equivalent SFSs.

Two practical examples were solved to illustrate the applicability and efficiency of the proposed method in solving MCDM problems. In the first example, the weights of the criteria and the ratings of the alternatives are SFSs. The best alternative obtained by the SF-PROMETHEE is the same as that obtained by the SF-TOPSIS and IF-TOPSIS.

The second example, ranking of the alternative energy exploitation projects, was previously solved by Goumas (1995) using conventional PROMETHEE with crisp values for the weights of the criteria and the performance of the scenarios, Goumas and Lygerou (2000) using F-PROMETHEE, Liao and Xu (2014) using IF-PROMETHEE, and Zhang et al. (2019) using PF-PROMETHEE. The SF-PROMETHEE is used to tackle this problem and the results are compared with the results of the previously used versions of the PROMETHEE. The ranking obtained by the SF-PROMETHEE coincides with the results of the conventional PROMETHEE, the IF-PROMETHEE and the PF- PROMETHEE. Meanwhile, the ranking differs from the ranking of the F-PROMETHEE.

It's worth noting that both the IF-PROMETHEE and the PF- PROMETHEE gave the same ranking despite the apparent difference in the hesitancy degrees of the used criteria weights. This can be attributed to dealing with only two degrees in the ranking process since the hesitancy degree depends on the membership degree and the non-membership degree.

When the SF-PROMETHEE is utilized to resolve the same problem using PF weights, to increase the hesitancy degree, the ranking obtained is different. The ranking is quite similar to the ranking of the F- PROMETHEE. Therefore, it can be concluded that the hesitancy degree of the used weights of the criteria plays a crucial role in ranking. This role was revealed when the SFSs was used in which the hesitancy degree is expressed as an independent variable.

For future research, a comparative study between the performance of various MCDM methods under IFSs, PFSs, and SFSs can be conducted to study the effect of the hesitancy degree on the solution of MCDM problems in diverse application areas.

References

Adali EA, Işik AT, Kundakci N (2016) An alternative approach based on fuzzy PROMETHEE method for the supplier selection problem. Uncertain Supply Chain Manag 4:183–194

Araz C, Ozfirat PM, Ozkarahan I (2007) An integrated multicriteria decision making methodology for outsourcing management. Comput Oper Res 34:3738–3756

Atanassov KT (1986) Intuitionistic fuzzy sets. Fuzzy Sets Syst 20(1):87–96

Behzadian M, Kazemzadeh RB, Albadvi A, Aghdasi M (2010) PROMETHEE: a comprehensive literature review on methodologies and applications. Eur J Oper Res 200(1):198–215

Bilsel RU, Büyüközkan G, Ruan D (2006) A fuzzy preference ranking model for a quality evaluation of hospital Websites. Int J Intell Syst 2:1181–1197

Bouyssou D (2001) Outranking Methods. In: Floudas CA, Pardalos PM (eds) Encyclopedia of Optimization. Springer, Boston, MA, pp 1919–1925

Brans JP (1982) L'ingénièrie de la decision; Elaboration d'instruments d'aide à la décision. La méthode PROMETHEE. In: Nadeau R, Landry M (eds) L'aide à la décision: Nature, Instruments et Perspectives d'Avenir. Canada. Presses de l'Université Laval, Québec, pp 183–213

Brans JP, Mareschal B (1992) PROMETHEE V-MCDM problems with segmentation constraints. INFOR 30(2):85–96

Brans JP, Mareschal B (1995) The PROMETHEE VI procedure. How to differentiate hard from soft multicriteria problems. J Decis Syst 4:213–223

Brans JP, Mareschal B (2005) Promethee methods. In: Greco S (ed) Multiple criteria decision analysis: state of the art surveys. International Series in Operations Research and Management Science, vol 78. Springer, New York, NY, pp 163–186
Brans JP, Mareschal B, Vincke P (1984) PROMETHEE: A new family of outranking methods in MCDM. In: Brans JP (ed) Operational Research'84 (IFORS'84). North-Holland, Amsterdam, pp 477–490
Brans JP, Vincke P (1985) A preference ranking organisation method: (the PROMETHEE method for multiple criteria decision-making). Manag Sci 31(6):647–656
Brans JP, Vincke P, Mareschal B (1986) How to select and how to rank projects: the Promethee method. Eur J Oper Res 24:228–238
Chen Y-H, Wang T-C, Wu C-Y (2011) Strategic decisions using fuzzy PROMETHEE for IS outsourcing. Expert Syst Appl 38:12216–13222
Dulmin R, Mininno V (2003) Supplier selection using a multi-criteria decision aid method. J Purch Supply Manag 9:177–187
Elevli B, Demirci A (2004) Multicriteria choice of ore transport system for an underground mine: application of PROMETHEE methods. J S Afr Inst Min Metall 104(5):251–256
Fernández-Castro AS, Jiménez M (2005) PROMETHEE: an extension through fuzzy mathematical programming. J Oper Res Soc 56:119–122
Garg H (2016a) A novel accuracy function under interval-valued Pythagorean fuzzy environment for solving multi-criteria decision making problem. J Intell Fuzzy Syst 31(1):529–540
Garg H (2016b) A new generalized pythagorean fuzzy information aggregation using einstein operations and its application to decision making. Int J Intell Syst 31(9):886–920
Geldermann J, Spengler T, Rentz O (2000) Fuzzy outranking for environmental assessment. Case study: iron and steel making industry. Fuzzy Sets Syst. 115(1):45–65
Goumas M (1995) Decision making for geothermal resources management by multi-criteria analysis. Ph.D. Dissertation, Chemical Engineering Department, National Technical University of Athens (in Greek)
Goumas M, Lygerou V (2000) An extension of the PROMETHEE method for decision making in fuzzy environment: ranking of alternative energy exploitation projects. Eur J Oper Res 123(3):606–613
Gul M (2018) Application of Pythagorean fuzzy AHP and VIKOR methods in occupational health and safety risk assessment: the case of a gun and rifle barrel external surface oxidation and colouring unit. Int J Occup Saf Ergon. https://doi.org/10.1080/10803548.2018.1492251
Gul M, Ak MF (2018) A comparative outline for quantifying risk ratings in occupational health and safety risk assessment. J Clean Prod 196:653–664
Gul M, Celik E, Gumus AT, Guneri AF (2018) A fuzzy logic based PROMETHEE method for material selection problems. Beni-Suef Univ J Basic Appl Sci 7:68–79
Gündoğdu FK, Kahraman C (2019) Spherical fuzzy sets and spherical fuzzy TOPSIS method. J Intell Fuzzy Syst 36(1):337–352
Gupta R, Sachdeva A, Bhardwaj A (2012) Selection of logistic service provider using fuzzy PROMETHEE for a cement industry. J Manuf Technol Manag 23(7):899–921
Halouani N, Chabchoub H, Martel JM (2009) PROMETHEE-MD-2T method for project selection. Eur J Oper Res 195:841–849
Krishankumar R, Ravichandran KS, Saeid AB (2017) A new extension to PROMETHEE under intuitionistic fuzzy environment for solving supplier selection problem with linguistic preferences. Appl Soft Comput 60:564–576
Le Téno JF, Mareschal B (1998) An interval version of PROMETHEE for the comparison of building products' design with ill-defined data on environmental quality. Eur J Oper Res 109(2):522–529
Leyva-López JC, Fernández-González E (2003) A new method for group decision support based on ELECTRE III methodology. Eur J Oper Res 148:14–27
Li W-X, Li B-Y (2010) An extension of the Promethee II method based on generalized fuzzy numbers. Expert Syst Appl 37(7):5314–5319

Liang D, Xu Z, Darko AP (2017) Projection model for fusing the information of Pythagorean fuzzy multicriteria group decision making based on geometric Bonferroni mean. Int J Intell Syst 32(9):966–987

Liao H, Xu Z (2014) multi-criteria decision making with intuitionistic fuzzy PROMETHEE. J Intell Fuzzy Syst 27:1703–1717

Radojevic D, Petrovic S (1997) A fuzzy approach to preference structure in multi-criteria ranking. Int Trans Oper Res 4(5/6):419–430

Rani P, Jain D (2017) Intuitionistic fuzzy PROMETHEE technique for multi-criteria decision making problems based on entropy measure. In: Singh M, Gupta P, Tyagi V, Sharma A, Ören T, Grosky W (eds) Advances in computing and data sciences. ICACDS 2016. Communications in Computer and Information Science, vol 721. Springer, Singapore, pp 290–301

Rao RV, Patel BK (2010) Decision making in the manufacturing environment using an improved PROMETHEE method. Int J Prod Res 48:4665–4682

Roy B (1968) Classement et choix en presence de points de vue multiples (la methode electre). Revue française d'informatique et de recherche opérationnelle 2(8):57–75

Roy B (1978) ELECTRE III: un algorithme de classement fonde sur une représentation floue des préférences en presence de critères multiples. Cahiers du Centre d'Études de Recherche Opérationnelle 20:3–24

Roy B (1986) Méthodologie multicritère d'aidé a la décision. Politiques et management public 4(3):138–140

Roy B (1991) The outranking approach and the foundations of ELECTRE methods. Theor Decis 31:49–73

Roy B, Bertier P (1973) La méthode ELECTRE II: une application au media-planning. In: Ross M (ed) Operational research 1972. North-Holland Publishing Company, pp 291–302

Senvar O, Tuzkaya G, Kahraman C (2014) Multi-criteria supplier selection using PROMETHEE method. In: Kahraman C, Öztayşi B (eds) Supply chain management under fuzziness. Studies in Fuzziness and Soft Computing, vol 313. Springer, Berlin, Heidelberg, pp 21–34

Shih H-S, Chang Y-T, Cheng C-P (2016) A generalized PROMETHEE III with risk preferences on losses and gains. Int J Inf Manage Sci 27:117–127

Shirinfar M, Haleh H (2011) Supplier selection and evaluation by fuzzy multi-criteria decision making methodology. Int J Industr Eng Prod Res 22(4):271–280

Tuzkaya G, Gulsun B, Kahraman C, Ozgen D (2010) An integrated fuzzy multi-criteria decision making methodology for material handling equipment selection problem and an application. Expert Syst Appl 37:2853–2863

Wang JJ, Yang DL (2007) Using a hybrid multi-criteria decision aid method for information systems outsourcing. Comput Oper Res 34:3691–3700

Wei G (2017) Pythagorean fuzzy interaction aggregation operators and their application to multiple attribute decision making. J Intell Fuzzy Syst 33(4):2119–2132

Yager RR (2014) Pythagorean membership grades in multicriteria decision making. IEEE Trans Fuzzy Syst 22(4):958–965

Zhang Z-X, Hao W-N, Yu X-H, Chen G, Zhang S-J, Chen J-Y (2019) Pythagorean fuzzy preference ranking organization method of enrichment evaluations. Int J Intell Syst 34(7):1416–1439

Zhang K, Kluck C, Achari G (2009) A comparative approach for ranking contaminated sites based on the risk assessment paradigm using fuzzy PROMETHEE. Environ Manag 44(5):952–967

Delivery Drone Design Using Spherical Fuzzy Quality Function Deployment

Elif Haktanır, Cengiz Kahraman, and Fatma Kutlu Gündoğdu

Abstract Quality Function Deployment (QFD) is a well-known structured approach meeting customer needs through technical applications. Customer requirements (CRs) for a delivery drone design are first determined and they are tried to be satisfied by design requirements (DRs) in this chapter. Uncertainty in this process is handled by fuzzy set theory. One of the extensions of ordinary fuzzy sets is spherical fuzzy sets (SFSs) theory proposed by Kutlu Gündoğdu and Kahraman (J Intell Fuzzy Syst 36(1):337–352, 2019), which is composed of those independent parameters: *membership degree*, *non-membership degree*, and *hesitancy degree*, satisfying that their squared sum is equal to at most 1. We employ Spherical fuzzy QFD (SF-QFD) in the design of delivery drones. The importance ratings and global weights of CRs and improvement directions of DRs are represented by SFSs. Spherical fuzzy aggregation operators are used to aggregate the opinions of different decision makers.

1 Introduction

Product development is the process of creating a new product that covers all stages from the concept development to placing the product on the market shelf. The product development process involves working with experts in different disciplines. The key to product success is that the entire product development process has been carefully and meticulously planned from the beginning of the idea creation stage. In this way,

E. Haktanır (✉)
Department of Industrial Engineering, Altınbas University, 34217 Bagcilar, Istanbul, Turkey
e-mail: elif.haktanir@altinbas.edu.tr

C. Kahraman
Industrial Engineering Department, Istanbul Technical University, Besiktas, Istanbul 34367, Turkey
e-mail: kahramanc@itu.edu.tr

F. Kutlu Gündoğdu
Turkish Air Force Academy, National Defence University, Yesilyurt, Istanbul 34367, Turkey
e-mail: fatmakutlugundogdu@gmail.com

© Springer Nature Switzerland AG 2021

C. Kahraman and F. Kutlu Gündoğdu (eds.), *Decision Making with Spherical Fuzzy Sets*, Studies in Fuzziness and Soft Computing 392,
https://doi.org/10.1007/978-3-030-45461-6_17

possible risks that may be experienced during the product development process can be estimated and kept under control and the products that are most suitable for customer and market needs can be introduced to the market.

Nowadays, the technical capability of a product is not enough alone for successful results in the market. In addition to the durable technical features of the product, users are looking for aesthetic elements that are pleasing to the eye. Therefore, industrial design has an important place in the product development process. Industrial design must include not only aesthetic concerns but also other engineering realities. This is because a trendy concept design may not actually be produced, or it may involve very expensive production methods. It is necessary to define the needs right from the beginning. The solutions that make a difference in order to take the lead in an environment where competition is very tough are only revealed through systematic work and new product development. In general, new product development consists of seven main stages as shown in Fig. 1. These stages are briefly explained in the following.

1. Idea generation: Idea generation is the systematic search for new product ideas. Those ideas could come from internal and/or external resources. Internal resources refer to the company's own R&D department, management and employees. External resources are such as customers, competitors, distributors and suppliers.
2. Screening: All ideas should be examined one by one in order to remove those who are faulty and problematic. Products that can be easily rejected as a result of this review:

   - Those that are impossible to manufacture or are technically very difficult,
   - Those who have been tried before and failed,
   - The same as an existing product,

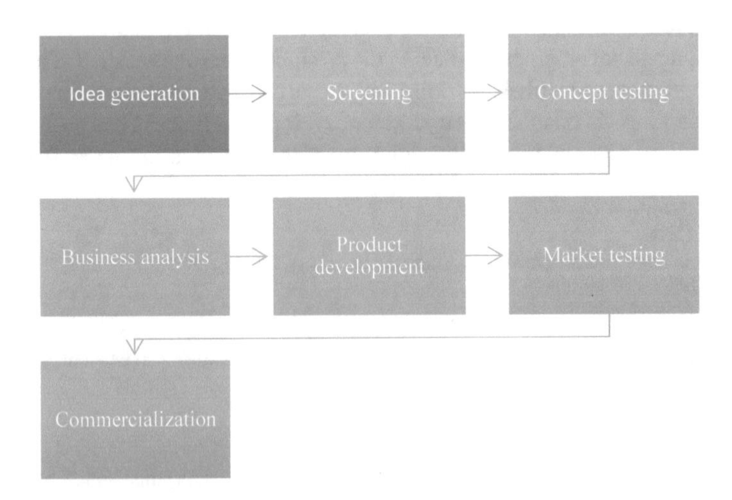

Fig. 1 Stages of new product development

- Those who require knowledge and experience that the organization does not have,
- Those that do not fit the existing operational structure.

3. Concept testing: At this stage, technical evaluation of ideas is performed. The applicability of the product is checked by the organization. There are typically two types of questions. A prototype can be developed and tested to answer some of these questions.

   (i) General questions: Can the product be implemented? Is the idea based on sound principles? Is it completely new or a variation of old ideas? Are there any patent problems or problems with competitors?
   (ii) Product specific questions: Is the proposed design technically feasible? Can the product be produced with existing technology? Does it fit the existing operational structure? Does the organization have the necessary experience and knowledge?

4. Business Analysis: If the product passes the technical evaluation stage, it will continue with the commercial evaluation stage where the profit analysis will be performed. At this stage, examinations such as how much the product will be sold, what is the competitive situation, how much investment is needed and what kind of return it will be generated are evaluated from the market and financial point of view. Market research that analyzes the customer response can be done at this stage. The technical evaluation and commercial analysis together constitute the feasibility study.
5. Product development: If the product passes the feasibility study, it will come to the final design and testing stage. This is the stage where the product is transformed from a prototype or conceptual model into a product form to be sold to customers. Data obtained from technical and commercial evaluation are used to complete the design.

The final design of the product is often the most important step. The design should be functional in order to work effectively, attractive for customers to like, and easy to produce to lower production costs. It can be difficult to design in accordance with all three criteria and a team of people from different functions must work to achieve the best solution. In this step, product design is completed including production processes. Then production starts and the product is put on the market.

6. Market Testing: If the product passes functional tests, the next step is market testing, the stage in which the product and the marketing program are introduced into a more realistic market environment. Market testing gives the marketer the opportunity to change the marketing mix before the product is released.

The amount of market testing varies according to the type of product. Costs can be tremendous and can also allow competitors to launch competitive product or sabotage the tester to get distorted results. Therefore, from time to time management may decide to skip this stage and move directly to the next one.

7. Commercialization: The last step in new product development is commercialization. Introducing the product to the market will cause high costs for production, advertising and promotion. The company should decide the timing of the launch (seasonality) and location (regional, national or international). This largely depends on the company's ability to take risks and access to the distribution network.

Customers get unsatisfied when their expectations do not meet the intended use of the products. Therefore, in order to meet the needs and desires of the existing and potential customers, companies apply various methods and models. One of the best knowns is the quality function deployment (QFD) model.

QFD is a process that enables the company to use all the competencies related to the existing products or services in the most efficient way according to customer requirements (CRs). It turns those demands into design goals by developing quality functions and enables this understanding to be used at every stage of the process. It is a multidisciplinary team process, which allows to divide each process of product or service development into stages and translate them into design requirements (DR).

The concept of QFD emerged as a result of total quality control technique applications between 1960–1965. The use of process diagrams in the case studies conducted on the subject has led to the development of the idea of functional deployment of jobs. QFD concept has been shaped by the application of process schemes. Later, Japanese scientist Akao adopted this approach and introduced the idea of transforming design characteristics into quality control points in order to take advantage of the power of this application in the product design process. In 1971, the problems experienced at the Kobe shipyard, which produces tankers of Mitsubishi Heavy Industries, were solved by applying the QFD method, thus the awareness of QFD has increased. After several applications, Akao named this process as "Quality Deployment" in his related study in 1972 (Akao 1972). Later, Mizuno and Akao developed the QFD matrix in their book which adapted CRs to all channels from design stage to production processes (Mizuno and Akao 1978).

The basis of the QFD is the transformation of CRs (What) into appropriate DRs (How) for each stage of product development and production. The process is carried out with a matrix called the House of Quality (HOQ). Main components of a classical HOQ are as follows:

1. Determination of CRs

The first step of HOQ is to identify who the customers are. The better the target audience is determined, the better their needs are analyzed. CRs can be determined in various ways. Conducting surveys, taking requests, complaints and suggestions from customers via phone and mail are the most common methods.

2. Determination of DRs

There are some requirements that cannot be determined by the customer, which must be considered during the production processes. These may be technical or legal requirements such as safety requirement, regulations, quality levels, product standards.

3. Relationship Matrix

The relationship matrix is the matrix between CRs and the firm's ability to meet these needs (DRs). For each DR that corresponds to each CR, the following question is asked: What is the severity of the relationship between the CR and the DR? It is placed in the matrix using the severity symbols and scores (linguistic terms and their corresponding SF values in this study).

4. The Roof of the HOQ (Correlation Matrix)

The correlation matrix, created on the roof of the HOQ, examines the relationship between DRs. The symbols used, present the type of the relationships:

- ↑sign for positive relationships
- ↓sign for negative relationships

5. Customer Rating

This evaluation made by the customers is very critical in determining which CR is more important while considering the DRs and deciding which CR should be given priority before starting production. As a result of the calculations made, CRs are ranked among themselves according to their importance evaluations.

Besides, the company has the opportunity to compare itself with its competitors in this section. It is seen which competitor meets which CRs better and how the competitiveness of the firm can be increased.

6. Technical Evaluation

The technical assessment indicates which DRs should be prioritized by the company in order to meet the CRs and gain advantage against its competitors. As in the customer assessment, some calculations are made, and the weights of the DRs are determined.

Customers generally tend to define their expectations by using linguistic terms rather than exact numerical values. Similarly, the importance scores of CRs, the relationship between CRs and DRs and relationship between the DRs themselves are usually expressed by linguistic terms. This causes a vagueness and impreciseness in these definitions. To cope with the vague nature of product development processes, fuzzy sets have been frequently used after its introduction to the literature by Zadeh (1965). Fuzzy sets extend the classical set theory by using membership degrees to represent the vagueness of the systems. Various extensions of fuzzy sets have been developed since its introduction. (Atanassov 1986) proposed intuitionistic fuzzy sets which assign both membership, non-membership, and hesitancy degrees to each element in the set. (Torra 2010) developed hesitant fuzzy sets which define the membership degree of a given element as a set of possible values. (Yager 2014) extended intuitionistic fuzzy sets to Pythagorean fuzzy sets which define elements with membership and non-membership degrees where their squared sum is smaller than or equal to 1. (Kutlu Gündoğdu and Kahraman 2019) introduced the spherical fuzzy sets (SFSs). The SFSs allow the hesitancy of a decision maker (DM)

Fig. 2 A delivery drone

to be assigned independently from membership and non-membership degrees. The advantage of the SFSs is that they gather the positive sides of other fuzzy sets extensions in a unique theory. SFSs present wider domain to define membership degrees than Pythagorean and intuitionistic fuzzy sets so they can provide more flexibility to DMs under imprecise and uncertain cases for real life problems. In this study all the linguistic evaluations of DMs are represented by SFSs.

The main objective of this study is to develop a SF QFD based product design and development approach. While dealing with linguistic data, SFSs based QFD system is utilized to rank the DRs for delivery drone design as an illustrative example.

Delivery drones are such unmanned aerial vehicles (UAVs) that can deliver lightweight packages. They generally use 4–8 propellers and rechargeable batteries and can be operated autonomously or remotely. They are mostly preferred for delivering time-sensitive items, such as medicine, or deliveries that would be difficult with traditional delivery vehicles. An illustrative delivery drone is given in Fig. 2. Delivery drones have the potential of reducing costs and time for deliveries, improving safety due to less heavy traffic, reducing gas emissions, providing greater route flexibility, and reducing roadway and bridge maintenance costs. Nevertheless, it still has such limitations that needs to be improved in time: constrained flight times and ranges due to limited battery capacity, limited package weights, being highly dependent on the weather conditions.

The rest of the paper is organized as follows: Sect. 2 gives a literature review on product design and development QFD under fuzziness. Section 3 presents the preliminaries of SFSs. Section 4 introduces the proposed SF-QFD method. Section 5 demonstrates the application of the proposed product development method using SF-QFD with a delivery drone design example. The last section concludes the paper with future directions.

2 Literature Review

A literature review on product design and development using QFD method based on Scopus database gave a list of 833 publications when the keyword search is limited into article titles, abstracts, and keywords. Figure 3 shows the distribution of the related publications with respect to years.

Although QFD is one of the long-standing product and quality planning methods in the literature, its usage has significantly increased in the 2000s. The year with the most publications on this topic is 2010 with 56 studies in total.

As it is given in Fig. 4, most of the product design and development studies applying QFD method are in conference paper type which is followed by articles, reviews, book chapters and other types of publications.

QFD method has been applied for product design and development studies within many subject areas and Fig. 5 shows their frequencies. Engineering, computer science

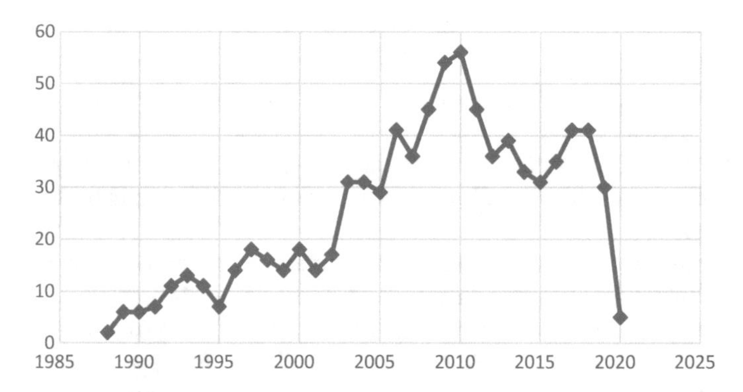

Fig. 3 Distribution of the product design and development publications applying QFD method with respect to years

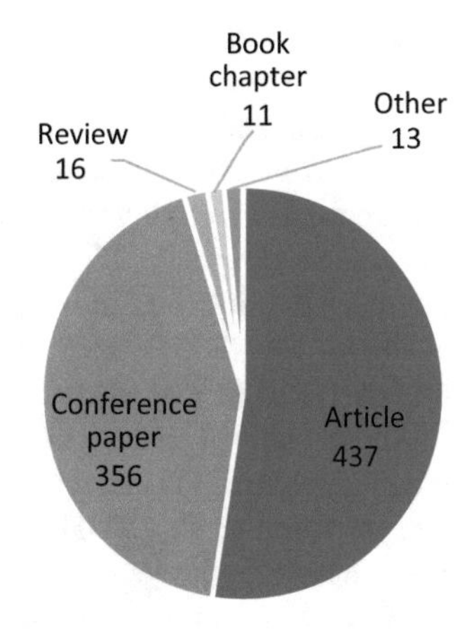

Fig. 4 Document types distributions of product design and development studies applying QFD method

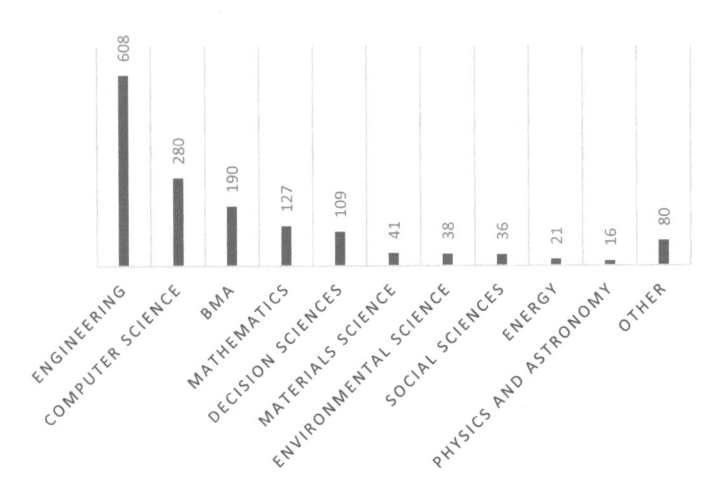

Fig. 5 Subject areas distribution of product design and development publications applying QFD method

and Business, Management and Accounting (BMA) are the most frequently applied subject areas respectively.

143 of overall QFD studies about product design and development applies fuzzy logic. Some representative studies are given below considering their development over the years.

Kalargeros and Gao (1998) aimed to simplify and rationalize application of QFD by utilizing fuzzy logic principles and applied to decision making during the product development stage. Verma et al. (1998) presented an expert system-based extension to the fuzzy QFD methodology to facilitate strategic product planning, early design decision-making and parameter target setting. Wang (1999) proposed a new fuzzy outranking approach to prioritize DRs recognized in QFD and used an example of a car design to illustrate the approach developed. Kwong and Bai (2002) presented a fuzzy AHP approach to the determine the importance weights of CRs in QFD and illustrated the proposed approach with an example of bicycle splashguard design. Yang and Fang (2003) integrated fuzzy logic into QFD to analyze product features and conduct product positioning. Chen et al. (2004) presented a fuzzy regression-based mathematical programming approach for QFD product planning and applied to a quality improvement problem for an emulsification dynamite packing machine. Karsak (2004) developed a fuzzy multiple objective decision-making approach to prioritize DRs in QFD and illustrated the application with a pencil design example. Liu (2005) rated DRs in fuzzy QFD via a mathematical programming approach and gave an example of a flexible manufacturing system design. Fung et al. (2005) developed a fuzzy expected value-based goal programing model for product planning using QFD and gave an illustrative example on dynamite-packing machine design. Chen et al. (2005) presented a fuzzy expected value modelling approach for determining target values of engineering characteristics in QFD and provided an example of a motor car design. Kahraman et al. (2006) proposed an integrated framework based on fuzzy QFD and a fuzzy optimization model to determine the product technical requirements to be considered in designing a product and presented an application in a Turkish Company producing PVC window and door systems. Chen and Weng (2006) developed an evaluation approach to engineering design in QFD processes using fuzzy goal programming models. They illustrated the proposed model with a writing instrument example. Lee et al. (2008) presented an integrative approach by incorporating the Kano model with fuzzy mode into the matrix of QFD and adjusting CR weights. They illustrated their approach on Taiwan's high-tech information electronic industry. Kuo et al. (2009) integrated environmental considerations in QFD by using fuzzy logic, used a real-life case for the toner cartridge design of the printer. Chen and Ko (2009) presented fuzzy linear programming models for new product design using QFD with failure modes and effect analysis and used a semiconductor packing case of the turbo thermal ball grid array package to exemplify the proposed models. Zhai et al. (2009) proposed a rough set based QFD approach to the management of imprecise design information in product development with a QFD case study on a bicycle design. Liu (2009) developed an extended fuzzy QFD approach which expands the research scope, from product planning to part deployment and gave an example on a small-medium scale manufacturing company located in the southern part of Taiwan that produces double-layered stainless vacuums for household use. Nepal et al. (2010) developed a fuzzy-AHP approach to prioritize the customer satisfaction attributes in target planning for automotive product development. Lee et al. (2010) proposed an evaluation framework for product planning using FANP, QFD and multi-choice goal programming, and carried out a case study of the product design process of backlight unit in thin film transistor liquid crystal display industry in Taiwan. Luo et al. (Luo et al. 2010) aimed to determine optimal levels of engineering characteristics in QFD under multi-segment market. They expressed the weights of importance of market segments and development costs in the model as triangular fuzzy numbers and illustrated their methodology by an example of an industrial pincers, which is a product of an oil equipment corporation in China. Liu (2011) proposed a product design and selection method using fuzzy QFD and fuzzy multi criteria decision making approaches and gave an example of an aluminum window design. Kavosi and Mavi (2011) developed a fuzzy QFD approach using TOPSIS and AHP methods. The methodology was applied for the Aras pen company in Iran. Hsu et al. (2011) presented an integrative approach by incorporating green product lifecycle management with fuzzy analysis into the matrix of QFD. Chen and Ko (2011) developed fuzzy nonlinear models for new product development using four-phase of QFD processes and illustrated the proposed model with a semiconductor packing case of the turbo thermal ball grid array package example. Wu (2011) explored an intelligent method to evaluate the product design time via fuzzy measurable HOQ and QFD for fuzzy regression estimation problem and illustrated the procedure of with a case of injection mold design. Lee et al. (2012) analyzed CRs and technology in new product development via fuzzy QFD and Delphi methods. Vinodh and Rathod (2012) presented an application of fuzzy eco-QFD to aid product design team in considering environmental concern and a case study describing the evolution of eco-friendly model of rotary switch. Bevilacqua et al. (2012) developed and tested a new fuzzy-QFD approach for characterizing customers rating of extra virgin olive oil. Karimi et al. (2012) applied a hybrid fuzzy QFD-TOPSIS method to design product in the industry with a case study in sum service company in Iran. Wang (2012) developed a fuzzy-normalization-based group decision-making approach for prioritizing engineering DRs in QFD under uncertainty and examined a real design case of a flexible manufacturing system. Wang and Chen (2012) proposed a fuzzy multi-criteria decision making based QFD to assist an enterprise in fulfilling collaborative product design and optimal selection of module mix. They demonstrated a real case study on developing various types of sport and water digital cameras. Li (2013) developed a method for product design selection with incomplete linguistic weight information based on QFD in a fuzzy environment and used a fully automatic washing machine development case to illustrate the proposed approach. Delice and Güngör (2013) proposed a fuzzy mixed-integer goal programming model that determines a combination of optimal DR values in QFD and presented a real-world application in the Turkish white goods industry. Cebi et al. (2014) evaluated design parameters for vessel engine room by using a modified QFD technique including fuzzy AHP and DEMATEL. Wu and Ho (2015) integrated green QFD and fuzzy theory and illustrated a case study on green mobile phone design. Hundal and Kant (2017) proposed a product design development method integrating QFD approach with heuristics-AHP, ANN and fuzzy logics and provided a case study in miniature circuit breaker. Liu et al. (2017) developed a two-phase approach for process optimization of customer collaborative design based on fuzzy QFD and design structure matrix and illustrated the development of a new type of automotive part as a numerical example. Vinodh et al. (2017) applied fuzzy QFD for sustainable design of consumer electronics products. Kang et al. (2018) integrated evaluation grid method and fuzzy QFD to new product development and used minicars as an example to illustrate the proposed method. Alptekin and Alptekin (2018) presented a fuzzy QFD approach for differentiating cloud products and presented a real-life cloud product design scenario with three customer profiles. Peng et al. (2018) presented a systematical decision-making approach for QFD based on hesitant linguistic term sets and conducted an empirical study on the research project called vortex recoil hydraulic retarder. Wang et al. (2019) integrated fuzzy based QFD and AHP for the design and implementation of a hand training device. Wang (2019) integrated a novel intuitive fuzzy method with QFD for product design and presented a case study on touch panels. Zheng et al. (2019) presented a weighted interval rough number based method to determine relative importance ratings of CRs in QFD product planning and gave an illustrative example on a mountain bicycle frameset from a local bicycle company in New Zealand. Huang et al. (2019) developed an approach for QFD based on proportional hesitant fuzzy linguistic term sets and prospect theory. They provided an example of product development of electric vehicles in a manufacturing company located in Shanghai, China. Haktanır and Kahraman (2019) presented a novel interval-valued Pythagorean fuzzy QFD method and its application to solar photovoltaic technology development. Li et al. (2019) developed a probabilistic language method based on fuzzy QFD to rate engineering characteristics in open design and presented an example of the smartphone. Kutlu Gündoğdu and Kahraman (2020) presented a novel SF-QFD method and its application to the linear delta robot technology development.

3 Spherical Fuzzy Sets: Preliminaries

SFSs are inspired from neutrosophic sets and picture fuzzy sets which are the extensions of intuitionistic fuzzy sets. They provide a larger preference domain for DMs to assign membership degrees since the squared sum of membership parameters can be at most 1.0. DMs can define their hesitancy information independently under SF environment like neutrosophic sets and picture fuzzy sets (Gündoğdu and Kahraman 2019a, b, c, d; Gündoğdu 2019; Kahraman et al. 2019; Kutlu Gündoğdu Preprint). In the following, preliminaries of SFS are presented:

Definition 3.1 Single valued SFSs of the universe of discourse U is given by

$$\tilde{A}_S = \left\{u, \left(\mu_{\tilde{A}_S}(u), \nu_{\tilde{A}_S}(u), \pi_{\tilde{A}_S}(u)\right) \middle| u \in U\right\} \tag{1}$$

where

$$\mu_{\tilde{A}_S}(u) : U \to [0, 1], \nu_{\tilde{A}_S}(u) : U \to [0, 1], \pi_{\tilde{A}_S}(u) : U \to [0, 1]$$

and

$$0 \le \mu^2_{\tilde{A}_S}(u) + \nu^2_{\tilde{A}_S}(u) + \pi^2_{\tilde{A}_S}(u) \le 1 \forall u \in U \tag{2}$$

For each u, the numbers $\mu_{\tilde{A}_S}(u)$, $\nu_{\tilde{A}_S}(u)$ and $\pi_{\tilde{A}_S}(u)$ are the degree of membership, non-membership and hesitancy of u to $\tilde{A}_S$, respectively.

Definition 3.2 Basic operators of Single-valued SFS;

$$\text{i. } \tilde{A}_S \oplus \tilde{B}_S = \left\{ \begin{array}{l} \left(\mu^2_{\tilde{A}_S} + \mu^2_{\tilde{B}_S} - \mu^2_{\tilde{A}_S}\mu^2_{\tilde{B}_S}\right)^{1/2}, v_{\tilde{A}_S}v_{\tilde{B}_S}, \\ \left(\left(1-\mu^2_{\tilde{B}_S}\right)\pi^2_{\tilde{A}_S} + \left(1-\mu^2_{\tilde{A}_S}\right)\pi^2_{\tilde{B}_S} - \pi^2_{\tilde{A}_S}\pi^2_{\tilde{B}_S}\right)^{1/2} \end{array} \right\} \tag{3}$$

$$\text{ii. } \tilde{A}_S \otimes \tilde{B}_S = \left\{ \begin{array}{l} \mu_{\tilde{A}_S}\mu_{\tilde{B}_S}, \left(v^2_{\tilde{A}_S} + v^2_{\tilde{B}_S} - v^2_{\tilde{A}_S}v^2_{\tilde{B}_S}\right)^{1/2}, \\ \left(\left(1-v^2_{\tilde{B}_S}\right)\pi^2_{\tilde{A}_S} + \left(1-v^2_{\tilde{B}_S}\right)\pi^2_{\tilde{B}_S} - \pi^2_{\tilde{A}_S}\pi^2_{\tilde{B}_S}\right)^{1/2} \end{array} \right\} \tag{4}$$

$$\text{iii. } \lambda \cdot \tilde{A}_S = \left\{ \begin{array}{l} \left(1-\left(1-\mu^2_{\tilde{A}_S}\right)^{\lambda}\right)^{1/2}, v^{\lambda}_{\tilde{A}_S}, \\ \left(\left(1-\mu^2_{\tilde{A}_S}\right)^{\lambda} - \left(1-\mu^2_{\tilde{A}_S} - \pi^2_{\tilde{A}_S}\right)^{\lambda}\right)^{1/2} \end{array} \right\} \tag{5}$$

iv. $$\tilde{A}_S^{\lambda} = \left\{ \begin{array}{l} \mu_{\tilde{A}_S}^{\lambda}, \left(1 - \left(1 - v_{\tilde{A}_S}^2\right)^{\lambda}\right)^{1/2}, \\ \left(\left(1 - v_{\tilde{A}_S}^2\right)^{\lambda} - \left(1 - v_{\tilde{A}_S}^2 - \pi_{\tilde{A}_S}^2\right)^{\lambda}\right)^{1/2} \end{array} \right\} \tag{6}$$

Definition 3 Single-valued Spherical Weighted Arithmetic with respect to, $w = (w_1, w_2, \ldots, w_n)$; $w_i \in [0, 1]$; $\sum_{i=1}^{n} w_i = 1$, SWAM is defined as;

$$SWAM_w\left(\tilde{A}_{S1}, \ldots, \tilde{A}_{Sn}\right) = w_1\tilde{A}_{S1} + w_2\tilde{A}_{S2} + \cdots + w_n\tilde{A}_{Sn}$$
$$= \left\{ \begin{array}{l} \left[1 - \prod_{i=1}^{n}(1 - \mu_{\tilde{A}_{Si}}^2)^{w_i}\right]^{1/2}, \\ \prod_{i=1}^{n} v_{\tilde{A}_{Si}}^{w_i}, \left[\prod_{i=1}^{n}(1 - \mu_{\tilde{A}_{Si}}^2)^{w_i} - \prod_{i=1}^{n}(1 - \mu_{\tilde{A}_{Si}}^2 - \pi_{\tilde{A}_{Si}}^2)^{w_i}\right]^{1/2} \end{array} \right\} \tag{7}$$

Definition 3.4 Single-valued Spherical Weighted Geometric Mean (SWGM) with respect to, $w = (w_1, w_2 \ldots, w_n)$; $w_i \in [0, 1]$; $\sum_{i=1}^{n} w_i = 1$, SWGM is defined as;

$$SWGM_w\left(\tilde{A}_1, \ldots, \tilde{A}_n\right) = \tilde{A}_{S1}^{w_1} + \tilde{A}_{S2}^{w_2} + \cdots + \tilde{A}_{Sn}^{w_n}$$
$$= \left\{ \begin{array}{l} \prod_{i=1}^{n} \mu_{\tilde{A}_{Si}}^{w_i}, \left[1 - \prod_{i=1}^{n}(1 - v_{\tilde{A}_{Si}}^2)^{w_i}\right]^{1/2}, \\ \left[\prod_{i=1}^{n}(1 - v_{\tilde{A}_{Si}}^2)^{w_i} - \prod_{i=1}^{n}(1 - v_{\tilde{A}_{Si}}^2 - \pi_{\tilde{A}_{Si}}^2)^{w_i}\right]^{1/2} \end{array} \right\} \tag{8}$$

Definition 3.5 Defuzzification functions and Accuracy functions of sorting SFS are defined by;

$$Score\left(\tilde{A}_s\right) = \left(2\mu_{\tilde{A}_s} - \frac{\pi_{\tilde{A}_s}}{2}\right)^2 - \left(v_{\tilde{A}_s} - \frac{\pi_{\tilde{A}_s}}{2}\right)^2 \tag{9}$$

$$Accuracy\left(\tilde{A}_S\right) = \mu_{\tilde{A}_S}^2 + v_{\tilde{A}_S}^2 + \pi_{\tilde{A}_S}^2 \tag{10}$$

Note that: $\tilde{A}_S < \tilde{B}_S$ if and only if;

i. $Score\left(\tilde{A}_S\right) < Score\left(\tilde{B}_S\right)$ or

ii. $Score\left(\tilde{A}_S\right) = Score\left(\tilde{B}_S\right)$ and $Accuracy\left(\tilde{A}_S\right) < Accuracy\left(\tilde{B}_S\right)$

4 Spherical Fuzzy QFD Method

The proposed SF-QFD method is composed of seventeen steps as presented in this section (Kutlu Gündoğdu and Kahraman 2020).

Step 1. Assign linguistic CRs and constitute pairwise comparison matrices using SF judgment matrices based on the linguistic terms given in Table 1.

Step 2. Check for the consistency of each pairwise comparison matrix. To do that, convert the linguistic terms in the pairwise comparison matrix to their corresponding score indices. Then, use the classical consistency check. The critical value of the consistency ratio is 10%. If the consistency ratio (cr) of crisp pairwise comparison matrix is less than 10%, we can continue with its corresponding SF pairwise comparison matrix.

Step 3. Compute the SF global weight of each CR. Let more than one DM rate CRs by using the scale in Table 1. Then, determine the importance evaluation ($I\tilde{E}_i$) of each CR using SWGM operator given in Eq. (8).

Step 4. Define the DRs and determine the direction of improvement of DRs. Then, fill in the correlation matrix among DRs as given in Fig. 2 by using the linguistic scale given in Table 1. Positive correlations (↑) and negative correlations (↓) are

Table 1 Linguistic and corresponding spherical scale

Linguistic terms	(μ, v, π)	Score Index (SI)
Absolutely More Importance (AMI), Absolutely More Relation (AMR), Absolutely More Satisfactory (AMS), Absolutely More Correlation (AMC)	(0.9, 0.1, 0.1)	9
Very High Importance (VHI), Very High Relation (VHR), Very High Satisfactory (VHS), Very High Correlation (VHC)	(0.8, 0.2, 0.2)	7
High Importance (HI), High Relation (HR), High Satisfactory (HS), High Correlation (HC)	(0.7, 0.3, 0.3)	5
Slightly More Importance (SMI), Slightly More Relation (SMR), Slightly More Satisfactory (SMS), Slightly More Correlation (SMC)	(0.6, 0.4, 0.4)	3
Equally Importance (EI), Medium Relation (MR), Medium Satisfactory (MS), Medium Correlation (MC)	(0.5, 0.5, 0.5)	1
Slightly Low Importance (SLI), Slightly Low Relation (SLR), Slightly Low Satisfactory (SLS), Slightly Low Correlation (SLC)	(0.4, 0.6, 0.4)	1/3
Low Importance (LI), Low Relation (LR), Low Satisfactory (LS), Low Correlation (LC)	(0.3, 0.7, 0.3)	1/5
Very Low Importance (VLI), Very Low Relation (VLR), Very Low Satisfactory (VLS), Very low Correlation (VLC)	(0.2, 0.8, 0.2)	1/7
Absolutely Low Importance (ALI), Absolutely Low Relation (ALR), Absolutely Low Satisfactory (ALS), Absolutely Low Correlation (ALC)	(0.1, 0.9, 0.1)	1/9

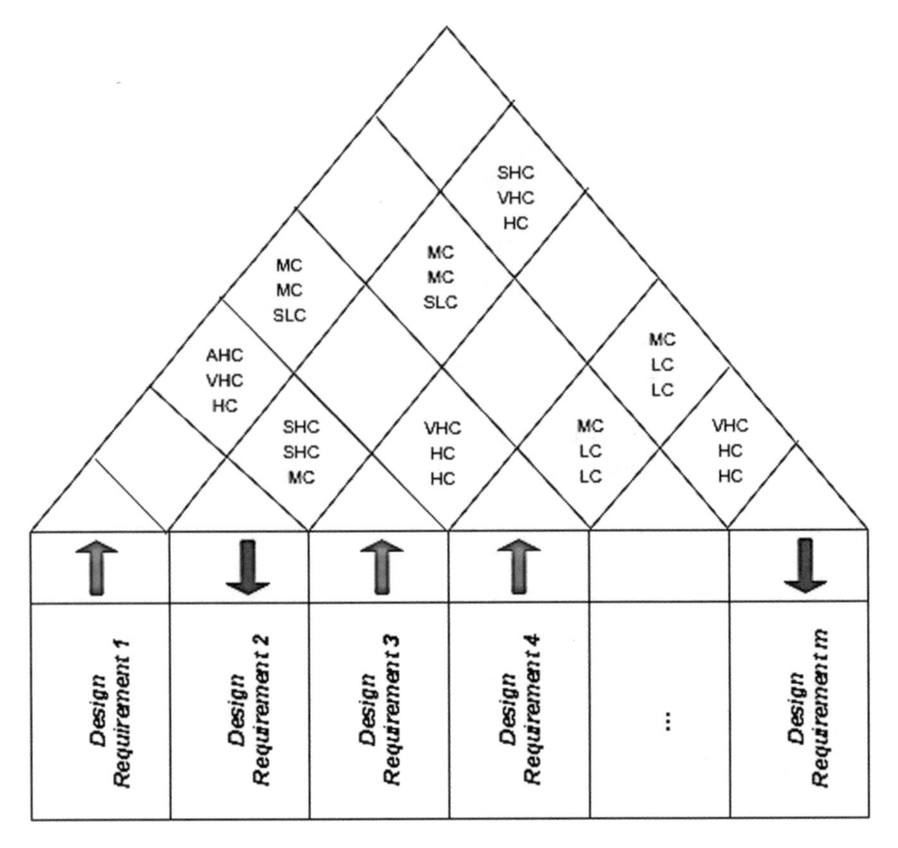

Fig. 6 DRs and their directions of improvement

represented with $po\bar{z}_{DR_j}$ and $ne\bar{g}_{DR_j}$, respectively. If there is no correlation, the cell is left empty. HOQ is filled by three experts who are experienced in drone technologies. In Fig. 6, each cell shows three evaluations from three experts. We aggregate these linguistic terms for getting aggregated positive and negative correlations of each DR based on SWGM operator given in Eq. (8).

Step 5. Aggregate all positive ($po\bar{z}_{DR_j}$) and negative ($ne\bar{g}_{DR_j}$) correlations separately by using SWAM operator given in Eq. (7). Defuzzify the aggregated positive and negative correlations using Eq. (11).

$$defuzz\left(\tilde{A}_s\right) = \left(2\mu_{\tilde{A}_s}\right)^2 - v_{\tilde{A}_s}^2 - \frac{\pi_{\tilde{A}_s}^2}{2} \tag{11}$$

After defuzzifying step, Correlation Impact Factor (CIF_{DR_j}) of DRs can be obtained from Eq. (12).

Direction of improvement			↓	↑	↑		↑
WHATS \ HOWS		Design Requirements	Design Requirement 1	Design Requirement 2	Design Requirement 3	....	Design Requirement m
	Main Criteria	Importance Evaluations (IE)					
Customer Requirements	Customer Requirement 1	SMI, LI, SLI	SMR, LR, SLR				
	Customer Requirement 2	VHI, HI, HI		VHR, HR, HR		MR, AMR, AMR	
	Customer Requirement 3	AMI, VHI, HI			MR, MR, AMR		VHR, MR, AMR
	...						
	Customer Requirement n	SMI, LI, SLI	SLR, LR, VLR				HR, VHR, HR

Fig. 7 The relationship matrix between DRs and CRs

$$CIF_{DR_j} = \beta + \left(poz\bar{c}_{DR_j} - neg\bar{c}_{DR_j}\right) * \left(\frac{n_j}{(m-1)}\right) \quad j = 1, 2, \ldots, m \tag{12}$$

where n_j is the number of correlations in one DR and β value can be used as the smallest integer to make CIF_{DR_j} value positive. In our study, β value is 1.0.

Step 6. Construct the relationship matrix $\left(\tilde{R}_{ij}\right)$ between DRs and CRs as given in Fig. 7. Aggregate the linguistic terms of each relationship between DRs and CRs using SWGM operator given in Eq. (8). Instead of SWGM operator, SWAM operator is also used for this calculation.

Step 7. To get the SF overall priorities of each DR ($A\tilde{I}_{DR_j}$), multiply the aggregated correlation impact factor (CIF_{DR_j}) of DRs by the summed aggregated linguistic terms in the relationship matrix ($\oplus_{i=1}^{n} \tilde{R}_{ij}$) based on Eq. (13).

$$A\tilde{I}_{DR_j} = \left(CIF_{DR_j} \cdot \oplus_{i=1}^{n} \tilde{R}_{ij}\right)$$
$$= \left\{ \begin{array}{l} \left(1 - \left(1 - \mu^2_{\oplus_{i=1}^{n} \tilde{R}_{ij}}\right)^{CIF_{DR_j}}\right)^{1/2}, v^{CIF_{DR_j}}_{\oplus_{i=1}^{n} \tilde{R}_{ij}}, \\ \left(\left(1 - \mu^2_{\oplus_{i=1}^{n} \tilde{R}_{ij}}\right)^{CIF_{DR_j}} - \left(1 - \mu^2_{\oplus_{i=1}^{n} \tilde{R}_{ij}} - \pi^2_{\oplus_{i=1}^{n} \tilde{R}_{ij}}\right)^{CIF_{DR_j}}\right)^{1/2} \end{array} \right\} \tag{13}$$

where $i = 1, 2, \ldots, n$ and $j = 1, 2, \ldots, m$.
Step 8. To get the SF overall priorities of each CR ($A\tilde{I}_{CR_i}$), multiply the aggregated linguistic importance evaluations ($I\tilde{E}_i$) of CRs by the summed aggregated linguistic terms in the relationship matrix ($\oplus_{j=1}^{m}\tilde{R}_{ij}$) based on Eq. (14).

$$A\tilde{I}_{CR_i} = I\tilde{E}_i \otimes \left(\oplus_{j=1}^{m}\tilde{R}_{ij}\right) = \left\{ \begin{array}{l} \mu_{I\tilde{E}_i}\mu_{\oplus_{j=1}^{m}\tilde{R}_{ij}}, \left(v_{I\tilde{E}_i}^2 + v_{\oplus_{j=1}^{m}\tilde{R}_{ij}}^2 - v_{I\tilde{E}_i}^2 v_{\oplus_{j=1}^{m}\tilde{R}_{ij}}^2\right)^{1/2}, \\ \left(\left(1 - v_{\oplus_{j=1}^{m}\tilde{R}_{ij}}^2\right)\pi_{I\tilde{E}_i}^2 + \left(1 - v_{I\tilde{E}_i}^2\right)\pi_{\oplus_{j=1}^{m}\tilde{R}_{ij}}^2 - \pi_{I\tilde{E}_i}^2\pi_{\oplus_{j=1}^{m}\tilde{R}_{ij}}^2\right)^{1/2} \end{array} \right\} \tag{14}$$

where $i = 1, 2, \ldots, n$ and $j = 1, 2, \ldots, m$.
Step 9. Determine the linguistic customer/expert ratings of the alternatives in terms of DRs and CRs, separately, by using the SF scale given in Table 1.
Step 10. Aggregate the evaluations of each DM using SWGM or SWAM operators. Construct aggregated SF decision matrix based on the opinions of experts in terms of DRs. Denote the evaluation values of each alternative $X_k (k = 1, 2 \ldots s)$ with respect to design requirement $A\tilde{I}_{DR_j}$ by $A\tilde{I}_{DR_j}\left(\tilde{X}_k\right) = \left(\mu_{kj}, v_{kj}, \pi_{kj}\right)$. $\tilde{D}^{DR_j} = \left(A\tilde{I}_{DR_j}\left(\tilde{X}_k\right)\right)_{mxs}$ is a spherical fuzzy decision matrix. For a SF-MCDM problem, decision matrix should be constructed as in Eq. (15).

$$\tilde{D}^{DR_j} = \left(A\tilde{I}_{DR_j}\left(\tilde{X}_k\right)\right)_{mxs} = \begin{pmatrix} (\mu_{11}, v_{11}, \pi_{11}) & (\mu_{12}, v_{12}, \pi_{12}) & \ldots (\mu_{1s}, v_{1s}, \pi_{1s}) \\ (\mu_{21}, v_{21}, \pi_{11}) & (\mu_{22}, v_{22}, \pi_{22}) & \ldots (\mu_{2s}, v_{2s}, \pi_{2s}) \\ . & . & . \\ . & . & . \\ (\mu_{m1}, v_{m1}, \pi_{m1}) & (\mu_{m2}, v_{m2}, \pi_{m2}) & \ldots (\mu_{ms}, v_{ms}, \pi_{ms}) \end{pmatrix} \tag{15}$$

Construct the aggregated SF decision matrix based on the opinions of DMs in terms of CRs. Denote the evaluation values of each alternative $X_k (k = 1, 2 \ldots s)$ with respect to customer requirement $A\tilde{I}_{CR_i}$ by $A\tilde{I}_{CR_i}\left(\tilde{X}_k\right) = \left(\mu_{ki}, v_{ki}, \pi_{ki}\right)$. $\tilde{D}^{CR_i} = \left(A\tilde{I}_{CR_i}\left(\tilde{X}_k\right)\right)_{mxs}$ is a spherical fuzzy decision matrix. For a SF-MCDM problem, decision matrix should be constructed as in Eq. (16).

$$\tilde{D}^{CR_i} = \left(A\tilde{I}_{CR_i}\left(\tilde{X}_k\right)\right)_{mxs}$$

$$= \begin{pmatrix} (\mu_{11}, v_{11}, \pi_{11}) & (\mu_{12}, v_{12}, \pi_{12}) & \ldots (\mu_{1s}, v_{1s}, \pi_{1s}) \\ (\mu_{21}, v_{21}, \pi_{11}) & (\mu_{22}, v_{22}, \pi_{22}) & \ldots (\mu_{2s}, v_{2s}, \pi_{2s}) \\ . & . & . \\ . & . & . \\ (\mu_{m1}, v_{m1}, \pi_{m1}) & (\mu_{m2}, v_{m2}, \pi_{m2}) & \ldots (\mu_{ms}, v_{ms}, \pi_{ms}) \end{pmatrix} \quad (16)$$

Step 11. Construct the weighted aggregated SF decision matrix for DRs $\left(\tilde{D}^{DR_j} = \left(A\tilde{I}_{DR_j}\left(\tilde{X}_k\right)\right)_{mxs}\right)$ by multiplying the SF overall priorities of each DR $\left(A\tilde{I}_{DR_j}\right)$, and aggregated SF decision matrix based on Eq. (7). Then, defuzzify the values of the weighted aggregated decision matrix for DRs by using Eq. (11).
Step 12. Construct the weighted aggregated SF decision matrix for CRs $\left(\tilde{D}^{CR_i} = \left(A\tilde{I}_{CR_i}\left(\tilde{X}_k\right)\right)_{mxs}\right)$ by multiplying the SF overall priorities of each CR $(A\tilde{I}_{CR_i})$ and aggregated SF decision matrix. Then, defuzzify the values of the weighted aggregated decision matrix for CRs based on Eq. (11).
Step 13. Conclude the SF Positive Ideal Solution (SF-PIS) and the SF Negative Ideal Solution (SF-NIS) based on the defuzzified values in the weighted aggregated SF decision matrix for CRs and DRs obtained in Step 11 and Step 12, respectively.

For the SF-PIS, Eqs. (17) and (18) are used to find the maximum scores in the CRs and DRs decision matrices. Based on the crisp maximum scores, their corresponding SF numbers are determined.

$$X^*_{DR_j} = \left\{ A\tilde{I}_{DR_j}, \max_k \left\langle defuzz\left(A\tilde{I}_{wDR_j}\left(\tilde{X}_k\right)\right)\right\rangle \middle| j = 1, \ldots, m \right\} \quad (17)$$

$$X^*_{CR_i} = \left\{ A\tilde{I}_{CR_i}, \max_k \left\langle defuzz\left(A\tilde{I}_{wCR_i}\left(\tilde{X}_k\right)\right)\right\rangle \middle| i = 1, \ldots, n \right\} \quad (18)$$

For the SF-NIS, Eqs. (19) and (20) are used to find the minimum scores in the decision matrix. Based on the crisp minimum scores, their corresponding SF numbers are determined.

$$X^-_{DR_j} = \left\{ A\tilde{I}_{DR_j}, \min_k \left\langle defuzz\left(A\tilde{I}_{wDR_j}\left(\tilde{X}_k\right)\right)\right\rangle \middle| j = 1, \ldots, m \right\} \quad (19)$$

$$X^-_{CR_i} = \left\{ A\tilde{I}_{CR_i}, min_k \left\langle defuzz\left(A\tilde{I}_{wCR_i}\left(\tilde{X}_k\right)\right)\right\rangle \middle| i = 1, \ldots, n \right\} \quad (20)$$

Step 14. Calculate the distances between alternative X_k and the SF-PIS as well as SF-NIS obtained from Eqs. (17) and (19), respectively. Modified Zhang and Xu's (Ejegwa 2019) distance can be used for this step.

For the SF-PIS:

$$D\left(\tilde{X}_{kj}, \tilde{X}^{*}_{DR_j}\right) = \frac{1}{2m}\sum_{j=1}^{m}\left(\left|\mu^2_{\tilde{X}_{kj}} - \mu^2_{\tilde{X}^{*}_{DR_j}}\right| + \left|v^2_{\tilde{X}_{kj}} - v^2_{\tilde{X}^{*}_{DR_j}}\right| + \left|\pi^2_{\tilde{X}_{kj}} - \pi^2_{\tilde{X}^{*}_{DR_j}}\right|\right), \quad \forall k (k = 1, 2 \ldots s) \tag{21}$$

For the SF-NIS:

$$D\left(\tilde{X}_{kj}, \tilde{X}^{-}_{DR_j}\right) = \frac{1}{2m}\sum_{j=1}^{m}\left(\left|\mu^2_{\tilde{X}_{kj}} - \mu^2_{\tilde{X}^{-}_{DR_j}}\right| + \left|v^2_{\tilde{X}_{kj}} - v^2_{\tilde{X}^{-}_{DR_j}}\right| + \left|\pi^2_{\tilde{X}_{kj}} - \pi^2_{\tilde{X}^{-}_{DR_j}}\right|\right), \quad \forall k (k = 1, 2 \ldots s) \tag{22}$$

Step 15. Calculate the distances between alternative X_k and the SF-PIS as well as SF-NIS obtained from Eqs. (18) and (20), respectively.

For the SF-PIS:

$$D\left(\tilde{X}_{ki}, \tilde{X}^{*}_{CR_i}\right) = \frac{1}{2n}\sum_{i=1}^{n}\left(\left|\mu^2_{\tilde{X}_{ki}} - \mu^2_{\tilde{X}^{*}_{CR_i}}\right| + \left|v^2_{\tilde{X}_{ki}} - v^2_{\tilde{X}^{*}_{CR_i}}\right| + \left|\pi^2_{\tilde{X}_{ki}} - \pi^2_{\tilde{X}^{*}_{CR_i}}\right|\right), \quad \forall k (k = 1, 2 \ldots s) \tag{23}$$

For the SF-NIS:

$$D\left(\tilde{X}_{ki}, \tilde{X}^{-}_{CR_i}\right) = \frac{1}{2n}\sum_{i=1}^{n}\left(\left|\mu^2_{\tilde{X}_{ki}} - \mu^2_{\tilde{X}^{-}_{CR_i}}\right| + \left|v^2_{\tilde{X}_{ki}} - v^2_{\tilde{X}^{-}_{CR_i}}\right| + \left|\pi^2_{\tilde{X}_{ki}} - \pi^2_{\tilde{X}^{-}_{CR_i}}\right|\right), \quad \forall k (k = 1, 2 \ldots s) \tag{24}$$

Step 16. Determine the closeness ratios of the alternatives based on Eqs. (25) and (26) in terms of CRs and DRs, respectively.

$$\xi\left(X_k^{DR}\right) = \frac{D\left(\tilde{X}_{kj}, \tilde{X}^{-}_{DR_j}\right)}{D\left(\tilde{X}_{kj}, \tilde{X}^{-}_{DR_j}\right) + D\left(\tilde{X}_{kj}, \tilde{X}^{*}_{DR_j}\right)} \tag{25}$$

and

$$\xi\left(X_k^{CR}\right) = \frac{D\left(\tilde{X}_{ki}, \tilde{X}^{-}_{CR_i}\right)}{D\left(\tilde{X}_{ki}, \tilde{X}^{-}_{CR_i}\right) + D\left(\tilde{X}_{ki}, \tilde{X}^{*}_{CR_i}\right)} \tag{26}$$

Step 17. Gain the joint performance rating value (JPR_k) of the alternatives in order to define our company position among the competitors by considering both customer evaluations and engineering assessments as given in Eq. (27). We put the alternatives into orders with respect to the decreasing values of JPR_k.

$$JPR_k = \alpha\left(\xi\left(X_k^{CR}\right)\right) + (1-\alpha)\left(\xi\left(X_k^{DR}\right)\right) \tag{27}$$

where α and $(1-\alpha)$ are the importance coefficients of CRs and DRs, respectively.

5 An Application: Product Development Using SFS

To demonstrate our proposed SF-QFD methodology, we select a case study that focuses on conceptual design and dimensional synthesis of delivery drones.

In order to design of drones, many studies have been focused on this topic. Many CRs and DRs which have major and minor effects on drone design are listed in the literature. This literature review shows that most important CRs are the safe/easy to operate (CR_1), precise navigation and data transmission (CR_2), high endurance (CR_3), long flight/battery time (CR_4), all weather capable (CR_5), and low life cycle cost (CR_6). According to the mentioned paper, the most important DRs are strong structure (DR_1), large power supply (DR_2), lightweight (DR_3), stable aerodynamic configuration (DR_4), simple design (DR_5), high endurance (DR_6), water resistant (DR_7). In our study, the selected CRs and the DRs of delivery drone design have been arranged on the QFD matrix. There are three decision-makers (DM1, DM2, and DM3), which are composed of experienced engineers and customers on drones at each evaluation stage. The weights of the DMs DM1, DM2, and DM3 who have different experience levels are 0.5, 0.3 and 0.2, respectively.

CRs have been rated as in Table 2 by using the scale in Table 1. We determine the importance evaluation ($I\tilde{E}_i$) of each CR using SWGM operator given in Eq. (8). To check the consistency of each pairwise comparison matrix, we converted the linguistic terms in the pairwise comparison matrix to their corresponding score indices. After applying the classical consistency test of each pairwise comparison matrix, the consistency of the CR matrices has been set to be less than 10%.

DRs and the direction of improvement of DRs have been defined in Table 3. Positive correlations and negative correlations are represented with $poz\bar{c}_{DR_j}$ and $neg\bar{c}_{DR_j}$, respectively. Three experts evaluate the DRs. We aggregate these linguistic terms for getting aggregated positive and negative correlations of each DR based on SWGM operator given in Table 3. In addition, by using Eq. (12), CIF_{DR_j} values are obtained as in Table 3.

Relationship matrix ($\tilde{R}_{ij}$) between DRs and CRs is constructed as given in Table 4. Then, we aggregate the linguistic terms of each relationship between delivery drone DRs and CRs using SWGM operator as in Table 5.

SF overall priorities of each DR ($A\tilde{I}_{DR_j}$) can be found by using Eq. (13) and are given in Tables 6 and 7.

SF overall priorities of each DR ($A\tilde{I}_{CR_i}$) can be found by using Eq. (14) and are given in Table 8.

Table 8 presents the linguistic terms of design engineering evaluation assigned by three DMs.

Table 2 CRs' importance ratings and global weights

DMs	Customer requirements (CRs)	CR1	CR2	CR3	CR4	CR5	CR6	SF CRs global weights
DM1	CR_1	EI	VLI	SLI	LI	ALI	SMI	(0.44, 0.58, 0.24)
DM2		EI	HI	SMI	SLI	AMI	VHI	
DM3		EI	SMI	HI	VHI	AMI	VHI	
DM1	CR_2	VHI	EI	SMI	SMI	SLI	AMI	(0.58, 0.43, 0.27)
DM2		LI	EI	SMI	LI	VHI	HI	
DM3		SLI	EI	SMI	HI	AMI	VHI	
DM1	CR_3	SMI	SLI	EI	SLI	SLI	HI	(0.48, 0.52, 0.29)
DM2		SLI	SLI	EI	VLI	HI	SMI	
DM3		LI	SLI	EI	SMI	VHI	HI	
DM1	CR_4	HI	SLI	SMI	EI	SLI	AMI	(0.54, 0.48, 0.26)
DM2		SMI	HI	VHI	EI	AMI	VHI	
DM3		VLI	ALI	SLI	EI	HI	SMI	
DM1	CR_5	AMI	SMI	SMI	HI	EI	AMI	(0.40, 0.63, 0.21)
DM2		ALI	VLI	LI	ALI	EI	SLI	
DM3		ALI	ALI	VLI	HI	EI	SLI	
DM1	CR_6	SLI	ALI	LI	ALI	ALI	EI	(0.26, 0.75, 0.19)
DM2		VLI	LI	SLI	VLI	SMI	EI	
DM3		VLI	VLI	LI	SLI	SMI	EI	

All engineering evaluation linguistic terms are aggregated based on Eq. (15) as given in Table 9.

Based on Table 9 and Eqs. (17) and (19), the defuzzified values are obtained to determine maximum scores in the weighted aggregated decision matrix of DRs. In each column, the largest scores represent PIS while the lowest scores represent NIS. According to the best and worst scores, the corresponding SF-PIS and SF-NIS for DRs are given in Table 10.

The distances between each alternative and the SF-PIS for DRs as well as SF-NIS are given in Table 11.

The closeness ratios indicate that the best alternative is Company A and overall ranking is $Alternative\ A > Our\ Company > Alternative\ C > Alternative\ B > Alternative\ D$ as seen in Table 11.

Table 12 shows the linguistic evaluations assigned by three DMs for CRs.

Linguistic evaluations of all CRs are aggregated based on Eq. (16) and are given in Table 13.

According to the best and worst scores in Table 13, the corresponding SF-PIS and SF-NIS for CRs are given in Table 14.

The distances between each alternative and the SF-PIS for CRs as well as SF-NIS are given in Table 15.

Table 3 Correlation impact factor calculations

DR_j	Direction of improvement	Aggregated correlations of each DR with respect to other DRs (μ, v, π)	Aggregated positive and negative correlations separately (μ, v, π)	Defuzzified values of aggregated positive and negative correlations	$\left(poz\bar{c}_{DR_j} - neg\bar{c}_{DR_j}\right)*\left(\frac{n_j}{(m-1)}\right)$	CIF_{DR_j}
DR_1	Positive with DR_2 (AMR, HR, VHR)	(0.82, 0.20, 0.12)	(0.81, 0.20, 0.13)	2.603	−0.432	0.568
	Positive with DR_6 (AMR, VHR, SMR)	(0.80, 0.23, 0.16)				
	Positive with DR_7 (AMR, VHR, HR)	(0.83, 0.19, 0.11)				
	Negative with DR_3 (AMR, AMR, AMR)	(0.90, 0.10, 0.00)	(0.89, 0.12, 0.04)	3.122		
	Negative with DR_5 (AMR, VHR, AMR)	(0.87, 0.14, 0.06)				
DR_2	Positive with DR_1 (AMR, HR, VHR)	(0.82, 0.20, 0.12)	(0.82, 0.20, 0.12)	2.610	0.756	1.756
	Negative with DR_3 (MR, HR, SMR)	(0.57, 0.37, 0.34)	(0.64, 0.37, 0.23)	1.476		
	Negative with DR_4 (LR, LR,LR)	(0.30, 0.70, 0.20)				

(continued)

Table 3 (continued)

DR_j	Direction of improvement	Aggregated correlations of each DR with respect to other DRs (μ, v, π)	Aggregated positive and negative correlations separately (μ, v, π)	Defuzzified values of aggregated positive and negative correlations	$\left(poz\bar{c}_{DR_j} - neg\bar{c}_{DR_j}\right)*\left(\frac{n_j}{(m-1)}\right)$	CIF_{DR_j}
	Negative with DR_5 (AMR, HR, VHR)	(0.82, 0.20, 0.12)				
DR_3	Positive with DR_4 (VHR, VHR, HR)	(0.78, 0.22, 0.13)	(0.82, 0.18, 0.09)	1.476	0.115	1.115
	Positive with DR_6 (AMR, VHR, AMR)	(0.87, 0.14, 0.06)				
	Positive with DR_7 (VHR, VMR, VHR)	(0.80, 0.20, 0.10)				
	Negative with DR_1 (AMR, AMR, AMR)	(0.90, 0.10, 0.00)	(0.80, 0.19, 0.18)	2.519		
	Negative with DR_2 (MR, HR, SMR)	(0.57, 0.37, 0.34)				
DR_4	Positive with DR_3 (VHR, VHR, HR)	(0.78, 0.22, 0.13)	(0.74, 0.27, 0.18)	2.103	0.315	1.315
	Positive with DR_5 (VHR, SMR, SMR)	(0.69, 0.32, 0.23)				

(continued)

Table 3 (continued)

DR_j	Direction of improvement	Aggregated correlations of each DR with respect to other DRs (μ, v, π)	Aggregated positive and negative correlations separately (μ, v, π)	Defuzzified values of aggregated positive and negative correlations	$\left(poz\bar{c}_{DR_j} - neg\bar{c}_{DR_j}\right)*\left(\frac{n_j}{(m-1)}\right)$	CIF_{DR_j}
	Negative with DR_2 (LR, LR, LR)	(0.30, 0.70, 0.20)	(0.30, 0.70, 0.20)	0.300		
DR_5	Positive with DR_4 (VHR, SMR, SMR)	(0.69, 0.32, 0.23)	(0.69, 0.32, 0.23)	1.791	−0.296	0.704
	Negative with DR_1 (AMR, VHR, AMR)	(0.87, 0.14, 0.06)	(0.76, 0.27, 0.12)	2.235		
	Negative with DR_2 (AMR, HR, VHR)	(0.82, 0.20, 0.12)				
	Negative with DR_7 (LR, LR, LR)	(0.30, 0.70, 0.20)				
DR_6	Positive with DR_1 (AMR, VHR, SMR)	(0.80, 0.23, 0.16)	(0.82, 0.19, 0.12)	2.654	1.327	2.327
	Positive with DR_3 (AMR, VHR, AMR)	(0.87, 0.14, 0.06)				
	Positive with DR_7 VHR, VHR, VHR)	(0.78, 0.22, 0.13)				

(continued)

Table 3 (continued)

DR_j	Direction of improvement	Aggregated correlations of each DR with respect to other DRs (μ, v, π)	Aggregated positive and negative correlations separately (μ, v, π)	Defuzzified values of aggregated positive and negative correlations	$\left(poz\bar{c}_{DR_j} - neg\bar{c}_{DR_j}\right) * \left(\frac{n_j}{(m-1)}\right)$	CIF_{DR_j}
DR_7	Positive with DR_1 (AMR, VHR, HR)	(0.83, 0.19, 0.11)	(0.80, 0.20, 0.11)	2.530	1.487	2.487
	Positive with DR_3 (VHR, VHR, VHR)	(0.80, 0.20, 0.10)				
	Positive with DR_6 VHR, VHR, VHR)	(0.78, 0.22, 0.13)				
	Negative with DR_5 (LR, LR, LR)	(0.30, 0.70, 0.20)	(0.30, 0.70, 0.20)	0.3		

Table 4 Relationship matrix ($\tilde{R}_{ij}$)

$\tilde{R}_{ij}$	DR_1	DR_2	DR_3	DR_4	DR_5	DR_6	DR_7
CR_1	VHR, VHR, HR	AMR, AMR, AMR	MR, HR, SMR		VHR, LR, VHR		AMR, VHR, HR
CR_2	VHR, AMR, AMR		VHR, VHR, SMR	LR, HR, HR	HR, SMR, MR	AMR, VHR, AMR	AMR, AMR, VHR
CR_3		MR, HR, SMR		MR, VHR, VHR	VHR, SMR, SMR	HR, SMR, VHR	AMR, HR, AMR
CR_4	LR, SMR, SMR	AMR, VHR, MR	AMR, AMR, VHR			HR, HR, HR	HR, HR, SMR
CR_5	MR, MR, VHR	VHR, VHR, HR	VHR, VHR, VHR	VHR, VHR, VHR	AMR, VHR, AMR		MR, MR, SMR
CR_6		MR, MR, MR	SMR, AMR, AMR	AMR, AMR, AMR	AMR, VHR, AMR		

Table 5 The aggregated relationship matrix ($\tilde{R}_{ij}$)

$\tilde{R}_{ij}$	DR_1	DR_2	DR_3	DR_4	DR_5	DR_6	DR_7
CR_1	(0.78, 0.22, 0.13)	(0.90, 0.10, 0.00)	(0.57, 0.37, 0.34)		(0.60, 0.45, 0.16)		(0.83, 0.19, 0.11)
CR_2	(0.85, 0.16, 0.07)		(0.76, 0.26, 0.17)	(0.46, 0.56, 0.20)	(0.62, 0.35, 0.29)	(0.87, 0.14, 0.06)	(0.88, 0.13, 0.05)
CR_3		(0.57, 0.37, 0.34)		(0.63, 0.32, 0.31)	(0.69, 0.32, 0.23)	(0.69, 0.32, 0.23)	(0.90, 0.10, 0.00)
CR_4	(0.42, 0.59, 0.25)	(0.77, 0.23, 0.21)	(0.88, 0.13, 0.05)			(0.70, 0.30, 0.20)	(0.68, 0.32, 0.23)
CR_5	(0.55, 0.37, 0.37)	(0.78, 0.22, 0.13)	(0.80, 0.20, 0.10)	(0.80, 0.20, 0.10)	(0.87, 0.14, 0.06)		(0.52, 0.40, 0.38)
CR_6		(0.50, 0.40, 0.40)	(0.65, 0.36, 0.27)	(0.90, 0.10, 0.00)	(0.87, 0.14, 0.06)		

Table 6 SF overall priorities of each delivery drone DR ($A\tilde{I}_{DR_j}$)

DR_j	DR_1	DR_2	DR_3	DR_4	DR_5	DR_6	DR_7
$A\tilde{I}_{DR_j}$	(0.56, 0.50, 0.18)	(0.88, 0.08, 0.21)	(0.78, 0.21, 0.19)	(0.87, 0.11, 0.14)	(0.68, 0.38, 0.15)	(0.93, 0.03, 0.13)	(0.98, 0.00, 0.08)

Table 7 SF overall priorities of each delivery drone CR ($A\tilde{I}_{CR_i}$)

CR_i	CR_1	CR_2	CR_3	CR_4	CR_5	CR_6
$A\tilde{I}_{CR_i}$	(0.34, 0.61, 0.26)	(0.47, 047, 0.29)	(0.35, 0.57, 0.33)	(0.40, 0.54, 0.30)	(0.30, 0.66, 0.22)	(0.20, 0.76, 0.22)

Table 8 Evaluations of DRs with respect to alternatives

Companies	DR_1	DR_2	DR_3	DR_4	DR_5	DR_6	DR_7
Our company	HS, VHS, LS	MS, VHS, AMS	VHS, SLS, SLS	LS, VHS, VLS	HS, SLS, HS	VHS, AMS, AMS	AMS, VHS, VHS
Alternative A	AMS, MS, VLS	HS, LS, VHS	HS, HS, VHS	HS, HS, SLS	AMS, AMS, VHS	HS, HS, LS	HS, VHS, VHS
Alternative B	HS, HS, MS	MS, VHS, AMS	AMS, SLS, SLS	LS, HS, VLS	HS, SLS, HS	MS, VHS, AMS	AMS, HS, SLS
Alternative C	VHS, MS, SMS	SMS, HS, SMS	HS, HS, SLS	VLS, HS, SMS	VHS, SMS, MS	SMS, HS, SMS	HS, AMS, HS
Alternative D	LS, VHS, VLS	MS, LS, SLS	AMS, HS, VLS	HS, SLS, LS	LS, MS, SLS	MS, VHS, SLS	LS, MS, VHS

The closeness ratios indicate that the best alternative is Company A and overall ranking is $Alternative\,A > Our\,Company > Alternative\,B > Alternative\,C > Alternative\,D$. as seen in Table 15.

Joint performance rating values (JPR_k) of the alternatives are given in Table 16. According to the JPR_k values, the best alternative is Alternative A.

We applied sensitivity analysis by changing the threshold value (α) observed similar results of the given decisions. If the threshold value (α) is less and equal to 0.14, the final ranking is

$$Alternative\,A > Our\,Company > Alternative\,C > Alternative\,B > \\ > Alternative\,D.$$

Table 9 Weighted aggregated decision matrix for engineering evaluations

Companies	DR_1	DR_2	DR_3	DR_4	DR_5	DR_6	DR_7
Our company	(0.35, 0.62, 0.23)	(0.57, 0.32, 0.36)	(0.44, 0.50, 0.29)	(0.32, 0.66, 0.20)	(0.43, 0.49, 0.30)	(0.80, 0.16, 0.15)	(0.83, 0.16, 0.11)
Alternative A	(0.32, 0.56, 0.31)	(0.49, 0.48, 0.26)	(0.56, 0.35, 0.25)	(0.54, 0.40, 0.26)	(0.55, 0.42, 0.19)	(0.55, 0.44, 0.23)	(0.73, 0.26, 0.18)
Alternative B	(0.37, 0.56, 0.24)	(0.57, 0.32, 0.36)	(0.47, 0.49, 0.29)	(0.31, 0.66, 0.21)	(0.43, 0.49, 0.30)	(0.61, 0.31, 0.33)	(0.70, 0.34, 0.21)
Alternative C	(0.37, 0.57, 0.31)	(0.55, 0.38, 0.33)	(0.49, 0.43, 0.28)	(0.32, 0.67, 0.20)	(0.48, 0.46, 0.26)	(0.59, 0.38, 0.30)	(0.74, 0.26, 0.19)
Alternative D	(0.21, 0.69, 0.21)	(0.36, 0.57, 0.35)	(0.49, 0.50, 0.20)	(0.43, 0.33, 0.27)	(0.25, 0.68, 0.28)	(0.52, 0.42, 0.35)	(0.42, 0.57, 0.27)

Table 10 SF-PIS and SF-NIS for engineering evaluations

	DR_1	DR_2	DR_3	DR_4	DR_5	DR_6	DR_7
$X^*_{DR_j}$	(0.37, 0.56, 0.24)	(0.57, 0.32, 0.36)	(0.56, 0.35, 0.25)	(0.54, 0.40, 0.26)	(0.55, 0.42, 0.19)	(0.80, 0.16, 0.15)	(0.83, 0.16, 0.11)
$X^-_{DR_j}$	(0.21, 0.69, 0.21)	(0.36, 0.57, 0.35)	(0.49, 0.50, 0.29)	(0.31, 0.66, 0.21)	(0.25, 0.68, 0.28)	(0.52, 0.42, 0.35)	(0.42, 0.57, 0.27)

Table 11 Distances to SF-PIS, SF-NIS, and ranking

Companies	$D\left(\tilde{X}_{kj}, \tilde{X}^*_{DR_j}\right)$	$D\left(\tilde{X}_{kj}, \tilde{X}^-_{DR_j}\right)$	$\xi(X_k^{DR})$	Ranking
Our company	0.0777	0.1758	0.6935	2
Alternative A	0.0775	0.1904	0.7108	1
Alternative B	0.1232	0.1297	0.5130	4
Alternative C	0.1159	0.1448	0.5555	3
Alternative D	0.2290	0.0262	0.1027	5

While the threshold value is greater than 0.14, Company C and Company B have just replaced each other in the final ranking.

Table 12 Evaluations of CRs with respect to alternatives

Companies	CR_1	CR_2	CR_3	CR_4	CR_5	CR_6
Our company	VHS, HS, VHS	VHS, MS, VHS	SLS, AMS, AMS	HS, LS, MS	AMS, HS, HS	SLS, VHS, HS
Alternative A	MS, AMS, AMS	HS, HS, MS	HS, HS, AMS	HS, HS, LS	HS, AMS, AMS	HS, HS, AMS
Alternative B	VHS, HS, HS	MS, MS, AMS	AMS, AMS, SMS	LS, LS, VLS	AMS, HS, SLS	AMS, MS, AMS
Alternative C	SMS, VHS, MS	VLS, SMS, MS	LS, HS, SLS	HS, VLS, SMS	HS, VHS, SMS	AMS, SMS, AMS
Alternative D	HS, LS, SLS	ALS, MS, SLS	HS, AMS, VLS	SMS, HS, HS	LS, LS, MS	SMS, MS, SLS

Table 13 Weighted aggregated decision matrix for CRs' assessments

Companies	CR_1	CR_2	CR_3	CR_4	CR_5	CR_6
Our company	(0.26, 0.64, 0.27)	(0.33, 0.53, 0.35)	(0.21, 0.68, 0.35)	(0.20, 0.68, 0.33)	(0.24, 0.68, 0.27)	(0.11, 0.82, 0.25)
Alternative A	(0.23, 0.65, 0.34)	(0.31, 0.55, 0.35)	(0.26, 0.61, 0.35)	(0.24, 0.65, 0.31)	(0.24, 0.68, 0.27)	(0.15, 0.78, 0.24)
Alternative B	(0.26, 0.64, 0.28)	(0.27, 0.57, 0.41)	(0.29, 0.59, 0.34)	(0.11, 0.81, 0.25)	(0.21, 0.70, 0.28)	(0.15, 0.78, 0.26)
Alternative C	(0.22, 0.67, 0.33)	(0.16, 0.75, 0.28)	(0.14, 0.76, 0.32)	(0.19, 0.72, 0.29)	(0.21, 0.69, 0.28)	(0.15, 0.78, 0.24)
Alternative D	(0.17, 0.74, 0.28)	(0.10, 0.84, 0.23)	(0.21, 0.69, 0.31)	(0.26, 0.61, 0.34)	(0.10, 0.82, 0.26)	(0.11, 0.81, 0.28)

Table 14 SF-PIS and SF-NIS for CRs' assessments

CR_i	CR_1	CR_2	CR_3	CR_4	CR_5	CR_6
$X^*_{CR_i}$	(0.26, 0.64, 0.27)	(0.33, 0.53, 0.35)	(0.29, 0.59, 0.34)	(0.26, 0.61, 0.34)	(0.24, 0.68, 0.27)	(0.16, 0.78, 0.24)
$X^-_{CR_i}$	(0.17, 0.74, 0.28)	(0.10, 0.84, 0.23)	(0.14, 0.76, 0.32)	(0.11, 0.81, 0.25)	(0.10, 0.82, 0.26)	(0.11, 0.82, 0.25)

Table 15 Distances to SF-PIS, SF-NIS, and ranking

Companies	$D\left(\tilde{X}_{ki}, \tilde{X}^*_{CR_i}\right)$	$D\left(\tilde{X}_{ki}, \tilde{X}^-_{CR_i}\right)$	$\xi\left(X_k^{CR}\right)$	Ranking
Our company	0.0261	0.1055	0.8015	2
Alternative A	0.0172	0.1195	0.8739	1
Alternative B	0.0427	0.0965	0.6930	3
Alternative C	0.0762	0.0603	0.4416	4
Alternative D	0.0955	0.0386	0.2880	5

Table 16 Joint performance rating value (JPR_k) and ranking ($\alpha = 0.5$)

Companies	$\xi(X_k^{CR})$	$\xi(X_k^{DR})$	JPR_k	Ranking
Our company	0.8015	0.6935	0.7475	2
Alternative A	0.8739	0.7108	0.7924	1
Alternative B	0.6930	0.5130	0.6030	3
Alternative C	0.4416	0.5555	0.4986	4
Alternative D	0.2880	0.1027	0.1954	5

6 Conclusions

Type-2 fuzzy sets, neutrosophic sets, fuzzy multisets, intuitionistic fuzzy sets, hesitant fuzzy sets, Pythagorean fuzzy sets, SFSs, and picture fuzzy sets are the recent extensions of ordinary fuzzy sets. In this chapter, SFSs have been employed in the design of delivery drones. For this aim, SF-QFD model has been proposed since SFSs let a larger preference domain to be independently define for membership, non-membership and hesitancy degrees. Aggregation operators for SFSs have been employed since the QFD model of delivery drones is based on multi DMs' opinions. The distances between alternative companies using SF-PIS and SF-NIS for CRs and DRs have been determined. Joint performance rating values have been also calculated for alternative companies.

Our suggestions for future research are the usage of neutrosophic sets or picture fuzzy sets to be used in the QFD model of delivery drones. A comparison between our results and suggested works are also recommended.

References

Akao Y (1972) New product development and quality assurance deployment system (in Japanese). Standard Q Control 25(4):243–246

Alptekin SE, Alptekin GI (2018) A fuzzy quality function deployment approach for differentiating cloud products. Int J Comput Intell Syst 11(1):1041–1055

Atanassov KT (1986) Intuitionistic fuzzy sets. Fuzzy Sets Syst 20(1):87–96

Bevilacqua M, Ciarapica FE, Marchetti B (2012) Development and test of a new fuzzy-QFD approach for characterizing customers rating of extra virgin olive oil. Food Qual Prefer 24(1):75–84

Cebi S, Ozkok M, Demirci E (2014) Evaluation of design parameters for vessel engine room by using a modified QFD technique. J Multiple-Valued Logic Soft Comput 23(5–6):559–587

Chen L, Ko W (2009) Fuzzy linear programming models for new product design using QFD with FMEA. Appl Math Model 33(2):633–647

Chen L, Ko W (2011) Fuzzy nonlinear models for new product development using four-phase quality function deployment processes. IEEE Trans Syst Man Cybern Part A Syst Hum 41(5):927–945

Chen L, Weng M (2006) An evaluation approach to engineering design in QFD processes using fuzzy goal programming models. Eur J Oper Res 172(1):230–248

Chen Y, Tang J, Fung RYK, Ren Z (2004) Fuzzy regression-based mathematical programming model for quality function deployment. Int J Prod Res 42(5):1009–1027
Chen Y, Fung RYK, Tang J (2005) Fuzzy expected value modelling approach for determining target values of engineering characteristics in QFD. Int J Prod Res 43(17):3583–3604
Delice EK, Güngör Z (2013) Determining design requirements in QFD using fuzzy mixed-integer goal programming: application of a decision support system. Int J Prod Res 51(21):6378–6396
Ejegwa PA (2019) Modified Zhang and Xu's distance measure for Pythagorean fuzzy sets and its application to pattern recognition problems. Neural Comput Appl 1–10
Fung RYK, Chen Y, Chen LI, Tang J (2005) A fuzzy expected value-based goal programing model for product planning using quality function deployment. Eng Optim 37(6):633–647
Gündoğdu FK (2019) Principals of spherical fuzzy sets. In: International conference on intelligent and fuzzy systems. Springer, Cham, pp 15–23
Gündoğdu FK, Kahraman C (2019a) A novel fuzzy TOPSIS method using emerging interval-valued spherical fuzzy sets. Eng Appl Artif Intell 85:307–323
Gündoğdu FK, Kahraman C (2019) A novel spherical fuzzy analytic hierarchy process and its renewable energy application. Soft Comput 1–15
Haktanır E, Kahraman C (2019) A novel interval-valued Pythagorean fuzzy QFD method and its application to solar photovoltaic technology development. Comput Ind Eng 132:361–372
Hsu C, Chang A, Kuo H (2011) Green supply implementation based on fuzzy QFD: An application in GPLM system. WSEAS Trans Syst 10(6):183–192
Huang J, You X, Liu H, Si S (2019) New approach for quality function deployment based on proportional hesitant fuzzy linguistic term sets and prospect theory. Int J Prod Res 57(5):1283–1299
Hundal GPS, Kant S (2017) Product design development by integrating QFD approach with heuristics-AHP, ANN and fuzzy logics-a case study in miniature circuit breaker. Int J Prod Q Manage 20(1):1–28
Kahraman C, Ertay T, Büyüközkan G (2006) A fuzzy optimization model for QFD planning process using analytic network approach. Eur J Oper Res 171(2):390–411
Kahraman C, Gundogdu FK, Onar SC, Oztaysi B (2019) Hospital location selection using spherical fuzzy TOPSIS. In: 2019 Conference of the international fuzzy systems association and the European society for fuzzy logic and technology (EUSFLAT 2019). Atlantis Press
Kalargeros N, Gao JX (1998) QFD: focusing on its simplification and easy computerization using fuzzy logic principles. Int J Veh Des 19(3):315–325
Kang X, Yang M, Wu Y, Ni B (2018) Integrating evaluation grid method and fuzzy quality function deployment to new product development. Math Probl Eng 2018
Karimi BH, Mozafari MM, Asli MN (2012) Applying a hybrid QFD-TOPSIS method to design product in the industry (case study in sum service company). Res J Appl Sci Eng Technol 4(18):3283–3288
Karsak EE (2004) Fuzzy multiple objective decision making approach to prioritize design requirements in quality function deployment. Int J Prod Res 42(18):3957–3974
Kavosi M, Mavi RK (2011) Fuzzy quality function deployment approach using TOPSIS and analytic hierarchy process methods. Inter J Prod Q Manage 7(3):304–324
Kuo T, Wu H, Shieh J (2009) Integration of environmental considerations in quality function deployment by using fuzzy logic. Expert Syst Appl 36(3 PART 2):7148–7156
Kutlu Gundogdu F, Kahraman C (2019) Extension of WASPAS with spherical fuzzy sets. Informatica 30(2):269–292
Kutlu Gündoğdu F, Kahraman C (2019a) Spherical fuzzy sets and spherical fuzzy TOPSIS method. J Intell Fuzzy Syst 36(1):337–352
Kutlu Gündoğdu F, Kahraman C (2019b) A novel VIKOR method using spherical fuzzy sets and its application to warehouse site selection. J Intell Fuzzy Syst 37(1):1197–1211
Kutlu Gündoğdu F, Kahraman C (2020) A novel spherical fuzzy QFD method and its application to the linear delta robot technology development. Eng Appl Artif Intell 87

Kwong CK, Bai H (2002) A fuzzy AHP approach to the determination of importance weights of customer requirements in quality function deployment. J Intell Manuf 13(5):367–377
Lee Y, Sheu L, Tsou Y (2008) Quality function deployment implementation based on fuzzy kano model: an application in PLM system. Comput Ind Eng 55(1):48–63
Lee AHI, Kang H, Yang C, Lin C (2010) An evaluation framework for product planning using FANP, QFD and multi-choice goal programming. Int J Prod Res 48(13):3977–3997
Lee Z, Pai C, Yang C (2012) Customer needs and technology analysis in new product development via fuzzy QFD and delphi. WSEAS Trans Bus Econ 9(1):1–15
Li M (2013) The method for product design selection with incomplete linguistic weight information based on quality function deployment in a fuzzy environment. Math Probl Eng 2013(7)
Li S, Tang D, Wang Q (2019) Rating engineering characteristics in open design using a probabilistic language method based on fuzzy QFD. Comput Ind Eng 135:348–358
Liu S (2005) Rating design requirements in fuzzy quality function deployment via a mathematical programming approach. Int J Prod Res 43(3):497–513
Liu H (2009) The extension of fuzzy QFD: From product planning to part deployment. Expert Syst Appl 36(8):11131–11144
Liu H (2011) Product design and selection using fuzzy QFD and fuzzy MCDM approaches. Appl Math Model 35(1):482–496
Liu A, Hu H, Zhang X, Lei D (2017) Novel two-phase approach for process optimization of customer collaborative design based on fuzzy-QFD and DSM. IEEE Trans Eng Manage 64(2):193–207
Luo XG, Kwong CK, Tang JF (2010) Determining optimal levels of engineering characteristics in quality function deployment under multi-segment market. Comput Ind Eng 59(1):126–135
Mizuno S, Akao Y (1978) Quality function deployment: a company-wide quality approach (in Japanese). JUSE Press, Tokyo
Nepal B, Yadav OP, Murat A (2010) A fuzzy-AHP approach to prioritization of CS attributes in target planning for automotive product development. Expert Syst Appl 37(10):6775–6786
Peng J, Xia G, Sun B, Wang S (2018) Systematical decision-making approach for quality function deployment based on uncertain linguistic term sets. Int J Prod Res 56(18):6183–6200
Torra V (2010) Hesitant fuzzy sets. Int J Intell Syst 25(6):529–539
Verma D, Chilakapati R, Fabrycky WJ (1998) Analyzing a quality function deployment matrix: An expert system-based approach to identify inconsistencies and opportunities. J Eng Des 9(3):250–261
Vinodh S, Rathod G (2012) Application of fuzzy logic-based environmental conscious QFD to rotary switch: a case study. Clean Technol Environ Policy 14(2):319–332
Vinodh S, Manjunatheshwara KJ, Karthik Sundaram S, Kirthivasan V (2017) Application of fuzzy quality function deployment for sustainable design of consumer electronics products: a case study. Clean Technol Environ Policy 19(4):1021–1030
Wang J (1999) Fuzzy outranking approach to prioritize design requirements in quality function deployment. Int J Prod Res 37(4):899–916
Wang Y (2012) A fuzzy-normalisation-based group decision-making approach for prioritising engineering design requirements in QFD under uncertainty. Int J Prod Res 50(23):6963–6977
Wang C (2019) Integrating a novel intuitive fuzzy method with quality function deployment for product design: case study on touch panels. J Intell Fuzzy Syst 37(2):2819–2833
Wang C, Chen J (2012) Using quality function deployment for collaborative product design and optimal selection of module mix. Comput Ind Eng 63(4):1030–1037
Wang D, Yu H, Wu J, Meng Q, Lin Q (2019) Integrating fuzzy based QFD and AHP for the design and implementation of a hand training device. J Intell Fuzzy Syst 36(4):3317–3331
Wu Q (2011) Fuzzy measurable house of quality and quality function deployment for fuzzy regression estimation problem. Expert Syst Appl 38(12):14398–14406
Wu Y, Ho CC (2015) Integration of green quality function deployment and fuzzy theory: a case study on green mobile phone design. J Clean Prod 108:271–280
Kutlu Gündoğdu F (Preprint) A spherical fuzzy extension of MULTIMOORA method. J Intell Fuzzy Syst 1–16

Yager RR (2014) Pythagorean membership grades in multicriteria decision making. IEEE Trans Fuzzy Syst 22(4):958–965
Yang CL, Fang HH (2003) Integrating fuzzy logic into quality function deployment for product positioning. J Chin Inst Ind Eng 20(3):275–281
Zadeh LA (1965) Fuzzy Sets. Information and Control 338–353
Zhai L, Khoo L, Zhong Z (2009) A rough set based QFD approach to the management of imprecise design information in product development. Adv Eng Inform 23(2):222–228
Zheng P, Xu X, Xie SQ (2019) A weighted interval rough number based method to determine relative importance ratings of customer requirements in QFD product planning. J Intell Manuf 30(1):3–16

Analysis of Usability Test Parameters Affecting the Mobile Application Designs by Using Spherical Fuzzy Sets

Fatma Kutlu Gündoğdu, Ezgi Cotari, Selcuk Cebi, and Cengiz Kahraman

Abstract Investments have been drastically increasing on mobile phone applications based on the increase in mobile internet usage. The fact that these investments should be utilized by many users has become an important agenda of business plans for many brands. It's obvious that the differentiate value proposals of the applications are no longer just a function or a new idea, but the usability of the whole system is become more and more critical. For this purpose, published academic studies about the usability of mobile applications in the literature have been examined. From this research, it's found out that researchers have been used various and differentiate criteria for the same objective. Thus, in this chapter, a new method has been proposed to decide which design parameters are affective for the related mobile application and to determine the importance degree of the design parameters to be taken within the scope of the usability evaluation of the applications. The methods including Kano Model, Quality Function Deployment (QFD), and Spherical Fuzzy Sets have been integrated into the proposed method. For this reason, the design requirements are categorized by using the Kano Model while QFD is used to integrate design requirements with design parameters to increase the usability dimension of the product. Spherical Fuzzy Sets were used to handle uncertainties and vagueness in the evaluation process of the approach.

C. Kahraman
Department of Industrial Engineering, Istanbul Technical University, Besiktas, Istanbul, Turkey

E. Cotari · S. Cebi (✉)
Department of Industrial Engineering, Yildiz Technical University, Besiktas, Istanbul, Turkey
e-mail: scebi@yildiz.edu.tr

F. Kutlu Gündoğdu
Industrial Engineering Department, National Defence University, Turkish Air Force Academy, 34149 Istanbul, Turkey
e-mail: fatmakutlugundogdu@gmail.com

© Springer Nature Switzerland AG 2021

C. Kahraman and F. Kutlu Gündoğdu (eds.), *Decision Making with Spherical Fuzzy Sets*, Studies in Fuzziness and Soft Computing 392,
https://doi.org/10.1007/978-3-030-45461-6_18

1 Introduction

Achieving the desired performance from a product and promoting a product requirement in the market is directly related to the usability of the product. The usability concept, which emerged in the late 1980s, is widely used to evaluate the performance and acceptance of a system, application, or product (Butler 1996; Wichansky 2000). Jakob Nielsen has defined the term usability by five criteria: learnability, efficiency, memorability, errors, and satisfaction (Nielsen 1994a). The International Organization for Standardization describes the usability in ISO 9241-11 as "the extent to which a system, product, or service can be used by specified users to achieve specified goals with effectiveness, efficiency and satisfaction in a specified context of use" (Bevan 1995; ISO 1998). Dumas and Redish (1999) described usability as the ability of individuals using the product to perform tasks quickly and easily. Krug (2000) defined usability as making sure that products work well, and that a person of average (or even below average) ability and experience can use the product for its purpose without frustration, regardless of the product (whether a website or a revolving door) (Krug 2000). Today, usability tests are widely used in the development of products ranging from household appliances to chairs, from websites to the advertising industry.

The number of mobile applications has increased rapidly with the development of smartphone technology in the last 10 years and it continues to increase day by day. Usability tests have an important role in the development and promotion of mobile applications in the market. Although various usability studies have been conducted on websites in the literature, studies on mobile applications are quite limited. The usability of mobile applications is more complex than the usability analysis of websites, due to the unique limitations of mobile devices. The main reasons for this challenge are (i) mobile content, (ii) internet connectivity, (iii) small screens, (iv) different screen resolutions, (v) limited processor capacity and battery power, and (vi) data entry methods (Zhang and Adipat 2005). Another difficulty in the promotion of mobile applications is that, unlike desktop computer users, mobile application users go directly for a different application when they do not get the performance they expect from the application (Chincholle et al. 2002). Therefore, in mobile applications, the correct information should be presented to the user within 3 min, otherwise, the user will give up the application (Bortenschlager et al. 2010). Hence, there are many different methods or approaches in the literature for evaluating the usability of mobile applications and the number of these studies continues to increase day by day. When the studies in the literature are examined, it is seen that there is no clear methodology for the standardization of the criteria utilized in the usability evaluation of mobile applications. The main objective of this study is to provide an approach to how design parameters that play an important role in the product design phase can be defined in terms of the level of significance in the design and which design parameters should be taken into consideration in usability tests. For this, an integrated approach that consists of the Kano Model, Quality Function Deployment and Spherical Fuzzy Sets has been proposed. In the scope of the study, firstly the design parameters affecting the product design will be compiled with the literature review and the design parameters will be categorized by using the Kano Model method. The Kano Model method is used to determine the importance degree of design parameters, while QFD is used to identify the design parameters that should be considered in the usability tests for the related design.

The rest of this chapter is organized as follows; Sect. 2 presents the methods used in the usability evaluation process. The usability evaluation criteria are given in Sect. 3. The proposed approach and its application have been given in Sects. 4 and 5, respectively. Finally, concluding remarks are presented in Sect. 6.

2 Methods Used in Usability Evaluation

In the literature, *Expert Analysis* and *User Participatory Evaluation* are commonly used methods in the usability studies (Dix et al. 2004).

2.1 Expert Analysis

The usability evaluation should be performed throughout the entire development process of the product, process, or system. However, periodic tests during the product development process may be expensive or it may fail to evaluate the actual interaction experience of users with designs and prototypes that do not fully reflect the experience. Therefore, several methods have been proposed to evaluate products by expert analysis. These are based on either the designer or the human criteria expertise, and the design is assessed in terms of its impact on a typical user. The main objective is to identify areas that violate known cognitive principles or that may cause difficulties for users as they ignore accepted empirical results. Since the expert analysis does not require user participation or an empirical experiment, they are relatively inexpensive methods. There are two approaches for expert analysis; these are *cognitive evaluation* and *intuitive evaluation* (Web 12012).

2.1.1 Cognitive Evaluation

Cognitive evaluation is an approach where one or more experts walk through each stage of the design step by step by putting themselves in place of users and trying to identify the points where users may experience problems (Wharton et al. 1994).

2.1.2 Intuitive Evaluation

Intuitive evaluation is an approach where experts evaluate product availability based on a specific checklist or a specific guideline. For example, the 10 control points described by Nielsen (1994b) are as follows:

- Visibility of system status
- Converging the real world with the system
- User control and freedom
- Consistency and standards
- Error prevention
- Recognition instead of remembering
- Flexibility and efficiency of use
- Aesthetic and minimalist design
- Help users recognize, understand and get rid of errors
- Assistance and documentation.

2.2 *User Participatory Evaluation*

The user participatory evaluation is divided into usability tests and survey methods.

2.2.1 Usability Tests

There are specific requirements to perform usability tests. These can be listed as follows;

- **User**: A user profile is a system user with a realistic definition that meets a specific purpose and needs through the relevant system (Cooper 1999; Zimmermann and Vanderheiden 2008).
- **Task**: Task defines the targets that users try to achieve by using the system/application/product.
- **Context**: All the environment and environmental conditions in which users will use the system/application/product. Usability tests are generally performed in two different contexts (Cooper 1999). These are the laboratory environment and fieldwork. Fieldwork is more preferred than the laboratory environment. Because the fieldwork reflects the reality of the tasks performed by the user to meet the needs of the user. Thus, the real user experience obtained in the real environment is more reliable and realistic than the results obtained in the laboratory (Wei et al. 2015). However, there are three main challenges mentioned in the literature to perform field studies (Kjeldskov and Stage 2004); Conducting realistic studies far from established usability evaluation techniques is quite complex and data collection is difficult. Unlike fieldwork, the laboratory environment has several advantages,

including full control over tests, easy measurement of usability features, and the ability to use video to capture user behavior (Cooper 1999). It is stated that it is very difficult to use an external camera to display the screen of the mobile device since the screens of the mobile phones related to the video capability to be used in the testing of mobile applications in laboratory studies are small and generally users close the page (Balagtas-Fernandez and Hussmann 2009).

- **Use Case**: Use cases were first defined by Jacobson in 1992 (Zimmermann and Vanderheiden 2008). A user scenario is a sequence of system actions that yields an observable value to a particular user (Kruchten 2004).
- **Scenario**: Scenarios are useful to show usage scenarios in more detail. Use cases reflect a generalized view of a user and a task, while scenarios define a specific use case as a concrete workflow with specific data, specific events, and a specific user interface (Zimmermann and Vanderheiden 2008). Generally, creating three to four scenarios to cover the standard users of a website provides a good starting point (Brinck et al. 2002).
- **Usability Testing Facilitator**: People who monitor the users when they use the system/application/product over a specific scenario to perform a specific task in a specific context. Depending on the preferred usability test method, it may not be required to have a test manager for each test. However, for the accuracy and consistency of the test, the test manager should not be too directive. Another risky situation is that different test managers manage the process in different ways for the same test.

2.2.2 Surveys

Kim et al. (2012) stated that the Human-Computer Interaction (HCI) Research Society developed various usability surveys as one of the usability evaluation methods. Because subjective ratings tend to have a high correlation with performance measures, they are sufficient to support comparable usability evaluation (Chin et al. 1988; Kirakowski and Corbett 1993; Annett 2002; Salvendy 2002; Ryu 2005). The survey types used in the studies are given in Table 1 (Filippi and Barattin 2012; Seffah et al. 2006).

2.3 Hybrid Studies

In some studies, both expert analysis (EA) and user participatory evaluation (UPE) are utilized together. This process is called as hybrid evaluation (HE). Table 2 summarizes the hybrid studies that include both expert analysis and user participatory evaluation simultaneously (Filippi and Barattin 2012).

Table 1 Survey types used in usability studies

Name of the survey	# of questions
Software usability measurement inventory (SUMI)	50
Website analysis and measurement inventory (WAMI)	20
Questionnaire for user interaction satisfaction (QUIS)	27
Purdue usability testing questionnaire (PUTQ)	100
System usability scale (SUS)	10
After scenario questionnaire (ASQ)	3
Post study system usability questionnaire (PSSUQ)	16
Computer system usability questionnaire (CSUQ)	16
Perceived usefulness and ease of use (PUEU)	12
Usability metric for user experience (UMUX)	4

3 Usability Evaluation Criteria

3.1 Evaluation Criteria Used in Usability Studies

A variety of usability evaluation methods is also present for usability evaluation criteria. However, the integrity and distinction of methods for the selection of evaluation methods are quite out of the question for the evaluation criteria. Although the preferred method is the same in many studies and articles in the literature, the evaluation criteria are quite different.

It is stated that five evaluation criteria are sufficient for usability testing (Nielsen 1994a): Learnability, Efficiency, Memorability, Errors (less error rate), Satisfaction. Usability evaluation criteria are also specified by the International Standardization Organization in ISO 9126-4 (ISO 1998) as; Understandability, Learnability, Operability, Attractiveness, Usability Compliance. Harrison et al. (2013) used the criteria which consists of combination of the International Standardization Organization (ISO) and Nielsen's evaluation criteria. Zhang and Adipat (2005) identified nine criteria for usability evaluation: Learnability, Efficiency, Memorability, Errors, Satisfaction, Effectiveness, Simplicity, Readability, and Learning Performance (Learning Performance). The evaluation criteria used by Abrahao and Insfran (2006) are as follows; Learnability, Understandability, Operability, Attractiveness, and Compliance. The usability criteria proposed by Shackel (2009) are Effectiveness, Learnability, Flexibility, and Attitude. Coursaris and Kim (2006) defined 9 criteria for usability: Effectiveness, Efficiency, Satisfaction, Learnability, Accessibility, Operability, Memorability, Acceptability, and Flexibility.

Table 2 EA, UPE, and HE methods applied in user experience studies and proposed approaches

The research study	UPE	EA	HE	Proposed approach
Smith and Mosier (1986)		✓		The guideline-based evaluation
Smith and Mosier (1986)		✓		Model based evaluation
Wharton et al. (1994)		✓		Cognitive walkthrough
Nielsen (1994a)	✓			Focus group
Nielsen (1994a)	✓			Post-talk walkthrough/retrospective testing
Nielsen (1994a)		✓		Consistency inspection
Nielsen (1994a)		✓		The feature inspection
Nielsen and Mack (1994)		✓		Formal usability inspection
Nielsen and Mack (1994)		✓		Heuristic evaluation
Nielsen and Mack (1994)			✓	Competitive/comparative analysis
Nielsen and Mack (1994)			✓	The pluralistic walkthrough
Dumas and Redish (1999)	✓			Query techniques
Constantine and Lockwood (1999)			✓	The collaborative usability inspection
Kamper (2002)		✓		Heuristic evaluation
Thompson (2003)	✓			Remote usability testing
Dix et al. (2004)	✓			Controlled experiment
Dix et al. (2004)	✓			Psychological monitoring
Huart et al. (2004)		✓		Cognitive walkthrough
Nielsen and Mack (1994)	✓			Coaching method
Sarodnick and Brau (2006)	✓			Coaching method
Sarodnick and Brau (2006)			✓	The participatory heuristic evaluation
Downey (2007)	✓			Co-discovery learning
Rubin and Chisnell (2008)	✓			Think aloud
Rubin and Chisnell (2008)			✓	Competitive/comparative analysis
Shneiderman et al. (2009)		✓		Heuristic evaluation

In addition to these, some researchers are used usability criteria for mobile phone evaluation. For instance, the usability evaluation criteria used by Ryu and Smith-Jackson for mobile telephone usability survey are (Ryu and Smith-Jackson 2006) Satisfaction, Affect, Mental Effort, Frustration, Perceived Usefulness, Flexibility, Ease of Use, Learnability, Controllability, Task Accomplishment, Temporal Efficiency, Helpfulness, Compatibility, Accuracy, Clarity of Presentation, Understandability, Installation, Documentation, Pleasurability, Specific Tasks, and Feedback. Hussain et al. (2013) proposed a metric-based evaluation model for mobile telephone applications given in Table 3.

Table 3 Usability evaluation model of mobile applications proposed by Hussain et al. (2013)

Availability criteria	Purpose	Guidelines
Effectiveness	Simplicity	Ease of data entry
		Easy of using output
		Ease of installation
		Ease of learning
	Accuracy	Correct
		Error-free
		Successful
Efficiency	Elapsed time	Response
		Completing a task
	Features	Help/support
		Touch screen capabilities
		Voice command
		System resources information
		Automatic update
Satisfaction	Safety	When using the application
		When driving
	Attractiveness	Interface

3.2 Usability Criteria to Be Considered in the Proposed Approach

The following parameters, which are widely used in the literature, will be considered within the scope of this study;

- ***Effectiveness*** indicates the level of users' ability to complete certain objectives and tasks accurately and completely (Zhang and Adipat 2005).
- ***Efficiency*** defines how fast users can perform a given task by using a mobile application (Zhang and Adipat 2005). In other words, it is the measure of time, money, mental effort, etc. resources needed to be spent to achieve the specified objectives (ISO 1998). According to Park and Lim (1999), the efficiency of the criteria can be measured by the following metrics: (i) The required time to complete the given task and (ii) Number of tasks completed per unit time.
- ***Satisfaction*** is the acceptable degree of the mobile application by users (Zhang and Adipat 2005). Satisfaction criteria are measured by the following metrics (Park and Lim 1999): (i) user satisfaction rating scale, (ii) proportion of users who say they prefer to total, (iii) the ratio of the positive opinions to the negative opinions, and (iv) complaint frequency.

- ***Learnability*** examines how easily users can accomplish a task when they first use an application/system/product, and how quickly they can improve their performance levels (Zhang and Adipat 2005).
- ***Errors*** examines the number of errors, the severity of errors, and how easily users can correct them that users make when using a system/application/product (Zhang and Adipat 2005). According to Zhang and Adipat, the following metrics can be used to measure errors criteria (Zhang and Adipat 2005): (i) the number of errors made by users, (ii) the average accuracy of completed tasks, and (iii) the ratio between successful interactions and errors.
- ***Usefulness*** is an indication of the belief that increases the ability to perform a task or to achieve a goal (Davis 1989).
- ***Memorability*** indicates the level of users' recognition of how to use a mobile application again after disusing it for a period of time (Zhang and Adipat 2005).
- ***Attractiveness*** presents the appeal of a mobile application to users (e.g. colour usage or graphic design) (ISO 1998; Seffah et al. 2006).
- ***Ease of Use*** refers to the effort capacity of the users in using a particular mobile application. The effort is a limited resource that one can allocate to the various activities (Radner and Rothschild 1975). It is claimed that if everything else is equal, users are more likely to adopt a mobile application that is easier to perceive than others (Davis 1989).
- ***Flexibility*** refers to the adaptability of a mobile application to the personal preferences of the user (Seffah et al. 2006).
- ***Feedback*** is the response of the product to user inputs or events in a meaningful manner (Seffah et al. 2006). This allows users to achieve their task.
- ***Accessibility*** presents the usability of a mobile application for people with any kind of disability (e.g., visual, auditory, psychomotor) (Seffah et al. 2006).
- ***Helpfulness*** presents the capacity of providing all kinds of useful information that a user needs when needed by the user (Kim et al. 2012).

4 Methodology

In this section, components of the proposed approach, Kano model and spherical fuzzy sets are studied.

4.1 Preliminaries of Spherical Fuzzy Sets

Fuzzy sets were introduced by Zadeh (1965) to present vagueness and impreciseness of human opinions. After the presentation of ordinary fuzzy sets, they have been very popular in almost all branches of science. Various researchers have developed several extensions of ordinary fuzzy sets. One of the extensions is spherical fuzzy sets proposed by Kutlu Gündoğdu and Kahraman (2019). The concept of SFS (Spherical Fuzzy Sets) provides a larger preference domain for decision makers to assign membership degrees since the squared sum of the spherical parameters is allowed to be at most 1.0. DMs can define their hesitancy information independently under spherical fuzzy environment. In recent years, several researchers have utilized this extension in the solution of multi-criteria decision making problems.

In the following, definition and preliminaries of SFS is presented:

Definition 1 Single valued Spherical Fuzzy Sets (SFS) $\tilde{A}_S$ of the universe of discourse U is given by

$$\tilde{A}_S = \left\{\left\langle u, (\mu_{\tilde{A}_S}(u), \nu_{\tilde{A}_S}(u), \pi_{\tilde{A}_S}(u))\right| u \in U\right\} \tag{1}$$

where

$$\mu_{\tilde{A}_S}(u) : U \to [0, 1],\ \nu_{\tilde{A}_S}(u) : U \to [0, 1],\ \pi_{\tilde{A}_S}(u) : U \to [0, 1]$$

and

$$0 \le \mu^2_{\tilde{A}_S}(u) + \nu^2_{\tilde{A}_S}(u) + \pi^2_{\tilde{A}_S}(u) \le 1 \quad \forall u \in U \tag{2}$$

For each u, the numbers $\mu_{\tilde{A}_S}(u)$, $\nu_{\tilde{A}_S}(u)$ and $\pi_{\tilde{A}_S}(u)$ are the degree of membership, non-membership and hesitancy of u to $\tilde{A}_S$, respectively.

Definition 2 Basic operators of Single-valued SFS;

1. $$\tilde{A}_S \oplus \tilde{B}_S = \left\{ \begin{array}{l} \left(\mu^2_{\tilde{A}_S} + \mu^2_{\tilde{B}_S} - \mu^2_{\tilde{A}_S}\mu^2_{\tilde{B}_S}\right)^{1/2}, \upsilon_{\tilde{A}_S}\upsilon_{\tilde{B}_S}, \\ \left(\left(1 - \mu^2_{\tilde{B}_S}\right)\pi^2_{\tilde{A}_S} + \left(1 - \mu^2_{\tilde{A}_S}\right)\pi^2_{\tilde{B}_S} - \pi^2_{\tilde{A}_S}\pi^2_{\tilde{B}_S}\right)^{1/2} \end{array} \right\} \tag{3}$$

2. $$\tilde{A}_S \otimes \tilde{B}_S = \left\{ \begin{array}{l} \mu_{\tilde{A}_S}\mu_{\tilde{B}_S}, \left(\upsilon^2_{\tilde{A}_S} + \upsilon^2_{\tilde{B}_S} - \upsilon^2_{\tilde{A}_S}\upsilon^2_{\tilde{B}_S}\right)^{1/2}, \\ \left(\left(1 - \upsilon^2_{\tilde{B}_S}\right)\pi^2_{\tilde{A}_S} + \left(1 - \upsilon^2_{\tilde{A}_S}\right)\pi^2_{\tilde{B}_S} - \pi^2_{\tilde{A}_S}\pi^2_{\tilde{B}_S}\right)^{1/2} \end{array} \right\} \tag{4}$$

3. $$\lambda \cdot \tilde{A}_S = \left\{ \begin{array}{l} \left(1 - \left(1 - \mu^2_{\tilde{A}_S}\right)^{\lambda}\right)^{1/2}, \upsilon^{\lambda}_{\tilde{A}_S}, \\ \left(\left(1 - \mu^2_{\tilde{A}_S}\right)^{\lambda} - \left(1 - \mu^2_{\tilde{A}_S} - \pi^2_{\tilde{A}_S}\right)^{\lambda}\right)^{1/2} \end{array} \right\} \quad for\ \lambda > 0 \tag{5}$$

4. $$\tilde{A}^{\lambda}_S = \left\{ \begin{array}{l} \mu^{\lambda}_{\tilde{A}_S}, \left(1 - \left(1 - \upsilon^2_{\tilde{A}_S}\right)^{\lambda}\right)^{1/2}, \\ \left(\left(1 - \upsilon^2_{\tilde{A}_S}\right)^{\lambda} - \left(1 - \upsilon^2_{\tilde{A}_S} - \pi^2_{\tilde{A}_S}\right)^{\lambda}\right)^{1/2} \end{array} \right\} \quad for\ \lambda > 0 \tag{6}$$

Definition 3 Single-valued Spherical Weighted Arithmetic Mean (SWAM) with respect to, $w = (w_1, w_2 \ldots, w_n)$; $w_i \in [0, 1]$; $\sum_{i=1}^{n} w_i = 1$, SWAM is defined as;

$$SWAM_w(\tilde{A}_{S1}, \ldots, \tilde{A}_{Sn}) = w_1\tilde{A}_{S1} + w_2\tilde{A}_{S2} + \cdots + w_n\tilde{A}_{Sn}$$
$$= \left\{ \begin{array}{l} \left[1 - \prod_{i=1}^{n}(1 - \mu^2_{\tilde{A}_{Si}})^{w_i}\right]^{1/2}, \prod_{i=1}^{n} v^{w_i}_{\tilde{A}_{Si}}, \\ \left[\prod_{i=1}^{n}(1 - \mu^2_{\tilde{A}_{Si}})^{w_i} - \prod_{i=1}^{n}(1 - \mu^2_{\tilde{A}_{Si}} - \pi^2_{\tilde{A}_{Si}})^{w_i}\right]^{1/2} \end{array} \right\} \quad (7)$$

Definition 4 Single-valued Spherical Weighted Geometric Mean (SWGM) with respect to,$w = (w_1, w_2 \ldots, w_n)$; $w_i \in [0, 1]$; $\sum_{i=1}^{n} w_i = 1$, SWGM is defined as;

$$SWGM_w(\tilde{A}_1, \ldots, \tilde{A}_n) = \tilde{A}^{w_1}_{S1} + \tilde{A}^{w_2}_{S2} + \cdots + \tilde{A}^{w_n}_{Sn}$$
$$= \left\{ \begin{array}{l} \prod_{i=1}^{n} \mu^{w_i}_{\tilde{A}_{Si}}, \left[1 - \prod_{i=1}^{n}(1 - v^2_{\tilde{A}_{Si}})^{w_i}\right]^{1/2}, \\ \left[\prod_{i=1}^{n}(1 - v^2_{\tilde{A}_{Si}})^{w_i} - \prod_{i=1}^{n}(1 - v^2_{\tilde{A}_{Si}} - \pi^2_{\tilde{A}_{Si}})^{w_i}\right]^{1/2} \end{array} \right\} \quad (8)$$

Definition 5 Defuzzification functions and Accuracy functions of sorting SFS are defined by;

$$defuzz\left(\tilde{A}_s\right) = \left(2\mu_{\tilde{A}_s}\right)^2 - v^2_{\tilde{A}_s} - \pi^2_{\tilde{A}_s}/2 \quad (9)$$

$$Accuracy\left(\tilde{A}_S\right) = \mu^2_{\tilde{A}_S} + v^2_{\tilde{A}_S} + \pi^2_{\tilde{A}_S} \quad (10)$$

Note that: $\tilde{A}_S < \tilde{B}_S$ if and only if

1. $defuzz\left(\tilde{A}_s\right) < defuzz(\tilde{B}_S)$ or
2. $defuzz\left(\tilde{A}_s\right) = defuzz(\tilde{B}_S)$ and $Accuracy(\tilde{A}_S) < Accuracy(\tilde{B}_S)$.

Since its development, fuzzy sets have been widely utilized as integrated to various methods such as quality function deployment (QFD) and Kano model.

4.2 Kano Model

Kano model analyze customers' expectations by taking into consideration the functional and dysfunctional queries. Generally, Kano's model classifies product attributes into six categories as follows (Wu and Wang 2012):

- ***"Must-be attributes (M)"***: These attributes should be granted in the design. Their presence does not lead to customer satisfaction, but their absence or poor fulfillment leads to customer dissatisfaction.
- ***"Attractive attributes (A)"***: These attributes usually are not expected by customers. Their presence lead to customer satisfaction, but their absence or poor fulfillment does not lead to customer dissatisfaction.
- ***"One-dimensional attributes (P)"***: There is a one-dimensional relationship between customer satisfaction and product characteristics. The better the feature, the higher the satisfaction.
- ***"Indifferent attributes (I)"***: Customers do not care about these features, so their presence or absence does not affect their level of satisfaction or dissatisfaction.
- ***"Reverse attributes (R)"***: Their presence lead to customer dissatisfaction, but their absence result in customer satisfaction.
- ***"Questionable (Q)"***: There is another category called "questionable" (Q), which demonstrates that either the respondent provides an illogical answer or the question is described incorrectly.

The classifications of the attributes given above are conducted after analyzing Kano's questionnaire which contains a functional and dysfunctional pair of questions. The functional question presents the customer's perception if the product has a certain attribute while the dysfunctional question presents the customer's perception if the product does not have that attribute (Ilbahar and Cebi 2017). The final classification of the product attribute is made based on the total responses. In other words, the product attribute is determined one of the six categories according to the Kano evaluation table as given in Table 4 (Pouliot 1993).

Table 4 Evaluation table for Kano survey

			Dysfunctional (X)				
			Enjoy	Expect it	Neutral	Live with	Dislike
			−2	−1	0	2	4
Functional (Y)	Enjoy	4	Q	A	A	A	P
	Expect it	2	R	Q	I	I	M
	Neutral	0	R	I	I	I	M
	Live with	−1	R	I	I	Q	M
	Dislike	−2	R	R	R	R	Q

4.3 Proposed Approach

The proposed method consists of Kano model, QFD and spherical fuzzy sets. The Steps 1–3 are considered as Kano Model application while the Steps 4–10 are in QFD application. The main steps of the proposed integrated method are as follows:
Step 1. To identify main customer requirements ($i = 1, 2, \ldots n$), a research is conducted.
Step 2. Prepare a questionnaire with respect to the identified customer requirements and conduct for analysis. This questionnaire consists of a pair of questions for each customer requirement attribute we would like to evaluate:

1. How the user feels if this attribute exists?
2. How the user feels in the absence of this attribute?

The first question is positive and the second question is negative and these questions are not open-ended. After asking these two questions to users, the category of each customer requirement can be determined.
Step 3. Obtain A, O, M, I, R, Q categories based on Table 4 as a results of the questionnaire and determine the crisp weights of the customer requirements ($\bar{w}_i^s$) by using average formula by utilizing functional and dysfunctional attributes.
Step 4. To identify usability criteria ($j = 1, 2, \ldots m$), thorough research is conducted.
Step 5. Determine the relationship matrix $\left(\tilde{R}_{S_{ij}}\right)$ between customer requirements and usability criteria based on spherical fuzzy linguistic scale given in Table 5. In this step, the project team asks "which usability criteria are associated with which customer requirement and what is the power of the relationship between them" to experts.
Step 6. Aggregate the judgments of each expert using Spherical Weighted Arithmetic Mean (SWAM) by Eq. (11) or Spherical Weighted Geometric Mean (SWGM) by Eq. (12). Based on their experiences, each expert may have a different importance weight.

Table 5 Linguistic terms and their corresponding spherical fuzzy numbers (Kutlu Gündoğdu and Kahraman 2019)

Linguistic terms	(μ, v, π)
Absolutely more relation (AMR)	(0.9, 0.1, 0.1)
Very high relation (VHR)	(0.8, 0.2, 0.2)
High relation (HR)	(0.7, 0.3, 0.3)
Slightly more relation (SMR)	(0.6, 0.4, 0.4)
Medium relation (MR)	(0.5, 0.5, 0.5)

$$SWAM_w(\tilde{A}_{S1},\ldots,\tilde{A}_{Sn}) = w_1\tilde{A}_{S1} + w_2\tilde{A}_{S2} + \cdots + w_n\tilde{A}_{Sn} = \left\{ \begin{array}{l} \left[1-\prod_{i=1}^{n}(1-\mu_{\tilde{A}_{Si}}^2)^{w_i}\right]^{1/2}, \prod_{i=1}^{n} v_{\tilde{A}_{Si}}^{w_i}, \\ \left[\prod_{i=1}^{n}(1-\mu_{\tilde{A}_{Si}}^2)^{w_i} - \prod_{i=1}^{n}(1-\mu_{\tilde{A}_{Si}}^2-\pi_{\tilde{A}_{Si}}^2)^{w_i}\right]^{1/2} \end{array} \right\} \quad (11)$$

or

$$SWGM_w(\tilde{A}_1,\ldots,\tilde{A}_n) = \tilde{A}_{S1}^{w_1} + \tilde{A}_{S2}^{w_2} + \cdots + \tilde{A}_{Sn}^{w_n} = \left\{ \begin{array}{l} \prod_{i=1}^{n}\mu_{\tilde{A}_{Si}}^{w_i}, \left[1-\prod_{i=1}^{n}(1-v_{\tilde{A}_{Si}}^2)^{w_i}\right]^{1/2}, \\ \left[\prod_{i=1}^{n}(1-v_{\tilde{A}_{Si}}^2)^{w_i} - \prod_{i=1}^{n}(1-v_{\tilde{A}_{Si}}^2-\pi_{\tilde{A}_{Si}}^2)^{w_i}\right]^{1/2} \end{array} \right\} \quad (12)$$

Step 7. Construct the aggregated weighted spherical fuzzy relationship matrix. After the weights of the customer requirements and the aggregated relationship matrix are determined, the aggregated weighted spherical fuzzy relationship matrix $\left(\tilde{R}_{S_{ij}}^{w}\right)$ is constructed by utilizing Eq. (13).

$$\tilde{R}_{S_{ij}}^{w} = \bar{w}_i^s \cdot \tilde{R}_{S_{ij}} = \left\langle \left(1-\left(1-\mu_{\tilde{R}_{S_{ij}}}^2\right)^{\bar{w}_i^s}\right)^{1/2}, v_{\tilde{R}_{S_{ij}}}^{\bar{w}_i^s}, \left(\left(1-\mu_{\tilde{R}_{S_{ij}}}^2\right)^{\bar{w}_i^s} - \left(1-\mu_{\tilde{R}_{S_{ij}}}^2-\pi_{\tilde{R}_{S_{ij}}}^2\right)^{\bar{w}_i^s}\right)^{1/2} \right\rangle \quad \forall i, \forall j \quad (13)$$

Step 8. Calculate the spherical fuzzy weight of the each usability criterion $\left(\tilde{F}_{S_j}\right)$ based on spherical fuzzy addition operator is given in Eq. (14).

$$\tilde{F}_{S_j} = \frac{\sum_{i=1}^{n}\tilde{R}_{S_{ij}}^{w}}{n} = \frac{\tilde{R}_{S_{1j}}^{w} \oplus \tilde{R}_{S_{2j}}^{w} \cdots \oplus \tilde{R}_{S_{nj}}^{w}}{n} \quad \forall j$$

$$i.e. \ \frac{\tilde{R}_{S_{11}}^{w} \oplus \tilde{R}_{S_{21}}^{w}}{2}$$

$$= \left\langle \left[1-(1-\mu^2_{\tilde{R}^w_{S_{11}}})^{1/2}(1-\mu^2_{\tilde{R}^w_{S_{21}}})^{1/2}\right]^{1/2}, \upsilon^{1/2}_{\tilde{R}^w_{S_{11}}}\upsilon^{1/2}_{\tilde{R}^w_{S_{21}}}, \left[(1-\mu^2_{\tilde{R}^w_{S_{11}}})^{1/2}(1-\mu^2_{\tilde{R}^w_{S_{21}}})^{1/2}-(1-\mu^2_{\tilde{R}^w_{S_{11}}}-\pi^2_{\tilde{R}^w_{S_{11}}})^{1/2}(1-\mu^2_{\tilde{R}^w_{S_{21}}}-\pi^2_{\tilde{R}^w_{S_{21}}})^{1/2}\right]^{1/2} \right\rangle \tag{14}$$

Step 9. Defuzzify the spherical fuzzy weight of the each usability criterion $\left(\tilde{F}_{S_j}\right)$ based on Eq. (15) is defined in Definition 5.

$$defuzz\left(\tilde{F}_{S_j}\right) = \left(2\mu_{\tilde{F}_{S_j}}\right)^2 - \upsilon^2_{\tilde{F}_{S_j}} - \pi^2_{\tilde{F}_{S_j}} \Big/ 2 \tag{15}$$

Step 10. Determine which usability criterion is more important than others based on defuzzified values calculated in Step 9. The largest value provides the more important usability criterion with respect to customer requirements.

5 A Case Study: Usability of Travel Mobile Application

A questionnaire on the usability of a travel mobile application is prepared and conducted with 30 respondents. 50% of the participants are male and rests of them are female. The average age of respondents is 37.

As a result of the interviews, it was determined that the customers have requests for the following functions besides the basic functions for a travel mobile application:

F1: Keeping calls, recognizing the customer,
F2: Keeping the applied filtering, recognizing the customer,
F3: Offline city map arrangement,
F4: Marking of desired points on the map, density information about desired points and day suggestion information,
F5: To be able to provide personalized place recommendations on the map,
F6: Recording and sharing the travel route on the map,
F7: Earn points or money as you buy travel from the application,
F8: Gaining discounts on following trips,
F9: Follows the information of who spent how much for crowded trips,
F10: Total package recommendation: Airplane, hotel, car rental, trip plan,
F11: Access to 48-h city tours information.

Impacts of determined eleven parameters on the usability of mobile applications are investigated through functional and dysfunctional questions. According to the results of the questionnaire, A, O, M, I, R, Q categories are obtained, and the weights of the customer requirements are determined according to the traditional Kano model (Table 6).

Table 6 Priorities of travel mobile application features

Customer requirements	Dysfunctional (X)	Functional (Y)	Category
F3	1.93	3.40	A
F8	1.83	3.38	A
F5	1.27	3.27	A
F6	1.33	3.27	A
F7	0.89	3.07	A
F4	1.87	2.73	A
F2	2.17	2.69	P
F1	2.47	2.63	P
F11	0.93	2.00	I
F10	1.15	1.77	I
F9	0.48	1.07	I

According to the results of the Kano analysis, F4, F11, F10 and F9 are determined as indifferent attributes, F2, F7, F6, F8, and F3 are determined as attractive attributes and F1 is determined as one-dimensional attribute.

In the following step, the usability criteria are determined as follows: Effectiveness, efficiency, satisfaction, learnability, errors, usefulness, memorability, attractiveness ease of use, flexibility, feedback, accessibility, helpfulness.

The relationship matrix $\left(\tilde{R}_{S_{ij}}\right)$ between customer requirements and usability criteria based on spherical fuzzy linguistic scale is gathered from the four experts. Four user experience experts with 4–6 years of experience have identified the relationship spherical fuzzy linguistic assessments between customer requirements and usability criteria by consensus. Determined relationship spherical fuzzy linguistic evaluations are shown in Table 7.

The aggregated weighted spherical fuzzy relationship matrix is constructed by utilizing Eq. (13) as given in Table 8. The spherical fuzzy weight of the each usability criterion $\left(\tilde{F}_{S_j}\right)$ is calculated based on spherical fuzzy addition operator is given in Eq. (14). $\tilde{F}_{S_j}$ spherical fuzzy weights are defuzzified by using Eq. (15) as given in Table 9.

It is seen that the errors and memorability criteria are less important than the other criteria in terms of usability criteria. The most important usability criteria are satisfaction, ease of use, and accessibility with respect to customer requirements.

6 Conclusions

When determining priorities according to the design parameters of a product or system, the technical and organizational capabilities of the relevant company are taken

Table 7 The relationship matrix between customer requirements and usability criteria

		Usability criteria												
Customer requirements	Weights	Effectiveness	Efficiency	Satisfaction	Learnability	Errors	Usefulness	Memorability	Attractiveness	Ease of use	Flexibility	Feedback	Accessibility	Helpfulness
F1: Keeping calls, recognizing the customer	2.63	SMR	AMR	AMR	SMR	MR	AMR	MR	MR	AMR	AMR	SMR	AMR	SMR
F2: Keeping the applied filtering, recognizing the customer	2.69	SMR	AMR	AMR	SMR	MR	AMR	MR	MR	AMR	SMR	SMR	AMR	SMR
F3: Offline city map arrangement	3.4	AMR	AMR	AMR	SMR	SMR	AMR	SMR	AMR	AMR	SMR	AMR	AMR	AMR
F4: Marking of desired points on the map, density information about desired points and day suggestion information	2.73	AMR	AMR	AMR	AMR	AMR	AMR	SMR	AMR	AMR	AMR	AMR	AMR	AMR
F5: To be able to provide personalized place recommendations on the map	3.27	SMR	AMR	AMR	SMR	MR	AMR	SMR	AMR	AMR	AMR	SMR	AMR	SMR
F6: Recording and sharing the travel route on the map	3.27	AMR	AMR	AMR	AMR	AMR	AMR	AMR	AMR	AMR	AMR	AMR	AMR	AMR
F7: Earn points or money as you buy travel from the application	3.07	MR	MR	AMR	AMR	MR	SMR	MR	SMR	AMR	SMR	AMR	AMR	AMR
F8: Gaining discounts on following trips	3.38	MR	MR	AMR	AMR	MR	SMR	MR	SMR	AMR	SMR	SMR	AMR	AMR

(continued)

Table 7 (continued)

		Usability criteria												
Customer requirements	Weights	Effectiveness	Efficiency	Satisfaction	Learnability	Errors	Usefulness	Memorability	Attractiveness	Ease of use	Flexibility	Feedback	Accessibility	Helpfulness
F9: Follows the information of who spent how much for crowded trips	1.07	AMR	AMR	AMR	AMR	AMR	AMR	SMR	SMR	AMR	MR	SMR	AMR	AMR
F10: Total package recommendation: airplane, hotel, car rental, trip plan	1.77	SMR	MR	AMR	SMR	MR	AMR	SMR	SMR	AMR	AMR	SMR	AMR	AMR
F11: Access to 48-h city tours information	2	AMR	AMR	AMR	SMR	AMR	AMR	AMR	SMR	AMR	AMR	SMR	AMR	AMR

Table 8 Aggregated weighted relationship matrix

Customer requirements	Effectiveness	Efficiency	Satisfaction	Learnability	Errors	Usefulness	Memorability	Attractiveness	Ease of use	Flexibility	Feedback	Accessibility	Helpfulness
F1	(0.83, 0.09, 0.32)	(0.99, 0.00, 0.00)	(0.99, 0.00, 0.00)	(0.83, 0.09, 0.32)	(0.73, 0.09, 0.47)	(0.99, 0.00, 0.00)	(0.73, 0.09, 0.47)	(0.73, 0.09, 0.47)	(0.99, 0.00, 0.00)	(0.99, 0.00, 0.00)	(0.83, 0.09, 0.32)	(0.99, 0.00, 0.00)	(0.83, 0.09, 0.32)
F2	(0.84, 0.09, 0.32)	(0.9, 0.00, 0.00)	(0.99, 0.00, 0.00)	(0.84, 0.09, 0.32)	(0.73, 0.09, 0.47)	(0.99, 0.00, 0.00)	(0.73, 0.09, 0.47)	(0.73, 0.09, 0.47)	(0.99, 0.00, 0.00)	(0.84, 0.09, 0.32)	(0.84, 0.09, 0.32)	(0.99, 0.00, 0.00)	(0.84, 0.09, 0.32)
F3	(1.00, 0.00, 0.00)	(1.00, 0.00, 0.00)	(1.00, 0.00, 0.00)	(0.88, 0.04, 0.30)	(0.88, 0.04, 0.30)	(1.00, 0.00, 0.00)	(0.88, 0.04, 0.30)	(1.00, 0.00, 0.00)	(1.00, 0.00, 0.00)	(0.88, 0.04, 0.30)	(1.00, 0.00, 0.00)	(1.00, 0.00, 0.00)	(1.00, 0.00, 0.00)
F4	(0.99, 0.00, 0.00)	(0.99, 0.00, 0.00)	(0.99, 0.00, 0.00)	(0.99, 0.00, 0.00)	(0.99, 0.00, 0.00)	(0.99, 0.00, 0.00)	(0.84, 0.08, 0.32)	(0.99, 0.00, 0.00)	(0.99, 0.00, 0.00)	(0.99, 0.00, 0.00)	(0.99, 0.00, 0.00)	(0.99, 0.00, 0.00)	(0.99, 0.00, 0.00)
F5	(0.88, 0.05, 0.30)	(1.00, 0.00, 0.00)	(1.00, 0.00, 0.00)	(0.88, 0.05, 0.30)	(0.78, 0.05, 0.46)	(1.00, 0.00, 0.00)	(0.88, 0.05, 0.30)	(1.00, 0.00, 0.00)	(1.00, 0.00, 0.00)	(1.00, 0.00, 0.00)	(0.88, 0.05, 0.30)	(1.00, 0.00, 0.00)	(0.88, 0.05, 0.30)
F6	(1.00, 0.00, 0.00)	(1.00, 0.00, 0.00)	(1.00, 0.00, 0.00)	(1.00, 0.00, 0.00)	(1.00, 0.00, 0.00)	(1.00, 0.00, 0.00)	(1.00, 0.00, 0.00)	(1.00, 0.00, 0.00)	(1.00, 0.00, 0.00)	(1.00, 0.00, 0.00)	(1.00, 0.00, 0.00)	(1.00, 0.00, 0.00)	(1.00, 0.00, 0.00)
F7	(0.77, 0.0, 0.46)	(0.77, 0.06, 0.46)	(1.00, 0.00, 0.00)	(1.00, 0.00, 0.00)	(0.77, 0.06, 0.46)	(0.86, 0.06, 0.31)	(0.77, 0.06, 0.46)	(0.86, 0.06, 0.31)	(1.00, 0.00, 0.00)	(0.86, 0.06, 0.31)	(1.00, 0.00, 0.00)	(1.00, 0.00, 0.00)	(1.00, 0.00, 0.00)
F8	(0.79, 0.05, 0.46)	(0.79, 0.05, 0.46)	(1.00, 0.00, 0.00)	(1.00, 0.00, 0.00)	(0.79, 0.05, 0.46)	(0.88, 0.05, 0.30)	(0.79, 0.05, 0.46)	(0.88, 0.05, 0.30)	(1.00, 0.00, 0.00)	(0.88, 0.05, 0.30)	(0.88, 0.05, 0.30)	(1.00, 0.00, 0.00)	(1.00, 0.00, 0.00)
F9	(0.91, 0.09, 0.00)	(0.91, 0.09, 0.00)	(0.91, 0.09, 0.00)	(0.91, 0.09, 0.00)	(0.91, 0.09, 0.00)	(0.91, 0.09, 0.00)	(0.62, 0.38, 0.30)	(0.62, 0.38, 0.30)	(0.91, 0.09, 0.00)	(0.51, 0.38, 0.41)	(0.62, 0.38, 0.30)	(0.91, 0.09, 0.00)	(0.91, 0.09, 0.00)
F10	(0.74, 0.20, 0.33)	(0.74, 0.20, 0.33)	(0.97, 0.02, 0.00)	(0.74, 0.20, 0.33)	(0.74, 0.20, 0.33)	(0.97, 0.02, 0.00)	(0.74, 0.20, 0.33)	(0.74, 0.20, 0.33)	(0.97, 0.02, 0.00)	(0.97, 0.02, 0.00)	(0.74, 0.20, 0.33)	(0.97, 0.02, 0.00)	(0.97, 0.02, 0.00)
F11	(0.98, 0.01, 0.00)	(0.98, 0.01, 0.00)	(0.98, 0.01, 0.00)	(0.77, 0.16, 0.33)	(0.98, 0.01, 0.00)	(0.98, 0.01, 0.00)	(0.98, 0.01, 0.00)	(0.77, 0.16, 0.33)	(0.98, 0.01, 0.00)	(0.98, 0.01, 0.00)	(0.77, 0.16, 0.33)	(0.98, 0.01, 0.00)	(0.98, 0.01, 0.00)

Table 9 Importance levels of usability criteria

Usability criteria	$\tilde{F}_{S_j}$	$defuzz\left(\tilde{F}_{S_j}\right)$
Effectiveness	(0.88, 0.08, 0.28)	3.042
Efficiency	(0.92, 0.07, 0.23)	3.339
Satisfaction	(0.98, 0.03, 0.00)	3.879
Learnability	(0.89, 0.09, 0.24)	3.127
Errors	(0.84, 0.08, 0.35)	2.753
Usefulness	(0.96, 0.03, 0.13)	6.687
Memorability	(0.81, 0.14, 0.35)	2.517
Attractiveness	(0.84, 0.15, 0.30)	2.736
Ease of use	(0.98, 0.03, 0.00)	3.879
Flexibility	(0.89, 0.12, 0.23)	3.116
Feedback	(0.86, 0.15, 0.25)	2.891
Accessibility	(0.98, 0.03, 0.00)	3.879
Helpfulness	(0.94, 0.05, 0.17)	3.538

into consideration. However, if a high-tech product is designed in today's digitalizing world conditions, the usability parameter is the most important component to be considered in prioritization. In particular, the usability parameter is very important in the development of mobile applications and in determining the performance of the developed features. When the usability parameters are analyzed in the literature, it is seen that many different and various parameters are used for the same purpose. This indicates that no specific standard has been established for the measurement of product availability. Therefore, within the scope of the proposed study, an integrated method has been developed to assist the designers in deciding the design parameters that should be taken into consideration in the design and usability tests of mobile applications and determining the importance of the design parameters. For this purpose, the method including Kano Model, QFD, and Spherical Fuzzy Sets has been developed. Thus, the developed method will suggest to the designers which design parameters should be kept in the foreground according to the application to be developed and which parameters should be considered at first when performing the usability test of this product. A mobile travel application was chosen to illustrate the application steps of the developed method.

In this chapter, all design parameters are assumed to be independent of each other. In some applications and some real-life problems, there may be interactions between design parameters. Therefore, the proposed method in future studies can be developed to take into account the interdependencies and interactions between design parameters.

References

Abrahao S, Insfran E (2006) Early usability evaluation in model driven architecture environments. In: 2006 sixth international conference on quality software (QSIC'06). IEEE, pp 287–294

Annett J (2002) Subjective rating scales: science or art? Ergonomics 45(14):966–987

Balagtas-Fernandez F, Hussmann H (2009) A methodology and framework to simplify usability analysis of mobile applications. In: Proceedings of the 2009 IEEE/ACM international conference on automated software engineering. IEEE Computer Society, pp 520–524

Bevan N (1995) Human-computer interaction standards. Adv Hum Factors Ergon 20:885–890

Bortenschlager M, Häusler E, Schwaiger W, Egger R, Jooss M (2010) Evaluation of the concept of early acceptance tests for touristic mobile applications. In: Information and communication technologies in tourism 2010, pp 149–158

Brinck T, Gergle D, Wood SD (2002) Designing web sites that work: usability for the web. Morgan Kaufmann Publishers

Butler KA (1996) Usability engineering turns 10. Interactions 3(1):58–75

Chin JP, Diehl VA, Norman KL (1988) Development of an instrument measuring user satisfaction of the human-computer interface. In: Proceedings of the SIGCHI conference on human factors in computing systems. ACM, pp 213–218

Chincholle D, Goldstein M, Nyberg M, Eriksson M (2002) Lost or found? A usability evaluation of a mobile navigation and location-based service. In: International conference on mobile human-computer interaction, Berlin, Heidelberg, pp 211–224

Constantine LL, Lockwood LA (1999) Software for use: a practical guide to the models and methods of usage-centered design. Pearson Education

Cooper A (1999) The inmates are running the asylum. SAMS, Indianapolis, pp 123–124

Coursaris C, Kim D (2006) A qualitative review of empirical mobile usability studies. In: AMCIS 2006 proceedings, p 352

Davis FD (1989) Perceived usefulness, perceived ease of use, and user acceptance of information technology. MIS Q 319–340

Dix A, Finlay J, Abowd GD, Beale R (2004) Human–computer interaction, 3rd edn. Pearson Prentice Hall

Downey LL (2007) Group usability testing: evolution in usability techniques. J Usability Stud 2(3):133–144

Dumas JS, Redish J (1999) A practical guide to usability testing. Intellect Books

Filippi S, Barattin D (2012) Generation, adoption and tuning of usability evaluation multimethods. Int J Hum Comput Interact 28(6):406–422

Harrison R, Flood D, Duce D (2013) Usability of mobile applications: literature review and rationale for a new usability model. J Interact Sci 1(1):1

Huart J, Kolski C, Sagar M (2004) Evaluation of multimedia applications using inspection methods: the cognitive walkthrough case. Interact Comput 16(2):183–215

Hussain A, Hashim NL, Nordin N, Tahir HM (2013) A metric-based evaluation model for applications on mobile phones. J ICT 12:55–71

Ilbahar E, Cebi S (2017) Classification of design parameters for E-commerce websites: a novel fuzzy Kano approach. Telematics Inform 34(8):1814–1825

ISO (1998) ISO 9241-11: ergonomic requirements for office work with visual display terminals (VDTs)—part 11: guidance on usability. International Organization for Standardization

Kamper RJ (2002) Extending the usability of heuristics for design and evaluation: lead, follow, and get out of the way. Int J Hum Comput Interact 14(3–4):447–462

Kim K, Proctor RW, Salvendy G (2012) The relation between usability and product success in cell phones. Behav Inform Technol 31(10):969–982

Kirakowski J, Corbett M (1993) SUMI: the software usability measurement inventory. Br J Educ Technol 24(3):210–212

Kjeldskov J, Stage J (2004) New techniques for usability evaluation of mobile systems. Int J Hum Comput Stud 60(5–6):599–620

Kruchten P (2004) The rational unified process: an introduction. Addison-Wesley Professional
Krug S (2000) Don't make me think!: a common sense approach to web usability. Pearson Education India
Kutlu Gündoğdu F, Kahraman C (2019) Spherical fuzzy sets and spherical fuzzy TOPSIS method. J Intell Fuzzy Syst 36(1):337–352
Nielsen J (1994a) Usability engineering. Elsevier
Nielsen J (1994b) Heuristic evaluation in usability inspection methods. Wiley, pp 25–62
Nielsen J, Mack RL (1994) Usability inspection methods, vol 1. Wiley, New York
Park KS, Lim CH (1999) A structured methodology for comparative evaluation of user interface designs using usability criteria and measures. Int J Ind Ergon 23(5–6):379–389
Pouliot F (1993) Theoretical issues of Kano's methods. Cent Qual Manag J 2(4):28–36
Radner R, Rothschild M (1975) On the allocation of effort. J Econ Theory 10(3):358–376
Rubin J, Chisnell D (2008) Handbook of usability testing: how to plan, design and conduct effective tests. Wiley
Ryu YS (2005) Development of usability questionnaires for electronic mobile products and decision making methods. Doctoral dissertation, Virginia Tech
Ryu YS, Smith-Jackson TL (2006) Reliability and validity of the mobile phone usability questionnaire (MPUQ). J Usability Stud 2(1):39–53
Salvendy G (2002) Use of subjective rating scores in ergonomics research and practice. Ergonomics 45(14):1005–1007
Sarodnick F, Brau H (2006) Methoden der Usability Evaluation: Wissenschaftliche Grundlagen und Praktische Anwendung. Huber
Seffah A, Donyaee M, Kline RB, Padda HK (2006) Usability measurement and metrics: a consolidated model. Software Qual J 14(2):159–178
Shackel B (2009) Usability—context, framework, definition, design and evaluation. Interact Comput 21(5–6):339–346
Shneiderman B, Plaisant C, Cohen M, Jacobs S (2009) Designing the user interface: strategies for effective human-computer interaction, 5th edn. Addison-Wesley, Boston, MA
Smith SL, Mosier JN (1986) Guidelines for designing user interface software. No. MTR-10090. Mitre Corporation, Bedford, MA
Thompson SM (2003) Remote observation strategies for usability testing. Inform Technol Libr 22(1):22
Web 1 (2012) www.usabilitybok.org/cognitive-walkthrough, 15 Oct 2019
Wei Q, Chang Z, Cheng Q (2015) Usability study of the mobile library app: an example from Chongqing University. Libr Hi Tech 33(3):340–355
Wharton C, Rieman J, Lewis C, Polson P (1994) Usability inspection methods. In: Nielsen J, Mack R (eds), pp 105–140
Wichansky AM (2000) Usability testing in 2000 and beyond. Ergonomics 43(7):998–1006
Wu M, Wang L (2012) A continuous fuzzy Kano's model for customer requirements analysis in product development. Proc IMechE Part B J Eng Manuf 226(3):535–546
Zadeh LA (1965) Fuzzy sets. Inf Control 338–353
Zhang D, Adipat B (2005) Challenges, methodologies and issues in the usability testing of mobile applications. Int J Hum Comput Interact 18(3):293–308
Zimmermann G, Vanderheiden G (2008) Accessible design and testing in the application development process: considerations for an integrated approach. Univ Access Inf Soc 7(1–2):117–128

Mathematical Programming with Spherical Fuzzy Sets

Spherical Fuzzy Linear Programming Problem

Firoz Ahmad and Ahmad Yusuf Adhami

Abstract The new extension of the uncertain set is presented by Kutlu Gündogdu and Kahraman (J Intell Fuzzy Syst (Preprint):1–16, 2019) and named as a spherical fuzzy set (SFS). The SFS is the superset of fuzzy, intuitionistic fuzzy, and Pythagorean fuzzy sets, respectively (Yager in Pythagorean fuzzy subsets, vol 2. IEEE, pp 57–61, 2013). The SFS inherently involves three membership functions, namely; positive, neutral, and negative membership degrees of the element into the SFS. In this chapter, we present the spherical fuzzy linear programming problem (SFLPP) in which the different parameters are represented by spherical fuzzy numbers (SFNs). The crisp version of the SFLPP is obtained with the aid of positive, neutral, and negative membership degrees. Furthermore, the spherical fuzzy optimization model is presented to solve the SFLPP. A numerical example and case study are presented to show the working efficiency of the proposed research. At last, the conclusion and future research scope are also discussed.

1 Introduction

In mathematical programming models, the well-renowned, simplest, and most extensively used model is linear programming problem (LPP). The nature of the LPP model is trivial and easily applicable to various real-life applications such as transportation problems, supplier selections, assignment problems, production planning problems, supply chain management, etc. The extension and advancement in conventional LPP are presented over several decades. Goal linear programming problem, multiobjective linear programming problem, bi-level or multilevel linear programming problem, etc. are some examples of extended LPP.

Uncertainty is also introduced in the LPP, which is widely adopted by many researchers. Firstly, Bellman and Zadeh (1970) proposed a fuzzy set (FS) and, based on the FS theory, the concept of fuzzy linear programming problem (FLPP) came

F. Ahmad (✉) · A. Y. Adhami
Department of Statistics and Operations Research, Aligarh Muslim University, Aligarh, India
e-mail: firoz.ahmad02@gmail.com

© Springer Nature Switzerland AG 2021

C. Kahraman and F. Kutlu Gündoğdu (eds.), *Decision Making with Spherical Fuzzy Sets*, Studies in Fuzziness and Soft Computing 392,
https://doi.org/10.1007/978-3-030-45461-6_19

into existence. Zimmermann (1978) was the first who introduced the fuzzy programming technique to solve the multiobjective LPP under fuzzy environment. The fuzzy optimization technique is based on the maximization of the marginal satisfaction (membership functions, degree of belongingness) of each element into the fuzzy decision set.

Later on, it has been realized that only the membership degrees are not well enough to represent the marginal attainment of the element into the fuzzy decision set. To extend or explore the fuzzy set, Atanassov (2017) was the first to introduce the intuitionistic fuzzy set (IFS) theory. The IFS theory deals with the membership function (degree of belongingness) as well as non-membership function (degree of non-belongingness) of the element into the intuitionistic fuzzy decision set. It was Angelov (1997) who first proposed the intuitionistic fuzzy optimization techniques based on intuitionistic fuzzy decision set in decision-making problems.

A tremendous amount of work has been presented in the literature by various authors in the fuzzy and intuitionistic fuzzy environment. Furthermore, in the multicriteria decision making (MCDM) problem, sometimes the performance of intuitionistic fuzzy set theory is found to be confined in many cases. For example, if the decision maker(s) assign the degree to which the alternatives satisfies the criteria is 0.6 and alternatives that dissatisfies the criteria is 0.7. Since both the membership and non-membership degrees are independent to each other and can any values between 0 and 1. Hence this situation is beyond the scope of intuitionistic fuzzy decision set as there must be the sum of membership and non-membership degrees less than or equal to 1. To overcome the aforementioned drawbacks or limitations, Yager (2013) propounded Pythagorean fuzzy set (PFS) which comply over the sum of squares of membership and non-membership functions must be less than or equal to 1.

Moreover, the neutrosophic set (NS) is also proposed by Smarandache (1999), which is inspired by indeterminacy or neutral thoughts in the decision-making process. Ahmad and Adhami (2019) presented a study on nonlinear transportation problems under the neutrosophic environment. Ahmad et al. (2018) proposed a new algorithm, named neutrosophic hesitant fuzzy programming approach, based on the single-valued neutrosophic hesitant fuzzy decision set and applied it to the manufacturing system. Ahmad et al. (2019) also discussed the neutrosophic optimization technique for the optimal shale gas water management system under uncertainty. Furthermore, Ahmad et al. (2020) also suggested the modified neutrosophic programming technique for multiobjective supply chain planning problem under uncertainty. Recently, the generalization of PFS is presented by Kutlu Gündogdu and Kahraman (2019) and named as a spherical fuzzy set (SFS). The SFS contemplates over the three different membership functions, namely; positive, neutral, and negative membership functions, along with the condition that the square sum of these membership functions must be less than or equal to 1. Thus the SFS provides a more generalized framework in comparisons to FS, IFS, and PFS while dealing with uncertainty.

In this chapter, we have discussed the LPP under a spherical fuzzy environment. The parameters are taken as SF numbers, and the corresponding deterministic version is obtained based on SFS theory. Different SF optimization models are presented to get the optimal solution of the SFLPP. Numerical examples and two real-life problems, such as production planning problems and purchasing strategy problems, are presented to show the validity and applicability of the SF optimization models. Conclusions and future scope are also discussed based on the given work.

The remainder of the chapter is organized as follows: Sect. 2 discusses the basic definitions regarding SFS, whereas, in Sect. 3, the proposed SFLPP is presented. The spherical fuzzy optimization model is discussed in Sect. 4. Numerical illustration and a case study are presented in Sect. 5, whereas conclusion and future research are revealed in Sect. 6.

2 Preliminaries

The basics definitions and ideas about the spherical fuzzy set is described in the following definitions.

Definition 1 (Rafiq et al. 2019) (*Spherical fuzzy set*) Let W represents the universe of discourse then a SFS $\tilde{X}_S$ can be defined with the aid of ordered triplets given as follows:

$$\tilde{X}_S = \left\{ \langle w, t_{\tilde{X}_S}(w), i_{\tilde{X}_S}(w), f_{\tilde{X}_S}(w) \rangle | w \in W \right\}$$

Such that $t_{\tilde{X}_S} : W \to [0, 1], i_{\tilde{X}_S} : W \to [0, 1], f_{\tilde{X}_S} : W \to [0, 1]$ and $0 \leq t^2_{\tilde{X}_S}(w) + i^2_{\tilde{X}_S}(w) + f^2_{\tilde{X}_S}(w) \leq 1 \quad \forall w \in W$.
Where $t_{\tilde{X}_S}(w), i_{\tilde{X}_S}(w)$ and $f_{\tilde{X}_S}(w)$ represents the positive membership degree, neutral membership degree and negative membership degree for each element $w \in W$ to $\tilde{X}_S$ respectively.

Definition 2 (Kutlu Gündogdu and Kahraman 2019) (*Union*) Let $\tilde{X}_S$ and $\tilde{Y}_S$ be two spherical fuzzy set then their union can be defined as follows:

$$\tilde{X}_S \cup \tilde{Y}_S = \left\{ \begin{array}{l} \max\left\{t_{\tilde{X}_S}, t_{\tilde{Y}_S}\right\}, \min\left\{f_{\tilde{X}_S}, f_{\tilde{Y}_S}\right\}, \\ \min\left\{\left(1 - \left(\left(\max\left\{t_{\tilde{X}_S}, t_{\tilde{Y}_S}\right\}\right)^2 + \left(\min\left\{f_{\tilde{X}_S}, f_{\tilde{Y}_S}\right\}\right)^2\right)\right)^{1/2}, \max\left\{i_{\tilde{X}_S}, i_{\tilde{Y}_S}\right\}\right\} \end{array} \right\}$$

Definition 3 (Kutlu Gündogdu and Kahraman 2019) (*Intersection*) Let $\tilde{X}_S$ and $\tilde{Y}_S$ be two spherical fuzzy set then their intersection can be defined as follows:

$$\tilde{X}_S \cap \tilde{Y}_S = \left\{ \begin{array}{l} \min\left\{t_{\tilde{X}_S}, t_{\tilde{Y}_S}\right\}, \max\left\{f_{\tilde{X}_S}, f_{\tilde{Y}_S}\right\}, \\ \max\left\{\left(1 - \left(\left(\min\left\{t_{\tilde{X}_S}, t_{\tilde{Y}_S}\right\}\right)^2 + \left(\max\left\{f_{\tilde{X}_S}, f_{\tilde{Y}_S}\right\}\right)^2\right)\right)^{1/2}, \min\left\{i_{\tilde{X}_S}, i_{\tilde{Y}_S}\right\}\right\} \end{array} \right\}$$

Definition 4 (Rafiq et al. 2019) (*Spherical fuzzy number*) In a SFS $\tilde{X}_S = \{\langle w, t_{\tilde{X}_S}(w), i_{\tilde{X}_S}(w), f_{\tilde{X}_S}(w)\rangle | w \in W\}$, the triple component $\langle t_{\tilde{X}_S}(w), i_{\tilde{X}_S}(w), f_{\tilde{X}_S}(w)\rangle$ are said to be a spherical fuzzy number (SFN) and each SFN can be represented by $a = \langle t_a, i_a, f_a\rangle$, where t_a, i_a and $f_a \in [0, 1]$, along with the restrictions $0 \leq t_a^2 + i_a^2 + f_a^2 \leq 1$.

Definition 5 (Rafiq et al. 2019) Let us consider that $a_r = \langle t_{a_r}, i_{a_r}, f_{a_r}\rangle$ and $a_s = \langle t_{a_s}, i_{a_s}, f_{a_s}\rangle$ be any two SFNs and $\lambda \geq 0$, then some basic operations can be defined as follows:

1. $a_r \oplus a_s = \left\langle \sqrt{t_{a_r}^2 + t_{a_s}^2 - t_{a_r}^2 \cdot t_{a_s}^2}, i_{a_r} \cdot i_{a_s}, f_{a_r} \cdot f_{a_s} \right\rangle$

2. $a_r \otimes a_s = \left\langle t_{a_r} \cdot t_{a_s}, i_{a_r} \cdot i_{a_s}, \sqrt{f_{a_r}^2 + f_{a_s}^2 - f_{a_r}^2 \cdot f_{a_s}^2} \right\rangle$

3. $\lambda a_r = \left\langle \sqrt{1 - \left(1 - t_{a_r}^2\right)^{\lambda}}, \left(i_{a_r}\right)^{\lambda}, \left(f_{a_r}\right)^{\lambda} \right\rangle$

Definition 6 (Rafiq et al. 2019) (*Ranking function*) Suppose that $a_s = \langle t_{a_s}, i_{a_s}, f_{a_s}\rangle$ be any SFNs, then the ranking function R and accuracy function A of SFNs can be defined as follows:

$$R(a_s) = \frac{\left(t_{a_s} + 1 - i_{a_s} + 1 - f_{a_s}\right)}{3} = \frac{1}{3}\left(2 + t_{a_s} - i_{a_s} - f_{a_s}\right) \tag{1}$$

$$A(a_s) = \left(t_{a_s} - f_{a_s}\right)$$

3 Spherical Fuzzy Linear Programming Problem

A conventional and most extensively used mathematical programming is the linear programming problem (LPP). The different extensions of linear programming problem such as fuzzy LPP, intuitionistic fuzzy LPP and neutrosophic LPP have been elaborately discussed by many researchers (Ahmad and Adhami 2019; Ahmad et al. 2018, 2019; Smarandache 1999; Rizk-Allahet al. 2018). The further extension of LPP is being discussed by introducing spherical fuzzy concept named as spherical fuzzy linear programming problem (SFLPP).

The first model presents the SFLPP in which only the co-efficient of the objective function is represented by the spherical fuzzy number, but all other parameters are assumed to be real numbers and can be described as follows:

Model I
Optimize $Z = \sum_{k=1}^{K} \tilde{c}_k^{sf} x_k$

Subject to

$$\sum_{k=1}^{K} a_{jk}x_k \leq, =, \geq b_j, \quad \forall j = 1, 2, \ldots, J$$

$$x_k \geq 0, \quad \forall k = 1, 2, \ldots, K$$

where $\tilde{c}_k^{sf}$ denotes a spherical fuzzy number. The parameters a_{jk}, b_j is real numbers.

In the second model of SFLPP, the co-efficient of constraints variables and right hand sides are represented by spherical fuzzy numbers whereas co-efficient of objective function is represented by real number. Therefore the equivalent model can be given as follows:

Model II
Optimize $Z = \sum_{k=1}^{K} c_k x_k$
Subject to

$$\sum_{k=1}^{K} \tilde{a}_{jk}^{sf} x_k \leq, =, \geq \tilde{b}_j^{sf}, \quad \forall j = 1, 2, \ldots, J$$

$$x_k \geq 0, \quad \forall k = 1, 2, \ldots, K$$

where $\tilde{a}_{jk}^{sf}$ and $\tilde{b}_j^{sf}$ denote the spherical fuzzy number. The parameter c_k is real number.

The third model represents the fully SFLPP in which all the parameters are represented by spherical fuzzy number respectively.

Model III
Optimize $Z = \sum_{k=1}^{K} \tilde{c}_k^{sf} x_k$
Subject to

$$\sum_{k=1}^{K} \tilde{a}_{jk}^{sf} x_k \leq, =, \geq \tilde{b}_j^{sf}, \quad \forall j = 1, 2, \ldots, J$$

$$x_k \geq 0, \quad \forall k = 1, 2, \ldots, K$$

where $\tilde{c}_k^{sf}$, $\tilde{a}_{jk}^{sf}$ and $\tilde{b}_j^{sf}$ are spherical fuzzy number.

The spherical fuzzy number $\tilde{a}^{sf} = \langle t, i, f \rangle$ such that $t, i, f \in [0, 1]$.

The positive membership function for spherical fuzzy number ($\tilde{a}^{sf}$) can be defined as follows:

$$t_{\tilde{a}^{sf}}(x) = \begin{cases} \frac{x-t_1}{t_2-t_1}, & t_1 \le x \le t_2 \\ \frac{t_2-x}{t_3-t_2}, & t_2 \le x \le t_3 \\ 0 & otherwise \end{cases} \tag{2}$$

The neutral membership degree of spherical fuzzy number can be defined as follows:

$$i_{\tilde{a}^{sf}}(x) = \begin{cases} \frac{x-i_1}{i_2-i_1}, & i_1 \le x \le i_2 \\ \frac{i_2-x}{i_3-i_2}, & i_2 \le x \le i_3 \\ 0 & otherwise \end{cases} \tag{3}$$

The negative membership function for spherical fuzzy number can be defined as follows:

$$f_{\tilde{a}^{sf}}(x) = \begin{cases} \frac{x-f_1}{f_2-f_1}, & f_1 \le x \le f_2 \\ \frac{f_2-x}{f_3-f_2}, & f_2 \le x \le f_3 \\ 0 & otherwise \end{cases} \tag{4}$$

To define the positive, neutral and negative membership functions for the objective function under spherical fuzzy concept, first we determine the upper and lower bound for each membership functions with the help of following expressions:

$$Z_U^t = \max\{Z(x_k)\} \text{ and } Z_L^t = \min\{Z(x_k)\} \text{ (Positive membership)} \tag{5}$$

$$Z_U^i = Z_L^t + P\left(Z_U^t - Z_L^t\right) \text{ and } Z_L^i = Z_L^t \text{ (Neutral membership)} \tag{6}$$

$$Z_U^f = Z_U^t \text{ and } Z_L^f = Z_L^t + Q\left(Z_U^t - Z_L^t\right) \text{ (Negative membership)} \tag{7}$$

where P and $Q \in [0, 1]$ are the predetermined real positive numbers.

With the aid of Eqs. (5)–(7), the positive, neutral and negative membership functions for objective function can be given as follows:

$$T(Z(x)) = \begin{cases} 1 & if\ z \ge z_U^t \\ \frac{z-z_L^t}{z_U^t-z_L^t} & if\ z_L^t \le z \le z_U^t \\ 0 & if\ z \le z_L^t \end{cases} \tag{8}$$

$$I(Z(x)) = \begin{cases} 1 & if\ z \ge z_U^i \\ \frac{z-z_L^i}{z_U^i-z_L^i} & if\ z_L^i \le z \le z_U^i \\ 0 & if\ z \le z_L^i \end{cases} \tag{9}$$

$$F(Z(x)) = \begin{cases} 1 & if\ z \geq z_U^f \\ \frac{z_U^f - z}{z_U^f - z_L^f} & if\ z_L^f \leq z \leq z_U^f \\ 0 & if\ z \leq z_L^f \end{cases} \tag{10}$$

where $T(Z(x)), I(Z(x))$ and $F(Z(x))$ are the positive, neutral and negative membership degrees for object functions.

The characterization of membership function for C_j^{th} constraints under spherical fuzzy set is given as follows:

$$T(C_j(x)) = \begin{cases} 1 & if\ b_j \geq \sum_{k=1}^{K} (a_{jk} + d_{jk})x_k \\ \frac{b_j - \sum_{k=1}^{K} a_{jk}x_k}{\sum_{k=1}^{K} d_{jk}x_k} & if\ \sum_{k=1}^{K} a_{jk}x_k \leq b_j \leq \sum_{k=1}^{K} (a_{jk} + d_{jk})x_k \\ 0 & if\ b_j \leq \sum_{k=1}^{K} a_{jk}x_k \end{cases} \tag{11}$$

$$I(C_j(x)) = \begin{cases} 1 & if\ b_j \geq \sum_{k=1}^{K} (a_{jk} + d_{jk})x_k \\ \frac{b_j - \sum_{k=1}^{K} d_{jk}x_k}{\sum_{k=1}^{K} a_{jk}x_k} & if\ \sum_{k=1}^{K} a_{jk}x_k \leq b_j \leq \sum_{k=1}^{K} (a_{jk} + d_{jk})x_k \\ 0 & if\ b_j \leq \sum_{k=1}^{K} a_{jk}x_k \end{cases} \tag{12}$$

$$F(C_j(x)) = \begin{cases} 0 & if\ b_j \geq \sum_{k=1}^{K} (a_{jk} + d_{jk})x_k \\ \frac{\sum_{k=1}^{K} (a_{jk} + d_{jk})x_k - b_j}{\sum_{k=1}^{K} d_{jk}x_k} & if\ \sum_{k=1}^{K} a_{jk}x_k \leq b_j \leq \sum_{k=1}^{K} (a_{jk} + d_{jk})x_k \\ 1 & if\ b_j \leq \sum_{k=1}^{K} a_{jk}x_k \end{cases} \tag{13}$$

where $d_{jk} \in [0, 1]$ is predetermined tolerance limit of the jth constraint.

4 Spherical Fuzzy Optimization Model

The above discussed spherical fuzzy linear programming model (Sect. 3) can be solved using spherical fuzzy optimization techniques. In the proposed SF optimization model, we have considered the maximization of positive membership degree, minimization of neutral and negative membership degrees under the spherical fuzzy decision set. Hence the SF optimization model can be expressed as follows:

$$\text{Max}\, T(Z(x)),\ \text{Min}\, I(Z(x)),\ \text{Min}\, F(Z(x))$$

Subject to

$$T(Z(x)) \geq I(Z(x)),\ T(Z(x)) \geq F(Z(x)),$$
$$0 \leq T(Z(x))^2 + I(Z(x))^2 + F(Z(x))^2 \leq 1,$$
$$T(Z(x)),\ I(Z(x)),\ F(Z(x)) \geq 0,\ x \geq 0.$$

where $T(Z(x))$, $(Z(x))$ and $F(Z(x))$ represent the positive, neutral and negative membership degrees of the spherical fuzzy objective function and constraints.

The above problem can be transformed into the following optimization model as follows:

$$\text{Max}\,\alpha,\ \text{Min}\,\beta,\ \text{Min}\,\gamma$$

Subject to

$$T(Z(x)) \geq \alpha,\ \ I(Z(x)) \leq \beta,\ \ F(Z(x)) \leq \gamma,$$

$$\alpha \geq \beta,\ \beta \geq \gamma$$

$$0 \leq \alpha^2 + \beta^2 + \gamma^2 \leq 1,\quad x \geq 0.$$

where α, β and γ represents the minimal degree of acceptance of positive membership, maximal degree of acceptance of neutral degree and maximal degree of acceptance of negative membership degree respectively.

Equivalently, the SF optimization model can be transformed into the following model:

$$\text{Max}\,(\alpha - \beta - \gamma)$$

Subject to,

$$T(Z(x)) \geq \alpha,\ I(Z(x)) \leq \beta,\ F(Z(x)) \leq \gamma,$$

$$\alpha \geq \beta,\ \beta \geq \gamma,\ \alpha,\ \beta,\ \gamma \geq 0$$

$$0 \leq \alpha^2 + \beta^2 + \gamma^2 \leq 1,\ x \geq 0.$$

The above SF optimization model can solve the SFLPP with different spherical fuzzy parameters and captures the more real aspects of uncertainty among parameters.

5 Numerical Examples

To illustrate the validity and applicability of the spherical fuzzy optimization model, the following numerical examples have been presented. The different spherical fuzzy parameters are converted into the crisp version by using Eqs. (2)–(4) for each membership degree assigned by the decision makers. The obtained crisp version of all the numerical examples are written in AMPL language and solved by using Knitro solvers online facility provided by University of Wisconsin (Czyzyk et al. 1998).

Example 1 (Model I) Maximize $Z = \tilde{c}_1^{sf} x_1 + \tilde{c}_2^{sf} x_2 + \tilde{c}_3^{sf} x_3$

Subject to

$$
\begin{aligned}
&0.2x_1 + 0.4x_2 + 0.6x_3 \leq 12 \\
&0.5x_1 + 0.4x_2 + 0.7x_3 \leq 15 \\
&0.3x_1 + 0.8x_2 + 0.2x_3 \leq 16 \\
&x_1,\ x_2,\ x_3 \geq 0
\end{aligned}
$$

where

$$
\tilde{c}_1^{sf} = (0.6, 0.6, 0.2),\ \tilde{c}_2^{sf} = (0.8, 0.4, 0.1),\ \tilde{c}_3^{sf} = (0.8, 0.2, 0.3)
$$

The crisp version of the SFLPP (Model I) can be obtained as follows:
Maximize $Z = 0.6x_1 + 0.76x_2 + 0.76x_3$
Subject to

$$
\begin{aligned}
&0.2x_1 + 0.4x_2 + 0.6x_3 \leq 12 \\
&0.5x_1 + 0.4x_2 + 0.7x_3 \leq 15 \\
&0.3x_1 + 0.8x_2 + 0.2x_3 \leq 16 \\
&x_1,\ x_2,\ x_3 \geq 0
\end{aligned}
$$

The solution results are as follows: $x_1 = 25,\ x_2 = 10,\ x_3 = 0$ and Max $Z = 25$.

Example 2 (Model I) Maximize $Z = \tilde{c}_1^{sf} x_1 + \tilde{c}_2^{sf} x_2 + \tilde{c}_3^{sf} x_3$

Subject to

$$
\begin{aligned}
&0.2x_1 + 0.4x_2 + 0.6x_3 \leq 12 \\
&0.5x_1 + 0.4x_2 + 0.7x_3 \leq 15 \\
&0.3x_1 + 0.8x_2 + 0.2x_3 \leq 16 \\
&x_1,\ x_2,\ x_3 \geq 0
\end{aligned}
$$

where

$$\tilde{c}_1^{sf} = (0.86, 0.52, 0.27),\ \tilde{c}_2^{sf} = (0.68, 0.42, 0.51),\ \tilde{c}_3^{sf} = (0.48, 0.62, 0.53)$$

The crisp version of the SFLPP (Model I) can be obtained as follows:
Maximize $Z = 0.28x_1 + 0.56x_2 + 0.49x_3$
Subject to

$$\begin{aligned}
&0.2x_1 + 0.4x_2 + 0.6x_3 \le 12\\
&0.5x_1 + 0.4x_2 + 0.7x_3 \le 15\\
&0.3x_1 + 0.8x_2 + 0.2x_3 \le 16\\
&x_1,\ x_2,\ x_3 \ge 0
\end{aligned}$$

The solution results are as follows: $x_1 = 16,\ x_2 = 19,\ x_3 = 07$ and Max $Z = 18.55$.

Example 3 (Model II) Maximize $Z = 60x_1 + 40x_2 + 30x_3$

Subject to

$$\begin{aligned}
&\tilde{a}_{11}^{sf}x_1 + \tilde{a}_{12}^{sf}x_2 + \tilde{a}_{13}^{sf}x_3 \le \tilde{b}_1^{sf}\\
&\tilde{a}_{21}^{sf}x_1 + \tilde{a}_{22}^{sf}x_2 + \tilde{a}_{23}^{sf}x_3 \le \tilde{b}_2^{sf}\\
&\tilde{a}_{31}^{sf}x_1 + \tilde{a}_{32}^{sf}x_2 + \tilde{a}_{33}^{sf}x_3 \ge \tilde{b}_3^{sf}\\
&x_1,\ x_2,\ x_3 \ge 0
\end{aligned}$$

where

$$\begin{aligned}
&\tilde{a}_{11}^{sf} = (0.8, 0.3, 0.2),\ \tilde{a}_{12}^{sf} = (0.5, 0.4, 0.3),\ \tilde{a}_{13}^{sf} = (0.4, 0.3, 0.4)\\
&\tilde{a}_{21}^{sf} = (0.4, 0.8, 0.2),\ \tilde{a}_{22}^{sf} = (0.6, 0.5, 0.2),\ \tilde{a}_{23}^{sf} = (0.5, 0.3, 0.6)\\
&\tilde{a}_{31}^{sf} = (0.7, 0.1, 0.7),\ \tilde{a}_{32}^{sf} = (0.7, 0.5, 0.1),\ \tilde{a}_{33}^{sf} = (0.7, 0.3, 0.5)\\
&\tilde{b}_1^{sf} = (0.9, 0.3, 0.2),\ \tilde{b}_2^{sf} = (0.8, 0.4, 0.2),\ \tilde{b}_3^{sf} = (0.7, 0.2, 0.3)
\end{aligned}$$

The equivalent crisp formulation of SFLPP (Model II) can be given as follows:
Maximize $Z = 60x_1 + 40x_2 + 30x_3$
Subject to

$$\begin{aligned}
&0.76x_1 + 0.60x_2 + 0.56x_3 \le 0.8\\
&0.46x_1 + 0.63x_2 + 0.53x_3 \le 0.73\\
&0.63x_1 + 0.70x_2 + 0.63x_3 \ge 0.73\\
&x_1,\ x_2,\ x_3 \ge 0
\end{aligned}$$

The above crisp problem is solved using the same solvers and solution results are summarized as follows: $x_1 = 0.79,\ x_2 = 0.32,\ x_3 = 0$ and Max $Z = 60.72$.

Example 4 (Model II) Maximize $Z = 60x_1 + 40x_2 + 30x_3$

Subject to

$$\begin{aligned} &\tilde{a}_{11}^{sf} x_1 + \tilde{a}_{12}^{sf} x_2 + \tilde{a}_{13}^{sf} x_3 \leq \tilde{b}_1^{sf} \\ &\tilde{a}_{21}^{sf} x_1 + \tilde{a}_{22}^{sf} x_2 + \tilde{a}_{23}^{sf} x_3 \leq \tilde{b}_2^{sf} \\ &\tilde{a}_{31}^{sf} x_1 + \tilde{a}_{32}^{sf} x_2 + \tilde{a}_{33}^{sf} x_3 \geq \tilde{b}_3^{sf} \\ &x_1,\ x_2,\ x_3 \geq 0 \end{aligned}$$

where

$$\begin{aligned} &\tilde{a}_{11}^{sf} = (0.82, 0.53, 0.62),\ \tilde{a}_{12}^{sf} = (0.56, 0.74, 0.83),\ \tilde{a}_{13}^{sf} = (0.74, 0.93, 0.44) \\ &\tilde{a}_{21}^{sf} = (0.48, 0.85, 0.26),\ \tilde{a}_{22}^{sf} = (0.56, 0.58, 0.52),\ \tilde{a}_{23}^{sf} = (0.15, 0.93, 0.36) \\ &\tilde{a}_{31}^{sf} = (0.74, 0.51, 0.27),\ \tilde{a}_{32}^{sf} = (0.57, 0.54, 0.81),\ \tilde{a}_{33}^{sf} = (0.57, 0.63, 0.45) \\ &\tilde{b}_1^{sf} = (0.89, 0.93, 0.42),\ \tilde{b}_2^{sf} = (0.68, 0.84, 0.62),\ \tilde{b}_3^{sf} = (0.57, 0.82, 0.39) \end{aligned}$$

The equivalent crisp formulation of SFLPP (Model II) can be given as follows:
Maximize $Z = 60x_1 + 40x_2 + 30x_3$
Subject to

$$\begin{aligned} &0.46x_1 + 0.34x_2 + 0.61x_3 \leq 0.79 \\ &0.61x_1 + 0.32x_2 + 0.23x_3 \leq 0.54 \\ &0.13x_1 + 0.40x_2 + 0.33x_3 \geq 0.87 \\ &x_1,\ x_2,\ x_3 \geq 0 \end{aligned}$$

The above crisp problem is solved using the same solvers and solution results are summarized as follows: $x_1 = 0.19,\ x_2 = 0.12,\ x_3 = 0.14$ and Max $Z = 20.40$.

Example 5 (Model III) Maximize $Z = \tilde{c}_1^{sf} x_1 + \tilde{c}_2^{sf} x_2 + \tilde{c}_3^{sf} x_3$

Subject to

$$\begin{aligned} &\tilde{a}_{11}^{sf} x_1 + \tilde{a}_{12}^{sf} x_2 + \tilde{a}_{13}^{sf} x_3 \leq \tilde{b}_1^{sf} \\ &\tilde{a}_{21}^{sf} x_1 + \tilde{a}_{22}^{sf} x_2 + \tilde{a}_{23}^{sf} x_3 \leq \tilde{b}_2^{sf} \\ &\tilde{a}_{31}^{sf} x_1 + \tilde{a}_{32}^{sf} x_2 + \tilde{a}_{33}^{sf} x_3 \geq \tilde{b}_3^{sf} \\ &x_1,\ x_2,\ x_3 \geq 0 \end{aligned}$$

where

$$\begin{aligned} &\tilde{c}_1^{sf} = (0.7, 0.7, 0.2),\ \tilde{c}_2^{sf} = (0.9, 0.4, 0.2),\ \tilde{c}_3^{sf} = (0.7, 0.2, 0.4) \\ &\tilde{a}_{11}^{sf} = (0.6, 0.3, 0.1),\ \tilde{a}_{12}^{sf} = (0.6, 0.4, 0.4),\ \tilde{a}_{13}^{sf} = (0.2, 0.3, 0.5) \end{aligned}$$

$$\tilde{a}_{21}^{sf} = (0.3, 0.9, 0.2),\ \tilde{a}_{22}^{sf} = (0.7, 0.5, 0.3),\ \tilde{a}_{23}^{sf} = (0.6, 0.3, 0.5)$$
$$\tilde{a}_{31}^{sf} = (0.8, 0.2, 0.7),\ \tilde{a}_{32}^{sf} = (0.8, 0.5, 0.4),\ \tilde{a}_{33}^{sf} = (0.7, 0.4, 0.5)$$
$$\tilde{b}_{1}^{sf} = (0.8, 0.3, 0.3),\ \tilde{b}_{2}^{sf} = (0.6, 0.4, 0.3),\ \tilde{b}_{3}^{sf} = (0.7, 0.6, 0.3)$$

The crisp formulation of the fully SFLPP (Model III) can be represented as follows:

Maximize $Z = 0.62x_1 + 0.43x_2 + 0.53x_3$

Subject to

$$\begin{aligned} &0.72x_1 + 0.61x_2 + 0.57x_3 \leq 0.80\\ &0.45x_1 + 0.63x_2 + 0.51x_3 \leq 0.79\\ &0.61x_1 + 0.72x_2 + 0.43x_3 \geq 0.77\\ &x_1,\ x_2,\ x_3 \geq 0 \end{aligned}$$

The solution result is obtained as follows: $x_1 = 0$, $x_2 = 0.64$, $x_3 = 0.72$ and Max $Z = 0.65$.

Example 6 (Model III) Maximize $Z = \tilde{c}_1^{sf} x_1 + \tilde{c}_2^{sf} x_2 + \tilde{c}_3^{sf} x_3$

Subject to

$$\begin{aligned} &\tilde{a}_{11}^{sf} x_1 + \tilde{a}_{12}^{sf} x_2 + \tilde{a}_{13}^{sf} x_3 \leq \tilde{b}_1^{sf}\\ &\tilde{a}_{21}^{sf} x_1 + \tilde{a}_{22}^{sf} x_2 + \tilde{a}_{23}^{sf} x_3 \leq \tilde{b}_2^{sf}\\ &\tilde{a}_{31}^{sf} x_1 + \tilde{a}_{32}^{sf} x_2 + \tilde{a}_{33}^{sf} x_3 \geq \tilde{b}_3^{sf}\\ &x_1,\ x_2,\ x_3 \geq 0 \end{aligned}$$

where

$$\tilde{c}_{1}^{sf} = (0.71, 0.76, 0.42),\ \tilde{c}_{2}^{sf} = (0.92, 0.94, 0.82),\ \tilde{c}_{3}^{sf} = (0.47, 0.42, 0.64)$$
$$\tilde{a}_{11}^{sf} = (0.76, 0.38, 0.41),\ \tilde{a}_{12}^{sf} = (0.68, 0.45, 0.49),\ \tilde{a}_{13}^{sf} = (0.62, 0.73, 0.85)$$
$$\tilde{a}_{21}^{sf} = (0.38, 0.99, 0.82),\ \tilde{a}_{22}^{sf} = (0.78, 0.8, 0.73),\ \tilde{a}_{23}^{sf} = (0.6, 10.39, 0.58)$$
$$\tilde{a}_{31}^{sf} = (0.88, 0.72, 0.77),\ \tilde{a}_{32}^{sf} = (0.58, 0.65, 0.74),\ \tilde{a}_{33}^{sf} = (0.87, 0.47, 0.85)$$
$$\tilde{b}_{1}^{sf} = (0.85, 0.36, 0.37),\ \tilde{b}_{2}^{sf} = (0.68, 0.46, 0.39),\ \tilde{b}_{3}^{sf} = (0.79, 0.64, 0.34)$$

The crisp formulation of the fully SFLPP (Model III) can be represented as follows:

Maximize $Z = 0.24x_1 + 0.73x_2 + 0.13x_3$

Subject to

$$\begin{aligned} &0.32x_1 + 0.23x_2 + 0.17x_3 \leq 0.60\\ &0.15x_1 + 0.13x_2 + 0.11x_3 \leq 0.89 \end{aligned}$$

$$0.21x_1 + 0.22x_2 + 0.31x_3 \geq 0.57$$
$$x_1,\ x_2,\ x_3 \geq 0$$

The solution result is obtained as follows: $x_1 = 0.32,\ x_2 = 0.14,\ x_3 = 0.52$ and Max $Z = 0.86$.

Example 7 Production Planning Problem A cell phone manufacturing company produces four different types of phones based on their appearance. Each type of cell phone passed through four various service facilities, namely; Soldering, Assembling, Polishing, and Inspection, respectively. The finished products are delivered to the market after completion of these service facilities. The uncertain data has been provided by the manager of the company and represented by a spherical fuzzy number according to the nature of the data. The stipulated time for each product at each service facility and its expected profit is presented in Table 1. The restriction over the monthly production quantity is also depicted in Table 2. The decision-maker(s) or manager(s) of the company intends to manufacture the products in such a manner that the total monthly expected profit is maximized. To capture the uncertainty, some parameters, such as expected profit, time, and production level, are represented by the spherical fuzzy number.

The degree of confirmation based on the previous knowledge of decision maker is given as $(0.9, 0.7, 0.6)$ for the positive, neutral and negative memberships of spherical fuzzy numbers respectively.

Let the number of P_1 type cell phone $= x_1$
Let the number of P_2 type cell phone be $= x_2$
Let the number of P_3 type cell phone be $= x_3$
Let the number of P_4 type cell phone be $= x_4$

Table 1 Service facility

Types of product	Soldering	Assembling	Polishing	Inspection	Expected profit
P_1	9	12	4	11	$(22)^{sf}$
P_2	5	14	1	10	$(28)^{sf}$
P_3	7	13	2	8	$(23)^{sf}$
P_4	6	16	3	15	$(27)^{sf}$

Table 2 Time and production limit level

Service facility	Time (in h)	Types of product	Minimum production level
Soldering	$(1200)^{sf}$	P_1	$(125)^{sf}$
Assembling	$(1450)^{sf}$	P_2	$(165)^{sf}$
Polishing	$(1100)^{sf}$	P_3	$(130)^{sf}$
Inspection	$(1500)^{sf}$	P_4	$(115)^{sf}$

Maximize $Z = (22)^{sf} x_1 + (28)^{sf} x_2 + (23)^{sf} x_3 + (27)^{sf} x_4$
Subject to

$$
\begin{aligned}
&9x_1 + 12x_2 + 4x_3 + 11x_4 \le (1200)^{sf} \\
&5x_1 + 14x_2 + 1x_3 + 10x_4 \le (1450)^{sf} \\
&7x_1 + 13x_2 + 2x_3 + 8x_4 \le (1100)^{sf} \\
&6x_1 + 16x_2 + 3x_3 + 15x_4 \le (1500)^{sf} \\
&x_1 \ge (125)^{sf} \\
&x_2 \ge (165)^{sf} \\
&x_3 \ge (130)^{sf} \\
&x_4 \ge (115)^{sf} \\
&x_1,\ x_2,\ x_3,\ x_4 \ge 0
\end{aligned}
$$

where

$$
\begin{aligned}
&(22)^{sf} = (24, 22, 20),\ \ (28)^{sf} = (30, 28, 26) \\
&(23)^{sf} = (26, 23, 20),\ \ (27)^{sf} = (29, 27, 25) \\
&(1200)^{sf} = (1250, 1200, 1150),\ \ (1450)^{sf} = (1500, 1450, 1400) \\
&(1100)^{sf} = (1150, 1100, 1050),\ \ (1500)^{sf} = (1550, 1500, 1450) \\
&(125)^{sf} = (130, 125, 120),\ \ (165)^{sf} = (170, 165, 160) \\
&(130)^{sf} = (135, 130, 125),\ \ (115)^{sf} = (120, 115, 110)
\end{aligned}
$$

Using the positive, neutral and negative membership functions of the spherical fuzzy number, the crisp formulation of the SFLPP can be obtained as follows:
Maximize $Z = 8x_1 + 12x_2 + 9x_3 + 11x_4$
Subject to

$$
\begin{aligned}
&9x_1 + 12x_2 + 4x_3 + 11x_4 \le 1750 \\
&5x_1 + 14x_2 + 1x_3 + 10x_4 \le 2300 \\
&7x_1 + 13x_2 + 2x_3 + 8x_4 \le 1950 \\
&6x_1 + 16x_2 + 3x_3 + 15x_4 \le 2250 \\
&x_1 \ge 26 \\
&x_2 \ge 32 \\
&x_3 \ge 28 \\
&x_4 \ge 18
\end{aligned}
$$

The optimal quantities of different types of cell-phones that must be produced are $x_1 = 26,\ x_2 = 32,\ x_3 = 233,\ x_4 = 18$ with the expected total profit Max $Z = 2891$.

Table 3 Purchasing cost and demand and supply capacities

Farm	Price (\$/ton) of fruits type					Maximum (of all 1 2 3 4 5 types combined they can supply)
	1 (Apple)	2 (Mango)	3 (Pears)	4 (Orange)	5 (Grapes)	
1	$(200)^{sf}$	$(600)^{sf}$	$(1600)^{sf}$	$(800)^{sf}$	$(1200)^{sf}$	$(180)^{sf}$
2	$(300)^{sf}$	$(550)^{sf}$	$(1400)^{sf}$	$(850)^{sf}$	$(1100)^{sf}$	$(200)^{sf}$
3	$(250)^{sf}$	$(600)^{sf}$	$(1500)^{sf}$	$(700)^{sf}$	$(1000)^{sf}$	$(100)^{sf}$
4	$(150)^{sf}$	$(500)^{sf}$	$(1700)^{sf}$	$(900)^{sf}$	$(1300)^{sf}$	$(120)^{sf}$
Minimum amount required (tons)	$(100)^{sf}$	$(60)^{sf}$	$(20)^{sf}$	$(80)^{sf}$	$(40)^{sf}$	

Example 8 Purchasing Problem A fruit seller wants to buy five different types of fruits from four farms in a month. Based on the imprecise information provided by fruit sellers, the best representation of the data has been taken in the form of spherical fuzzy parameters. The prices of the vegetables at different farms, the capacities of the farms, and the minimum requirements of the grocery store are summarized in Table 3. The fruit seller intends to adopt the optimal purchasing policy that minimizes the total purchasing cost without affecting the minimum requirements at the grocery. To capture the uncertainty, all the parameters such as expected price, supply, and demand level are represented by the spherical fuzzy number.

The degree of confirmation based on the previous knowledge of decision maker is given as $(0.7, 0.9, 0.8)$ for the positive, neutral and negative memberships of spherical fuzzy numbers respectively. Let x_{ij} be the amount of each types of fruits purchased from different firms; where i $(i = 1, 2, 3, 4)$ and j $(j = 1, 2, 3, 4, 5)$ represents the types of fruits and different firms respectively.

Minimize $Z = \sum_{i=1}^{4} \sum_{j=1}^{5} (c_{ij})^{sf} x_{ij}$

Subject to

$$\sum_{j=1}^{5} (c_{1j})^{sf} x_{1j} \leq (180)^{sf}$$

$$\sum_{j=1}^{5} (c_{2j})^{sf} x_{2j} \leq (200)^{sf}$$

$$\sum_{j=1}^{5} (c_{3j})^{sf} x_{3j} \leq (100)^{sf}$$

$$\sum_{j=1}^{5} (c_{4j})^{sf} x_{4j} \leq (120)^{sf}$$

$$\sum_{i=1}^{4} (c_{i1})^{sf} x_{i1} \geq (100)^{sf}$$
$$\sum_{i=1}^{4} (c_{i2})^{sf} x_{i2} \geq (60)^{sf}$$
$$\sum_{i=1}^{4} (c_{i3})^{sf} x_{i3} \geq (20)^{sf}$$
$$\sum_{i=1}^{4} (c_{i4})^{sf} x_{i4} \geq (80)^{sf}$$
$$\sum_{i=1}^{4} (c_{i5})^{sf} x_{i5} \geq (40)^{sf}$$
$$x_{ij} \geq 0$$

where

$$\begin{aligned}
&(200)^{sf} = (150, 200, 250),\ (300)^{sf} = (270, 300, 330),\\
&(250)^{sf} = (240, 250, 260),\ (150)^{sf} = (140, 150, 160)\\
&(100)^{sf} = (70, 90, 110),\ (600)^{sf} = (550, 600, 650),\\
&(550)^{sf} = (550, 600, 650),\ (20)^{sf} = (10, 20, 30)\\
&(500)^{sf} = (450, 550, 650),\ (60)^{sf} = (50, 60, 70),\\
&(1600)^{sf} = (1550, 1600, 1650)\\
&(1400)^{sf} = (1300, 1400, 1500),\ (1500)^{sf} = (1400, 1500, 1600),\\
&(1700)^{sf} = (1600, 1700, 1800)\\
&(120)^{sf} = (110, 120, 130),\ (800)^{sf} = (700, 800, 900),\\
&(850)^{sf} = (800, 850, 900),\ (700)^{sf} = (600, 700, 800)\\
&(900)^{sf} = (800, 900, 1000),\ (80)^{sf} = (70, 80, 90),\\
&(1200)^{sf} = (1100, 1200, 1300),\ (1100)^{sf} = (1000, 1100, 1200)\\
&(1000)^{sf} = (950, 1000, 1050),\ (1300)^{sf} = (1200, 1300, 1400),\\
&(40)^{sf} = (35, 40, 45),\ (180)^{sf} = (170, 180, 190)
\end{aligned}$$

Using the positive, neutral and negative membership functions of the spherical fuzzy number, the crisp formulation of the SFLPP can be obtained as follows:

Minimize $Z = \sum_{i=1}^{4} \sum_{j=1}^{5} c_{ij} x_{ij}$

Subject to

$$\sum_{j=1}^{5} c_{1j}x_{1j} \leq 113$$

$$\sum_{j=1}^{5} c_{2j}x_{2j} \leq 126$$

$$\sum_{j=1}^{5} c_{3j}x_{3j} \leq 75$$

$$\sum_{j=1}^{5} c_{4j}x_{4j} \leq 83$$

$$\sum_{i=1}^{4} c_{i1}x_{i1} \geq 52$$

$$\sum_{i=1}^{4} c_{i2}x_{i2} \geq 23$$

$$\sum_{i=1}^{4} c_{i3}x_{i3} \geq 16$$

$$\sum_{i=1}^{4} c_{i4}x_{i4} \geq 47$$

$$\sum_{i=1}^{4} c_{i5}x_{i5} \geq 19$$

$$x_{ij} \geq 0$$

The optimal quantities of different types of fruits that must be purchased are
$x_{11} = 17,\ x_{12} = 09,\ x_{13} = 13,\ x_{14} = 12,\ x_{15} = 18,$

$x_{21} = 17, x_{22} = 19,\ x_{23} = 03,\ x_{24} = 02,\ x_{25} = 08,$

$x_{31} = 15,\ x_{32} = 19,\ x_{33} = 16,\ x_{34} = 22,\ x_{35} = 08,$

$x_{41} = 13,\ x_{42} = 09,\ x_{43} = 18,\ x_{44} = 23,\ x_{45} = 11,$

$x_{51} = 02,\ x_{52} = 05,\ x_{53} = 03,\ x_{54} = 07,\ x_{55} = 03$ with the expected total purchasing costs Min $Z = 1362$.

6 Conclusions

Recently, the spherical fuzzy set is introduced, and it is the first time when LPP is discussed in the spherical fuzzy environment. The ample opportunity to capture the imprecise and inconsistent information by SFS leads to a significant contribution to decision-making problems. Thus, in this chapter, we introduce SFLPP under a spherical fuzzy environment, which comprises the maximization of positive and minimization of neutral and negative membership functions into the spherical fuzzy decision set. The deterministic version of SFLPP is obtained with the help of positive, neutral, and negative membership degrees, respectively. Furthermore, the various numerical illustrations are presented to show the applicability of the proposed SFLPP solution approach. The SFLPP is also applied to two different real-life problems, such as production planning problems and purchasing strategy problems.

In the future, many research works can be explored in the spherical fuzzy domain by introducing the duality theory of LPP under a spherical fuzzy environment. Furthermore, the presented approach can be extended for nonlinear programming problems, fractional programming problems, etc. The implication of SFLPP in real-life problems is also an open door for researchers such as transportation problems, supplier selection problems, supply chain, inventory control, and portfolio management, etc.

References

Ahmad F, Adhami AY (2019) Neutrosophic programming approach to multiobjective nonlinear transportation problem with fuzzy parameters. Int J Manag Sci Eng Manag 14(3):218–229

Ahmad F, Adhami AY, Smarandache F (2018) Single valued neutrosophic hesitant fuzzy computational algorithm for multiobjective nonlinear optimization problem. Neutrosophic Sets Syst 22:76–86

Ahmad F, Adhami AY, Smarandache F (2019) Neutrosophic optimization model and computational algorithm for optimal shale gas water management under uncertainty. Symmetry 11(4):544

Ahmad F, Adhami AY, Smarandache F (2020) Modified neutrosophic fuzzy optimization model for optimal closed-loop supply chain management under uncertainty. In: Optimization theory based on neutrosophic and plithogenic sets. Academic Press, pp 343–403

Angelov PP (1997) Optimization in an intuitionistic fuzzy environment. Fuzzy Sets Syst 86(3):299–306

Atanassov KT (2017) Intuitionistic fuzzy logics. Springer International Publishing

Bellman RE, Zadeh LA (1970) Decision-making in a fuzzy environment. Manage Sci 17(4):140–164

Czyzyk J, Mesnier MP, Moré JJ (1998) The NEOS server. IEEE J Comput Sci Eng 5(3):68–75

Kutlu Gündoğdu F, Kahraman C (2019) Spherical fuzzy sets and spherical fuzzy TOPSIS method. J Intell Fuzzy Syst (Preprint):1–16

Rafiq M, Ashraf S, Abdullah S, Mahmood T, Muhammad S (2019) The cosine similarity measures of spherical fuzzy sets and their applications in decision making. J Intell Fuzzy Syst (Preprint):1–15

Rizk-Allah RM, Hassanien AE, Elhoseny M (2018) A multi-objective transportation model under neutrosophic environment. Comput Electr Eng 69:705–719

Smarandache F (1999) A unifying field in logics: neutrosophic logic. Philosophy. American Press

Yager RR (2013) Pythagorean fuzzy subsets, vol 2. IEEE, pp 57–61

Zimmermann HJ (1978) Fuzzy programming and linear programming with several objective functions. Fuzzy Sets Syst 1(1):45–55

Spherical Fuzzy Multiobjective Linear Programming Problem

Firoz Ahmad and Ahmad Yusuf Adhami

Abstract Multiobjective optimization techniques are an essential and burning topic for the last few decades. The significant development can be realized by having enormous researches in this field, such as the fuzzy technique, an intuitionistic fuzzy optimization technique, different goal programming techniques, etc. Thus this chapter investigates a new algorithm based on the spherical fuzzy set (SF) named as the spherical fuzzy multiobjective programming problem (SFMOLPP) under the spherical fuzzy environment. The SFMOLPP inevitably involves the degree of neutrality along with positive and negative membership degrees of the element into the feasible solution set. It also generalizes the decision set by imposing the restriction that the sum of squares of each membership function must be less than or equal to one. The attainment of achievement function is determined by maximizing the positive membership function and minimization of neutral and negative membership function of each objective function under the spherical fuzzy decision set. At last, numerical examples and the conclusion are presented to reveal the applicability and future research scope in the SF domain.

1 Introduction

In mathematical programming problems, the multiobjective linear programming problem (MOLPP) is the most encountered problem in real-life. In a multiobjective programming problem, more than one conflicting objective is to be optimized under a given set of well-defined constraints. It may not always be possible to obtain a single solution that satisfies each objective function efficiently; however, a compromise solution can be determined. Literature suggests various approaches for MOLPP under different conditions. Zimmermann (1978) proposed a fuzzy programming technique to solve MOLPP in a fuzzy decision set. Later on, Angelov (1997) also investigated the intuitionistic fuzzy programming approach for MOLPP, which considers

F. Ahmad (✉) · A. Y. Adhami
Department of Statistics and Operations Research, Aligarh Muslim University, Aligarh, India
e-mail: firoz.ahmad02@gmail.com

© Springer Nature Switzerland AG 2021

C. Kahraman and F. Kutlu Gündoğdu (eds.), *Decision Making with Spherical Fuzzy Sets*, Studies in Fuzziness and Soft Computing 392,
https://doi.org/10.1007/978-3-030-45461-6_20

the degree of belongingness and non-belongingness of the solution set into the intuitionistic fuzzy decision set. In the past few years, lots of efforts are being made in the domain of different robust techniques for MOLPP. Various real-life problems such as transportation problems, supplier selection problems, inventory control, etc. inherently takes the form of the multiobjective optimization problem.

Based on the different uncertain sets, such as fuzzy set (Bellman and Zadeh (1970)), intuitionistic fuzzy set (Atanassov (1986)), neutrosophic set (Smarandache (1999), Ahmad and Adhami (2019), Ahmad et al. (2019), Ahmad et al. (2018)) etc. different multiobjective optimization techniques have been developed.. Moreover, the neutrosophic set (NS) is also proposed by Smarandache (1999), which is inspired by indeterminacy or neutral thoughts in the decision-making process. Ahmad and Adhami (2019) presented a study on nonlinear transportation problems under the neutrosophic environment. Ahmad et al. (2018) proposed a new algorithm named neutrosophic hesitant fuzzy programming approach based on the single-valued neutrosophic hesitant fuzzy decision set and applied it to the manufacturing system. Ahmad et al. (2019) also discussed the neutrosophic optimization technique for the optimal shale gas water management system under uncertainty. The decision-making problem can be classified into two broad cases, such as discrete and continuous. In discrete case, multi-attribute and multi-criteria decision-making problem are dealt with whereas multiobjective optimization problem falls in continuous case. Recently, Ahmad et al. (2020) suggested modified neutrosophic optimization technique under uncertainty and successfully applied to supply chain planning problems.

The generalization of fuzzy set, intuitionistic fuzzy set and Pythagorean fuzzy set (Yager 2013) is presented by Kutlu Gündoğdu and Kahraman (2019) and termed as spherical fuzzy set (SFS) which is based on three different membership degrees named as positive, neutral and negative membership degrees of the element into the feasible solution set. The limitations of fuzzy, intuitionistic fuzzy, Pythagorean fuzzy sets can be described by the following explanations. Suppose that in multi-attribute decision-making problems, the acceptance degree of selecting an alternative is 0.6, the rejection of the options is 0.7, and the indeterminacy or neutral degree of choosing the alternative is found to be 0.8 then this situation is beyond the coverage of above discussed uncertain sets. Hence to deal with such situations, the spherical fuzzy set is a powerful decision-making tool under the spherical fuzzy environment (Rafiq et al. 2019). It also allows the decision-makers to incorporate indeterminacy and neutral thoughts in decision-making processes.

The remaining part of the chapter is summarized as follows: In Sect. 2, basic definition regarding spherical fuzzy set is presented whereas Sect. 3 describes the spherical fuzzy multiobjective programming problem and solution algorithm. Section 4 presents the numerical examples whereas a conclusion along with the future research scope is described in Sect. 5.

2 Basic Definitions

In this section, some important definitions regarding spherical fuzzy set are discussed.

Definition 1 (*Kutlu Gündoğdu and Kahraman (*2019)) (Spherical fuzzy set) A spherical fuzzy set $\tilde{X}_s$ defined over the universe of discourse Y under the ordered

$$\tilde{X}_S = \left\{ \left\langle y,\ \left(\alpha_{\tilde{X}_S}(y),\ \beta_{\tilde{X}_S}(y),\ \gamma_{\tilde{X}_S}(y) \right) \mid y \in Y \right\rangle \right\}$$

Such that $\alpha_{\tilde{X}_S} : Y \to [0, 1],\ \beta_{\tilde{X}_S} : Y \to [0, 1],\ \gamma_{\tilde{X}_S} : Y \to [0, 1]$ and $0 \le \alpha^2_{\tilde{X}_S}(y) + \beta^2_{\tilde{X}_S}(y) + \gamma^2_{\tilde{X}_S}(y) \le 1 \quad \forall\, y \in Y$

where $\alpha_{\tilde{X}_S}(y),\ \beta_{\tilde{X}_S}(y)$ and $\gamma_{\tilde{X}_S}(y)$ represents the positive membership degree, neutral membership degree and negative membership degree for each element $y \in Y$ to $\tilde{X}_S$ respectively.

Definition 2 (*Kutlu Gündoğdu and Kahraman (*2019)) For any two SFSs $\tilde{X}_S = \left(\alpha_{\tilde{X}_S},\ \beta_{\tilde{X}_S},\ \gamma_{\tilde{X}_S} \right)$ and $\tilde{W}_S = \left(\alpha_{\tilde{W}_S},\ \beta_{\tilde{W}_S},\ \gamma_{\tilde{W}_S} \right)$, the followings are the condition that holds with $\lambda,\ \lambda_1,\ \lambda_2 > 0$.

i. $\tilde{X}_S \oplus \tilde{W}_S = \tilde{W}_S \oplus \tilde{X}_S$

ii. $\tilde{X}_S \otimes \tilde{W}_S = \tilde{W}_S \otimes \tilde{X}_S$

iii. $\lambda\left(\tilde{X}_S \oplus \tilde{W}_S \right) = \lambda\, \tilde{W}_S \oplus \lambda\, \tilde{X}_S$

iv. $\lambda_1 \tilde{X}_S \oplus \lambda_2\, \tilde{X}_S = (\lambda_1 + \lambda_2)\tilde{X}_S$

v. $\left(\tilde{X}_S \otimes \tilde{W}_S \right)^{\lambda} = \tilde{X}_s^{\lambda} \otimes \tilde{W}_s^{\lambda}$

vi. $\tilde{X}_s^{\lambda_1} \otimes \tilde{W}_s^{\lambda_2} = \tilde{X}_s^{\lambda_1 + \lambda_2}$

Definition 3 (*Optimal Solution*) A solution x^* is said to be an optimal solution to the multiobjective linear programming problem if and only if there exists $x^* \in X$ such that $Z_o(x^*) \le$ *or* $\ge Z_o(x)$ (for minimization or maximization case), for $o = 1, 2, \ldots, O$ and for all $x \in X$.

Definition 4 (*Pareto-Optimal Solution*) A solution x^* is said to be a Pareto-optimal solution to the multiobjective linear programming problem if and only if there does not exist another $x \in X$ such that $Z_o(x^*) \le$ *or* $\ge Z_o(x)$ (for minimization or maximization case) for $o = 1, 2, \ldots, O$ and $Z_o(x^*) \ne Z_o(x)$ for at least one $o,\ o \in (1, 2, \ldots, O)$.

3 Spherical Fuzzy Multiobjective Linear Programming Model

Based on the SFS (Kutlu Gündoğdu and Kahraman 2019), we have developed the spherical fuzzy multiobjective optimization techniques which characterize three memberships function, such as maximization of positive membership degree and minimization of neutral and negative membership degrees under spherical fuzzy environment. The spherical fuzzy decision set also provides an ample opportunity to express the neutral thoughts of the decision-maker(s) and consequently captures the behavior of indeterminacy in the decision-making problem. Fuzzy multiobjective programming approach considers the only maximization of membership function, and the intuitionistic fuzzy optimization technique deals with the maximization of membership grades and minimization of non-membership grades under the respective decision set. Furthermore, the situation arises when membership and non-membership degrees are somehow not a well representative of decision-makers (s) thoughts, and indeterminacy condition exists in some cases. Such a situation could be dealt with SFS efficiently. Therefore the spherical fuzzy programming approach assists in the decision-making process when there is the existence of some indeterminacy thoughts in the decision-making process.

The mathematical formulation of multiobjective linear programming problem can be summarized as follows:

$$\begin{aligned} &\text{Maximize } Z_o(x) = (Z_1, Z_2, Z_3, \ldots, Z_o) \ \ \forall o = 1, 2, \ldots, O_1 \\ &\text{Minimize } Z_o(x) = (Z_1, Z_2, Z_3, \ldots, Z_o) \ \ \forall o = O_1 + 1, \ \ O_1 + 2, \ldots, O \end{aligned}$$

Subject to

$$\begin{aligned} &g_i(x) \le b_i, \quad \forall\, i = 1, 2, \ldots, I_1, \\ &g_i(x) \ge b_i, \quad \forall\, i = I_1 + 1, I_1 + 2, \ldots, I_2, \\ &g_i(x) = b_i, \quad \forall\, i = I_2 + 1, I_2 + 2, \ldots, I. \\ &x = \left(x_1, x_2, \ldots, x_j\right) \in X, \ x \ge 0. \end{aligned} \tag{1}$$

where $Z_o(x)$ is the Oth objective function. $g_i(x)$, b_i are the real valued function and numbers respectively. $x = \left(x_1, x_2, \ldots, x_j\right)$ represents the set of decision variables.

According to (Bellman and Zadeh 1970), the fuzzy decision set (D) consists of fuzzy goals (Z) and fuzzy constraints (C) under fuzzy environment. The decision set has been widely used in decision making processes and can be stated as follows:

$$D = (Z \cap C) \tag{2}$$

Hence, based on the fuzzy decision set theory (Eq. 2), the spherical fuzzy decision set can be developed under spherical fuzzy environment. Consequently, the spherical fuzzy decision set (D_{sf}) (Eq. 3), spherical fuzzy goals (Z_o) and spherical fuzzy constraints (C_o) can be given as follows:

$$D_{sf} = \left(\bigcap_{o=1}^{O} Z_o\right)\left(\bigcap_{j=1}^{J} C_j\right) = (x,\ T_D(x),\ I_D(x),\ F_D(x)) \tag{3}$$

where

$$T_D(x) = \begin{Bmatrix} T_{D_1}(x), T_{D_2}(x), T_{D_3}(x), \ldots, T_{D_O}(x) \\ T_{C_1}(x), T_{C_2}(x), T_{C_3}(x), \ldots, T_{C_J}(x) \end{Bmatrix} \quad \forall\ x \in X$$

$$I_D(x) = \begin{Bmatrix} I_{D_1}(x), I_{D_2}(x), I_{D_3}(x), \ldots, I_{D_O}(x) \\ I_{C_1}(x), I_{C_2}(x), I_{C_3}(x), \ldots, I_{C_J}(x) \end{Bmatrix} \quad \forall\ x \in X$$

$$F_D(x) = \begin{Bmatrix} F_{D_1}(x), F_{D_2}(x), F_{D_3}(x), \ldots, F_{D_O}(x) \\ F_{C_1}(x), F_{C_2}(x), F_{C_3}(x), \ldots, F_{C_J}(x) \end{Bmatrix} \quad \forall\ x \in X \tag{4}$$

where $T_D(x)$, $I_D(x)$ and $F_D(x)$ are the positive membership function, neutral membership function and negative membership function under spherical fuzzy decision set D_{sf}.

To determine the marginal evaluation of each objective (Eq. 1), first we calculate the upper and lower bound for each objective function. Suppose that the obtained decision variables are $X^1, X^1, X^1, \ldots, X^o$. By substituting these values in each objective function, the mathematical expressions of upper and lower bound can be given as follows (Eq. 5):

$$U_o = \max\{Z_o(X^o)\} \text{ and } L_o = \min\{Z_o(X^o)\} \quad \forall o = 1, 2, \ldots, O \tag{5}$$

where U_o and L_o represents the upper and lower bound of Oth objectives. The upper and lower bounds for positive, neutral and negative membership functions for each objective under spherical fuzzy environment can be derived by using the following mathematical expressions:

$$\begin{aligned} &U_o^T = U_o, \quad L_o^T = L_o \\ &U_o^I = L_o^T + p_o(U_o^T - L_o^T), \quad L_o^I = L_o^T \\ &U_o^F = U_o^T, \quad L_o^F = L_o^T + q_o(U_o^T - L_o^T) \end{aligned} \tag{6}$$

where p_o and q_o are the predetermined real numbers such that $p_o,\ q_o \in (0,\ 1)$ assigned by decision maker.

Case I When the objective function are of minimization type then positive, neutral and negative membership functions for each objective under spherical fuzzy environment can be obtained as follows:

$$T_D(x) = \begin{cases} 1, & if \;\; Z_o(x) \leq L_o^T \\ 1 - \frac{Z_o(x) - L_o^T}{U_o^T - L_o^T}, & if \;\; L_o^T \leq Z_o(x) \leq U_o^T \\ 0, & if \;\; Z_o^T \geq U_o^T \end{cases}$$

$$I_D(x) = \begin{cases} 1, & if \;\; Z_o(x) \leq L_o^I \\ 1 - \frac{Z_o(x) - L_o^I}{U_o^I - L_o^I}, & if \;\; L_o^I \leq Z_o(x) \leq U_o^I \\ 0, & if \;\; Z_o^I \geq U_o^I \end{cases}$$

$$F_D(x) = \begin{cases} 1, & if \;\; Z_o(x) \geq U_o^F \\ 1 - \frac{U_o^F - Z_o(x)}{U_o^F - L_o^F}, & if \;\; L_o^F \leq Z_o(x) \leq U_o^F \\ 0, & if \;\; Z_o^T \leq L_o^T \end{cases} \tag{7}$$

Case II When the objective function are of minimization type then positive, neutral and negative membership functions for each objective under spherical fuzzy environment can be obtained as follows:

$$T_D(x) = \begin{cases} 0, & if \;\; Z_o(x) \leq L_o^T \\ 1 - \frac{U_o^T - Z_o(x)}{U_o^T - L_o^T}, & if \;\; L_o^T \leq Z_o(x) \leq U_o^T \\ 1, & if \;\; Z_o^T \geq U_o^T \end{cases}$$

$$I_D(x) = \begin{cases} 0, & if \;\; Z_o(x) \leq L_o^I \\ 1 - \frac{U_o^I - Z_o(x)}{U_o^I - L_o^I}, & if \;\; L_o^I \leq Z_o(x) \leq U_o^I \\ 1, & if \;\; Z_o^I \geq U_o^I \end{cases}$$

$$F_D(x) = \begin{cases} 0, & if \;\; Z_o(x) \geq U_o^F \\ 1 - \frac{Z_o(x) - L_o^F}{U_o^F - L_o^F}, & if \;\; L_o^F \leq Z_o(x) \leq U_o^F \\ 1, & if \;\; Z_o^T \leq L_o^T \end{cases} \tag{8}$$

where $U_o^{(.)} \neq L_o^{(.)}$ must hold for all objectives. If for any objective $U_o^{(.)} = L_o^{(.)}$ then the membership value will be one. The spherical fuzzy decision making model based on (Bellman and Zadeh 1970) concept for MOPP can be given as follows (Eq. 9):

$$Max \quad \min_{o=1,2,\ldots,O} \quad T_o(Z_o(x))^2$$

$$Min \quad \max_{o=1,2,\ldots,O} \quad I_o(Z_o(x))^2$$

$$Min \quad \max_{o=1,2,\ldots,O} \quad F_o(Z_o(x))^2$$

Subject to

$$g_i(x) \leq b_i, \quad \forall \; i = 1, 2, \ldots, I_1,$$

$$g_i(x) \geq b_i, \quad \forall \; i = I_1 + 1, I_1 + 2, \ldots, I_2,$$

$$\begin{aligned}
&g_i(x) = b_i, \quad \forall\ i = I_2 + 1, I_2 + 2, \ldots, I \\
&x = \left(x_1, x_2, \ldots, x_j\right) \in X, \quad x \geq 0 \\
&T_o(Z_o(x))^2 \geq I_o(Z_o(x))^2, \quad T_o(Z_o(x))^2 \geq F_o(Z_o(x))^2 \\
&0 \leq T_o(Z_o(x))^2 + I_o(Z_o(x))^2 + F_o(Z_o(x))^2 \leq 1
\end{aligned} \tag{9}$$

Using the auxiliary variables, the above mathematical model (Eq. 9) can be reformulated as follows (Eq. 10):

$$\begin{aligned}
&Max\ \ \alpha^2 \\
&Min\ \ \beta^2 \\
&Min\ \ \gamma^2
\end{aligned}$$

Subject to

$$\begin{aligned}
&T_o(Z_o(x))^2 \geq \alpha^2, \quad I_o(Z_o(x))^2 \leq \beta^2, \quad F_o(Z_o(x))^2 \leq \gamma^2 \\
&g_i(x) \leq b_i, \quad \forall\ i = 1, 2, \ldots, I_1, \\
&g_i(x) \geq b_i, \quad \forall\ i = I_1 + 1, I_1 + 2, \ldots, I_2, \\
&g_i(x) = b_i, \quad \forall\ i = I_2 + 1, I_2 + 2, \ldots, I. \\
&x = \left(x_1, x_2, \ldots, x_j\right) \in X, \quad x \geq 0 \\
&\alpha^2 \geq \beta^2, \quad \alpha^2 \geq \gamma^2, \quad 0 \leq \alpha^2 + \beta^2 + \gamma^2 \leq 1
\end{aligned} \tag{10}$$

After simplifying the above model (Eq. 10), the following mathematical spherical fuzzy multiobjective optimization model can be obtained as follows:

$$Max\left(\alpha^2 - \beta^2 - \gamma^2\right)$$

Subject to

$$\begin{aligned}
&T_o(Z_o(x))^2 \geq \alpha^2, \quad I_o(Z_o(x))^2 \leq \beta^2, \quad F_o(Z_o(x))^2 \leq \gamma^2 \\
&g_i(x) \leq b_i, \quad \forall\ i = 1, 2, \ldots, I_1, \\
&g_i(x) \geq b_i, \quad \forall\ i = I_1 + 1, I_1 + 2, \ldots, I_2, \\
&g_i(x) = b_i, \quad \forall\ i = I_2 + 1, I_2 + 2, \ldots, I. \\
&x = \left(x_1, x_2, \ldots, x_j\right) \in X, \quad x \geq 0 \\
&\alpha^2 \geq \beta^2, \quad \alpha^2 \geq \gamma^2, \quad 0 \leq \alpha^2 + \beta^2 + \gamma^2 \leq 1
\end{aligned} \tag{11}$$

Table 1 Pay-off matrix

	Z_1	Z_2	...	Z_o
X^1	$Z_1(X^1)$	$Z_2(X^1)$	...	$Z_o(X^1)$
X^2	$Z_1(X^2)$	$Z_2(X^2)$	...	$Z_o(X^2)$
...	...	...	...	...
...	...	...	...	...
...	...	...	...	...
X^o	$Z_1(X^o)$	$Z_2(X^o)$	...	$Z_o(X^o)$

Hence the above optimization model (Eq. 11) provides the optimal solution for the MOLPP under spherical fuzzy environment.

Solution steps

The stepwise algorithm to solve the MOLPP under spherical fuzzy environment can be summarized as follows:

Step I. Determine the pay-off matrix as given in Table 1 by solving each objective function individually and define the upper and lower bound for each objective.

Step II. Construct the upper and lower bound for positive, neutral and negative membership functions by using upper and lower bound given in Eq. (6).

Step III. Transform the different membership functions into linear membership function under spherical fuzzy environment.

Step IV. Formulate the spherical multiobjective programming problem of each objective under spherical fuzzy environment.

Step V. Solve the proposed optimization model by using either some suitable software or robust optimization techniques to obtain the compromise solution of MOLPP.

4 Numerical Examples

The following are the numerical example to show the validity and applicability of the proposed approach. The obtained spherical fuzzy model of all the numerical example has been coded in AMPL language and solved by using the Knitro optimization software online facility provided by University of Wisconsin (Dolan 2001).

Example 1 (Maximization type objectives)

$$\begin{aligned}\text{Maximize } Z_1 &= 50x_1 + 100x_2 + 17.5x_3\\ \text{Maximize } Z_2 &= 92x_1 + 75x_2 + 50x_3\\ \text{Maximize } Z_3 &= 25x_1 + 100x_2 + 75x_3\end{aligned}$$

Table 2 Individual best and worst value of objective functions

	Z_1	Z_2	Z_3	X
Maximum Z_1	8041	10,020.33	9319.25	X^1
Maximum Z_2	5452.63	10,950.59	5903.00	X^2
Maximum Z_3	7983.60	10,056.99	9355.90	X^3

Table 3 Individual best and worst values of objective functions

	Z_1	Z_2	Z_3	X
Minimum Z_1	468.21	625.95	625.95	X^1
Minimum Z_2	2375.62	1856.67	1856.67	X^2
Minimum Z_3	2077.77	927.85	927.85	X^3

Subject to

$$\begin{aligned}
&12x_1 + 17x_2 \le 1400\\
&3x_1 + 9x_2 + 8x_3 \le 1000\\
&10x_1 + 13x_2 + 15x_3 \le 1750\\
&6x_1 + 16x_3 \le 1325\\
&x_1, x_2, x_3 \ge 0.
\end{aligned}$$

On solving each objective function individually, we have obtained the following best and worst solution to define the upper and lower bound for each membership functions respectively (Table 2 and 3).

The upper and lower bound for the positive, neutral and negative membership degrees have been obtained as follows:

$$\begin{aligned}
&U_1^T = 8041, \quad L_1^T = 5452.63\\
&U_1^I = 5452.63 + p_1(8041 - 5452.63), \quad L_1^I = 5452.63\\
&U_1^F = 8041, \quad L_1^F = 5452.63 + q_1(8041 - 5452.63)
\end{aligned}$$

The different membership function for the first objective function is defined as follows:

$$T_D(Z_1(x)) = \begin{cases} 0, & if\ Z_1(x) \le L_1^T \\ 1 - \frac{U_1^T - Z_1(x)}{U_1^T - L_1^T}, & if\ L_1^T \le Z_1(x) \le U_1^T \\ 1, & if\ Z_1(x) \ge U_1^T \end{cases}$$

$$I_D(Z_1(x)) = \begin{cases} 0, & if\ Z_1(x) \le L_1^I \\ 1 - \frac{U_1^I - Z_1(x)}{U_1^I - L_1^I}, & if\ L_1^I \le Z_1(x) \le U_1^I \\ 1, & if\ Z_1(x) \ge U_1^I \end{cases}$$

$$F_D(Z_1(x)) = \begin{cases} 0, & if\ Z_1(x) \geq U_1^F \\ 1 - \frac{Z_1(x) - L_1^F}{U_1^F - L_1^F}, & if\ L_1^F \leq Z_1(x) \leq U_1^F \\ 1, & if\ Z_1(x) \leq L_1^T \end{cases}$$

The upper and lower bound for the positive, neutral and negative membership degrees have been obtained as follows:

$$U_2^T = 10950.59, \quad L_2^T = 10020.33$$
$$U_2^I = 10020.33 + p_2(10950.59 - 10020.33), \quad L_2^I = 10020.33$$
$$U_2^F = 10950.59, \quad L_2^F = 10020.33 + q_2(10950.59 - 10020.33)$$

The different membership function for the second objective function is defined as follows:

$$T_D(Z_2(x)) = \begin{cases} 0, & if\ Z_2(x) \leq L_2^T \\ 1 - \frac{U_2^T - Z_2(x)}{U_2^T - L_2^T}, & if\ L_2^T \leq Z_2(x) \leq U_2^T \\ 1, & if\ Z_2(x) \geq U_2^T \end{cases}$$

$$I_D(Z_2(x)) = \begin{cases} 0, & if\ Z_2(x) \leq L_2^I \\ 1 - \frac{U_2^I - Z_2(x)}{U_2^I - L_2^I}, & if\ L_2^I \leq Z_2(x) \leq U_2^I \\ 1, & if\ Z_2(x) \geq U_2^I \end{cases}$$

$$F_D(Z_2(x)) = \begin{cases} 0, & if\ Z_2(x) \geq U_2^F \\ 1 - \frac{Z_2(x) - L_2^F}{U_2^F - L_2^F}, & if\ L_2^F \leq Z_2(x) \leq U_2^F \\ 1, & if\ Z_2(x) \leq L_2^T \end{cases}$$

The upper and lower bound for the positive, neutral and negative membership degrees have been obtained as follows:

$$U_3^T = 9355.90, \quad L_3^T = 5903$$
$$U_3^I = 5903 + p_3(9355.90 - 5903), \quad L_3^I = 5903$$
$$U_3^F = 9355.90, \quad L_3^F = 5903 + q_3(9355.90 - 5903)$$

The different membership function for the third objective function is defined as follows:

$$T_D(Z_3(x)) = \begin{cases} 0, & if\ Z_3(x) \leq L_3^T \\ 1 - \frac{U_3^T - Z_3(x)}{U_3^T - L_3^T}, & if\ L_3^T \leq Z_3(x) \leq U_3^T \\ 1, & if\ Z_3(x) \geq U_3^T \end{cases}$$

$$I_D(Z_3(x)) = \begin{cases} 0, & if\ Z_3(x) \le L_3^I \\ 1 - \frac{U_3^I - Z_3(x)}{U_3^I - L_3^I}, & if\ L_3^I \le Z_3(x) \le U_3^I \\ 1, & if\ Z_3(x) \ge U_3^I \end{cases}$$

$$F_D(Z_3(x)) = \begin{cases} 0, & if\ Z_3(x) \ge U_3^F \\ 1 - \frac{Z_3(x) - L_3^F}{U_3^F - L_3^F}, & if\ L_3^F \le Z_3(x) \le U_3^F \\ 1, & if\ Z_3(x) \le L_3^T \end{cases}$$

Now the formulation of the spherical fuzzy optimization model is given as follows:

$$Max\left(\alpha^2 - \beta^2 - \gamma^2\right)$$

Subject to

$$\begin{aligned} & T_1(Z_1(x))^2 \ge \alpha^2,\ \ T_2(Z_2(x))^2 \ge \alpha^2,\ \ T_3(Z_3(x))^2 \ge \alpha^2 \\ & I_1(Z_1(x))^2 \le \beta^2,\ \ I_2(Z_2(x))^2 \le \beta^2,\ \ I_3(Z_3(x))^2 \le \beta^2 \\ & F_1(Z_1(x))^2 \le \gamma^2,\ \ F_2(Z_2(x))^2 \le \gamma^2,\ \ F_3(Z_3(x))^2 \le \gamma^2 \\ & 12x_1 + 17x_2 \le 1400 \\ & 3x_1 + 9x_2 + 8x_3 \le 1000 \\ & 10x_1 + 13x_2 + 15x_3 \le 1750 \\ & 6x_1 + 16x_3 \le 1325 \\ & x_1, x_2, x_3 \ge 0,\ \ \alpha^2 \ge \beta^2,\ \ \alpha^2 \ge \gamma^2,\ \ 0 \le \alpha^2 + \beta^2 + \gamma^2 \le 1 \end{aligned}$$

The optimal solution of the problem is $x_1 = 58.48, x_2 = 34.59, x_3 = 47.69$ and the value of each objective is obtained as $Z_1 = 7217.97, Z_2 = 10359.72, Z_3 = 8498.59$ respectively.

Example 2 (Minimization type objectives)

$$\begin{aligned} \text{Minimize}\ \ Z_1 &= 10x_1 + 13x_2 + 18x_3 \\ \text{Minimize}\ \ Z_2 &= 45x_1 + 84x_2 + 42x_3 \\ \text{Minimize}\ \ Z_3 &= 57x_1 + 18x_2 + 27x_3 \end{aligned}$$

Subject to

$$\begin{aligned} & 2x_1 + 7x_2 + 3x_3 \ge 140 \\ & 3x_1 + 9x_2 + 8x_3 \ge 100 \\ & 4x_1 + 3x_2 + 5x_3 \ge 150 \\ & 6x_1 - 7x_2 + 6x_3 \ge 125 \\ & x_1, x_2, x_3 \ge 0. \end{aligned}$$

Table 4 Individual best and worst values of objective functions

	Z_1	Z_2	Z_3	X
Minimum Z_1	657.68	1425.87	1298.23	X^1
Minimum Z_2	2152.66	1754.36	2254.56	X^2
Maximum Z_3	1125.77	1327.85	1927.85	X^3

On solving each objective function individually, we have obtained the following best and worst solution to define the upper and lower bound for each membership functions respectively (Table 4).

The upper and lower bound for the positive, neutral and negative membership degrees have been obtained as follows:

$$U_1^T = 2375.62, \quad L_1^T = 468.21$$
$$U_1^I = 468.21 + p_1(2375.62 - 468.21), \quad L_1^I = 468.21$$
$$U_1^F = 2375.62, \quad L_1^F = 468.21 + q_1(2375.62 - 468.21)$$

The different membership function for the first objective function is defined as follows:

$$T_D(Z_1(x)) = \begin{cases} 1, & if\ Z_1(x) \leq L_1^T \\ 1 - \frac{Z_1(x)-L_1^T}{U_1^T-L_1^T}, & if\ L_1^T \leq Z_2(x) \leq U_1^T \\ 0, & if\ Z_2(x) \geq U_1^T \end{cases}$$

$$I_D(Z_1(x)) = \begin{cases} 1, & if\ Z_2(x) \leq L_1^I \\ 1 - \frac{Z_1(x)-L_1^I}{U_1^I-L_1^I}, & if\ L_1^I \leq Z_2(x) \leq U_1^I \\ 0, & if\ Z_2(x) \geq U_1^I \end{cases}$$

$$F_D(Z_1(x)) = \begin{cases} 1, & if\ Z_2(x) \geq U_1^F \\ 1 - \frac{U_1^F-Z_1(x)}{U_1^F-L_1^F}, & if\ L_1^F \leq Z_2(x) \leq U_1^F \\ 0, & if\ Z_2(x) \leq L_1^F \end{cases}$$

The upper and lower bound for the positive, neutral and negative membership degrees have been obtained as follows:

$$U_2^T = 1856.67, \quad L_2^T = 625.95$$
$$U_2^I = 625.95 + p_2(1856.67 - 625.95), \quad L_2^I = 625.95$$
$$U_2^F = 1856.67, \quad L_2^F = 625.95 + q_2(1856.67 - 625.95)$$

The different membership function for the second objective function is defined as follows:

$$T_D(Z_2(x)) = \begin{cases} 1, & if\ Z_2(x) \leq L_2^T \\ 1 - \frac{Z_2(x)-L_2^T}{U_2^T-L_2^T}, & if\ L_o^T \leq Z_2(x) \leq U_2^T \\ 0, & if\ Z_2(x) \geq U_2^T \end{cases}$$

$$I_D(Z_2(x)) = \begin{cases} 1, & if\ Z_2(x) \leq L_2^I \\ 1 - \frac{Z_2(x)-L_2^I}{U_2^I-L_2^I}, & if\ L_2^I \leq Z_2(x) \leq U_2^I \\ 0, & if\ Z_2(x) \geq U_2^I \end{cases}$$

$$F_D(Z_2(x)) = \begin{cases} 1, & if\ Z_2(x) \geq U_2^F \\ 1 - \frac{U_2^F-Z_2(x)}{U_2^F-L_2^F}, & if\ L_o^F \leq Z_2(x) \leq U_2^F \\ 0, & if\ Z_2(x) \leq L_2^F \end{cases}$$

The upper and lower bound for the positive, neutral and negative membership degrees have been obtained as follows:

$$U_3^T = 1856.67, \quad L_3^T = 625.95$$
$$U_3^I = 625.95 + p_3(1856.67 - 625.95), \quad L_3^I = 625.95$$
$$U_3^F = 1856.67, \quad L_3^F = 625.95 + q_3(1856.67 - 625.95)$$

The different membership function for the third objective function is defined as follows:

$$T_D(Z_3(x)) = \begin{cases} 1, & if\ Z_3(x) \leq L_3^T \\ 1 - \frac{Z_3(x)-L_3^T}{U_3^T-L_3^T}, & if\ L_3^T \leq Z_3(x) \leq U_3^T \\ 0, & if\ Z_3(x) \geq U_3^T \end{cases}$$

$$I_D(Z_3(x)) = \begin{cases} 1, & if\ Z_3(x) \leq L_3^I \\ 1 - \frac{Z_3(x)-L_3^I}{U_3^I-L_3^I}, & if\ L_3^I \leq Z_3(x) \leq U_3^I \\ 0, & if\ Z_3(x) \geq U_3^I \end{cases}$$

$$F_D(Z_3(x)) = \begin{cases} 1, & if\ Z_3(x) \geq U_3^F \\ 1 - \frac{U_3^F-Z_3(x)}{U_3^F-L_3^F}, & if\ L_3^F \leq Z_3(x) \leq U_3^F \\ 0, & if\ Z_3(x) \leq L_3^F \end{cases}$$

Now the formulation of the spherical fuzzy optimization model is given as follows:

$$Max\left(\alpha^2 - \beta^2 - \gamma^2\right)$$

Subject to

$$T_1(Z_1(x))^2 \geq \alpha^2, \quad T_2(Z_2(x))^2 \geq \alpha^2, \quad T_3(Z_3(x))^2 \geq \alpha^2$$
$$I_1(Z_1(x))^2 \leq \beta^2, \quad I_2(Z_2(x))^2 \leq \beta^2, \quad I_3(Z_3(x))^2 \leq \beta^2$$
$$F_1(Z_1(x))^2 \leq \gamma^2, \quad F_2(Z_2(x))^2 \leq \gamma^2, \quad F_3(Z_3(x))^2 \leq \gamma^2$$

$$
\begin{aligned}
&2x_1 + 7x_2 + 3x_3 \geq 140 \\
&3x_1 + 9x_2 + 8x_3 \geq 100 \\
&4x_1 + 3x_2 + 5x_3 \geq 150 \\
&6x_1 - 7x_2 + 6x_3 \geq 125 \\
&x_1, x_2, x_3 \geq 0, \ \ \alpha^2 \geq \beta^2, \ \ \alpha^2 \geq \gamma^2, \ \ 0 \leq \alpha^2 + \beta^2 + \gamma^2 \leq 1
\end{aligned}
$$

The above model has been coded in AMPL language and solved by using the Knitro optimization software online facility provided by University of Wisconsin (Dolan (2001)). The optimal solution of the problem is $x_1 = 3.02, x_2 = 7.38, x_3 = 29.44$ and the value of each objective is obtained as $Z_1 = 625.95, Z_2 = 1856.67, Z_3 = 927.86$, respectively.

Example 3 (Mixed type objectives)

$$
\begin{aligned}
\text{Minimize } Z_1 &= 19x_1 + 33x_2 + 58x_3 \\
\text{Minimize } Z_2 &= 25x_1 + 68x_2 + 72x_3 \\
\text{Maximize } Z_3 &= 17x_1 + 35x_2 + 63x_3
\end{aligned}
$$

Subject to

$$
\begin{aligned}
&7x_1 + 2x_2 + 5x_3 \geq 125 \\
&4x_1 + 5x_2 + 3x_3 \geq 116 \\
&6x_1 + 9x_2 + 5x_3 \geq 214 \\
&x_1, x_2, x_3 \geq 0.
\end{aligned}
$$

On solving each objective function individually, we have obtained the following best and worst solution to define the upper and lower bound for each membership functions respectively.

The upper and lower bound for the positive, neutral and negative membership degrees have been obtained as follows:

$$
\begin{aligned}
&U_1^T = 1425.87, \quad L_1^T = 657.68 \\
&U_1^I = 657.68 + p_1(1425.87 - 657.68), \quad L_1^I = 657.68 \\
&U_1^F = 1425.87, \quad L_1^F = 657.68 + q_1(1425.87 - 657.68)
\end{aligned}
$$

The different membership function for the first objective function is defined as follows:

$$T_D(Z_1(x)) = \begin{cases} 1, & if\ Z_1(x) \le L_1^T \\ 1 - \frac{Z_1(x)-L_1^T}{U_1^T-L_1^T}, & if\ L_1^T \le Z_2(x) \le U_1^T \\ 0, & if\ Z_2(x) \ge U_1^T \end{cases}$$

$$I_D(Z_1(x)) = \begin{cases} 1, & if\ Z_2(x) \le L_1^I \\ 1 - \frac{Z_1(x)-L_1^I}{U_1^I-L_1^I}, & if\ L_1^I \le Z_2(x) \le U_1^I \\ 0, & if\ Z_2(x) \ge U_1^I \end{cases}$$

$$F_D(Z_1(x)) = \begin{cases} 1, & if\ Z_2(x) \ge U_1^F \\ 1 - \frac{U_1^F-Z_1(x)}{U_1^F-L_1^F}, & if\ L_1^F \le Z_2(x) \le U_1^F \\ 0, & if\ Z_2(x) \le L_1^F \end{cases}$$

The upper and lower bound for the positive, neutral and negative membership degrees have been obtained as follows:

$$U_2^T = 2254.56, \quad L_2^T = 1754.36$$
$$U_2^I = 1754.36 + p_2(2254.56 - 1754.36), \quad L_2^I = 1754.36$$
$$U_2^F = 2254.56, \quad L_2^F = 1754.36 + q_2(2254.56 - 1754.36)$$

The different membership function for the second objective function is defined as follows:

$$T_D(Z_2(x)) = \begin{cases} 1, & if\ Z_2(x) \le L_2^T \\ 1 - \frac{Z_2(x)-L_2^T}{U_2^T-L_2^T}, & if\ L_o^T \le Z_2(x) \le U_2^T \\ 0, & if\ Z_2(x) \ge U_2^T \end{cases}$$

$$I_D(Z_2(x)) = \begin{cases} 1, & if\ Z_2(x) \le L_2^I \\ 1 - \frac{Z_2(x)-L_2^I}{U_2^I-L_2^I}, & if\ L_2^I \le Z_2(x) \le U_2^I \\ 0, & if\ Z_2(x) \ge U_2^I \end{cases}$$

$$F_D(Z_2(x)) = \begin{cases} 1, & if\ Z_2(x) \ge U_2^F \\ 1 - \frac{U_2^F-Z_2(x)}{U_2^F-L_2^F}, & if\ L_o^F \le Z_2(x) \le U_2^F \\ 0, & if\ Z_2(x) \le L_2^F \end{cases}$$

The upper and lower bound for the positive, neutral and negative membership degrees have been obtained as follows:

$$U_3^T = 1927.85, \quad L_3^T = 1125.77$$
$$U_3^I = 1125.77 + p_3(1927.85 - 1125.77), \quad L_3^I = 1125.77$$
$$U_3^F = 1927.85, \quad L_3^F = 1125.77 + q_3(1927.85 - 1125.77)$$

The different membership function for the third objective function is defined as follows:

$$T_D(Z_3(x)) = \begin{cases} 1, & if\ Z_3(x) \leq L_3^T \\ 1 - \frac{Z_3(x)-L_3^T}{U_3^T-L_3^T}, & if\ L_3^T \leq Z_3(x) \leq U_3^T \\ 0, & if\ Z_3(x) \geq U_3^T \end{cases}$$

$$I_D(Z_3(x)) = \begin{cases} 1, & if\ Z_3(x) \leq L_3^I \\ 1 - \frac{Z_3(x)-L_3^I}{U_3^I-L_3^I}, & if\ L_3^I \leq Z_3(x) \leq U_3^I \\ 0, & if\ Z_3(x) \geq U_3^I \end{cases}$$

$$F_D(Z_3(x)) = \begin{cases} 1, & if\ Z_3(x) \geq U_3^F \\ 1 - \frac{U_3^F-Z_3(x)}{U_3^F-L_3^F}, & if\ L_3^F \leq Z_3(x) \leq U_3^F \\ 0, & if\ Z_3(x) \leq L_3^F \end{cases}$$

Now the formulation of the spherical fuzzy optimization model is given as follows:

$$Max\left(\alpha^2 - \beta^2 - \gamma^2\right)$$

Subject to

$$\begin{aligned}
&T_1(Z_1(x))^2 \geq \alpha^2,\ \ T_2(Z_2(x))^2 \geq \alpha^2,\ \ T_3(Z_3(x))^2 \geq \alpha^2 \\
&I_1(Z_1(x))^2 \leq \beta^2,\ \ I_2(Z_2(x))^2 \leq \beta^2,\ \ I_3(Z_3(x))^2 \leq \beta^2 \\
&F_1(Z_1(x))^2 \leq \gamma^2,\ \ F_2(Z_2(x))^2 \leq \gamma^2,\ \ F_3(Z_3(x))^2 \leq \gamma^2 \\
&7x_1 + 2x_2 + 5x_3 \geq 125 \\
&4x_1 + 5x_2 + 3x_3 \geq 116 \\
&6x_1 + 9x_2 + 5x_3 \geq 214 \\
&x_1, x_2, x_3 \geq 0,\ \ \alpha^2 \geq \beta^2,\ \ \alpha^2 \geq \gamma^2,\ \ 0 \leq \alpha^2 + \beta^2 + \gamma^2 \leq 1
\end{aligned}$$

The above model has been coded in AMPL language and solved by using the Knitro optimization software online facility provided by University of Wisconsin (Dolan (2001)). The optimal solution of the problem is $x_1 = 12.83$, $x_2 = 07.38$, $x_3 = 15.29$ and the value of each objective is obtained as $Z_1 = 1374.13$, $Z_2 = 1923.47$, $Z_3 = 1439.68$ respectively.

Example 4 (Production Planning Problem)

A well-known leather company manufactures three different leather products (such as a belt, wallet, etc.). The various departments provide the manufacturing services to the raw materials that result in the finished goods. The data related to the total time availability and the time devoted to offering various services to different products are summarized in Table 3. Furthermore, the customer satisfaction parameters over the different products are 36, 43, and 67, along with the expected profit of \$57, \$62, and \$74, respectively. The quality assurance parameters are 52, 28, and 75 for each product (Table 5).

Suppose that, x_1, x_2 and x_3 represent the number of three different products. Then mathematical production planning model for the given data can be given as follows:

Table 5 Values of different parameters

Departments	Service time (in hrs.)	Unit price	Types of products		
			x_1	x_2	x_3
Milling	1600	1.85	15	11	09
Lather	1250	1.55	06	09	12
Grinder	1900	2.25	10	07	15
Jig saw	1455	1.75	00	14	14
Drill press	1100	1.35	02	00	09
Band saw	1175	0.85	09	10	13
Total capacity cost	478,562				

$$\begin{aligned}
\text{Maximize } Z_1 &= 36x_1 + 43x_2 + 67x_3 \text{ (Customer satisfaction)} \\
\text{Maximize } Z_2 &= 57x_1 + 62x_2 + 74x_3 \text{ (Profit function)} \\
\text{Maximize } Z_3 &= 52x_1 + 28x_2 + 75x_3 \text{ (Quality assurance)}
\end{aligned}$$

Subject to

$$\begin{aligned}
&15x_1 + 11x_2 + 9x_3 \leq 1600 \\
&6x_1 + 9x_2 + 12x_3 \leq 1250 \\
&10x_1 + 7x_2 + 15x_3 \leq 1900 \\
&14x_2 + 14x_3 \leq 1455 \\
&2x_1 + 9x_3 \leq 1100 \\
&9x_1 + 10x_2 + 13x_3 \leq 1175 \\
&x_1, x_2, x_3 \geq 0
\end{aligned}$$

On solving each objective function individually, we have obtained the following best and worst solution to define the upper and lower bound for each membership functions respectively (Table 6).

The upper and lower bound for the positive, neutral and negative membership degrees have been obtained as follows:

$$U_1^T = 6778.85, \quad L_1^T = 6055.77$$

Table 6 Individual best and worst values of objective functions

	Z_1	Z_2	Z_3	X
Maximum Z_1	6055.77	4889.71	5124.34	X^1
Maximum Z_2	6688.46	7357.35	7205.92	X^2
Maximum Z_3	6778.85	4905.88	6785.75	X^3

$$U_1^I = 6055.77 + p_1(6778.85 - 6055.77), \quad L_1^I = 6055.77$$
$$U_1^F = 6778.85, \quad L_1^F = 6055.77 + q_1(6778.85 - 6055.77)$$

The upper and lower bound for the positive, neutral and negative membership degrees have been obtained as follows:

$$U_2^T = 7357.35, \quad L_2^T = 4889.71$$
$$U_2^I = 4889.71 + p_2(7357.35 - 4889.71), \quad L_2^I = 4889.71$$
$$U_2^F = 7357.35, \quad L_2^F = 4889.71 + q_2(7357.35 - 4889.71)$$

The upper and lower bound for the positive, neutral and negative membership degrees have been obtained as follows:

$$U_3^T = 7205.92, \quad L_3^T = 5124.34$$
$$U_3^I = 5124.34 + p_3(7205.92 - 5124.34), \quad L_3^I = 5124.34$$
$$U_3^F = 7205.92, \quad L_3^F = 5124.34 + q_3(7205.92 - 5124.34)$$

The different membership function for the first objective function is defined as follows:

$$T_D(Z_1(x)) = \begin{cases} 0, & if\ Z_1(x) \leq L_1^T \\ 1 - \frac{U_1^T - Z_1(x)}{U_1^T - L_1^T}, & if\ L_1^T \leq Z_1(x) \leq U_1^T \\ 1, & if\ Z_1(x) \geq U_1^T \end{cases}$$

$$I_D(Z_1(x)) = \begin{cases} 0, & if\ Z_1(x) \leq L_1^I \\ 1 - \frac{U_1^I - Z_1(x)}{U_1^I - L_1^I}, & if\ L_1^I \leq Z_1(x) \leq U_1^I \\ 1, & if\ Z_1(x) \geq U_1^I \end{cases}$$

$$F_D(Z_1(x)) = \begin{cases} 0, & if\ Z_1(x) \geq U_1^F \\ 1 - \frac{Z_1(x) - L_1^F}{U_1^F - L_1^F}, & if\ L_1^F \leq Z_1(x) \leq U_1^F \\ 1, & if\ Z_1(x) \leq L_1^T \end{cases}$$

The different membership function for the second objective function is defined as follows:

$$T_D(Z_2(x)) = \begin{cases} 0, & if\ Z_2(x) \leq L_2^T \\ 1 - \frac{U_2^T - Z_2(x)}{U_2^T - L_2^T}, & if\ L_2^T \leq Z_2(x) \leq U_2^T \\ 1, & if\ Z_2(x) \geq U_2^T \end{cases}$$

$$I_D(Z_2(x)) = \begin{cases} 0, & if\ Z_2(x) \leq L_2^I \\ 1 - \frac{U_2^I - Z_2(x)}{U_2^I - L_2^I}, & if\ L_2^I \leq Z_2(x) \leq U_2^I \\ 1, & if\ Z_2(x) \geq U_2^I \end{cases}$$

$$F_D(Z_2(x)) = \begin{cases} 0, & if\ Z_2(x) \geq U_2^F \\ 1 - \frac{Z_2(x)-L_2^F}{U_2^F-L_2^F}, & if\ L_2^F \leq Z_2(x) \leq U_2^F \\ 1, & if\ Z_2(x) \leq L_2^T \end{cases}$$

The different membership function for the third objective function is defined as follows:

$$T_D(Z_3(x)) = \begin{cases} 0, & if\ Z_3(x) \leq L_3^T \\ 1 - \frac{U_3^T-Z_3(x)}{U_3^T-L_3^T}, & if\ L_3^T \leq Z_3(x) \leq U_3^T \\ 1, & if\ Z_3(x) \geq U_3^T \end{cases}$$

$$I_D(Z_3(x)) = \begin{cases} 0, & if\ Z_3(x) \leq L_3^I \\ 1 - \frac{U_3^I-Z_3(x)}{U_3^I-L_3^I}, & if\ L_3^I \leq Z_3(x) \leq U_3^I \\ 1, & if\ Z_3(x) \geq U_3^I \end{cases}$$

$$F_D(Z_3(x)) = \begin{cases} 0, & if\ Z_3(x) \geq U_3^F \\ 1 - \frac{Z_3(x)-L_3^F}{U_3^F-L_3^F}, & if\ L_3^F \leq Z_3(x) \leq U_3^F \\ 1, & if\ Z_3(x) \leq L_3^T \end{cases}$$

Here the aim is to maximize the positive membership degree and minimize the neutral and negative membership degrees under spherical fuzzy environment. Now the formulation of the spherical fuzzy optimization model is given as follows:

$$Max\left(\alpha^2 - \beta^2 - \gamma^2\right)$$

Subject to;

$$T_1(Z_1(x))^2 \geq \alpha^2,\ \ T_2(Z_2(x))^2 \geq \alpha^2,\ \ T_3(Z_3(x))^2 \geq \alpha^2$$
$$I_1(Z_1(x))^2 \leq \beta^2,\ \ I_2(Z_2(x))^2 \leq \beta^2,\ \ I_3(Z_3(x))^2 \leq \beta^2$$
$$F_1(Z_1(x))^2 \leq \gamma^2,\ \ F_2(Z_2(x))^2 \leq \gamma^2,\ \ F_3(Z_3(x))^2 \leq \gamma^2$$
$$15x_1 + 11x_2 + 9x_3 \leq 1600$$
$$6x_1 + 9x_2 + 12x_3 \leq 1250$$
$$10x_1 + 7x_2 + 15x_3 \leq 1900$$
$$14x_2 + 14x_3 \leq 1455$$
$$2x_1 + 9x_3 \leq 1100$$
$$9x_1 + 10x_2 + 13x_3 \leq 1175$$
$$x_1, x_2, x_3 \geq 0,\ \ \alpha^2 \geq \beta^2,\ \ \alpha^2 \geq \gamma^2,\ \ 0 \leq \alpha^2 + \beta^2 + \gamma^2 \leq 1$$

The optimal solution of the problem is $x_1 = 4, x_2 = 0, x_3 = 90$ and the value of each objective is obtained as $Z_1 = 6055$, $Z_2 = 6688$, $Z_3 = 6778$ respectively.

5 Conclusions

In this chapter, we have discussed the multiobjective optimization problem under a spherical fuzzy environment. The presented multiobjective optimization technique is based on a spherical fuzzy decision set. The spherical fuzzy set includes the positive, neutral, and negative membership grades of the element into the spherical fuzzy decision set. The advanced modeling and optimization approach is presented efficiently along with the numerical examples to show the validity and applicability of the model. The spherical fuzzy multiobjective programming model provides more flexibility in the decision making process as compared to fuzzy, intuitionistic fuzzy, and Pythagorean fuzzy optimization techniques. The spherical fuzzy optimization technique exhibits the degree of neutral thoughts or indeterminacy with the sum of squares of each membership degrees to one (1).

The aim of this chapter is to introduce the spherical fuzzy multiobjective programming problem under the SF concept and to explore the efficient utility in continuous optimization cases such as multiobjective optimization problems. Hence, it may be applied to a wide range of real-life applications in a different field such as transportation problems, supplier selection problem, inventory control, supply chain management, etc.

In the future, the proposed spherical fuzzy multiobjective programming problem may be applied to a non-linear programming problem, fractional programming problem, bi-level, and multilevel programming problem, etc. It is also possible to extend the spherical fuzzy concept with the stochastic and uncertain programming domain.

References

Ahmad F, Adhami AY (2019) Neutrosophic programming approach to multiobjective nonlinear transportation problem with fuzzy parameters. Int J Manag Sci Eng Manag 14(3):218–229

Ahmad F, Adhami AY, Smarandache F (2018) Single valued neutrosophic hesitant fuzzy computational algorithm for multiobjective nonlinear optimization problem. Neutrosophic Sets Syst 22:76–86

Ahmad F, Adhami AY, Smarandache F (2020) Modified neutrosophic fuzzy optimization model for optimal closed-loop supply chain management under uncertainty. In: Optimization theory based on neutrosophic and plithogenic sets. Academic Press, pp 343–403

Ahmad F, Adhami AY, Smarandache F (2019) Neutrosophic optimization model and computational algorithm for optimal shale gas water management under uncertainty. Symmetry 11(4):544

Angelov PP (1997) Optimization in an intuitionistic fuzzy environment. Fuzzy Sets Syst 86(3):299–306

Atanassov KT (1986) Studies in fuzziness and soft computing intuitionistic fuzzy logics. Fuzzy Sets Syst

Bellman RE, Zadeh LA (1970) Decision-making in a fuzzy environment. Manage Sci 17(4):140–164

Dolan E (2001) The NEOS Server 4.0 administrative guide. Technical memorandum ANL/MCS-TM-250. Mathematics and Computer Science Division, Argonne National Laboratory

Kutlu Gündoğdu F, Kahraman C (2019) Spherical fuzzy sets and spherical fuzzy TOPSIS method. J Intell Fuzzy Syst, (Preprint): 1–16

Rafiq M, Ashraf S, Abdullah S, Mahmood T (2019) The cosine similarity measures of spherical fuzzy sets and their applications in decision making, 1–15

Smarandache F (1999) A unifying field in logics: neutrosophic logic. Philosophy. American Press
Yager RR (2013) Pythagorean fuzzy subsets, vol 2, IEEE, pp 57–61
Zimmermann HJ (1978) Fuzzy programming and linear programming with several objective functions. Fuzzy Sets Syst 1(1):45–55

Spherical Fuzzy Goal Programming Problem

Firoz Ahmad and Ahmad Yusuf Adhami

Abstract Goal programming optimization techniques are an essential and exciting topic since the 1970s. The significant development can be realized by having enormous researches in this field, such as fuzzy goal optimization technique, intuitionistic fuzzy goal optimization technique, different goal programming techniques, etc. Thus this chapter investigates a new algorithm based on the spherical fuzzy set (SF) named as spherical fuzzy goal programming problem (SFGP) under the spherical fuzzy environment. The SFGP inevitably involves the degree of neutrality along with truth and a falsity membership degree of the element into the feasible decision set. It also generalizes the decision set by imposing the restriction that the sum of squares of each membership function must be less than or equal to one. The attainment of achievement function is determined by reducing the deviations from ideal solutions for the truth, neutral and negative membership goals of each objective function under the spherical fuzzy decision set. At last, numerical examples and conclusions are presented to reveal the applicability and future research scope in the SF domain.

1 Introduction

In mathematical programming, Goal Programming (GP) is one of the most popular optimization techniques for the multiobjective programming problem. Goal-programming models are being used increasingly in decision-making problems where the alternatives cannot be compared on the basis of a single performance criterion. In a typical goal-programming formulation, the goals and constraints are defined precisely. The GP technique is basically concerned with the achievement of different pre-determined goals. While dealing with multiple objectives, the tendencies of GP are to minimize the negative and positive deviations from their respective goals simultaneously. Many real-life problems have been formulated and solved using GP

F. Ahmad (✉) · A. Y. Adhami
Department of Statistics and Operations Research, Aligarh Muslim University, Aligarh, India
e-mail: firoz.ahmad02@gmail.com

© Springer Nature Switzerland AG 2021

C. Kahraman and F. Kutlu Gündoğdu (eds.), *Decision Making with Spherical Fuzzy Sets*, Studies in Fuzziness and Soft Computing 392,
https://doi.org/10.1007/978-3-030-45461-6_21

techniques. Firstly, the concept of Goal Programming technique was introduced by (Charnes and Cooper, 1977) and later on the original name was used by (Charnes and Cooper, 1977) and (Dharmar et al. 1987). The opportunity of setting pre-determined goals makes the wide range of applicability in real-life problems having multiple objectives. In the GP model, deviations from the targets are the essential aspect that provides the dynamic optimization framework for a multiobjective programming problem. The GP model reduces the negative (overachievement) and positive (underachievement) from their respective goals of each objective function.

Moreover, the GP problem is extended into the fuzzy environment. Apart from its theoretical interest, decision making in a fuzzy environment is of practical interest since much of the decision making in the real world takes place in an imprecise environment, in that both the goals and their importance are not stated with precision. Hence the concept of fuzzy set (FS) (Bellman and Zadeh 1970) came into existence. The FS deals with the degree of belongingness (membership function) of the element into the feasible decision set. Based on the fuzzy set (FS), (Zimmermann 1978) proposed the fuzzy goal programming (FGP) problem, which is widely accepted in almost every domain of real-life issues. The FGP model consists of membership goals with the maximum attainment degree unity (1). The FGP optimizes the unwanted deviation from the ideal solution of each objective simultaneously.

Many real-life decision-making problems have vagueness as well as hesitation degree associated with the element. Hence, FS theory fails to capture the hesitation degree of the component into the feasible set. To include the hesitancy degree, (Atanassov 1986) investigated the intuitionistic fuzzy set (IFS), which is based on more intuition than the FS. The IFS deals with the degree of belongingness as well as the non-belongingness of the element into the feasible decision set.

Firstly, Angelov (1997) proposed the intuitionistic fuzzy programming approach (IFPA) for the multiobjective optimization problem. In IFPA, the overall satisfaction level is achieved by maximizing the membership degree and minimizing the non-membership degree simultaneously. Based on IFS, the concept of intuitionistic fuzzy goal programming (IFGP) came into existence. The deviations from ideal solutions under each membership goals are defined in such a way that the overall achievement of satisfaction degree reaches to its maximum. Hence the degree of attainment of membership goal is set to unity (1), whereas the attainment degree of non-membership goal is assigned by zero (0). Therefore, IFGP is also a powerful optimization technique for different decision-making problems.

Further extension of FS and IFS is introduced in the 1990s. (Smarandache 1999) propounded the concept of neutrosophic logic and neutrosophic set (NS) with the fact that indeterminacy/neutral exists in real-life scenarios. The NS contains the knowledge of neutral thoughts which makes it differ from FS and IFS. Thus NS contemplate over three different membership function such as truth, indeterminacy and falsity membership functions respectively. Based on the NS, neutrosophic goal programming problem (NGPP) is developed and widely been used for multiobjective optimization problem (Smarandache (1999), Ahmad and Adhami (2019), Ahmad et al. (2019), Ahmad et al. (2018)). The truth, indeterminacy, and falsity membership goals are defined with their attainment degrees unity (1), half (0.5) and zero

(0) respectively and achieved by minimizing the deviations from each membership goals (see Ahmad et al. (2020)).

Recently, Kutlu Gündoğdu and Kahraman (2019) addressed the modified version of NS and presented spherical fuzzy set (SFS). The SFS discuss the situation when the sum of truth, indeterminacy, and falsity membership function is not unity (1). In such typical circumstances, the SFS based decision–making technique is one of the most effective and convenient tools. In the SFS, the sum of the square of each type of membership function is considered, and their sum is restricted to lies between zero (0) to unity (1). Thus, SF decision set is much beneficial to deal with the above-discussed situation effectively. Hence, this chapter addresses the spherical fuzzy goal programming (SFGP) problem based on the SF decision set.

The remaining part of the chapter is summarized as follows: In Sect. 2, basic definition regarding spherical fuzzy set is presented whereas Sect. 3 describes the spherical fuzzy goal programming problem and solution algorithm. Section 4 presents the numerical examples whereas a conclusion along with the future research scope is described in Sect. 5.

2 Preliminaries

Some basic definitions and related terms of the fuzzy, intuitionistic fuzzy, neutrosophic and spherical fuzzy sets have been discussed in this section.

Definition 1 (*Fuzzy Set (FS)*) Let W be the universe of discourse with its generic element w, then a fuzzy set X in W can be defined by a function $X :\rightarrow [0, 1]$

$$X = \{w, \mu_X(w)|w \in W\}$$

where

$$\mu_X(w) : W \rightarrow [0, 1]$$

With conditions

$$0 \leq \mu_X(w) \leq 1$$

where $\mu_X(w)$ denotes the membership function of the element $w \in W$, into the set X respectively.

Definition 2 (*Intuitionistic Fuzzy Set (IFS)*) Let W denote a universe of discourse; then an IFS X in W is given by the ordered triplets as follows:

$$X = \{w, \mu_X(w), \nu_X(w) | w \in W\}$$

where

$$\mu_X(w) : W \to [0, 1]; \quad \nu_X(w) : W \to [0, 1]$$

Such that,

$$0 \leq \mu_X(w) + \nu_X(w) \leq 1$$

$\mu_X(w)$ and $\nu_X(w)$ denotes the membership and non-membership function of the element w into the set X.

Definition 3 (*Neutrosophic Set (NS)*) Let W be the universe of discourse with its generic element w, then a neutrosophic set X in W can be defined by a function $X :\to [0, 1]$.

$$X = \{w, \mu_X(w), \lambda_X(w), \nu_X(w) | w \in W\}$$

where

$$\mu_X(w) : W \to [0, 1], \lambda_X(w) : W \to [0, 1] \text{ and } \nu_X(w) : W \to [0, 1]$$

With the conditions that

$$0 \leq \mu_X(w) + \lambda_X(w) + \nu_X(w) :\leq 3$$

where $\mu_X(w), \lambda_X(w)$ and $\nu_X(w)$ denotes the truth, indeterminacy and a falsity membership functions of the element $w \in W$, into the set X respectively.

Definition 4 (*Over achievement*) Let $Z_o(x^*)$ be the ideal solution of a goal programming problem then the positive deviation from this ideal point $d_o^+(Z_o(x))$ for all $o = 1, 2, \ldots, O$ is termed as over achievement and it should be always minimized.

Definition 5 (*Under achievement*) Let $Z_o(x^*)$ be the ideal solution of a goal programming problem then the negative deviation from this ideal point $d_o^-(Z_o(x))$ for all $o = 1, 2, \ldots, O$ is termed as under achievement and it should be always minimized.

Definition 6 (*Target values*) For each $Z_o(x)$, an individual optimal solution $Z_o(x^*)$ is associated which are to be achieved is termed as target values in the goal programming problem.

3 Spherical Fuzzy Goal Programming Model

Based on the spherical fuzzy decision set, spherical fuzzy goal programming (SFGP) problem is discussed. The spherical fuzzy set contains three different membership functions, namely truth, indeterminacy, and falsity, along with the restrictions that sum of the squares of all three membership degrees must lies between the intervals 0 to 1. Consequently, the SFGP has also three different fuzzy goals with varying levels of attainment. The truth spherical fuzzy goal is defined for the acceptance degree with attainment value unity (1), the indeterminacy goal is set with achievement degree half (0.5), and the falsity spherical fuzzy goal is defined for the rejection degree with attainment value zero (0) respectively. Thus SFGP is a more convenient optimization tool than FGP and IFGP because of the indeterminacy goal while dealing with multiobjective decision-making problems.

The mathematical formulation of multiobjective programming problem can be summarized as follows:

Maximize $Z_o(x) = (Z_1, Z_2, Z_3, \ldots, Z_o) \;\; \forall\, o = 1, 2, \ldots, O_1$

Minimize $Z_o(x) = (Z_1, Z_2, Z_3, \ldots, Z_o) \;\; \forall\, o = O_1 + 1, O_1 + 2, \ldots, O$

Subject to

$$\begin{aligned} &g_i(x) \le b_i, \quad \forall\, i = 1, 2, \ldots, I_1, \\ &g_i(x) \ge b_i, \quad \forall\, i = I_1 + 1, I_1 + 2, \ldots, I_2, \\ &g_i(x) = b_i, \quad \forall\, i = I_2 + 1, I_2 + 2, \ldots, I. \\ &x = \left(x_1, x_2, \ldots, x_j\right) \in X, \;\; x \ge 0. \end{aligned} \tag{1}$$

where $Z_o(x)$ is the Oth objective function. The function $g_i(x)$ and b_i are the real valued function and numbers respectively. $x = \left(x_1, x_2, \ldots, x_j\right)$ represents the set of decision variables.

According to (Bellman and Zadeh 1970), the fuzzy decision set (D) consists of fuzzy goals (Z) and fuzzy constraints (C) under fuzzy environment. The decision set has been widely used in decision making processes and can be stated as follows:

$$D = (Z \cap C)$$

Hence, based on the fuzzy decision set theory, the spherical fuzzy decision set can be developed under spherical fuzzy environment. Consequently, the spherical fuzzy decision set (D_{sf}), spherical fuzzy goals (Z_o) and spherical fuzzy constraints (C_o) can be given as follows:

Table 1 Pay-off matrix

	Z_1	Z_2	…	Z_o
X^1	$Z_1(X^1)$	$Z_2(X^1)$	…	$Z_o(X^1)$
X^2	$Z_1(X^2)$	$Z_2(X^2)$	…	$Z_o(X^2)$
…	…	…	…	…
…	…	…	…	…
…	…	…	…	…
X^o	$Z_1(X^o)$	$Z_2(X^o)$	…	$Z_o(X^o)$

$$D_{sf} = \left(\bigcap_{o=1}^{O} Z_o\right)\left(\bigcap_{j=1}^{J} C_j\right) = (x, T_D(x), I_D(x), F_D(x))$$

where

$$T_D(x) = \begin{Bmatrix} T_{D_1}(x), T_{D_2}(x), T_{D_3}(x), \ldots, T_{D_O}(x) \\ T_{C_1}(x), T_{C_2}(x), T_{C_3}(x), \ldots, T_{C_J}(x) \end{Bmatrix} \quad \forall\ x \in X$$

$$I_D(x) = \begin{Bmatrix} I_{D_1}(x), I_{D_2}(x), I_{D_3}(x), \ldots, I_{D_O}(x) \\ I_{C_1}(x), I_{C_2}(x), I_{C_3}(x), \ldots, I_{C_J}(x) \end{Bmatrix} \quad \forall\ x \in X$$

$$F_D(x) = \begin{Bmatrix} F_{D_1}(x), F_{D_2}(x), F_{D_3}(x), \ldots, F_{D_O}(x) \\ F_{C_1}(x), F_{C_2}(x), F_{C_3}(x), \ldots, F_{C_J}(x) \end{Bmatrix} \quad \forall\ x \in X \tag{2}$$

where $T_D(x)$, $I_D(x)$ and $F_D(x)$ are the truth membership function, indeterminacy membership function and a falsity membership function under spherical fuzzy decision set D_{sf}.

To determine the marginal evaluation of each objective, first we calculate the upper and lower bound for each objective function. Suppose that the obtained decision variables are $X^1, X^1, X^1, \ldots, X^o$. The pay-off matrix has been shown in Table 1. By substituting these values in each objective function, the mathematical expressions of upper and lower bound can be given as follows (Eq. 3):

$$U_o = \max\{Z_o(X^o)\} \text{ and } L_o = \min\{Z_o(X^o)\} \ \ \forall\, o = 1, 2, \ldots, O \tag{3}$$

where U_o and L_o represents the upper and lower bound of Oth objectives. The upper and lower bounds for truth, indeterminacy and a falsity membership functions for each objective under spherical fuzzy environment can be derived by using the following mathematical expressions (Eq. 4):

$$U_o^T = U_o, \quad L_o^T = L_o$$
$$U_o^I = L_o^T + p_o(U_o^T - L_o^T), \quad L_o^I = L_o^T$$

$$U_o^F = U_o^T, \quad L_o^F = L_o^T + q_o(U_o^T - L_o^T) \tag{4}$$

where p_o and q_o are the predetermined real numbers such that $p_o, q_o \in (0, 1)$ assigned by decision maker.

Case I When the objective function are of minimization type then truth, indeterminacy and a falsity membership functions for each objective under spherical fuzzy environment can be obtained as follows:

$$T_D(x) = \begin{cases} 1, & if\ Z_o(x) \leq L_o^T \\ 1 - \frac{Z_o(x) - L_o^T}{U_o^T - L_o^T}, & if\ L_o^T \leq Z_o(x) \leq U_o^T \\ 0, & if\ Z_o^T \geq U_o^T \end{cases}$$
$$I_D(x) = \begin{cases} 1, & if\ Z_o(x) \leq L_o^I \\ 1 - \frac{Z_o(x) - L_o^I}{U_o^I - L_o^I}, & if\ L_o^I \leq Z_o(x) \leq U_o^I \\ 0, & if\ Z_o^I \geq U_o^I \end{cases}$$
$$F_D(x) = \begin{cases} 1, & if\ Z_o(x) \geq U_o^F \\ 1 - \frac{U_o^F - Z_o(x)}{U_o^F - L_o^F}, & if\ L_o^F \leq Z_o(x) \leq U_o^F \\ 0, & if\ Z_o^T \leq L_o^T \end{cases} \tag{5}$$

Case II When the objective function are of minimization type then truth, indeterminacy and a falsity membership functions for each objective under spherical fuzzy environment can be obtained as follows:

$$T_D(x) = \begin{cases} 0, & if\ Z_o(x) \leq L_o^T \\ 1 - \frac{U_o^T - Z_o(x)}{U_o^T - L_o^T}, & if\ L_o^T \leq Z_o(x) \leq U_o^T \\ 1, & if\ Z_o^T \geq U_o^T \end{cases}$$
$$I_D(x) = \begin{cases} 0, & if\ Z_o(x) \leq L_o^I \\ 1 - \frac{U_o^I - Z_o(x)}{U_o^I - L_o^I}, & if\ L_o^I \leq Z_o(x) \leq U_o^I \\ 1, & if\ Z_o^I \geq U_o^I \end{cases}$$
$$F_D(x) = \begin{cases} 0, & if\ Z_o(x) \geq U_o^F \\ 1 - \frac{Z_o(x) - L_o^F}{U_o^F - L_o^F}, & if\ L_o^F \leq Z_o(x) \leq U_o^F \\ 1, & if\ Z_o^T \leq L_o^T \end{cases} \tag{6}$$

where $U_o^{(.)} \neq L_o^{(.)}$ must hold for all objectives. If for any objective $U_o^{(.)} = L_o^{(.)}$ then the membership value will be one. The spherical fuzzy decision making model based on (Bellman and Zadeh 1970) concept for MOPP can be given as follows:

Moreover, all the above three discussed membership degrees can be transformed into membership goals according to their respective degrees of attainment. The highest degree of truth membership function that can be achieved is unity (1), the indeterminacy membership function is neutral and independent with the most upper attainment degree half (0.5), and the falsity membership function can be achieved with the highest attainment degree zero (0). Now the transformed membership goals under a spherical fuzzy environment can be expressed as follows:

$$T_o(Z_o(x))^2 + d_{oT}^- - d_{oT}^+ = 1 \tag{7}$$

$$I_o(Z_o(x))^2 + d_{oI}^- - d_{oI}^+ = 0.5 \tag{8}$$

$$F_o(Z_o(x))^2 + d_{oF}^- - d_{oF}^+ = 0 \tag{9}$$

where $d_{oT}^-, d_{oT}^+, d_{oI}^-, d_{oI}^+, d_{oF}^-$ and d_{oF}^+ are over and under deviational variable for truth, indeterminacy and a falsity membership functions respectively.

Intuitionally, the motive is to maximize the truth and indeterminacy membership degrees of spherical fuzzy objectives and constraints and minimize the falsity membership degree of spherical fuzzy objectives and constraints. The general formulation of the spherical fuzzy goal programming (SFGP) model for the multiobjective optimization problem is represented as follows:

Model-I represents the general SFGP model under spherical fuzzy environment where the satisfaction level is defined by the simply sum of the unwanted deviations from each spherical fuzzy goals.

Model-I

$$Minimize\ \alpha = \sum_{0=1}^{O} d_{oT}^- + \sum_{0=1}^{O} d_{oI}^- + \sum_{0=1}^{O} d_{oF}^+$$

Subject to

$$\begin{aligned}
&T_o(Z_o(x))^2 + d_{oT}^- - d_{oT}^+ \geq 1 \\
&I_o(Z_o(x))^2 + d_{oI}^- - d_{oI}^+ \geq 0.5 \\
&F_o(Z_o(x))^2 + d_{oF}^- - d_{oF}^+ \leq 0 \\
&g_i(x) \leq b_i, \quad \forall\, i = 1, 2, \ldots, I_1, \\
&g_i(x) \geq b_i, \quad \forall\, i = I_1 + 1, I_1 + 2, \ldots, I_2, \\
&g_i(x) = b_i, \quad \forall\, i = I_2 + 1, I_2 + 2, \ldots, I \\
&x = \left(x_1, x_2, \ldots, x_j\right) \in X, x \geq 0 \\
&T_o(Z_o(x))^2 \geq I_o(Z_o(x))^2, \;\; T_o(Z_o(x))^2 \geq F_o(Z_o(x))^2
\end{aligned}$$

$$0 \leq T_o(Z_o(x))^2 + I_o(Z_o(x))^2 + F_o(Z_o(x))^2 \leq 1$$

Model-II represents the Weighted SFGP (WSFGP) model under spherical fuzzy environment where the satisfaction level is defined by the weighted sum of the unwanted deviations from each spherical fuzzy goal.

Model-II

$$Minimize\ \alpha = \sum_{0=1}^{O} w_{oT} \cdot d_{oT}^{-} + w_{oI} \cdot d_{oI}^{-} + w_{oF} \cdot d_{oF}^{+}$$

Subject to

$$\begin{aligned}
&T_o(Z_o(x))^2 + d_{oT}^{-} - d_{oT}^{+} \geq 1 \\
&I_o(Z_o(x))^2 + d_{oI}^{-} - d_{oI}^{+} \geq 0.5 \\
&F_o(Z_o(x))^2 + d_{oF}^{-} - d_{oF}^{+} \leq 0 \\
&g_i(x) \leq b_i, \quad \forall\, i = 1, 2, \ldots, I_1, \\
&g_i(x) \geq b_i, \quad \forall\, i = I_1 + 1, I_1 + 2, \ldots, I_2, \\
&g_i(x) = b_i, \quad \forall\, i = I_2 + 1, I_2 + 2, \ldots, I \\
&x = \left(x_1, x_2, \ldots, x_j\right) \in X, x \geq 0 \\
&T_o(Z_o(x))^2 \geq I_o(Z_o(x))^2, \ \ T_o(Z_o(x))^2 \geq F_o(Z_o(x))^2 \\
&0 \leq T_o(Z_o(x))^2 + I_o(Z_o(x))^2 + F_o(Z_o(x))^2 \leq 1
\end{aligned}$$

where w_{oT}, w_{oI} and w_{oF} are the weight parameters assigned to deviational variables according to the decision makers preferences.

Hence the above optimization models (Models I and II) provide the optimal solution for the MOOP problem under spherical fuzzy environment.

4 Numerical Examples

To show the applicability and validity of the proposed SFGP models, following numerical examples are presented. All the three numerical problems are written in AMPL language and solved by using the global optimization solvers Knitro 5.0 online facility provided by University of Wisconsin (Dolan et al 2002).

Example 1 (Maximization type objectives)

$$\begin{aligned}
&\text{Maximize } Z_1 = 50x_1 + 100x_2 + 17.5x_3 \\
&\text{Maximize } Z_2 = 92x_1 + 75x_2 + 50x_3 \\
&\text{Maximize } Z_3 = 25x_1 + 100x_2 + 75x_3
\end{aligned}$$

Table 2 Individual best and worst value of objective functions

	Z_1	Z_2	Z_3	X
Maximum Z_1	8041	10,020.33	9319.25	X^1
Maximum Z_2	5452.63	10,950.59	5903.00	X^2
Maximum Z_3	7983.60	10,056.99	9355.90	X^3

Subject to

$$\begin{aligned}
&12x_1 + 17x_2 \leq 1400\\
&3x_1 + 9x_2 + 8x_3 \leq 1000\\
&10x_1 + 13x_2 + 15x_3 \leq 1750\\
&6x_1 + 16x_3 \leq 1325\\
&x_1, x_2, x_3 \geq 0.
\end{aligned}$$

On solving each objective function individually, we have obtained the following best and worst solution to define the upper and lower bound for each membership functions respectively. The best and worst solution is summarized in Table 2.

The upper and lower bound for the truth, indeterminacy and falsity membership degrees have been obtained as follows:

$$\begin{aligned}
&U_1^T = 8041, \quad L_1^T = 5452.63\\
&U_1^I = 5452.63 + p_1(8041 - 5452.63), \quad L_1^I = 5452.63\\
&U_1^F = 8041, \quad L_1^F = 5452.63 + q_1(8041 - 5452.63)
\end{aligned}$$

The different membership function for the first objective function is defined as follows:

$$T_D(Z_1(x)) = \begin{cases} 0, & if\ Z_1(x) \leq L_1^T \\ 1 - \frac{U_1^T - Z_1(x)}{U_1^T - L_1^T}, & if\ L_1^T \leq Z_1(x) \leq U_1^T \\ 1, & if\ Z_1(x) \geq U_1^T \end{cases}$$

$$I_D(Z_1(x)) = \begin{cases} 0, & if\ Z_1(x) \leq L_1^I \\ 1 - \frac{U_1^I - Z_1(x)}{U_1^I - L_1^I}, & if\ L_1^I \leq Z_1(x) \leq U_1^I \\ 1, & if\ Z_1(x) \geq U_1^I \end{cases}$$

$$F_D(Z_1(x)) = \begin{cases} 0, & if\ Z_1(x) \geq U_1^F \\ 1 - \frac{Z_1(x) - L_1^F}{U_1^F - L_1^F}, & if\ L_1^F \leq Z_1(x) \leq U_1^F \\ 1, & if\ Z_1(x) \leq L_1^T \end{cases}$$

The upper and lower bound for the truth, indeterminacy and a falsity membership degree have been obtained as follows:

$$U_2^T = 10950.59, \quad L_2^T = 10020.33$$
$$U_2^I = 10020.33 + p_2(10950.59 - 10020.33), \quad L_2^I = 10020.33$$
$$U_2^F = 10950.59, \quad L_2^F = 10020.33 + q_2(10950.59 - 10020.33)$$

The different membership function for the second objective function is defined as follows:

$$T_D(Z_2(x)) = \begin{cases} 0, & if\ Z_2(x) \leq L_2^T \\ 1 - \frac{U_2^T - Z_2(x)}{U_2^T - L_2^T}, & if\ L_2^T \leq Z_2(x) \leq U_2^T \\ 1, & if\ Z_2(x) \geq U_2^T \end{cases}$$

$$I_D(Z_2(x)) = \begin{cases} 0, & if\ Z_2(x) \leq L_2^I \\ 1 - \frac{U_2^I - Z_2(x)}{U_2^I - L_2^I}, & if\ L_2^I \leq Z_2(x) \leq U_2^I \\ 1, & if\ Z_2(x) \geq U_2^I \end{cases}$$

$$F_D(Z_2(x)) = \begin{cases} 0, & if\ Z_2(x) \geq U_2^F \\ 1 - \frac{Z_2(x) - L_2^F}{U_2^F - L_2^F}, & if\ L_2^F \leq Z_2(x) \leq U_2^F \\ 1, & if\ Z_2(x) \leq L_2^T \end{cases}$$

The upper and lower bound for the truth, indeterminacy and a falsity membership degree have been obtained as follows:

$$U_3^T = 9355.90, \quad L_3^T = 5903$$
$$U_3^I = 5903 + p_3(9355.90 - 5903), \quad L_3^I = 5903$$
$$U_3^F = 9355.90, \quad L_3^F = 5903 + q_3(9355.90 - 5903)$$

The different membership function for the third objective function is defined as follows:

$$T_D(Z_3(x)) = \begin{cases} 0, & if\ Z_3(x) \leq L_3^T \\ 1 - \frac{U_3^T - Z_3(x)}{U_3^T - L_3^T}, & if\ L_3^T \leq Z_3(x) \leq U_3^T \\ 1, & if\ Z_3(x) \geq U_3^T \end{cases}$$

$$I_D(Z_3(x)) = \begin{cases} 0, & if\ Z_3(x) \leq L_3^I \\ 1 - \frac{U_3^I - Z_3(x)}{U_3^I - L_3^I}, & if\ L_3^I \leq Z_3(x) \leq U_3^I \\ 1, & if\ Z_3(x) \geq U_3^I \end{cases}$$

$$F_D(Z_3(x)) = \begin{cases} 0, & if\ Z_3(x) \geq U_3^F \\ 1 - \frac{Z_3(x) - L_3^F}{U_3^F - L_3^F}, & if\ L_3^F \leq Z_3(x) \leq U_3^F \\ 1, & if\ Z_3(x) \leq L_3^T \end{cases}$$

Now the formulations of the spherical fuzzy optimization models are given as follows:

Model-I

$$Minimize\ \alpha = \sum_{0=1}^{3} d_{oT}^{-} + \sum_{0=1}^{3} d_{oI}^{-} + \sum_{0=1}^{3} d_{oF}^{+}$$

Subject to

$$\begin{aligned}
&T_o(Z_o(x))^2 + d_{oT}^{-} - d_{oT}^{+} \geq 1\\
&I_o(Z_o(x))^2 + d_{oI}^{-} - d_{oI}^{+} \geq 0.5\\
&F_o(Z_o(x))^2 + d_{oF}^{-} - d_{oF}^{+} \leq 0\\
&T_o(Z_o(x))^2 \geq I_o(Z_o(x))^2,\ \ T_o(Z_o(x))^2 \geq F_o(Z_o(x))^2\\
&0 \leq T_o(Z_o(x))^2 + I_o(Z_o(x))^2 + F_o(Z_o(x))^2 \leq 1\\
&12x_1 + 17x_2 \leq 1400\\
&3x_1 + 9x_2 + 8x_3 \leq 1000\\
&10x_1 + 13x_2 + 15x_3 \leq 1750\\
&6x_1 + 16x_3 \leq 1325\\
&x_1, x_2, x_3 \geq 0
\end{aligned}$$

The optimal solution of the problem is $x_1 = 58.48$, $x_2 = 34.59$, $x_3 = 47.69$ and the value of each objective is obtained as $Z_1 = 7217.97$, $Z_2 = 10359.72$, $Z_3 = 8498.59$ respectively.

Model-II

$$Minimize\ \alpha = \sum_{0=1}^{3} w_{oT} \cdot d_{oT}^{-} + w_{oI} \cdot d_{oI}^{-} + w_{oF} \cdot d_{oF}^{+}$$

Subject to

$$\begin{aligned}
&T_o(Z_o(x))^2 + d_{oT}^{-} - d_{oT}^{+} \geq 1\\
&I_o(Z_o(x))^2 + d_{oI}^{-} - d_{oI}^{+} \geq 0.5\\
&F_o(Z_o(x))^2 + d_{oF}^{-} - d_{oF}^{+} \leq 0\\
&12x_1 + 17x_2 \leq 1400\\
&3x_1 + 9x_2 + 8x_3 \leq 1000\\
&10x_1 + 13x_2 + 15x_3 \leq 1750
\end{aligned}$$

$$
\begin{aligned}
&6x_1 + 16x_3 \leq 1325 \\
&x_1, x_2, x_3 \geq 0, \\
&T_o(Z_o(x))^2 \geq I_o(Z_o(x))^2, \quad T_o(Z_o(x))^2 \geq F_o(Z_o(x))^2 \\
&0 \leq T_o(Z_o(x))^2 + I_o(Z_o(x))^2 + F_o(Z_o(x))^2 \leq 1
\end{aligned}
$$

where w_{oT}, w_{oI} and w_{oF} are the weight parameters and are equally assigned to each deviational variables.

The optimal solution of the problem is $x_1 = 62.78, x_2 = 36.94, x_3 = 51.19$ and the value of each objective is obtained as $Z_1 = 7412.52, Z_2 = 11025.72, Z_3 = 8368.59$ respectively.

Example 2 (Minimization type objectives)

$$
\begin{aligned}
\text{Minimize } Z_1 &= 10x_1 + 13x_2 + 18x_3 \\
\text{Minimize } Z_2 &= 45x_1 + 84x_2 + 42x_3 \\
\text{Minimize } Z_3 &= 57x_1 + 18x_2 + 27x_3
\end{aligned}
$$

Subject to

$$
\begin{aligned}
&2x_1 + 7x_2 + 3x_3 \geq 140 \\
&3x_1 + 9x_2 + 8x_3 \geq 100 \\
&4x_1 + 3x_2 + 5x_3 \geq 150 \\
&6x_1 - 7x_2 + 6x_3 \geq 125 \\
&x_1, x_2, x_3 \geq 0.
\end{aligned}
$$

On solving each objective function individually, we have obtained the following best and worst solution to define the upper and lower bound for each membership functions respectively. The best and worst solution is summarized in Table 3.

The upper and lower bound for the positive, neutral and negative membership degrees have been obtained as follows:

$$
\begin{aligned}
&U_1^T = 2375.62, \quad L_1^T = 468.21 \\
&U_1^I = 468.21 + p_1(2375.62 - 468.21), \quad L_1^I = 468.21
\end{aligned}
$$

Table 3 Individual best and worst values of objective functions

	Z_1	Z_2	Z_3	X
Minimum Z_1	468.21	625.95	625.95	X^1
Minimum Z_2	2375.62	1856.67	1856.67	X^2
Minimum Z_3	2077.77	927.85	927.85	X^3

$$U_1^F = 2375.62, \quad L_1^F = 468.21 + q_1(2375.62 - 468.21)$$

The different membership function for the first objective function is defined as follows:

$$T_D(Z_1(x)) = \begin{cases} 1, & if \ Z_1(x) \leq L_1^T \\ 1 - \frac{Z_1(x) - L_1^T}{U_1^T - L_1^T}, & if \ L_1^T \leq Z_2(x) \leq U_1^T \\ 0, & if \ Z_2(x) \geq U_1^T \end{cases}$$

$$I_D(Z_1(x)) = \begin{cases} 1, & if \ Z_2(x) \leq L_1^I \\ 1 - \frac{Z_1(x) - L_1^I}{U_1^I - L_1^I}, & if \ L_1^I \leq Z_2(x) \leq U_1^I \\ 0, & if \ Z_2(x) \geq U_1^I \end{cases}$$

$$F_D(Z_1(x)) = \begin{cases} 1, & if \ Z_2(x) \geq U_1^F \\ 1 - \frac{U_1^F - Z_1(x)}{U_1^F - L_1^F}, & if \ L_1^F \leq Z_2(x) \leq U_1^F \\ 0, & if \ Z_2(x) \leq L_1^F \end{cases}$$

The upper and lower bound for the truth, indeterminacy and a falsity membership degree have been obtained as follows:

$$U_2^T = 1856.67, \quad L_2^T = 625.95$$
$$U_2^I = 625.95 + p_2(1856.67 - 625.95), \quad L_2^I = 625.95$$
$$U_2^F = 1856.67, \quad L_2^F = 625.95 + q_2(1856.67 - 625.95)$$

The different membership function for the second objective function is defined as follows:

$$T_D(Z_2(x)) = \begin{cases} 1, & if \ Z_2(x) \leq L_2^T \\ 1 - \frac{Z_2(x) - L_2^T}{U_2^T - L_2^T}, & if \ L_o^T \leq Z_2(x) \leq U_2^T \\ 0, & if \ Z_2(x) \geq U_2^T \end{cases}$$

$$I_D(Z_2(x)) = \begin{cases} 1, & if \ Z_2(x) \leq L_2^I \\ 1 - \frac{Z_2(x) - L_2^I}{U_2^I - L_2^I}, & if \ L_2^I \leq Z_2(x) \leq U_2^I \\ 0, & if \ Z_2(x) \geq U_2^I \end{cases}$$

$$F_D(Z_2(x)) = \begin{cases} 1, & if \ Z_2(x) \geq U_2^F \\ 1 - \frac{U_2^F - Z_2(x)}{U_2^F - L_2^F}, & if \ L_o^F \leq Z_2(x) \leq U_2^F \\ 0, & if \ Z_2(x) \leq L_2^F \end{cases}$$

The upper and lower bound for the truth, indeterminacy and a falsity membership degree have been obtained as follows:

$$U_3^T = 1856.67, \quad L_3^T = 625.95$$

$$U_3^I = 625.95 + p_3(1856.67 - 625.95), \quad L_3^I = 625.95$$
$$U_3^F = 1856.67, \quad L_3^F = 625.95 + q_3(1856.67 - 625.95)$$

The different membership function for the third objective function is defined as follows:

$$T_D(Z_3(x)) = \begin{cases} 1, & if\ Z_3(x) \leq L_3^T \\ 1 - \frac{Z_3(x) - L_3^T}{U_3^T - L_3^T}, & if\ L_3^T \leq Z_3(x) \leq U_3^T \\ 0, & if\ Z_3(x) \geq U_3^T \end{cases}$$

$$I_D(Z_3(x)) = \begin{cases} 1, & if\ Z_3(x) \leq L_3^I \\ 1 - \frac{Z_3(x) - L_3^I}{U_3^I - L_3^I}, & if\ L_3^I \leq Z_3(x) \leq U_3^I \\ 0, & if\ Z_3(x) \geq U_3^I \end{cases}$$

$$F_D(Z_3(x)) = \begin{cases} 1, & if\ Z_3(x) \geq U_3^F \\ 1 - \frac{U_3^F - Z_3(x)}{U_3^F - L_3^F}, & if\ L_3^F \leq Z_3(x) \leq U_3^F \\ 0, & if\ Z_3(x) \leq L_3^F \end{cases}$$

Now the formulations of the spherical fuzzy optimization model are given as follows:

Model-I

$$Minimize\ \alpha = \sum_{0=1}^{3} d_{oT}^{-} + \sum_{0=1}^{3} d_{oI}^{-} + \sum_{0=1}^{3} d_{oF}^{+}$$

Subject to

$$T_o(Z_o(x))^2 + d_{oT}^{-} - d_{oT}^{+} \geq 1$$
$$I_o(Z_o(x))^2 + d_{oI}^{-} - d_{oI}^{+} \geq 0.5$$
$$F_o(Z_o(x))^2 + d_{oF}^{-} - d_{oF}^{+} \leq 0$$
$$T_o(Z_o(x))^2 \geq I_o(Z_o(x))^2, \quad T_o(Z_o(x))^2 \geq F_o(Z_o(x))^2$$
$$0 \leq T_o(Z_o(x))^2 + I_o(Z_o(x))^2 + F_o(Z_o(x))^2 \leq 1$$
$$2x_1 + 7x_2 + 3x_3 \geq 140$$
$$3x_1 + 9x_2 + 8x_3 \geq 100$$
$$4x_1 + 3x_2 + 5x_3 \geq 150$$
$$6x_1 - 7x_2 + 6x_3 \geq 125$$
$$x_1, x_2, x_3 \geq 0.$$

The optimal solution of the problem is $x_1 = 3.02, x_2 = 7.38, x_3 = 29.44$ and the value of each objective is obtained as $Z_1 = 625.95, Z_2 = 1856.67, Z_3 = 927.86$ respectively.

Model-II

$$Minimize\ \alpha = \sum_{0=1}^{3} w_{oT} \cdot d_{oT}^{-} + w_{oI} \cdot d_{oI}^{-} + w_{oF} \cdot d_{oF}^{+}$$

Subject to

$$\begin{aligned}
&T_o(Z_o(x))^2 + d_{oT}^{-} - d_{oT}^{+} \geq 1 \\
&I_o(Z_o(x))^2 + d_{oI}^{-} - d_{oI}^{+} \geq 0.5 \\
&F_o(Z_o(x))^2 + d_{oF}^{-} - d_{oF}^{+} \leq 0 \\
&2x_1 + 7x_2 + 3x_3 \geq 140 \\
&3x_1 + 9x_2 + 8x_3 \geq 100 \\
&4x_1 + 3x_2 + 5x_3 \geq 150 \\
&6x_1 - 7x_2 + 6x_3 \geq 125 \\
&x_1, x_2, x_3 \geq 0, \\
&T_o(Z_o(x))^2 \geq I_o(Z_o(x))^2, \ \ T_o(Z_o(x))^2 \geq F_o(Z_o(x))^2 \\
&0 \leq T_o(Z_o(x))^2 + I_o(Z_o(x))^2 + F_o(Z_o(x))^2 \leq 1
\end{aligned}$$

where w_{oT}, w_{oI} and w_{oF} are the weight parameters and are equally assigned to each deviational variables.

The optimal solution of the problem is $x_1 = 4.29, x_2 = 9.23, x_3 = 31.42$ and the value of each objective is obtained as $Z_1 = 636.58, Z_2 = 1902.52, Z_3 = 963.28$ respectively.

Example 3 (Industrial Application)

The problem discussed in this section is taken from (Smarandache and Zhou 2017) which is an industrial problem. Suppose that the Manager or Decision maker intends to reduce approx. 98.5% biological oxygen demand (BOD) and the tolerance limits of acceptance, indeterminacy and rejection on this goal are 0.1, 0.2 and 0.3 respectively. Moreover the Decision maker also intends to remove the quantity of BODS5 within 300 (thousand \$) tolerances of acceptance, indeterminacy and rejection 200, 250, 300 (thousand \$) respectively. Thus the spherical fuzzy goal programming problem can be formulated as follows:

$$\begin{aligned}
&Minimize\ \ Z_1 = 19.4x_1^{-1.47} + 16.8x_2^{-1.66} + 91.5x_3^{-0.3} + 120x_4^{-0.33} \\
&Minimize\ \ Z_2 = x_1 \cdot x_2 \cdot x_3 \cdot x_4
\end{aligned}$$

$$Subject\ to,$$
$$x_i \geq 0,\ i = 1, 2, 3, 4.$$

The target values assigned to first and second objectives are 300 and 0.015 respectively. For the first objective, the acceptance, indeterminacy and the rejection tolerance level are assigned as 200, 100 and 300 respectively. The acceptance, indeterminacy and the rejection tolerance level for the second objective are determined as 0.1, 0.05 and 0.2 respectively. The decision variables $x_i^{\prime}s$ represent the percentage of BOD5 (to remove 5 days BOD) in the consecutive step. After four processes, the left % of BOD5 will be $x_i,\ i = 1, 2, 3, 4$. The motive is to reduce the remaining of BOD5 with minimum yearly gross cost at its lowest level. The first objective (Z_1) represents the removal cost of BOD5 by different treatments such as primary clarifier, trickling filter, activated sludge, carbon adsorption. Similarly the second objective (Z_2) represents the annual cost incurred over removal of wastewater.

Now the truth membership, indeterminacy membership and falsity membership functions have been defined under spherical fuzzy environment.

The truth membership functions of the spherical fuzzy goals are obtained as follows:

$$T_D(Z_1(x)) = \begin{cases} 1, & if\ Z_1(x) \leq 300 \\ 1 - \frac{Z_1(x)-300}{200}, & if\ 300 \leq Z_1(x) \leq 500 \\ 0, & if\ Z_1(x) \geq 500 \end{cases}$$

$$T_D(Z_2(x)) = \begin{cases} 1, & if\ Z_2(x) \leq 0.015 \\ 1 - \frac{Z_2(x)-0.015}{0.1}, & if\ 0.015 \leq Z_2(x) \leq 0.115 \\ 0, & if\ Z_2(x) \geq 0.115 \end{cases}$$

The indeterminacy membership functions of the spherical fuzzy goals are obtained as follows:

$$I_D(Z_1(x)) = \begin{cases} 0, & if\ Z_1(x) \leq 300 \\ 1 - \frac{600-Z_1(x)}{300}, & if\ 300 \leq Z_1(x) \leq 600 \\ 1, & if\ Z_1(x) \geq 600 \end{cases}$$

$$I_D(Z_2(x)) = \begin{cases} 0, & if\ Z_2(x) \leq 0.015 \\ 1 - 0.05, & if\ 0.015 \leq Z_2(x) \leq 0.065 \\ 1, & if\ Z_2(x) \geq 0.065 \end{cases}$$

The falsity membership functions of the spherical fuzzy goals are obtained as follows:

$$F_D(Z_1(x)) = \begin{cases} 1, & if\ Z_1(x) \geq 600 \\ 1 - \frac{600-Z_1(x)}{300}, & if\ 300 \leq Z_1(x) \leq 600 \\ 0, & if\ Z_1(x) \leq 300 \end{cases}$$

Table 4 Comparison of optimal solution based on different methods

Methods	Z_1	Z_2	x_1	x_2	x_3	x_4
FGGPP (Smarandache and Zhou 2017))	363.8048	0.04692	0.705955	0.7248393	0.1598653	0.5733523
IFGGPP (Smarandache and Zhou 2017)	422.1483	0.01504	0.638019	0.662717	0.09737155	0.3653206
NGPP (Smarandache and Zhou 2017)	317.666	0.1323	0.774182	0.7865418	0.2512332	0.8647621
WNGPP (Smarandache and Zhou 2017)	417.6666	0.2150	2.628853	3.087266	0.181976E-01	1.455760
SFGP (Model-I)	317.257	0.1323	0.754182	0.76522	0.248992	0.853892
WSFGP (Model-II)	363.8128	0.04589	0.702133	0.719845	0.1588324	0.5724323

$$F_D(Z_2(x)) = \begin{cases} 1, & if\ Z_2(x) \geq 0.215 \\ 1 - \frac{0.215 - Z_{21}(x)}{0.2}, & if\ 0.015 \leq Z_2(x) \leq 0.215 \\ 0, & if\ Z_2(x) \leq 0.015 \end{cases}$$

The solution results of the above problem is summarized in Table 4. The same problem has been also solved by using four different methods (Smarandache and Zhou 2017) namely; fuzzy goal geometric programming problem (FGGPP), intuitionistic fuzzy goal geometric programming problem (IFGGPP), neutrosophic goal programming problem (NGPP) and weighted neutrosophic goal programming problem (WNGPP) respectively. From Table 4, it is clear that the performance of proposed SFGP (Model-I) and WSFGP (Model-II) is better as compared to other existing approaches.

5 Conclusions

In this chapter, we have addressed the spherical fuzzy goal programming problem under a spherical fuzzy environment. The discussed spherical fuzzy goal programming optimization technique is based on a spherical fuzzy decision set. The spherical fuzzy set includes the positive, neutral, and negative membership grades of the element into the spherical fuzzy decision set. The advanced modeling and optimization framework is presented efficiently along with the numerical examples to show the validity and applicability of the model. The propounded SFGP reduces the unwanted deviations from the ideal solution by satisfying the different spherical fuzzy goals under the SF decision set. The spherical fuzzy goal programming model provides more flexibility in the decision making process as compared to fuzzy, intuitionistic fuzzy, and Pythagorean fuzzy goal optimization techniques. The SFGP technique exhibits the degree of neutral thoughts or indeterminacy with the sum of squares of truth and falsity membership degrees to one (1). Both (Models-I and II) the SFGP model outperform while applied to the industrial problems and yield better results than other existing approaches.

The aim of this chapter is to introduce the spherical fuzzy goal programming model under the SF concept and to explore the efficient utility in continuous optimization cases such as multiobjective optimization problems. Hence, it may be applied to a wide range of real-life applications in a different field such as transportation problems, supplier selection problem, inventory control, supply chain management, etc.

In the future, the proposed spherical fuzzy goal programming problem may be applied to a non-linear programming problem, fractional programming problem, bi-level, and multilevel programming problem, etc. It is also possible to extend the spherical fuzzy concept with the stochastic and uncertain programming domain.

References

Ahmad F, Adhami AY (2019) Neutrosophic programming approach to multiobjective nonlinear transportation problem with fuzzy parameters. Int J Manag Sci Eng Manag 14(3):218–229

Ahmad F, Adhami AY, Smarandache F (2018) Single valued neutrosophic hesitant fuzzy computational algorithm for multiobjective nonlinear optimization problem. Neutrosophic Sets Syst 22:76–86

Ahmad F, Adhami AY, Smarandache F (2019) Neutrosophic optimization model and computational algorithm for optimal shale gas water management under uncertainty. Symmetry 11(4):544

Ahmad F, Adhami AY, Smarandache F (2020) Modified neutrosophic fuzzy optimization model for optimal closed-loop supply chain management under uncertainty. In: Optimization theory based on neutrosophic and plithogenic sets. Academic Press, pp 343–403

Angelov PP (1997) Optimization in an intuitionistic fuzzy environment. Fuzzy Sets Syst 86(3):299–306

Atanassov KT (1986) Studies in fuzziness and soft computing intuitionistic fuzzy logics. Fuzzy Sets Syst

Bellman RE, Zadeh LA (1970) Decision-making in a fuzzy environment. Manage Sci 17(4):140–164

Charnes A, Cooper WW (1977) Goal programming and multiple objective optimizations. Eur J Oper Res 1(1):39–54

Dharmar S, Tiwari RN, Rao JR (1987) Goal Program 24:27–34

Dolan E, Fourer R, Moré JJ, Munson TS (2002) The NEOS server for optimization: version 4 and beyond. Preprint ANL/MCS-TM-253, Mathematics and Computer Science Division, Argonne National Laboratory

Kutlu Gündoğdu F, Kahraman C (2019) Spherical fuzzy sets and spherical fuzzy TOPSIS method. J Intell Fuzzy Syst, (Preprint), 1–16

Smarandache F (1999) A unifying field in logics: neutrosophic logic. In: Philosophy. American Research Press, pp 1–141

Smarandache F, Zhou Y (2017) Neutrosophic Operational Research, vol I
Zimmermann HJ (1978) Fuzzy programming and linear programming with several objective functions. Fuzzy Sets Syst 1(1):45–55

Spherical Fuzzy Geometric Programming Problem

Firoz Ahmad and Ahmad Yusuf Adhami

Abstract Geometric programming (GP) optimization techniques are an important and interesting topic since the 1970s. The significant development can be realized by having enormous researches in this field, such as fuzzy geometric optimization technique, multiobjective goal geometric programming, different goal geometric programming techniques, etc. Thus this chapter investigates a new algorithm based on the spherical fuzzy set (SFS) named as spherical fuzzy geometric programming problem (SFGPP) under the spherical fuzzy environment. The SFGPP inevitably involves the degree of neutrality along with truth and a falsity membership degree of the element into the feasible decision set. It also generalizes the decision set by imposing the restriction that the sum of squares of each membership function must be less than or equal to one. The attainment of achievement function is determined by maximizing the truth and minimizing the neutral and negative membership degree of each objective function under the spherical fuzzy decision set. At last, numerical examples and conclusions are presented to reveal the applicability and future research scope in the SF domain.

1 Introduction

The concept of geometric programming (GP) was first introduced by Duffin et al. in the book "Geometric Programming—Theory and Applications" in 1967. Later on, the development of a numerical approach for the solution of GPs was first introduced by Duffin (1970). The basic concepts and theoretical aspects of GPs were discussed by Zener (1971); after that, Beightler and Philips (1976) made a significant contribution in their famous book on GP and its applications. Bazaraa et al. (2013), Floudas (2013), and Peterson (2001) presented the origin and brief history of the development of GP problems.

F. Ahmad (✉) · A. Y. Adhami
Department of Statistics and Operations Research, Aligarh Muslim University, Aligarh, India
e-mail: firoz.ahmad02@gmail.com

© Springer Nature Switzerland AG 2021

C. Kahraman and F. Kutlu Gündoğdu (eds.), *Decision Making with Spherical Fuzzy Sets*, Studies in Fuzziness and Soft Computing 392,
https://doi.org/10.1007/978-3-030-45461-6_22

Based on the different uncertain sets, such as fuzzy set (Bellman and Zadeh 1970), intuitionistic fuzzy set (Atanassov 1986), neutrosophic set (see Smarandache (1999), Ahmad and Adhami (2019a), Ahmad et al. (2019b), Ahmad et al. (2018)) etc.; different geometric programming optimization techniques have been developed. Recently, Ahmad et al. (2020) presented the modified neutrosophic programming approach for multiobjective programming problems which is based on the neutrosophic decision set. The decision-making problem can be classified into two broad cases, such as discrete and continuous. In the discrete case, multi-attribute and multi-criteria decision-making problems are dealt with, whereas the geometric programming problem falls in the continuous case. The generalization of fuzzy set, intuitionistic fuzzy set and Pythagorean fuzzy set (Yager 2013) is presented by Kutlu Gündoğdu and Kahraman (2019) and termed as spherical fuzzy set (SFS) which is based on three different membership degrees named as positive, neutral and negative membership degrees of the element into the feasible solution set. The limitations of fuzzy, intuitionistic fuzzy, Pythagorean fuzzy sets can be described by the following explanations. Suppose that in multi-attribute decision-making problems, the acceptance degree of selecting an alternative is 0.6, the rejection of the alternatives is 0.7, and the indeterminacy or neutral degree of selecting the alternative is found to be 0.8 then this situation is beyond the coverage of above discussed uncertain sets. Hence to deal with such cases, the spherical fuzzy set is a robust decision-making tool under the spherical fuzzy environment (see Rafiq et al. 2019). It also allows the decision-makers to incorporate indeterminacy and neutral thoughts in decision-making processes.

The remaining part of the chapter is summarized as follows: In Sect. 2, the spherical fuzzy geometric programming problem and solution algorithm have been studied. Section 3 presents the numerical examples, and the system reliability optimization problem is discussed, whereas a conclusion along with the future research scope is described in Sect. 4.

2 Spherical Fuzzy Geometric Programming Model

The spherical fuzzy set is the generalization of various classical sets such as fuzzy, intuitionistic fuzzy, Pythagorean fuzzy sets. The SFS reflects reality as compared to others by introducing the degree of neutrality or indeterminacy in decision-making processes. The marginal evaluation of the objective functions are represented by three different membership function, namely; the truth, indeterminacy or neutral and the falsity membership functions respectively. Therefore SFS holds the most critical aspects of decision making, where it may be indispensable to adopt the indeterminacy degree. The GP under the spherical fuzzy environment is quite apparent as many engineering and management problems may not be easy to solve without considering the indeterminacy aspects of the decision-making scenario. Thus the formulation of geometric programming problems under the spherical fuzzy environment is quite essential and much needed to obtain the optimal solutions in different fields of optimizations such as engineering and management problems, etc.

2.1 Geometric Programming Problem

The geometric programming problem is a particular case of nonlinear programming problems. The structure of the geometric programming problem depends on the representation or involvement of the decision variables in their product form. To specify the geometric type of objective functions, the following expressions are presented. With the aid of these technical terms, the representation of the geometric programming problem can be done easily.

Monomial The term "monomial" is derived from Latin word, ***mono*** meaning only one and ***mial*** meaning term. Hence monomial is an "algebraic expression which contains only one term." The definition of the monomial used in geometric programming is same but only difference is that in algebra, exponent of the variable cannot have a negative or fractional. Thus, in GP the exponent can be any real number, including fractional and negative.

Therefore, if $x_1, x_2, \ldots, x_n$ represents n real positive variables, then real valued function G of x, can be represented as follows (Eq. 1):

$$G(x) = cx_1^{a_1} x_2^{a_2} \ldots x_n^{a_n} \tag{1}$$

where $c > 0$ and $a_n \in \Re$ is known as monomial function.

Polynomial The term "polynomial" is derived from Latin word, ***polyno*** meaning many and ***mial*** meaning term. Therefore, polynomial is an "algebraic expression which contains many terms." Hence sum of one or more monomials, that is; any real valued function G of x of the following form (Eq. 2)

$$G(x) = \sum_{i=1}^{m} c_i x_1^{a_{i1}} x_2^{a_{i2}} \ldots x_n^{a_{in}} \tag{2}$$

represents the polynomial function.

Posynomial Areal valued function $G(x)$ is said to be posynomial, if all the coefficients $c_i > 0$. Thus, the sum of one or more monomials, that is; any real valued function of the following form (Eq. 3)

$$G(x) = \sum_{i=1}^{m} c_i x_1^{a_{i1}} x_2^{a_{i2}} \ldots x_n^{a_{in}} \tag{3}$$

where $c_i > 0$, represents the posynomial function.

By using the above terminology, the formulation of geometric programming problem can be given as follows (Eq. 4):

$$Minimize\ G_0(x) = \sum_{l=1}^{K_0} c_{0l} \prod_{j=1}^{J} x_j^{\lambda_{0lj}}$$

subject to

$$g_r(x) = \sum_{l=1+K_{r-1}}^{K_r} c_{rl} \prod_{j=1}^{J} x_j^{\lambda_{rlj}} \leq 1 \quad x_j > 0,\ j = 1, 2, \ldots, J \tag{4}$$

where $c_{rl}(> 0)$ and $\lambda_{rlj}(l = 1, 2, \ldots, 1 + K_{r-1}, \ldots, K_r; r = 0, 1, 2, \ldots, 1; j = 1, 2, \ldots, J)$ are real numbers. The problem in (1) is a constrained posynomial GP problem. Each posynomial constraints contains different number of terms and it is represented by K_r for all $r = 0, 1, 2, \ldots, 1$.

2.2 Multiobjective Geometric Programming Problem

In the present scenario, real-life optimization problems seldom contain a single objective. The existence of multiple objectives is very common in day to day life. For example, in the transportation problem, the cost, time, profit, etc. are the conflicting objectives that are to be optimized simultaneously. Thus, the need for multiobjective optimization techniques is of prime concern. The search for the global solution that satisfies all the different objective functions efficiently is quite challenging; however, a compromise solution is acceptable up to some extent. Therefore the continuous efforts in the development of different multiobjective optimization techniques laid down the base for future research in this domain. The multiobjective geometric programming problems (MOGPP) are very trivial in nature and exist in different real-life issues such as gravel-box problem, bar truss-region allocation, system reliability, inventory control, etc. with the multiple objectives such as volume, time, costs, scrap, reliability, etc. Many researchers have revealed a significant research contribution to MOGPP. Das et al. (2000) discussed the multi-item inventory model in which the quantity-dependent inventory costs and demand-dependent unit cost have been taken as uncertain and solved using geometric programming techniques. Mahapatra and Roy (2009) addressed a single and multi-container maintenance model under fuzzy uncertainty and solved by using a multiobjective geometric programming technique. Islam and Roy (2006) also investigated a new fuzzy multi-objective geometric programming problem and presented the application of transportation problems.

The mathematical formulation of multiobjective geometric programming problem can be given as follows (Eq. 5):

$$Minimize\ G_k(x) = \sum_{i=1}^{T_k^0} c_{ki}^0 \prod_{r=1}^{n} x_x^{\lambda_{kir}^0}$$

subject to

$$g_k(x) = \sum_{s=1}^{T_k} c_{ks} \prod_{r=1}^{n} x_r^{\lambda_{ksr}} \leq 1 \quad x > 0,\ k = 1, 2, \ldots, m \tag{5}$$

where $c_{ji}^0(> 0)$, $c_{ks}(> 0)$, a_{jir}^0, a_{jir} are real numbers for all $j = 1, 2, \ldots, k; i = 1, 2, \ldots, T_j^0; k = 1, 2, \ldots, m; s = 1, 2, \ldots, T_k$. $G_k(x)$ is the kth objective function. The function $g_k(x)$ is the real valued function and $x = (x_1, x_2, \ldots, x_r)$ represents the set of decision variables respectively.

Based on the spherical fuzzy decision set, a spherical fuzzy geometric programming (SFGP) problem is discussed. The spherical fuzzy set contains three different membership functions, namely truth, indeterminacy, and falsity, along with the restrictions that sum of the squares of all three membership degrees must lies between the intervals 0–1. Consequently, SFGP is a more convenient optimization tool than other geometric programming optimization techniques because of the existence of indeterminacy membership while dealing with multiobjective decision-making problems.

According to Bellman and Zadeh (1970), the fuzzy decision set (D) consists of fuzzy goals (Z) and fuzzy constraints (C) under fuzzy environment. The decision set has been widely used in decision making processes and can be stated as follows:

$$D = (Z \cap C)$$

Hence, based on the fuzzy decision set theory, the spherical fuzzy decision set can be developed under spherical fuzzy environment. Consequently, the spherical fuzzy decision set (D_{sf}), spherical fuzzy goals (Z_o) and spherical fuzzy constraints (C_o) can be given as follows (Eq. 6):

$$D_{sf} = \left(\bigcap_{o=1}^{O} Z_o\right)\left(\bigcap_{j=1}^{J} C_j\right) = (x, T_D(x), I_D(x), F_D(x)) \tag{6}$$

where

$$T_D(x) = \begin{Bmatrix} T_{D_1}(x), T_{D_2}(x), T_{D_3}(x), \ldots, T_{D_O}(x) \\ T_{C_1}(x), T_{C_2}(x), T_{C_3}(x), \ldots, T_{C_J}(x) \end{Bmatrix} \quad \forall x \in X$$

$$I_D(x) = \begin{Bmatrix} I_{D_1}(x), I_{D_2}(x), I_{D_3}(x), \ldots, I_{D_O}(x) \\ I_{C_1}(x), I_{C_2}(x), I_{C_3}(x), \ldots, I_{C_J}(x) \end{Bmatrix} \quad \forall x \in X$$

$$F_D(x) = \begin{Bmatrix} F_{D_1}(x), F_{D_2}(x), F_{D_3}(x), \ldots, F_{D_O}(x) \\ F_{C_1}(x), F_{C_2}(x), F_{C_3}(x), \ldots, F_{C_J}(x) \end{Bmatrix} \quad \forall x \in X$$

where $T_D(x)$, $I_D(x)$ and $F_D(x)$ are the truth membership function, indeterminacy membership function and a falsity membership function under spherical fuzzy decision set D_{sf}.

Table 1 Pay-off matrix

	Z_1	Z_2	…	Z_o
X^1	$Z_1(X^1)$	$Z_2(X^1)$	…	$Z_o(X^1)$
X^2	$Z_1(X^2)$	$Z_2\ (X^2)$	…	$Z_o(X^2)$
…	…	…	…	…
…	…	…	…	…
…	…	…	…	…
X^o	$Z_1(X^o)$	$Z_2\ (X^o)$	…	$Z_o(X^o)$

To determine the marginal evaluation of each objective, first we calculate the upper and lower bound for each objective function. Suppose that the obtained decision variables are $X^1, X^1, X^1, \ldots, X^o$. The pay-off matrix has been shown in Table 1. By substituting these values in each objective function, the mathematical expressions of upper and lower bound can be given as follows (Eq. 7):

$$U_o = \max\{Z_o(X^o)\} \text{ and } L_o = \min\{Z_o(X^o)\} \quad \forall o = 1, 2, \ldots, O \tag{7}$$

where U_o and L_o represents the upper and lower bound of Oth objectives. The upper and lower bounds for truth, indeterminacy and a falsity membership functions for each objective under spherical fuzzy environment can be derived by using the following mathematical expressions (Eq. 8):

$$\begin{aligned} &U_o^T = U_o,\ L_o^T = L_o \\ &U_o^I = L_o^T + p_o(U_o^T - L_o^T),\ L_o^I = L_o^T \\ &U_o^F = U_o^T,\ L_o^F = L_o^T + q_o(U_o^T - L_o^T) \end{aligned} \tag{8}$$

where p_o and q_o are the predetermined real numbers such that $p_o,\ q_o \in (0,\ 1)$ assigned by decision maker.

Case-I When the objective function are of minimization type then truth, indeterminacy and a falsity membership functions for each objective under spherical fuzzy environment can be obtained as follows (Eq. 9):

$$T_D(x) = \begin{cases} 1 & if\ Z_o(x) \le L_o^T \\ 1 - \frac{Z_o(x) - L_o^T}{U_o^T - L_o^T}, & if\ L_o^T \le Z_o(x) \le U_o^T \\ 0 & if\ Z_o^T \ge U_o^T \end{cases}$$

$$I_D(x) = \begin{cases} 1 & if\ Z_o(x) \le L_o^I \\ 1 - \frac{Z_o(x) - L_o^I}{U_o^I - L_o^I}, & if\ L_o^I \le Z_o(x) \le U_o^I \\ 0 & if\ Z_o^I \ge U_o^I \end{cases}$$

$$F_D(x) = \begin{cases} 1 & if\ Z_o(x) \geq U_o^F \\ 1 - \frac{U_o^F - Z_o(x)}{U_o^F - L_o^F}, & if\ L_o^F \leq Z_o(x) \leq U_o^F \\ 0 & if\ Z_o^T \leq L_o^T \end{cases} \tag{9}$$

Case-II When the objective function are of minimization type then truth, indeterminacy and a falsity membership functions for each objective under spherical fuzzy environment can be obtained as follows (Eq. 10):

$$T_D(x) = \begin{cases} 0, & if\ Z_o(x) \leq L_o^T \\ 1 - \frac{U_o^T - Z_o(x)}{U_o^T - L_o^T}, & if\ L_o^T \leq Z_o(x) \leq U_o^T \\ 1, & if\ Z_o^T \geq U_o^T \end{cases}$$

$$I_D(x) = \begin{cases} 0, & if\ Z_o(x) \leq L_o^I \\ 1 - \frac{U_o^I - Z_o(x)}{U_o^I - L_o^I}, & if\ L_o^I \leq Z_o(x) \leq U_o^I \\ 1, & if\ Z_o^I \geq U_o^I \end{cases}$$

$$F_D(x) = \begin{cases} 0, & if\ Z_o(x) \geq U_o^F \\ 1 - \frac{Z_o(x) - L_o^F}{U_o^F - L_o^F}, & if\ L_o^F \leq Z_o(x) \leq U_o^F \\ 1, & if\ Z_o^T \leq L_o^T \end{cases} \tag{10}$$

where $U_o^{(.)} \neq L_o^{(.)}$ must hold for all objectives. If for any objective $U_o^{(.)} = L_o^{(.)}$ then the membership value will be one. Intuitionally, the motive is to maximize the truth and indeterminacy membership degrees of spherical fuzzy objectives and constraints, and minimize the falsity membership degree of spherical fuzzy objectives and constraints. The general formulation of the spherical fuzzy geometric programming (SFGP) model for multiobjective optimization problem is represented as follows (Eq. 11):

$$\begin{aligned} &Max \min_{o=1,2,\ldots,O} T_o(Z_o(x))^2 \\ &Min \max_{o=1,2,\ldots,O} I_o(Z_o(x))^2 \\ &Min \max_{o=1,2,\ldots,O} F_o(Z_o(x))^2 \end{aligned}$$

Subject to

$$\begin{aligned} &g_i(x) \leq b_i, \forall\, i = 1, 2, \ldots, I_1, \\ &g_i(x) \geq b_i, \forall\, i = I_1 + 1, I_1 + 2, \ldots, I_2, \\ &g_i(x) = b_i, \forall\, i = I_2 + 1, I_2 + 2, \ldots, I \\ &x = (x_1, x_2, \ldots, x_j) \in X, x \geq 0 \\ &T_o(Z_o(x))^2 \geq I_o(Z_o(x))^2, T_o(Z_o(x))^2 \geq F_o(Z_o(x))^2 \\ &0 \leq T_o(Z_o(x))^2 + I_o(Z_o(x))^2 + F_o(Z_o(x))^2 \leq 1 \end{aligned} \tag{11}$$

Using the auxiliary variables, the above mathematical model (Eq. 11): can be reformulated as follows (Eq. 12):

$$Max\ \alpha^2$$
$$Min\ \beta^2$$
$$Min\ \gamma^2$$

Subject to

$$\begin{aligned}
&T_o(Z_o(x))^2 \geq \alpha^2,\ I_o(Z_o(x))^2 \leq \beta^2,\ F_o(Z_o(x))^2 \leq \gamma^2 \\
&g_i(x) \leq b_i, \forall\, i = 1, 2, \ldots, I_1, \\
&g_i(x) \geq b_i, \forall\, i = I_1 + 1, I_1 + 2, \ldots, I_2, \\
&g_i(x) = b_i, \forall\, i = I_2 + 1, I_2 + 2, \ldots, I. \\
&x = \left(x_1, x_2, \ldots, x_j\right) \in X, x \geq 0 \\
&\alpha^2 \geq \beta^2, \alpha^2 \geq \gamma^2, 0 \leq \alpha^2 + \beta^2 + \gamma^2 \leq 1
\end{aligned} \tag{12}$$

After simplifying the above model (Eq. 12), the spherical fuzzy geometric optimization model can be obtained as follows (Eq. 13):

$$Max\left(\alpha^2 - \beta^2 - \gamma^2\right)$$

Subject to

$$\begin{aligned}
&T_o(Z_o(x))^2 \geq \alpha^2,\ I_o(Z_o(x))^2 \leq \beta^2,\ F_o(Z_o(x))^2 \leq \gamma^2 \\
&g_i(x) \leq b_i, \forall\, i = 1, 2, \ldots, I_1, \\
&g_i(x) \geq b_i, \forall\, i = I_1 + 1, I_1 + 2, \ldots, I_2, \\
&g_i(x) = b_i, \forall\, i = I_2 + 1, I_2 + 2, \ldots, I. \\
&x = \left(x_1, x_2, \ldots, x_j\right) \in X, x \geq 0 \\
&\alpha^2 \geq \beta^2, \alpha^2 \geq \gamma^2, 0 \leq \alpha^2 + \beta^2 + \gamma^2 \leq 1
\end{aligned} \tag{13}$$

Hence the above optimization model (Eq. 13) provides the optimal solution for the SFGPP under spherical fuzzy environment.

Solution Steps

The stepwise algorithm to solve the SFGPP under spherical fuzzy environment can be summarized as follows:

Step-I. Determine the pay-off matrix as given in Table 1 by solving each objective function individually and define the upper and lower bound for each objective.

Step-II. Construct the upper and lower bound for positive, neutral and negative membership functions by using upper and lower values given in Eq. (8).

Step-III. Transform the different membership functions into linear membership function under spherical fuzzy environment.
Step-IV. Formulate the spherical fuzzy geometric programming problem for each type membership function (Eq. 11) under spherical fuzzy environment.
Step-V. Solve the proposed optimization model by using either some suitable software or robust optimization techniques to obtain the compromise solution of SFGPP (Eq. 13).

3 Numerical Examples

To show the applicability and validity of the proposed SFGP model, following numerical examples are presented. All the three numerical problems are written in AMPL language and solved by using the global optimization solvers Knitro 0.5.0 online facility provided by University of Wisconsin (Dolan et al. 2002).

Example-1

$$Minimize\ G_1(x) = \frac{80}{x_1x_2x_3} + 40x_2x_3$$

$$Minimize\ G_2(x) = \frac{80}{x_1x_2x_3}$$

$$Subject\ to,$$

$$\frac{1}{2}x_1x_3 + \frac{1}{4}x_1x_2 \leq 1;$$

$$x_1, x_2, x_3 > 0$$

On solving each objective function individually, we have obtained the following best and worst solution to define the upper and lower bound for each membership functions respectively. The best and worst solution is summarized in Table 2.

The upper and lower bound for the truth, indeterminacy and a falsity membership degree have been obtained as follows:

$$U_1^T = 114.365,\ L_1^T = 52.346$$

$$U_1^I = 52.346 + p_1(114.365 - 52.346),\ L_1^I = 52.346$$

$$U_1^F = 114.365,\ L_1^F = 52.346 + q_1(114.365 - 52.346)$$

Table 2 Individual best and worst value of objective functions

	Z_1	Z_2
Minimum Z_1	52.346	13.928
Minimum Z_2	114.365	1.238

The different membership function for the first objective function is defined as follows:

$$T_D(G_1(x)) = \begin{cases} 1, & if\ G_1(x) \le L_1^T \\ 1 - \frac{G_1(x)-L_1^T}{U_1^T-L_1^T}, & if\ L_1^T \le G_1(x) \le U_1^T \\ 0, & if\ G_1(x) \ge U_1^T \end{cases}$$

$$I_D(G_1(x)) = \begin{cases} 1, & if\ G_1(x) \le L_1^I \\ 1 - \frac{G_1(x)-L_1^I}{U_1^I-L_1^I}, & if\ L_1^I \le G_1(x) \le U_1^I \\ 0, & if\ G_1(x) \ge U_1^I \end{cases}$$

$$F_D(G_1(x)) = \begin{cases} 1, & if\ G_1(x) \ge U_1^F \\ 1 - \frac{U_1^F-G_1(x)}{U_1^F-L_1^F}, & if\ L_1^F \le G_1(x) \le U_1^F \\ 0, & if\ G_1(x) \le L_1^F \end{cases}$$

The upper and lower bound for the truth, indeterminacy and a falsity membership degree have been obtained as follows:

$$U_2^T = 13.928,\ L_2^T = 1.238$$
$$U_2^I = 1.238 + p_2(13.928 - 1.238),\ L_2^I = 1.238$$
$$U_2^F = 13.928,\ L_2^F = 1.238 + q_2(13.928 - 1.238)$$

The different membership function for the second objective function is defined as follows:

$$T_D(G_2(x)) = \begin{cases} 1, & if\ G_2(x) \le L_2^T \\ 1 - \frac{G_2(x)-L_2^T}{U_2^T-L_2^T}, & if\ L_o^T \le G_2(x) \le U_2^T \\ 0, & if\ G_2(x) \ge U_2^T \end{cases}$$

$$I_D(G_2(x)) = \begin{cases} 1, & if\ G_2(x) \le L_2^I \\ 1 - \frac{G_2(x)-L_2^I}{U_2^I-L_2^I}, & if\ L_2^I \le G_2(x) \le U_2^I \\ 0, & if\ G_2(x) \ge U_2^I \end{cases}$$

$$F_D(G_2(x)) = \begin{cases} 1, & if\ G_2(x) \ge U_2^F \\ 1 - \frac{U_2^F-G_2(x)}{U_2^F-L_2^F}, & if\ L_o^F \le G_2(x) \le U_2^F \\ 0, & if\ G_2(x) \le L_2^F \end{cases}$$

Now the formulations of the spherical fuzzy geometric optimization models are given as follows:

$$Max\left(\alpha^2 - \beta^2 - \gamma^2\right)$$

Subject to

Table 3 Individual best and worst values of objective functions

	$Z_1(X^1)$	$Z_2(X^2)$
Minimum Z_1	6.75	60.75
Minimum Z_2	6.94	57.87

$$T_1(G_1(x))^2 \geq \alpha^2, I_1(G_1(x))^2 \leq \beta^2, F_1(G_1(x))^2 \leq \gamma^2$$
$$T_2(G_2(x))^2 \geq \alpha^2, I_2(G_2(x))^2 \leq \beta^2, F_2(G_2(x))^2 \leq \gamma^2$$
$$\frac{1}{2}x_1x_3 + \frac{1}{4}x_1x_2 \leq 1;$$
$$x_1, x_2, x_3 > 0$$
$$\alpha^2 \geq \beta^2, \alpha^2 \geq \gamma^2, 0 \leq \alpha^2 + \beta^2 + \gamma^2 \leq 1$$

The optimal solution of the problem is $t_1 = 2.93, t_2 = 1.17, t_3 = 0.43$ and the value of each objective is obtained as $G_1 = 86.78$, $G_2 = 3.3$ respectively.

Example-2

$$Minimize\ G_1(x) = x_1^{-1}x_2^{-2}$$
$$Minimize\ G_2(x) = 2x_1^{-2}x_2^{-3}$$
$$Subject\ to,$$
$$x_1 + x_2 \leq 1;$$
$$x_1,\ x_2 > 0$$

On solving each objective function individually, we have obtained the following best and worst solution to define the upper and lower bound for each membership functions respectively. The best and worst solution is summarized in Table 3.

The upper and lower bound for the truth, indeterminacy and a falsity membership degree have been obtained as follows:

$$U_1^T = 6.94, L_1^T = 6.75$$
$$U_1^I = 6.75 + p_1(6.95 - 6.75), L_1^I = 6.75$$
$$U_1^F = 6.95, L_1^F = 6.75 + q_1(6.95 - 6.75)$$

The different membership function for the first objective function is defined as follows:

$$T_D(G_1(x)) = \begin{cases} 1, & if\ G_1(x) \leq L_1^T \\ 1 - \frac{G_1(x) - L_1^T}{U_1^T - L_1^T}, & if\ L_1^T \leq G_1(x) \leq U_1^T \\ 0, & if\ G_1(x) \geq U_1^T \end{cases}$$

$$I_D(G_1(x)) = \begin{cases} 1, & if\ G_1(x) \le L_1^I \\ 1 - \frac{G_1(x)-L_1^I}{U_1^I-L_1^I}, & if\ L_1^I \le G_1(x) \le U_1^I \\ 0, & if\ G_1(x) \ge U_1^I \end{cases}$$

$$F_D(G_1(x)) = \begin{cases} 1, & if\ G_1(x) \ge U_1^F \\ 1 - \frac{U_1^F-G_1(x)}{U_1^F-L_1^F}, & if\ L_1^F \le G_1(x) \le U_1^F \\ 0, & if\ G_1(x) \le L_1^F \end{cases}$$

The upper and lower bound for the truth, indeterminacy and a falsity membership degree have been obtained as follows:

$$U_2^T = 60.75,\ L_2^T = 57.87$$
$$U_2^I = 57.87 + p_2(60.75 - 57.87),\ L_2^I = 57.87$$
$$U_2^F = 60.75,\ L_2^F = 57.87 + q_2(60.75 - 57.87)$$

The different membership function for the second objective function is defined as follows:

$$T_D(G_2(x)) = \begin{cases} 1, & if\ G_2(x) \le L_2^T \\ 1 - \frac{G_2(x)-L_2^T}{U_2^T-L_2^T}, & if\ L_o^T \le G_2(x) \le U_2^T \\ 0, & if\ G_2(x) \ge U_2^T \end{cases}$$

$$I_D(G_2(x)) = \begin{cases} 1, & if\ G_2(x) \le L_2^I \\ 1 - \frac{G_2(x)-L_2^I}{U_2^I-L_2^I}, & if\ L_2^I \le G_2(x) \le U_2^I \\ 0, & if\ G_2(x) \ge U_2^I \end{cases}$$

$$F_D(G_2(x)) = \begin{cases} 1, & if\ G_2(x) \ge U_2^F \\ 1 - \frac{U_2^F-G_2(x)}{U_2^F-L_2^F}, & if\ L_o^F \le G_2(x) \le U_2^F \\ 0, & if\ G_2(x) \le L_2^F \end{cases}$$

Now the formulations of the spherical fuzzy geometric optimization models are given as follows:

$$Max\left(\alpha^2 - \beta^2 - \gamma^2\right)$$

Subject to

$$T_1(G_1(x))^2 \ge \alpha^2, I_1(G_1(x))^2 \le \beta^2, F_1(G_1(x))^2 \le \gamma^2$$
$$T_2(G_2(x))^2 \ge \alpha^2, I_2(G_2(x))^2 \le \beta^2, F_2(G_2(x))^2 \le \gamma^2$$
$$x_1 + x_2 \le 1;$$
$$x_1, x_2 > 0$$
$$\alpha^2 \ge \beta^2, \alpha^2 \ge \gamma^2, 0 \le \alpha^2 + \beta^2 + \gamma^2 \le 1$$

Table 4 Individual best and worst values of objective functions

	$G_1(X^1)$	$G_2(X^2)$
Minimum G_1	130	290
Maximum G_2	23	86

The optimal solution of the problem is $x_1 = 0.36577$, $x_2 = 0.63422$ and the value of each objective is obtained as $G_1 = 6.796$, $G_2 = 58.599$ respectively.

Example-3

$$\begin{aligned}
&Minimize\ G_1(x) = 80t_1^{-1}t_2^{-1}t_3^{-1} + 40t_2t_3 + 20t_1t_3 + 80t_1t_2 \\
&Maximize\ G_2(x) = 22t_2^{-1}t_3^{-1} + 36t_2t_3 - 20t_1t_2 \\
&Subject\ to, \\
&t_1, t_2, t_3 > 0
\end{aligned}$$

On solving each objective function individually, we have obtained the following best and worst solution to define the upper and lower bound for each membership functions respectively. The best and worst solution is summarized in Table 4.

The upper and lower bound for the truth, indeterminacy and a falsity membership degree have been obtained as follows:

$$\begin{aligned}
&U_1^T = 290,\ L_1^T = 130 \\
&U_1^I = 290 + p_1(290 - 130),\ L_1^I = 130 \\
&U_1^F = 130,\ L_1^F = 130 + q_1(290 - 130)
\end{aligned}$$

The different membership function for the first objective function is defined as follows:

$$T_D(G_1(x)) = \begin{cases} 1, & if\ G_1(x) \le L_1^T \\ 1 - \frac{G_1(x) - L_1^T}{U_1^T - L_1^T}, & if\ L_1^T \le G_1(x) \le U_1^T \\ 0, & if\ G_1(x) \ge U_1^T \end{cases}$$

$$I_D(G_1(x)) = \begin{cases} 1, & if\ G_1(x) \le L_1^I \\ 1 - \frac{G_1(x) - L_1^I}{U_1^I - L_1^I}, & if\ L_1^I \le G_1(x) \le U_1^I \\ 0, & if\ G_1(x) \ge U_1^I \end{cases}$$

$$F_D(G_1(x)) = \begin{cases} 1, & if\ G_1(x) \ge U_1^F \\ 1 - \frac{U_1^F - G_1(x)}{U_1^F - L_1^F}, & if\ L_1^F \le G_1(x) \le U_1^F \\ 0, & if\ G_1(x) \le L_1^F \end{cases}$$

The upper and lower bound for the truth, indeterminacy and a falsity membership degree have been obtained as follows:

$$U_2^T = 86,\ L_2^T = 23$$
$$U_2^I = 23 + p_2(86 - 23),\ L_2^I = 23$$
$$U_2^F = 86,\ L_2^F = 23 + q_2(86 - 23)$$

The different membership function for the second objective function is defined as follows:

$$T_D(G_2(x)) = \begin{cases} 1, & if\ G_2(x) \le L_2^T \\ 1 - \frac{G_2(x) - L_2^T}{U_2^T - L_2^T}, & if\ L_o^T \le G_2(x) \le U_2^T \\ 0, & if\ G_2(x) \ge U_2^T \end{cases}$$

$$I_D(G_2(x)) = \begin{cases} 1, & if\ G_2(x) \le L_2^I \\ 1 - \frac{G_2(x) - L_2^I}{U_2^I - L_2^I}, & if\ L_2^I \le G_2(x) \le U_2^I \\ 0, & if\ G_2(x) \ge U_2^I \end{cases}$$

$$F_D(G_2(x)) = \begin{cases} 1, & if\ G_2(x) \ge U_2^F \\ 1 - \frac{U_2^F - G_2(x)}{U_2^F - L_2^F}, & if\ L_o^F \le G_2(x) \le U_2^F \\ 0, & if\ G_2(x) \le L_2^F \end{cases}$$

Now the formulations of the spherical fuzzy geometric optimization models are given as follows:

$$Max\left(\alpha^2 - \beta^2 - \gamma^2\right)$$

Subject to

$$T_1(G_1(x))^2 \ge \alpha^2,\ I_1(G_1(x))^2 \le \beta^2,\ F_1(G_1(x))^2 \le \gamma^2$$
$$T_2(G_2(x))^2 \ge \alpha^2,\ I_2(G_2(x))^2 \le \beta^2,\ F_2(G_2(x))^2 \le \gamma^2$$
$$x_1 + x_2 \le 1;$$
$$x_1, x_2 > 0$$
$$\alpha^2 \ge \beta^2, \alpha^2 \ge \gamma^2, 0 \le \alpha^2 + \beta^2 + \gamma^2 \le 1$$

The optimal solution of the problem is $t_1 = 1, t_2 = 1/2, t_2 = 2$ and the value of each objective is obtained as $G_1 = 200,\ G_2 = 48$ respectively.

Example-4 (System Reliability Optimization) The problem discussed in this section is taken from Kundu and Islam (2018) which is a system reliability optimization problem. Consider that a series system of reliability with three components. Assume that $R_i (i = 1, 2, 3)$ represents the individual reliability of the ith component of the series system. Similarly, $R_S(R_1, R_2, R_3)$ and $C_S(C_1, C_2, C_3)$ are the system reliability and costs of all three component of the series system. Manager or Decision maker intends to maximize the reliability of the series system and minimize the total

Table 5 Input data

C_1	C_2	C_3	C_0	a_C	r_C	d_C	a_R	r_R	d_R	R_0
40	40	45	100	24	40	18	0.3	0.5	0.24	0.3

cost associated with all three components. The degree of acceptance (a_R), indeterminacy (d_R) and rejection (r_R) are the tolerance limits for the system whereas degree of acceptance (a_C), indeterminacy (d_C) and rejection (r_C) are the tolerance limits for the system cost respectively. The satisfaction target values for system reliability and cost are represented by R_0 and C_0. The related data are summarized in Table 5. Thus the geometric programming problem can be formulated as follows:

$$\begin{aligned}
&Maximize\ G_1 = \prod_{i=1}^{3} R_i = R_1 R_2 R_3 (System\ reliability)\\
&Minimize\ G_2 = \sum_{i=1}^{3} C_i R_i^{a_i} (System\ \text{Cost})\\
&Subject\ to,\\
&0 < R_i \leq 1; i = 1, 2, 3
\end{aligned}$$

The truth membership functions of the spherical fuzzy objective functions are obtained as follows:

$$T_D(G_1(x)) = \begin{cases} 1, & if\ G_1(x) \leq 0.3 \\ 1 - \frac{G_1(x)-0.3}{0.3}, & if\ 0.3 \leq G_1(x) \leq 0.3 + 0.3 \\ 0, & if\ G_1(x) \geq 0.3 + 0.3 \end{cases}$$

$$T_D(G_2(x)) = \begin{cases} 1, & if\ G_2(x) \leq 100 \\ 1 - \frac{G_2(x)-100}{24}, & if\ 100 \leq G_2(x) \leq 100 + 24 \\ 0, & if\ G_2(x) \geq 100 + 24 \end{cases}$$

The indeterminacy membership functions of the spherical fuzzy objective functions are obtained as follows:

$$I_D(G_1(x)) = \begin{cases} 0, & if\ G_1(x) \leq 0.24 \\ 1 - \frac{0.3-G_1(x)}{0.24}, & if\ 0.3 \leq G_1(x) \leq 0.3 + 0.24 \\ 1, & if\ G_1(x) \geq 0.3 + 0.24 \end{cases}$$

$$I_D(G_2(x)) = \begin{cases} 0, & if\ G_2(x) \leq 100 \\ 1 - \frac{100-G_2(x)}{18}, & if\ 100 \leq G_2(x) \leq 100 + 18 \\ 1, & if\ G_2(x) \geq 100 + 18 \end{cases}$$

The falsity membership functions of spherical fuzzy objective functions are obtained as follows:

$$F_D(G_1(x)) = \begin{cases} 1, & if\ G_1(x) \geq 0.3 \\ 1 - \frac{0.3 - G_1(x)}{0.5}, & if\ 0.3 \leq G_1(x) \leq 0.3 + 0.5 \\ 0, & if\ G_1(x) \leq 0.3 + 0.5 \end{cases}$$

$$F_D(G_2(x)) = \begin{cases} 1, & if\ G_2(x) \geq 100 \\ 1 - \frac{100 - G_2(x)}{40}, & if\ 100 \leq G_2(x) \leq 100 + 40 \\ 0, & if\ G_2(x) \leq 100 + 40 \end{cases}$$

Now the formulations of the spherical fuzzy geometric optimization models are given as follows:

$$Max\left(\alpha^2 - \beta^2 - \gamma^2\right)$$

Subject to

$$\begin{aligned} & T_1(G_1(x))^2 \geq \alpha^2,\ I_1(G_1(x))^2 \leq \beta^2,\ F_1(G_1(x))^2 \leq \gamma^2 \\ & T_2(G_2(x))^2 \geq \alpha^2,\ I_2(G_2(x))^2 \leq \beta^2,\ F_2(G_2(x))^2 \leq \gamma^2 \\ & 0 < R_i \leq 1; i = 1, 2, 3 \\ & \alpha^2 \geq \beta^2, \alpha^2 \geq \gamma^2, 0 \leq \alpha^2 + \beta^2 + \gamma^2 \leq 1 \end{aligned}$$

The solution results of the above problem is summarized in the Table 6. The same problem has been also solved by using two different methods (Kundu and Islam 2018) namely; intutionitic fuzzy goal geometric programming problem (IFGGPP), neutrosophic goal geometric programming problem (NGGPP) respectively. From Table 6, it is clear that the performance of proposed SFGPP approach is better as compared to other existing approaches.

Table 6 Results and comparison of optimal solution based on different methods

Method	Objective functions		Decision variables		
	R_s	C_s	R_1	R_2	R_3
IFGGPP (Kundu and Islam 2018)	0.523472	78.766	0.771292	0.875617	0.775105
NGGPP (Kundu and Islam 2018)	0.613664	80.372	0.824809	0.863345	0.861773
Proposed SFGPP	**0.618454**	**81.254**	**0.835402**	**0.862134**	**0.858712**

4 Conclusions

The geometric programming problem is a particular class of nonlinear programming problems. The structure of the GP problem is quite different from general mathematical programming problems. It inevitably involves posynomial terms into its objective functions. The formulation of GP in a different uncertain environment is quite beneficial in decision-making processes. The ample opportunity to capture the imprecise and inconsistent information by SFS leads to a significant contribution to decision-making problems. Thus, this chapter investigates SFGPP under a spherical fuzzy environment, which comprises with the maximization of positive and minimization of neutral and negative membership functions into the spherical fuzzy decision set. The restrictions over the sum of squares of all the membership functions have imposed to be less than or equal to unity. Different numerical examples have been presented to show the applicability of SFGPP. In system reliability optimization, the performance of SFGPP over others is quite better and provides the strength to the decision-making process by incorporating spherical fuzzy uncertainty.

Many real-life problems such as multi gravel-box optimization, system reliability, bar truss region optimization problem and inventory control, etc. take the form of the geometric programming model. Furthermore, the solution algorithm based on SFS also provides flexibility while solving the GP problem with multiple posynomial objectives. The designed optimization framework can also assist the decision-makers in incorporating the indeterminacy degree while solving the different real-life problems. In the future, many engineering problems can be formulated as SFGPP problems such as bar truss region, gravel box designing, product designing, etc.

References

Ahmad F, Adhami AY (2019) Neutrosophic programming approach to multiobjective nonlinear transportation problem with fuzzy parameters. Int J Manag Sci Eng Manag 14(3):218–229

Ahmad F, Adhami AY, Smarandache F (2018) Single valued neutrosophic hesitant fuzzy computational algorithm for multiobjective nonlinear optimization problem. Neutrosophic sets and systems 22:76–86

Ahmad F, Adhami AY, Smarandache F (2019) Neutrosophic optimization model and computational algorithm for optimal shale gas water management under uncertainty. Symmetry 11(4):544

Ahmad F, Adhami AY, Smarandache F (2020) Modified neutrosophic fuzzy optimization model for optimal closed-loop supply chain management under uncertainty. In: Optimization theory based on neutrosophic and plithogenic sets. Academic Press, pp 343–403

Atanassov KT (1986) Studies in fuzziness and soft computing intuitionistic fuzzy logics. In: Fuzzy sets and systems

Bazaraa MS, Sherali HD, Shetty CM (2013) Nonlinear programming: theory and algorithms. Wiley

Bellman RE, Zadeh LA (1970) Decision-making in a fuzzy environment. Manag Sci 17(4):B-141

Beightler CS, Phillips DT (1976) Applied geometric programming. Wiley

Das K, Roy TK, Maiti M (2000) Multi-item inventory model with quantity-dependent inventory costs and demand-dependent unit cost under imprecise objective and restrictions: a geometric programming approach. Prod Plan Control 11(8):781–788

Dolan BED, Fourer R, Moré JJ, Munson TS (2002) Optimization on the NEOS Server. SIAM News 35(6):1–5
Duffin RJ (1970) Linearizing geometric programs. SIAM review 12(2):211–227
Floudas CA (2013) Deterministic global optimization: theory, methods and applications, vol 37. Springer Science and Business Media
Islam S, Roy TK (2006) Continuous optimization a new fuzzy multi-objective programming: entropy based geometric programming and its application of transportation problems 173:387–404
Kundu T, Islam S (2018) Neutrosophic goal geometric programming problem and its application to multi-objective reliability optimization model. Int J Fuzzy Syst 20(6):1986–1994
Kutlu Gündoğdu F, Kahraman C (2019) Spherical fuzzy sets and spherical fuzzy TOPSIS method. J Intell Fuzzy Syst (Preprint), 1–16
Mahapatra GS, Roy TK (2009) Single and multi-container maintenance model: a fuzzy geometric programming approach. J Math Res 1(2):47
Peterson EL (2001) The origins of geometric programming. Ann Oper Res 105(1):15
Rafiq M, Ashraf S, Abdullah S, Mahmood T (2019) The cosine similarity measures of spherical fuzzy sets and their applications in decision making, (Preprint) 1–15
Smarandache F (1999) A unifying field in logics: neutrosophic logic. Philosophy. https://doi.org/10.5281/zenodo.49174
Yager RR (2013) Pythagorean fuzzy subsets. IEEE 2:57–61
Zener C (1971) Engineering design by geometric programming. Wiley